Systems Analysis and Design for the Global Enterprise

Jeffrey L. Whitten

Professor

Both at Purdue University
West Lafayette, IN

With contributions by
Gary Randolph
Purdue University

SEVENTH EDITION

Mc Graw Hill ■ **McGraw-Hill Irwin**

Boston   Burr Ridge, IL   Dubuque, IA   Madison, WI   New York   San Francisco   St. Louis
Bangkok   Bogotá   Caracas   Kuala Lumpur   Lisbon   London   Madrid   Mexico City
Milan   Montreal   New Delhi   Santiago   Seoul   Singapore   Sydney   Taipei   Toronto

McGraw-Hill
Irwin

SYSTEMS ANALYSIS AND DESIGN FOR THE GLOBAL ENTERPRISE
Published by McGraw-Hill/Irwin, a business unit of The McGraw-Hill Companies, Inc., 1221
Avenue of the Americas, New York, NY, 10020. Copyright © 2007 by The McGraw-Hill
Companies, Inc. All rights reserved. No part of this publication may be reproduced or distributed
in any form or by any means, or stored in a database or retrieval system, without the prior written
consent of The McGraw-Hill Companies, Inc., including, but not limited to, in any network or
other electronic storage or transmission, or broadcast for distance learning.

Some ancillaries, including electronic and print components, may not be available to customers
outside the United States.

This book is printed on acid-free paper.

1 2 3 4 5 6 7 8 9 0 QPD/QPD 0 9 8 7 6

ISBN-13: 978-0-07-110766-2
ISBN-10: 0-07-110766-5

www.mhhe.com

> Intended Audience

Systems Analysis and Design for the Global Enterprise, seventh edition, is intended to support one or more practical courses in information systems development. These courses are normally taught to both information systems and business majors at the sophomore, junior, senior, or graduate level.

We recommend that students take a computer- and information systems–literacy course before using this text. While not required or assumed, a programming course can significantly enhance the learning experience provided by this textbook.

> Why We Wrote This Book

More than ever, today's students are "consumer-oriented," due in part to the changing world economy, which promotes quality, competition, and professional currency. They expect to walk away from a course with more than a grade and a promise that they'll someday appreciate what they've learned. They want to "practice" the application of concepts, not just study applications of concepts. We wrote this book (1) to balance the coverage of concepts, tools, techniques, and their application, (2) to provide the most examples of system analysis and design deliverables available in any book, and (3) to balance the coverage of classic methods (such as *structured analysis* and *information engineering*) and emerging methods (e.g., *object-oriented analysis, agile development,* and *rapid application development*). Additionally, our goal is to serve the reader by providing a postcourse, professional reference for the best current practices.

We have written the book using a lively, conversational tone. This approach (and the numerous examples) delivers a comprehensive text that still connects with the student throughout the learning process.

> Changes for the Seventh Edition

- **Reorganization for Better Clarity:** The object-oriented analysis chapter has become Chapter 10 to better position it alongside the structured analysis chapters (Chapters 8 and 9). Other chapters have been reorganized internally. For example, Chapter 9, in response to reviewer comments, has undergone extensive reorganization. Also, the discussion of sequential versus iterative development has been moved to Chapter 3 to place it with related methodology concepts.
- **Expanded Object-Oriented Coverage:** As object-oriented analysis and design grows in importance, coverage continues to increase. The seventh edition more fully explains the object-oriented approach and tracks both where it follows the same path as the traditional, structured approach and where the two approaches part ways. The object-oriented analysis chapter (Chapter 10) features expanded coverage of activity diagrams. New to this edition in Chapter 10 is coverage of system sequence diagrams. Chapter 18 features expanded coverage of object-oriented design. Persistence and system design classes are discussed as well as entity, controller, and interface design classes. The discussion of sequence diagrams and CRC cards has been expanded, and their role in the design process explained more fully. Coverage of design patterns has been greatly expanded with a discussion of the Gang of Four patterns and an examination of two of the patterns.
- **UML 2.0:** Both Chapter 10 and Chapter 18 have been revised to cover the UML 2.0 specification. Each UML 2.0 diagram is listed with an explanation of its purpose. In Chapters 7, 10, and 18, five of the thirteen UML 2.0 diagrams are developed in depth and three more are shown and discussed.
- **Expanded Discussion of Feasibility:** The discussion of feasibility now includes legal feasibility and cultural (or political) feasibility as well as our traditional four tests of feasibility (operational, economic, schedule, and technical).
- **Use of Context Diagrams:** Even as the move away from data flow diagrams and to UML diagrams continues, the context diagram continues to be important as a

tool for understanding system scope. It has been added to the tools used in Chapter 5 and can be employed in the classroom as a first modeling assignment.
- **Updated Technology References:** The extensive references to example technologies has been continued in the seventh edition and updated to reflect technological changes, version updates, and mergers and acquisitions of technology companies.
- **Revision of the SoundStage Running Case:** The SoundStage case has been condensed, changed from a dialogue format to a narrative format, and integrated into the opening of each chapter. Featuring the perspective of a just-graduated systems analyst in his first assignment, SoundStage briefly introduces the concepts taught in each chapter and underscores their importance in a real systems project.

> Pedagogical Use of Color

The seventh edition continues the use of color applied to an adaptation of Zachman's *Framework for Information Systems Architecture.* The color mappings are displayed in the inside front cover of the textbook.

The information systems building blocks matrix uses these colors to introduce recurring concepts. System models then reinforce those concepts with a consistent use of the same colors.

> Organization

Systems Analysis and Design for the Global Enterprise, seventh edition, is divided into four parts. The text's organization is flexible enough to allow instructors to omit and resequence chapters according to what they feel is important to their audience. Every effort has been made to decouple chapters from one another as much as possible to assist in resequencing the material—even to the extent of reintroducing selected concepts and terminology.

Part One, "Developing Systems," presents the information systems development scenario and process. Chapters 1 through 4 introduce the student to systems analysts, other project team members (including users and management), information systems building blocks (based on the Zachman framework), a contemporary systems

> **Information Systems Framework**
> Color is used consistently throughout the text's framework to introduce recurring concepts.

represents methods

represents data and/or knowledge

represents process

represents communication/interface

represents people

development life cycle, and project management. Part One can be covered relatively quickly. Some readers may prefer to omit project management or delay it until the end of the book.

Part Two, "Systems Analysis Methods," covers the front-end life-cycle activities, tools, and techniques for analyzing business problems, specifying business requirements for an information system, and proposing a business and system solution. Coverage in Chapters 5 through 11 includes requirements gathering, use cases, data modeling with entity-relationship diagrams, process modeling with data flow diagrams, object-oriented analysis, and solution identification and the system proposal.

Part Three, "Systems Design Methods," covers the middle life-cycle activities, tools, and techniques. Chapters 12 through 18 include coverage of both general and detailed design, with a particular emphasis on application architecture, rapid development and prototyping, external design (inputs, outputs, and interfaces), internal design (e.g., database and software engineering), and object-oriented design.

Part Four, "Beyond Systems Analysis and Design," is a capstone unit that places systems analysis and design into perspective by surveying the back-end life-cycle activities. Specifically, Chapters 19 and 20 examine system implementation, support, maintenance, and reengineering.

> Supplements and Instructional Resources

It has always been our intent to provide a complete course, not just a textbook. We are especially excited about this edition's comprehensive support package. It includes software bundles and other resources for both the student and the instructor. The supplements for the seventh edition include the following components.

For the Instructor

Instructor's Manual with PowerPoint Presentations

The instructor's manual is offered on the *Instructor's CD-ROM.* This manual includes course planning materials, teaching guidelines and PowerPoint slides, templates, and answers to end-of-chapter problems, exercises, and minicases.

The PowerPoint presentations on the CD-ROM include over 400 slides. All slides are complete with instructor notes that provide teaching guidelines and tips. Instructors can (1) pick and choose the slides they wish to use, (2) customize slides to their own preferences, and (3) add new slides. Slides can be organized into electronic presentations or be printed as transparencies or transparency masters.

Test Bank

The *Instructor's CD-ROM* also includes an electronic test bank covering all the chapters. Computerized/Network Testing with Brownstone Diploma software is fully networkable for LAN test administration. Each chapter offers 75 questions in the following formats: true/false, multiple choice, sentence completion, and matching. The test bank and answers are cross-referenced to the page numbers in the textbook. A level-of-difficulty rating is also assigned to each question.

> Packages

System Architect Student Edition Version 8

An optional package combines the textbook, Student Resource CD, and a student version of System Architect. System Architect is a powerful, repository-based enterprise modeling tool which supports a comprehensive set of diagramming techniques and features, including all nine UML diagram types, business enterprise modeling, data modeling, business modeling with IDEFO and IDEF3 notations, plus many more.

Visible Analyst Workbench

Another optional package combines the textbook, Student Resource CD, and Visible Analyst Workbench. This tool integrates business function analysis, data modeling and database design, process modeling, and object modeling in one easy-to-use package. Print versions of each case can be ordered through McGraw-Hill's Custom Publishing group by visiting www.primiscontentcenter.com. A *build your own project* model is retained for instructors and students who want to maximize value by leveraging students' past and current work experience or for use with a live-client project.

Primis Content Center

Primis Online

Print versions of projects and cases, as well as other MIS content, can be ordered through McGraw-Hill's Custom Publishing Group.

Acknowledgements

We are indebted to many individuals who contributed to the development of this edition:

Grant Alexander, *Northeastern Oklahoma State University*
Richard J. Averbeck, *DeVry Institutes*
Emerson (Bill) Bailey, *Park University*
Jack Briner, *Charleston Southern University*
Jimmie Carraway, *Old Dominion University*
Casey Cegielski, *Auburn University*
Minder Chen, *George Mason University*
Glenn Dietrich, *University of Texas-San Antonio*
Dorothy Dologite, *Baruch College, CUNY*
Tom Erickson, *University of Virginia's Virginia Center for Continuing and Professional Education*
Bob Kilmer, *Messiah College*
Avram Malkin, *DeVry College of Technology*
Dat-Dao Nguyen, *California State University-Northridge*
Parag C. Pendharkar, *Penn State University*
Leah Pietron, *University of Nebraska-Omaha*
Charlene Riggle, *University of South Florida-Sarasota/Manatee*

A special thank-you is extended to the following focus group participants:

Jeffrey Parsons, *Memorial University of Newfoundland*
Parag C. Pendharkar, *Penn State University*
Carl Scott, *University of Houston*
Ron Thompson, *Wake Forest University*
Steve Walczak, *Colorado University-Denver*

We also are indebted to many individuals who contributed to the development of the previous editions of this text.

Jeanne M. Alm, *Moorhead State University*
Charles P. Bilbrey, *James Madison University*
Ned Chapin, *California State University-Hayward*
Carol Clark, *Middle Tennessee State University*
Gail Corbitt, *California State University-Chico*
Larry W. Cornwell, *Bradley University*
Barbara B. Denison, *Wright State University*
Linda Duxbury, *Carleton University*
Dana Edberg, *University of Nevada-Reno*
Craig W. Fisher, *Marist College*
Raoul J. Freeman, *California State University-Dominguez Hills*
Dennis D. Gagnon, *Santa Barbara City College*
Abhijit Gopal, *University of Calgary*

Patricia J. Guinan, *Boston University*
Bill C. Hardgrave, *University of Arkansas-Fayetteville*
Alexander Hars, *University of Southern California*
Richard C. Housley, *Golden Gate University*
Constance Knapp, *Pace University*
Riki S. Kuchek, *Orange Coast College*
Thom Luce, *Ohio University*
Charles M. Lutz, *Utah State University*
Ross Malaga, *University of Maryland-Baltimore County*
Chip McGinnis, *Park College*
William H. Moates, *Indiana State University*
Ronald J. Norman, *San Diego State University*
Charles E. Paddock, *University of Nevada-Las Vegas*
June A. Parsons, *Northern Michigan University*
Harry Reif, *James Madison University*
Gail L. Rein, *SUNY-Buffalo*
Rebecca H. Rutherfoord, *Southern College of Technology*
Craig W. Slinkman, *University of Texas-Arlington*
John Smiley, *Holy Family College*
Mary Thurber, *Northern Alberta Institute of Technology*
Jerry Tillman, *Appalachian State University*
Jonathan Trower, *Baylor University*
Margaret S. Wu, *University of Iowa*
Jacqueline E. Wyatt, *Middle Tennessee State University*
Vincent C. Yen, *Wright State University*
Ahmed S. Zaki, *College of William and Mary*

Finally, we acknowledge the contributions, encouragement, and patience of the staff at McGraw-Hill. Special thanks to Brent Gordon, publisher; Paul Ducham, sponsoring editor; Trina Hauger, developmental editor; Greta Kleinert, marketing manager; Kristin Bradley, project manager; and Kami Carter, designer. We also thank Judy Kausal, photo research coordinator; Michael McCormick, production supervisor; Greg Bates, media producer; and Rose Range, supplement coordinator.

To those of you who used our previous editions, thank you for your continued support. For those using the text for the first time, we hope you see a difference in this text. We eagerly await your reactions, comments, and suggestions.

Jeffrey L. Whitten
Lonnie D. Bentley

Brief Contents

Contents

14 DATABASE DESIGN 517

15 OUTPUT DESIGN AND PROTOTYPING 549

16 INPUT DESIGN AND PROTOTYPING 581

17 USER INTERFACE DESIGN 613

18 OBJECT-ORIENTED DESIGN AND MODELING USING THE UML 647

PART FOUR
Beyond Systems Analysis and Design 681

19 SYSTEMS CONSTRUCTION AND IMPLEMENTATION 683

20 SYSTEMS OPERATIONS AND SUPPORT 701

Systems Analysis and Design for the Global Enterprise

Part One

Developing Systems

This is a practical book about information systems development methods. All businesses and organizations develop information systems. You can be assured that you will play some role in the systems analysis and design for those systems—either as a customer or user of those systems or as a developer of those systems. Systems analysis and design is about business problem solving and computer applications. The methods you will learn in this book can be applied to a wide variety of problem domains, not just those involving the computer.

Before we begin, we assume you've completed an introductory course in computer-based information systems. Many of you have also completed one or more programming courses (using technologies such as *Access, Java,* C/C++, or *Visual Basic*). That will prove helpful, since systems analysis and design precedes and/or integrates with those activities. But don't worry—we'll review all the necessary principles on which systems analysis and design is based.

Part One focuses on the big picture. Before you learn about specific activities, tools, techniques, methods, and technology, you need to understand this big picture. As you explore the context of systems analysis and design, we will introduce many ideas, tools, and techniques that are not explored in great detail until later in the book. Try to keep that in mind as you explore the big picture.

Systems development isn't magic. There are no secrets for success, no perfect tools, techniques, or methods. To be sure, there are skills that can be mastered. But the complete and consistent application of those skills is still an art.

We start in Part One with fundamental concepts, philosophies, and trends that provide the context of systems analysis and design methods— in other words, the basics! If you understand these basics, you will be better able to apply, with confidence, the practical tools and techniques you will learn in Parts Two through Four. You will also be able to adapt to new situations and methods.

Four chapters make up this part. Chapter 1, "The Value of Systems Analysis and Design," introduces you to the *participants* in systems analysis and design with special emphasis on the modern systems analyst as the facilitator of systems work. You'll also learn about the relationships between systems analysts, end users, managers, and other information systems professionals. Finally, you'll learn to prepare yourself for a career as an analyst (if that is your goal). Regardless, you will understand how you will interact with this important professional.

Chapter 2, "The Components of Information Systems," introduces the

product we will teach you how to build—*information systems.* Specifically, you will learn to examine information systems in terms of common building blocks, KNOWLEDGE, PROCESSES, and COMMUNICATIONS— each from the perspective of different participants or stakeholders. A visual matrix framework will help you organize these building blocks so that you can see them applied in the subsequent chapters.

Chapter 3, "Developing Information Systems," introduces a high-level (meaning general) process for information systems development. This is called a *systems development life cycle.* We will present the life cycle in a form in which most of you will experience it—a *systems development methodology.* This methodology will be the context in which you will learn to use and apply the systems analysis and design methods taught in the remainder of the book.

Chapter 4, "Project Management," introduces project management techniques. All systems projects are dependent on the principles that are surveyed. This chapter introduces two modeling techniques for project management: *Gantt* and *PERT.* These tools help you schedule activities, evaluate progress, and adjust schedules.

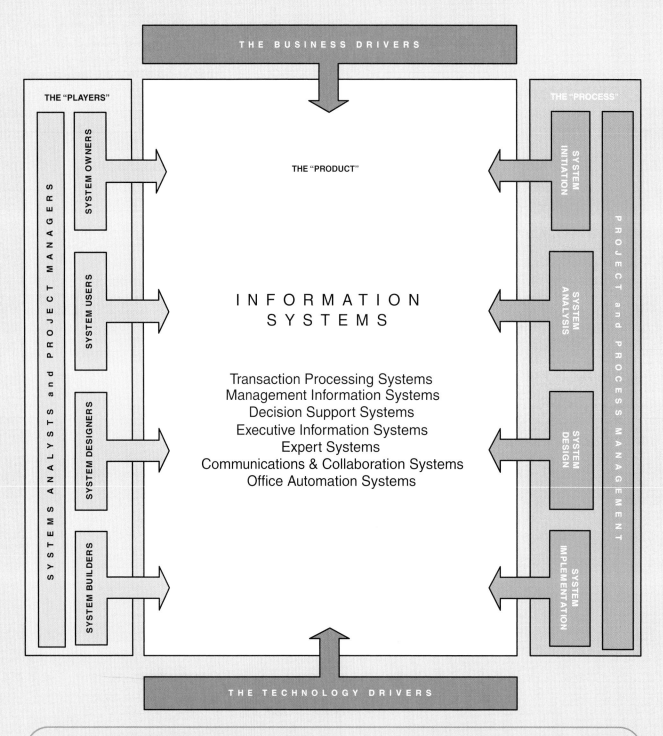

THE BUSINESS DRIVERS

THE "PLAYERS"

THE "PROCESS"

SYSTEMS ANALYSTS and PROJECT MANAGERS

SYSTEM OWNERS

SYSTEM USERS

SYSTEM DESIGNERS

SYSTEM BUILDERS

THE "PRODUCT"

INFORMATION SYSTEMS

Transaction Processing Systems
Management Information Systems
Decision Support Systems
Executive Information Systems
Expert Systems
Communications & Collaboration Systems
Office Automation Systems

SYSTEM INITIATION

SYSTEM ANALYSIS

SYSTEM DESIGN

SYSTEM IMPLEMENTATION

PROJECT and PROCESS MANAGEMENT

THE TECHNOLOGY DRIVERS

CHAPTER 1 HOME PAGE Each chapter in this book begins with a "home page" similar to the one above. The home page is something of a chapter map, a visual framework for systems thinking applicable to that chapter. Chapter 1 focuses on (1) the players in the systems game, (2) business drivers of interest to business players, (3) technology drivers and enablers of interest to the technical players, and (4) the process used to develop systems. We will also examine the critical role played by systems analysts in facilitating an understanding of how all four perspectives must come together.

1

The Value of Systems Analysis and Design

Chapter Preview and Objectives

This is a book about systems analysis and design as applied to information systems and computer applications. No matter what your chosen occupation or position in any business, you will likely participate in systems analysis and design. Some of you will become *systems analysts*, the key players in systems analysis and design activities. The rest of you will work *with* systems analysts as projects come and go in your organizations. This chapter introduces you to information systems from four different perspectives. You will understand the context for systems analysis and design methods when you can:

▌ Define *information system* and name seven types of information system applications.

▌ Identify different types of *stakeholders* who use or develop information systems, and give examples of each.

▌ Define the unique role of *systems analysts* in the development of information systems.

▌ Identify those *skills* needed to successfully function as an information systems analyst.

▌ Describe current *business drivers* that influence information systems development.

▌ Describe current *technology drivers* that influence information systems development.

▌ Briefly describe a simple *process* for developing information systems.

Introduction

It is Bob Martinez's first week at work as an analyst/programmer. Fresh out of college with a degree in computer information systems technology, Bob is eager to work with information systems in the real world. His employer is SoundStage Entertainment Club, one of the fastest-growing music and video clubs in America. SoundStage is just beginning systems analysis and design work on a reengineering of their member services information system. Bob has been appointed to the project team.

This morning was the kickoff meeting for the project, a meeting that included the vice president of member services, director of the audio club, director of the game club, director of marketing, director of customer services, and director of warehouse operations. With that lineup Bob was glad to mainly keep silent at the meeting and rely on his boss, Sandra Shepherd, a senior systems analyst. He was amazed at how well Sandra was able to speak the language of each of the participants and to explain the plans for the new information system in terms they could understand and with benefits they could appreciate. Bob had thought that being just out of college he would know more about cutting-edge technology than most of his co-workers. But Sandra seemed to understand everything about e-commerce and using mobile technologies plus many things of which Bob was only vaguely aware. He made a note to read up on ERP systems as that had come up in the discussion. By the end of the meeting Bob had a new appreciation for the job of systems analyst and of all the things he had yet to learn.

system a group of interrelated components that function together to achieve a desired result.

A Framework for Systems Analysis and Design

information system (IS) an arrangement of people, data, processes, and **information technology** that interact to collect, process, store, and provide as output the information needed to support an organization.

information technology (IT) a contemporary term that describes the combination of computer technology (hardware and software) with telecommunications technology (data, image, and voice networks).

transaction processing system (TPS) an information system that captures and processes data about business transactions.

management information system (MIS) an information system that provides for management-oriented reporting based on transaction processing and operations of the organization.

As its title suggests, this is a book about *systems analysis and design methods*. In this chapter, we will introduce the subject using a simple but comprehensive visual framework. Each chapter in this book begins with a *home page* (see page 4) that quickly and visually shows which aspects of the total framework we will be discussing in the chapter. We'll build this visual framework slowly over the first four chapters to avoid overwhelming you with too much detail too early. Thereafter, each chapter will highlight those aspects of the full framework that are being taught in greater detail in that chapter.

Ultimately, this is a book about "analyzing" business requirements for information systems and "designing" *information systems* that fulfill those business requirements. In other words, the *product* of systems analysis and design is an information system. That product is visually represented in the visual framework as the large rectangle in the center of the picture.

A **system** is a group of interrelated components that function together to achieve a desired result. For instance, you may own a home theater system made up of a DVD player, receiver, speakers, and display monitor.

Information systems (IS) in organizations capture and manage data to produce useful information that supports an organization and its employees, customers, suppliers, and partners. Many organizations consider information systems to be essential to their ability to compete or gain competitive advantage. Most organizations have come to realize that *all* workers need to participate in the development of information systems. Therefore, information systems development is a relevant subject to you regardless of whether or not you are studying to become an information systems professional.

Information systems come in all shapes and sizes. They are so interwoven into the fabric of the business systems they support that it is often difficult to distinguish between business systems and their support information systems. Suffice it to say that information systems can be classified according to the functions they serve. **Transaction processing systems (TPSs)** process business transactions such as orders, time cards, payments, and reservations. **Management information systems (MISs)** use the transaction data to produce information needed by managers to run the business.

Decision support systems (DSSs) help various decision makers identify and choose between options or decisions. **Executive information systems (EISs)** are tailored to the unique information needs of executives who plan for the business and assess performance against those plans. **Expert systems** capture and reproduce the knowledge of an expert problem solver or decision maker and then simulate the "thinking" of that expert. **Communication and collaboration systems** enhance communication and collaboration between people, both internal and external to the organization. Finally, **office automation systems** help employees create and share documents that support day-to-day office activities.

As illustrated in the chapter home page, information systems can be viewed from various perspectives, including:

- The players in the information system (the "team").
- The business drivers influencing the information system.
- The technology drivers used by the information system.
- The process used to develop the information system.

Let's examine each of these perspectives in the remaining sections of the chapter.

The Players—System Stakeholders

Let's assume you are in a position to help build an information system. Who are the **stakeholders** in this system? Stakeholders for information systems can be broadly classified into the five groups shown on the left-hand side of Figure 1-1. Notice that each stakeholder group has a different perspective of the same information system. The *systems analyst* is a unique stakeholder in Figure 1-1. The systems analyst serves as a facilitator or coach, bridging the communications gap that can naturally develop between the nontechnical system owners and users and the technical system designers and builders.

All the above stakeholders have one thing in common—they are what the U.S. Department of Labor calls **information workers.** The livelihoods of information workers depend on decisions made based on information. Today, more than 60 percent of the U.S. labor force is involved in producing, distributing, and using information. Let's examine the five groups of information workers in greater detail.

Let's briefly examine the perspectives of each group. But before we do so, we should point out that these groups actually define "roles" played in systems development. In practice, any individual person may play more than one of these roles. For example, a system owner might also be a system user. Similarly, a systems analyst may also be a system designer, and a system designer might also be a system builder. Any combination may work.

> Systems Owners

For any information system, large or small, there will be one or more **system owners.** System owners usually come from the ranks of management. For medium to large information systems, system owners are usually middle or executive managers. For smaller systems, system owners may be middle managers or supervisors. System owners tend to be interested in the bottom line—how much will the system cost? How much value or what benefits will the system return to the business? Value and benefits can be measured in different ways, as noted in the margin checklist.

> Systems Users

System users make up the vast majority of the information workers in any information system. Unlike system owners, system users tend to be less concerned with costs and benefits of the system. Instead, as illustrated in Figure 1-1, they are concerned with the functionality the system provides to their jobs and the system's ease of learning and ease of use. Although users have become more technology-literate over the years,

---

decision support system (DSS) an information system that either helps to identify decision-making opportunities or provides information to help make decisions.

executive information system (EIS) an information system that supports the planning and assessment needs of executive managers.

expert system an information system that captures the expertise of workers and then simulates that expertise to the benefit of nonexperts.

communications and collaboration system an information system that enables more effective communications between workers, partners, customers, and suppliers to enhance their ability to collaborate.

office automation system an information system that supports the wide range of business office activities that provide for improved work flow between workers.

stakeholder any person who has an interest in an existing or proposed information system. Stakeholders may include both technical and nontechnical workers. They may also include both internal and external workers.

information worker any person whose job involves creating, collecting, processing, distributing, and using information.

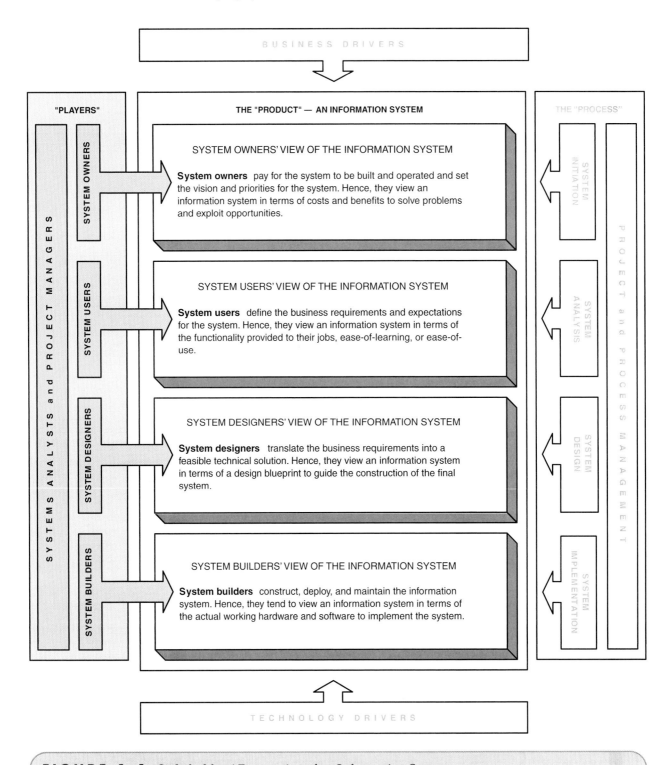

FIGURE 1-1 Stakeholders' Perspective of an Information System

---

their primary concern is to get the job done. Consequently, discussions with most
users need to be kept at the business requirements level as opposed to the technical
requirements level. Much of this book is dedicated to teaching you how to effectively
identify and communicate business requirements for an information system.

There are many classes of system users. Each class should be directly involved in
any information system development project that affects them. Let's briefly examine
these classes.

Internal System Users   Internal system users are employees of the businesses for which most information systems are built. Internal users make up the largest percentage of information system users in most businesses. Examples include:

- *Clerical and service workers*—perform most of the day-to-day transaction processing in the average business. They process orders, invoices, payments, and the like. They type and file correspondence. They fill orders in the warehouse. And they manufacture goods on the shop floor. Most of the fundamental data in any business is captured or created by these workers, many of whom perform manual labor in addition to processing data. Information systems that target these workers tend to focus on transaction processing speed and accuracy.
- *Technical and professional staff*—consists largely of business and industrial specialists who perform highly skilled and specialized work. Examples include lawyers, accountants, engineers, scientists, market analysts, advertising designers, and statisticians. Because their work is based on well-defined bodies of knowledge, they are sometimes called **knowledge workers.** Information systems that target technical and professional staff focus on data analysis as well as generating timely information for problem solving.
- *Supervisors, middle managers, and executive managers*—are the decision makers. Supervisors tend to focus on day-to-day problem solving and decision making. Middle managers are more concerned with tactical (short-term) operational problems and decision making. Executive managers are concerned with strategic (long-term) planning and decision making. Information systems for managers tend to focus entirely on information access. Managers need the right information at the right time to identify and solve problems and make good decisions.

External System Users   The Internet has allowed traditional information system boundaries to be extended to include other businesses or direct consumers as system users. These external system users make up an increasingly large percentage of system users for modern information systems. Examples include:

- *Customers*—any organizations or individuals that purchase our products and services. Today, our customers can become direct users of our information systems when they can directly execute orders and sales transactions that used to require intervention by an internal user. For example, if you purchased a company's product via the Internet, you became an external user of that business's sales information system. (There was no need for a separate internal user of the business to input your order.)
- *Suppliers*—any organizations from which our company may purchase supplies and raw materials. Today, these suppliers can interact directly with our company's information systems to determine our supply needs and automatically create orders to fill those needs. There is no longer always a need for an internal user to initiate those orders to a supplier.
- *Partners*—any organizations from which our company purchases services or with which it partners. Most modern businesses contract or outsource a number of basic services such as grounds maintenance, network management, and many others. And businesses have learned to partner with other businesses to more quickly leverage strengths to build better products more rapidly.
- *Employees*—those employees who work on the road or who work from home. For example, sales representatives usually spend much of their time on the road. Also, many businesses permit workers to telecommute (meaning "work from home") to reduce costs and improve productivity. As mobile or remote users, these employees require access to the same information systems as those needed by internal users.

POSSIBLE VALUES AND BENEFITS OF INFORMATION SYSTEMS

Increased Business Profit
Reduced Business Costs
Costs and Benefits of the System
Increased Market Share
Improved Customer Relations
Increased Efficiency
Improved Decision Making
Better Compliance with Regulations
Fewer Mistakes
Improved Security
Greater Capacity

---

system user a "customer" who will use or is affected by an information system on a regular basis—capturing, validating, entering, responding to, storing, and exchanging data and information.

---

knowledge worker any worker whose responsibilities are based on a specialized body of knowledge.

External system users are increasingly referred to as **remote users** and **mobile users.** They connect to our information systems through laptop computers, handheld computers, and smart phones—either wired or wireless. Designing information systems for these devices presents some of the most contemporary of challenges that we will address in this book.

> Systems Designers

System designers are technology specialists for information systems. As Figure 1-1 shows, system designers are interested in information technology choices and in the design of systems that use chosen technologies. Today's system designers tend to focus on technical specialties. Some of you may be educating yourselves to specialize in one of these technical specialties, such as:

- *Database administrators*—specialists in database technologies who design and coordinate changes to corporate databases.
- *Network architects*—specialists in networking and telecommunications technologies who design, install, configure, optimize, and support local and wide area networks, including connections to the Internet and other external networks.
- *Web architects*—specialists who design complex Web sites for organizations, including public Web sites for the Internet, internal Web sites for organizations (called *intranets*), and private business-to-business Web sites (called *extranets*).
- *Graphic artists*—relatively new in today's IT worker mix, specialists in graphics technology and methods used to design and construct compelling and easy-to-use interfaces to systems, including interfaces for PCs, the Web, handhelds, and smart phones.
- *Security experts*—specialists in the technology and methods used to ensure data and network security (and privacy).
- *Technology specialists*—experts in the application of specific technologies that will be used in a system (e.g., a specific commercial software package or a specific type of hardware).

> Systems Builders

System builders (again, see Figure 1-1) are another category of technology specialists for information systems. Their role is to construct the system according to the system designers' specifications. In small organizations or with small information systems, systems designers and systems builders are often the same people. But in large organizations and information systems they are often separate jobs. Some of you may be educating yourselves to specialize in one of their technical specialties, such as:

- *Applications programmers*—specialists who convert business requirements and statements of problems and procedures into computer languages. They develop and test computer programs to capture and store data and to locate and retrieve data for computer applications.
- *Systems programmers*—specialists who develop, test, and implement operating systems-level software, utilities, and services. Increasingly, they also develop reusable software "components" for use by applications programmers (above).
- *Database programmers*—specialists in database languages and technology who build, modify, and test database structures and the programs that use and maintain them.
- *Network administrators*—specialists who design, install, troubleshoot, and optimize computer networks.
- *Security administrators*—specialists who design, implement, troubleshoot, and manage security and privacy controls in a network.

---

remote user a user who is not physically located on the premises but who still requires access to information systems.

mobile user a user whose location is constantly changing but who requires access to information systems from any location.

system designer a technical specialist who translates system users' business requirements and constraints into technical solutions. She or he designs the computer databases, inputs, outputs, screens, networks, and software that will meet the system users' requirements.

system builder a technical specialist who constructs information systems and components based on the design specifications generated by the system designers.

- *Webmasters*—specialists who code and maintain Web servers.
- *Software integrators*—specialists who integrate software packages with hardware, networks, and other software packages.

Although this book is not directly intended to educate the system builder, it is intended to teach system designers how to better communicate design specifications to system builders.

> Systems Analysts

As you have seen, system owners, users, designers, and builders often have very different perspectives on any information system to be built and used. Some are interested in generalities, while others focus on details. Some are nontechnical, while others are very technical. This presents a communications gap that has always existed between those who need computer-based business solutions and those who understand information technology. The **systems analyst** bridges that gap. You can (and probably will) play a role as either a systems analyst or someone who works with systems analysts.

As illustrated in Figure 1-1, their role intentionally overlaps the roles of all the other stakeholders. For the system owners and users, systems analysts identify and validate business problems and needs. For the system designers and builders, systems analysts ensure that the technical solution fulfills the business needs and integrate the technical solution into the business. In other words, systems analysts *facilitate* the development of information systems through interaction with the other stakeholders.

There are several legitimate, but often confusing, variations on the job title we are calling "systems analyst." A *programmer/analyst* (or *analyst/programmer*) includes the responsibilities of both the computer programmer and the systems analyst. A *business analyst* focuses on only the nontechnical aspects of systems analysis and design. Other synonyms for "systems analyst" are systems consultant, business analyst, systems architect, systems engineer, information engineer, information analyst, and systems integrator.

Some of you will become systems analysts. The rest of you will routinely work with systems analysts who will help you solve your business and industrial problems by creating and improving your access to the data and information needed to do your job. Let's take a closer look at systems analysts as the key facilitators of information systems development.

The Role of the Systems Analyst   Systems analysts understand both business and computing. They study business problems and opportunities and then transform business and information requirements into specifications for information systems that will be implemented by various technical specialists including computer programmers. Computers and information systems are of value to a business only if they help solve problems or effect improvements.

Systems analysts initiate *change* within an organization. Every new system changes the business. Increasingly, the very best systems analysts literally change their organizations—providing information that can be used for competitive advantage, finding new markets and services, and even dramatically changing and improving the way the organization does business.

The systems analyst is basically a *problem solver.* Throughout this book, the term *problem* will be used to describe many situations, including:

- Problems, either real or anticipated, that require corrective action.
- Opportunities to improve a situation despite the absence of complaints.
- Directives to change a situation regardless of whether anyone has complained about the current situation.

The systems analyst's job presents a fascinating and exciting challenge to many individuals. It offers high management visibility and opportunities for important

systems analyst a specialist who studies the problems and needs of an organization to determine how people, data, processes, and information technology can best accomplish improvements for the business.

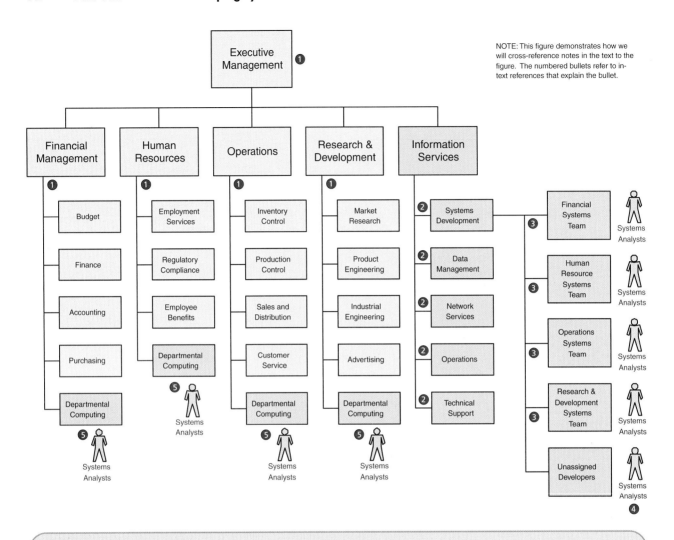

NOTE: This figure demonstrates how we will cross-reference notes in the text to the figure. The numbered bullets refer to in-text references that explain the bullet.

FIGURE 1-2 Systems Analysts in a Typical Organization

decision making and creativity that may affect an entire organization. Furthermore, this job can offer these benefits relatively early in your career (compared to other entry-level jobs and careers).

Where Do Systems Analysts Work? Every business organizes itself uniquely. But certain patterns of organization seem to reoccur. Figure 1-2 is a representative organization chart. The following numbered bullets cross-reference and emphasize key points in the figure:

❶ System owners and system users are located in the functional units and sub-units of the business, as well as in the executive management.

❷ System designers and builders are usually located in the information systems unit of the business. Most systems analysts work also for the information services unit of an organization.

❸ As shown in the figure, systems analysts (along with systems designers and builders) may be permanently assigned to a team that supports a specific business function (e.g., financial systems).

Numbers 2 and 3 above represent a traditional approach to organizing systems analysts and other developers. Numbers 4 and 5 below represent strategies intended to emphasize either efficiency or business expertise. All of the strategies can be combined in a single organization.

The Next Generation:
Career Prospects for Systems Analysts

Many of you are considering or preparing for a career as a systems analyst. The life of a systems analyst is both challenging and rewarding. But what are the prospects for the future? Do organizations need systems analysts? Will they need them in the foreseeable future? Is the job changing for the future, and if so, how? These questions are addressed in this box.

According to the U.S. Department of Labor, computer-related jobs account for 5 out of the 20 fastest-growing occupations in the economy. What's more, these fastest-growing computer-related occupations pay better than many other jobs.

In 2002, 468,000 workers were classified as systems analysts. By 2012, that number will grow to 653,000, an increase of 39%. This means that at least 185,000 new systems analysts must be educated and hired (not including those needed to replace the ones who retire or move into managerial positions or other occupations). The need is increasing because industry needs systems analysts to meet the seemingly endless demand for more information systems and software applications. As some programming jobs are being out-sourced to independent contractors and other countries, the need grows even greater for skilled systems analysts, who can create solid design specifications for remote development teams. Opportunities for success will be the greatest for the most educated, qualified, skilled, and experienced analysts.

What happens to the successful systems analyst? Does a position as a systems analyst lead to any other careers? Indeed, there are many career paths. Some analysts leave the information systems field and join the user community. Their experience with developing business applications, combined with their total systems perspective, can make experienced analysts unique business specialists. Alternatively, analysts can become project managers, information systems managers, or technical specialists (for databases, telecommunications, microcomputers, and so forth). Finally, skilled systems analysts are often recruited by the consulting and outsourcing industries. The career path opportunities are virtually limitless.

As with any profession, systems analysts can expect change. While it is always dangerous to predict changes, we'll take a shot at it. We believe that organizations will become increasingly dependent on external sources for their systems analysts—consultants and outsourcers. This will be driven by such factors as the complexity and rapid change of technology, the desire to accelerate systems development, and the continued difficulty in recruiting, retaining, and retraining skilled systems analysts (and other information technology professionals). In many cases, internally employed systems analysts will manage projects through consulting or outsourcing agreements.

We believe that an increasing percentage of tomorrow's systems analysts will not work in the information systems department. Instead, they will work directly for a business unit within an organization. This will enable them to better serve their users. It will also give users more power over what systems are built and supported.

Finally, we also believe that a greater percentage of systems analysts will come from noncomputing backgrounds. At one time most analysts were computer specialists. Today's computer graduates are becoming more business-literate. Similarly, today's business and noncomputing graduates are becoming more computer-literate. Their full-time help and insight will be needed to meet demand and to provide the business background necessary for tomorrow's more complex applications.

❹ Systems analysts (along with system designers and builders) may also be pooled and temporarily assigned to specific projects for any business function as needed. (Some organizations believe this approach yields greater efficiency because analysts and other developers are always assigned to the highest-priority projects regardless of business area expertise.)

❺ Some systems analysts may work for smaller, departmental computing organizations that support and report to their own specific business functions. (Some organizations believe this structure results in systems analysts that develop greater expertise in their assigned business area to complement their technical expertise.)

All of the above strategies can, of course, be reflected within a single organization.

Regardless of where systems analysts are assigned within the organization, it is important to realize that they come together in *project teams*. Project teams are usually created and disbanded as projects come and go. Project teams must also include appropriate representation from the other stakeholders that we previously discussed (system owners, system users, system designers, and system builders). Accordingly, we will emphasize team building and teamwork throughout this book.

Skills Needed by the Systems Analyst For those of you with aspirations of becoming a systems analyst, this section describes the skills you will need to develop. This book introduces many systems analysis and design concepts, tools, and techniques. But you will also need skills and experiences that neither this book nor your systems analysis and design course can fully provide.

When all else fails, the systems analyst who remembers the basic concepts and principles of "systems thinking" will still succeed. No tool, technique, process, or methodology is perfect in all situations! But concepts and principles of systems thinking will always help you adapt to new and different situations. This book emphasizes systems thinking.

Not too long ago, it was thought that the systems analyst's only real tools were paper, pencil, and a flowchart template. Over the years, several tools and techniques have been developed to help the systems analyst. Unfortunately, many books emphasize a specific class of tools that is associated with one methodology or approach to systems analysis and design. In this book, we propose a "toolbox" approach to systems analysis and design. As you read this book, your toolbox will grow to include many tools from different methodologies and approaches to systems analysis and design. Subsequently, you should pick and use tools based on the many different situations you will encounter as an analyst—the right tool for the right job!

In addition to having formal systems analysis and design skills, a systems analyst must develop or possess other skills, knowledge, and traits to complete the job. These include:

- *Working knowledge of information technologies*—The analyst must be aware of both existing and emerging information technologies. Such knowledge can be acquired in college courses, professional development seminars and courses, and in-house corporate training programs. Practicing analysts also stay current through disciplined reading and participation in appropriate professional societies. (To get started, see the Suggested Readings at the end of this and subsequent chapters.)
- *Computer programming experience and expertise*—It is difficult to imagine how systems analysts could adequately prepare business and technical specifications for a programmer if they didn't have some programming experience. Most systems analysts need to be proficient in one or more high-level programming languages.
- *General knowledge of business processes and terminology*—Systems analysts must be able to communicate with business experts to gain an understanding of their problems and needs. For the analyst, at least some of this knowledge comes only by way of experience. At the same time, aspiring analysts should avail themselves of every opportunity to complete basic business literacy courses available in colleges of business. Relevant courses may include financial accounting, management or cost accounting, finance, marketing, manufacturing or operations management, quality management, economics, and business law.
- *General problem-solving skills*—The systems analyst must be able to take a large business problem, break down that problem into its parts, determine problem causes and effects, and then recommend a solution. Analysts must avoid the tendency to suggest the solution before analyzing the problem. For aspiring analysts, many colleges offer philosophy courses that teach

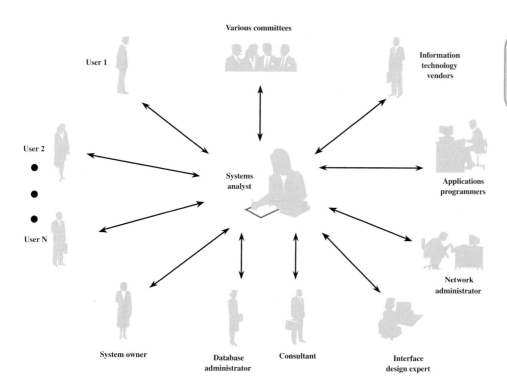

FIGURE 1-3

The Systems Analyst as a Facilitator

problem-solving skills, critical thinking, and reasoning. These "soft skills" will serve an analyst well.

- *Good interpersonal communication skills*—An analyst must be able to communicate effectively, both orally and in writing. Almost without exception, your communications skills, not your technical skills, will prove to be the single biggest factor in your career success or failure. These skills are learnable, but most of us must force ourselves to seek help and work hard to improve them. Most schools offer courses such as business and technical writing, business and technical speaking, interviewing, and listening—all useful skills for the systems analyst. These skills are taught in Chapter 6.

- *Good interpersonal relations skills*—As illustrated in Figure 1-3, systems analysts interact with all stakeholders in a systems development project. These interactions require effective interpersonal skills that enable the analyst to deal with group dynamics, business politics, conflict, and change. Many schools offer valuable interpersonal-skills development courses on subjects such as teamwork, principles of persuasion, managing change and conflict, and leadership.

- *Flexibility and adaptability*—No two projects are alike. Accordingly, there is no single, magical approach or standard that is equally applicable to all projects. Successful systems analysts learn to be flexible and to adapt to unique challenges and situations. Our aforementioned toolbox approach is intended to encourage flexibility in the use of systems analysis and design tools and methods. But you must develop an attitude of adaptability to properly use any box of tools.

- *Character and ethics*—The nature of the systems analyst's job requires a strong character and a sense of right and wrong. Analysts often gain access to sensitive or confidential facts and information that are not meant for public disclosure. Also, the products of systems analysis and design are usually considered the intellectual property of the employer. There are several standards for computer ethics. One such standard, from the Computer Ethics Institute, is called "The Ten Commandments of Computer Ethics" and is shown in Figure 1-4.

FIGURE 1-4 Ethics for Systems Analysts

The Ten Commandments of Computer Ethics

1. Thou shalt not use a computer to harm other people.
2. Thou shalt not interfere with other people's computer work.
3. Thou shalt not snoop around in other people's computer files.
4. Thou shalt not use a computer to steal.
5. Thou shalt not use a computer to bear false witness.
6. Thou shalt not copy or use proprietary software for which you have not paid.
7. Thou shalt not use other people's computer resources without authorization or proper compensation.
8. Thou shalt not appropriate other people's intellectual output.
9. Thou shalt think about the social consequences of the program you are writing or the system you are designing.
10. Thou shalt always use a computer in ways that insure consideration and respect for your fellow humans.

Source: Computer Ethics Institute.

> External Service Providers

external service provider (ESP) a systems analyst, system designer, or system builder who sells his or her expertise and experience to other businesses to help those businesses purchase, develop, or integrate their information systems solutions; may be affiliated with a consulting or services organization.

Those of you with some computing experience may be wondering where consultants fit in our taxonomy of stakeholders. They are not immediately apparent in our visual framework. But they are there! Any of our stakeholder roles may be filled by internal or external workers. Consultants are one example of an **external service provider (ESP).** Most ESPs are systems analysts, designers, or builders who are contracted to bring special expertise or experience to a specific project. Examples include technology engineers, sales engineers, systems consultants, contract programmers, and systems integrators.

> The Project Manager

project manager an experienced professional who accepts responsibility for planning, monitoring, and controlling projects with respect to schedule, budget, deliverables, customer satisfaction, technical standards, and system quality.

We've introduced most of the key players in modern information systems development—systems owners, users, designers, builders, and analysts. We should conclude by emphasizing the reality that these individuals must work together as a team to successfully build information systems and applications that will benefit the business. Teams require leadership. For this reason, usually one or more of these stakeholders takes on the role of **project manager** to ensure that systems are developed on time, within budget, and with acceptable quality. As Figure 1-1 indicates, most project managers are experienced systems analysts. But in some organizations, project managers are selected from the ranks of what we have called "system owners." Regardless, most organizations have learned that project management is a specialized role that requires distinctive skills and experience.

Business Drivers for Today's Information Systems

Another way to look at our information system product is from the perspective of business drivers. Using Figure 1-5, let's now briefly examine the most important business trends that are impacting information systems. Many trends quickly become fads, but here are some business trends we believe will influence systems development in the

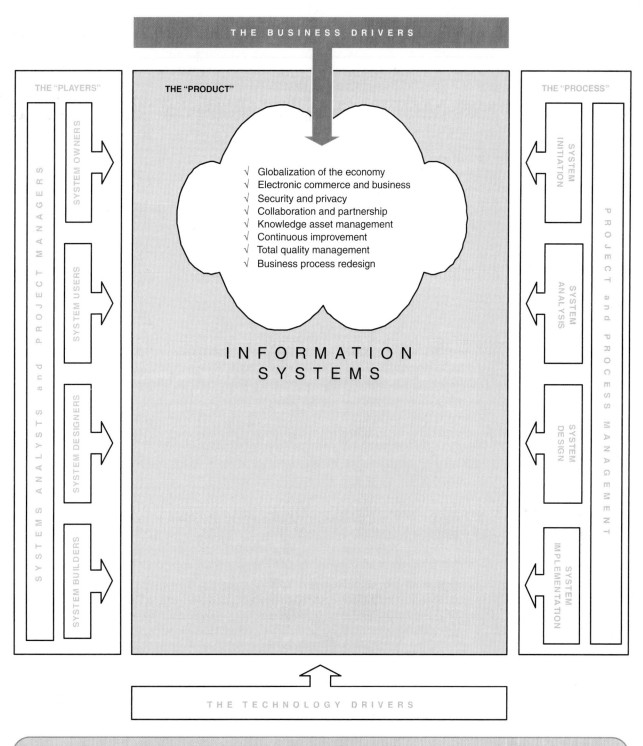

FIGURE 1-5 Business Drivers for an Information System

coming years. Many of these trends are related and integrated such that they form a new business philosophy that will impact the way everyone works in the coming years.

> Globalization of the Economy

Since the 1990s, there has been a significant trend of economic globalization. Competition is global, with emerging industrial nations offering lower-cost or higher-quality alternatives to many products. American businesses find themselves with new international competitors. On the other hand, many American businesses have discovered new and expanded international markets for their own goods and services. The bottom line is that most businesses were forced to reorganize to operate in this global economy.

How does economic globalization affect the players in the systems game? First, information systems and computer applications must be internationalized. They must support multiple languages, currency exchange rates, international trade regulations, and different business cultures and practices. Second, most information systems ultimately require information consolidation for performance analysis and decision making. The aforementioned language barriers, currency exchange rates, transborder information regulations, and the like, complicate such consolidation. Finally, there exists a demand for players who can communicate, orally and in writing, with management and users that speak different languages, dialects, and slang. Opportunities for international employment of systems analysts should continue to expand.

> Electronic Commerce and Business

In part due to the globalization of the economy, and in part because of the pervasiveness of the Internet, businesses are changing or expanding their business model to implement **electronic commerce (e-commerce)** and **electronic business (e-business).** The Internet is fundamentally changing the rules by which business is conducted. We live in a world where consumers and businesses will increasingly expect to conduct commerce (business transactions) using the Internet. But the impact is even more substantive. Because people who work in the business world have become so comfortable with "surfing the Web," organizations are increasingly embracing the Web interface as a suitable architecture for conducting day-to-day business *within* the organization.

There are three basic types of e-commerce- and e-business-enabled information systems applications:

electronic commerce (e-commerce) the buying and selling of goods and services by using the Internet.

electronic business (e-business) the use of the Internet to conduct and support day-to-day business activities.

- Marketing of corporate image, products, and services is the simplest form of electronic commerce application. The Web is used merely to "inform" customers about products, services, and policies. Most businesses have achieved this level of electronic commerce.
- *Business-to-consumer (B2C)* electronic commerce attempts to offer new, Web-based channels of distribution for traditional products and services. You, as a typical consumer, can research, order, and pay for products directly via the Internet. Examples include Amazon.com (for books and music) and E-trade.com (for stocks and bonds). Both companies are businesses that were created on the Web. Their competition, however, includes traditional businesses that have added Web-based electronic commerce front ends as an alternative consumer option (such as Barnes and Noble and Merrill Lynch). Figure 1-6 illustrates a typical B2C Web storefront.
- *Business-to-business (B2B)* electronic commerce is the real future. This is the most complex form of electronic commerce and could ultimately evolve into electronic business—the complete, paperless, and digital processing of virtually all business transactions that occur within and between businesses.

One example of B2B electronic commerce is electronic procurement. All businesses purchase raw materials, equipment, and supplies—frequently tens or hundreds of millions of dollars worth per year. B2B procurement allows employees to browse electronic storefronts and catalogs, initiate purchase requisitions and work orders, route requisitions and work orders electronically for expenditure approvals, order the goods and services, and pay for the delivered goods and completed services—all

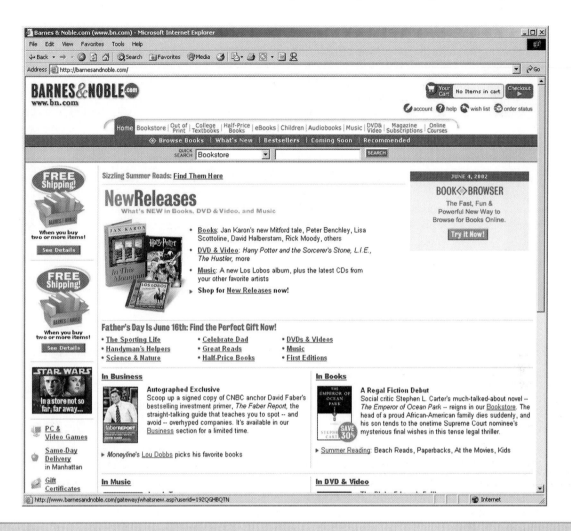

FIGURE 1-6   An Electronic Commerce Storefront

without the traditional time-consuming and costly paper flow and bureaucracy. Figure 1-7 illustrates a sample Web-based procurement storefront.

Largely due to the trend toward these e-business and e-commerce applications, most new information systems applications are being designed for an Internet architecture. Not that long ago, we were redesigning most applications to operate within a *Windows* user interface. Today, we increasingly see applications designed to run within an Internet browser such as *Internet Explorer* or *Netscape*. The choice of a desktop operating system, such as *Windows, Macintosh,* or *Linux,* is becoming less important than the availability of the browser itself.

> Security and Privacy

As the digital economy continues to evolve, citizens and organizations alike have developed a heightened awareness of the security and privacy issues involved in today's economy. Security issues tend to revolve around business continuity; that is, "How will the business continue in the event of a breach or disaster—any event that causes a disruption of business activity?" Additionally, businesses must ask themselves, "How can the business protect its digital assets from outside threats?" It is true that these questions ultimately come down to technology; however, the concerns have become fundamental business concerns.

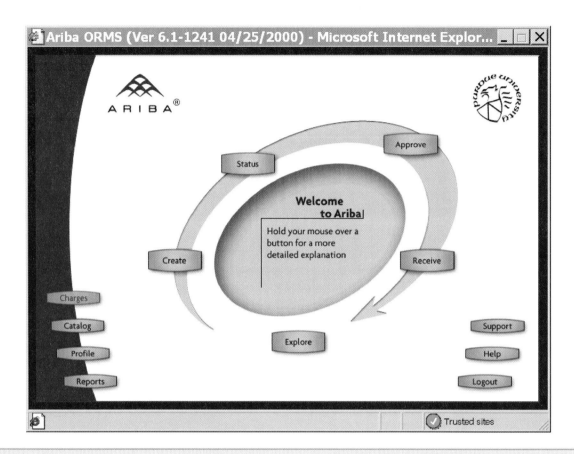

FIGURE 1-7 An Electronic Commerce Procurement Storefront

Related to security is the issue of privacy. Consumers are increasingly demanding privacy in the digital economy. Governments are regulating privacy issues, and the regulations will likely become more stringent as the digital economy continues to evolve. Go to your favorite commercial Web sites. Almost every business now has a privacy policy. Consumer groups are beginning to analyze and monitor such privacy policies, holding companies accountable and lobbying governments for stricter regulations and enforcement.

As information systems are developed and changed, you will increasingly be expected to incorporate more stringent security and privacy controls. In the global economy, you will need to become sensitive to a wide array of regulations that vary considerably from one country to another. Certainly, security and privacy mechanisms will be subject to the same internal audits that have become routine in systems that support or interact with financial systems.

> Collaboration and Partnership

Collaboration and partnership are significant business trends that are influencing information systems applications. Within organizations, management is emphasizing the need to break down the walls that separate organization departments and functions. Management speaks of "cross-functional" teams that collaborate to address common business goals from interdisciplinary perspectives. For example, new product design used to be the exclusive domain of engineers. Today, new product design typically involves a cross-functional team of representatives from many organizational units, such as engineering, marketing, sales, manufacturing, inventory control, distribution, and, yes, information systems.

Similarly, the trend toward collaboration extends beyond the organization to include other organizations—sometimes even competitors. Organizations choose to directly collaborate as partners in business ventures that make good business sense. Microsoft and Oracle sell competitive database management systems. But Microsoft and Oracle also partner to ensure that Oracle applications will operate on a Microsoft database. Both companies benefit financially from such cooperation.

In a similar vein, businesses have learned that it can be beneficial for their information systems to interoperate with one another. For example, while Wal-Mart could generate its own restocking orders for merchandise and send them to its suppliers, it makes more sense to integrate their respective inventory control systems. Suppliers can monitor Wal-Mart's inventory levels directly and can automatically initiate business-to-business transactions to keep the shelves stocked with their merchandise. Both companies benefit. (Of course, this also raises the aforementioned issue of requirements for good security.)

> Knowledge Asset Management

What is knowledge? *Knowledge* is the result of a continuum of how we process raw data into useful information. Information systems collect raw **data** by capturing business facts (about products, employees, customers, and the like) and processing business transactions. Data gets combined, filtered, organized, and analyzed to produce **information** to help managers plan and operate the business. Ultimately, information is refined by people to create **knowledge** and expertise. Increasingly, organizations are asking themselves, "How can the company manage and share knowledge for competitive advantage? And as workers come and go, how can the workers' knowledge and expertise be preserved within the organization?"

Thirty years of data processing and information systems have resulted in an enormous volume of data, information, and knowledge. All three are considered critical *business* resources, equal in importance to the classic economic resources of land, labor, and capital.

The need for knowledge asset management impacts information systems on a variety of fronts. Although we have captured (and continue to capture) a great amount of data and information in information systems, they are loosely integrated in most organizations—indeed, redundant and contradictory data and information are common in information systems. As new information systems are built, we will increasingly be expected to focus on integration of the data and information that can create and preserve knowledge in the organizations for which we work. This will greatly complicate systems analysis and design. In this book, we plan to introduce you to many tools and techniques that can help you integrate systems for improved knowledge management.

> Continuous Improvement and Total Quality Management

Information systems automate and support **business processes.** In an effort to continuously improve a business process, **continuous process improvement (CPI)** examines a business process to implement a series of small changes for improvement. These changes can result in cost reductions, improved efficiencies, or increased value and profit. Systems analysts are both affected by continuous process improvements and expected to initiate or suggest such improvements while designing and implementing information systems.

Another ongoing business driver is **total quality management (TQM).** Businesses have learned that quality has become a critical success factor in competition. They have also learned that quality management does not begin and end with the products and services sold by the business. Instead, it begins with a culture that recognizes that

---

data raw facts about people, places, events, and things that are of importance in an organization. Each fact is, by itself, relatively meaningless.

information data that has been processed or reorganized into a more meaningful form for someone. Information is formed from combinations of data that hopefully have meaning to the recipient.

knowledge data and information that are further refined based on the facts, truths, beliefs, judgments, experiences, and expertise of the recipient. Ideally information leads to *wisdom*.

business processes tasks that respond to business events (e.g., an order). Business processes are the work, procedures, and rules required to complete the business tasks, independent of any information technology used to automate or support them.

continuous process improvement (CPI) the continuous monitoring of business processes to effect small but measurable improvements in cost reduction and value added.

total quality management (TQM) a comprehensive approach to facilitating quality improvements and management within a business.

everyone in the business is responsible for quality. TQM commitments require that every business function, including information services, identify quality indicators, measure quality, and make appropriate changes to improve quality.

Information systems, and hence systems analysts, are part of the TQM requirement. Our discussions with college graduate recruiters suggest that an "obsessive" attitude toward quality management will become an essential characteristic of successful systems analysts (and all information technology professionals). Throughout this book, continuous process improvement and total quality management will be an underlying theme.

> Business Process Redesign

As stated earlier, many information systems support or automate business processes. Many businesses are learning that those business processes have not changed substantially in decades and that those business processes are grossly inefficient and/or costly. Many business processes are overly bureaucratic, and all their steps do not truly contribute value to the business. Unfortunately, information systems have merely automated many of these inefficiencies. Enter business process redesign!

business process redesign (BPR) the study, analysis, and redesign of fundamental business processes to reduce costs and/or improve value added to the business.

Business process redesign (BPR) involves making substantive changes to business processes across a larger system. In effect, BPR seeks to implement more substantial changes and improvements than does CPI. In a BPR, business processes are carefully documented and analyzed for timeliness, bottlenecks, costs, and whether or not each step or task truly adds value to the organization (or, conversely, adds only bureaucracy). Business processes are then redesigned for maximum efficiency and lowest possible costs.

So how does BPR affect information systems? There are two basic ways to implement any information system—build it or buy it. In other words, you can write the software yourself, or you can purchase and implement a commercial software package. In both cases, BPR figures prominently. In writing your own software, it is useful to redesign business processes before writing the software to automate them. This way, you avoid automating underlying inefficiencies. Alternatively, in purchasing software packages, most businesses have discovered it is easier to redesign the business processes to work with the software package than to attempt to force (and even cripple) the software package to work with existing business processes.

Technology Drivers for Today's Information Systems

Advances in information technology can also be drivers for information systems (as suggested in Figure 1-8). In some cases, outdated technologies can present significant problems that drive information system development projects. In other cases, newer technologies present business opportunities. Let's examine several technologies that are influencing today's information systems.

> Networks and the Internet

Scott McNealy, Sun Computer's charismatic CEO, is often cited as stating, "The network has become the computer." Few would argue that today's information systems are installed on a network architecture consisting of local and wide area networks. These networks include mainframe computers, network servers, and a variety of desktop, laptop, and handheld client computers. But today, the most pervasive networking technologies are based on the Internet. Some of the more relevant Internet technologies that you need to become aware of, if not develop some basic skill with, are described in the following list. (For now, don't be intimidated by these terms—we

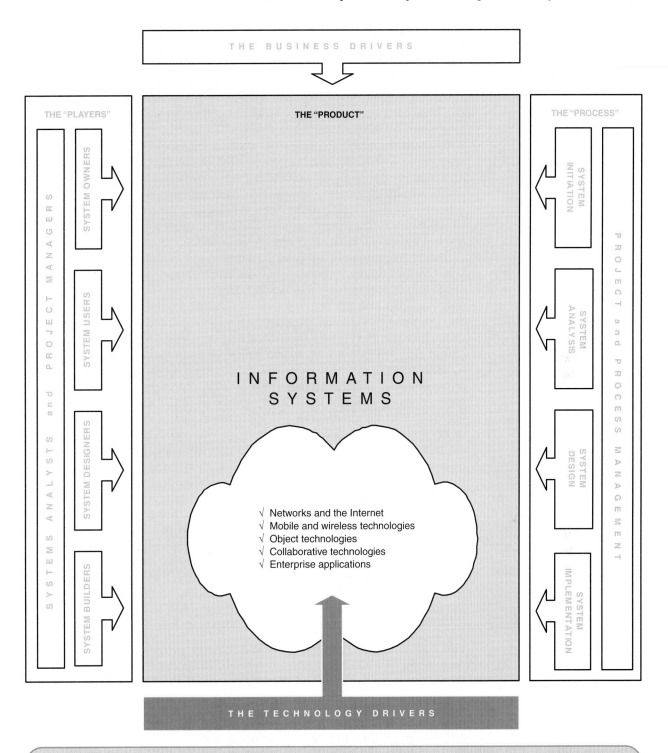

FIGURE 1-8 Technology Drivers for an Information System

will be teaching you more about these technologies and how to design systems that use them throughout this textbook.)

- *xHTML* and *XML* are the fundamental languages of Web page authoring and Internet application development. Extensible Hypertext Markup Language (xHTML) is the emerging second-generation version of HTML, the language used to construct Web pages. Extensible Markup Language (XML) is the language used to effectively transport data content along with its proper interpretation over the Internet. Introductory xHTML and XML courses have become core requirements in most information systems and information technology college curricula.

- *Scripting* languages are simple programming languages designed specifically for Internet applications. Examples include *Perl, VBScript,* and *JavaScript.* These languages are increasingly taught in college Web development and programming courses.

- Web-specific programming languages such as *Java* and *Cold Fusion* have emerged to specifically address construction of complex, Web-based applications that involve multiple servers and Web browsers. These languages are also becoming prevalent in college programming curricula.

- *Intranets* are essentially private Internets designed for use by employees of an organization. They offer the look and feel of the Internet; however, security and firewalls restrict their use to employees.

- *Extranets,* like intranets, are private Internets. But extranets are for use between specific organizations. Only the employees of those identified businesses can access and use the extranet. For example, an automotive manufacturer such as Chevrolet might set up an extranet for the sole use of its dealers. Through this extranet, the manufacturer can communicate information about parts, problems, sales incentives, and the like.

- *Portals* (in corporations) are "home pages" that can be customized to the specific needs of different individuals who use them. For example, portal technology can define Web pages that provide appropriate information and applications for different roles in the same company. Each individual's role determines which information and applications that person can use from her or his Web page. Examples of roles include "customer," "supplier," and different types of "employee." Portals can also effectively integrate public Internet, private intranet, and extranet content into each individual user's home page.

- *Web services* are the latest rage. Web services are reusable, Web-based programs that can be called from any other Internet program. For example, let's say you need to write a program to accept credit card payments over the Web. Sure, you could write, debug, and test the credit card validation program yourself. But an alternative approach would be to purchase the right to use a credit card validation program over the Web. By doing so, you need not maintain responsibility for the credit card validation code. You need only "call" the Web service from your program, much as you would call an internal subroutine. Of course, you will pay for the privilege of using Web services since somebody had to write the original Web service program.

These are but a few of the network and Internet technologies that you should seek out during your education. But you must recognize the volatility of the Internet and accept that these and other technologies will emerge and disappear frequently in the near future.

> Mobile and Wireless Technologies

Wireless Handheld

Mobile and wireless technologies are poised to significantly change the next generation of information systems. Handheld computers, or *personal data assistants* (*PDAs,* such as the HP *iPaq,* Palm, and RIM *BlackBerry*®), have become common in the ranks of information workers. These devices are increasingly including wireless capabilities (see margin photo) that provide Web access and e-mail. *Cell phones* are also increasingly featuring Internet and e-mail capabilities. And now, integrated devices such as *smart phones* are emerging that integrate the capabilities of PDAs and cell phones into a single device (see margin photo). For those who prefer separate devices, technologies like *Bluetooth* are emerging to allow the separate devices to interoperate as one logical device while preserving each one's form factors and advantages.

Additionally, laptop computers are increasingly equipped with wireless and mobile capabilities to allow information workers to more easily move between locations while preserving connectivity to information systems. All of these technical trends will significantly impact the analysis and design of new information systems. Increasingly, wireless access must be assumed. And the limitations of mobile devices and screen sizes must be accommodated in an information system's design. This textbook will teach and demonstrate tools and techniques to deal with the design of emerging mobile applications.

Smart Phone

> Object Technologies

Today, most contemporary information systems are built using **object technologies.** Today's pervasive programming languages are object-oriented. They include *C++, Java, Smalltalk,* and *Visual Basic .NET.* Object technologies allow programmers to build software from software parts called *objects.* (We will get into more specifics about objects later in this book.) Object-oriented software offers two fundamental advantages over nonobject software. First, objects are reusable. Once they are designed and built, objects can be reused in multiple information systems and applications. This reduces the time required to develop future software applications. Second, objects are extensible. They can be changed or expanded easily without adversely impacting any previous applications that used them. This reduces the lifetime costs of maintaining and improving software.

The impact of object technology is significant in the world of systems analysis and design. Prior to object technologies, most programming languages were based on so-called *structured methods.* Examples include *COBOL* (the dominant language), *C, FORTRAN, Pascal,* and *PL/1.* It is not important at this time for you to be able to differentiate between structured and object technologies and methods. Suffice to say, structured methods are inadequate to the task of analyzing and designing systems that will be built using object technologies. Accordingly, **object-oriented analysis and design** methods have emerged as the preferred approach for building *most* contemporary information systems. For this reason, we will integrate object-oriented analysis and design tools and techniques throughout this book to give you a competitive advantage in tomorrow's job market.

At the same time, structured tools and techniques are still important. Databases, for example, are still commonly designed using structured tools. And structured tools are still preferred by many systems analysts for analyzing and designing work flows and business processes. Thus, we will also teach you several popular structured tools and techniques for systems analysis and design.

It is easy to become a devout disciple of one analysis and design strategy, such as structured analysis and design or object-oriented analysis and design. You should avoid doing so! We will advocate both and teach you when and how to combine structured and object-oriented tools and techniques for systems analysis and design. As we write this chapter, this approach—called **agile development**—is gaining favor among experienced analysts who have become weary of overly prescriptive methods that usually insist that you use only *one* methodology's tools and processes. At the risk of oversimplifying agile methods, think of it as assembling a toolbox of different tools and techniques—structured, object-oriented, and others—and then selecting the best tool or technique for whatever problems or need you encounter as a systems analyst.

> Collaborative Technologies

Another significant technology trend is the use of collaborative technologies. Collaborative technologies are those that enhance interpersonal communications and teamwork. Four important classes of collaborative technologies are e-mail, instant messaging, groupware, and work flow.

Everybody knows what e-mail is. But e-mail's importance in information systems development is changing. Increasingly, modern information systems are *e-mail-enabled;*

object technology a software technology that defines a system in terms of objects that consolidate data and behavior (into objects). Objects become reusable and extensible components for the software developers.

object-oriented analysis and design a collection of tools and techniques for systems development that will utilize object technologies to construct a system and its software.

agile development a systems development strategy wherein the system developers are given the flexibility to select from a variety of appropriate tools and techniques to best accomplish the tasks at hand. Agile development is believed to strike an optimal balance between productivity and quality for systems development.

that is, e-mail capabilities are built right into the application software. There is no need to switch to a dedicated e-mail program such as *Outlook.* The application merely invokes the user's or organization's default e-mail program to enable relevant messages to be sent or received.

Related to e-mail technology is instant messaging (e.g., AOL's *Instant Messenger* and Microsoft's *MSN Messenger Service*). Instant messaging was popularized in public and private "chat rooms" on the Internet. But instant messaging is slowly being incorporated into enterprise information systems applications as well. For example, instant messaging can implement immediate response capabilities into a help system for a business application. Imagine being able to instantly send and receive messages with the corporate help desk when using a business application. The productivity and service-level implications are significant.

Finally, groupware technology allows teams of individuals to collaborate on projects and tasks regardless of their physical location. Examples of groupware technologies include Lotus's *SameTime* and Microsoft's *NetMeeting.* Using such groupware allows multiple individuals to participate in meetings and share software tools across a network. As with e-mail and instant messaging, groupware capabilities can be built into appropriate business applications.

Clearly, systems analysts and system designers must build these innovative collaborative technologies into their applications.

> Enterprise Applications

Virtually all organizations, large and small, require a core set of enterprise applications to conduct business. As shown in Figure 1-9, for most businesses the core applications include financial management, human resource management, marketing and sales, and operations management (inventory and/or manufacturing control). At one time, most organizations custom-built most or all of these core enterprise applications. But today, these enterprise applications are frequently purchased, installed, and configured for the business and integrated into the organization's business processes. Why? Because these core enterprise applications in different organizations or industries tend to be more alike than they are different.

Today, these "internal" core applications are being supplemented with other enterprise applications that integrate an organization's business processes with those of its suppliers and customers. These applications, called *customer relationship management* and *supply chain management,* are also illustrated in Figure 1-9.

The trend toward the use of purchased enterprise applications significantly impacts systems analysis and design. Purchased and installed enterprise applications are never sufficient to meet all of the needs for information systems in any organization. Thus, systems analysts and other developers are asked to develop value-added applications to meet additional needs of the business. But the purchased and installed enterprise applications become a technology constraint. Any custom application must properly integrate with and interface to the purchased enterprise applications. This is often called **systems integration,** and this is the business and systems environment into which most of you will graduate. Let's briefly explore some of the more common enterprise applications and describe their implications for systems analysis and design.

Enterprise Resource Planning (ERP)   As previously noted, the core business information system applications in most businesses were developed in-house incrementally over many years. Each system had its own files and databases with loose and awkward integration of all applications. During the 1990s, businesses tried very hard to integrate these legacy information systems, usually with poor results. Organizations would have probably preferred to redevelop these core business applications (see Figure 1-9 again) from scratch as a single *integrated* information system. Unfortunately, few if any businesses had enough resources to attempt this. Recognizing that the basic applications needed by most businesses were more similar than different, the software industry developed a solution—**enterprise resource planning (ERP)**

---

systems integration the process of building a unified information system out of diverse components of purchased software, custom-built software, hardware, and networking.

---

enterprise resource planning (ERP) a software application that fully integrates information systems that span most or all of the basic, core business functions (including transaction processing and management information for those business functions).

REPRESENTATIVE ERP VENDORS

SSA

Oracle/PeopleSoft

SAP AG (the Market Leader)

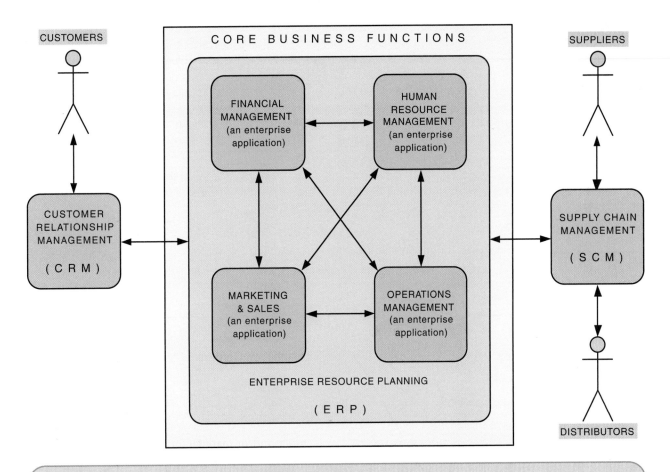

FIGURE 1-9   Enterprise Applications

applications. An ERP solution is built around a common database shared by common business functions. Examples of ERP software vendors are listed in the margin.

An ERP solution provides the core information system functions for the entire business. But usually an organization must redesign its business processes to fully exploit and use an ERP solution. Most organizations must still supplement the ERP solution with custom software applications to fulfill business requirements that are unique to the industry or business. For most organizations, an ERP implementation and integration represents the single largest information system project ever undertaken by the organization. It can cost tens of millions of dollars and require a small army of managers, users, analysts, technical specialists, programmers, and consultants.

ERP applications are significant to systems analysts for several reasons. First, systems analysts may be involved in the decision to select and purchase an ERP solution. Second, and more common, systems analysts are frequently involved in the customization of the ERP solution, as well as the redesign of business processes to use the ERP solution. Third, if custom-built applications are to be developed within an organization that uses an ERP core solution, the ERP system's architecture significantly impacts the analysis and design of the custom application that must coexist and interoperate with the ERP system.

Supply Chain Management   Today, many organizations are expending effort on enterprise applications that extend support beyond their core business functions. Companies are extending their core business applications to interoperate with their suppliers and distributors to more efficiently manage the flow of raw materials and products between their respective organizations. These **supply chain management (SCM)** applications utilize the Internet as a means for integration and communications.

REPRESENTATIVE SCM VENDORS

i2 Technologies
Manugistics
SAP
SCT

supply chain management (SCM) a software application that optimizes business processes for raw material procurement through finished product distribution by directly integrating the logistical information systems of organizations with those of their suppliers and distributors.

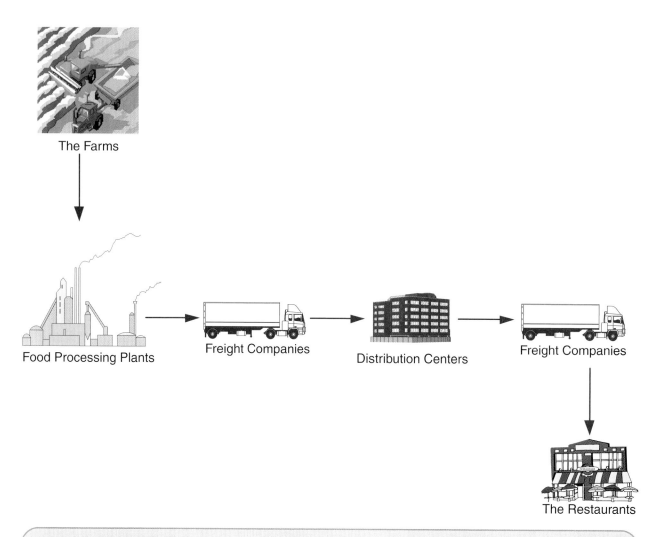

The Farms

Food Processing Plants

Freight Companies

Distribution Centers

Freight Companies

The Restaurants

FIGURE 1-10 Supply Chain

REPRESENTATIVE CRM VENDORS

BroadVision

E.piphany

Kana

Amdocs

Oracle/PeopleSoft

Siebel (the Market Leader)

SAP

customer relationship management (CRM) a software application that provides customers with access to a business's processes from initial inquiry through postsale service and support.

For example, Figure 1-10 demonstrates a logical supply chain ending at restaurants belonging to a franchise (e.g., Outback, Red Lobster, Wendy's). Notice that this supply chain includes many businesses and carriers to achieve its final result—ensuring that the restaurants have adequate food supplies to do business. Any delays or problems in any single link of this supply chain will adversely affect one and all. For that reason, several of these businesses will implement supply chain management using SCM software technology to plan, implement, and manage the chain. Examples of supply chain management vendors are listed in the margin. (It should be noted that several ERP application vendors are extending ERP software applications to include SCM capabilities. The SCM market is due for a shakeout because there are too many vendors for all to succeed.)

SCM applications are significant to systems analysts for the same reasons as stated for ERP applications. As an analyst, you may be involved in the evaluation and selection of an SCM package. Or you may be expected to implement and perhaps customize such packages to meet the organization's needs. And again, you may expect to participate in redesigning existing business processes to work appropriately with the SCM solution.

Customer Relationship Management   Many companies have discovered that highly focused customer relationship management can create loyalty that results in increased sales. Thus, many businesses are implementing **customer relationship management (CRM)** solutions that enable customer self-service via the Internet.

The theme of all CRM solutions is a focus on the "customer." CRM is concerned with not only providing effective customer inquiry responses and assistance but also helping the business better profile its customer base for the purpose of improving customer relations and marketing. Examples of CRM vendors are listed in the margin. As was the case with SCM technologies, many ERP vendors are developing or acquiring CRM capabilities to complement and extend their ERP solutions. And as with SCM, the larger number of players will likely be reduced through acquisition and attrition.

CRM technology impacts systems analysts in precisely the same ways as those we described for ERP and SCM technology. In many businesses, new applications must interface with a core, CRM enterprise application.

Enterprise Application Integration   Many companies face the significant challenge of integrating their existing legacy systems with new applications such as ERP, SCM, and CRM solutions. Any company that wants to do business across the Internet will also have to meet the challenge of integrating its systems with those of other organizations and their different systems and technologies. To meet this challenge, many organizations are looking at enterprise application integration software. **Enterprise application integration (EAI)** involves linking applications, whether purchased or developed in-house, so that they can transparently interoperate with one another. This is illustrated conceptually in Figure 1-11. Some vendors offering EAI tools are listed in the margin.

> **REPRESENTATIVE EAI VENDORS**
>
> BEA Systems
> IBM (MQSeries)
> Mercator Software
> TIBCO Software

enterprise application integration (EAI) the process and technologies used to link applications to support the flow of data and information between those applications. EAI solutions are usually based on **middleware.**

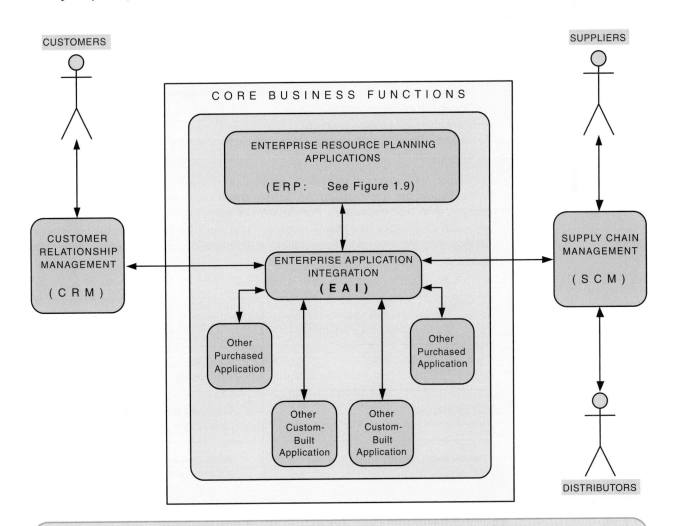

FIGURE 1-11 Enterprise Application Integration

middleware software (usually purchased) used to translate and route data between different applications.

Again, this market is rapidly expanding and contracting. The tools are used to define and construct communication pipelines between differing applications and technologies.

Today, as any new information system is developed, it must be integrated with all the information systems that preceded it. These "legacy" information systems may have been purchased or built in-house. Regardless, systems analysts and other developers must consider application integration for any new information system to be developed. And EAI technologies are at the core of the integration requirements.

A Simple System Development Process

Thus far you have learned about different types of information systems, the players involved in developing those systems, and several business and technology drivers that influence the development of information systems. In this section you will learn about another information system perspective, the "process" for developing an information system.

system development process a set of activities, methods, best practices, deliverables, and automated tools that stakeholders use to develop and maintain information systems and software.

Most organizations have a formal **system development process** consisting of a standard set of processes or steps they expect will be followed on any system development project. While these processes may vary greatly for different organizations, a common characteristic can be found: Most organizations' system development process follows a problem-solving approach. That approach typically incorporates the following general problem-solving steps:

1. Identify the problem.
2. Analyze and understand the problem.
3. Identify solution requirements and expectations.
4. Identify alternative solutions and choose the best course of action.
5. Design the chosen solution.
6. Implement the chosen solution.
7. Evaluate the results. (If the problem is not solved, return to step 1 or 2 as appropriate.)

Figure 1-12 adds a system development process perspective that we will use (with appropriate refinements) throughout this book as we study the development process, tools, and techniques. For the sake of simplicity our initial problem-solving approach establishes four stages or phases that must be completed for any system development project—system initiation, system analysis, system design, and system

Our Simplified System Development Process	General Problem-Solving Steps
System initiation	1. Identify the problem. (Also plan for the solution of the problem.)
System analysis	2. Analyze and understand the problem. 3. Identify solution requirements and expectations.
System design	4. Identify alternative solutions and choose the best course of action. 5. Design the chosen solution.
System implementation	6. Implement the chosen solution. 7. Evaluate the results. (If the problem is not solved, return to step 1 or 2 as appropriate.)

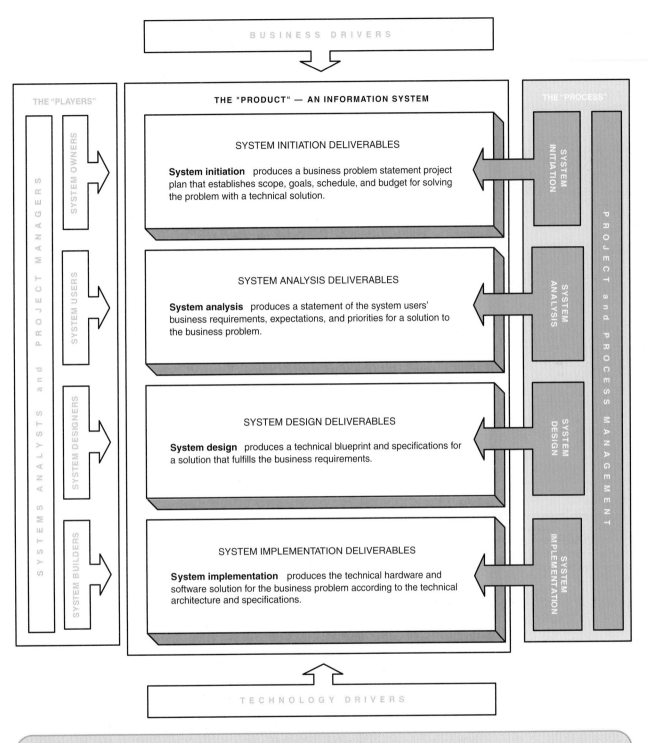

FIGURE 1-12 Systems Development and Problem Solving

implementation. The table on the previous page shows the correlation between the above general problem-solving steps and our process.

It is important to note that any system development process must be managed on a project-by-project basis. You learned earlier that at least one stakeholder accepts responsibility as the project manager for ensuring that the system is developed on time, within budget, and with acceptable quality. The activity of managing a project is referred to as **project management.** Accordingly, in Figure 1-12 we have added a process for project management. Also, to ensure that all projects are

project management the activity of defining, planning, directing, monitoring, and controlling a project to develop an acceptable system within the allotted time and budget.

process management
the ongoing activity that
defines, improves, and
coordinates the use of an
organization's chosen
methodology (the "process")
and standards for all system
development projects.

managed according to the same development process, we have included **process management** as an ongoing activity. Notice that project and process management overlap all of the process phases.

Let's *briefly* examine our system development process in Figure 1-12 to expand your understanding of each phase and activity in the process. Given a problem to be solved or a need to be fulfilled, what will we do during system initiation, analysis, design, and implementation? Also, who will be involved in each phase?

> System Initiation

system initiation the initial
planning for a project to define
initial business scope, goals,
schedule, and budget.

Information system projects are usually complicated. They require a significant time, effort, and economic investment. The problems to be solved are frequently stated vaguely, which means that the initial envisioned solution may be premature. For these reasons, system projects should be carefully planned. System initiation establishes project scope and the problem-solving plan. Thus, as shown in Figure 1-12, we see that **system initiation** establishes the project scope, goals, schedule, and budget required to solve the problem or opportunity represented by the project. Project scope defines the area of the business to be addressed by the project and the goals to be achieved. Scope and goals ultimately impact the resource commitments, namely, schedule and budget, that must be made to successfully complete the project. By establishing a project schedule and budget against the *initial* scope and goals, you also establish a *baseline* against which all stakeholders can accept the reality that any future changes in scope or goals *will* impact the schedule and budget.

Figure 1-12 also shows that project managers, system analysts, and system owners are the primary stakeholders in a system analysis. This book will teach you many tools and techniques for initiating a system project and establishing a suitable project plan.

> System Analysis

system analysis the study
of a business problem domain
to recommend improvements
and specify the business
requirements and priorities
for the solution.

The next step in our system development process is **system analysis.** System analysis is intended to provide the project team with a more thorough understanding of the problems and needs that triggered the project. As such, the business area (scope of the project—as defined during system initiation) may be studied and analyzed to gain a more detailed understanding of what works, what doesn't, and what's needed. As depicted in Figure 1-12, the system analysis requires working with system users to clearly define business requirements and expectations for any new system that is to be purchased or developed. Also, business priorities may need to be established in the event that schedule and budget are insufficient to accomplish all that is desired.

Recall the business drivers discussed earlier in the chapter. These (and future) business drivers most closely affect system analysis, which often defines business requirements in response to the business drivers. For example, we discussed a current trend toward e-business and e-commerce. This business driver may influence the business requirement for any information system, leading us to establish project goals to conduct all business transactions on the Web.

The completion of a system analysis often results in the need to update many of the deliverables produced earlier, during system initiation. The analysis may reveal the need to revise the business scope or project goals—perhaps we now feel the scope of the project is too large or too small. Accordingly, the schedule and budget for the project may need to be revised. Finally, the feasibility of the project itself becomes questionable. The project could be canceled or could proceed to the next phase.

As shown in Figure 1-12, project managers, system analysts, and system users are the primary stakeholders in a system analysis. Typically, results must be summarized and defended to the system owners, who will pay to design and implement an information system to fulfill the business requirements. This book will teach you many tools and techniques for performing a system analysis and documenting user requirements.

> System Design

Given an understanding of the business requirements for an information system, we can now proceed to **system design.** During system design we will initially need to explore alternative technical solutions. Rarely is there only one solution to any problem. For example, today most companies need to choose between purchasing a solution that is good enough and building a custom solution in-house. (We'll explore options such as this throughout this book.)

Once a technical alternative is chosen and approved, the system design phase develops the technical blueprints and specifications required to implement the final solution. These blueprints and specifications will be used to implement required databases, programs, user interfaces, and networks for the information system. In the case where we choose to purchase software instead of build it, the blueprints specify how the purchased software will be integrated into the business and with other information systems.

Recall the technology drivers discussed in the last section of the chapter. These (and future) technology drivers most closely impact the system design process and decisions. Many organizations define a common information technology architecture based on these technology drivers. Accordingly, all system designs for new information systems must conform to the standard IT architecture.

As depicted in Figure 1-12, project managers, system analysts, and system designers are the primary stakeholders in a system design. This book will teach you many tools and techniques for performing a system design.

system design the specification or construction of a technical, computer-based solution for the business requirements identified in a system analysis. (Note: Increasingly, the design takes the form of a working prototype.)

> System Implementation

The final step in our simple system development process is **system implementation.** As shown in Figure 1-12, system implementation constructs the new information system and puts it into operation. It is during system implementation that any new hardware and system software are installed and tested. Any purchased application software and databases are installed and configured. And any custom software and databases are constructed using the technical blueprints and specifications developed during system design.

As system components are constructed or installed, they must be individually tested. And the complete system must also be tested to ensure that it works properly and meets user requirements and expectations. Once the system has been fully tested, it must be placed into operation. Data from the previous system may have to be converted or entered into start-up databases, and system users must be trained to properly use the system. Finally, some sort of transition plan from older business processes and information systems may have to be implemented.

And once again, as depicted in Figure 1-12, project managers, system analysts, and system builders are the primary stakeholders in a system implementation. While this book will teach you some of the tools and techniques for performing a system implementation, most of these methods are taught in programming, database, and networking courses. This book emphasizes system initiation, analysis, and design skills, but it will also teach you the unique system implementation tools and techniques that are most commonly performed by systems analysts and, therefore, are not typically covered in these other information technology courses.

system implementation the construction, installation, testing, and delivery of a system into production (meaning day-to-day operation).

> System Support and Continuous Improvement

We would be remiss not to briefly acknowledge that implemented information systems face a lifetime of support and continuous improvement. But where is that shown in Figure 1-12? It is there! But it is subtle.

Implemented information systems are rarely perfect. Your users will find errors (bugs) and you will discover, on occasion, design and implementation flaws that require attention and fixes. Also, business and user requirements constantly change. Thus, there will be a need to continuously improve any information system until the time it becomes obsolete. So where does system support and change fit into our development process?

A change made for system support or improvement is merely another project, sometimes called a *maintenance* or *enhancement* project. Such a project should follow the exact same problem-solving approach defined for any other project. The only difference is the effort and budget required to complete the project. Many of the phases will be completed much more quickly, especially if the original stakeholders properly documented the system as initially developed. Of course, if they did not, a system improvement project can quickly consume much greater time, effort, and money. Much of what we will teach you in this book is intended to help you appropriately document information systems to significantly reduce lifetime costs of supporting and improving your information systems.

Learning Roadmap

Each chapter will provide guidance for self-paced instruction under the heading "Learning Roadmap." Recognizing that different students and readers have different backgrounds and interests, we will propose appropriate learning paths—most within this book, but some beyond the scope of this book.

Most readers should proceed directly to Chapter 2 because the first four chapters provide much of the context for the remainder of the book. Several recurring themes, frameworks, and terms are introduced in those chapters to allow you to define your own learning path from that point forward. This chapter focused on information systems from four different perspectives:

- The *players*—both developers and users of information systems.
- The *business drivers* that currently influence information systems.
- The *technology drivers* that currently influence information systems.
- The *process* of developing information systems.

Chapter 2 will take a closer look at the *product* itself—information systems—from an architectural perspective appropriate for systems development. We will define how different players and development stages view an information system.

Looking further ahead, Chapter 3 more closely examines the *process* of systems development. Chapter 4 completes the introduction to systems analysis and design methods by examining the *management* of systems development.

Summary

1. Information systems in organizations capture and manage data to produce useful information that supports an organization and its employees, customers, suppliers, and partners.
2. Information systems can be classified according to the functions they serve, including:

   a. Transaction processing systems that process business transactions such as orders, time cards, payments, and reservations.
   b. Management information systems that use transaction data to produce information needed by managers to run the business.

c. Decision support systems that help various decision makers identify and choose between options or decisions.

d. Executive information systems that are systems tailored to the unique information needs of executives who plan for the business and assess performance against the plans.

e. Expert systems that are systems that capture and reproduce the knowledge of an expert problem solver or decision maker and then simulate the "thinking" of that expert.

f. Communication and collaboration systems that enhance communication and collaboration between people, both internal and external to the organization.

g. Office automation systems that help employees create and share documents that support day-to-day office activities.

3. Information systems can be viewed from various perspectives, including from the perspective of the "players," the "business drivers" influencing the information system, the "technology drivers" used by the information system, and the "process" used to develop the information system.

4. Information workers are the stakeholders in information systems. Information workers include those people whose jobs involve the creation, collection, processing, distribution, and use of information. They include:

a. System owners, the sponsors and chief advocates of information systems.

b. System users, the people who use or are impacted by the information system on a regular basis. Geographically, system users may be internal or external.

c. System designers, technology specialists who translate system users' business requirements and constraints into technical solutions.

d. System builders, technology specialists who construct the information system based on the design specifications.

e. Systems analysts, who facilitate the development of information systems and computer applications. They coordinate the efforts of the owners, users, designers, and builders. Frequently, they may play one of those roles as well. Systems analysts perform systems analysis and design.

5. In addition to having formal systems analysis and design skills, a systems analyst must develop or

possess the following skills, knowledge, and traits:

a. Working knowledge of information technologies.

b. Computer programming experience and expertise.

c. General knowledge of business processes and terminology.

d. General problem-solving skills.

e. Good interpersonal communication skills.

f. Good interpersonal relations skills.

g. Flexibility and adaptability.

h. Character and ethics.

6. Any stakeholder role may be filled by an internal or external worker referred to as an external service provider (ESP). Most ESPs are systems analysts, designers, or builders who are contracted to bring special expertise or experience to a specific project.

7. Most information systems projects involve working as a team. Usually one or more of the stakeholders (team members) takes on the role of project manager to ensure that the system is developed on time, within budget, and with acceptable quality. Most project managers are experienced systems analysts.

8. Business drivers influence information systems. Current business drivers that will continue to influence the development of information systems include:

a. Globalization of the economy.

b. Electronic commerce and business.

c. Security and privacy.

d. Collaboration and partnership.

e. Knowledge asset management.

f. Continuous improvement and total quality management.

g. Business process redesign.

9. Information technology can be a driver of information systems. Outdated technologies can present problems that drive the need to develop new systems. Newer technologies such as the following are influencing today's information systems:

a. Networks and the Internet:

i) *xHTML* and *XML* are the fundamental languages of Web page authoring and Internet application development. Extensible Hypertext Markup Language (xHTML) is the emerging second-generation version of HTML, the language used to construct Web pages. Extensible Markup Language (XML) is the language used to effectively transport data content along with its proper interpretation over the Internet.

ii) *Scripting* languages are simple programming languages designed specifically for Internet applications.

iii) Web-specific programming languages such as *Java* and *Cold Fusion* have emerged to specifically address construction of complex, Web-based applications that involve multiple servers and Web browsers.

iv) *Intranets* are essentially private Internets designed for use by employees of an organization. They offer the look and feel of the Internet; however, security and firewalls restrict their use to employees.

v) *Extranets,* like intranets, are private Internets. But extranets are for use between specific organizations. Only the employees of those identified businesses can access and use the extranet.

vi) *Portals* (in corporations) are "home pages" that can be customized to the specific needs of different individuals who use them. For example, portal technology can define Web pages that provide appropriate information and applications for different roles in the same company. Each individual's role determines which information and applications that person can use from her or his Web page.

vii) Web services are reusable, Web-based programs that can be called from any other Internet program.

b. Mobile and wireless technologies—Increasingly, wireless access must be assumed. And the limitations of mobile devices and screen sizes must be accommodated in an information system's design. All of the following technical trends will significantly impact the analysis and design of new information systems:

i) Handheld computers, or *personal data assistants* (such as the HP *iPaq,* Palm, and RIM *BlackBerry*) have become common in the ranks of information workers. These devices are increasingly including wireless capabilities that provide Web access and e-mail

ii) *Cell phones* are also increasingly featuring Internet and e-mail capabilities.

iii) Integrated devices such as *smart phones* are emerging that integrate the capabilities

of PDAs and cell phones into a single device.

iv) Technologies like *Bluetooth* are emerging to allow separate devices to interoperate as one logical device while preserving each one's form factors and advantages.

c. Object technologies—Most contemporary information systems are built using object technologies. Object technologies allow programmers to build software from software parts called objects. Object-oriented software offers the advantage of reusability and extensibility.

d. Collaborative technologies—Collaborative technologies are those that enhance interpersonal communications and teamwork. Four important classes of collaborative technologies are e-mail, instant messaging, groupware, and work flow.

e. Enterprise applications—Virtually all organizations, large and small, require a core set of enterprise applications to conduct business. For most businesses the core applications include financial management, human resource management, marketing and sales, and operations management (inventory and/or manufacturing control). At one time, most organizations custom-built most or all of these core enterprise applications. But today, these enterprise applications are frequently purchased, installed, and configured for the business and integrated into the organization's business processes. These "internal" core applications are being supplemented with other enterprise applications that integrate an organization's business processes with those of its suppliers and customers. These applications are called customer relationship management (CRM) and supply chain management (SCM). Enterprise application integration (EAI) involves linking applications, whether purchased or developed in-house, so that they can transparently interoperate with one another.

10. Many organizations have a formal systems development process consisting of a standard set of processes or steps they expect will be followed on any systems development project. Systems development processes tend to mirror general problem-solving approaches. This chapter presented a simplified system development process that is composed of the following phases:

a. System initiation—the initial planning for a project to define initial business scope, goals, schedule, and budget.

b. System analysis—the study of a business process domain to recommend improvements and specify the business requirements and priorities for the solution.

c. System design—the specification or construction of a technical, computer-based solution for the business requirements identified in system analysis.

d. System implementation—the construction, installation, testing, and delivery of a system into operation.

11. Information systems face a lifetime of support and continuous improvement. A change made for system support or improvement is merely another project, sometimes called a maintenance or enhancement project. These projects follow the exact same problem-solving approach defined

for any other project, but they require less effort and budget.

12. Sequential development requires that each development process (phase) be completed—one after the other. This approach is referred to as the waterfall approach. An alternative development approach is iterative (or incremental) development. This approach requires completing enough analysis, design, and implementation as is necessary to fully develop a part of the new system. Once that version of the system is implemented, the strategy is to then perform some additional analysis, design, and implementation in order to release the next version of the system. These iterations continue until all parts of the entire information system have been developed.

 ## Review Questions

1. Why are information systems (IS) essential in organizations?
2. Why do systems analysts need to know who the stakeholders are in the organization?
3. Who are the typical stakeholders in an information system? What are their roles?
4. Please explain what the consequences are if an information system lacks a system owner.
5. What are the differences between internal users and external users? Give examples.
6. What are the differences between the role of system analysts and the role of the rest of the stakeholders?
7. What kind of knowledge and skills should a system analyst possess?
8. In addition to the business and computing knowledge that system analysts should possess, what

are the other essential skills that they need to effectively complete their jobs?
9. Why are good interpersonal communication skills essential for system analysts?
10. What are some of the business drivers for today's information systems?
11. What are the differences between electronic commerce (e-commerce) and electronic business (e-business)?
12. What are the differences between information and knowledge?
13. What are the most important technology drivers for today's information systems?
14. What are the four steps in a system development process? What happens in each step?
15. Why is system initiation essential in the system development process?

 ## Problems and Exercises

1. Assume you are a systems analyst who will be conducting a requirements analysis for an individually owned brick-and-mortar retail store with a point-of-sale system. Identify who the typical internal and external users might include.
2. Assume you are a systems analyst for a consulting company and have been asked to assist the chief executive officer (CEO) of a regional bank. The bank recently implemented a plan to reduce the

number of staff, including loan officers, as a strategy to maintain profitability. Subsequently, the bank has experienced chronic problems with backlogged loan requests because of the limited number of loan officers who are able to review and approve or disapprove loans. The CEO of the bank is interested in solutions that would allow the approval process to move faster without increasing the number of loan officers, and has engaged your company to come up with suggestions. What is one type of system that you might recommend to the bank?

3. How do communication and collaboration systems improve efficiency and effectiveness? What are some of the communication and collaboration systems that are being used by an increasing number of organizations?

4. Identify the type of information system that clerical workers in an organization would typically use and why.

5. As information systems increase in complexity and comprehensiveness, ethical issues regarding accessing and using data from these systems are also increasing. What are some of these ethical issues?

6. What are business to consumer (B2C) and business to business (B2B) Web applications, and what are some examples of each type?

7. While system development processes and methodologies can vary greatly, identify and briefly explain the "generic" phases of the system development process that are described in the textbook and which must be completed for any project. You are a contractor with a systems integration company.

8. Your company has a contract with a local firm to link all of their systems so they can transparently work together. Their applications include a number of existing legacy systems, which were built at different times by different developers using a variety of languages and platforms, as well as several newer contemporary applications. What is the term for this type of linking? What type of tool would you most likely use, and what are some examples of these tools?

9. Your company has asked you to develop a new Web-based system to replace its existing legacy system. There will be very little change in business requirements and functionality from the existing legacy system. Suggest which system development process you might use and why.

10. You recently joined a retail sales company which has recently bought out and assimilated a commercial industrial supply house. You have been asked to lead a project to develop a consolidated inventory-tracking system. Suggest which system development process you might use and why.

11. Your company president sits down beside you just before a meeting is to begin and tells you that people keep saying the customer needs to install a CRM, but doesn't really know what it is. The company president then asks you to explain it in nontechnical terms in the next 30 seconds.

12. Industry studies indicate that mobile and wireless technology has become one of the major technology drivers for designing new information systems. Why is this the case and what is the impact?

13. Briefly explain the impact of Web services on Web development. Give some examples of Web services.

14. Identify in which phase of the development process the following activities belong:

   a. Development of the technical blueprint or design document.
   b. Project scheduling.
   c. Integration testing.
   d. Interviewing system users to define business requirements.

15. What are the two most important advantages of object-oriented software technologies over structured software technologies?

Projects and Research

1. Research the average and/or median salaries for IT professionals. You can use a variety of methods to find this information, such as searching the Web for online sites that publish the results of salary surveys for IT professionals. You can also look at classified ads in newspapers, trade magazines, and/or online.

   a. Is there a significant difference between typical salaries for system analysts, designers and developers?

   b. Roughly, what is the difference in the typical salaries for these different groups?
   c. What do you think are the reasons for the difference?
   d. Is there a gender gap in the salaries of IT professionals? Discuss any trends that you found, and the implications.

2. Contact the chief information officers (CIOs) or top IT managers of several local or regional organizations. Ask them about the process or

methodology they use for system development in their organizations, and why they utilize that particular approach.

   a. Describe and compare the different approaches that you have found.

   b. Which approach do you believe to be the most effective approach?

   c. Why?

3. Career choices and personal skills:

   a. At this point in your education, if you had to choose between becoming a systems analyst, systems designer, or systems builder, which one would you choose?

   b. Why?

   c. Now divide a piece of paper into two columns. On one side, list the personal skills and traits you think are most important for each of these three groups of systems analysts, designers, and builders. In the second column, list at least five skills and traits that you feel to be your strongest ones, then map them to the skills and traits you listed for each of the three groups. With which group do you have the most skills and traits in common?

   d. Is this group the same one as the one you would choose in Question 3a? Why do you think this is (or is not) the case?

4. Your school library should have journals and periodicals dating back at least several decades, or may subscribe to online research services which do. Look at several recent articles in information technology journals regarding systems analysis, as well as several articles from at least 25 years ago.

   a. Compare the recent articles to the older ones. Do there appear to be significant differences in the typical knowledge, skills, abilities, and/or experience that systems analysts needed 25 years ago compared to now?

   b. If you found some differences, what are the ones that you consider most important?

   c. What do you think are some of the reasons for these changes?

   d. Now get out your crystal ball and look into the future 25 years from now. What differences do you predict between the systems analysts of today and those in 25 years?

5. Search the Web for several articles and information on ethical issues related to information technology professionals.

   a. What articles did you find?

   b. Based on your research, which ethical issues do you think systems analysts might face during their careers?

   c. Pick a particular ethical issue and explain the steps you would take if you were confronted with this issue.

   d. What would you do if you found your best friend and co-worker had committed a serious ethical violation? Facts to consider: The violation may never be discovered, but it will cost your company many thousands of dollars in higher costs over the next several years. Your company has a stringent policy of firing employees who commit serious ethical violations. Make sure to explain your reasons for the action(s) you would take, if any.

6. Search the Web or business periodicals in your library such as Forbes Magazine for information on three or four chief information officers of large companies or organizations.

   a. Which industry sector, companies, and CIOs did you find?

   b. For each CIO that you researched, what was their predominant experience prior to becoming a CIO; that is, did they have an information technology background, a business background, or both?

   c. For each CIO, what is their level of education?

   d. How many years has each been a CIO, and for approximately how many different companies has each one worked?

   e. Based upon your research, what knowledge and skills does a CIO need in order to be successful? Why?

Minicases

1. What do you think will be possible technologically 10 years from now? How about 20 or 30 years from now? Research a new and interesting technology that is in the research and development stage. Prepare a presentation using a movie clip and PowerPoint on this technology and present it to the class. Submit a short paper on the impacts this new technology might have on society and/or businesses.

2. Consider outsourcing: It is many times the case that at least part of the development process is outsourced. In fact, project leaders today must be

capable of handling geographically diverse teams as well as timeline and resource constraints. Outsourcing brings to the table increased efficiency and economic gains to the societies that are interacting. However, these gains are not quickly realized, and the negative impacts on a society that is outsourcing can be significant from a jobs perspective. Dr. Mankiw, as an economic advisor to President Bush, publicly touted the benefits of outsourcing and was deeply criticized for his stance. Do you think that it is good or bad for a business to outsource work? Do you think there are ethical dilemmas for companies who outsource? Find at least two articles on the impacts of outsourcing, and bring them to share with the class.

3. You are a network administrator, and as part of your job, you monitor employee e-mails. You discover that your boss is cutting corners on a system that your company is developing in order to finish the project more quickly and to stay under budget. There is a flaw in the system as a result, and this flaw will cause a network crash if more than 20 people are on the network at a time. The client expects approximately 12 people on the network at any given time. You are sure, as apparently your boss is, that the customer will not find out until well after the project is accepted (if ever). What do you do?

4. A systems analyst must be both technically proficient and capable of successful customer communication. Developing a good system requires a complete understanding of user requirements. Many times, users don't know what is available (technologically) or even what they would like from a system. What are characteristics of good communication?

Suggested Readings

Ambler, Scott. *Agile Modeling: Effective Practices for eXtreme Programming and the Unified Process.* New York: John Wiley & Sons, 2002. This book has significantly shaped our thinking about the software development process. Those of you who are critical of the "extreme-programming" movement need not fear that our enthusiasm for this suggested reading indicates an endorsement of extreme programming. We simply like the sanity that Scott brings to the process of systems and software development through the use of flexible methods within the context of an iterative process. We will reference this book in several chapters.

Ernest, Kallman; John Grillo; and James Linderman. *Ethical Decision Making and Information Technology: An Introduction with Cases,* 2nd ed. Burr Ridge, IL: McGraw-Hill/Irwin, 1995. This is an excellent textbook for teaching ethics in an MIS curriculum. It is a collection of case studies that can complement a systems analysis and design course.

Gartner Group IT Symposium and Expo (annual). Our university's management information unit has long subscribed to the Gartner Group's service that reports on industry trends, the probabilities for success of trends and technologies, and suggested strategies for information technology transfer. Gartner research has played a significant, ongoing role in helping us to chart business and technology drivers as described in this chapter. We have also been fortunate to be able to attend Gartner's annual IT symposium. Gartner Group reports and symposiums have influenced each edition of the book. For more information about the Gartner Group, see www.gartner.com.

Gause, Donald, and Gerald Weinberg. *Are Your Lights On? How to Figure Out What the Problem REALLY Is.* New York: Dorset House Publishing, 1990. Yes, this is not a recent book, but neither are the fundamentals of problem solving. Here's a short and easy-to-read book about general problem solving. You can probably read the entire book in one night, and it could profoundly improve your problem-solving potential as a systems analyst (or, for that matter, any other profession).

Levine, Martin. *Effective Problem Solving,* 2nd ed. Englewood Cliffs, NJ: Prentice Hall, 1994. This is another older book, but as we stated before, problem-solving methods are timeless. At only 146 pages, this title can serve as an excellent professional reference.

Weinberg, Gerald. *Rethinking Systems Analysis and Design.* New York: Dorset House Publishing, 1988. Don't let the date fool you. This is one of the best and most important books on this subject ever written. This book may not teach any of the popular systems analysis and design methods of our day, but it challenges the reader to leap beyond those methods to consider something far more important—the people side of systems work. Dr. Weinberg's theories and concepts are presented in the context of dozens of delightful fables and short experiential stories. We are grateful to him for our favorite systems analysis fable of all time, "The Three Ostriches."

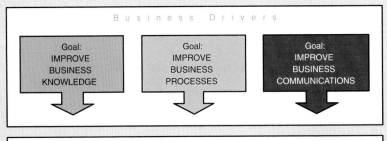
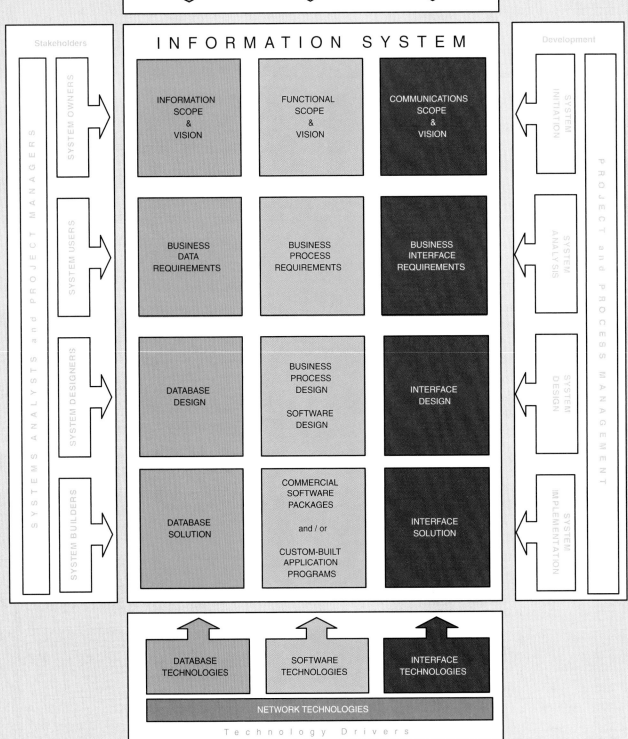

2

The Components of Information Systems

Chapter Preview and Objectives

Systems analysis and design methods are used to develop information systems for organizations. Before learning the *process* of building systems, you need a clear understanding of the *product* you are trying to build. This chapter takes an architectural look at information systems and applications. We will build a framework for information systems architecture that will subsequently be used to organize and relate all of the chapters in this book. The chapter will address the following areas:

▌ Differentiate between *front-* and *back-office* information systems.

▌ Describe the different classes of information system applications (*transaction processing, management information, decision support, expert, communication and collaboration,* and *office automation systems*) and how they interoperate to supplement one another.

▌ Describe the role of information systems architecture in systems development.

▌ Identify three high-level goals that provide system owners and system users with a perspective of an information system.

▌ Name three goal-oriented perspectives for any information systems.

▌ Identify three technologies that provide system designers and builders with a perspective of an information system.

▌ Describe four building blocks of the KNOWLEDGE goal for an information system.

▌ Describe four building blocks of the PROCESS goal for an information system.

▌ Describe four building blocks of the COMMUNICATIONS goal for an information system.

▌ Describe the role of network technologies as it relates to KNOWLEDGE, PROCESSES, and COMMUNICATIONS building blocks.

Introduction

The SoundStage member services system project is getting underway. Bob Martinez has been assigned the task of conducting initial meetings with groups of system users to gain their perspective on the system and what it must accomplish. He quickly discovered that everyone had a different perspective and expressed that perspective in a different language. In college, majoring in computer information systems technology, Bob tended to think about information systems in terms of programming languages, networking technologies, and databases. He found that the others didn't think in those terms. The system users talked about manual forms and how they were routed. They talked about policies and procedures and reports they needed. As he met with managers he heard them talk about strategic plans and how the system could give the organization a competitive edge.

It reminded Bob of the old story he had heard about three blind men who came upon an elephant. One felt the trunk and concluded that an elephant was like a snake. Another felt a leg and concluded that an elephant was like a tree. The third felt an ear and concluded that an elephant was like a fan. It didn't take Bob long to realize that the owners' and users' perspectives were just as valid as his. An information system is more than technology. It is mainly a tool that serves the goals of the organization.

The Product—Information Systems

front-office information system an information system that supports business functions that extend out to the organization's customers.

In Chapter 1 you were introduced to information systems from four different perspectives, including stakeholders, business drivers, technology drivers, and the process of systems development. As suggested by the *home page* (see p. 42), this chapter will more closely examine the information system "product."

Organizations are served not by a single information system but, instead, by a federation of information systems that support various business functions. This idea is illustrated in Figure 2-1. Notice that most businesses have both **front-office**

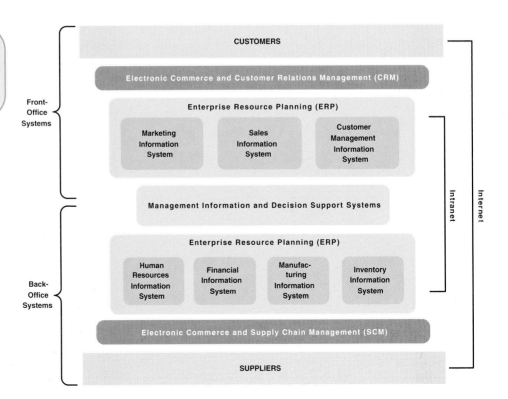

FIGURE 2-1

A Federation of Information Systems

information systems that support business functions that reach out to customers (or constituents) and **back-office information systems** that support internal business operations as well as interact with suppliers. These front- and back-office information systems feed data to management information systems and decision support systems that support management needs of the business. Contemporary information systems are interfacing with customers and suppliers using electronic commerce technology, customer relations management (CRM), and supply chain management (SCM) applications (see descriptions in Chapter 1) over the Internet. Finally, most companies have some sort of intranet (internal to the business) to support communications between employees and the information systems.

In Chapter 1 you learned that there are several classes of information system applications (see opposite page). Each class serves the needs of different types of users. In practice, these classes overlap such that it isn't always easy to differentiate one from another. The various applications should ideally interoperate to complement and supplement one another. Take a few moments to study Figure 2-2, which illustrates typical roles of information systems in an organization. The rounded rectangles represent various information systems. Notice that an organization can and will have multiple transaction-processing systems, office automation systems, and the like. The "drum" shapes represent stored data. Notice that an organization has multiple

> **back-office information system** an information system that supports internal business operations of an organization, as well as reaches out to suppliers.

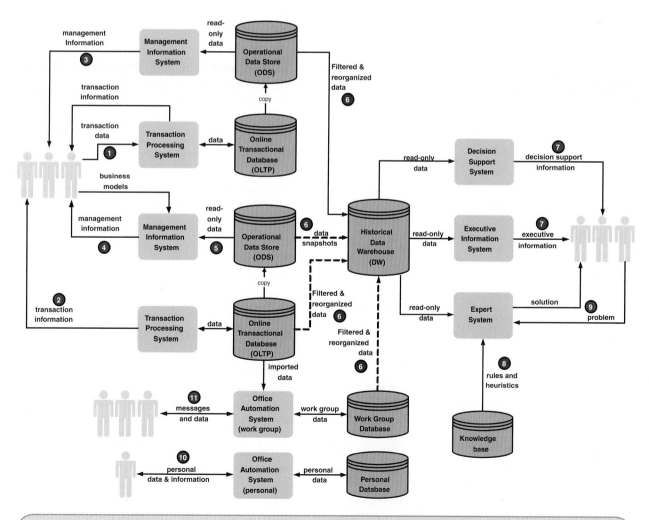

FIGURE 2-2 Information Systems Applications

sets of stored data, and only some of them work together. We call your attention to the following number annotations on the diagram:

❶ The first transaction processing system responds to an input transaction's data (e.g., an order). It produces transaction information to verify the correct processing of the input transaction.

❷ The second transaction processing system merely produces an output transaction (e.g., an invoice). Such a system may respond to something as simple as the passage of time (e.g., it is the end of the month; therefore, generate all invoices).

❸ The first management information system simply produces reports or information (e.g., sales analysis reports) using data stored in transactional databases (maintained by the aforementioned transaction processing systems).

❹ The second management information system uses business models (e.g., MRP) to produce operational management information (e.g., a production schedule).

❺ Notice that an MIS may use data from more than one transactional database.

❻ Notice that snapshots of data from the transactional databases populate a data warehouse. The snapshots may be taken at various time intervals, and different subsets of data may be included in various snapshots. The data in the warehouse will be organized to ensure easy access and inquiry by managers.

❼ Decision support and executive information systems applications will typically provide read-only access to the data warehouses to produce decision support and executive management information.

❽ An expert system requires a special database that stores the expertise in the form of rules and heuristics.

❾ An expert system either accepts problems as inputs (e.g., Should we grant credit to a specific customer?), or senses problems in the environment (e.g., Is the lathe producing parts within acceptable specifications?), and then responds to a problem with an appropriate solution based on the system's expertise.

❿ Personal office automation systems tend to revolve around the data and business processing needs of an individual. Such systems are typically developed by the users themselves (and run on personal computers).

⓫ Work group office automation systems are frequently message-based (e.g., e-mail-based) and are smaller-scale solutions to departmental needs. As shown in the figure, they can access or import data from larger, transaction processing systems.

In the average business, there will be many instances of each of these different applications.

A Framework for Information Systems Architecture

It has become fashionable to deal with the complexity of modern information systems by using the term *architecture.* Information technology professionals speak of data architectures, application architectures, network architectures, software architectures, and so forth. An **information systems architecture** serves as a higher-level framework for understanding different views of the fundamental building blocks of an information system. Essentially, information systems architecture provides a foundation for organizing the various components of any information system you care to develop.

Different stakeholders have different perspectives on or views of an information system. System owners and system users tend to focus on three common business goals of any information system. These goals are typically established in response to

information systems architecture a unifying framework into which various stakeholders with different perspectives can organize and view the fundamental building blocks of information systems.

one or more of the business drivers you read about in Chapter 1. These goal-oriented perspectives of an information system include:

- The goal to improve *business knowledge.* Knowledge is a product of information and data.
- The goal to improve *business processes* and services.
- The goal to improve *business communications* and people collaboration.

The role of the system designers and builders is more technical. As such, their focus tends to be placed more on the technologies that may be used by the information system in order to achieve the business goal. The system designers' and builders' perspectives of an information system tend to focus more on:

- The *database technologies* that support business accumulation and use of business knowledge.
- The *software technologies* that automate and support business processes and services.
- The *interface technologies* that support business communications and collaboration.

As shown in Figure 2-3, the intersection of these perspectives (rows and columns) defines *building blocks* for an information system. In the next section, we will describe all these information system building blocks.

NOTE: Throughout this book, we use a consistent color scheme for both the framework and the various tools that relate to, or document, the building blocks. The color scheme is based on the building blocks as follows:

represents something to do with KNOWLEDGE
represents something to do with PROCESSES
represents something to do with COMMUNICATIONS

The information system building blocks do not exist in isolation. They must be carefully synchronized to avoid inconsistencies and incompatibilities within the system. For example, a database designer (a *system designer*) and a programmer (a *system builder*) have their own architectural views of the system; however, these views must be compatible and consistent if the system is going to work properly. Synchronization occurs both horizontally (across any given row) and vertically (down any given column).

In the remainder of this chapter, we'll briefly examine each focus and perspective—the building blocks of information systems.

> KNOWLEDGE Building Blocks

Improving business knowledge is a fundamental goal of an information system. As you learned in Chapter 1, business knowledge is derived from data and information. Through processing, data is refined to produce information that results in knowledge. Knowledge is what enables a company to achieve its mission and vision.

The KNOWLEDGE column of your framework is illustrated in Figure 2-4. Notice at the bottom of the KNOWLEDGE column that our goal is to capture and store business data using DATABASE TECHNOLOGIES. Database technology (such as *Access, SQL Server, DB2,* or *Oracle*) will be used to organize and store data for all information systems. Also, as you look down the KNOWLEDGE column, each of our different stakeholders has different perspectives of the information system. Let's examine those views and discuss their relevance to the KNOWLEDGE column.

System Owners' View of KNOWLEDGE   The average system owner is not interested in raw data. The system owner is interested in information that adds new business knowledge. Business knowledge and information help managers make

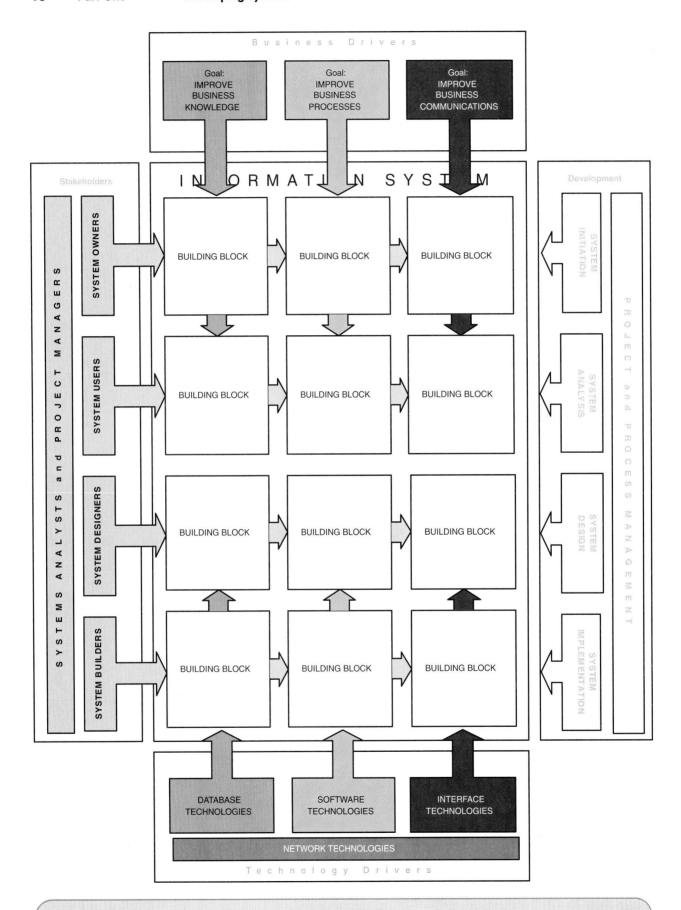

FIGURE 2-3 Information System Perspectives and Focuses

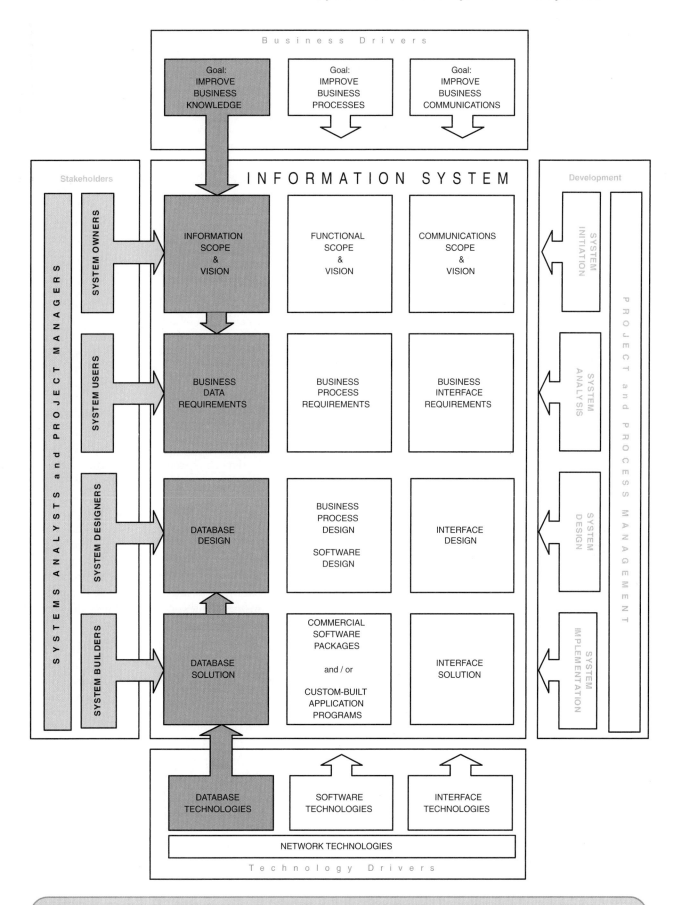

FIGURE 2-4   A BUSINESS KNOWLEDGE Perspective of Information Systems

intelligent decisions that support the organization's mission, goals, objectives, and competitive edge.

Business knowledge may initially take the form of a simple list of business entities and business rules. Examples of business entities might include CUSTOMERS, PRODUCTS, EQUIPMENT, BUILDINGS, ORDERS, and PAYMENTS. What do business entities have to do with knowledge? Information is produced from raw data that describe these business entities. Therefore, it makes sense that we should quickly identify relevant business entities about which we need to capture and store data.

It is also useful to understand simple business associations or rules that describe how the business entities interact. Examples of useful business rules for a sales system might include the following:

- A CUSTOMER <u>can place</u> ORDERS—an ORDER <u>must be placed by</u> a CUSTOMER.
- An ORDER <u>sells</u> PRODUCTS—a PRODUCT <u>may be sold on</u> an ORDER.

Intuitively, a system's database needs to track these business entities and rules in order to produce useful information (for example, "Has CUSTOMER 2846 placed any unfilled ORDERS?").

System owners are concerned with the big picture. They are generally not interested in details (such as what fields describe a CUSTOMER or an ORDER). The primary role of system owners in a systems development project should be to define the scope and vision for the project. For KNOWLEDGE, scope can be defined in simple terms such as the aforementioned business entities and rules. System owners define project vision and expectations in terms of their insight into problems, opportunities, and constraints as they relate to the business entities and rules.

System Users' View of KNOWLEDGE   Information system users are knowledgeable about the data that describe the business. As information workers, they capture, store, process, edit, and use that data every day. They frequently see the data only in terms of how data are currently stored or how they think data should be stored. To them, the data are recorded on forms, stored in file cabinets, recorded in books and binders, organized into spreadsheets, or stored in computer files and databases. The challenge in systems development is to correctly identify and verify users' business data requirements. **Data requirements** are an extension of the business entities and rules that were initially identified by the system owners. System users may identify additional entities and rules because of their greater familiarity with the data. More importantly, system users must specify the exact data attributes to be stored and the precise business rules for maintaining that data. Consider the following example:

data requirement a representation of users' data in terms of entities, attributes, relationships, and rules.

> A system owner may identify the need to store data about a business entity called CUSTOMER. System users might tell us that we need to differentiate between PROSPECTIVE CUSTOMERS, ACTIVE CUSTOMERS, and INACTIVE CUSTOMERS because they know that slightly different types of data describe each type of customer. System users can also tell us precisely what data must be stored about each type of customer. For example, an ACTIVE CUSTOMER might require such data attributes as CUSTOMER NUMBER, NAME, BILLING ADDRESS, CREDIT RATING, and CURRENT BALANCE. Finally, system users are also knowledgeable about the precise rules that govern entities and relationships. For example, they might tell us that the credit rating for an ACTIVE CUSTOMER must be PREFERRED, NORMAL, or PROBATIONARY and that the default for a new customer is NORMAL. They might also specify that only an ACTIVE CUSTOMER can place an ORDER, but an ACTIVE CUSTOMER might not necessarily have any current ORDERS at any given time.

Notice from the above example that the system user's data requirements can be identified independently of the DATABASE TECHNOLOGY that can or will be used to store the data. System users tend to focus on the "business" issues as they pertain to the data. It is important that the system users provide data requirements that are consistent with and complementary to the information scope and vision provided by the system owners.

System Designers' View of KNOWLEDGE   The system designer's KNOWLEDGE perspective differs significantly from the perspectives of system owners and system users. The system designer is more concerned with the DATABASE TECHNOLOGY that will be used by the information system to support business knowledge. System designers translate the system users' business data requirements into database designs that will subsequently be used by system builders to develop computer databases that will be made available via the information system. The system designers' view of data is constrained by the limitations of whatever database management system (DBMS) is chosen. Often, the choice has already been made and the developers must use that technology. For example, many businesses have standardized on an enterprise DBMS (such as *Oracle, DB2,* or *SQL Server*) and a work group DBMS (such as *Access*).

In any case, the system designer's view of KNOWLEDGE consists of data structures, database schemas, fields, indexes, and other technology-dependent components. Most of these technical specifications are too complex to be reasonably understood by system users. The systems analyst and/or database specialists design and document these technical views of the data. This book will teach tools and techniques for transforming business data requirements into database schemas.

System Builders' View of KNOWLEDGE   The final view of KNOWLEDGE is relevant to the system builders. In the KNOWLEDGE column of Figure 2-4, system builders are closest to the actual database management system technology. They must represent data in very precise and unforgiving languages. The most commonly encountered database language is *SQL (Structured Query Language)*. Alternatively, many database management systems, such as *Access* and *Visual FoxPro* include proprietary languages or facilities for constructing a new database.

Not all information systems use database technology to store their business data. Many older legacy systems were built with *flat-file* technologies such as VSAM. These flat-file data structures were constructed directly within the programming language used to write the programs that use those files. For example, in a *COBOL* program the flat-file data structures are expressed as PICTURE clauses in a DATA DIVISION. It is not the intent of this book to teach either database or flat-file construction languages, but only to place them in the context of the KNOWLEDGE building block of information systems.

> PROCESS Building Blocks

Improving business and services processes is another fundamental goal of an information system. Processes deliver the desired functionality of an information system. Processes represent the *work* in a system. People may perform some processes, while computers and machines perform others.

The PROCESS building blocks of information systems are illustrated in Figure 2-5. Notice at the bottom of the PROCESS column that SOFTWARE TECHNOLOGIES will be used to automate selected processes. As you look down the PROCESS column, each of our different stakeholders has different perspectives of the information system. Let's examine those views and discuss their relevance to the PROCESS column.

System Owners' View of PROCESSES   As usual, system owners are generally interested in the big picture. They tend to focus not so much on work flow and procedures as on high-level **business functions,** such as those listed in the margin of page 53. Organizations are often organized around these business fuctions with a vice president

business function a group of related processes that support the business. Functions can be decomposed into other subfunctions and eventually into processes that do specific tasks.

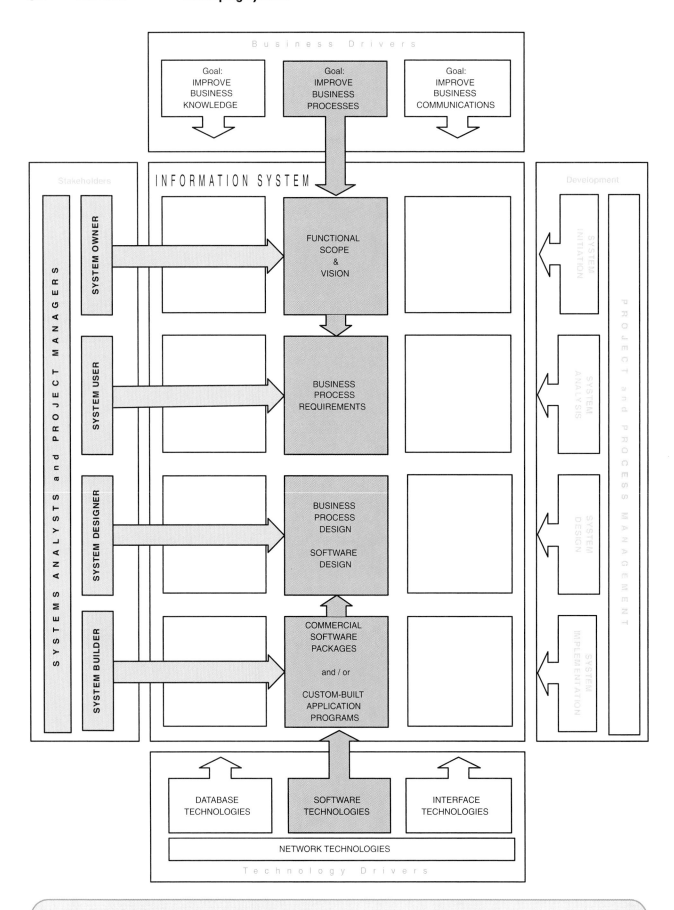

FIGURE 2-5 A BUSINESS PROCESS Perspective of Information Systems

overseeing each function. Unlike business events (such as CUSTOMER SUBMITS ORDER) that have a definite beginning and end, a business function has no starting time or stopping time.

Historically, most information systems were (or are) *function-centered.* That means the system supported one business function. An example would be a SALES INFORMATION SYSTEM that supports only the initial processing of customer orders. Today, many of these single-function information systems are being redesigned as **cross-functional information systems** that support several business functions. As a contemporary alternative to the traditional SALES INFORMATION SYSTEM, a cross-functional ORDER FULFILLMENT INFORMATION SYSTEM would also support all relevant processes subsequent to the processing of the customer order. This would include filling the order in the warehouse, shipping the products to the customer, billing the customer, and providing any necessary follow-up service to the customer—in other words, all business processes required to ensure a complete and satisfactory response to the customer order, regardless of which departments are involved.

As shown in Figure 2-5, the system owners view a system's business PROCESSES with respect to the functional scope being supported by the systems and to a vision or expectation for improvements. The system's business functions are frequently documented by systems analysts in terms of simple lists of business events and responses to those events. Some examples of business events and responses are as follows:

- Event: CUSTOMER SUBMITS ORDER
  Response: CUSTOMER RECEIVES ORDERED PRODUCTS

- Event: EMPLOYEE SUBMITS PURCHASE REQUISITION FOR SUPPLIES
  Response: EMPLOYEE RECEIVES REQUESTED SUPPLIES

- Event: END OF MONTH
  Response: INVOICE CUSTOMERS AGAINST ACCOUNTS

With respect to each event and response identified, system owners would identify perceived problems, opportunities, goals, objectives, and constraints. The costs and benefits of developing information systems to support business functions would also be discussed. As was the case with KNOWLEDGE, system owners are not concerned with PROCESS details. That level of detail is identified and documented as part of the system users' view of processes.

System owners also frequently identify services and levels of service that they seek to provide to customers, suppliers, and employees. A popular example is customer, supplier, or employee *self*-service. Human resource systems, for example, increasingly provide employees with the ability to enter their own transactions such as change of address, medical claims, and training requests. System owners also identify needs for information systems to improve service by reducing errors and improving service.

This book will teach you how to identify and document project scope in terms of relevant business functions, business events, and responses.

System Users' View of PROCESSES

Returning again to Figure 2-5, we are ready to examine the system users' view of processes. Users are concerned with the business processes, or "work," that must be performed in order to provide the appropriate responses to business events. System users specify the business process in terms of **process requirements** for a new system. Process requirements are often documented in terms of activities, data flows, or work flow.

These process requirements must be precisely specified, especially if they are to be automated or supported by software technology. Business process requirements are frequently defined in terms of **policies** and **procedures.** Policies are explicit rules that must be adhered to when completing a business process. Procedures are the precise steps to be followed in completing the business process. Consider the following example:

TYPICAL BUSINESS FUNCTIONS

Sales

Service

Manufacturing

Shipping

Receiving

Accounting

---

cross-functional information system a system that supports relevant business processes from several business functions without regard to traditional organizational boundaries such as divisions, departments, centers, and offices.

---

process requirements a user's expectation of the processing requirements for a business process and its information systems.

---

policy a set of rules that govern a business process.

---

procedure step-by-step set of instructions and logic for accomplishing a business process.

CREDIT APPROVAL is a policy. It establishes a set of rules for determining whether or not to extend credit to a customer. That credit approval policy is usually applied within the context of a specific CREDIT CHECK procedure that established the correct steps for checking credit against the credit policy.

---

work flow the flow of transactions through business processes to ensure appropriate checks and approvals are implemented.

Process requirements are also frequently specified in terms of **work flow.** Most businesses are very dependent on checks and balances to implement work flow. For example, a purchase requisition may be initiated by any employee. But that requisition follows a specific work flow of approvals and checks before it becomes a purchase order transaction that is entered into an information processing system. Of course, these checks and balances can become cumbersome and bureaucratic. Systems analysts and users seek an approropriate balance between checks and balances and service and performance.

Once again, the challenge in systems development is to identify, express, and analyze business process requirements exclusively in business terms that can be understood by system users. Tools and techniques for process modeling and documentation of policies and procedures are taught extensively in this book.

System Designers' View of PROCESSES As was the case with the KNOWLEDGE building block, the system designer's view of business processes is constrained by the limitations of specific application development technologies such as *Java, Visual Basic.NET, C++,* and *C#.* Sometimes the analyst is able to choose the software technology; however, often the choices are limited by software architecture standards that specify which software and hardware technologies must be used. In either case, the designer's view of processes is technical.

Given the business processes from the system users' view, the designer must first determine which processes to automate and how to best automate those processes. Models are drawn to document and communicate how selected business processes are, or will be, implemented using the software and hardware.

Today, many businesses purchase commercial off-the-shelf (COTS) software instead of building that software in-house. In fact, many businesses prescribe that software that can be purchased should never be built—or that only software that provides true competitive advantage should be built. In the case of purchasing software, business processes must usually be changed or adapted to work with the software. Hence, in this scenario the business process design specifications must document how the software package will be integrated into the enterprise.

---

software specifications the technical design of business processes to be automated or supported by computer programs to be written by system builders.

Alternatively, in the case of building software in-house, business processes are usually designed first. And the business process specifications must then be supplemented with **software specifications** that document the technical design of computer programs to be written. You may have encountered some of these software specifications in a programming course. As was the case with KNOWLEDGE, some of these technical views of PROCESSES can be understood by users but most cannot. The designers' intent is to prepare software specifications that (1) fulfill the business process requirements of system users and (2) provide sufficient detail and consistency for communicating the software design to system builders. The systems design chapters in this book teach tools and techniques for transforming business process requirements into both business process design and software design specifications.

System Builders' View of PROCESSES System builders represent PROCESSES using precise computer programming languages or application development environments (ADEs) that describe inputs, outputs, logic, and control. Examples include *C++, Visual Basic .NET, C#* (part of the Microsoft *Visual Studio .NET* ADE), and *Java* (available in ADEs such as IBM *WebSphere* and BEA *WebLogic*). Additionally, some applications and database management systems provide their own internal languages for programming. Examples include *Visual Basic for Applications* (in *Access*) and

PL-SQL (in *Oracle*). All these languages are used to write custom-built **application programs** that automate business processes.

This book does not teach application programming. We will, however, demonstrate how some of these languages provide an excellent environment for rapidly developing a system using prototyping software. **Prototyping** has become the design technique of choice for many system designers and builders. Prototypes typically evolve into the final version of the system or application.

As mentioned earlier, sometimes decisions may involve purchasing a commercial software package as a system solution. In this scenario, the system builder may need to focus on customization that must be done to the software package. The system builder may also be expected to develop application programs that must be integrated with the commercial package to extend the package's functional capabilities. Finally, the system builder must also focus on program utilities that must be written to help with the conversion and integration of the commercial program and existing systems.

> **application program** a language-based, machine-readable representation of what a software process is supposed to do or how a software process is supposed to accomplish its task.

> **prototyping** a technique for quickly building a functioning but incomplete model of the information system using rapid application development tools.

> COMMUNICATIONS **Building Blocks**

Let's examine our final building block—COMMUNICATIONS. A common goal of most organizations is to improve business communications and collaboration between employees and other constituents. Communication improvements in information systems are typically directed toward two critical interface goals for an information system:

- Information systems must provide effective and efficient communication interfaces to the system's users. These interfaces should promote teamwork and coordination of activities.
- Information systems must interface effectively and efficiently with *other* information systems—both with those within the business and increasingly with other businesses' information systems.

The COMMUNICATIONS building blocks of information systems are illustrated in our framework in Figure 2-6. Notice at the bottom of the COMMUNICATION column that it utilizes INTERFACE TECHNOLOGIES to implement the communication interfaces. And once again, as you look down the COMMUNICATION column, each of our different stakeholders has different views of the system. Let's examine those views and discuss their relevance to systems development.

System Owners' View of COMMUNICATION   The system owners' view of COMMUNICATION is relatively simple. Early in a systems development project, system owners need to specify:

- With which business units, employees, customers, and external businesses must the new system interface?
- Where are these business units, employees, customers, and external businesses located?
- Will the system have to interface with any other information, computer, or automated systems?

Answers to these questions help to define the communications scope of an information systems development project. Minimally, a suitable system owners' view of information system communication scope and vision might be expressed as a simple list of business locations or systems with which the information system must interface. Again, relevant problems, opportunities, or constraints may be identified and analyzed.

System Users' View of COMMUNICATION   System users' view of COMMUNICATION is more in terms of the information system's inputs and outputs. Those inputs and outputs can take many forms; however, the business interface requirements are more

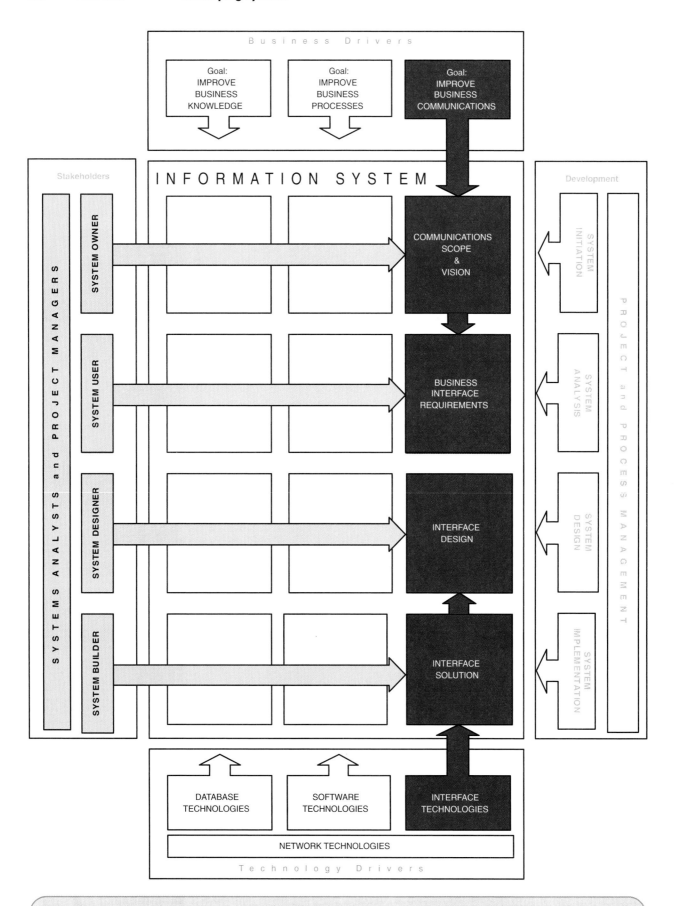

important than the technical format. The inputs and outputs represent how the proposed system would interact with users, employees, business units, customers, and other businesses.

The details of those inputs and outputs are important. System users might specify the details in the form of a list of fields (and their values) that make up the inputs or outputs. Alternatively, and because system users have become comfortable with the graphical user interface (e.g., *Windows* or Web browsers) for the system, the details might be specified in the form of prototypes. System users are increasingly demanding that their custom-built information system applications have the same "look and feel" as their favorite PC tools such as word processors and spreadsheets. This common graphical user interface makes each new application easier to learn and use.

Both list and prototype approaches to documenting the system users' view of COMMUNICATION will be addressed in various chapters of this book.

System Designers' View of COMMUNICATION   System designers must be concerned with the technical design of both the user and the system-to-system communication interfaces. We call these **interface specifications.** Let's begin with the user interface.

Users and designers can be involved in interface design. But whereas system users are interested in requirements and format, system designers have other interests such as consistency, compatibility, completeness, and user dialogues. The **user dialogue** (sometimes called *interface navigation*) specifies how the user will navigate through an application to perform useful work.

The trend toward graphical user interfaces (GUIs) such as *Windows* and Web browsers has simplified life for system users but complicated the design process for system designers. In a typical *Windows* application, there are many different things users can do at any given time—type something, click the left mouse button on a menu item or toolbar icon, press the F1 key for help, maximize the current window, minimize the current window, switch to a different program, and many others. Accordingly, the system designer views the interface in terms of various system states, events that change the system from one state to another, and responses to those events. Today, there are many more design decisions and considerations the system designer must address to document the dialogue of a graphical user interface solution. Tools used to document user dialogues will be discussed in the design unit of this book.

Web interfaces have further complicated the designer's activities. Society has come to expect more glitz in Web interfaces. For that reason, it is not at all uncommon for the design team to include graphical design specialists and human–computer interface specialists to ensure that the interface for a Web server is both compelling and easy to use.

Although not depicted in Figure 2-6, modern system designers may also design *keyless interfaces* such as bar coding, optical character recognition, pen, and voice or handwriting recognition. These alternatives reduce errors by eliminating the keyboard as a source of human error. However, these interfaces, like graphical user interfaces, must be carefully designed to both exploit the underlying technology and maximize the return on what can be a sizable investment.

Finally, and as suggested earlier, system designers are also concerned with system-to-system interfaces. Increasingly, system interfaces are the most difficult to design and implement. For instance, consider a procurement information system that is used to initiate and purchase everything from supplies to equipment. A procurement system must interface with other information systems such as human resources (to determine authority to purchase and approve orders), accounting (to determine if funds are available against an account), receiving (to determine if ordered goods were received, and accounts payable (to initiate payment). These interfacing systems may use very different software and databases. This can greatly complicate system interface design. System interfaces become even more complex when the interface is between information systems in different businesses. For example, in the aforementioned system, we might want to enable our procurement system to directly interface with a supplier's order fulfillment system.

interface specifications technical designs that document how system users are to interact with a system and how a system interacts with other systems.

user dialogue a specification of how the user moves from window to window or page to page, interacting with the application programs to perform useful work.

Legacy information systems in most businesses were each built with the technologies and techniques that represented the best practices at the time when they were developed. Some systems were built in-house. Others were purchased from software vendors or developed with consultants. As a result, the integration of these heterogeneous systems can be difficult. Consequently, the need for different systems to interoperate is pervasive. Accordingly, the time system designers spend on system-to-system integration is frequently as much as or more than the time they spend on system development. The system designer's mission is to find or build interfaces between these systems that (1) do not create maintenance projects for the legacy systems, (2) do not compromise the superior technologies and design of the new systems, and (3) are ideally transparent to the system users.

System Builders' View of COMMUNICATION   System builders construct, install, test, and implement both user and system-to-system interface solutions using INTERFACE TECHNOLOGY (see Figure 2-6). For user interfaces, the interface technology is frequently embedded into the *application development environment (ADE)* used to construct software for the system. For example, ADEs such as those for *Visual Studio .NET,* and *Powerbuilder* include all the interface technology required to construct a *Windows* graphical user interface (GUI). ADEs such as those for *Java* and *Cold Fusion* provide similar functionality for Web interfaces. Alternatively, the user interface could be constructed with a stand-alone interface technology that supports *xHTML* (e.g., Macromedia's *Dreamweaver*).

System-to-system interfaces are considerably more complex than user interfaces to construct or implement. One system-to-system interfacing technology that is currently popular is middleware. **Middleware** is a layer of utility software that sits in between application software and systems software to transparently integrate differing technologies so that they can interoperate.

middleware utility software that allows application software and systems software that utilize differing technologies to interoperate.

One common example of middleware is the open database connectivity (ODBC) tools that allow application programs to work with different database management systems without having to be rewritten to take into consideration the nuances and differences of those database management systems. Programs written with ODBC commands can, for the most part, work with any ODBC-compliant database (which includes dozens of different database management systems). Similar middleware products exist for each of the columns in our information system framework. System designers help to select and apply these products to integrate systems.

At the time of this writing, *XML (eXtensible Markup Language)* has emerged as an evolving standard for system-to-system communication. *XML* is unique in its ability to share data between systems through data streams that not only include the data but also include the meaning and structural definitions for that data. *XML* capabilities are the new frontier for software that implements electronic data exchange over the Web.

Once again, this book is not about system construction; however, we present the system builder's view because the other COMMUNICATION views lead to the construction of the actual communication interfaces.

Network Technologies and the IS Building Blocks

In this chapter, we unveiled a framework for information systems architecture that was initially inspired by the work of John Zachman.[1] The Zachman "Framework for Information Systems Architecture" achieved international recognition and use. The Zachman framework is a matrix (similar to the chapter map at the beginning of this chapter). The rows correspond to what Zachman calls *perspectives* of different

---

[1]John A. Zachman, "A Framework for Information Systems Architecture," *IBM Systems Journal* 26, no. 3 (1987), pp. 276–292.

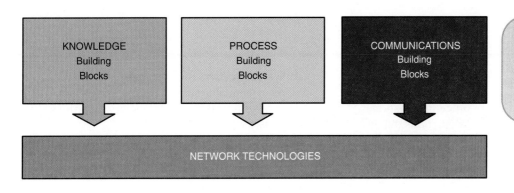

FIGURE 2-7

The Role of the Network in Information Systems

people involved in systems development and use. The columns correspond to *focuses* on different aspects of the information system. Zachman's architecture includes a separate column that closely equates to what our framework recognizes as NETWORK TECHNOLOGIES. (We have chosen to omit that column because network frameworks are more typically covered in data communications and networking textbooks—and those textbooks tend to focus on the Open Systems Interconnect (OSI) framework as opposed to Zachman's.)

But unquestionably, today's information systems are built on networks. Figure 2-7 shows a modern high-level information systems framework that demonstrates the contemporary layering of an information system's KNOWLEDGE, PROCESSES, and COMMUNICATIONS building blocks on NETWORK TECHNOLOGIES. Today's best-designed information systems tend to separate these layers and force them to communicate across the network. This *clean-layering* approach allows any one building block to be replaced with another while having little or no impact on the other building blocks. For example, the DATABASE TECHNOLOGY, SOFTWARE TECHNOLOGY, or INTERFACE TECHNOLOGY could be changed without impacting the other building blocks.

It is not the intent of this book to teach network technology. Most information systems and technology programs offer courses that can expand your understanding of network technology.

Learning Roadmap

So where are we now? If you have already read Chapter 1, you learned about information systems development projects with a focus on the stakeholders, the process, and the business and technical drivers that influence the need for new systems. If you haven't already done so, you should at least skim Chapter 1 to learn about the *context of systems analysis and design methods.*

In Chapter 2, you learned about the product itself—information systems—in terms of basic building blocks. This architectural perspective focused on the different information system views of the various stakeholders. You learned that system owners and users view information systems from the standpoint of achieving goals—improving business knowledge, processes, and communications—whereas system designers and builders view information systems in terms of technology that supports the achievement of goals.

Most readers should proceed directly to Chapter 3, which introduces you to the process of information system development. You'll learn about information systems problem solving, methodologies, and development technology as you expand your education in the fundamentals for systems analysis and design methods.

Summary

1. Organizations are served by a federation of information systems that support various business functions. Businesses have front-office information systems that support business functions that extend out to their customers and back-office information systems that support internal business operations and interact with suppliers.

2. The many classes of information system applications overlap and interoperate to complement and supplement one another.

3. Information systems architecture provides a unifying framework into which various stakeholders with different perspectives can organize and view the fundamental building blocks of information systems:

   a. System owners and system users tend to focus on three common business goals of any information system—improvements in business knowledge, business processes, and business communications.

   b. System designers and builders tend to focus on technologies used by the information system in order to achieve the business goals. They focus on the database technologies that support business knowledge, software technologies that support business processes, and interface technologies that support business communications.

4. The three views represented in the model are:

   a. KNOWLEDGE—the business knowledge that helps managers make intelligent decisions.

   b. PROCESSES—the activities (including management) that carry out the mission of the business.

   c. COMMUNICATIONS—how the system interfaces with its users and other information systems.

5. Improving business knowledge is a fundamental goal of an information system:

   a. The system owner is interested in information that adds new business knowledge.

   b. Information system users are knowledgeable about the data that describes the business. This data is used to create information and subsequent business knowledge.

   c. System designers are concerned with the database technology that will be used by the information system to support business knowledge.

   d. System builders focus on the actual database management system technology used to store the business data that will support business knowledge.

6. Improving business processes is a fundamental goal of an information system:

   a. System owners are interested in the business functions the groups of related processes, that support a business.

   b. System users specify the business process in terms of process requirements for a new system. Business process requirements are frequently defined in terms of policies and procedures. Policies are explicit rules that must be adhered to when completing business processes. Procedures are the precise steps to be followed in completing business processes.

   c. System designers view business processes in terms of the application development environment and the software technology used to develop the system. Many businesses purchase commercial off-the-shelf software solutions instead of building the software in-house.

   d. System builders focus on custom-built applications programs that automate business processes.

7. A common goal of most organizations is to improve business communications:

   a. System owners define the communications scope of an information system development project.

   b. System users view communications in terms of the information system's inputs and outputs.

   c. System designers are concerned with the technical design of both user and system-to-system communication interfaces.

   d. System builders are concerned with the interface technology they use to implement user and system-to-system communication interfaces.

8. Today's information systems are built on networks. Network technology allows properly designed information systems to separate the KNOWLEDGE, PROCESS, and COMMUNICATION building blocks and force them to communicate across the network.

Review Questions

1. What is the difference between front-office information systems and back-office information systems?
2. How do transaction processing systems (TPSs), management information systems (MISs), and decision support systems (DSSs) interact with each other?
3. Why do we need to identify the information system architecture?
4. What are the three business goal–oriented perspectives or views of an information system that systems owners and system users tend to focus on? What are the three technological perspectives that system designers and builders tend to focus on?
5. How are the business perspectives and the technology perspectives of an information system related?
6. In any given building blocks of an information system, the views of four groups of stakeholders need to be taken into account during the development of the system. What are these four stakeholder groups?

7. Briefly describe how system designers and system builders tend to view KNOWLEDGE in a system.
8. Understanding business functions is essential in the process building block of an information system. What are six high-level business functions typical of many companies?
9. If you were the system owner of an online CD store, list two business functions of your online store in terms of business events and responses to those events.
10. Give an example of a policy and the procedures needed to implement the policy.
11. What is prototyping? Why do we need such a technique?
12. What are the two most critical goals in the communication building blocks?
13. What is user dialogue?
14. Why has the increasing use of graphical user interfaces (GUI) complicated the design process for system designers?
15. What role does network technology play in developing an information system?

 Problems and Exercises

1. Companies generally need to use more than one information system to support all their different business functions. These functions are frequently referred to as either front-office information systems or back-office systems. Define each of these two types of systems and identify some of the typical business functions supported by them.
2. As a systems analyst, designer, or builder, you will frequently be involved with your organization's information systems architecture. What is an information systems architecture, and what is its purpose?
3. Although system owners and system users generally have different perspectives of their organization's information system, both groups tend to focus on three business goals that are common to any information system. What are these goal-oriented perspectives, and what is their importance?
4. In an information system, the process building blocks represent the work that occurs in a sys-

tem, which may be performed by people or by computers and machinery. Stakeholders tend to have different views or perspectives of these building blocks. What are these different stakeholder perspectives regarding processes, and how they differ from each other?
5. Assume you are a systems designer and your organization is building a new inventory management system. In reviewing the requirements documentation, it appears that an error was made and some additional data elements were left out that are needed to meet the business or technical objectives of the inventory management system. What should you **not** do at this point?
6. Assume you are designing a retail point-of-sale (POS) system for your company. What are the typical system interfaces of a point-of-sale system that need to be taken into account in designing the POS system?
7. As business technology becomes more powerful and sophisticated, many businesses are redesigning

their single-function information systems, such as sales, into cross-functional information systems that provide integrated support for separate, but related, business functions. Assume that you are designing an order management system that will integrate all business functions triggered by the submission of a sales order. What typical business functions would be included in a cross-functional information system?

8. Middleware is frequently used in systems integration projects when different information systems are tied together to exchange data via system-to-system interfaces. Briefly define middleware, explain its benefits, and provide an example.

9. In identifying and documenting business requirements, systems analysts need to be able to distinguish between laws, policies, and procedures. Why is this important?

10. It is common for system owners and system users to have very different views of the same business processes used in an information system. Why do you think this is? Consider an airline that is developing a customer self-check-in system at airports. What do you think the perspective of the system owners is? What about the system users? Give examples of how the business processes for an airline check-in system will be viewed by the system owners and system users.

11. System designers and system builders also tend to have very different views of system building blocks. Explain the different ways that designers and builders might view the communication building blocks using the customer self check-in system scenario described in the last question.

12. System designers frequently have a number of technical design options to choose from when designing interfaces between different systems and applications. What should designers always keep in mind when designing these interfaces?

13. The framework for information systems architecture used in this textbook is derived from the pioneering framework developed by John Zachman. What is one of the advantages of designing systems based upon this or similar frameworks?

14. At times, an organization may choose to purchase a commercial off-the-shelf (COTS) software package. What do you think are the pros and cons of using off-the-shelf applications compared to custom-built applications?

15. If an organization chose a COTS package as their solution, would the view of the system builder be the same as for a custom-built application? If not, how would it be different?

Projects and Research

1. Select a medium to large organization. The organization can be in the public or private sector, and it can be one with which you are personally familiar or one for which information is readily available.

   a. Describe the organization you have selected (type of organization, mission, products or services, size, annual sales or revenues, etc.).

   b. Select one of the major information systems the organization uses and/or is developing, and describe it.

   c. In the organization you selected, who would typically be the owner of this system?

   d. Describe, from the viewpoint of the owner, the information produced by this system.

   e. If the organization initiated a project to replace or modify this system, how might the system owner define the scope and vision of the project within the context of the organization you selected

   f. Who are the typical users of this system?

   g. Describe, from the perspective of the users, the information produced by this system.

   h. What is an essential difference in how system owners and users view the information produced by the system?

2. Contact and interview two or three systems analysts, in different organizations if possible, regarding this chapter's subtopic on communication building blocks.

   a. Describe the nature of each analyst's company or organization, its mission, and current business issues or needs.

   b. How important does each of the system analysts consider communications with system users versus system builders; that is, which is more important and why?

   c. Do they find it more difficult to communicate with the system users or with the system builders? Why?

d. If they were CIOs for a day, what would each of them change about the way their designers communicate with system users and builders?

e. Which viewpoints do you agree with, or do you have a totally different one than the people you interviewed? Justify your answer.

3. Select an information system used by a medium to large organization. It can be one with which you are personally familiar or one whose organizational structure and information system you have researched.

   a. What is the nature of the organization you have selected, its mission, and the high-level purpose of their information system?

   b. Who is the owner of the system?

   c. If you were the owner of the system, describe how you would see the system processes from that viewpoint.

   d. Who are the users of the system?

   e. If you were one of the system users, describe how you would see the system processes from that viewpoint.

   f. If you were the system designer, describe the system processes from that viewpoint.

   g. What are the essential differences in viewpoints?

4. Imagine that you are the owner of a small business and are searching on the Web for a company that can supply the products or services needed by your business. Find several business-to-business (B2B) Web sites that offer the products or services for which your business is looking. Familiarize yourself with their Web sites from the viewpoint of a typical business customer who is visiting these sites for the first time.

   a. What is the nature of your organization, and for what type of goods or services are you looking?

   b. Which B2B sites did you find on the Web?

   c. Compare the different sites. If all other things were equal (price, availability, brands offered, etc.), would you be more likely to purchase goods or services from one than the other(s), solely because of differences in their Web sites? Why or why not?

   d. From the viewpoint of a business customer, do you think design or usability is more important for a Web site? Explain your answer.

   e. From the viewpoint of a consumer, would your answer be the same as in the preceding question? Explain.

5. Research several articles published in the last few years in your library and/or on the Web that discuss ethical issues related to systems design.

   a. What articles were you able to find?

   b. Describe some of the situations and ethical issues that might arise from time to time in systems design.

   c. Pick one of the situations described in (b) and describe what you believe to be the system designer's ethical obligation, if any.

   d. (Extra credit) Do you think that requiring IT professionals, that is, systems analysts, designers, and builders, to be licensed or certified would increase professionalism and/or reduce unethical behavior? Why or why not?

6. The textbook uses a framework for describing information systems architecture that is based upon John Zachman's "Framework for Information Systems Architecture" model. Using the Web or your school library, research other frameworks for describing IS architectures, and select one, such as Open Systems Interconnect (OSI).

   a. Which frameworks did you find, and which did you select?

   b. Describe its approach to communicating systems architecture. Include a diagram if applicable.

   c. What are its similarities to the framework used in the textbook?

   d. What are its differences?

   e. If you were a systems owner, which one would you find easier to understand?

   f. If you were a systems analyst, which one would you find easier to understand?

Minicases

1. An IT manager requests an amount of funds to upgrade the e-mail server. Without the necessary upgrade, the server will be burdened by the sheer amount of e-mail and will run the risk of crashing. The business manager denies the request, citing the past reliability of the server, and expresses concern at the recent large IT expenditures. The business manager leaves the conversation wondering what IT investments are really necessary, and if the IT manager is just creating "job security."

The IT manager, likewise, leaves the meeting frustrated at not having the tools he/she needs to do the job properly. The IT manager knows that when the server crashes, it will be his/her responsibility to fix.

a. Do you think this happens often in business?
b. What perspectives do you think each are taking on the problem?
c. How could each have communicated those perspectives and business needs better?

2. Interview at least one person in marketing, customer service, and accounting/payroll in the same company. What types of information do they handle? Do they share information across departments? Do you notice overlap in information or in data entry?

3. Government service departments are deeply burdened by the amount of data that they hold and process. Interview someone from a service department and draft a short essay. Service departments that must sift through vast amounts of data are those that deal with, for example, missing persons, child protective services, DMV, and tracking of persons on probation following a

crime. You should include, but are not limited to, topics such as:

- What is the department's (or person's) job?
- What kind of data do they collect and analyze?
- What kind of analyses do they do on the data?
- How much information do they collect, from whom, and what programs do they use?

4. Your neighborhood grocery store, Wow Grocery, always seems to be running out of your favorite ice cream. In frustration, you ask the store manager why they always seem to be out. The store manager, Bob, tells you that the small store cannot afford an inventory management system, so inventory is updated manually. This means that oftentimes the store must either stock extra quantities of well-liked items or risk running out. Unfortunately, Wow Grocery does not have a large enough freezer to store additional stocks of ice cream. As a result, the store runs out of the ice cream quite frequently.

5. What can Wow Grocery do to automate or manage its inventory system without spending much money? Draft your solution into a short paper.

Suggested Readings

Galitz, Wilbert O. *The Essential Guide to User Interface Design: An Introduction to GUI Design Principles and Techniques,* 2nd ed. New York: John Wiley & Sons, 2002.

Goldman, James E.; Phillip T. Rawles; and Julie R. Mariga. *Client/Server Information Systems: A Business-Oriented Approach.* New York: John Wiley & Sons, 1999. For students who are looking for a student-oriented introduction to information technology architecture and data communications, we recommend our colleagues' book because it was written for business and information systems majors to provide a comprehensive survey of the technology that supports today's information systems.

Inmon, W. H. *Building the Data Warehouse,* 3rd ed. New York: John Wiley & Sons, 2002.

Sethi, Vikram, and William R. King. *Organizational Transformation through Business Process Reengineering.* Upper Saddle River, NJ: Prentice Hall, 1998.

Taylor, David, and Alyse D. Terhune. *Doing E-Business: Strategies for Thriving in an Electronic Marketplace.* New York: John Wiley & Sons, 2000.

Zachman, John A. "A Framework for Information System Architecture." *IBM Systems Journal* 26, no. 3 (1987). We adapted the matrix model for information system building blocks from Mr. Zachman's conceptual framework. We first encountered John Zachman on the lecture circuit, where he delivers a remarkably informative and entertaining talk on the same subject as this article. Mr. Zachman's framework has drawn professional acclaim and inspired at least one conference on his model. His framework is based on the concept that architecture means different things to different people. His framework suggests that information systems consist of three distinct "product-oriented" views—data, processes, and technology (which we renamed *communications*)—to which we added a fourth view, "interface." The Zachman framework offers six different audience-specific views—for each of those product views—the ballpark and owner's views (which we renamed as *owner's* and *user's views,* respectively), the designer's and builder's views (which we combined into our *designer's view*), and an out-of-context view (which we called the *builder's view*).

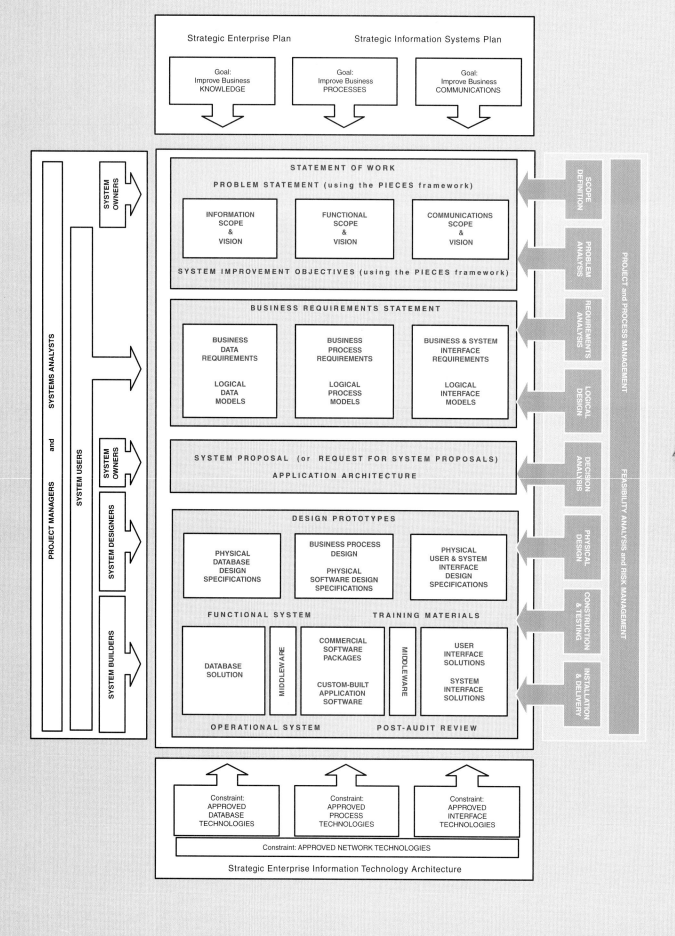

Developing Information Systems

Chapter Preview and Objectives

This chapter more closely examines the systems development process that was first introduced in Chapter 1. Successful systems development is governed by some fundamental, underlying principles that we introduce in this chapter. We also introduce a basic, representative systems development methodology as a disciplined approach to developing information systems. Although such an approach will not guarantee success, it will greatly improve the chances of success. You will know that you understand information systems development when you can:

▌ Describe the motivation for a standard systems development process in terms of the Capability Maturity Model (CMM) for quality management.

▌ Differentiate between the system life cycle and a system development methodology.

▌ Describe 10 basic principles of systems development.

▌ Define problems, opportunities, and directives—the triggers for systems development projects.

▌ Describe the PIECES framework for categorizing problems, opportunities, and directives.

▌ Describe the essential phases of systems development. For each phase, describe its purpose, inputs, and outputs.

▌ Describe cross life-cycle activities that overlap multiple system development phases.

▌ Describe typical, alternative "routes" through the essential phases of systems development. Describe how routes may be combined or customized for different types of projects.

▌ Describe various automated tools for systems development.

Introduction

Work is getting underway at SoundStage Entertainment Club for the systems analysis and design of their member services information system. But the more Bob Martinez learns about the system, the more confused he gets. Bob can recall some of his programming assignments in college. Most of them were just a page or two of bulleted points listing required features. It was pretty easy to get your head around that. But the new SoundStage system will involve tracking member contacts and purchase requirements, promotions, sales, shipments, inventory, multiple warehouses, Web sites, and more. Bob wonders how they will even list all the requirements, let alone keep them straight. How will they know what data they need to track? How will they know what every piece of programming needs to do? He mentioned that to his boss, Sandra. She said it was all about following "the methodology." He remembered something about methodology from a systems analysis class. At the time it seemed like a lot of unnecessary work. But he is starting to see now that on a large project, following an established method may be the only path that is safe to travel.

The Process of Systems Development

systems development process a set of activities, methods, best practices, deliverables, and automated tools that stakeholders (from Chapter 1) use to develop and continuously improve information systems and software (from Chapters 1 and 2).

This chapter introduces a focus on information systems development. We will examine a **systems development process.** Notice we did not say "the" process—there are as many variations on the process as there are experts and authors. We will present one representative process and use it consistently throughout this book. The chapter home page shows an expanded number of phases compared to the home page of Chapter 1. This is because we have factored the high-level phases such as system analysis and system design from Chapter 1 into multiple phases and activities. We have also refined the size and place of the stakeholder roles to reflect "involvement" as opposed to emphasis or priority. And we have edited and enhanced the building blocks to indicate deliverables and artifacts of system development. All of these modifications will be explained in due time.

Why do organizations embrace standardized processes to develop information systems? Information systems are a complex product. Recall from Chapter 2 that an information system includes data, process, and communications building blocks and technologies that must serve the needs of a variety of stakeholders. Perhaps this explains why as many as 70 percent or more of information system development projects have failed to meet expectations, cost more than budgeted, and are delivered much later than promised. The Gartner Group suggests that "consistent adherence to moderately rigorous methodology guidelines can provide 70 percent of [systems development] organizations with a productivity improvement of at least 30 percent within two years."[1]

Increasingly, organizations have no choice but to adopt and follow a standardized systems development process. First, using a consistent process for systems development creates efficiencies that allow management to shift resources between projects. Second, a consistent methodology produces consistent documentation that reduces lifetime costs to maintain the systems. Finally, the U.S. government has mandated that any organization seeking to develop software for the government must adhere to certain quality management requirements. A consistent process promotes quality. And many other organizations have aggressively committed to total quality management goals in order to increase their competitive advantage. In order to realize quality and productivity improvements, many organizations have turned to project and process management frameworks such as the Capability Maturity Model, discussed in the next section.

---

[1]Richard Hunter, "AD Project Portfolio Management," *Proceedings of the Gartner Group IT98 Symposium/Expo* (CD-ROM). The Gartner Group is an industry watchdog and research group that tracks and projects industry trends for IT managers.

> The Capability Maturity Model

It has been shown that as an organization's information system development process matures, project timelines and cost decrease while productivity and quality increase. The Software Engineering Institute at Carnegie Mellon University has observed and measured this phenomenon and developed the **Capability Maturity Model (CMM)** to assist all organizations to achieve these benefits. The CMM has developed a wide following, both in industry and government. Software evaluation based on CMM is being used to qualify information technology contractors for most U.S. federal government projects.

The CMM framework for systems and software is intended to help organizations improve the maturity of their systems development processes. The CMM is organized into five maturity levels (see Figure 3-1):

- *Level 1—Initial:* This is sometimes called anarchy or chaos. At this level, system development projects follow no consistent process. Each development team uses its own tools and methods. Success or failure is usually a function of the skill and experience of the team. The process is unpredictable and not repeatable. A project typically encounters many crises and is frequently over budget and behind schedule. Documentation is sporadic or not consistent from one project to the next, thus creating problems for those who must maintain a system over its lifetime. Almost all organizations start at Level 1.
- *Level 2—Repeatable:* At this level, project management processes and practices are established to track project costs, schedules, and functionality. The focus is on project management. A system development process is always followed, but it may vary from project to project. Success or failure is still a function of the skill and experience of the project team; however, a concerted effort is made to repeat earlier project successes. Effective project management practices lay the foundation for standardized processes in the next level.

Capability Maturity Model (CMM) a standardized framework for assessing the maturity level of an organization's information systems development and management processes and products. It consists of five levels of maturity.

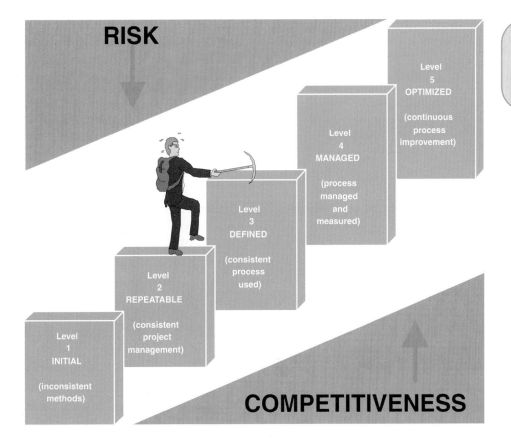

FIGURE 3-1

The Capability Maturity Model (CMM)

TABLE 3-1 Impact of System Development "Process" on Quality

CMM Project Statistics for a Project Resulting in 200,000 Lines of Code						
Organization's CMM Level	Project Duration (months)	Project Person-Months	Number of Defects Shipped	Median Cost ($ millions)	Lowest Cost ($ millions)	Highest Cost ($ millions)
1	30	600	61	5.5	1.8	100+
2	18.5	143	12	1.3	.96	1.7
3	15	80	7	.728	.518	.933

Source: Master Systems, Inc.

system development methodology a formalized approach to the systems development process; a standardized process that includes the activities, methods, best practices, deliverables, and automated tools to be used for information systems development.

- *Level 3—Defined:* In this level, a standard **system development** process (sometimes called a *methodology*) is purchased or developed. All projects use a tailored version of this process to develop and maintain information systems and software. As a result of using the standardized process for all projects, each project results in consistent and high-quality documentation and deliverables. The process is stable, predictable, and repeatable.
- *Level 4—Managed:* In this level, measurable goals for quality and productivity are established. Detailed measures of the standard system development process and product quality are routinely collected and stored in a database. There is an effort to improve individual project management based on this collected data. Thus, management seeks to become more proactive than reactive to systems development problems (such as cost overruns, scope creep, schedule delays, etc.). Even when a project encounters unexpected problems or issues, the process can be adjusted based on predictable and measurable impacts.
- *Level 5—Optimizing:* In this level, the standardized system development process is continuously monitored and improved based on measures and data analysis established in Level 4. This can include changing the technology and best practices used to perform activities required in the standard system development process, as well as adjusting the process itself. Lessons learned are shared across the organization, with a special emphasis on eliminating inefficiencies in the system development process while sustaining quality. In summary, the organization has institutionalized continuous systems development process improvement.

It is very important to recognize that each level is a prerequisite for the next level.

Currently, many organizations are working hard to achieve at least CMM Level 3 (sometimes driven by a government or organizational mandate). A central theme to achieving Level 3 (Defined) is the use of a standard process or methodology to build or integrate systems. As shown in Table 3-1, an organization can realize significant improvements in schedule and cost by institutionalizing CMM Level 3 process improvements.[2]

system life cycle the factoring of the lifetime of an information system into two stages, (1) systems development and (2) systems operation and maintenance—first you build it; then you use and maintain it. Eventually, you cycle back to redevelopment of a new system.

> Life Cycle versus Methodology

The terms *system life cycle* and *system development methodology* are frequently and incorrectly interchanged. Most system development processes are derived from a natural **system life cycle.** The system life cycle just happens. Figure 3-2 illustrates two

---

[2]White Paper, "Rapidly Improving Process Maturity: Moving Up the Capability Maturity Model through Outsourcing" (Boston: Keane, 1997, 1998, p.11).

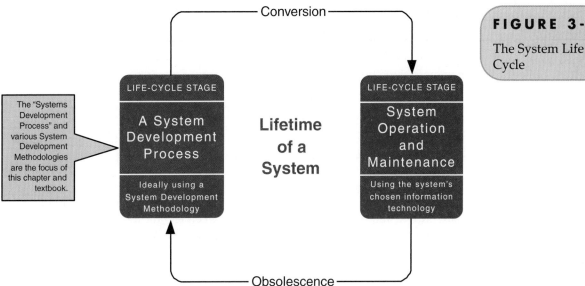

FIGURE 3-2
The System Life Cycle

life-cycle stages. Notice that there are two key events that trigger a change from one stage to the other:

- When a system cycles from development to operation and maintenance, a *conversion* must take place.
- At some point in time, *obsolescence* occurs (or is imminent) and a system cycles from operation and maintenance to redevelopment.

Actually, a system may be in more than one stage at the same time. For example, one version may be in *operation and support* while the next version is in *development.*

So how does this contrast with a systems development methodology? A systems development methodology "executes" the systems development stage of the system life cycle. Each individual information system has its own system life cycle. The methodology is the standard process to build and maintain that system and all other information systems through their life cycles. Consistent with the goals of the CMM, methodologies ensure that:

- A consistent, reproducible approach is applied to all projects.
- There is reduced risk associated with shortcuts and mistakes.
- Complete and consistent documentation is produced from one project to the next.
- Systems analysts, designers, and builders can be quickly reassigned between projects because all use the same process.
- As development teams and staff constantly change, the results of prior work can be easily retrieved and understood by those who follow.

Methodologies can be purchased or homegrown. Why purchase a methodology? Many information system organizations can't afford to dedicate staff to the development and continuous improvement of a homegrown methodology. Methodology vendors have a vested interest in keeping their methodologies current with the latest business and technology trends. Homegrown methodologies are usually based on generic methodologies and techniques that are well documented in books and on Web sites. Examples of system development methodologies are listed in the margin on the following page. You should be able to research most of them on the Web. Many of their underlying methods will be taught in this textbook.

Throughout this book, we will use a methodology called **FAST**, which stands for *F*ramework for the *A*pplication of *S*ystems *T*hinking. *FAST* is not a real-world commercial methodology. We developed it as a composite of the best practices we've

FAST a hypothetical methodology used throughout this book to demonstrate a representative systems development process. The acronym's letters stand for *F*ramework for the *A*pplication of *S*ystems *T*hinking.

REPRESENTATIVE SYSTEM DEVELOPMENT METHODOLOGIES

Architected Rapid Application Development (Architected RAD)

Dynamic Systems Development Methodology (DSDM)

Joint Application Development (JAD)

Information Engineering (IE)

Rapid Application Development (RAD)

Rational Unified Process (RUP)

Structured Analysis and Design (old, but still occasionally encountered)

eXtreme Programming (XP)

Note: There are many commercial methodologies and software tools (sometimes called *methodware*) based on the above general methodologies.

encountered in many commercial and reference methodologies. Unlike many commercial methodologies, *FAST* is not prescriptive. That is to say, *FAST* is an agile framework that is flexible enough to provide for different types of projects and strategies. Most important, *FAST* shares much in common with both the book-based and the commercial methodologies that you will encounter in practice.

> Underlying Principles for Systems Development

Before we examine the *FAST* methodology, let's introduce some general principles that should underlie all systems development methodologies.

Principle 1: Get the System Users Involved   Analysts, programmers, and other information technology specialists frequently refer to "my system." This attitude has, in part, created an "us versus them" conflict between technical staff and their users and management. Although analysts and programmers work hard to create technologically impressive solutions, those solutions often backfire because they don't address the real organization problems. Sometimes they even introduce new organization problems. For this reason, system user involvement is an absolute necessity for successful systems development. Think of systems development as a partnership between system users, analysts, designers, and builders. The analysts, designers, and builders are responsible for systems development, but they must engage their owners and users, insist on their participation, and seek agreement from all stakeholders concerning decisions that may affect them.

Miscommunication and misunderstandings continue to be a significant problem in many systems development projects. However, owner and user involvement and education minimize such problems and help to win acceptance of new ideas and technological change. Because people tend to resist change, information technology is often viewed as a threat. The best way to counter that threat is through constant and thorough communication with owners and users.

Principle 2: Use a Problem-Solving Approach   A system development methodology is, first and foremost, a problem-solving approach to building systems. The term *problem* is broadly used throughout this book to include (1) real problems, (2) opportunities for improvement, and (3) directives from management. The classical problem-solving approach is as follows:

1. Study and understand the problem, its context, and its impact.
2. Define the requirements that must be met by *any* solution.
3. Identify candidate solutions that fulfill the requirements, and select the best solution.
4. Design and/or implement the chosen solution.
5. Observe and evaluate the solution's impact, and refine the solution accordingly.

Systems analysts should approach all projects using some variation of this problem-solving approach.

Inexperienced or unsuccessful problem solvers tend to eliminate or abbreviate one or more of the above steps. For example, they fail to completely understand the problem, or they prematurely commit to the first solution they think of. The result can range from (1) solving the wrong problem, to (2) incorrectly solving the problem, (3) picking the wrong solution, or (4) picking a less-than-optimal solution. A methodology's problem-solving process, when correctly applied, can reduce or eliminate these risks.

Principle 3: Establish Phases and Activities   All methodologies prescribe phases and activities. The number and scope of phases and activities vary from author to author, expert to expert, methodology to methodology, and business to business. The chapter home page at the beginning of this chapter illustrates the eight phases of our *FAST* methodology in the context of your information systems framework. The phases

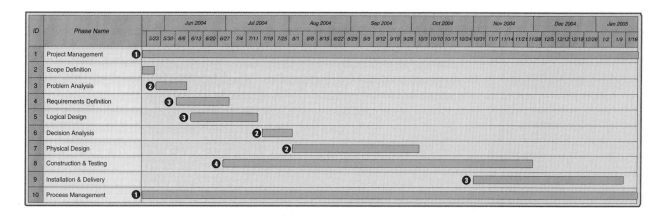

ID	Phase Name	Jun 2004					Jul 2004				Aug 2004				Sep 2004				Oct 2004				Nov 2004				Dec 2004				Jan 2005		
		5/23	5/30	6/6	6/13	6/20	6/27	7/4	7/11	7/18	7/25	8/1	8/8	8/15	8/22	8/29	9/5	9/12	9/19	9/26	10/3	10/10	10/17	10/24	10/31	11/7	11/14	11/21	11/28	12/5	12/12	12/19	12/26	1/2	1/9	1/16
1	Project Management																																			
2	Scope Definition																																			
3	Problem Analysis																																			
4	Requirements Definition																																			
5	Logical Design																																			
6	Decision Analysis																																			
7	Physical Design																																			
8	Construction & Testing																																			
9	Installation & Delivery																																			
10	Process Management																																			

FIGURE 3-3   Overlap of System Development Phases and Activities

are listed on the far right-hand side of the illustration. In each phase, the focus is on those building blocks and on stakeholders that are aligned to the left of that phase.

The phases are: scope definition, problem analysis, requirements analysis, logical design, decision analysis, physical design and integration, construction and testing, and installation and delivery. Each of these phases will be discussed later in this chapter. These phases are not absolutely sequential. The phases tend to overlap one another, as illustrated in Figure 3-3. Also, the phases may be customized to the special needs of a given project (e.g., deadlines, complexity, strategy, resources). In this chapter, we will describe each customization as alternative routes through the methodology and problem-solving process.

Principle 4: Document throughout Development   When do you document the programs you write? Be honest. We must confess that, like most students, we did our documentation after we wrote the programs. We knew better, but we postdocumented anyway. That just does not work in the business world. In medium to large organizations, system owners, users, analysts, designers, and builders come and go. Some will be promoted; some will have extended medical leaves; some will quit the organization; and still others will be reassigned. To promote good communication between constantly changing stakeholders, documentation should be a working byproduct of the entire systems development effort.

Documentation enhances communications and acceptance. Documentation reveals strengths and weaknesses of the system to multiple stakeholders. It stimulates user involvement and reassures management about progress. At the same time, some methodologies have been criticized for expecting too much documentation that adds little value to the process or resulting system. Our *FAST* methodology advocates a balance between the value of documentation and the effort to produce it. Experts call this *agile modeling*.

Principle 5: Establish Standards   In a perfect world, all information systems would be integrated such that they behave as a single system. Unfortunately, this never happens because information systems are developed and replaced over a very long period of time. Even organizations that purchase and install an enterprise resource planning (ERP) product usually discover that there are applications and needs that fall outside the ERP system. Systems integration has become critical to the success of any organization's information systems.

To achieve or improve systems integration, organizations turn to standards. In many organizations, these standards take the form of enterprise information technology architecture. An IT architecture sets standards that serve to direct technology solutions and information systems to a common technology vision or configuration.

An information technology architecture typically standardizes on the following (note: it is not important that you know what all these sample technologies are):

- *Database technology*—What database engine(s) will be used (e.g., *Oracle*, IBM *DB2*, Microsoft *SQL Server*)? On what platforms will they be operated (e.g., *UNIX, Linux, Windows XP, MVS*)? What technologies will be used to load data into online transaction processing (OLTP) databases, operational data stores, and data warehouses (i.e., Extract Transform and Load [ETL])?
- *Software technology*—What application development environment(s)/language(s) will be used to write software (e.g., IBM's *Websphere* with *Java*, Microsoft's *Visual Studio .NET* with *Visual Basic .NET, Visual C++*, and/or *Visual C#*; Syebase's *Powerbuilder*, Oracle's *Oracle Forms*)?
- *Interface technology*—How will user interfaces be developed—with *MS Windows* components or Web languages and components (e.g., an *xHTML* editor such as Macromedia's *Dreamweaver*, a portal engine such as IBM's *Websphere*)? How will data be exchanged between different information systems (e.g., a data broker such as IBM's *MQ Messaging*, an XML-based data exchange, or a custom programmed interface)?

Notice how these architectural questions closely correspond to the technology drivers in your information system model.

In the absence of an IT architecture, each information system and application may be built using radically different technologies. Not only does this make it difficult to integrate applications, but it creates resource management problems—IT organizations cannot as easily move developers between projects as priorities change or emergencies occur because different teams are staffed with technical competencies based on the various technologies used and being used to develop information systems. Creating an enterprise IT architecture and driving projects and teams to that architecture make more sense.

As new technologies emerge, an IT architecture must change. But that change can be managed. The chief technology officer (CTO) in an organization is frequently charged with technology exploration and IT architecture management. Given that architecture, all information systems projects are constrained to implement new systems that conform to the architecture (unless otherwise approved by the CTO).

Principle 6: Manage the Process and Projects   Most organizations have a system development process or methodology, but they do not always use it consistently on projects. Both the process and the projects that use it must be managed. **Process management** ensures that an organization's chosen process or management is used consistently on and across all projects. Process management also defines and improves the chosen process or methodology over time. **Project management** ensures that an information system is developed at minimum cost, within a specified time frame, and with acceptable quality (using the standard system development process or methodology). Effective project management is essential to achieving CMM Level 2 success. Use of a repeatable process gets us to CMM Level 3. CMM Levels 4 and 5 require effective process management. Project management can occur without a standard process, but in mature organizations all projects are based on a standardized and managed process.

Process management and project management are influenced by the need for quality management. Quality standards are built into a process to ensure that the activities and deliverables of each phase will contribute to the development of a high-quality information system. They reduce the likelihood of missed problems and requirements, as well as flawed designs and program errors (bugs). Standards also make the IT organization more agile. As personnel changes occur, staff can be relocated between projects with the assurance that every project is following an understood and accepted process.

process management an ongoing activity that documents, teaches, oversees the use of, and improves an organization's chosen methodology (the "process") for systems development. Process management is concerned with phases, activities, deliverables, and quality standards that should be consistently applied to all projects.

project management the process of scoping, planning, staffing, organizing, directing, and controlling a project to develop an information system at minimum cost, within a specified time frame, and with acceptable quality.

Principle 7: Justify Information Systems as Capital Investments   Information systems are capital investments, just like a fleet of trucks or a new building. System owners commit to this investment. Notice that the initial commitment occurs early in a project, when system owners agree to sponsor and fund the project. Later (during the phase called *decision analysis*), system owners recommit to the more costly technical decisions. In considering a capital investment, two issues must be addressed:

1. For any problem, there are likely to be several possible solutions. The systems analyst and other stakeholders should not blindly accept the first solution suggested. The analyst who fails to look at alternatives may be doing the business a disservice.
2. After identifying alternative solutions, the systems analyst should evaluate each possible solution for feasibility, especially for **cost-effectiveness.** Cost-effectiveness is measured using a technique called *cost-benefit analysis.*

Like project and process management, cost-benefit analysis is performed throughout the system development process.

A significant advantage of the phased approach to systems development is that it provides several opportunities to reevaluate cost-effectiveness, risk, and feasibility. There is often a temptation to continue with a project only because of the investment already made. In the long run, canceled projects are usually much less costly than implemented disasters. This is extremely important for young analysts to remember.

Most system owners want more from their systems than they can afford or are willing to pay for. Furthermore, the scope of most information system projects increases as the analyst learns more about the business problems and requirements as the project progresses. Unfortunately, most analysts fail to adjust estimated costs and schedules as the scope increases. As a result, the analyst frequently and needlessly accepts responsibility for cost and schedule overruns.

Because information systems are recognized as capital investments, system development projects are often driven by enterprise planning. Many contemporary information technology business units create and maintain a **strategic information systems plan.** Such a plan identifies and prioritizes information system development projects. Ideally, a strategic information systems plan is driven by a **strategic enterprise plan** that charts a course for the entire business.

Principle 8: Don't Be Afraid to Cancel or Revise Scope   There is an old saying: "Don't throw good money after bad." In other words, don't be afraid to cancel a project or revise scope, regardless of how much money has been spent so far—cut your losses. To this end, we advocate a **creeping commitment** approach to systems development.[3] With the creeping commitment approach, multiple feasibility checkpoints are built into any systems development methodology. At each checkpoint feasibility is reassessed. All previously expended costs are considered sunk (meaning not recoverable). They are, therefore, irrelevant to the decision. Thus, the project should be reevaluated at each checkpoint to determine if it remains feasible to continue investing time, effort, and resources into the project. At each checkpoint, the analyst should consider the following options:

- Cancel the project if it is no longer feasible.
- Reevaluate and adjust the costs and schedule if project scope is to be increased.
- Reduce the scope if the project budget and schedule are frozen and not sufficient to cover all project objectives.

The concept of sunk costs is more or less familiar to most financial analysts, but it is frequently forgotten or not used by the majority of systems analysts, most system users, and even many system owners.

cost-effectiveness the result obtained by striking a balance between the lifetime costs of developing, maintaining, and operating an information system and the benefits derived from that system. Cost-effectiveness is measured by cost-benefit analysis.

strategic information systems plan a formal strategic plan (3 to 5 years) for building and improving an information technology infrastructure and the information system applications that use that infrastructure.

strategic enterprise plan a formal strategic plan (3 to 5 years) for an entire business that defines its mission, vision, goals, strategies, benchmarks, and measures of progress and achievement. Usually, the strategic enterprise plan is complemented by strategic business unit plans that define how each business unit will contribute to the enterprise plan. The information systems plan (above) is one of those unit-level plans.

creeping commitment a strategy in which feasibility and risks are continuously reevaluated throughout a project. Project budgets and deadlines are adjusted accordingly.

---

[3]Thomas Gildersleeve, *Successful Data Processing Systems Analysis,* 2nd ed. (Englewood Cliffs, NJ: Prentice Hall, 1985), pp. 5–7.

risk management the process of identifying, evaluating, and controlling what might go wrong in a project before it becomes a threat to the successful completion of the project or implementation of the information system. Risk management is driven by risk analysis or assessment.

PRINCIPLES OF SYSTEMS DEVELOPMENT

Get the System Users Involved.

Use a Problem-Solving Approach.

Establish Phases and Activities.

Document throughout Development.

Establish Standards.

Manage the Process and Projects.

Justify Information Systems as Capital Investments.

Don't Be Afraid to Cancel or Revise Scope.

Divide and Conquer.

Design Systems for Growth and Change.

In addition to managing feasibility throughout the project, we must manage risk. **Risk management** seeks to balance risk and reward. Different organizations are more or less averse to risk, meaning that some are willing to take greater risks than others in order to achieve greater rewards.

Principle 9: Divide and Conquer   Whether you realize it or not, you learned the divide-and-conquer approach throughout your education. Since high school, you've been taught to outline a paper before you write it. Outlining is a divide-and-conquer approach to writing. A similar approach is used in systems development. We divide a system into subsystems and components in order to more easily conquer the problem and build the larger system. In systems analysis, we often call this *factoring*. By repeatedly dividing a larger problem (system) into more easily managed pieces (subsystems), the analyst can simplify the problem-solving process. This divide-and-conquer approach also complements communication and project management by allowing different pieces of the system to be communicated to different and the most appropriate stakeholders.

The building blocks of your information system framework provide one basis for dividing and conquering an information system's development. We will use this framework throughout the book.

Principle 10: Design Systems for Growth and Change   Businesses change over time. Their needs change. Their priorities change. Accordingly, information systems that support a business must change over time. For this reason, good methodologies should embrace the reality of change. Systems should be designed to accommodate both growth and changing requirements. In other words, well-designed information systems can both scale up and adapt to the business. But regardless of how well we design systems for growth and change, there will always come a time when they simply cannot support the business.

System scientists describe the natural and inevitable decay of all systems over time as *entropy*. As described earlier in this section, after a system is implemented it enters the *operations and maintenance* stage of the life cycle. During this stage the analyst encounters the need for changes that range from correcting simple mistakes, to redesigning the system to accommodate changing technology, to making modifications to support changing user requirements. Such changes direct the analyst and programmers to rework formerly completed phases of the life cycle. Eventually, the cost of maintaining the current system exceeds the costs of developing a replacement system—the current system has reached entropy and becomes obsolete.

But system entropy can be managed. Today's tools and techniques make it possible to design systems that can grow and change as requirements grow and change. This book will teach you many of those tools and techniques. For now, it's more important to simply recognize that flexibility and adaptability do not happen by accident—they must be built into a system.

We have presented 10 principles that should underlie any methodology. These principles are summarized in the margin and can be used to evaluate any methodology, including our *FAST* methodology.

A Systems Development Process

In this section we'll examine a logical process for systems development. (Reminder: *FAST* is a hypothetical methodology created for learning purposes.) We'll begin by studying types of system projects and how they get started. Then we'll introduce the eight *FAST* phases. Finally, we'll examine alternative variations, or "routes" through the phases, for different types of projects and development strategies.

> Where Do Systems Development Projects Come From?

System owners and system users initiate most projects. The impetus for most projects is some combination of **problems, opportunities,** and **directives.** To simplify this discussion, we will frequently use the term *problem* to collectively refer to problems, opportunities, and directives. Accordingly, *problem solving* refers to solving problems, exploiting opportunities, and fulfilling directives.

There are far too many potential system problems to list them all in this book. However, James Wetherbe developed a useful framework for classifying problems.[4] He calls it *PIECES* because the letters of each of the six categories, when put together, spell the word "pieces." The categories are:

P   the need to correct or improve *performance.*
I   the need to correct or improve *information* (and data).
E   the need to correct or improve *economics,* control costs, or increase profits.
C   the need to correct or improve *control* or security.
E   the need to correct or improve *efficiency* of people and processes.
S   the need to correct or improve *service* to customers, suppliers, partners, employees, and so on.

Figure 3-4 expands on each of the PIECES categories.

The categories of the PIECES framework are neither exhaustive nor mutually exclusive—they overlap. Any given project is usually characterized by one or more categories, and any given problem or opportunity may have implications with respect to more than one category. But PIECES is a practical framework (used in *FAST*), not just an academic exercise. We'll revisit PIECES several times in this book.

Projects can be either planned or unplanned. A *planned project* is the result of one of the following:

- An *information systems strategy plan* has examined the business as a whole to identify those system development projects that will return the greatest strategic (long-term) value to the business.
- A *business process redesign* has thoroughly analyzed a series of business processes to eliminate redundancy and bureaucracy and to improve efficiency and value added. Now it is time to redesign the supporting information system for those redesigned business processes.

The opposite of planned projects are *unplanned projects*—those that are triggered by a specific problem, opportunity, or directive that occurs in the course of doing business. Most organizations have no shortage of unplanned projects. Anyone can submit a proposed project based on something that is happening in the business. The number of unplanned-project proposals can easily overwhelm the largest information systems organization; therefore, they are frequently screened and prioritized by a **steering committee** of system owners and IT managers to determine which requests get approved. Those requests that are not approved are **backlogged** until resources become available (which sometimes never happens).

Both planned and unplanned projects go through the same essential system development process. Let's now examine the project phases in somewhat greater detail.

> The *FAST* Phases

FAST, like most methodologies, consists of phases. The number of phases will vary from one methodology to another. In Chapter 1 you were introduced to the four classic phases of the system development life cycle. The *FAST* methodology employs

---

[4]James Wetherbe and Nicholas P. Vitalari, *Systems Analysis and Design: Traditional, Best Practices,* 4th ed. (St. Paul, MN: West Publishing, 1994), pp. 196–199. James Wetherbe is a respected information systems educator, researcher, and consultant.

---

problem an undesirable situation that prevents the organization from fully achieving its mission, vision, goals, and/or objectives.

opportunity a chance to improve the organization even in the absence of an identified problem.

directive a new requirement that's imposed by management, government, or some external influence.

steering committee an administrative body of system owners and information technology executives that prioritizes and approves candidate system development projects.

backlog a repository of project proposals that cannot be funded or staffed because they are a lower priority than those that have been approved for system development. Note that priorities change over time; therefore, a backlogged project might be approved at some future date.

The PIECES Problem-Solving Framework and Checklist

The following checklist for problem, opportunity, and directive identification uses Wetherbe's PIECES framework. Note that the categories of PIECES are not mutually exclusive; some possible problems show up in multiple lists. Also, the list of possible problems is not exhaustive. The PIECES framework is equally suited to analyzing both manual and computerized systems and applications.

PERFORMANCE

  A. Throughput – the amount of work performed over some period of time.
  B. Response times – the average delay between a transaction or request, and a response to that transaction or request.

INFORMATION (and Data)

  A. Outputs
    1. Lack of any information
    2. Lack of necessary information
    3. Lack of relevant information
    4. Too much information – "information overload"
    5. Information that is not in a useful format
    6. Information that is not accurate
    7. Information that is difficult to produce
    8. Information is not timely to its subsequent use
  B. Inputs
    1. Data is not captured
    2. Data is not captured in time to be useful
    3. Data is not accurately captured – contains errors
    4. Data is difficult to capture
    5. Data is captured redundantly – same data captured more than once
    6. Too much data is captured
    7. Illegal data is captured
  C. Stored data
    1. Data is stored redundantly in multiple files and/or databases
    2. Same data items have different values in different files (poor data integration)
    3. Stored data is not accurate
    4. Data is not secure to accident or vandalism
    5. Data is not well organized
    6. Data is not flexible – not easy to meet new information needs from stored data
    7. Data is not accessible

ECONOMICS

  A. Costs
    1. Costs are unknown
    2. Costs are untraceable to source
    3. Costs are too high
  B. Profits

    1. New markets can be explored
    2. Current marketing can be improved
    3. Orders can be increased

CONTROL (and Security)

  A. Too little security or control
    1. Input data is not adequately edited
    2. Crimes (e.g., fraud, embezzlement) are (or can be) committed against data
    3. Ethics are breached on data or information – refers to data or information getting to unauthorized people
    4. Redundantly stored data is inconsistent in different files or databases
    5. Data privacy regulations or guidelines are being (or can be) violated
    6. Processing errors are occurring (either by people, machines, or software)
    7. Decision-making errors are occurring
  B. Too much control or security
    1. Bureaucratic red tape slows the system
    2. Controls inconvenience customers or employees
    3. Excessive controls cause processing delays

EFFICIENCY

  A. People, machines, or computers waste time
    1. Data is redundantly input or copied
    2. Data is redundantly processed
    3. Information is redundantly generated
  B. People, machines, or computers waste materials and supplies
  C. Effort required for tasks is excessive
  D. Material required for tasks is excessive

SERVICE

  A. The system produces inaccurate results
  B. The system produces inconsistent results
  C. The system produces unreliable results
  D. The system is not easy to learn
  E. The system is not easy to use
  F. The system is awkward to use
  G. The system is inflexible to new or exceptional situations
  H. The system is inflexible to change
  I. The system is incompatible with other systems

FIGURE 3-4 The PIECES Framework for Problem Identification

eight phases to better define periodic milestones and the deliverables. The grid below compares the *FAST* phases to the classic phases. As you can see, both sets of phases cover the same ground, but *FAST* is more detailed.

FAST Phases	Classic Phases			
	Project Initiation	System Analysis	System Design	System Implementation
Scope definition	X			
Problem analysis	X	X		
Requirements analysis		X		
Logical design		X		
Decision analysis	(a system analysis transition phase)			
Physical design and integration			X	
Construction and testing			X	X
Installation and delivery				X

Figure 3-5 illustrates the phases of the *FAST* methodology. Each phase produces deliverables that are passed to the next phase. And documentation accumulates as you complete each phase. Notice that we have included an iconic representation of the building blocks to symbolize this accumulation of knowledge and work-in-process artifacts during system development. Notice also that a project starts with some combination of PROBLEMS, OPPORTUNITIES, and DIRECTIVES from the user community (the green arrow) and finishes with a WORKING BUSINESS SOLUTION (the red arrow) for the user community.

Figure 3-6 shows the *FAST* methodology from the perspective of your information system building blocks that you learned in Chapters 1 and 2. The phases are on the right-hand side. The deliverables are built around the building blocks for knowledge, processes, and communications. The stakeholders are on the left-hand side. Notice how we have expanded and duplicated some stakeholders to reflect their involvement opposite the phases in which they primarily participate.

NOTE: The remainder of this section briefly describes each of the eight basic phases. Throughout this discussion, we will be referring to the process flowchart in Figure 3-5, as well as the building blocks view of the process in Figure 3-6. Also throughout the discussion, all terms printed in SMALL CAPS refer to phases, prerequisites (inputs), and deliverables (outputs) shown in Figures 3-5 and 3-6.

Scope Definition   The first phase of a typical project is SCOPE DEFINITION. The purpose of the scope definition phase is twofold. First, it answers the question, "Is this problem worth looking at?" Second, and assuming the problem *is* worth looking at, it establishes the size and boundaries of the project, the project vision, any constraints or limitations, the required project participants, and, finally, the budget and schedule.

In Figure 3-6, we see that the participants in the scope definition phase primarily include SYSTEM OWNERS, PROJECT MANAGERS, and SYSTEM ANALYSTS. System users are generally excluded because it is too early to get into the level of detail they will eventually bring to the project.

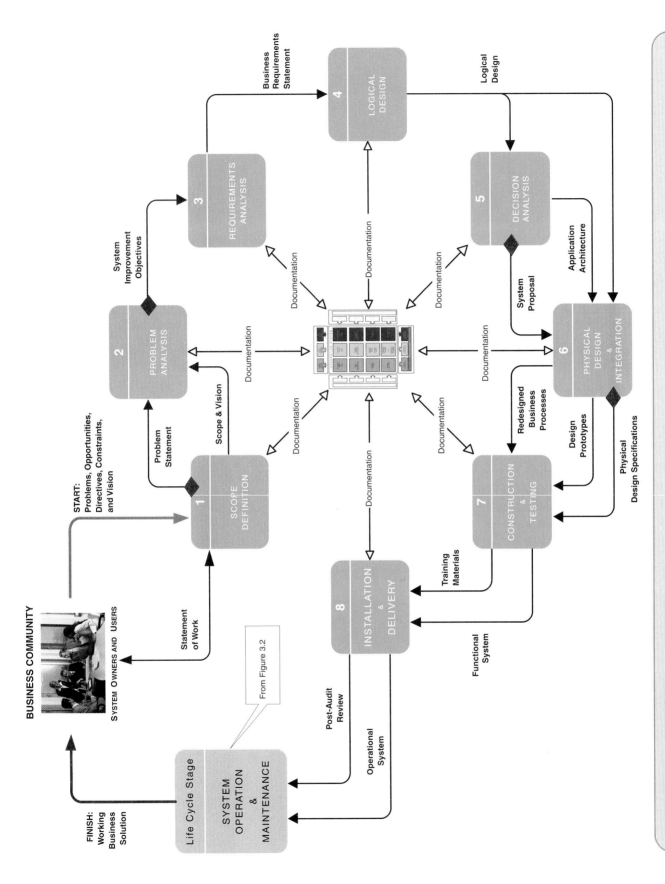

BUSINESS COMMUNITY

SYSTEM OWNERS AND USERS

START:
Problems, Opportunities,
Directives, Constraints,
and Vision

FINISH:
Working
Business
Solution

Statement
of Work

Life Cycle Stage

SYSTEM
OPERATION
&
MAINTENANCE

From Figure 3.2

Post-Audit
Review

Operational
System

1
SCOPE
DEFINITION

Scope & Vision

Problem
Statement

Documentation

2
PROBLEM
ANALYSIS

System
Improvement
Objectives

3
REQUIREMENTS
ANALYSIS

Business
Requirements
Statement

4
LOGICAL
DESIGN

Documentation

Documentation

Logical
Design

5
DECISION
ANALYSIS

System
Proposal

Application
Architecture

6
PHYSICAL
DESIGN
&
INTEGRATION

Documentation

Physical
Design Specifications

Redesigned
Business
Processes

Design
Prototypes

7
CONSTRUCTION
&
TESTING

Training
Materials

Functional
System

8
INSTALLATION
&
DELIVERY

Documentation

Documentation

Documentation

FIGURE 3-5 Process View of System Development

80

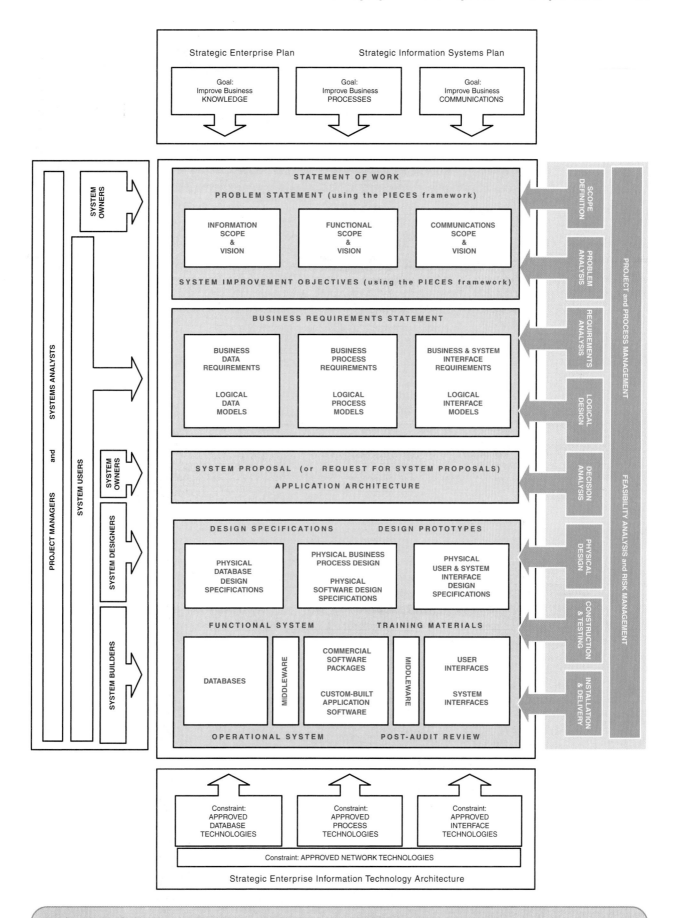

FIGURE 3-6 Building Blocks View of System Development

problem statement a statement and categorization of problems, opportunities, and directives; may also include constraints and an initial vision for the solution. Synonyms include *preliminary study and feasibility assessment*.

constraint any factor, limitation, or restraint that may limit a solution or the problem-solving process.

scope creep a common phenomenon wherein the requirements and expectations of a project increase, often without regard to the impact on budget and schedule.

statement of work a contract with management and the user community to develop or enhance an information system; defines vision, scope, constraints, high-level user requirements, schedule, and budget. Synonyms include *project charter, project plan,* and *service-level agreement*.

In Figure 3-5, we see that the scope definition phase is triggered by some combination of PROBLEMS, OPPORTUNITIES, and DIRECTIVES (to which we will add CONSTRAINTS and VISION). There are several deliverables or outcomes of a scope definition. One important outcome is a PROBLEM STATEMENT, a succinct overview of the problems, opportunities, and/or directives that triggered the project. The PIECES framework provides an excellent outline for a **problem statement.** The goal here is not to solve the problems, opportunities, and directives but only to catalog and categorize them. We should also identify any **constraints** that may impact the proposed project. Examples of constraints include budget limits, deadlines, human resources available or not available, business policies or government regulations, and technology standards. Finally, the system owners should be asked for at least a high-level vision for the system improvements they are seeking.

Given a basic understanding of problems, opportunities, directives, constraints, and vision, we need to establish initial scope. Thus, an initial SCOPE STATEMENT is another important outcome of this phase. Scope defines how big we think the project is. Your information system building blocks provide a useful framework for defining scope. Figure 3-6 illustrates that scope and vision can be defined in terms of INFORMATION, FUNCTIONS, and INTERFACES. Scope can, and frequently does, change during a project. But by documenting initial scope, you establish a baseline for controlling **scope creep** on both the budget and the schedule.

Given the initial problem and scope statements for the project, the analyst can staff the project team, estimate the budget for system development, and prepare a schedule for the remaining phases. Ultimately, this phase concludes with a "go or no-go" decision from system owners. Either the system owners agree with the proposed scope, budget, and schedule for the project, or they must reduce scope (to reduce costs and time) or cancel the project. This feasibility checkpoint is illustrated in Figure 3-5 as a diamond.

The final and most important deliverable is a STATEMENT OF WORK. A **statement of work** is a contract or agreement to develop the information system. It consolidates the problem statement, scope statement, and schedule and budget for all parties who will be involved in the project.

Problem Analysis   There is always an existing system, regardless of whether it currently uses information technology. The PROBLEM ANALYSIS phase studies the existing system and analyzes the findings to provide the project team with a more thorough understanding of the problems that triggered the project. The analyst frequently uncovers new problems and answers the most important question, "Will the benefits of solving these problems exceed the costs of building the system to solve these problems?"

Once again, Figure 3-6 provides a graphical overview of the problem analysis phase in terms of your information system building blocks. Notice that the participants still include the SYSTEM OWNERS but that this phase begins to actively involve the SYSTEM USERS as well. The system users are the business subject matter experts in any project. (Notice the intentional expansion of the system users' perspective to overlap many phases—remember principle 1: "Get the system users involved.") Of course, PROJECT MANAGERS and SYSTEM ANALYSTS are always involved in all phases of a project.

As shown in Figure 3-5, the prerequisites for the problem analysis phase are the SCOPE and PROBLEM STATEMENTS as defined and approved in the scope definition phase. The deliverable of the problem analysis phase is a set of SYSTEM IMPROVEMENT OBJECTIVES derived from a thorough understanding of the business problems. These objectives do not define inputs, outputs, or processes. Instead, they define the business criteria on which any new system will be evaluated. For instance, we might define a system improvement objective as any of the following:

Reduce the time between order processing and shipping by three days.
Reduce bad credit losses by 45 percent.
Comply with new financial aid federal qualification requirements by January 1.

Think of system improvement objectives as the *grading criteria* for evaluating any new system that you might eventually design and implement. System improvement objectives may be presented to system owners and users as a written recommendation or an oral presentation.

Depending on the complexity of the problem and the project schedule, the team may or may not choose to formally document the existing system. Such documentation frequently occurs when the business processes are considered dated or overly bureaucratic. Documentation of the existing system is sometimes called an "AS IS" BUSINESS MODEL. The as-is model may be accompanied by analysis demonstrating inefficiencies, bottlenecks, or other problems related to the business processes.

Every existing system has its own terminology, history, culture, and nuances. Learning those aspects of the system is an important by-product of this phase. From all of the information gathered, the project team gains a better understanding of the existing system's problems and opportunities. After reviewing the findings, the system owners will either agree or disagree with the recommended system improvement objectives. And consistent with the creeping commitment principle, we include another go or no-go feasibility checkpoint (the red diamond) at the end of the phase. The project can be either:

- Canceled if the problems are deemed no longer worth solving.
- Approved to continue to the next phase.
- Reduced or expanded in scope (with budget and schedule modifications) and then approved to continue to the next phase.

Requirements Analysis   Given system owner approval to continue from the problem analysis phase, now you can design a new system, right? No, not yet! What capabilities should the new system provide for its users? What data must be captured and stored? What performance level is expected? Careful! This requires decisions about *what* the system must do, *not how* it should do those things. The REQUIREMENTS ANALYSIS phase defines and prioritizes the *business* requirements. Simply stated, the analyst approaches the users to find out what they need or want out of the new system, carefully avoiding any discussion of technology or technical implementation. This is perhaps the most important phase of systems development. Errors and omissions in requirements analysis result in user dissatisfaction with the final system and costly modifications.

Returning again to Figure 3-6, notice that the participants primarily include both SYSTEM USERS (which may include owners who will actually *use* the system) and SYSTEMS ANALYSTS. PROJECT MANAGERS are also involved. SYSTEM DESIGNERS are omitted from this phase in order to prevent premature attention to technology solutions. The building blocks can themselves provide the framework for defining many business requirements, including BUSINESS DATA REQUIREMENTS, BUSINESS PROCESS REQUIREMENTS, and BUSINESS AND SYSTEM INTERFACE REQUIREMENTS. Because the business requirements are intended to solve problems, the PIECES framework can also provide a useful outline, this time for a requirements statement.

In Figure 3-5, we see that the SYSTEM IMPROVEMENT OBJECTIVES from the problem analysis phase are the prerequisite to the requirements analysis phase. The deliverable is a BUSINESS REQUIREMENTS STATEMENT. Again, this requirements statement does not specify any technical possibilities or solutions. The requirements statement may be a document as small as a few pages, or it may be extensive with a page or more of documentation per requirement.

To produce a business requirements statement, the systems analyst works closely with system users to identify needs and priorities. This information is collected by way of interviews, questionnaires, and facilitated meetings. The challenge to the team is to validate those requirements. The system improvement objectives provide the "grading key" for business requirements: *Does each requirement contribute to meeting one or more system improvement objectives?* Chapters 6 and 7 will introduce systems analysis tools and techniques for identifying and documenting user requirements.

Typically, requirements must also be prioritized. Priorities serve two purposes. First, if project timelines become stressed, requirements priorities can be used to rescope the project. Second, priorities can frequently be used to define iterations of design and construction to create staged releases or versions of the final product.

The requirements analysis phase should never be skipped or shortchanged. One of the most common complaints about new systems and applications is that they don't really satisfy the users' needs. This usually happens when system designers and builders become preoccupied with a technical solution before fully understanding the business needs. System designers and builders are dependent on competent systems analysts to work with users to define and document complete and accurate business requirements before applying any technology.

Logical Design   Business requirements (above) are usually expressed in words. Systems analysts have found it useful to translate those words into pictures called **system models** to validate the requirements for completeness and consistency. (Figure 3-5 is an example of a common system model called a *data flow diagram.*) System modeling implements a timeless concept: "A picture is worth a thousand words."

The LOGICAL DESIGN PHASE translates business requirements into system models. The term **logical design** should be interpreted as "technology independent," meaning the pictures illustrate the system independent of any possible technical solution—hence, they model business requirements that must be fulfilled by any technical solution we might want to consider.

Different methodologies require or recommend different amounts and degrees of system modeling or logical design. Prescriptive methodologies like *structured analysis and design, information engineering,* and the *Rational Unified Process (RUP)* usually require that many types and/or instances of system models be drawn in various levels of detail. Fortunately, computer-automated tools are available to assist the systems analyst in these drawing tasks. Alternatively, agile methodologies like *architected rapid application development* and *extreme programming* recommend "just enough modeling." This so-called *agile modeling* seeks to prevent the project from degenerating into a condition called **analysis paralysis.** This textbook leans toward agile methods but recognizes that complex problems may best be solved using more prescriptive approaches.

In Figure 3-6, we see that the participants include SYSTEM ANALYSTS (who draw the models) and SYSTEM USERS (who validate the models). PROJECT MANAGERS are always included to ensure that modeling meets standards and does not deter overall project progress. We can draw (1) LOGICAL DATA MODELS that depict data and information requirements, (2) LOGICAL PROCESS MODELS that depict business processes requirements, and (3) LOGICAL INTERFACE MODELS that depict business and system interface requirements.[5]

In Figure 3-5, we see that the prerequisite to logical design is the BUSINESS REQUIREMENTS STATEMENT from the previous phase. In practice, the requirements analysis and logical design phases almost always have considerable overlap. In other words, as business requirements are identified and documented, they can be modeled. The deliverables of logical design are the LOGICAL SYSTEM MODELS AND SPECIFICATIONS themselves. Depending on the methodology used, the level of detail in the specifications will vary. For example, we may define a business rule that specifies the legitimate values for a data attribute such as *Credit Rating* or a rule that specifies the business policy for a *Credit Check.*

---

system model a picture of a system that represents reality or a desired reality. System models facilitate improved communication between system users, system analysts, system designers, and system builders.

---

logical design the translation of business user requirements into a system model that depicts only the business requirements and not any possible technical design or implementation of those requirements. Common synonyms include *conceptual design* and *essential design,* the latter of which refers to modeling the "essence" of a system, or the "essential requirements" independent of any technology. The antonym of logical design is *physical design* (defined later in this chapter).

---

analysis paralysis a satirical term coined to describe a common project condition in which excessive system modeling dramatically slows progress toward implementation of the intended system solution.

---

[5]Those of you already familiar with *object-oriented* modeling should note that object models tend to blur the boundaries of our framework somewhat, but the framework can still be applied since the problem to be solved is still driven by the three fundamental business goals illustrated in our framework. This will be demonstrated in the object-oriented analysis and design chapters of this book.

Before we move on to the next phase, we should note that the SCOPE DEFINITION, PROBLEM ANALYSIS, REQUIREMENTS ANALYSIS, and LOGICAL DESIGN PHASES are collectively recognized by most experts as *system analysis*. Some experts would also include our next phase, DECISION ANALYSIS. But we consider it to be a system analysis to system design transition phase because it makes the transition from the business concerns of system owners and users to the technology concerns of system designers and builders. And of course, systems analysts are the common thread that ensures continuity as we make this transition. Let's examine the transition.

Decision Analysis   Given business requirements and the logical system models, there are usually numerous alternative ways to design a new information system to fulfill those requirements. Some of the pertinent questions include the following:

- How much of the system should be automated with information technology?
- Should we purchase software or build it ourselves (called the *make-versus-buy decision*)?
- Should we design the system for an internal network, or should we design a Web-based solution?
- What information technologies (possibly emerging) might be useful for this application?

These questions are answered in the DECISION ANALYSIS phase of the methodology. The purpose of this phase is to (1) identify candidate technical solutions, (2) analyze those candidate solutions for feasibility, and (3) recommend a candidate system as the target solution to be designed.

In Figure 3-6, we see that the decision analysis phase is positioned halfway through the development process. Half the building blocks are positioned higher, and half are positioned lower. This is consistent with the decision analysis phase's role as a transition from analysis to design—and from business concerns of SYSTEM USERS to those of SYSTEM DESIGNERS (and, ultimately, system builders). Designers (the technical experts in specific technologies) begin to play a role here along with system users and SYSTEM ANALYSTS. Analysts help to define and analyze the alternatives. Decisions are made regarding the technologies to be used as part of the application's architecture. Ultimately, SYSTEM OWNERS will have to approve or disapprove the approved decisions since they are paying for the project.

Figure 3-5 shows that a decision analysis is triggered by validated business requirements plus any logical system models and specifications that expand on those requirements. The project team solicits ideas and opinions for technical design and implementation from a diverse audience, possibly including IT software vendors. Candidate solutions are identified and characterized according to various criteria. It should be noted that many modern organizations have information technology and architecture standards that constrain the number of candidate solutions that might be considered and analyzed. (The existence of such standards is illustrated at the bottom of your information system building blocks model in Figure 3-6.) After the candidate solutions have been identified, each one is evaluated by the following criteria:

- *Technical feasibility*—Is the solution technically practical? Does our staff have the technical expertise to design and build this solution?
- *Operational feasibility*—Will the solution fulfill the user's requirements? To what degree? How will the solution change the user's work environment? How do users feel about such a solution?
- *Economic feasibility*—Is the solution cost-effective (as defined earlier in the chapter)?
- *Schedule feasibility*—Can the solution be designed and implemented within an acceptable time period?
- *Risk feasibility*—What's the probability of a successful implementation using the technology and approach?

The project team is usually looking for the *most* feasible solution—the solution that offers the best combination of technical, operational, economic, schedule, and risk feasibility. Different candidate solutions may be most feasible on a single criterion; however, one solution will usually prove *most* feasible based on all of the criteria.

The key deliverable of the decision analysis phase is a SYSTEM PROPOSAL. This proposal may be written and/or presented verbally. Several outcomes are possible. The creeping commitment feasibility checkpoint (again, the red diamond) may result in any one of the following options:

- Approve and fund the system proposal for design and construction (possibly including an increased budget and timetable if scope has significantly expanded).
- Approve or fund one of the alternative candidate solutions.
- Reject all the candidate solutions and either cancel the project or send it back for new recommendations.
- Approve a reduced-scope version of the proposed solution.

Optionally, the decision analysis phase may also produce an APPLICATION ARCHITECTURE for the approved solution. Such a model serves as a high-level blueprint (like a simple house floor plan) for the recommended or approved proposal.

Before we move on, you may have noticed in Figure 3-6 a variation on the SYSTEM PROPOSAL deliverable called a REQUEST FOR SYSTEM PROPOSALS (or RFP). This variation is for a recommendation to purchase the hardware and/or software solution as opposed to building it in-house. We'll defer any further discussion of this option until later in the chapter when we discuss the commercial package integration variation of our basic process.

Physical Design and Integration   Given approval of the SYSTEM PROPOSAL from the decision analysis phase, you can finally design the new system. The purpose of the PHYSICAL DESIGN AND INTEGRATION phase is to transform the business requirements (represented in part by the LOGICAL SYSTEM MODELS) into PHYSICAL DESIGN SPECIFICATIONS that will guide system construction. In other words, physical design addresses greater detail about *how* technology will be used in the new system. The design will be constrained by the approved ARCHITECTURAL MODEL from the previous phase. Also, design requires adherence to any internal technical design standards that ensure completeness, usability, reliability, performance, and quality.

physical design the translation of business user requirements into a system model that depicts a technical implementation of the users' business requirements. Common synonyms include *technical design* or, in describing the output, *implementation model*. The antonym of physical design is *logical design* (defined earlier in this chapter).

Physical design is the opposite of logical design. Whereas logical design dealt exclusively with business requirements independent of any technical solution, physical design represents a specific technical solution. Figure 3-6 demonstrates the physical design phase from the perspective of your building blocks. Notice that the design phase is concerned with technology-based views of the system: (1) PHYSICAL DATABASE DESIGN SPECIFICATIONS, (2) PHYSICAL BUSINESS PROCESS and SOFTWARE DESIGN SPECIFICATIONS, and (3) PHYSICAL USER AND SYSTEM INTERFACE SPECIFICATIONS. The SYSTEM DESIGNER and SYSTEM ANALYST (possibly overlapping roles for some of the same individuals) are the key participants; however, certain aspects of the design usually have to be shared with the SYSTEM USERS (e.g., screen designs and work flow). You may have already had some exposure to physical design specifications in either programming or database courses.

There are two extreme philosophies of physical design.

- *Design by specification*—Physical system models and detailed specifications are produced as a series of written (or computer-generated) blueprints for construction.
- *Design by prototyping*—Incomplete but functioning applications or subsystems (called *prototypes*) are constructed and refined based on feedback from users and other designers.

In practice, some combination of these extremes is usually performed.

No new information system exists in isolation from other existing information systems in an organization. Consequently, a design must also reflect system integration concerns. The new system must be integrated both with other information systems and with the business's processes themselves. Integration is usually reflected in physical system models and design specifications.

In summary, Figure 3-5 shows that the deliverables of the physical design and integration phase include some combination of PHYSICAL DESIGN MODELS AND SPECIFICATIONS, DESIGN PROTOTYPES, and REDESIGNED BUSINESS PROCESSES. Notice that we have included one final go or no-go feasibility checkpoint for the project (the red diamond). A project is rarely canceled after the design phase unless it is hopelessly over budget or behind schedule. On the other hand, scope could be decreased to produce a minimum acceptable product in a specified time frame. Or the schedule could be extended to build a more complete solution in multiple versions. The project plan (schedule and budget) would need to be adjusted to reflect these decisions.

It should be noted that in modern methodologies, there is a trend toward merging the design phase with our next phase, construction. In other words, the design and construction phases usually overlap.

Construction and Testing   Given some level of PHYSICAL DESIGN MODELS AND SPECIFICATIONS (and/or DESIGN PROTOTYPES), we can begin to construct and test system components for that design. Figure 3-5 shows that the primary deliverable of the CONSTRUCTION AND TESTING phase is a FUNCTIONAL SYSTEM that is ready for implementation. The purpose of the construction and testing phase is twofold: (1) to build and test a system that fulfills business requirements and physical design specifications, and (2) to implement the interfaces between the new system and existing systems. Additionally, FINAL DOCUMENTATION (e.g., help systems, training manuals, help desk support, production control instructions) will be developed in preparation for training and system operation. The construction phase may also involve installation of purchased software.

Your information system framework (Figure 3-6) identifies the relevant building blocks and activities for the construction phase. The focus is on the last row of building blocks. The project team must construct or install:

- DATABASES—Databases may include *online transaction processing (OLTP)* databases to support day-to-day business transactions, *operational data stores (ODS)* to support day-to-day reporting and queries, and *data warehouses* to support data analysis and decision support needs.
- COMMERCIAL SOFTWARE PACKAGES and/or CUSTOM-BUILT SOFTWARE—Packages are installed and customized as necessary. Application programs are constructed according to the physical design and/or prototypes from the previous phase. Both packages and custom software must be thoroughly tested.
- USER AND SYSTEM INTERFACES—User interfaces (e.g., Windows and Web interfaces) must be constructed and tested for usability and stability. System-to-system interfaces must be either constructed or implemented using application integration technologies. Notice that MIDDLEWARE (a type of system software) is often used to integrate disparate database, software, and interface technologies. We'll talk more about middleware in the design unit of this book.

Figure 3-6 also identifies the participants in this phase as SYSTEM BUILDERS, SYSTEM ANALYSTS, SYSTEM USERS, and PROJECT MANAGERS. SYSTEM DESIGNERS may also be involved to clarify design specifications.

You probably already have some experience with part of this activity—application programming. Programs can be written in many different languages, but the current trend is toward the use of visual and object-oriented programming languages such as *Java, C++,* or *Visual Basic.* As components are constructed, they are typically demonstrated to users in order to solicit feedback.

One of the most important aspects of construction is conducting tests of both individual system components and the overall system. Once tested, a system (or version of a system) is ready for INSTALLATION AND DELIVERY.

Installation and Delivery   What's left to do? New systems usually represent a departure from the way business is currently done; therefore, the analyst must provide for a smooth transition from the old system to the new system and help users cope with normal start-up problems. Thus, the INSTALLATION AND DELIVERY phase serves to deliver the system into operation (sometimes called *production*).

In Figure 3-5, the FUNCTIONAL SYSTEM from the construction and testing phase is the key input to the INSTALLATION AND DELIVERY phase. The deliverable is an OPERATIONAL SYSTEM. SYSTEM BUILDERS install the system from its development environment into the production environment. SYSTEM ANALYSTS must train SYSTEM USERS, write various user and production control manuals, convert existing files and databases to the new databases, and perform final system testing. Any problems may initiate rework in previous phases thought to be complete. System users provide continuous feedback as new problems and issues arise. Essentially, the installation and delivery phase considers the same building blocks as the construction phase.

To provide a smooth transition to the new system, a conversion plan should be prepared. This plan may call for an abrupt cutover, where the old system is terminated and replaced by the new system on a specific date. Alternatively, the plan may run the old and new systems in parallel until the new system has been deemed acceptable to replace the old system.

The installation and delivery phase also involves training individuals who will use the final system and developing documentation to aid the system users. The implementation phase usually includes some form of POST-AUDIT REVIEW to gauge the success of the completed systems project. This activity promotes continuous improvement of the process and future project management.

System Operation and Maintenance Once the system is placed into operation, it will require ongoing **system support** for the remainder of its useful, productive lifetime. System support consists of the following ongoing activities:

- *Assisting users*—Regardless of how well the users have been trained and how thorough and clear the end-user documentation is, users will eventually require additional assistance as unanticipated problems arise, new users are added, and so forth.
- *Fixing software defects (bugs)*—Software defects are errors that slipped through the testing of software. These are inevitable, but they can usually be resolved, in most cases, by knowledgeable support.
- *Recovering the system*—From time to time, a system failure may result in a program "crash" and/or loss of data. Human error or a hardware or software failure may cause this. The systems analyst or technical support specialists may then be called on to recover the system—that is, to restore a system's files and databases and to restart the system.
- *Adapting the system to new requirements*—New requirements may include new business problems, new business requirements, new technical problems, or new technology requirements.

Eventually, we expect that the user feedback and problems, or changing business needs, will indicate that it is time to start over and reinvent the system. In other words, the system has reached entropy, and a new project to create an entirely new system development process should be initiated.

> Cross Life-Cycle Activities

System development also involves a number of **cross life-cycle activities.** These activities, listed in the margin definition, are not explicitly depicted in Figure 3-5, but they are vital to the success of any project. Let's briefly examine each of these activities.

Fact-Finding There are many occasions for **fact-finding** during a project. Fact-finding is most crucial to the early phases of a project. It is during these phases that

---

system support the ongoing technical support for users of a system, as well as the maintenance required to deal with any errors, omissions, or new requirements that may arise.

cross life-cycle activity any activity that overlaps multiple phases of the system development process. Examples include *fact-finding, documentation, presentation, estimation, feasibility analysis, project and process management, change management,* and *quality management.*

fact-finding the formal process of using research, interviews, meetings, questionnaires, sampling, and other techniques to collect information about system problems, requirements, and preferences. It is also called *information gathering* or *data collection.*

the project team learns about a business's vocabulary, problems, opportunities, constraints, requirements, and priorities. But fact-finding is also used during the decision analysis, physical design, construction and testing, and installation and delivery phases—only to a lesser extent. It is during these latter phases that the project team researches technical alternatives and solicits feedback on technical designs, standards, and working components.

Documentation and Presentation   Communication skills are essential to the successful completion of any project. In fact, poor communication is frequently cited as the cause of project delays and rework. Two forms of communication that are common to systems development projects are **documentation** and **presentation.**

Clearly, documentation and presentation opportunities span all the phases. In Figure 3-7, the black arrows represent various instances of documentation of a phase. The red arrows represent instances where presentations are frequently required. Finally, the green arrows represent the storage of documentation and other artifacts of systems development in a **repository.** A repository saves documentation for reuse and rework as necessary.

Feasibility Analysis   Consistent with our creeping commitment approach to systems development, **feasibility analysis** is a cross life-cycle activity. Different measures of **feasibility** are applicable in different phases of the methodology. These measures include technical, operational, economic, schedule, and risk feasibility, as described when we introduced the decision analysis phase. Feasibility analysis requires good **estimation** techniques.

Process and Project Management   Recall that the CMM considers systems development to be a *process* that must be managed on a project-by-project basis. For this reason and others, process management and project management are ongoing, cross life-cycle activities. Both types of management were introduced earlier, but their definitions are repeated in the margin on page ••• for your convenience. **Process management** defines the methodology to be used on every project—think of it as the *recipe* for building a system. **Project management** is concerned with administering a single instance of the process as applied to a single project.

Failures and limited successes of systems development projects often outnumber successful projects. Why is that? One reason is that many systems analysts are unfamiliar with, or undisciplined in how to properly apply, tools and techniques of systems development. But most failures are attributed to poor leadership and management. This mismanagement results in unfulfilled or unidentified requirements, cost overruns, and late delivery.

> Sequential versus Iterative Development

The above discussion of phases might lead you to assume that systems development is a naturally sequential process, moving in a one-way direction from phase to phase. Such sequential development is, in fact, one alternative. This approach is depicted in part (a) of Figure 3-8. In the figure we have used the four classic phases rather than the eight *FAST* phases in the interest of simplicity. This strategy requires that each phase be "completed" one after the other until the information system is finished. In reality, the phases may somewhat overlap one another in time. For example, some system design can be started prior to the completion of system analysis. Given its waterfall-like visual appearance, this approach is often called the **waterfall development approach.**

The waterfall approach has lost favor with most modern system developers. A more popular strategy, shown in part (b) of Figure 3-8, is commonly referred to as the **iterative development approach,** or incremental development process. This

documentation the ongoing activity of recording facts and specifications for a system for current and future reference.

presentation the ongoing activity of communicating findings, recommendations, and documentation for review by interested users and managers. Presentations may be either written or verbal.

repository a database and/or file directory where system developers store all documentation, knowledge, and artifacts for one or more information systems or projects. A repository is usually automated for easy information storage, retrieval, and sharing.

feasibility analysis the activity by which feasibility is measured and assessed.

feasibility a measure of how beneficial the development of an information system would be to an organization.

estimation the calculated prediction of the costs and effort required for system development. A somewhat facetious synonym is *guesstimation,* usually meaning that the estimation is based on experience or empirical evidence but is lacking in rigor—in other words, a *guess.*

process management an ongoing activity that documents, teaches, oversees the use of, and improves an organization's chosen methodology (the "process") for systems development. Process management is concerned with phases, activities, deliverables, and quality standards that should be consistently applied to all projects.

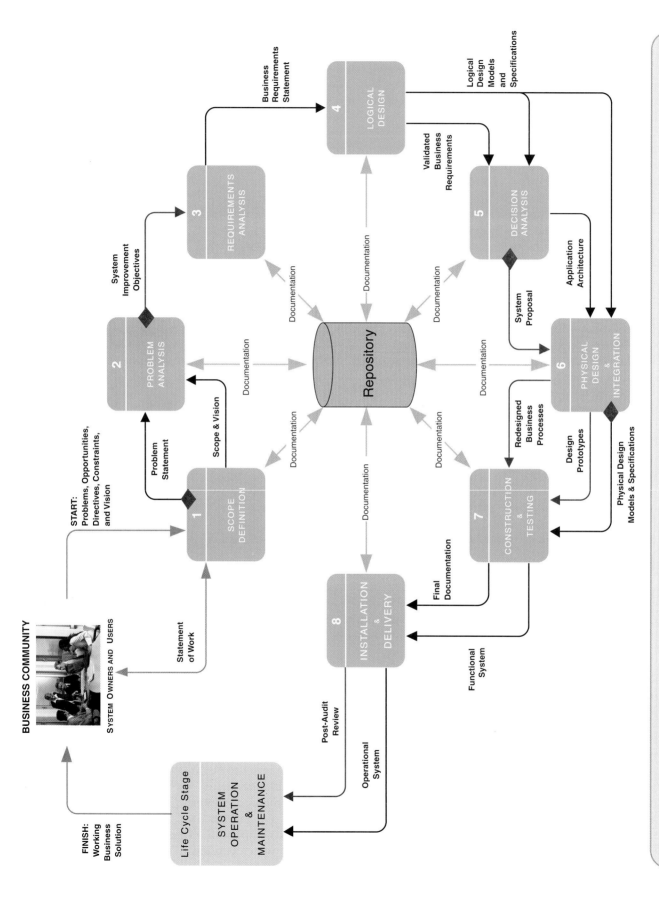

FIGURE 3-7 System Development Documentation, Repository, and Presentation

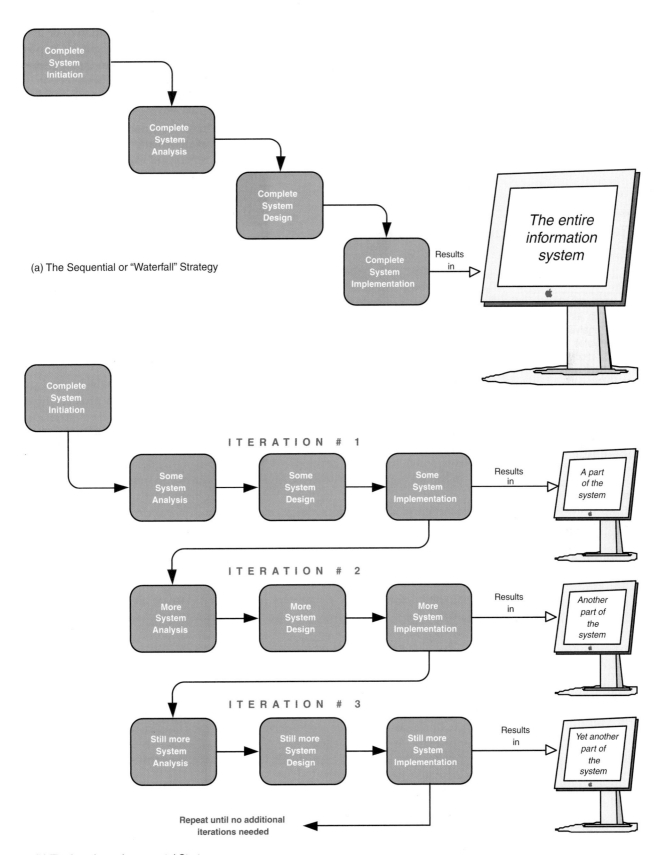

(a) The Sequential or "Waterfall" Strategy

(b) The Iterative or Incremental Strategy

FIGURE 3-8   Sequential versus Iterative Systems Development Approach

project management the process of scoping, planning, staffing, organizing, directing, and controlling a project to develop an information system at minimum cost, within a specified time frame, and with acceptable quality.

approach requires completing enough analysis, design, and implementation to be able to fully develop a *part* of the new system and place it into operation as quickly as possible. Once that version of the system is implemented, the strategy is to then perform some additional analysis, design, and implementation to release the next version of the system. These iterations continue until all parts of the entire information system have been implemented. The popularity of this iterative and incremental process can be explained simply: System owners and users have long complained about the excessive time required to develop and implement information systems using the waterfall approach. The iterative approach allows versions of useable information to be delivered in regular and shorter time frames. This results in improved customer (system owner and user) satisfaction.

Alternative Routes and Strategies

waterfall development approach an approach to systems analysis and design that completes each phase one after another and only once.

iterative development approach an approach to systems analysis and design that completes that entire information system in successive iterations. Each iteration does some analysis, some design, and some construction. Synonyms include incremental and spiral.

Given any destination, there are many routes to that destination and many modes of transport. You could take the superhighway, highways, or back roads, or you could fly. Deciding which route is best depends on your goals and priorities. Do you want to get there fast, or do you want to see the sights? How much are you willing to spend? Are you comfortable with the mode of travel? Just as you would pick your route and means to a travel destination, you can and should pick a route and means for a systems development destination.

So far, we've described a basic set of phases that comprise our *FAST* methodology. At one time, a "one size fits all" methodology was common for most projects; however, today a variety of types of projects, technologies, and development strategies exist—one size no longer fits all projects! Like many contemporary methodologies, *FAST* provides alternative routes and strategies to accommodate different types of projects, technology goals, developer skills, and development paradigms.

In this section, we will describe several *FAST* routes and strategies. Before we do so, examine Figure 3-9 on the following page. The figure illustrates a taxonomy or classification scheme for methodological strategies. Notice the following:

- Methodologies and routes can support the option of either *building software solutions* in-house or *buying a commercial software solution* from a software vendor. Generally, many of the same methods and techniques are applicable to both options.
- Methodologies may be either very *prescriptive* ("Touch all the bases; follow all the rules") or relatively *adaptive* ("Change as needed within certain guidelines").
- Methodologies can also be characterized as *model-driven* ("Draw pictures of the system") or *product-driven* ("Build the product and see how the users react").
- Model-driven methodologies are rapidly moving to a focus on the *object-oriented* technologies being used to construct most of today's systems (more about this later). Earlier model-driven approaches emphasized either process modeling *or* data modeling.
- Finally, product-driven approaches tend to emphasize either rapid *prototyping* or writing program *code* as soon as possible (perhaps you've heard the term *extreme programming*).

So many strategies! Which should you choose? A movement is forming known as *agile methods*. In a nutshell, advocates of agile methods suggest that system analysts and programmers should have a tool box of methods that include tools and techniques from all of the above methodologies. They should choose their tools and techniques based on the problem and situation. *FAST* is an agile methodology. It advocates the integrated use of tools and techniques from many methodologies, applied in the context of repeatable processes (as in CMM Level 3). That said, let's examine some of the route variations and strategies for the *FAST* process. As we

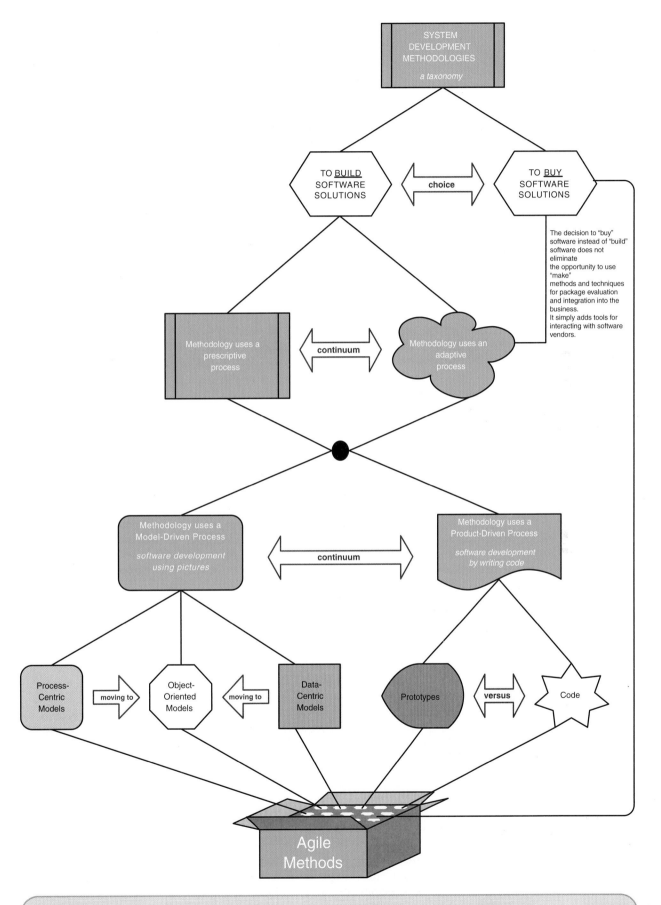

SYSTEM
DEVELOPMENT
METHODOLOGIES

a taxonomy

TO <u>BUILD</u>
SOFTWARE
SOLUTIONS

choice

TO <u>BUY</u>
SOFTWARE
SOLUTIONS

The decision to "buy"
software instead of "build"
software does not
eliminate
the opportunity to use
"make"
methods and techniques
for package evaluation
and integration into the
business.
It simply adds tools for
interacting with software
vendors.

Methodology uses a
prescriptive
process

continuum

Methodology uses an
adaptive
process

Methodology uses a
Model-Driven Process

*software development
using pictures*

continuum

Methodology uses a
Product-Driven Process

*software development
by writing code*

Process-
Centric
Models

moving to

Object-
Oriented
Models

moving to

Data-
Centric
Models

Prototypes

versus

Code

Agile
Methods

FIGURE 3-9   A Taxonomy for System Development Methodologies and Strategies

navigate through each route, we will use red typefaces and arrows to highlight those aspects of the route that differ from the basic route you've already learned.

> The Model-Driven Development Strategy

One of the oldest and most commonly used approaches to analyzing and designing information systems is based on system modeling. As a reminder, a system model is a picture of a system that represents reality or a desired reality. System models facilitate improved communication between system users, system analysts, system designers, and system builders. In the *FAST* methodology, system models are used to illustrate and communicate the KNOWLEDGE, PROCESS, or INTERFACE building blocks of information systems. This approach is called **model-driven development.**

The model-driven development route for *FAST* is illustrated in Figure 3-10. The model-driven approach does not vary much from the basic phases we described earlier. We call your attention to the following notes that correspond to the numbered bullets:

model-driven development a system development strategy that emphasizes the drawing of system models to help visualize and analyze problems, define business requirements, and design information systems.

❶ System models may exist from the project that created the current system. Be careful! These models are notorious for being out of date. But they can still be useful as a point of departure.

❷ Earlier you learned that it is important to define scope for a project. One of the simplest ways to communicate scope is by drawing MODELS THAT SHOW SCOPE DEFINITION. Scope models show which aspects of a problem are within scope and which aspects are outside scope. This is sometimes called a *context diagram* or context model.

❸ Some system modeling techniques call for extensive MODELS OF THE EXISTING SYSTEM to identify problems and opportunities for system improvement. This is sometimes called the *as-is system model.* Modeling of the current system has waned in popularity today. Many project managers and analysts view it as counterproductive or of little value added. The exception is modeling of as-is business processes for the purpose of business process redesign.

❹ The requirements statement is one of the most important deliverables of system development. It sometimes includes MODELS THAT DEPICT HIGH-LEVEL BUSINESS REQUIREMENTS. One of the most popular modeling techniques today is called *use case* (introduced in Chapter 7). Use cases identify requirements and track their fulfillment through the life cycle.

logical model a pictorial representation that depicts *what* a system is or does. Synonyms include essential model, conceptual model, and business model.

❺ Most model-driven techniques require that analysts document business requirements with **logical models** (defined earlier). Business requirements are frequently expressed in LOGICAL MODELS THAT DEPICT MORE DETAILED USER REQUIREMENTS. They show only *what* a system must be or must do. They are implementation *in*dependent; that is, they depict the system independent of any possible technical implementation. Hence, they are useful for depicting and validating business requirements.

❻ As a result of the decision analysis phase, the analyst may produce system MODELS THAT DEPICT APPLICATION ARCHITECTURE. Such models illustrate the planned technical implementation of a system.

physical model a technical pictorial representation that depicts what a system is or does and *how* the system is implemented. Synonyms include implementation model and technical model.

❼ Many model-driven techniques require that analysts develop MODELS THAT DEPICT PHYSICAL DESIGN SPECIFICATIONS (defined earlier in this chapter). Recall that **physical models** show not only what a system is or does but also *how* the system is implemented with technology. They are implementation *de*pendent because they reflect technology choices and the limitations of those technology choices. Examples include database schemas, structure charts, and flowcharts. They serve as a blueprint for construction of the new system.

❽ New information systems must be interwoven into the fabric of an organization's business processes. Accordingly, the analyst and users may develop MODELS OF REDESIGNED BUSINESS PROCESSES.

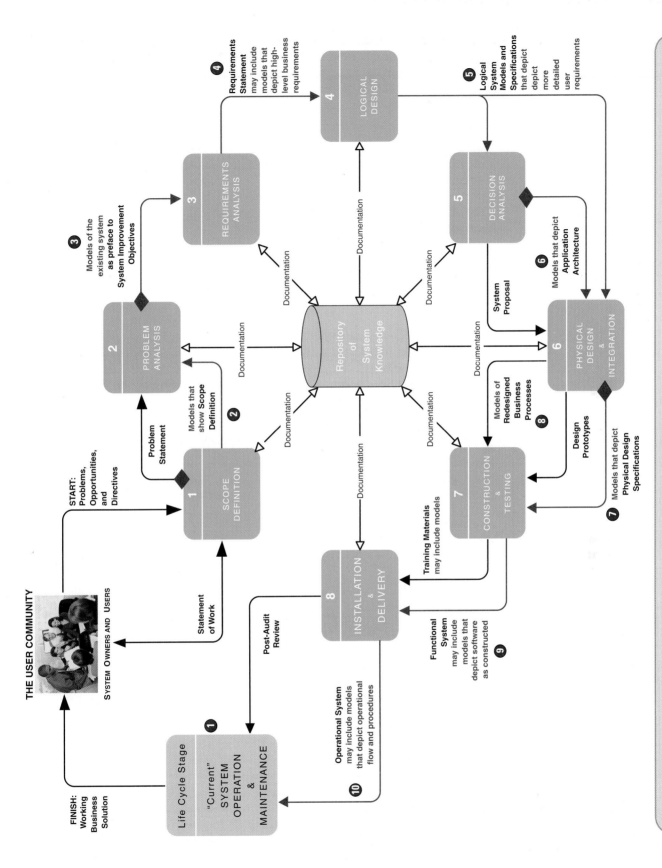

THE USER COMMUNITY

SYSTEM OWNERS AND USERS

START:
Problems,
Opportunities,
and
Directives

FINISH:
Working
Business
Solution

1
SCOPE
DEFINITION

2
PROBLEM
ANALYSIS

3
REQUIREMENTS
ANALYSIS

4
LOGICAL
DESIGN

5
DECISION
ANALYSIS

6
PHYSICAL
DESIGN
&
INTEGRATION

7
CONSTRUCTION
&
TESTING

8
INSTALLATION
&
DELIVERY

Life Cycle Stage

1
"Current"
SYSTEM
OPERATION
&
MAINTENANCE

Repository
of
System
Knowledge

Statement of Work

Problem Statement

2 Models that show Scope Definition

3 Models of the existing system as preface to System Improvement Objectives

4 Requirements Statement may include models that depict high-level business requirements

5 Logical System Models and Specifications that depict more detailed user requirements

6 Models that depict Application Architecture

System Proposal

7 Models that depict Physical Design Specifications

Design Prototypes

8 Models of Redesigned Business Processes

Training Materials may include models

9 Functional System may include models that depict software as constructed

10 Operational System may include models that depict operational flow and procedures

Post-Audit Review

Documentation

Documentation

Documentation

Documentation

Documentation

Documentation

Documentation

Documentation

Documentation

FIGURE 3-10 The Model-Driven System Development Strategy

95

9 Construction translates the physical system models into software. In some cases, automated tools exist to automatically translate software into PHYSICAL MODELS THAT DEPICT SOFTWARE CONSTRUCTED. This is called *reverse engineering*.

10 Finally, the operational system may include MODELS THAT DEPICT FLOW AND PROCEDURE. For example, system models may document backup and recovery procedures.

In summary, system models can be produced as a portion of the deliverables for most phases. Model-driven approaches emphasize system modeling. Once implemented, the system models serve as documentation for any changes that might be needed during the operation and support stage of the life cycle.

The model-driven approach is believed to offer several advantages and disadvantages, as listed below:

Advantages

- Requirements specification tends to be more thorough and better documented.
- Business requirements and system designs are easier to validate with pictures than words.
- It is easier to identify, conceptualize, and analyze alternative technical solutions.
- Design specifications tend to be more sound, stable, adaptable, and flexible because they are model based and more thoroughly analyzed *before* they are built.
- Systems can be constructed more correctly the first time when built from thorough and clear model based specifications. Some argue that code-generating software can automatically generate skeleton or near-complete code from good system models.

Disadvantages

- It is time-consuming. It takes time to collect the facts, draw the models, and validate those models. This is especially true if users are uncertain or imprecise about their system requirements.
- The models can only be as good as the users' understanding of those requirements.
- Pictures are not software—some argue that this reduces the users' role in a project to passive participation. Most users don't get excited about pictures. Instead, they want to see working software, and they gauge project progress by the existence of software (or its absence).
- The model-driven approach is considered by some to be inflexible—users must fully specify requirements before design, design must fully document technical specifications before construction, and so forth. Some view such rigidity as impractical.

process modeling a process-centered technique popularized by the *structured analysis and design* methodology that used models of business process requirements to derive effective software designs for a system. Structured analysis introduced a modeling tool called the *data flow diagram* to illustrate the flow of data through a series of business processes. Structured design converted data flow diagrams into a process model called *structure charts* to illustrate a top-down software structure that fulfills the business requirements.

Model-driven development is most effective for systems for which requirements are well understood and which are so complex that they require large project teams to complete. The approach also works well when fulfillment of user expectations and quality is more important than cost and schedule.

There are several different model-driven techniques. They differ primarily in terms of the types of models that they require the systems analyst to draw and validate. Let's briefly examine three of the most popular model-driven development techniques that will be taught in this book. Please note that we are introducing only the techniques here, not the models. We'll teach the models themselves later, in the "how to" chapters.

Process Modeling   **Process modeling** was founded in the structured analysis and design methodologies in 1978. While structured analysis and design has lost favor as a methodology, process modeling remains a viable and important technique. Recall that your information system building blocks include several possible focuses: KNOWLEDGE, PROCESSES, and INTERFACES. Process modeling focuses on the PROCESS column of building blocks. *Flowcharts* are one type of process model (used primarily by SYSTEM BUILDERS) that you may have encountered in a programming course. Process modeling has enjoyed something of a renaissance with the emergence of business process redesign (introduced in Chapter 1).

Data flow diagrams and structure charts have contributed significantly to reducing the communications gap that often exists between nontechnical system owners and users and technical system designers and builders. Process modeling is taught in this book.

Data Modeling   Recall that KNOWLEDGE improvement is a fundamental goal and set of building blocks in your framework. Knowledge is the product of *information,* which in turn is the product of *data.* **Data modeling** methods emphasize the knowledge building blocks, especially data. In the data modeling approach, emphasis is placed on diagrams that capture business data requirements and translate them into database designs. Arguably, data modeling is the most widely practiced system modeling technique. Hence, it will be taught in this book.

> **data modeling** a data-centered technique used to model business data requirements and design database systems that fulfill those requirements. The most frequently encountered data models are *entity relationship diagrams.*

Object Modeling   Object modeling is the result of technical advancement. Today, most programming languages and methods are based on the emergence of object technology. While the concepts of object technology are covered extensively throughout this book, a brief but oversimplified introduction is appropriate here.

For the past 30 years, techniques like process and data modeling deliberately separated the concerns of PROCESSES from those of DATA. In other words, process and data models were separate and distinct. Because virtually all systems included processes *and* data, the techniques were frequently used in parallel and the models had to be carefully synchronized. Object techniques are an attempt to eliminate the separation of concerns, and hence the need for synchronization of data and process concerns. This has given rise to **object modeling** methods.

Business objects correspond to real things of importance in the business such as *customers* and the *orders* they place for *products.* Each object consists of *both* the data that describes the object and the processes that can create, read, update, and delete that object. With respect to your information system building blocks, object-oriented analysis and design (OOAD) significantly changes the paradigm. The DATA and PROCESS columns (and, arguably, the INTERFACE column as well) are essentially merged into a single OBJECT column. The models then focus on identifying objects, building objects, and assembling appropriate objects, as with *Legos,* into useful information systems.

> **object modeling** a technique that attempts to merge the data and process concerns into singular constructs called *objects.* Object models are diagrams that document a system in terms of its objects and their interactions. Object modeling is the basis for *object-oriented analysis and design* methodologies.

The current popularity of object technology is driving the interest in object models and OOAD. For example, most of today's popular operating systems like Microsoft *Windows* and Apple *Mac/OS* have object-oriented user interfaces ("point and click," using objects such as windows, frames, drop-down menus, radio buttons, checkboxes, scroll bars, and the like). Web user interfaces like Microsoft *Internet Explorer* and Netscape *Navigator* are also based on object technology. Object programming languages such as *Java, C++, C#, Smalltalk,* and *Visual Basic .NET* are used to construct and assemble such object-oriented operating systems and applications. And those same languages have become the tools of choice for building next-generation information system applications. Not surprisingly, object modeling techniques have been created to express business and software requirements and designs in terms of objects. This edition of this book extensively integrates the most popular object modeling techniques to prepare you for systems analysis and design that ultimately produces today's object-based information systems and applications.

> The Rapid Application Development Strategy

rapid application development (RAD) a system development strategy that emphasizes speed of development through extensive user involvement in the rapid, iterative, and incremental construction of a series of functioning **prototypes** of a system that eventually evolves into the final system (or a version).

In response to the faster pace of the economy in general, **rapid application development (RAD)** has become a popular route for accelerating systems development. The basic ideas of RAD are:

- To more actively involve system users in the analysis, design, and construction activities.
- To organize systems development into a series of focused, intense workshops jointly involving SYSTEM OWNERS, USERS, ANALYSTS, DESIGNERS, and BUILDERS.
- To accelerate the requirements analysis and design phases through an iterative construction approach.
- To reduce the amount of time that passes before the users begin to see a working system.

prototype a small-scale, representative, or working model of the users' requirements or a proposed design for an information system. Any given prototype may omit certain functions or features until such time as the prototype has sufficiently evolved into an acceptable implementation of requirements.

The basic principle behind prototyping is that users know what they want when they see it working. In RAD, a **prototype** eventually evolves into the final information system. The RAD route for *FAST* is illustrated in Figure 3-11. Again, the red text and flows indicate the deviations from the basic *FAST* process. We call your attention to the following notes that correspond to the numbered bullets:

❶ The emphasis is on reducing time in developing applications and systems; therefore, the initial problem analysis, requirements analysis, and decision analysis phases are consolidated and accelerated. The deliverables are typically abbreviated, again in the interest of time. The deliverables are said to be INITIAL, meaning "expected to change" as the project progresses.

   After the above initial analysis, the RAD uses an iterative approach, as discussed earlier in the chapter. Each iteration emphasizes only enough new functionality to be accomplished within a few weeks.

❷ LOGICAL AND PHYSICAL DESIGN SPECIFICATIONS are usually significantly abbreviated and accelerated. In each iteration of the cycle, only some design specifications will be considered. While some system models may be drawn, they are selectively chosen and the emphasis continues to be on rapid development. The assumption is that errors can be caught and fixed in the next iteration.

❸ In some, but rarely all, iterations, some business processes may need to be redesigned to reflect the likely integration of the evolving software application.

❹ In each iteration of the cycle, SOME DESIGN PROTOTYPES or SOME PARTIAL FUNCTIONAL SYSTEM elements are constructed and tested. Eventually, the completed application will result from the final iteration through the cycle.

❺ After each prototype or partial functional subsystem is constructed and tested, system users are given the opportunity to experience working with that prototype. The expectation is that users will clarify requirements, identify new requirements, and provide BUSINESS FEEDBACK on design (e.g., ease of learning, ease of use) for the next iteration through the RAD cycle.

❻ After each prototype or functioning subsystem is constructed and tested, system analysts and designers will review the application architecture and design to provide TECHNICAL FEEDBACK and direction for the next iteration through the RAD cycle.

❼ Based on the feedback, systems analysts will identify REFINED SYSTEM IMPROVEMENT OBJECTIVES and/or BUSINESS REQUIREMENTS. This analysis tends to focus on revising or expanding objectives and requirements and identifying user concerns with the design.

❽ Based on the feedback, systems analysts and system designers will identify a REFINED APPLICATION ARCHITECTURE and/or DESIGN CHANGES.

❾ Eventually, the system (or a version of the system) will be deemed worthy of implementation. This CANDIDATE RELEASE VERSION OF THE FUNCTIONAL SYSTEM is system tested and placed into operation. The next version of the system may continue iterating through the RAD cycle.

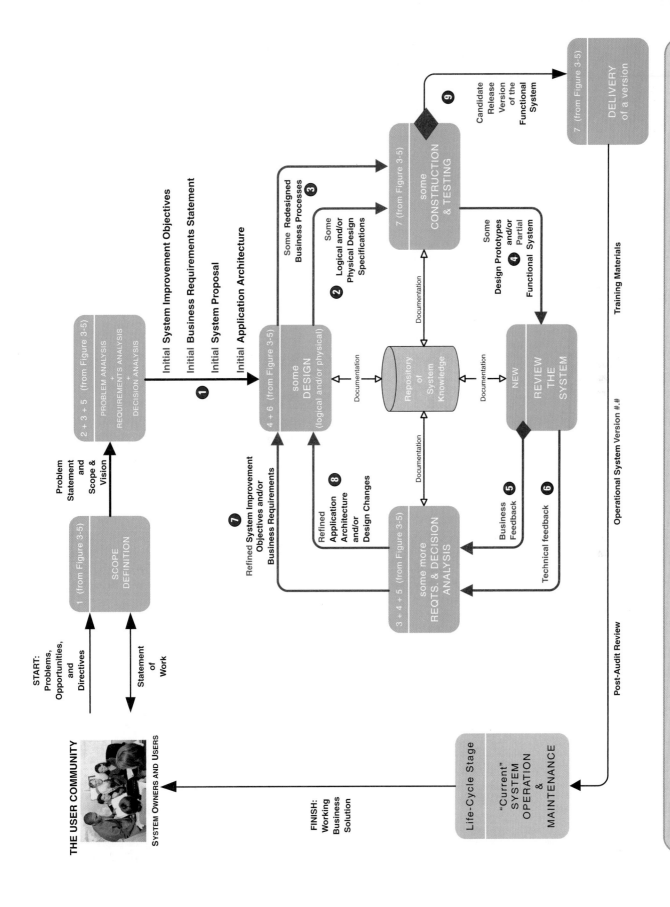

FIGURE 3-11 The Rapid Application Development (RAD) Strategy

timeboxing the imposition of a nonextendable period of time, usually 60 to 90 days, by which the first (or next) version of a system must be delivered into operation.

Although not a rigid requirement for RAD, the duration of the prototyping loop can be limited using a technique called **timeboxing.** Timeboxing seeks to deliver an operational system to users and management on a regular, recurring basis. Advocates of timeboxing argue that management and user enthusiasm for a project can be enhanced and sustained because a working version of the system is implemented on a regular basis.

The RAD approach offers several advantages and disadvantages:

Advantages

- It is useful for projects in which the user requirements are uncertain or imprecise.
- It encourages active user and management participation (as opposed to a passive reaction to nonworking system models). This increases end-user enthusiasm for the project.
- Projects have higher visibility and support because of the extensive user involvement throughout the process.
- Users and management see working, software-based solutions more rapidly than they do in model-driven development.
- Errors and omissions tend to be detected earlier in prototypes than in system models.
- Testing and training are natural by-products of the underlying prototyping approach.
- The iterative approach is a more natural process because change is an expected factor during development.

Disadvantages

- Some argue that RAD encourages a "code, implement, and repair" mentality that increases lifetime costs required to operate, support, and maintain the system.
- RAD prototypes can easily solve the wrong problems since problem analysis is abbreviated or ignored.
- A RAD-based prototype may discourage analysts from considering other, more worthy technical alternatives.
- Sometimes it is best to throw a prototype away, but stakeholders are often reluctant to do so because they see this as a loss of time and effort in the current product.
- The emphasis on speed can adversely impact quality because of ill-advised shortcuts through the methodology.

RAD is most popular for small- to medium-size projects. We will demonstrate prototyping and RAD techniques in appropriate chapters of this book.

> The Commercial Application Package Implementation Strategy

commercial application package a software application that can be purchased and customized (within limits) to meet the business requirements of a large number of organizations or a specific industry. A synonym is *commercial off-the-shelf (COTS) system.*

Sometimes it makes more sense to buy an information system solution than to build one in-house. In fact, many organizations increasingly expect to build software in-house only when there is a competitive advantage to be gained. And for many core applications such as human resources, financials, procurement, manufacturing, and distribution, there is little competitive value in building your own system—hence a **commercial application package** is purchased. Accordingly, our *FAST* methodology includes a commercial software package route.

The ultimate commercial solution is enterprise resource planning, or ERP (defined in Chapter 1). ERP solutions provide *all* of the core information system applications for an entire business. For most organizations, an ERP implementation represents the single largest information system project ever undertaken by the organization. It can cost tens or hundreds of millions of dollars and require a small army of managers, users, analysts, technical specialists, programmers, consultants, and contractors.

The *FAST* methodology's route for commercial application package integration is not really intended for ERP projects. Indeed, most ERP vendors provide their own implementation methodology (and consulting partners) to help their customers implement such a massive software solution. Instead, our *FAST* methodology provides a route for implementing all other types of information system solutions that may be purchased by a business. For example, an organization might purchase a commercial application package for a single business function such as accounting, human resources, or procurement. The package must be selected, installed, customized, and integrated into the business and its other existing information systems.

The basic ideas behind our commercial application package implementation route are:

- Packaged software solutions must be carefully selected to fulfill business needs—"You get what you ask and pay for."
- Packaged software solutions not only are costly to purchase but can be costly to implement. In fact, the package route can actually be more expensive to implement than an in-house development route.
- Software packages must usually be customized for and integrated into the business. Additionally, software packages usually require the redesign of existing business processes to adapt to the software.
- Software packages rarely fulfill all business requirements to the users' complete satisfaction. Thus, some level of in-house systems development is necessary in order to meet the unfulfilled requirements.

The commercial application package implementation route is illustrated in Figure 3-12. Once again, the red typeface and arrows indicate differences from the basic *FAST* process. We call your attention to the following notes that correspond to the numbered bullets:

❶ It should be noted that the decision to purchase a package is determined in the problem analysis phase. The red diamond represents the "make versus buy" decision. The remainder of this discussion assumes that a decision to buy has been approved.

❷ Problem analysis usually includes some initial TECHNOLOGY MARKET RESEARCH to identify what package solutions exist, what features are in the software, and what criteria should be used to evaluate such application packages. This research may involve software vendors, IT research services (such as the Gartner Group), or consultants.

❸ After defining business requirements, the requirements must be communicated to the software vendors who offer viable application solutions. The business (and technical) requirements are formatted and communicated to candidate software vendors as either a REQUEST FOR PROPOSAL (RFP) or a REQUEST FOR QUOTATION (RFQ). The double-ended arrow implies that there may need to be some clarification of requirements and criteria.

❹ Vendors submit PROPOSALS or QUOTATIONS for their application solutions. These proposals are evaluated against the business and technical requirements specified in the RFP. The double-ended arrow indicates that claimed features and capabilities must be validated and in some instances clarified. This occurs during the decision analysis phase.

❺ A CONTRACT AND ORDER is negotiated with the winning vendor for the software and possibly for services necessary to install and maintain the software.

❻ The vendor provides the BASELINE COMMERCIAL APPLICATION software and documentation. Services for installation and implementation of the software are frequently provided by the vendor or its service providers (certified consultants).

❼ When an application package is purchased, the organization must nearly always change its business processes and practices to work efficiently with the package. The need for REDESIGNED BUSINESS PROCESSES is rarely greeted with enthusiasm,

request for proposal (RFP) a formal document that communicates business, technical, and support requirements for an application software package to vendors that may wish to compete for the sale of that application package and services.

request for quotation (RFQ) a formal document that communicates business, technical, and support requirements for an application software package to a single vendor that has been determined as being able to supply that application package and services.

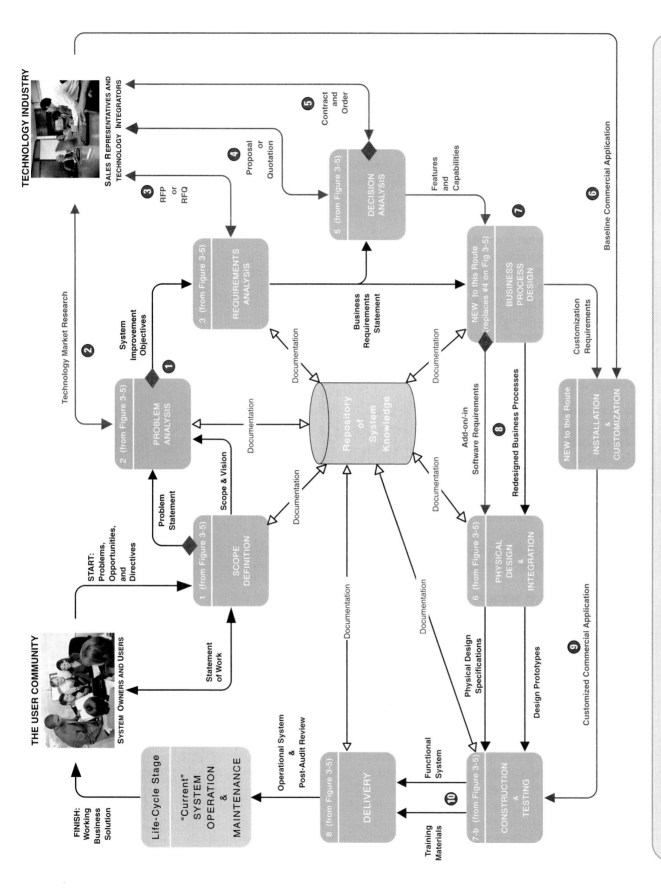

THE USER COMMUNITY

SYSTEM OWNERS AND USERS

TECHNOLOGY INDUSTRY

SALES REPRESENTATIVES AND TECHNOLOGY INTEGRATORS

START: Problems, Opportunities, and Directives

Statement of Work

1 (from Figure 3-5) SCOPE DEFINITION

Problem Statement

Scope & Vision

2 (from Figure 3-5) PROBLEM ANALYSIS

System Improvement Objectives

Technology Market Research

3 (from Figure 3-5) REQUIREMENTS ANALYSIS

Business Requirements Statement

③ RFP or RFQ

④ Proposal or Quotation

⑤ Contract and Order

5 (from Figure 3-5) DECISION ANALYSIS

Features and Capabilities

7 NEW to this Route (replaces #4 on Fig 3-5) BUSINESS PROCESS DESIGN

⑥ Baseline Commercial Application

Customization Requirements

NEW to this Route INSTALLATION & CUSTOMIZATION

Add-on/-in Software Requirements

⑧ Redesigned Business Processes

6 (from Figure 3-5) PHYSICAL DESIGN & INTEGRATION

Physical Design Specifications

Design Prototypes

⑨ Customized Commercial Application

Repository of System Knowledge

Documentation

7-b (from Figure 3-5) CONSTRUCTION & TESTING

Functional System

Training Materials

⑩

8 (from Figure 3-5) DELIVERY

Operational System & Post-Audit Review

Life-Cycle Stage

"Current" SYSTEM OPERATION & MAINTENANCE

FINISH: Working Business Solution

①

②

FIGURE 3-12 The Commercial Application Package Implementation Strategy

102

but they are usually necessary. In many cases, the necessary changes are not wrong—they are just different and unfamiliar. System users tend to be uncomfortable with changing the way they have always done something.

❽ An application package rarely meets all business requirements upon installation. Typically, a **gap analysis** must be performed to determine which business requirements are not fulfilled by the package's capabilities and features. For requirements that will not be fulfilled, the following options exist:

- Request customizations of the package within allowable limits as specified by the software vendor. Most commercial application packages allow the purchaser to set specific options, preferences, and defined values and ranges for certain parameters. Within limits, these customizations allow you to "personalize" the package to the business accounting and business practices. Such necessary CUSTOMIZATION REQUIREMENTS need to be specified.

- Define ADD-ON SOFTWARE REQUIREMENTS. Add-on software requirements specify programs that must be designed and constructed to augment the commercial application package and deliver additional functionality. It should be noted that add-on programs carry some risk that they may have to be modified in the future when a new version of the software becomes available. But this risk is nominal, and most organizations take the risk in order to provide additional functionality.

- Define ADD-IN SOFTWARE REQUIREMENTS. Add-in software requirements are very dangerous! They specify changes to the actual commercial application package to meet business requirements. In other words, users are requesting that changes be made to the purchased software, its database, or its user interfaces. At best, such changes can make version upgrades extremely difficult and prohibitively expensive. At worst, such changes can invalidate technical support from the vendor. (And most vendors encourage keeping versions relatively current by canceling technical support on older versions.) Changing program code and database structures should be discouraged. Insistence by users is often a symptom of unwillingness to adapt business processes to work with the application. Many organizations prohibit changes to application packages and force users to adapt in order to preserve their upgrade path.

❾ The BASELINE COMMERCIAL APPLICATION is installed and tested. Allowable changes based on options, preferences, and parameters are completed and tested. Note: These customizations within the limits specified by the software vendor will typically carry forward to version upgrades. In most instances the vendor has provided for this level of CUSTOMIZED COMMERCIAL APPLICATION.

❿ Any add-on (or add-in) software changes are designed and constructed to meet additional business requirements. The system is subsequently tested and placed into operation using the same activities described in the basic *FAST* process.

The commercial application package strategy offers its own advantages and disadvantages:

> **gap analysis** a comparison of business and technical requirements for a commercial application package against the capabilities and features of a specific commercial application package for the purpose of defining the requirements that cannot be met.

Advantages

- New systems can usually be implemented more quickly because extensive programming is not required.
- Many businesses cannot afford the staffing and expertise required to develop in-house solutions.
- Application vendors spread their development costs across all customers that purchase their software. Thus, they can invest in continuous

Disadvantages

- A successful COTS implementation is dependent on the long-term success and viability of the application vendor—if the vendor goes out of business, you can lose your technical support and future improvements.
- A purchased system rarely reflects the ideal solution that the business might achieve with an in-house-developed system that could be

improvements in features, capabilities, and usability that individual businesses cannot always afford.
- The application vendor assumes responsibility for significant system improvements and error corrections.
- Many business functions are more similar than dissimilar for all businesses in a given industry. For example, business functions across organizations in the health care industry are more alike than different. It does not make good business sense for each organization to "reinvent the wheel."

customized to the precise expectations of management and the users.
- There is almost always at least some resistance to changing business processes to adapt to the software. Some users will have to give up or assume new responsibilities. And some people may resent changes they perceive to be technology-driven instead of business-driven.

Regardless, the trend toward purchased commercial application packages cannot be ignored. Today, many businesses require that a package alternative be considered prior to engaging in any type of in-house development project. Some experts estimate that by the year 2005 businesses will purchase 75 percent of their new information system applications. For this reason, we will teach systems analysis tools and techniques needed by system analysts to function in this environment.

> Hybrid Strategies

The *FAST* routes are not mutually exclusive. Any given project may elect to or be required to use a combination of, or variation of, more than one route. The route to be used is always selected during the *scope definition* phase and is negotiated as part of the *statement of work*. One strategy that is commonly applied to both model-driven and rapid application development routes is an incremental strategy. Figure 3-13 illustrates one possible implementation of an incremental strategy in combination with rapid application development. The project delivers the information system into operation in four stages. Each stage implements a version of the final system using a RAD route. Other variations on routes are possible.

> System Maintenance

All routes ultimately result in placing a new system into operation. System maintenance is intended to guide projects through the operation and support stage of their life cycle—which could last decades! Figure 3-14 places system maintenance into perspective. System maintenance in *FAST* is not really a unique route. As illustrated in the figure, it is merely a smaller-scale version of the same *FAST* process (or route) that was used to originally develop the system. The figure demonstrates that the starting point for system maintenance depends on the problem to be solved. We call your attention to the following numbered bullets in the figure:

❶ Maintenance and reengineering projects are triggered by some combination of user and technical feedback. Such feedback may identify new problems, opportunities, or directives.

❷ The maintenance project is initiated by a SYSTEM CHANGE REQUEST that indicates the problems, opportunities, or directives.

❸ The simplest fixes are SOFTWARE BUGS (errors). Such a project typically jumps right into a RECONSTRUCTION AND RETESTING phase and is solved relatively quickly.

❹ Sometimes a DESIGN FLAW in the system becomes apparent after implementation. For example, users may frequently make the same mistake due to a confusing screen design. For this type of maintenance project, the PHYSICAL DESIGN AND INTEGRATION phase would need to be revisited, followed of course by the construction and delivery phases.

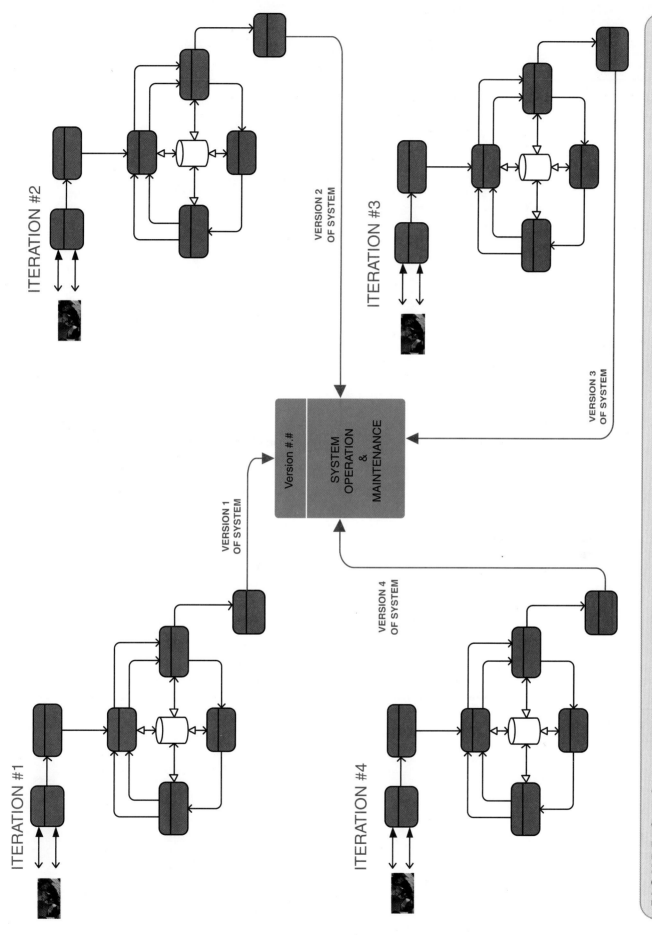

ITERATION #1

ITERATION #2

ITERATION #3

ITERATION #4

VERSION 1
OF SYSTEM

VERSION 2
OF SYSTEM

VERSION 3
OF SYSTEM

VERSION 4
OF SYSTEM

Version #.#

SYSTEM
OPERATION
&
MAINTENANCE

FIGURE 3-13 An Incremental Implementation Strategy

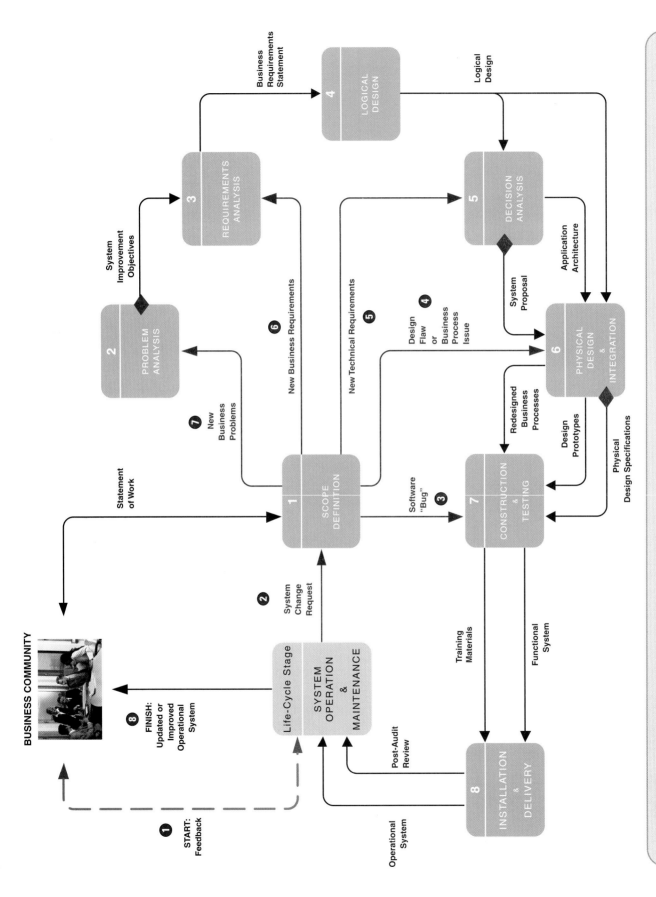

BUSINESS COMMUNITY

Statement of Work

1 START: Feedback

2 System Change Request

Life-Cycle Stage

SYSTEM OPERATION & MAINTENANCE

8 FINISH: Updated or Improved Operational System

Operational System

Post-Audit Review

8 INSTALLATION & DELIVERY

Training Materials

Functional System

7 CONSTRUCTION & TESTING

3 Software "Bug"

1 SCOPE DEFINITION

7 New Business Problems

6 New Business Requirements

2 PROBLEM ANALYSIS

System Improvement Objectives

3 REQUIREMENTS ANALYSIS

Business Requirements Statement

4 LOGICAL DESIGN

Logical Design

5 New Technical Requirements

5 DECISION ANALYSIS

System Proposal

Application Architecture

4 Design Flaw or Business Process Issue

6 PHYSICAL DESIGN & INTEGRATION

Redesigned Business Processes

Design Prototypes

Physical Design Specifications

FIGURE 3-14 A System Maintenance Perspective

106

In some cases a BUSINESS PROCESS ISSUE may become apparent. In this case, only the business process needs to be redesigned.

❺ On occasion, NEW TECHNICAL REQUIREMENTS might dictate a change. For example, an organization may standardize on the newest version of a particular database management system such as *SQL Server* or *Oracle*. For this type of project, the DECISION ANALYSIS phase may need to be revisited to first determine the risk and feasibility of converting the existing, operational database to the new version. As appropriate, such a project would subsequently proceed to the physical design, construction, and delivery phases as necessary.

❻ Businesses constantly change; therefore, business requirements for a system also change. One of the most common triggers for a reengineering project is a NEW (or revised) BUSINESS REQUIREMENT. Given the requirement, the REQUIREMENTS ANALYSIS phase must be revisited with a focus on the impact of the new requirement on the existing system. Based on requirements analysis, the project would then proceed to the logical design, decision analysis, physical design, construction, and delivery phases.

Time out! By now, you've noticed that maintenance and reengineering projects initiate in different phases of the basic methodology. You might be concerned that repeating these phases would require excessive time and effort. Keep in mind, however, that the scope of maintenance and reengineering projects is much, much smaller than the original project that created the operational system. Thus, the work required in each phase is much less than you might have guessed.

❼ Again, as businesses change, significant NEW BUSINESS PROBLEMS, opportunities, and constraints can be encountered. In this type of project, work begins with the PROBLEM ANALYSIS phase and proceeds as necessary to the subsequent phases.

❽ In all cases, the final result of any type of maintenance or reengineering project is an updated operational business system that delivers improved value to the system users and owners. Updates may include revised programs, databases, interfaces, or business processes.

As described earlier in the chapter, we expect all systems to eventually reach entropy. The business and/or technical problems and requirements have become so troublesome as to warrant "starting over."

That completes our introduction to the systems development methodology and routes. Before we end this chapter, we should introduce the role of automated tools that support systems development.

Automated Tools and Technology

You may be familiar with the old fable of the cobbler (shoemaker) whose own children had no shoes. That situation is not unlike the one faced by some systems developers. For years we've been applying information technology to solve our users' business problems; however, we've sometimes been slow to apply that same technology to our own problem—developing information systems. In the not-too-distant past, the principal tools of the systems analyst were paper, pencil, and flowchart template.

Today, entire suites of automated tools have been developed, marketed, and installed to assist systems developers. While system development methodologies do not always require automated tools, most methodologies do benefit from such technology. Some of the most commonly cited benefits include:

- Improved productivity—through automation of tasks.
- Improved quality—because automated tools check for completeness, consistency, and contradictions.
- Better and more consistent documentation—because the tools make it easier to create and assemble consistent, high-quality documentation.

- Reduced lifetime maintenance—because of the aforementioned system quality improvements combined with better documentation.
- Methodologies that really work—through rule enforcement and built-in expertise.

Chances are that your future employer is using or will be using this technology to develop systems. We will demonstrate the use of various automated tools throughout this textbook. There are three classes of automated tools for developers:

- Computer-aided systems modeling.
- Application development environments.
- Project and process managers.

Let's briefly examine each of these classes of automated tools.

> Computer-Assisted Systems Engineering

Systems developers have long aspired to transform information systems and software development into engineering-like disciplines. The terms *systems engineering* and *software engineering* are based on a vision that systems and software development can and should be performed with engineering-like precision and rigor. Such precision and rigor are consistent with the model-driven approach to systems development. To help systems analysts better perform system modeling, the industry developed automated tools called **computer-assisted software engineering (CASE)** tools. Think of CASE technology as software that is used to design and implement other software. This is very similar to the computer-aided design (CAD) technology used by most contemporary engineers to design products such as vehicles, structures, machines, and so forth. Representative modeling products are listed in the margin. Most modeling products run on personal computers, as depicted in Figure 3-15.

CASE Repositories   At the center of any true CASE tool's architecture is a developer's database called a **CASE repository.** The repository concept was introduced earlier (see Figure 3-7).

Around the CASE repository is a collection of tools, or facilities, for creating system models and documentation.

CASE Facilities   To use the repository, the CASE tools provide some combination of the following facilities, illustrated in Figure 3-16:

- *Diagramming tools* are used to draw the system models required or recommended in most system development methodologies. Usually, the shapes on one system model can be linked to other system models and to detailed descriptions (see next item below).
- *Dictionary tools* are used to record, delete, edit, and output detailed documentation and specifications. The descriptions can be associated with shapes appearing on system models that were drawn with the diagramming tools.
- *Design tools* can be used to develop mock-ups of system components such as inputs and outputs. These inputs and outputs can be associated with both the aforementioned system models and the descriptions.
- *Quality management tools* analyze system models, descriptions and specifications, and designs for completeness, consistency, and conformance to accepted rules of the methodologies.
- *Documentation tools* are used to assemble, organize, and report on system models, descriptions and specifications, and prototypes that can be reviewed by system owners, users, designers, and builders.
- *Design and code generator tools* automatically generate database designs and application programs or significant portions of those programs.
- *Testing tools* simulate transactions and data traffic, measure performance, and provide configuration management of test plans and test scripts.

Forward and Reverse Engineering   As previous stated, CASE technology automates system modeling. Today's CASE tools provide two distinct ways to develop system models—**forward engineering** and **reverse engineering.** Think of reverse

REPRESENTATIVE CASE TOOLS

Computer Associates' *Erwin*
Oracle's *Designer 2000*
Popkin's *System Architect*
Rational *ROSE*
Visible Systems' *Visible Analyst*

computer-assisted software engineering (CASE) the use of automated software tools that support the drawing and analysis of system models and associated specifications. Some CASE tools also provide prototyping and code generation capabilities.

CASE repository a system developers' database where developers can store system models, detailed descriptions and specifications, and other products of systems development. Synonyms data include data *dictionary* and *encyclopedia*.

forward engineering a CASE tool capability that can generate initial software or database code directly from system models.

reverse engineering a CASE tool capability that can automatically generate initial system models from software or database code.

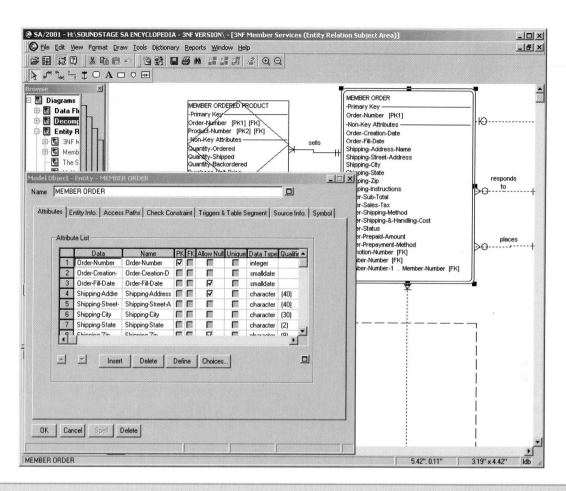

FIGURE 3-15   Screen Capture of *System Architect* CASE Tool

engineering as allowing you to generate a flowchart from an existing program and of forward engineering as allowing you to generate a program directly from a flowchart. CASE tools that allow for bidirectional, forward, and reverse engineering are said to provide for "round-trip engineering." For example, you reverse engineer a poorly designed system into a system model, edit and improve that model, and then forward engineer the new model into an improved system.

> Application Development Environments

The emphasis on speed and quality in software development has resulted in RAD approaches. The potential for RAD has been amplified by the transformation of programming language compilers into complete **application development environments (ADEs).** ADEs make programming simpler and more efficient. Indeed, most programming language compilers are now integrated into an ADE. Examples of ADEs (and the programming languages they support, where applicable) are listed in the margin.

Application development environments provide a number of productivity and quality management facilities. The ADE vendor provides some of these facilities. Third-party vendors provide many other facilities that can integrate into the ADE.

- *Programming languages* or *interpreters* are the heart of an ADE. Powerful debugging features and assistance are usually provided to help programmers quickly identify and solve programming problems.
- *Interface construction tools* help programmers quickly build the user interfaces using a component library.

REPRESENTATIVE ADEs

IBM's *Websphere (Java)*
Borland's *J Builder (Java)*
Macromedia's *Cold Fusion*
Microsoft's *Visual Studio.NET (VB .NET, C#, C++ .NET)*
Oracle's *Developer*
Sybase's *Powerbuilder*

application development environment (ADE) an integrated software development tool that provides all the facilities necessary to develop new application software with maximum speed and quality. A common synonym is *integrated development environment (IDE).*

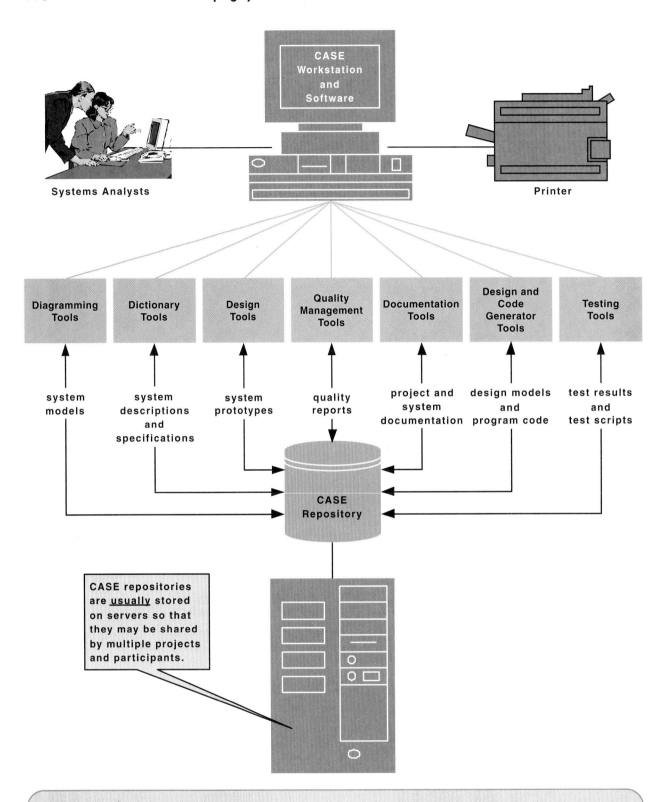

FIGURE 3-16 CASE Tool Architecture

- *Middleware* is software that helps programmers integrate the software being developed with various databases and computer networks.
- *Testing tools* are used to build and execute test scripts that can consistently and thoroughly test software.
- *Version control* tools help multiple programmer teams manage multiple versions of a program, both during development and after implementation.
- *Help authoring tools* are used to write online help systems, user manuals, and online training.
- *Repository links* permit the ADE to integrate with CASE tool products as well as other ADEs and development tools.

> Process and Project Managers

A third class of automated tools helps us manage the system development methodology and projects that use the methodology. While CASE tools and ADEs are intended to support analysis, design, and construction of new information systems and software, **process manager application** and **project manager application** tools are intended to support cross life-cycle activities. Microsoft's *Project* and Niku's *Open Workbench* and *Project Manager* are examples of automated project management tools. Because process and project management is the subject of the next chapter, you'll learn more about these automated tools in that chapter.

process manager application an automated tool that helps to document and manage a methodology and routes, its deliverables, and quality management standards. An emerging synonym is *methodware*.

project manager application an automated tool that helps to plan system development activities (preferably using the approved methodology), estimate and assign resources (including people and costs), schedule activities and resources, monitor progress against schedule and budget, control and modify schedule and resources, and report project progress.

Learning Roadmap

This chapter, along with the first two, completes the minimum context for studying systems analysis and design. We have described that context in terms of the three Ps—the participants (the stakeholders; Chapter 1), the product (the information system; Chapter 2), and the process (the system development; Chapter 3). Armed with this understanding, you are now ready to study systems analysis and/or design methods.

For many of you, Chapter 4, "Project Management," will provide a more complete context for studying systems analysis and design. It builds on Chapter 3 by emphasizing a variety of management issues related to the system development process. These include methodology management, resource management, management of expectations, change management, and others.

For those of you who skip the project and process management chapter, your next assignment will depend on whether your goal is:

- A comprehensive survey of systems development—We recommend you continue your sequential path to Chapter 5, "Systems Analysis." In that chapter you will study in greater depth the scope definition, problem analysis, requirements analysis, and logical design phases that were introduced in Chapter 3.
- The study of systems analysis techniques—Again, we recommend you continue your sequential path to Chapter 5, "Systems Analysis." Chapter 5 will provide a context for subsequently studying the tools and techniques of systems analysis.
- The study of systems design techniques—You might still want to quickly skim or review Chapter 5, "Systems Analysis," for context. Then continue your detailed study in Chapter 12, "System Design." In that chapter, you will learn more about the process and strategies for system design and construction of information systems.

Summary

1. A systems development process is a set of activities, methods, best practices, deliverables, and automated tools that stakeholders use to develop and continuously improve information systems and software.

2. The Capability Maturity Model (CMM) is a framework for assessing the maturity level of an organization's information systems development and management processes and products. It defines the need for a system development process.

3. A system life cycle divides the life of an information system into two stages, *systems development* and *systems operation and maintenance.*

4. A systems development methodology is a process for the system development stage. It defines a set of activities, methods, best practices, deliverables, and automated tools that systems developers and project managers are to use to develop and maintain information systems and software.

5. The following principles should underlie all systems development methodologies:

   a. Get the system users involved.
   b. Use a problem-solving approach.
   c. Establish phases and activities.
   d. Document throughout development.
   e. Establish standards.
   f. Manage the process and projects.
   g. Justify information systems as capital investments.
   h. Don't be afraid to cancel or revise scope.
   i. Divide and conquer.
   j. Design systems for growth and change.

6. System development projects are triggered by problems, opportunities, and directives:

   a. Problems are undesirable situations that prevent the organization from fully achieving its purpose, goals, and/or objectives.
   b. Opportunities are chances to improve the organization even in the absence of specific problems.
   c. Directives are new requirements that are imposed by management, government, or some external influence.

7. Wetherbe's PIECES framework is useful for categorizing problems, opportunities, and directives. The letters of the PIECES acronym correspond to Performance, Information, Economics, Control, Efficiency, and Service.

8. Traditional, basic systems development phases include:

   a. Scope definition
   b. Problem analysis
   c. Requirements analysis
   d. Logical design
   e. Decision analysis
   f. Physical design and integration
   g. Construction and testing
   h. Installation and delivery

9. Cross life-cycle activities are activities that overlap many or all phases of the methodology. They may include:

   a. Fact-finding, the formal process of using research, interviews, meetings, questionnaires, sampling, and other techniques to collect information about systems, requirements, and preferences.
   b. Documentation, the activity of recording facts and specifications for a system for current and future reference. Documentation is frequently stored in a repository, a database where systems developers store all documentation, knowledge, and products for one or more information systems or projects.
   c. Presentation, the activity of communicating findings, recommendations, and documentation for review by interested users and managers. Presentations may be either written or verbal.
   d. Feasibility analysis, the activity by which feasibility, a measure of how beneficial the development of an information system would be to an organization, is measured and assessed.
   e. Process management, the ongoing activity that documents, manages the use of, and improves an organization's chosen methodology (the "process") for systems development.
   f. Project management, the activity of defining, planning, directing, monitoring, and controlling a project to develop an acceptable system within the allotted time and budget.

10. There are different routes through the basic systems development phases. An appropriate route is selected during the scope definition phase. Typical routes include:

    a. Model-driven development strategies, which emphasize the drawing of diagrams to help visualize and analyze problems, define business requirements, and design information systems. Alternative model-driven strategies include:

    i) Process modeling

    ii) Data modeling

    iii) Object modeling

b. Rapid application development (RAD) strategies, which emphasize extensive user involvement in the rapid and evolutionary construction of working prototypes of a system to accelerate the system development process.

c. Commercial application package implementation strategies, which focus on the purchase and integration of a software package or solution to support one or more business functions and information systems.

d. System maintenance, which occurs after a system is implemented and lasts throughout the system's lifetime. Essentially, system maintenance executes a smaller-scale version of the development process with different starting points depending on the type of problem to be solved.

11. Automated tools support all systems development phases:

a. Computer-aided systems engineering (CASE) tools are software programs that automate or support the drawing and analysis of system models and provide for the translation of system models into application programs.

    i) A CASE repository is a system developers' database. It is a place where developers can store system models, detailed descriptions and specifications, and other products of systems development.

    ii) Forward engineering requires that the systems analyst draw system models, either from scratch or from templates. The resulting models are subsequently transformed into program code.

    iii) Reverse engineering allows a CASE tool to read existing program code and transform that code into a representative system model that can be edited and refined by the systems analyst.

b. Application development environments (ADEs) are integrated software development tools that provide all the facilities necessary to develop new application software with maximum speed and quality.

c. Process management tools help us document and manage a methodology and routes, its deliverables, and quality management standards.

d. Project management tools help us plan system development activities (preferably using the approved methodology), estimate and assign resources (including people and costs), schedule activities and resources, monitor progress against schedule and budget, control and modify schedule and resources, and report project progress.

Review Questions

1. Explain why having a standardized system development process is important to an organization.

2. How are system life cycle and system development methodology related?

3. What are the 10 underlying principles for systems development?

4. Why is documentation important throughout the development process?

5. Why are process management and project management necessary?

6. What is risk management? Why is it necessary?

7. Which stakeholders initiate most projects? What is the impetus for most projects?

8. Who are the main participants in the scope definition? What are their goals in the scope definition?

9. What are the three most important deliverables in scope definition?

10. Who are the main participants in the requirements analysis phase? Why are they the main participants?

11. What feasibility analyses are made in the decision analysis?

12. What is model-driven development?

13. Why is model-driven development popular?

14. What is rapid application development (RAD)?

15. What benefits can RAD bring to the system development process?

16. What is computer-assisted software engineering (CASE)? List some examples of CASE.

Problems and Exercises

1. The Capability Maturity Model (CMM) was developed by the Software Engineering Institute at Carnegie Mellon, and is widely used by both the private and public sectors. What is the purpose of the CMM framework and how does it achieve this?

2. List the five maturity levels, and briefly describe each of them.

3. Table 3-1 in the textbook illustrates the difference in a typical project's duration, person-months, quality, and cost, depending upon whether an organization's system development process is at CMM level 1, 2, or 3. Between which two CMM levels does an organization gain the greatest benefit in terms of percentage of improvement? What do you think is the reason for this?

4. *Systems development methodology* and *system life cycle* are two terms that are frequently used and just as frequently misused. What is the difference between the two terms?

5. Describe how using a systems development methodology is in line with CMM goals and can help an organization increase its maturity level.

6. A number of underlying principles are common to all systems development methodologies. Identify these underlying principles and explain them.

7. The PIECES framework was developed by James Wetherbe as a means to classify problems. Identify the categories, then categorize the following problems using the PIECES framework:

   a. Duplicate data is stored throughout the system.
   b. There is a need to port an existing application to PDA devices.
   c. Quarterly sales reports need to be generated automatically.
   d. Employees can gain access to confidential portions of the personnel system.
   e. User interfaces for the inventory system are difficult and confusing, resulting in a high frequency of incorrect orders.

8. Each phase of a project includes specific deliverables that must be produced and delivered to the next phase. Using the textbook's hypothetical *FAST* methodology, what are the deliverables for the requirements analysis, logical design, and physical design/integration phases?

9. Scope definition is the first phase of the *FAST* methodology, and it is either the first phase or part of the first phase of most methodologies. What triggers the scope phase, which stakeholders are involved in this phase, what two essential questions need to be answered, and what three important deliverables come out of this phase?

10. The requirements analysis phase is an essential part of a system development methodology. According to the *FAST* methodology, which stakeholders typically participate in this phase? What is the primary focus of requirements analysis? What is *not* the focus? How should each proposed requirement be evaluated? What critical error must be avoided?

11. In the *FAST* methodology, as well as most system methodologies, system owners and system designers do not participate in the requirements analysis phase. What do you think the reason is for this?

12. What is the essential purpose of the logical design phase? How does it accomplish this? How are technological solutions incorporated in this phase? What are some common synonyms for this phase used by other methodologies? Who are the typical participants in this phase? What is agile modeling and what is its purpose? What are the deliverables coming out of this phase? In terms of the development team, what critical transition takes place by the end of this phase?

13. What is the essential purpose of the physical design phase? Who must be involved in this phase, and who may be involved? What are the two philosophies of physical design on the different ends of the continuum, and how are they different? Is this a likely phase in which a project might be canceled? With what other phase is there likely to be overlap, and what do you think is the reason for this?

14. A customer has engaged your software development company to develop a new order-processing system. However, the time frames are very tight and inflexible for delivery of at least the basic part of the new system. Further, user requirements are sketchy and unclear. What are two system development strategies that might be advantageous to use in this engagement?

15. What is the potential downside to using the strategies described in the preceding question?

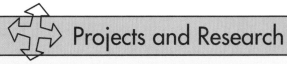

Projects and Research

1. The Software Engineering Institute (SEI) at Carnegie Mellon University has developed a series of related Capability Maturity Models (CMMs). You can read about these different CMM products at SEI's Web site (http://www.sei.cmu.edu).

   a. Identify the current CMM products being maintained or developed by SEI.
   b. What are their differences and similarities?
   c. If you were to rate your organization, or an organization with which you are familiar, using the CMM described in the textbook, at which level would it be? Why?
   d. What steps would you recommend that your organization take in order to advance to the next CMM level?
   e. Do you feel that the time, cost, and resources to advance to the next level would be worth the perceived benefits for your organization? Why or why not?

2. You are a new project manager and have been assigned responsibility for an enterprise information systems project that touches every division in your organization. The chief executive officer stated at project initiation that successfully implementing this project was the number 1 priority of your organization. The project is in midst of the requirements analysis phase. While it is on schedule, you notice that attendance of the system users and owners at meetings on requirements has been dropping. A more experienced project manager has told you not to worry, that this is normal. Should you be concerned?

3. There are many different systems development methodologies in use, each with its own terminology, and number and scope of phases. Search the Web for information on two or three of these other systems development methodologies, then do the following:

   a. Note the systems development methodologies that you found. When, by whom, and why were they developed?
   b. What phases and terminology do they employ?
   c. Draw a matrix comparing their phases to the textbook's *FAST* methodology.
   d. What significant differences did you find?
   e. Do you see any advantages or disadvantages in any of the methodologies that you found compared to the *FAST* methodology?

4. The *PIECES* framework, which was developed by James Wetherbe and is described in the textbook, is intended to be a framework for classifying problems, opportunities, and directives.

   Contact one of the systems analysts for your organization, your school, and/or another organization. Ask them about the information systems used in their organization and to describe what the problems are in general terms. Select three of these systems:

   a. Describe the systems you selected, their problems, and the organizations that use them.
   b. Use the *PIECES* framework in the textbook to categorize each system's problems.
   c. Describe the *PIECES* category or categories you found for each problem. Did each problem generally have one or more categories associated with it?
   d. For the systems used by different organizations, what commonality did you find in the categories of problems? If you found a great deal of commonality of categories, do you think this is significant or just coincidental?
   e. Where in the systems development life cycle do you think the PIECES framework would be of the greatest value?

5. Computer-assisted software engineering (CASE) tools can significantly help developers improve productivity, quality, and documentation. Conduct an informal survey of about a dozen information technology departments regarding whether they use CASE tools, and if so, what type. Also, find out how long they have been using the CASE tools and whether they are used for all or just some projects. Add any other questions you may find meaningful. Try to split your survey between public and private sector agencies, and/or large and small organizations.

   a. What types of organizations did you survey?
   b. What did you ask?
   c. What were the results?
   d. Given the limited and informal nature of this survey, were you able to find any patterns or trends?
   e. Based upon your readings and your survey, what are your feelings regarding the use of CASE tools?

6. Projects at times are canceled or abandoned, sometime by choice, sometimes not. Research the Web for articles on project abandonment strategies, and select two of them.

a. What articles did you select?
b. What are their central themes, findings, and recommendations regarding project abandonment?
c. Compare and contrast their findings and their recommendations. Which strategy would you choose, if any?

d. Do you think that abandoning a project is always avoidable and/or always represents a failure?

Minicases

1. Interview at least two project managers. What are their experiences with *scope creep?*
2. George is the CEO of a major corporation that has been trying to develop a program that captures the keystrokes of employees on their computers. The project is currently $100,000 over budget and behind schedule and will require at least another $50,000 and six months to complete. The CEO wants to continue the project because so much has already been invested in it. What is your recommendation? Why?
3. Beatrice is an excellent manager—she is very capable of managing the bureaucratic process and following the business rules in her corporation.

She is a "by the book" person who can always be counted on to do things "right." Will Beatrice make a good *project manager?* Why or why not?
4. A company is trying to decide between using an off-the-shelf program or developing a custom program for inventory management. The off-the-shelf product is less expensive than the custom solution and still has most of the needed functionality. The CEO believes that the missing capability can be addressed through tweaking the program once it is purchased. As the CIO of the company, what are your concerns and recommendations to the CEO?

Suggested Readings

Ambler, Scott. *Agile Modeling: Effective Practices for eXtreme Programming and the Unified Process.* New York: John Wiley & Sons, 2002. This is the definitive book on agile methods and modeling.

Application Development Trends (monthly periodical). Framingham, MA: Software Productivity Group, Inc. This is our favorite periodical for keeping up with the latest trends in methodology and automated tools. Each month features several articles on different topics and products.

DeMarco, Tom. *Structured Analysis and System Specification.* Englewood Cliffs, NJ: Prentice Hall, 1978. This is the classic book on structured systems analysis, a process-centered, model-driven methodology.

Gane, Chris. *Rapid Systems Development.* Englewood Cliffs, NJ: Prentice Hall, 1989. This book presents a nice overview of RAD that combines model-driven development and prototyping in the correct balance.

Gildersleeve, Thomas. *Successful Data Processing Systems Analysis,* 2nd ed. Englewood Cliffs, NJ: Prentice Hall, 1985. We are indebted to Gildersleeve for the creeping commitment approach. The classics never become obsolete.

Jacobson, Ivar; Grady Booch; and James Rumbaugh. *The Unified Software Development Process.* Reading, MA:

Addison-Wesley, 1999. The Rational Unified Process is a currently popular example of a model-driven, object-oriented methodology.

McConnell, Steve. *Rapid Development.* Redmond, WA: Microsoft Press, 1996. Chapter 7 of this excellent reference book provides what may be the definitive summary of system development life cycle and methodology variations that we call "routes" in our book.

Orr, Ken. *The One Minute Methodology.* New York: Dorsett House, 1990. Must reading for those interested in exploring the need for methodology. This very short book can be read in one sitting. It follows the satirical story of an analyst's quest for the development silver bullet, "the one minute methodology."

Paulk, Mark C.; Charles V. Weber; Bill Curtis; and Mary Beth Chrissis. *The Capability Maturity Model: Guidelines for Improving the Software Process.* Reading, MA: Addison-Wesley, 1995. This book fully describes version 1.1 of the Capability Maturity Model. Note that version 2.0 was under development at the time we were writing this chapter.

Wetherbe, James. *Systems Analysis and Design: Best Practices,* 4th ed. St. Paul, MN: West, 1994. We are indebted to Wetherbe for the PIECES framework.

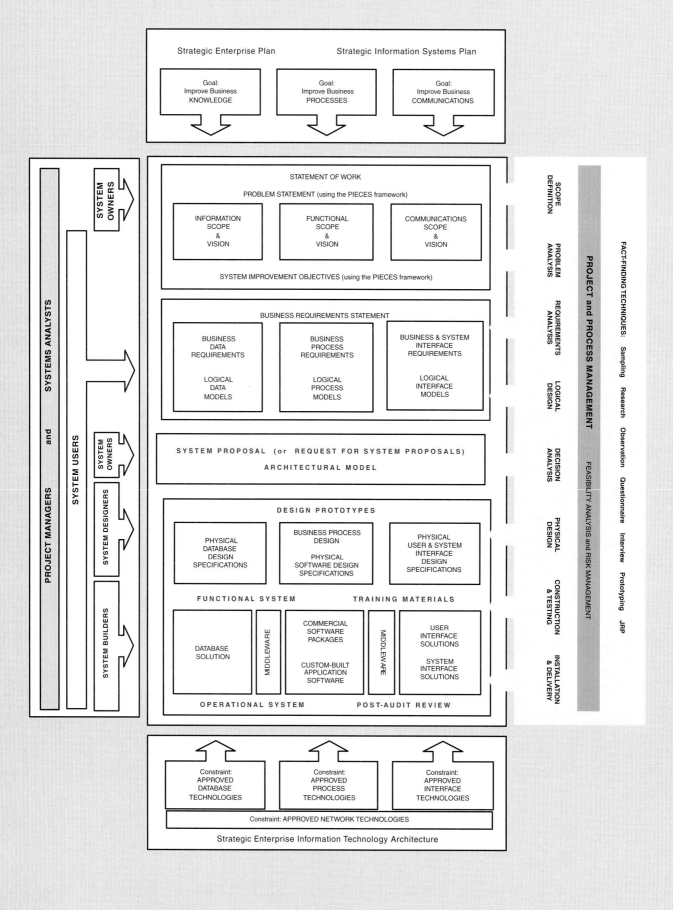

Project Management

Chapter Preview and Objectives

Project management skills are greatly in demand in the information technology community. Project management is a natural extension of the previous chapter's introduction to system development. This chapter provides a process-centric survey of key project management tools and techniques as they apply to systems analysis and design. You will know that you understand the basics of project management when you can:

▐ Define the terms *project* and *project management* and differentiate between project and process management.

▐ Describe the causes of failed information systems and technology projects.

▐ Describe the basic competencies required of project managers.

▐ Describe the basic functions of project management.

▐ Differentiate between PERT and Gantt charts as project management tools.

▐ Describe the role of project management software as it relates to project management tools.

▐ Describe eight activities in project management.

▐ Define *joint project planning* and its role in project management.

▐ Define *scope* and write a *statement of work* to document scope.

▐ Use a *work breakdown structure* to decompose a project into tasks.

▐ Estimate tasks' durations and specify intertask dependencies on a PERT chart.

▐ Assign resources to a project and produce a project schedule with a Gantt chart.

▐ Assign people to tasks and direct the team effort.

▐ Use critical path analysis to adjust schedule and resource allocations in response to schedule and budget deviations.

▐ Manage user expectations of a project and adjust project scope.

Introduction

Bob Martinez was in the office of his boss, Sandra Shepherd, discussing the Sound-Stage Member Services system project.

"This sure is a big project," Bob said, "bigger than anything I've ever worked on before. How will you make sure it stays on track?"

"Well, first we have to get consensus on the scope of the project and document assumptions and constraints," Sandra answered. "We also have to negotiate the project budget and schedule. Then we identify all the tasks that need to be performed. The *FAST* methodology is our template, but we always customize it for each project. We have to plan each task and analyze how its work and its own schedule fit in with the overall project. Then we assign people and other resources to each task. As the project goes on we have to manage the process to make sure everything stays on schedule."

"Wow!" Bob replied. "On some of the semester group projects I did in college, we just kind of dived right in with the work. If we got behind we just pulled a couple of all-nighters."

"Believe me," Sandra said, "you don't want to be around when the finger pointing starts on a real systems project that is behind schedule or over budget. That's something a couple of all-nighters won't fix. We have a long road ahead of us, and we want to plan this as carefully as possible."

What Is Project Management?

Most of you are familiar with Murphy's Law, which suggests, "If anything can go wrong, it will." Murphy has motivated numerous pearls of wisdom about projects, machines, people, and why things go wrong. This chapter will teach you strategies, tools, and techniques for project management as applied to information systems projects.

project manager the person responsible for supervising a systems project from initiation to conclusion. Successful project managers possess a wide range of technical, management, leadership, and communication skills.

The demand for **project managers** in the information systems community is strong. Typically, IS project managers come from the ranks of experienced IS developers such as systems analysts. While it is unlikely that your first job responsibilities will include project management, you should immediately become aware of project management processes, tools, and techniques. Eventually you will combine this knowledge with development experience plus your own observation of project managers to form the basis for your own career opportunities in project management.

Before we can define *project management,* we should first define *project.* There are as many definitions as there are authors, but we like the definition put forth by Wysocki, Beck, and Crane:

project a sequence of activities that must be completed on time, within budget, and according to specification.

> A **project** is a [temporary] <u>sequence</u> of <u>unique</u>, <u>complex</u>, and <u>connected activities</u> that have <u>one goal</u> or purpose and that must be completed by a <u>specific time</u>, <u>within budget</u>, and <u>according to specification</u>.[1]

The keywords are underlined to draw your attention to some key aspects of the definition. As applied to information system development, we note the following:

- A system development process or methodology, such as *FAST,* defines a sequence of <u>activities</u>, mandatory and optional.
- Every system development project is <u>unique</u>; that is, it is different from every other system development project that preceded it.
- The activities that comprise system development are relatively <u>complex</u>. They require the skills that you are learning in this book, and they require that you be able to adapt concepts and skills to changing conditions and unanticipated events.

---

[1]Robert K. Wysocki, Robert Beck, Jr., and David B. Crane, *Effective Project Management: How to Plan, Manage, and Deliver Projects on Time and within Budget* (New York: John Wiley & Sons, 1995), p. 38.

- By now, you've already learned that the phases and activities that make up a system development methodology are generally <u>sequential</u>. While some tasks may overlap, many tasks are dependent on the completion of other tasks.
- The development of an information system represents a <u>goal</u>. Several objectives may need to be met to achieve that goal.
- Although many information system development projects do not have absolute deadlines or <u>specified times</u> (there are exceptions), they are notoriously completed later than originally projected. This is becoming less acceptable to upper management given the organizationwide pressures to reduce cycle times for products and business processes.
- Few information systems are completed <u>within budget</u>. Again, upper management is increasingly rejecting this tendency.
- Information systems must satisfy the business, user, and management expectations <u>according to specifications</u> (which we call *requirements* throughout this book).

For any systems development project, effective **project management** is necessary to ensure that the project meets the deadline, is developed within an acceptable budget, and fulfills customer expectations and specifications. You learned in Chapter 3 that project management is a cross life-cycle activity because it overlaps all phases of any systems development methodology.

Corporate rightsizing has changed the structure and culture of most organizations and, hence, project management. More flexible and temporary interdepartmental teams that are given greater responsibility and authority for the success of organizations have replaced rigid hierarchical command structures and permanent teams. Contemporary system development methodologies depend on having teams that include both technical and nontechnical users, managers, and information technologists all directed to the project goal. These dynamic teams require leadership and project management.

Organizations take different approaches to project management. One approach is to appoint a project manager from the ranks of the team (once it has been formed). Alternatively, many organizations believe that successful project managers apply a unique body of knowledge and skills that must be learned. These organizations tend to hire and/or develop professional project managers who are then assigned to one or more projects.

The prerequisite for good project management is a well-defined system development process. In Chapter 3, we introduced the Capability Maturity Model (CMM) as a framework for quality management that is based on a sound systems development process. In Chapter 3 we differentiated between project and process management. Project management was defined above. **Process management** is an ongoing activity that documents, manages the use of, and improves an organization's chosen methodology (the "process") for systems development. Process management is concerned with the activities, deliverables, and quality standards to be applied to *all* projects. The scope of process management is all projects, whereas the scope of project management is a single project. This chapter will focus on project management.

> **project management** the process of scoping, planning, staffing, organizing, directing, and controlling the development of an acceptable system at a minimum cost within a specified time frame.

> **process management** the activity of documenting, managing, and continually improving the process of systems development.

> The Causes of Failed Projects

What causes a project to succeed or fail? Chapter 3 introduced 10 basic principles of systems development that are critical success factors for all projects. See Chapter 3 for a review of those principles. From a project management perspective, a project is considered a success if:

- The resulting information system is acceptable to the customer.
- The system is delivered "on time."
- The system is delivered "within budget."
- The system development process had a minimal impact on ongoing business operations.

Not all projects meet these criteria, and as a result, not all projects are successful. Failures and limited successes far outnumber successful information systems. Project mismanagement can undermine the best application of the systems analysis and design methods taught in this book. We can develop an appreciation for the importance of project management by studying the mistakes of some project managers. Let's examine some project mismanagement problems and consequences:

- *Failure to establish upper-management commitment to the project*—Sometimes commitment changes during a project.
- *Lack of organization's commitment to the system development methodology*—Many system development methodologies do little more than collect dust.
- *Taking shortcuts through or around the system development methodology*—Project teams often take shortcuts for one or more of the following reasons:

  — The project gets behind schedule, and the team wants to catch up.
  — The project is over budget, and the team wants to make up costs by skipping steps.
  — The team is not trained or skilled in some of the methodology's activities and requirements, so it skips them.

- *Poor expectations management*—All users and managers have expectations of the project. Over time, these expectations may change. This can lead to two undesirable situations:

  — **Scope creep** is the unexpected growth of user expectations and business requirements for an information system as the project progresses. The schedule and budget can be adversely affected by such changes.
  — **Feature creep** is the uncontrolled addition of technical features to a system under development without regard to schedule or budget.

- *Premature commitment to a fixed budget and schedule*—You can rarely make accurate estimates of project costs and schedule before completing a detailed problem analysis or requirements analysis. Premature estimates are inconsistent with the creeping commitment approach introduced in Chapter 3.
- *Poor estimating techniques*—Many systems analysts estimate by making a best-calculated estimate and then doubling that number. This is not a scientific approach.
- *Overoptimism*—Systems analysts and project managers tend to be optimists. As project schedules slip, they respond, "No big deal. We can make it up later." They fail to recognize that certain tasks are dependent on other tasks. Because of these dependencies, a schedule slip in one phase or activity will cause corresponding slips in many other phases and activities, thus contributing to cost overruns.
- *The mythical man-month*[2]—As the project gets behind schedule, project leaders frequently try to solve the problem by assigning more people to the team. It doesn't work! There is no linear relationship between time and number of personnel. The addition of personnel usually creates more communication problems, causing the project to get even further behind schedule.
- *Inadequate people management skills*—Managers tend to be thrust into management positions and are not prepared for management responsibilities. This problem is easy to identify. No one seems to be in charge; customers don't know the status of the project; teams don't meet regularly to discuss and monitor progress; team members aren't communicating with one another; the project is always said to be "95 percent complete."
- *Failure to adapt to business change*—If the project's importance changes during the project, or if the management or the business reorganizes,

scope creep the unexpected and gradual growth of requirements during an information systems project.

feature creep the uncontrolled addition of technical features to a system.

---

[2]Fred Brooks, *The Mythical Man-Month* (Reading, MA: Addison-Wesley, 1975).

projects should be reassessed for compatibility with those changes and their importance to the business.

- *Insufficient resources*—This could be due to poor estimating or to other priorities, or it could be that the staff resources assigned to a project do not possess the necessary skills or experience.
- *Failure to "manage to the plan"*—Various factors may cause the project manager to become sidetracked from the original project plan.

Ultimately, *the* major cause of project failure is that most project managers were not educated or trained to be project managers. Just as good programmers don't always go on to become good systems analysts, good systems analysts don't automatically perform well as project managers. To be a good project manager, you should be educated and skilled in the "art of project management."

> The Project Management Body of Knowledge

The Project Management Institute was created as a professional society to guide the development and certification of *professional* project managers. The institute created the Project Management Body of Knowledge (PMBOK) for the education and certification of professional project managers. This chapter's content was greatly influenced by the PMBOK.

Project Manager Competencies   Good project managers possess a core set of competencies. Table 4-1 summarizes these competencies. Some of these competencies can be taught, in courses, books, and professional workshops; however, some come only with professional experience in the field. There are two basic premises of project management competencies: First, individuals cannot manage a process they

TABLE 4-1   **Project Manager Competencies**

Competency	Explanation	How to Obtain?
Business Achievement Competencies		
Business awareness	Ties every systems project to the mission, vision, and goals of the organization.	Business experience
Business partner orientation	Keeps managers and users involved throughout a systems project.	Business experience
Commitment to quality	Ensures that every systems project contributes to the quality expectation of the organization as a whole.	Business experience
Problem-Solving Competencies		
Initiative	Demonstrates creativity, calculated risks, and persistence necessary to get the job done.	Business experience
Information gathering	Skillfully obtains the factual information necessary to analyze, design, and implement the information system.	Chapter 6 in this book plus business experience
Analytical thinking	Can assess and select appropriate system development processes and use project management tools to plan, schedule, and budget for system development.	This chapter
	Can solve problems through the analytical approach of decomposing systems into their parts and then reassembling the parts into improved systems.	Chapters 8, 9, and 11 in this book plus business experience
Conceptual thinking	Understands systems theory and applies it to systems analysis and design of information systems.	Chapters 2 and 5–11 in this book

continued

TABLE 4-1 (Continued)

Competency	Explanation	How to Obtain?
Influence Competencies		
Interpersonal awareness	Understands, recognizes, and reacts to interpersonal motivations and behaviors.	Can be learned in courses but requires business experience
Organizational awareness	Understands the politics of the organization and how to use them in a project.	Business experience
Anticipation of impact	Understands implications of project decisions and manages expectations and risk.	Introduced in this chapter but requires business experience
Resourceful use of influence	Skillfully obtains cooperation and consensus of managers, users, and technologists to solutions.	Business experience
People Management Competencies		
Motivating others	Coaches and directs individuals to overcome differences and achieve project goals as a team.	Business experience
Communication skills	Communicates effectively, both orally and in writing, in the context of meetings, presentations, memos, and reports.	Can be learned in courses but usually requires business experience
Developing others	Ensures that project team members receive sufficient training, assignments, supervision, and performance feedback required to complete projects.	Business experience
Monitoring and controlling	Develops the project plan, schedule, and budget and continuously monitors progress and makes adjustments when necessary.	Tools and techniques taught in this chapter, but requires project experience
Self-Management Competencies		
Self-confidence	Consistently makes and defends decisions with a strong personal confidence in the process and/or facts.	Business experience
Stress management	Works effectively under pressure or adversity.	Business experience
Concern for credibility	Consistently and honestly delivers on promises and solutions. Maintains technical or business currency in the field as appropriate.	Business experience
Flexibility	Capable of adjusting process, management style, or decision making based on situations and unanticipated problems.	Business experience

Source: Adapted from Robert K. Wysocki, Robert Beck, Jr., and David B. Crane, *Effective Project Management: How to Plan, Manage, and Deliver Projects on Time and within Budget* (New York: John Wiley & Sons, 1995).

have never used. Second, managers must have an understanding of the business and culture that provides a context for the project.

Project Management Functions   The basic functions of a project manager have been studied and refined by management theorists for many years. These functions include scoping, planning, staffing, organizing, scheduling, directing, controlling, and closing:

- *Scoping*—Scope defines the boundaries of the project. A project manager must scope project expectations and constraints in order to plan activities, estimate costs, and manage expectations.
- *Planning*—Planning identifies the tasks required to complete the project. This is based on the manager's understanding of the project scope and the methodology used to achieve the goal.
- *Estimating*—Each task that is required to complete the project must be estimated. How much time will be required? How many people will be needed?

What skills will be needed? What tasks must be completed before other tasks are started? Can some of the tasks overlap? How much will it cost? These are all estimating issues. Some of these issues can be resolved with the project modeling tools that will be discussed later in this chapter.

- *Scheduling*—Given the project plan, the project manager is responsible for scheduling all project activities. The project schedule should be developed with an understanding of the required tasks, task duration, and task prerequisites.
- *Organizing*—The project manager should make sure that members of the project team understand their own individual roles and responsibilities as well as their reporting relationship to the project manager.
- *Directing*—Once the project has begun, the project manager must direct the team's activities. Every project manager must demonstrate people management skills to coordinate, delegate, motivate, advise, appraise, and reward team members.
- *Controlling*—Perhaps the manager's most difficult and important function is controlling the project. Few plans will be executed without problems and delays. The project manager must monitor and report progress against goals, schedule, and costs and make appropriate adjustments when necessary.
- *Closing*—Good project managers always assess successes and failures at the conclusion of a project. They learn from their mistakes and plan for continuous improvement of the systems development process.

All the above functions are dependent on ongoing interpersonal *communication* among the project manager, the team, and other managers.

Project Management Tools and Techniques—PERT and Gantt Charts

The PMBOK includes tools and techniques that support project managers. Two such tools are PERT and Gantt charts.

PERT, which stands for *Project Evaluation and Review Technique,* was developed in the late 1950s to plan and control large weapons development projects for the U.S. Navy. A **PERT chart** is a graphical network model that depicts a project's tasks and the relationships between those tasks. A sample PERT chart is illustrated in Figure 4-1. PERT was developed to make clear the *interdependence* between project tasks before those tasks are scheduled. The boxes represent project tasks (we used phases from Chapter 3). (The content of the boxes can be adjusted to show various project attributes such as schedule and actual start and finish times.) The arrows indicate that one task is dependent on the start or completion of another task.

The Gantt chart, first conceived by Henry L. Gantt in 1917, is the most commonly used project scheduling and progress evaluation tool. A **Gantt chart** is a simple horizontal bar chart that depicts project tasks against a calendar. Each bar represents a named project task. The tasks are listed vertically in the left-hand column. The horizontal axis is a calendar time line. Figure 4-2 illustrates a phase-level Gantt chart, once again based on Chapter 3. We used the same project that was illustrated in Figure 4-1.

Gantt charts offer the advantage of clearly showing *overlapping* tasks, that is, tasks that can be performed at the same time. The bars can be shaded to clearly indicate percentage completion and project progress. The figure demonstrates which phases are ahead of and behind schedule at a glance. The popularity of Gantt charts stems from their simplicity—they are easy to learn, read, prepare, and use.

Gantt and PERT charts are not mutually exclusive. Gantt charts are more effective when you are seeking to communicate schedule. PERT charts are more effective when you want to study the relationships between tasks.

Project Management Software

Project management software is routinely used to help project managers plan projects, develop schedules, develop budgets, monitor progress and costs, generate reports, and effect change. Representative automated project management tools are listed in the margin.

PERT chart a graphical network model used to depict the interdependencies between a project's tasks.

Gantt chart a bar chart used to depict project tasks against a calendar.

PROJECT MANAGEMENT SOFTWARE

Niku's *Project Manager*

Artemis *International Solutions Corporation's 9000*

Computer Associates' *AllFusion Process Management Suite*

Microsoft's *Project*

Primavera's *Project Planner* and *Project Manager*

C/S Solutions' *Risk +*.

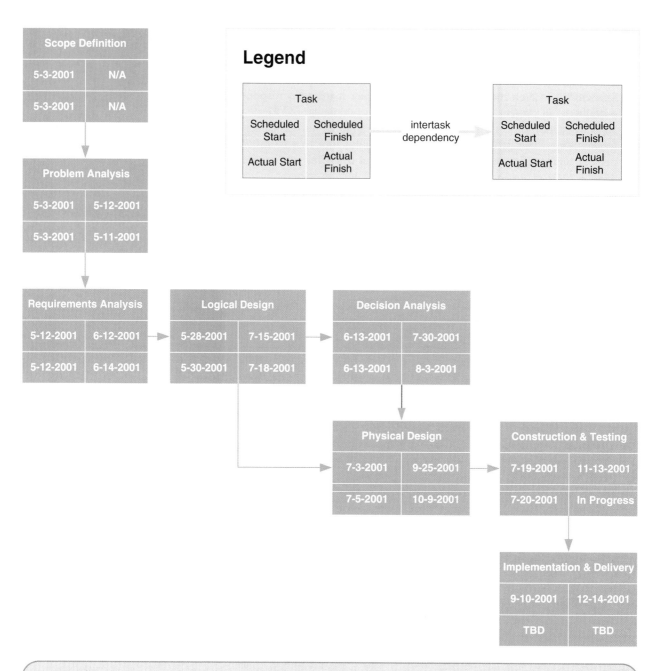

FIGURE 4-1 A PERT Chart

We will teach you project modeling and management techniques in the context of project management software. We used Microsoft *Project* because that tool is frequently available to students and institutions at special academic prices through their college bookstore. Microsoft *Project*, like most project management software tools, supports both PERT and Gantt charts.

Figure 4-3(a) illustrates one possible Microsoft *Project* Gantt chart for the Sound-Stage Member Services project. We call your attention to the following numbered bullets:

❶ The black bars are *summary tasks* that represent project *phases* that are further broken down into other tasks.

❷ The red bars indicate tasks that have been determined to be critical to the schedule, meaning that any extension to the duration of those tasks will delay other tasks and the project as a whole. We'll talk more about critical tasks later.

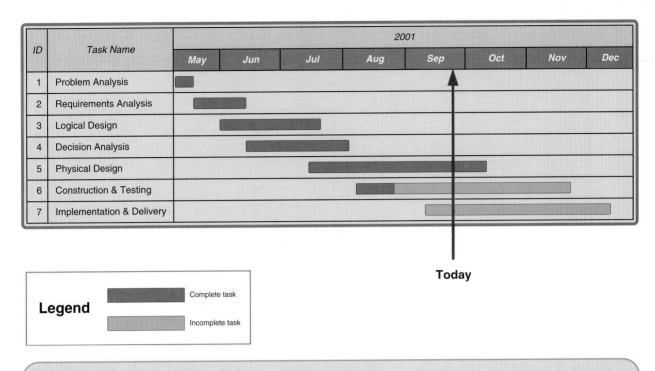

FIGURE 4-2 A Gantt Chart

❸ The blue bars indicate tasks that are not critical to the schedule, meaning they have some slack time during which delays will not affect other tasks and the project as a whole.

❹ The red arrows indicate prerequisites between two critical tasks. (The blue arrows indicate prerequisites between two noncritical tasks.)

❺ The teal diamonds indicate milestones—events that have no duration. They signify the end of some significant task or deliverable.

Figure 4-3(b) shows a Microsoft *Project* PERT chart based on the same project plan that was illustrated in the Gantt chart. The contents of each cell in the task rectangles are able to be customized in Microsoft *Project*.

The Project Management Life Cycle

Recall from Chapter 3 that the Capability Maturity Model defines a framework for assessing the quality of an organization's information systems development activities. CMM Level 1 is defined as "initial" and is characterized by the lack of any consistent project or process management function. The first stage of maturity improvement is to implement a consistent project management function—called CMM Level 2. In this section we introduce a project management life cycle representative of CMM Level 2 maturity.

Figure 4-4 illustrates a project management process or life cycle. Recall that project management is a cross life-cycle activity; that is, project management activities overlap all the system development phases that were introduced in Chapter 3. The illustrated project management activities correspond to classic management functions: scoping, planning, estimating, scheduling, organizing, directing, controlling, and closing.

The project management process shown in Figure 4-4 incorporates a joint project planning (JPP) technique.[3] **Joint project planning (JPP)** is a strategy wherein all

joint project planning (JPP) a strategy in which all stakeholders attend an intensive workshop aimed at reaching consensus on project decisions.

---

[3]Wysocki, Beck, and Crane, *Effective Project Management: How to Plan, Manage, and Deliver Projects on Time and within Budget*, p. 38.

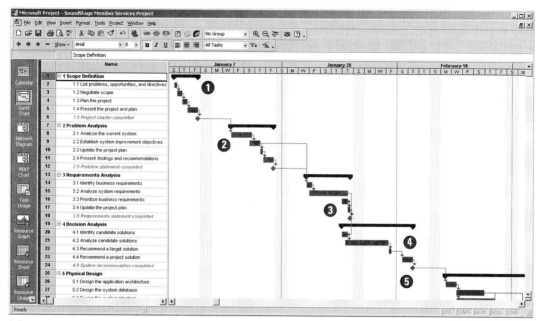

(a)

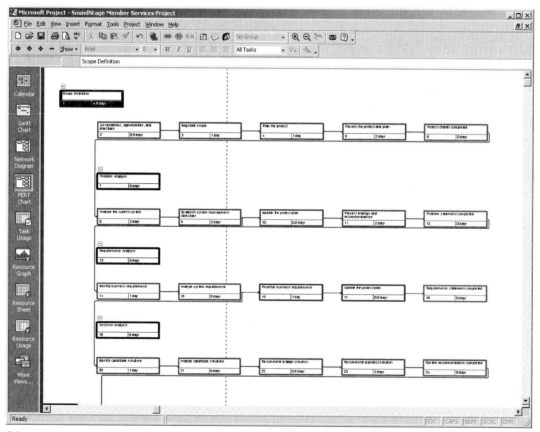

(b)

FIGURE 4-3 Microsoft *Project* Gantt and PERT Charts

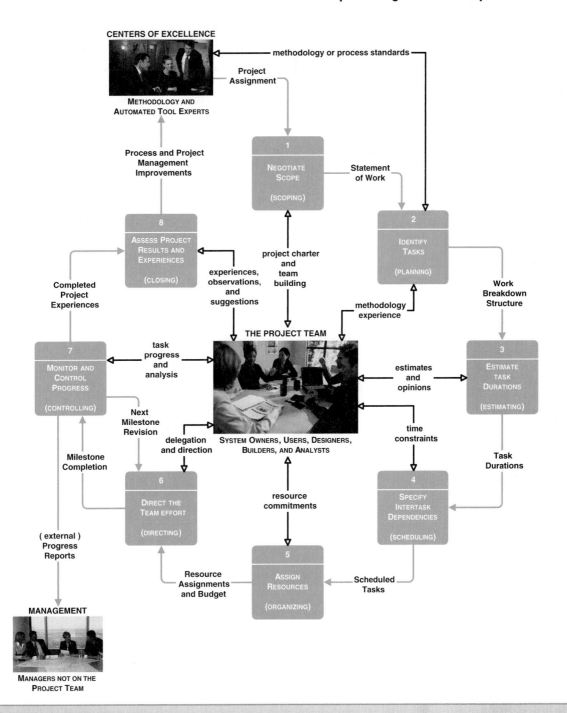

FIGURE 4-4 A Project Management Life Cycle

stakeholders in a project (meaning system owners, users, analysts, designers, and builders) participate in a one- to three-day project management workshop, the result of which is consensus on project scope, schedule, resources, and budget. (Subsequent workshops or meetings may be required to adjust scope, budget, and schedule.) Notice that in JPP, the project team is *actively* involved in all inputs and deliverables of all project management activities.

In the following subsections, we will review each of the illustrated project management activities and discuss how to use appropriate project management tools and techniques.

> Activity 1—Negotiate Scope

Perhaps the most important prerequisite to effective project management occurs at the beginning. All parties must agree to the project scope before any attempt is made to identify and schedule tasks or to assign resources (people) to those tasks. Scope defines the expectations of a project, and expectations ultimately determine satisfaction and degrees of success. Accordingly, the negotiation of project scope is a necessary activity in the project management life cycle. What is scope? **Scope** defines the boundaries of a project—the parts of the business that are to be studied, analyzed, designed, constructed, implemented, and ultimately improved. Scope also defines the aspects of a system that are considered outside the project. The answers to five basic questions influence the negotiation of project scope:

- *Product*—What do you want?
- *Quality*—How good do you want it to be?
- *Time*—When do you want it?
- *Cost*—How much are you willing to pay for it?
- *Resources*—What resources are you willing or able to bring to the table?

Negotiation of the above factors is a give-and-take activity that includes much iteration. The deliverable is an agreed-on **statement of work** that describes the work to be performed during the project. In consulting engagements, the statement of work has become a commonly used contract between the consultant and client. This approach works equally well for internal system development projects to establish a contract between business management and the project manager and team. According to Keane, Inc., a leading project management consulting firm,

> The statement of work affirms that the project manager understands who is really in charge of the effort, who is controlling the purse strings, what is the formal and informal organization within which the project will be developed, who are the "kings and queens" that have interest, and other similar but mainly nontechnical issues. It establishes a firm business relationship between the project manager and both the customer and the extended project team.[4]

An outline for a typical statement-of-work document is shown in Figure 4-5. The size of the document will vary in different organizations. It may be as small as one to two pages, or it may run several pages.

> Activity 2—Identify Tasks

Given the project scope, the next activity is to identify project tasks. Tasks identify the work to be done. Typically, this work is defined in a top-down, outline manner. In Chapter 3, you learned about system development routes and their phases. But phases are too large and complex for planning and scheduling a project. We need to break them down into activities and tasks until each task represents a manageable amount of work that can be planned, scheduled, and assigned. Some experts advocate decomposing tasks until the tasks represent an amount of work that can be completed in two weeks or less.

Ultimately, the project manager will determine the level of detail in the outline; however, most system development methodologies decompose phases for you—into suggested activities and tasks. These activities and tasks are not necessarily carved in stone; that is, most methodologies allow for some addition, deletion, and changing of activities and tasks based on the unique nature of each project. One popular tool used to identify and document project activities and tasks is a work breakdown structure. A **work breakdown structure (WBS)** is a hierarchical decomposition of the project into phases, activities, and tasks.

scope the boundaries of a project—the areas of a business that a project may (or may not) address.

statement of work a narrative description of the work to be performed as part of a project. Common synonyms include *scope statement, project definition, project overview,* and *document of understanding.*

work breakdown structure (WBS) a graphical tool used to depict the hierarchical decomposition of a project into phases, activities, and tasks.

---

[4]Updated and revised by Donald H. Plumber, *Productivity Management: Keane's Project Management Approach for Systems Development* (Boston: Keane, Inc., 1995), p. 5.

STATEMENT OF WORK

I.	Purpose
II.	Background
	A. Problem, opportunity, or directive statement
	B. History leading to project request
	C. Project goal and objectives
	D. Product description
III.	Scope
	(notice the use of the information system building blocks)
	A. Stakeholders
	B. Knowledge
	C. Processes
	D. Communications
IV.	Project Approach
	A. Route
	B. Deliverables
V.	Managerial Approach
	A. Team-building considerations
	B. Manager and experience
	C. Training requirements
	D. Meeting schedules
	E. Reporting methods and frequency
	F. Conflict management
	G. Scope management
VI.	Constraints
	A. Start date
	B. Deadlines
	C. Budget
	D. Technology
VII.	Ballpark Estimates
	A. Schedule
	B. Budget
VIII.	Conditions of Satisfaction
	A. Success criteria
	B. Assumptions
	C. Risks
IX.	Appendixes

FIGURE 4-5

An Outline for a Statement of Work

Work breakdown structures can be drawn using top-down hierarchy charts similar to organization charts (Figure 4-6). In Microsoft *Project*, a WBS is depicted using a simple outline style, indentation of activities and tasks on the Gantt chart "view" of the project. Microsoft *Project* also offers a military numbering scheme to represent hierarchical decomposition of a project as follows:

1. Phase 1 of the project
   1.1 Activity 1 of Phase 1
              1.1.1 Task 1 of Activity 1 in Phase 1
              1.1.2 Task 2 of Activity 1 in Phase 1
   1.2 Activity 2 of Phase 1 . . .
2. Phase 2 of the project . . .

If you reexamine Figure 4-3(a), you will notice that Microsoft *Project* provides a column for the WBS in the Gantt chart. Also notice its use of the indentation and numbering to differentiate between tasks and subtasks.

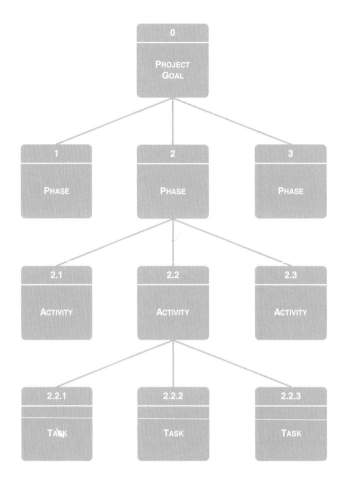

milestone an event signifying the completion of a major project deliverable.

We may want to include in a WBS special tasks called **milestones.** These are events that signify the accomplishment or completion of major deliverables during a project. In information systems projects, an example of a milestone might be the completion of all the tasks associated with producing a major project deliverable such as a requirements statement (see Chapter 3). It might be useful to distinguish milestones from other tasks in a WBS by using special formatting, such as italics.

> Activity 3—Estimate Task Durations

Given a work breakdown structure with a suitable level of detail, the project manager must estimate duration for each task. Duration of any task is a random variable subject to factors such as the size of the team, number of users, availability of users, aptitudes of users, complexity of the business system, information technology architecture, experience of team personnel, time committed to other projects, and experience with other projects.

Most system development methodologies not only define tasks but also provide baseline estimates for task duration. The project manager must adjust these baselines into reasonable estimates for each unique project.

In Microsoft *Project,* all phases, activities, and tasks of a methodology are simply called *tasks.* The work breakdown structure then consists of both summary and primitive tasks. A *summary task* is one that consists of other tasks (such as phases and activities). A *primitive task* is one that does not consist of any other tasks. It is the primitive tasks for which the project manager must estimate duration. (Like most project management software, Microsoft *Project* will automatically calculate the duration of all summary tasks based on the estimated durations of their component primitive tasks.)

For those primitive tasks that are not milestones, we must estimate duration. In estimating task duration, it is important to understand the concept of *elapsed time*. Elapsed time takes into consideration two important factors with respect to people:

- *Efficiency*—No worker performs at 100 percent efficiency. Most people take coffee breaks, lunch breaks, restroom breaks, and time to read their e-mail, check their calendars, participate in nonproject work, and even engage in idle conversation. Experts differ on just how productive the average worker is, but one commonly used figure is 75 percent.
- *Interruptions*—People experience phone calls, visitors, and other unplanned interruptions that increase the time required for project work. This is variable for different workers. Interruptions can consume as little as 10 percent of a worker's day or as much as 50 percent.

Why is this important? Given a task that could be completed in 10 hours with 100 percent efficiency and no interruptions, and assuming a worker efficiency of 75 percent and 15 percent interruptions, the true estimate for the task would be

$$10 \text{ hours} \div 0.75 \text{ efficiency} = 13.3 \text{ hours} \div (1.00 - 0.15 \text{ interruptions})$$

$$= 15.7 \text{ hours}$$

There are many techniques for estimating task duration. For the sake of demonstration, we offer the following classic technique:

1. *Estimate the minimum amount of time it would take to perform the task.* We'll call this the **optimistic duration (OD).** The optimistic duration assumes that even the most likely interruptions or delays, such as occasional employee illnesses, will *not* happen.
2. *Estimate the maximum amount of time it would take to perform the task.* We'll call this the **pessimistic duration (PD).** The pessimistic duration assumes that nearly anything that can go wrong will go wrong. All possible interruptions or delays, such as labor strikes, illnesses, inaccurate specification of requirements, equipment delivery delays, and underestimation of the system's complexity, are assumed to be inevitable.
3. *Estimate the* **expected duration (ED)** *that will be needed to perform the task.* Don't just take the median of the optimistic and pessimistic durations. Attempt to identify interruptions or delays that are most likely to occur, such as occasional employee illnesses, inexperienced personnel, and occasional training.
4. *Calculate the* **most likely duration (D),** as follows:

$$D = \frac{(1 \times OD) + (4 \times ED) + (1 \times PD)}{6}$$

where 1, 4, and 1 are default weights used to calculate a weighted average of the three estimates.

Developing OD, PD, and ED estimates can be tricky and require experience. Several techniques are used in estimating. Three of the most common techniques are:

- *Decomposition*—a simple technique wherein a project is decomposed into small, manageable pieces that can be estimated based on historical data of past projects and similarly complex pieces.
- *COCOMO* (pronounced like "Kokomo")—a model-based technique wherein standard parameters based on prior projects are applied to the new project to estimate duration of a project and its tasks.
- *Function points*—a model-based technique wherein the "end product" of a project is measured based on number and complexity of inputs, outputs, files, and queries. The number of function points is then compared to projects that had a similar number of function points to estimate duration.

optimistic duration (OD) the estimated minimum amount of time needed to complete a task.

pessimistic duration (PD) the estimated maximum amount of time needed to complete a task.

expected duration (ED) the estimated amount of time required to complete a task.

most likely duration (D) an estimated amount of time required to complete a task, based on a weighted average of optimistic, pessimistic, and expected durations.

Some automated project management tools, such as *CS/10000* and *Cost•Xpert,* provide expert system technology that makes these estimates for you based on your answers to specific questions.

Milestones (as defined in the previous subsection) have no duration. They simply happen. In Microsoft *Project,* milestones are designated by setting the duration to *zero.* (In the Gantt chart, those zero-duration tasks change from bars to diamonds.)

> Activity 4—Specify Intertask Dependencies

Given the duration estimates for all tasks, we can now begin to develop a project schedule. The project schedule depends not only on task durations but also on inter-task dependencies. In other words, the start or completion of individual tasks may depend on the start or completion of other tasks. There are four types of intertask dependencies:

- *Finish-to-start (FS)*—The finish of one task triggers the start of another task.
- *Start-to-start (SS)*—The start of one task triggers the start of another task.
- *Finish-to-finish (FF)*—Two tasks must finish at the same time.
- *Start-to-finish (SF)*—The start of one task signifies the finish of another task.

Intertask dependencies can be established and depicted in both Gantt and PERT charts. Figure 4-7 illustrates how to enter intertask dependencies in the Gantt chart view in Microsoft *Project.* We call your attention to the following annotated bullets:

❶ Intertask dependencies may be entered in the Gantt chart view in the *Predecessors* column by entering the dependent tasks' row numbers. Note that a task can have zero, one, or many predecessors.

❷ Intertask dependencies may also be entered (or modified) by opening the *Task Information* dialogue box for a given task.

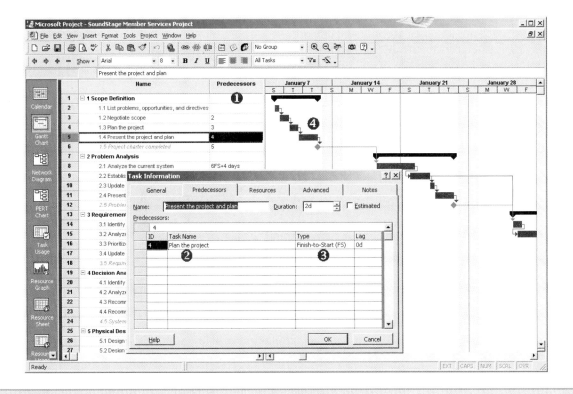

FIGURE 4-7 Entering Intertask Dependencies in Microsoft *Project*

❸ The type of dependency can be entered in the *Task Information* dialogue box for any given dependent task.

❹ Intertask dependencies are graphically illustrated in the Gantt chart as arrows between the bars that represent each task. Arrows may begin or terminate on the left side (to indicate a "start" dependency) or right side (to indicate a "finish" dependency).

Milestones (as defined earlier) almost always have several predecessors to signify those tasks that must be completed before the milestone has been achieved.

Given the start date for a project, the tasks to be completed, the task durations, and the intertask dependencies, the project can now be scheduled. There are two approaches to scheduling:

- **Forward scheduling** establishes a project start date and then schedules forward from that date. Based on the planned duration of required tasks, their interdependencies, and the allocation of resources to complete those tasks, a projected project completion date is calculated.

- **Reverse scheduling** establishes a project deadline and then schedules backward from that date. Tasks, their duration, interdependencies, and resources must be considered to ensure that the project can be completed by the deadline.

Each task can be given its own start and finish dates. Like most project management tools, Microsoft *Project* actually builds the schedule for you as you enter the task durations and intertask dependencies (predecessors). On the Gantt chart, the task bars are expanded to reflect duration and shifted left and right to reflect start and end dates. Microsoft *Project* can also produce a traditional calendar view of the final schedule, as shown in Figure 4-8.

forward scheduling a project scheduling approach that establishes a project start date and then schedules forward from that date.

reverse scheduling a project scheduling strategy that establishes a project deadline and then schedules backward from that date.

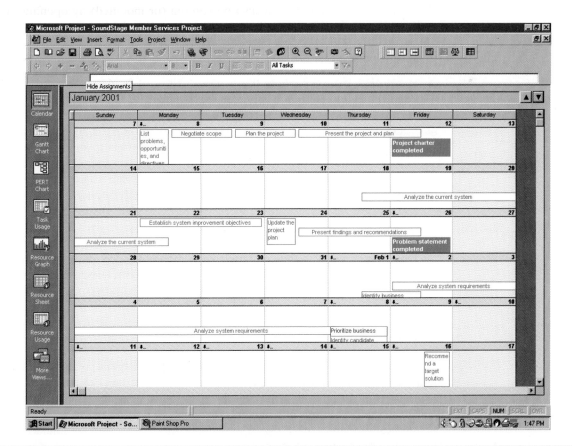

FIGURE 4-8 The Project Schedule in Calendar View

> Activity 5—Assign Resources

The previous steps resulted in "a" schedule, but not "the" schedule! We have yet to consider the allocation of resources to the project. Resources include the following categories:

- *People*—includes all the system owners, users, analysts, designers, builders, external agents, and clerical help that will be involved in the project in any way.
- *Services*—includes services such as a quality review that may be charged on a per-use basis.
- *Facilities and equipment*—includes all rooms and technology that will be needed to complete the project.
- *Supplies and materials*—includes everything from pencils, paper, and note-books to toner cartridges, and so on.
- *Money*—includes a translation of all of the above into budgeted dollars!

The availability of resources, especially people and facilities, can significantly alter the project schedule.

Most system development methodologies identify *people* resources required for each task in the form of roles. A role is not the same as a job title. Think of a role as a "hat" that someone wears because he or she possesses a certain skill(s). Any given individual may be capable of wearing many hats (thus playing many roles). Also, many people may possess the skills required to play a given role. The project manager's task is either to assign specific people to fill roles or to gain commitments from management to provide people to fill roles. Representative roles from the *FAST* methodology are listed in the margin.

In Microsoft *Project,* roles and assignments are specified in the *Resource Sheet* view, as shown in Figure 4.9(a). Predefined roles and resources may be available in the chosen methodology and route templates.

❶ The project manager enters the names or titles of people (roles) in the *Resource Name* column. Resources may also include specific services, facilities, equipment, supplies, materials, and so forth.

❷ Notice that the Resource Sheet provides a column for establishing what percentage of a resource will be allocated to the project. For example, a database administrator might be allocated one-quarter time (25 percent) to a project. Allocations greater than 100 percent indicate a need for more than one person to fill a given role in the project. By setting *Max. Units* to 250 percent for that resource, there would be a need for the equivalent of 2½ full-time programmers.

❸ *Project* also allows the project manager to estimate the cost of each resource. These costs can be estimated based on company history, consulting contracts, or internal cost accounting standards. Notice that both standard and overtime costs can be estimated. These costs are usually based on standards to protect information about anyone's actual salary.

❹ Each resource has a *calendar* that considers the standard workweek and holidays, as well as individual vacations and other commitments.

Given the resources, they now can be specifically assigned to tasks, as shown in Figure 4.9(b). As resources are assigned to the tasks, the project manager would specify the units of that resource that will be required to complete each assigned task. (This may be a *percentage* of a person's time needed for that task.)

As these resources are formally assigned, the schedule will be adjusted (which happens automatically in tools such as *Project*). If you enter the cost of resources, tools such as Microsoft *Project* will automatically calculate and maintain a budget based on the resources and schedule.

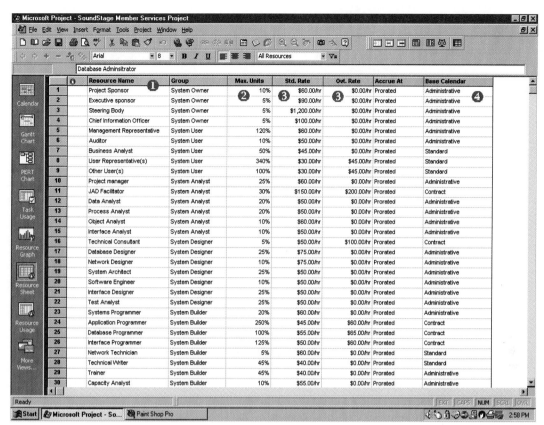

(a)

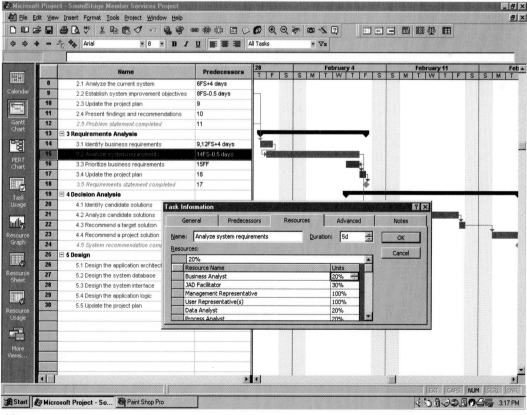

(b)

FIGURE 4-9 Defining and Assigning Project Resources

Assigning People to Tasks   Recruiting the right team members can make or break a project. The following are guidelines for selecting and recruiting the team:

- *Recruit talented, highly motivated people.* Highly skilled and motivated team members are more likely to overcome project obstacles unaided and are more likely to meet project deadlines and produce quality work.
- *Select the best task for each person.* All workers have strengths and weaknesses. Effective project managers learn to exploit the strengths of team members and avoid assigning tasks to team members not skilled in those areas.
- *Promote team harmony.* Project managers should select team members who will work well together.
- *Plan for the future.* Junior personnel with potential to be mentored by project leaders must be considered. Although junior personnel might not be as productive as the seasoned veterans, project managers will need them and have to rely on them on future projects.
- *Keep the team size small.* By limiting the team size, communication overhead and difficulties will be reduced. A 2-person team has only 1 communication path; a 4-person team has 6 communication paths; and a 50-person team has at least 1,200 communication paths. The more communication paths there are, the greater the probability that there will be increased communication problems. By the same token the teams should be large enough to provide adequate backup and coverage in key skills if a team member is lost.

Resource Leveling   So far, we have identified tasks, task durations, and intertask dependencies and assigned resources to each task to produce the project schedule. It is common to overallocate resources when assigning resources to tasks. *Overallocate* refers to the act of assigning more resources than are available.

For example, during a specific period in the project (day, week, etc.), a project manager may have assigned a specific person to work on multiple tasks that add up to more hours than the person has available to work during that period. This renders the overall schedule infeasible because the overallocated resource cannot reasonably complete all assigned tasks according to schedule. To correct this problem, project managers must use a technique called resource leveling. **Resource leveling** is a strategy used to correct resource overallocations by some combination of delaying or splitting tasks. Let's briefly explain both approaches.

Delaying tasks is based on the concepts of *critical path* and *slack time.* When it comes to the project schedule, some tasks are more sensitive to schedule delays than others. For this reason, project managers must become aware of the critical path for a project. The **critical path** for a project is the sequence of dependent tasks that have the largest sum of *most likely durations.* The critical path determines the earliest possible completion date of the project. (We previously described how to estimate *most likely duration* for a task.) The critical path tasks have no slack time available—thus, any delay in completion of any of the tasks on the critical path will cause an overall delay in the completion of the entire project. The opposite of a critical task is one that has some slack time. The **slack time** available for any noncritical task is the amount of delay that can be tolerated between the starting time and the completion time of a task without causing a delay in the completion date of the entire project. Tasks that have slack time can get behind schedule by an amount less than or equal to the slack time without having any impact on the project's final completion date. The availability of slack time in certain tasks provides an opportunity to delay the start of the tasks to level resources while not affecting the project completion date. Of course, it may be necessary to delay a critical path task to level resources, unless you can split the task.

Splitting tasks involves breaking a task into multiple tasks to assign alternate resources to the tasks. Thus, a single task for which a resource was overallocated is now apportioned to two or more resources that are (presumably) not overallocated. Splitting tasks requires identifying and assigning new resources such as analysts, contractors, or consultants.

resource leveling a strategy for correcting resource overallocations.

critical path the sequence of dependent tasks that determines the earliest completion date for a project.

slack time the amount of delay that can be tolerated between the starting time and the completion time of a task without causing a delay in the completion date of a project.

Resource leveling can be tedious to perform manually. For each resource, the project manager needs to know the total time available to the project for the resource, all task assignments made to the resource, and the sum of all durations of those task assignments over various time periods. All project management software tools, such as Microsoft *Project,* automatically determine critical *paths* and slack times. This enables those same software tools to track resource allocations and automatically perform resource leveling. It is extraordinarily rare for any modern project manager to manually level resource assignments.

Resource leveling will be an ongoing activity because the schedule and resource assignments are likely to change over the course of a project.

Schedule and Budget　　Given a schedule based on leveled resources and given the cost of each resource (e.g., cost per hour of a systems analyst or database administrator) the project manager can produce a printed (or Web-based) document that communicates the project plan to all concerned parties. Project management tools will provide multiple views of a project such as calendars, Gantt chart, PERT chart, resource and resource leveling reports, and budget reports. All that remains is to direct resources to the completion of project tasks and deliverables.

Communication　　The statement of work, timetable for major deliverables, and overall project schedule should be communicated to all parties involved in the project. This communication must also include a plan for reporting progress, both orally and in writing, the frequency of such communications, and a contact person and method for parties to submit feedback and suggestions. A corporate intranet can be an effective way to keep everyone informed of project progress and issues.

> Activity 6—Direct the Team Effort

All the preceding project management activities led to a master plan for the project. It's now time to execute that plan. There are several dimensions to directing the team effort. Tom Demarco states in his book *The Deadline: A Novel about Project Management* that the hardest job in management is people.

Few new project managers are skilled at supervising people. Most learn supervision through their own experiences as subordinates—things they liked and disliked about those who supervised them. This topic could easily take up an entire chapter. In the margin checklist, we provide a classic list of project supervision recommendations from *The People Side of Systems,* by Keith London.

As noted by Graham McLeod and Derek Smith, "Individuals brought together in a systems development team do not form a close-knit unit immediately." McLeod and Smith explain that teams go through stages of team development, as shown in Figure 4-10.

In *The One Minute Manager,* by Kenneth Blanchard and Spencer Johnson, a classic and indispensable aid to anyone managing people for the first time, the authors share the simple secrets of managing people and achieving success through the actions of subordinates.

Most young, and many experienced, managers have difficulty with the subtle arts of delegation and accountability. Worse still, they let subordinates reverse-delegate tasks back to the manager. This leads to poor time management and manager frustration. In *The One Minute Manager Meets the Monkey,* Kenneth Blanchard teams with William Oncken and Hal Burrows to help managers overcome this problem. The solution is based on Oncken's classic principle of "the care and feeding of monkeys." Monkeys are "problems" that managers delegate to their subordinates, who in turn attempt to reverse-delegate back to the manager. In this 125-page book the authors teach managers how to keep the monkeys on the subordinates' backs. Doing so increases the manager's available work time, accelerates task accomplishment by subordinates, and teaches subordinates how to take responsibility and solve their own problems.

10 HINTS FOR PROJECT LEADERSHIP

Be Consistent.

Provide Support.

Don't Make Promises You Can't Keep.

Praise in Public; Criticize in Private.

Be Aware of Morale Danger Points.

Set Realistic Deadlines.

Set Perceivable Targets.

Explain and Show, Rather Than Do.

Don't Rely Just on Status Reports.

Encourage a Good Team Spirit.

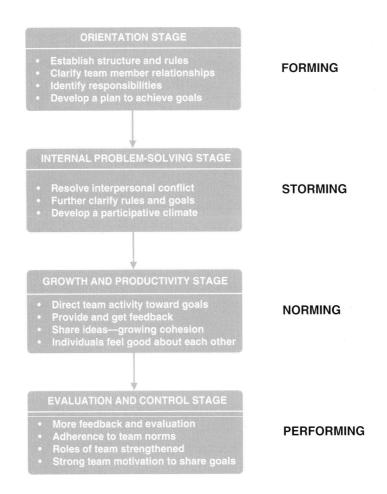

FIGURE 4-10

Stages of Team Maturity

Source: Adapted from Graham McLeod and Derek Smith, *Managing Information Technology Projects* (Cambridge, MA: Course Technology, 1996).

> Activity 7—Monitor and Control Progress

While executing the project, the project manager must control the project, that is, monitor its progress against the scope, schedule, and budget. The manager must report progress and, when necessary, adjust scope, schedule, and resources.

Progress Reporting Progress reporting should be frequent enough to establish accountability and control, but not so frequent as to become a burden and impediment to real project progress. For example, Keane, Inc., a consulting firm, recommends that progress reports or meetings occur every two weeks—consistent with the firm's project-planning strategy that decomposes projects into tasks that produce deliverables that require no more than 80 work hours.

Project progress reports can be verbal or written. Figure 4-11 illustrates a template for a written progress report. Project progress reports (or presentations) should be honest and accurate, even if the news is less than good. Project progress reports should report successes but should clearly identify problems and concerns such that they can be addressed before they escalate unto major issues or catastrophes.

As tasks are completed, progress can be recorded in Microsoft *Project* (see Figure 4-12). We call your attention to the following Gantt progress items:

❶ All the tasks in the preliminary investigation phase are complete as indicated by the yellow lines that run the full length of each task bar. Notice that because all these tasks are complete, they are no longer critical—the bars have changed from red to blue.

❷ In the problem analysis phase, only the first task, "Analyze the current system," is 100 percent complete.

PROJECT PROGRESS REPORT

I. Cover page
    A. Project name or identification
    B. Project manager
    C. Date of report
II. Summary of progress
    A. Schedule analysis
    B. Budget analysis
    C. Scope analysis
       (describe any changes that may have an impact on future progress)
    D. Process analysis
       (describe any problems encountered with strategy or methodology)
    E. Gantt progress chart(s)
III. Activity analysis
    A. Tasks completed since last report
    B. Current tasks and deliverables
    C. Short-term future tasks and deliverables
IV. Previous problems and issues
    A. Action item and status
    B. New or revised action items
       1. Recommendation
       2. Assignment of responsibility
       3. Deadline
V. New problems and issues
    A. Problems
       (actual or anticipated)
    B. Issues
       (actual or anticipated)
    C. Possible solutions
       1. Recommendation
       2. Assignment of responsibility
       3. Deadline
VI. Attachments
    (include relevant printouts from project management software)

FIGURE 4-11

Outline for a
Progress Report

❸ Notice that the "Establish system improvement objectives" task bar has a partial yellow line running 60 percent of its length. This indicates the task is about 60 percent complete. The task bar is still red because any delay in completing the task will threaten the project completion date.

❹ All remaining tasks shown in the displayed chart have not been started. Actual progress will be recorded when the task is started, in process, or completed.

❺ Progress for any given task is recorded in the task information dialogue box for that task. In this example, the project manager is recording 10 percent completion of the named task.

Microsoft *Project* also provides a number of preconfigured and customizable reports that can present useful project status information.

Change Management   It is not uncommon for scope to grow out of control even when a properly completed statement of work was agreed on early in the planning process. We refer to scope growth as "change." As noted by Keane, Inc., "Change is frequently a point of contention between the customer and the information systems organization, because they disagree on whether a particular function is a change or a part of the initial agreement." The inevitability of scope change necessitates that we have a formal strategy and process to deal with change and its impact on schedule

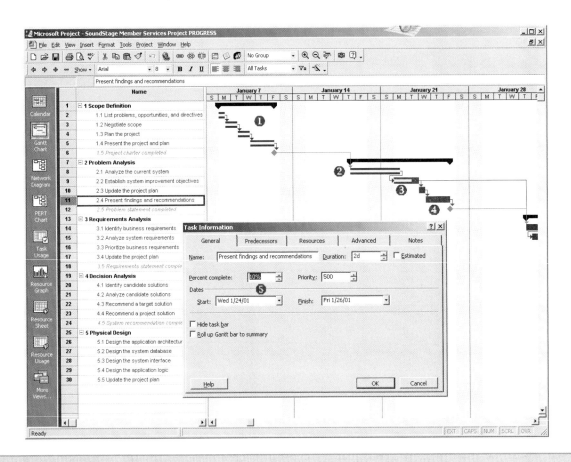

FIGURE 4-12 Progress Reporting on a Gantt Chart

change management a formal strategy wherein a process is established to facilitate changes that occur during a project.

and budget. **Change management** is intended to protect the project manager and team from being held accountable for schedule and budget overruns that were driven by changes in scope.

Changes can be the result of various events and factors, including:

- An omission in defining initial scope (as documented in the statement of work).
- A misunderstanding of the initial scope (the desired product is more complicated than originally communicated or perceived).
- An external event such as government regulations that create new requirements.
- Organizational changes, such as mergers, acquisitions, and partnerships, that create new business problems and opportunities (not to mention "players").
- Availability of better technology.
- Shifts in planned technology that force unexpected and significant changes to the business organization, culture, and/or processes.
- Management's desire to have the system do more than was originally requested or agreed to.
- Reduced funding for the project or imposition of an earlier deadline.

A change management system results in a collection of procedures for documenting a change request and defines the steps necessary to consider the change based on the expected impact of the change. Most change management systems require that a change request form be initiated by one or more project stakeholders (e.g., system owners, users, analysts, designers, or builders). Ideally, change requests are considered by a *change control board (CCB),* which is responsible for approving or rejecting all change requests. The CCB's composition typically includes members of the project team as well as outsiders who may have an interest or stake in the project. The CCB's decision should be based on impact analysis.

Feasibility impact analysis should assess the importance of the change to the business, the impact of the change on the project schedule, and the impact of the change on the project budget and long-term operating costs.

Ultimately, change management boils down to managing the expectations of the stakeholders. In the next section, we introduce a simple but conceptually sound framework for managing expectations and their impact on project schedule and budget.

Expectations Management   Experienced project managers often complain that managing system owners' and users' expectations of a project is more difficult than managing cost, schedule, people, or quality. In this section we introduce a simple tool that we'll call an *expectations management matrix* that can help project managers deal with this problem. We first learned about this tool from Dr. Phil Friedlander, a consultant and trainer then with McDonnell Douglas. He attributes the matrix to "folklore" but also credits Jerry Gordon, of Majer, and Ron Leflour, a project management educator/trainer. Dr. Friedlander's paper (listed in the Suggested Readings for this chapter) is adapted in this text for this presentation.

Every project has goals and constraints when it comes to cost, schedule, scope, and quality. In an ideal world, each of these parameters could be optimized. Management often has that expectation. Reality suggests, however, that you can't optimize them all—you must strike a balance that is both feasible and acceptable to management. That is the purpose of the expectations management matrix. An **expectations management matrix** is a rule-driven tool for helping management understand the dynamics and impact of changing project parameters such as cost, schedule, scope, and quality.

> **expectations management matrix** a tool used to understand the dynamics and impact of changing the parameters of a project.

The basic matrix, shown in Figure 4-13, consists of three rows and three columns (plus headings). The rows correspond to the measures of success in any project: cost, schedule, and scope and/or quality. The columns correspond to priorities: first, second, and third. To establish expectations, we assign names to the priorities as follows:

- *Maximize or minimize*—the measure of success that is determined to be the most important for a given project.
- *Constrain*—the second most important of the three measures of success in a project.
- *Accept*—the least important of the three measures in a project.

Most managers would, ideally, like to give equal priority to all three measures; experience suggests that the three measures tend to balance themselves naturally. For example, if you increase scope or quality requirements, the project will take more time and/or money. If you try to get any job done faster, you generally have to reduce scope or

PRIORITIES → ↓MEASURES OF SUCCESS	Max or Min	Constrain	Accept
Cost			
Schedule			
Scope and/or Quality			

FIGURE 4-13

A Management Expectations Matrix

FIGURE 4-14

Management of
Expectations for
the Lunar Landing
Project

↓MEASURES OF SUCCESS	PRIORITIES → Max or Min	Constrain	Accept
Cost • $20 billion (estimated)			X
Schedule • Dec 31, 1969 (deadline)		X	
Scope and/or Quality • Land a man on the moon • Get him back safely	X		

quality requirements or pay more money to compensate. The management expectations matrix helps (or forces) management to understand this through three simple rules:

1. For any project, you must record three Xs within the nine available cells.
2. No row may contain more than one X. In other words, a single measure of success must have one and only one priority.
3. No column may contain more than one X. In other words, there must be a first, second, and third priority.

Let's illustrate the tool using Dr. Friedlander's own example. In 1961 President John F. Kennedy established a major project—land a man on the moon and return him safely before the end of the decade. Figure 4-14 shows the realistic expectations of the project. Let's walk through the example:

1. The system owner (the public) had both scope and quality expectations. The scope (or requirement) was to successfully land a man on the moon. The quality measure was to return the man (or men) safely. Because the public would expect no less from the new space program, this had to be made the first priority. In other words, we had to maximize safety and minimize risk as a first priority. Hence, we record the X in column 1, row 3.
2. At the time of the project's inception, the Soviet Union was ahead in the race to space. This was a matter of national pride; therefore, the second priority was to get the job done by the end of the decade. We call this the project constraint—there is no need to rush the deadline, but we don't want to miss the deadline. Thus, we record the second X in column 2, row 2.
3. By default, the third priority had to be cost (estimated at $20 billion in 1961). By making cost the third priority, we are not stating that cost will not be controlled. We are merely stating that we may have to accept cost overruns to achieve the scope and quality requirement by the constrained deadline. We record the third X in column 3, row 1.

History records that we achieved the scope and quality requirement, and did so in 1969. The project actually cost well in excess of $30 billion, more than a 50 percent cost overrun. Did that make the project a failure? On the contrary, most people perceived the project as a grand success. The government managed the public's expectations of the project in realizing that maximum safety and minimum risk, plus meeting the deadline (beating the Soviets), was an acceptable trade-off for the cost overrun. The government brilliantly managed public opinion. Systems development project managers can learn a valuable lesson from this balancing act.

At the beginning of any project, the project manager should consider introducing the system owner to the expectations matrix concept and should work with the system owner to complete the matrix. For most projects, it would be difficult to record all the scope and quality requirements in the matrix. Instead, they would be listed in the statement of work. The estimated costs and deadlines could be recorded directly in the matrix.

The project manager doesn't establish the priorities; he or she merely enforces the rules of the matrix. This sounds easy, but it rarely is. Many managers are unwilling to be pinned down on the priorities—"Shouldn't we be able to maximize everything?" These managers need to be educated about the reason for the priorities. They need to understand the priorities if they cannot maximize all three measures. This leads to intelligent compromises instead of merely guesswork.

What if a system owner refuses to prioritize? The tool becomes less useful, except as a mechanism for documenting concerns before they become disasters. A system owner who refuses to set priorities may be setting the project manager up for a no-win performance review. And as Dr. Friedlander points out, "Those who do not 'believe' the principles [of the matrix] will eventually 'know' the truth. You do not have to believe in gravity, but you will hit the ground just as hard as the person who does."

Let's assume the management expectations matrix that conforms to the aforementioned rules. How does this help a project manager manage expectations? During the course of the average systems development project, priorities are not stable. Various factors such as the economy, government, and company politics can change the priorities. Budgets may become more or less constrained. Deadlines may become more or less important. Quality may become more important. And, most frequently, requirements increase. As already noted, these changing factors affect all the measures in some way. The trick is to manage expectations despite the ever-changing project parameters.

The technique is relatively straightforward. Whenever the "max/min measure" or the "constrain measure" begins to slip, it can result in a potential management expectations problem. For example, consider a project manager who is faced with the following priorities (see Figure 4-15):

1. Explicit requirements and quality expectations were established at the start of a project and given the highest priority.
2. An absolute maximum budget was established for the project.
3. The project manager agreed to shoot for the desired deadline, but the system owner(s) accepted the reality that if something must slip, it should be schedule.

Now suppose that during systems analysis, significant and unanticipated business problems are identified. The analysis of these problems has placed the project behind schedule. Furthermore, solving the new business problems substantially expands the user

PRIORITIES → ↓MEASURES OF SUCCESS	Max or Min	Constrain	Accept
Cost		X	
Schedule			X
Scope and/or Quality	X		

FIGURE 4-15

A Typical Initial Expectations Matrix

FIGURE 4-16

Adjusting
Expectations
(a sample)

PRIORITIES → MEASURES OF SUCCESS	Max or Min	Constrain	Accept
Cost • **Adjusted budget**		**X+** Increase budget	
Schedule • **Adjusted deadline**			**X-** Extend deadline
Scope and/or Quality • **Adjusted scope**	**X+** Accept expanded requirements		

requirements for the new system. How does the project manager react? There should be no overreaction to the schedule slippage—schedule slippage was the "accept" priority in the matrix. The scope increase (in the form of several new requirements) is the more significant problem because the added requirements will increase the cost of the project. Cost is the constrained measure of success. As it stands, an expectations problem exists. The project manager needs to review the matrix with the system owner.

First, the system owner needs to be made aware of which measure or measures are in jeopardy and why. Then together, the project manager and system owner can discuss courses of action. Several courses of action are possible:

- The resources (cost and/or schedule) can be reallocated. Perhaps the system owner can find more money somewhere. All priorities would remain the same (noting, of course, the revised deadline based on schedule slippages already encountered during systems analysis).
- The budget might be increased, but it would be offset by additional planned schedule slippages. For instance, by extending the project into a new fiscal year, additional money might be allocated without taking any money from existing projects or uses. This solution is shown in Figure 4-16.
- The user requirements (or quality) might be reduced through prioritizing those requirements and deferring some of those requirements until version 2 of the system. This alternative would be appropriate if the budget cannot be increased.
- Finally, measurement priorities can be changed.

Only the system owner may initiate priority changes. For example, the system owner may agree that the expanded requirements are worth the additional cost. He or she may allocate sufficient funds to cover the requirements but may migrate priorities such that minimizing cost now becomes the highest priority (see Figure 4-17, step 1). But now the matrix violates a rule—there are two Xs in column 1. To compensate, we must migrate the scope and/or quality criterion to another column, in this case, the constrain column (see Figure 4-17, step 2). Expectations have been adjusted. In effect, the system owner is freezing growth of requirements and still accepting schedule slippage.

There are three final comments about priority changes. First, priorities may change more than once during a project. Expectations can be managed through any number of changes as long as the matrix is balanced (meaning it conforms to our

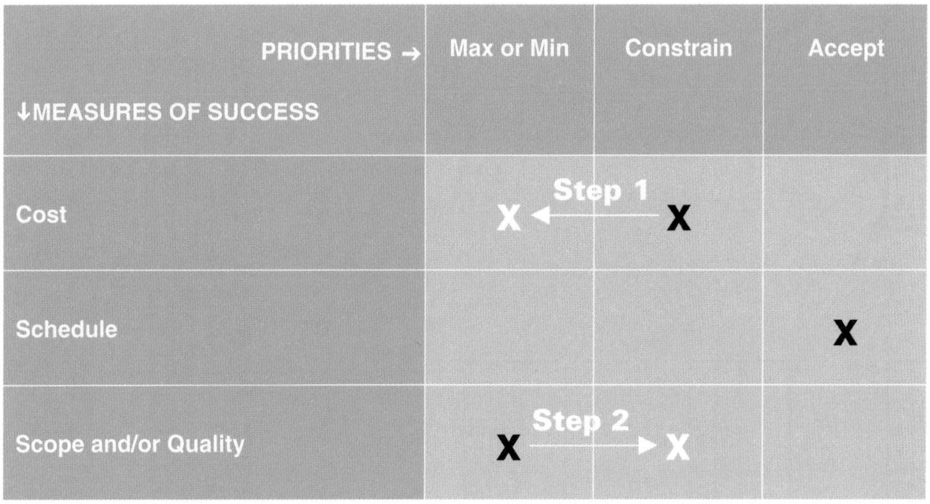

FIGURE 4-17
Changing Priorities

rules). Second, expectation management can be achieved through any combination of priority changes and resource adjustments. Finally, system owners can initiate priority changes even if the project is on schedule. For example, government regulation might force an uncompromising deadline on an existing project. That would suddenly migrate our "accept" schedule slippages to "max constraint." The other Xs would have to be migrated to rebalance the matrix.

While the management expectations matrix is a simple tool, it may be one of the most effective.

Schedule Adjustments—Critical Path Analysis   When it comes to the project schedule, some tasks are more sensitive to schedule delays than others. For this reason, project managers must become aware of the critical path and slack times for a project.

Understanding the critical path and slack time in a project is indispensable to the project manager. Knowledge of such project factors influences the people management decisions to be made by the project manager. Emphasis can and should be placed on the critical path tasks, and if necessary, resources might be temporarily diverted from tasks with slack time to help get critical tasks back on schedule.

The critical path and slack time for a project can be depicted on both Gantt and PERT charts; however, PERT charts are generally preferred because they more clearly depict intertask dependencies that define the critical path. Most project management software, including Microsoft *Project,* automatically calculates and highlights the critical path based on intertask dependencies combined with durations. It is useful, however, to understand how the critical path and slack times are calculated.

Consider the following hypothetical example. A project consists of the nine primitive tasks shown in Figure 4-18. The most likely duration (in days) for each task is recorded. There are four distinct sequences of tasks in a project. They are:

Path 1:	A → B → C → D → I
Path 2:	A → B → C → E → I
Path 3:	A → B → C → F → G → I
Path 4:	A → B → C → F → H → I

The total of most likely duration times for each path is calculated as follows:

Path 1:	3 + 2 + 2 + 7 + 5 = 19
Path 2:	3 + 2 + 2 + 6 + 5 = 18
Path 3:	3 + 2 + 2 + 3 + 2 + 5 = 17
Path 4:	3 + 2 + 2 + 3 + 1 + 5 = 16

In this example, path 1 is the *critical path* at 19 days. (Note: You can have multiple critical paths if they have the same total duration.)

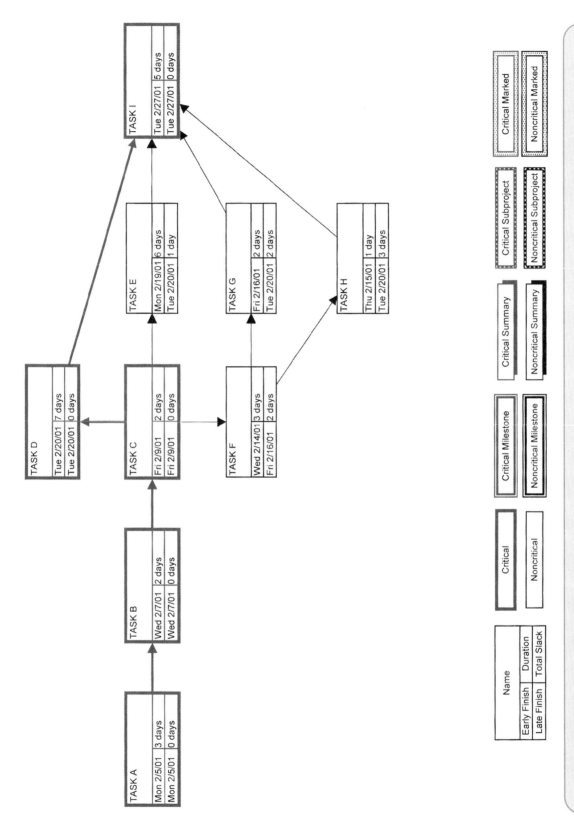

FIGURE 4-18 Critical Path Analysis

In this example, tasks E, F, and G are not on the critical path; they each have some slack time. For example, task E is included in a path that has one day less duration than the critical path; therefore, task E can get behind by as much as one day without adversely affecting the project completion date. Similarly, tasks F and G can *combine* for a maximum slack of two days without delaying the entire schedule.

In Figure 4-18, the critical path is shown in red. The tasks that have slack capacity are shown in black. Similarly, project management software also uses color to differentiate critical path tasks in a Gantt or PERT chart.

> Activity 8—Assess Project Results and Experiences

Project managers must learn from their mistakes! They should embrace continuous process improvement. This final activity involves soliciting feedback from project team members (including customers) concerning their project experiences and suggestions aimed at improving the project and process management of the organization. Project review(s) should be conducted to answer the following fundamental questions:

- Did the final product meet or exceed user expectations?
- Did the project come in on schedule?
- Did the project come in under budget?

The answers to these questions should be followed up with the basic question "Why or why not?" Subsequently, and based on the responses to the above questions, changes should be made to improve the system development and project management methods that will be used on future projects. Suggestions for improvements are communicated to "Centers for Excellence," which can modify standards and processes, as well as share useful ideas and experiences with other project teams that may solicit their help or expertise. Project assessments often contribute improvements to specific project deliverables (milestones), processes or tasks that created the deliverables, and the overall management of the project.

Where you go from here depends on where you are coming from and where you want to go. If you are reading through the chapters sequentially, you should probably move on to Chapter 5, "Systems Analysis," to expand your understanding of systems analysis tasks, tools, and techniques. Alternatively, if you are enrolled in a system design–focused course, you might skip ahead to either Chapter 11, "Feasibility Analysis and the System Proposal" (which marks the end of systems analysis), or Chapter 12, "Systems Design" (which provides an in-depth look at the activities of system design, prototyping, and rapid application development).

Some instructors have deferred this project management chapter to the end of your course. If so, you may be interested in expanding your knowledge of project management tools, techniques, and methods. Some schools offer a project management course. If not, you may find that your systems analysis and design instructor might supervise you to complete an independent study course on the subject. If so, we direct you to two specific references at the end of this chapter as possible texts: (1) the Wysocki et al. book is well organized around the Project Management Body of Knowledge that we presented in our chapter, and (2) the McLeod and Smith book is especially comprehensive in its coverage of project management dimensions that we could not cover fully in our chapter.

Learning Roadmap

Summary

1. A project is a (temporary) sequence of unique, complex, and connected activities that have one goal or purpose and that must be completed by a specific time, within budget, and according to specification.

2. Project management is the process of scoping, planning, staffing, organizing, directing, and controlling the development of an acceptable system at a minimum cost within a specified time frame.

3. Process management is an ongoing activity that documents, manages the use of, and improves an organization's chosen methodology (the "process") for systems development.

4. From a project management perspective, a project is considered a success if the resulting information system is acceptable to the customer, the system is delivered "on time" and "within budget," and the system development process had a minimal impact on ongoing business operations.

5. The Project Management Institute has created the Project Management Body of Knowledge (PMBOK) for the education and certification of professional project managers. It addresses:

   a. Project manager competencies.
   b. Project management functions.
   c. Tools and techniques such as:

      i) PERT charts, graphical network models that depict a project's tasks, and the relationships between those tasks.
      ii) Gantt charts, simple horizontal bar charts that depict project tasks against a calendar.

   d. Project management software.

6. Project management is a cross life-cycle activity; that is, project management tasks overlap all the system development phases. A project management process is essential to achieving CMM Level 2 maturity.

7. Joint project planning (JPP) is a strategy wherein all stakeholders in a project participate in a one- to three-day project management workshop, the result of which is consensus agreement on project scope, schedule, resources, and budget.

8. The tasks of project management include:

   a. Negotiate scope. Scope defines the boundaries of a project and is included in the statement of work, a narrative description of the work to be performed as part of a project.

   b. Identify tasks. A work breakdown structure (WBS) is a hierarchical decomposition of the project into its tasks and subtasks. Some tasks represent the completion of milestones or the completion of major deliverables during a project.

   c. Estimate task durations. There are many techniques and tools for estimating task durations.

   d. Specify intertask dependencies. The start or completion of individual tasks may be dependent on the start or completion of other tasks. These dependencies impact the completion of any project.

   e. Assign resources. The following resources may impact a project schedule: people, services, facilities and equipment, supplies and materials, and money.

      i) Such resources must be assigned to tasks to develop a schedule.
      ii) Resource leveling is a strategy used to correct resource overallocations by some combination of delaying or splitting tasks. Resource leveling requires knowledge of:

         (1) The critical path—that sequence of dependent tasks that have the largest sum of most likely durations. The critical path determines the earliest possible completion date of the project.
         (2) Slack time—the amount of delay that can be tolerated between the starting time and completion time of a task without causing a delay in the completion date of the entire project.

   f. Direct the team effort. One of the most important dimensions of directing the team effort is the supervision of people.

   g. Monitor and control progress. During the project, the project manager must monitor project progress against the scope, schedule, and budget and, when necessary, make adjustments to scope, schedule, and resources.

      i) Progress reporting is an essential control process that uses communication to keep a project within scope, on time, and within budget.
      ii) A complete project plan provides mechanisms and a process to manage requests for changes to scope. This is called change management.
      iii) Change management frequently requires that a project manager manage the expectations of management and users themselves. An expectations management matrix is a rule-driven tool for helping management understand the dynamics and impact of changing

project parameters such as cost, schedule, scope, and quality.

iv) Schedule adjustments are required when a project's scope changes or when other factors drive schedule or budget out of the projected range.

h. Assess project results and experiences. This final activity involves soliciting feedback from project team members (including customers) concerning their project experiences and suggestions aimed at improving the project and process management of the organization.

Review Questions

1. What is a project?
2. Of the many different reasons that projects fail, what is the major cause of project failure?
3. What is the difference between scope creep and feature creep?
4. What are the five main categories of competencies that a project manager should have?
5. Why are business achievement competencies important?
6. What are the basic project management functions?
7. What are PERT and Gantt charts? How do we decide which one to use?
8. What are the eight major activities in the project management life cycle?
   - Negotiate scope
   - Identify tasks
   - Estimate task durations
   - Specify intertask dependencies
   - Assign resources
   - Direct the team effort
   - Monitor and control progress
   - Assess project results and experiences
9. Why is negotiating scope important? What is the deliverable in the process of negotiating the scope?
10. What is a popular tool used to identify tasks in the project management life cycle?
11. What are the factors to consider in estimating task durations?
12. What are the differences between forward scheduling and reverse scheduling?
13. What are the categories of resources to be allocated to the project?
14. What should project managers do to manage changes that occur and/or are requested during a project?
15. Why is critical path analysis important?

Problems and Exercises

1. Assume you are a systems analyst and a proud member of a project team that has just completed a major project that spanned several years and that touched almost every business unit in your organization. The project was completed ahead of schedule and well within budget. Development and implementation went very smoothly with virtually no disruption of business operations. A postimplementation survey indicates that system users have been able to use the system with minimal training, although there have been some comments from the more vocal users that it wasn't quite what they expected and doesn't do some of the things they thought it would. Should the project be considered a success?
2. Executive management is concerned that some users are less than satisfied with the new system described in the preceding question and have assigned you to lead a postimplementation work group to determine the cause. Of the dozen project mismanagement problems described in the textbook, which ones do you think were most likely to have contributed to user dissatisfaction?

3. As a newly appointed project manager, you are eager to get started on your first project. What should your first activity be? How important is it? Who is typically involved? What questions do you need to make sure are answered? What's the ultimate outcome from this activity, and what is included in this deliverable?
4. You are the project manager of a medium-size project that is scheduled to take 10 months from project initiation on September 1st through delivery on June 30th. It is now April 1st, seven months since the project began, and the project is slightly behind schedule, by perhaps a week. Draw a Gantt chart (you may use the style shown in Figure 4.2 or another Gantt chart style if you prefer). Assume you are using the *FAST* methodology, and that project phases can overlap.
5. You are the project manager for a company that is building a behavioral health system for some of the counties in your state. The project is slightly ahead of schedule and there haven't been any significant problems to date. In reviewing some of the screens under construction, you are surprised to

find a number of features that were not part of the design. The system builder was one of your most talented and creative programmers. When you ask about these features, the builder proudly tells you that they add to the functionality of the system without taking any additional programming time, and that they were intended to be a surprise. You can see that the features definitely do add to the functionality of the system. The code has already been written for them—should you allow them to be included in the system, even though they were not part of the approved technical design?

6. The methodology used in your organization calls for change requests to be considered by a change control board (CCB). After some reflection and a discussion with the programmer, you have decided to submit a change request to the CCB to add the new features. In your presentation to the CCB, what reason might you give for the change request and what things should you take into consideration?

7. The CEO of your organization was so impressed with your last project that you have been given responsibility with a larger, even more important project. The CEO calls you in for a discussion regarding the importance of the project, and tells you that the very survival of the organization may hinge upon completing this project and rolling out the new system to customers before a certain date when a competitor is expected to complete a similar project. The company can afford to budget only up to a certain maximum, although if other, less critical projects-in-progress are delayed, there may be some additional funding available if absolutely necessary. Finally, in order to be a competitive product in the market, the new system must contain at least a certain minimum feature set, although more would be desirable, and the quality must be of the highest level. At the conclusion of this discussion, the CEO shakes your hand and wishes you good luck. Use the priorities set by the CEO to create an initial management expectations matrix.

8. Now suppose that during the course of this project, it becomes apparent that costs were significantly underestimated and the budget is rapidly becoming depleted. In addition, the head of marketing has picked up a trade magazine and read that your organization's main competitor is adding some really exciting features to their product without changing their release date. The budget overage is not the major problem; you know additional money can be allocated, although it may delay other projects. But you also know that your marketing stakeholders will be demanding that similar features be added to the system you are developing while keeping to the original

schedule. This presents an expectations conflict since scope is the constrained measure of success. What should you do at this point?

9. Suppose the CEO decides that no matter what, the new features absolutely must be added in order for the new system to be competitive. What issues does this raise, and how would this be reflected in the expectations matrix?

10. You are working on the schedule for the system design phase and are trying to estimate the duration of a complex design task. From breaking this task down into smaller tasks similar to ones that you've had experience with on other projects, you estimate the task should normally take an expected duration (ED) of three workdays, given a typical 75 percent worker efficiency rate and 15 percent interruption factor. But you also know of some instances where absolutely nothing went right, and it took up to two full workweeks, or a pessimistic duration (PD) of 80 hours, to complete the design task. Using the classic technique described in the textbook, calculate the most likely duration of the task.

11. In the preceding question, what technique did you use to estimate the expected duration of the design task? Describe some of the other techniques you could use to estimate task duration.

12. During one phase of the project, you review the project schedule and realize that a member of your project team has been assigned multiple tasks that add up to more hours than the person has available to work during that period. What technique could you use to resolve this?

13. You have been asked to complete a project in shortest time possible. The project tasks, most likely duration (in days), and predecessors are shown below. What are the different paths (sequence of tasks) and the number of days for each? What is the critical path, that is, the shortest time in which the project can be completed? Is it actually important in the business world for project managers to understand critical path analysis, or is this just theoretical knowledge?

Tasks	Duration	Predecessors
A	2	None
B	2	None
C	1	None
D	4	A
E	5	B
F	1	C, D
G	6	A, E
H	4	F
I	7	G,H

14. As a new project manager in a rapidly growing organization, you have been asked to lead a project team for an important project. The scope of the project is not too broad, project time frames are somewhat on the tight side but definitely doable, and the budget is more than generous. In fact, you have been given the authority to hire as many people as you want for your project team.

You estimate that 5 people would be about right for this type of project, 8 would provide a healthy amount of backup, and 10 could give you the resources to deliver an outstanding system in record time. What is something you might want to keep in mind before making your decision on how many people to hire?

 Projects and Research

1. Projects fail, sometimes spectacularly. Search the Web for articles on major project failures; numerous articles should be readily available. Find and review articles on approximately 10 major project failures during the past decade, then do the following:

   a. List the project failures that you found, and describe them.
   b. What was the cost of each project failure?
   c. What were the consequences of each project failure?
   d. Categorize the reason(s) for each project's failure based upon the causes listed in this chapter.
   e. What were the most common causes for the project failures?
   f. In hindsight, how many of the project failures were avoidable?
   g. What is the most important lesson that new systems analysts can learn from these project failures?

2. The Project Management Institute (PMI) is one of the leading and perhaps *the* leading project management organization in the world. PMI created and maintains the "Project Management Body of Knowledge" (PMBOK), which is a de facto standard for project managers. PMI also certifies a project manager who passes its knowledge and experience requirements as a "Project Management Professional" (PMP).

   a. Go to the PMI Web site (*www.pmi.org*). What are the requirements to become certified as a PMP?
   b. Based upon your readings and experience, how important do you think the PMP certification is for the project manager? Do you think it is worth the investment?
   c. What about the organization that employs a PMP? Is the certification an assurance that the organization's projects will be completed successfully. How much more should an organization pay a project manager with a PMP certification?

   d. Professionals in many fields, such as medicine, engineering, accountancy, and law, are required to be licensed or certified. Do you think professional certification should be required before someone can manage a large project?

3. You work in the information technology division of a large law firm with offices throughout the state. One of the vice presidents of the company has asked you to manage the development of an automated case-tracking system for your company. The project, which is just beginning, is the first large project you have been asked to manage. You take your duties very seriously and want to do an exemplary job on this project.

   a. You are meeting with the vice president of the company to discuss the scope of the project. In your meeting, what questions need to be answered and negotiated in order to be able to determine the scope of the project?
   b. Once you have finished negotiating scope, the vice president has asked you to write a Statement of Work. What does the Statement of Work represent in this situation? How long should it be?
   c. Write a Statement of Work, using the outline shown in Figure 4-5 as an example. Assume that the vice president has given you carte blanche (although that will probably never happen in real life).

4. Project management software, such as Microsoft *Project,* have become commonplace. Many of them incorporate traditional tools, such as PERT and Gantt charts, which were developed decades ago.

   a. Conduct an informal survey of about a dozen project managers in industry. How many of them use project management software?
   b. For those who don't use project management software, what are their reasons for not using it?
   c. For those who do use project management software, which ones do they use? What are their opinions regarding the software they use?

d. Search on the Web for different project management programs. Which ones did you find?

e. Review their features and specifications. Do any of them appear to have unique features? Which one do you think is the most popular, or at least the one most widely used? Which one would you pick if cost was not a consideration?

5. You are managing the development of a case-tracking system project for your large law firm. The requirements phase of the project is almost complete, and preliminary design work has begun. The project is running several days behind schedule, which you don't consider serious, and it is within budget, but barely. Quality, in terms of requirements analysis, has generally been acceptable so far in your opinion, but some of the project team members have mentioned that they are not sure if certain issues have been fully resolved. Based upon this information, write a Project Progress Report, using the outline in Figure 4-11 as an example and following the guidelines described in Activity 7 in this chapter.

6. As part of continuous improvement, it is important for project managers and project teams to assess the results and their experiences once a project has been completed. There are numerous methods and techniques for doing this. Search on the Web for pertinent articles, using phrases such as project assessment, project postimplementation reports, and the like.

a. What articles did you find?
b. Describe the methods and techniques they suggest.
c. Select the ones you feel are the most valuable, and explain why.
d. Do you think that assessing project results can make a significant difference in the quality of future projects?

Minicases

1. You are on a team that is developing a Web site for a local business, Custom Car Care. There is a set schedule of four months for requirements analysis, development, and successful deployment. The team is on schedule, in week 8, and has just shown Debbie, the CEO of Custom Car Care, the prototype. Debbie is very happy with your work so far, but has some additional capabilities she would like added to the site. Although the additions are not in the previous time or cost estimate, she requires that you stay on schedule and within current budget. What do you do?

2. Alicia and John are a team coding a difficult and sizable program in Java. They have some experience with the language, but will have to learn a significant amount "on the fly." They have estimated that the project will take two months as the optimistic estimate, three months as the expected estimate, and four months as the pessimistic estimate. You are their project manager and must develop a contract for completion with the client for the code development. How much time should you allow in the contract for this deliverable?

3. Develop both a forward and backward schedule of tasks and timelines for a major project you are completing for a class. If there is discrepancy between the two schedules, err on the side of front-loading your tasks. Monitor your project timeline and keep track of the milestones as you complete the project. At the end of the project, submit your timeline and project notes to your professor, along with a copy of the class project. Did the schedule development and management of the project help you? Share with your class your experience.

4. In an interview with a project manager, find out how often personnel issues affect the successful (and on-time) completion of a project. How does this project manager deal with personal or family problems that distract or remove key members of a team?

Suggested Readings

Blanchard, Kenneth, and Spencer Johnson. *The One Minute Manager.* New York: Berkley Publishing Group, 1981, 1982. Arguably, this is one of the best people management books ever written. Available in most bookstores, it can be read overnight and used for discussion material for the lighter side of project management (or any kind of management). This is must reading for all college students with management aspirations.

Blanchard, Kenneth; William Oncken, Jr.; and Hal Burrows. *The One Minute Manager Meets the Monkey.* New York: Simon & Schuster, 1988. A sequel to *The One Minute Manager,* this book effectively looks at the topic of delegation and time management. The monkey refers to Oncken's classic article, "Managing Management Time: Who's Got the Monkey?" as printed in the *Harvard Business Review* in 1974. The book teaches managers how to achieve results by helping their staff (their monkeys) solve their own problems.

Brooks, Fred. *The Mythical Man-Month.* Reading, MA.: Addison-Wesley, 1975. This is a classic set of essays on software engineering (also known as systems analysis, design, and implementation). Emphasis is on managing complex projects.

Catapult, Inc. *Microsoft Project 98: Step by Step.* Redmond, WA: Microsoft Press, 1997. An update for *Project 2000* is expected.

Demarco, Tom. *The Deadline: A Novel about Project Management.* New York: Dorset House Publishing, 1997. This would be an excellent companion to a project management text, especially for a graduate-level course. It demonstrates the "good, bad, and ugly" of project management, told as a story.

Duncan, William R., Director, and Standards Committee. *A Guide to the Project Management Body of Knowledge.* Upper Darby, PA: Project Management Institute, 1996. This is a concise overview of the generally accepted Project Management Body of Knowledge and practices used for certification of project managers.

Friedlander, Phillip. "Ensuring Software Project Success with Project Buyers," *Software Engineering Tools, Techniques, and Practices 2,* no. 6 (March/April 1992), pp. 26–29. We adapted our expectations management matrix from Dr. Friedlander's work.

Kernzer, Harold. *Project Management: A Systems Approach to Planning, Scheduling, and Controlling,* 4th ed. New York: Van Nostrand Reinhold, 1989. Many experts consider this book to be the definitive work in the field of project management. Dr. Kernzer's seminars and courses on the subject are renowned.

London, Keith. *The People Side of Systems.* New York: McGraw-Hill, 1976. This is a timeless classic about various people aspects of systems work. Chapter 8, "Handling a Project Team," does an excellent job of teaching the leadership aspects of project management.

McLeod, Graham, and Derek Smith. *Managing Information Technology Projects.* Cambridge, MA: Course Technology, 1996. If you are looking for a good academic book for a course or independent study project to expand your knowledge of IT project and process management, this is it! The book provides a comprehensive treatment of virtually all dimensions of IT project and process management.

Roetzheim, William H., and Reyna A. Beasley. *Software Project Cost and Schedule Estimating: Best Practices.* Upper Saddle River, NJ: Prentice Hall, 1998. This is one of the more complete books on the subject of estimating techniques. Better still, the book includes evaluation copies of *Cost•Xpert, Risk•Xpert,* and *Strategy•Xpert* (™of Marotz, Inc.), software tools for estimating.

Wysocki, Robert K; Robert Beck, Jr; and David B. Crane. *Effective Project Management: How to Plan, Manage, and Deliver Projects on Time and within Budget,* 2nd ed. New York: John Wiley & Sons, 2000. Buy this book! This is our new benchmark for introducing project management. It is easy to read and worth its weight in gold. We were surprised how compatible the book is with past editions of our book, and our project management directions continue to be influenced by this work.

Part Two

Systems Analysis Methods

The five chapters in Part Two introduce you to systems analysis activities and methods. Chapter 5, "Systems Analysis," provides the context for all the subsequent chapters by introducing the activities of *systems analysis*. Systems analysis is the most critical phase of a project. During systems analysis we learn about the existing business system, come to understand its problems, define objectives for improvement, and define the detailed business requirements that must be fulfilled by *any* subsequent technical solution. Clearly, any subsequent system design and implementation of a new system depends on the quality of the preceding systems analysis. Systems analysis is often shortchanged in a project because (1) many analysts are not skilled in the concepts and logical modeling techniques to be used, and (2) many analysts do not understand the significant impact of those shortcuts. Chapter 5 introduces you to systems analysis and its overall importance in a project. Subsequent chapters teach you specific systems analysis skills with an emphasis on logical system modeling.

Chapter 6, "Requirements Gathering," teaches various fact-finding techniques and strategies used to solicit user requirements for a new system.

In Chapter 7, "Use Cases," you will learn about the tools and techniques necessary to perform use-case modeling to document system requirements.

In Chapter 8, "Data Modeling and Analysis," we teach you *data modeling,* a technique for organizing and documenting the stored data requirements for a system. You will learn to draw entity relationship diagrams as a tool for structuring business data that will eventually be designed as a database. These models will capture the business associations and rules that must govern the data.

Chapter 9, "Process Modeling," introduces *process modeling.* It explains how data flow diagrams can be used to depict the essential business processes in a system, the flow of data through a system, and policies and procedures to be implemented by processes. If you've done any programming, you recognize the importance of understanding the business processes for which you are trying to write the programs.

Chapter 10, "Object-Oriented Analysis and Modeling with UML," teaches you about the object-oriented approach to performing systems analysis using UML tools.

Chapter 11, "Feasibility Analysis and the System Proposal," teaches you how to brainstorm possible system solutions, analyze those solutions for feasibility, select the best overall solution, and then present your recommendation in the form of a written and oral proposal to management.

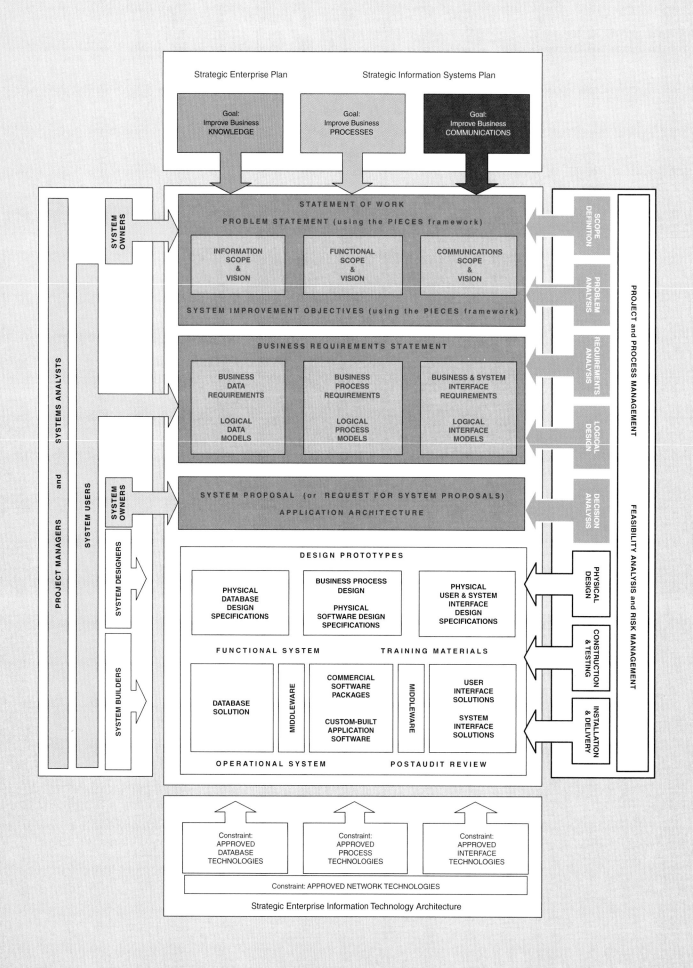

Systems Analysis

Chapter Preview and Objectives

In this chapter you will learn more about the systems analysis phases in a systems development project—namely, the scope definition, problem analysis, requirements analysis, and decision analysis phases. The first three phases are collectively referred to as *systems analysis*. The latter phase provides transition between systems analysis and systems design. You will know that you understand the process of systems analysis when you can:

▌ Define systems analysis and relate the term to the scope definition, problem analysis, requirements analysis, logical design, and decision analysis phases of this book's systems development methodology.

▌ Describe a number of systems analysis approaches for solving business system problems.

▌ Describe the scope definition, problem analysis, requirements analysis, logical design, and decision analysis phases in terms of your information system building blocks.

▌ Describe the scope definition, problem analysis, requirements analysis, logical design, and decision analysis phases in terms of purpose, participants, inputs, outputs, techniques, and steps.

▌ Identify the chapters in this textbook that can help you learn specific systems analysis tools and techniques.

NOTE: Although some of the tools and techniques of systems analysis are previewed in this chapter, it is *not* the intent of this chapter to teach those tools and techniques. This chapter teaches only the *process* of systems analysis. The tools and techniques will be taught in the subsequent six chapters.

Introduction

Bob Martinez remembers learning in college that systems analysis defines what an information system needs to do while system design defines how it needs to do it. At the time, it sounded like a simple two-step process. Now, as he begins working on the SoundStage Member Services system project, he sees that there are multiple phases and several steps within systems analysis and system design.

The SoundStage project is at the beginning of systems analysis, in what Sandra, his boss, calls the scope definition phase. After that they'll do problem analysis, requirements analysis, and decision analysis. It sounds like a lot of work just to understand *what* the system needs to do. But this is a complicated system. As Sandra says, would you build a house without a good set of plans?

What Is Systems Analysis?

systems analysis a problem-solving technique that decomposes a system into its component pieces for the purpose of studying how well those component parts work and interact to accomplish their purpose.

systems design a complementary problem-solving technique (to systems analysis) that reassembles a system's component pieces back into a complete system—hopefully, an improved system. This may involve adding, deleting, and changing pieces relative to the original system.

information systems analysis those development phases in an information systems development project that primarily focus on the business problem and requirements, independent of any technology that can or will be used to implement a solution to that problem.

repository a location (or set of locations) where systems analysts, systems designers, and system builders keep all of the documentation associated with one or more systems or projects.

In Chapter 3 you learned about the systems development process. In that chapter we purposefully limited our discussion to only briefly examining each phase. In this chapter, we take a much closer look at those phases that are collectively referred to as **systems analysis.** Formally defined in the margin, systems analysis is the study of a system and its components. It is a prerequisite to **systems design,** the specification of a new and improved system. This chapter will focus on systems analysis. Chapter 12 will do the same for systems design.

Moving from this classic definition of systems analysis to something a bit more contemporary, we see that *systems analysis* is a term that collectively describes the early phases of systems development. Figure 5-1 uses color to identify the systems analysis phases in the context of the full classic route for our *FAST* methodology (from Chapter 3). There has never been a universally accepted definition of systems analysis. In fact, there has never been universal agreement on when information systems analysis ends and when information systems design begins. For the purpose of this book, **information systems analysis** emphasizes *business* issues, *not* technical or implementation concerns.

Systems analysis is driven by the business concerns of SYSTEM OWNERS and SYSTEM USERS. Hence, it addresses the KNOWLEDGE, PROCESS, and COMMUNICATIONS building blocks from SYSTEM OWNERS' and SYSTEM USERS' perspectives. The SYSTEMS ANALYSTS serve as facilitators of systems analysis. This context is illustrated in the chapter home page that preceded the objectives for this chapter.

The documentation and deliverables produced by systems analysis tasks are typically stored in a repository. A repository may be created for a single project or shared by all projects and systems. A repository is normally implemented as some combination of the following:

- A network directory of word processing, spreadsheet, and other computer-generated files that contain project correspondence, reports, and data.
- One or more CASE tool dictionaries or encyclopedias (as discussed in Chapter 3).
- Printed documentation (such as that stored in binders and system libraries).
- An *intranet* Web site interface to the above components (useful for communication).

Hereafter, we will refer to these components collectively as the **repository.**

This chapter examines each of our five systems analysis phases in greater detail. But first, let's examine some overall strategies for systems analysis.

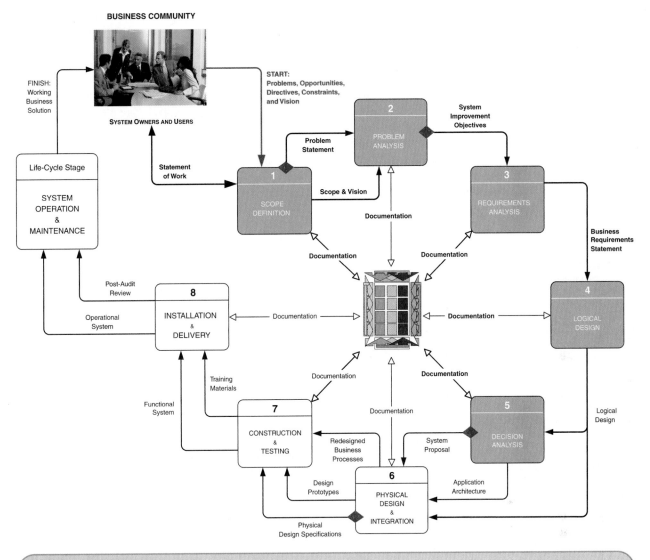

FIGURE 5-1 The Context of Systems Analysis

Systems Analysis Approaches

Fundamentally, systems analysis is about *problem solving*. There are many approaches to problem solving; therefore, it shouldn't surprise you that there are many approaches to systems analysis. These approaches are often viewed as *competing* alternatives. In reality, certain combinations can and should actually complement one another. This was characterized in Chapter 3 as *agile methods*. Let's briefly examine the varied approaches.

NOTE: The intent here is to develop a high-level understanding only. Subsequent chapters in this unit will actually teach you the underlying techniques.

> Model-Driven Analysis Approaches

Structured analysis, information engineering, and object-oriented analysis are examples of **model-driven analysis.** Model-driven analysis uses pictures to communicate

model-driven analysis a problem-solving approach that emphasizes the drawing of pictorial system models to document and validate existing and/or proposed systems. Ultimately, the system model becomes the blueprint for designing and constructing an improved system.

model a representation of either reality or vision. Since "a picture is worth a thousand words," most models use pictures to represent the reality or vision.

business problems, requirements, and solutions. Examples of **models** with which you may already be familiar include flowcharts, structure or hierarchy charts, and organization charts.

Today, model-driven approaches are almost always enhanced by the use of automated tools. Some analysts draw system models with general-purpose graphics software such as Microsoft *Visio.* Other analysts and organizations require the use of repository-based CASE or modeling tools such as *System Architect, Visible Analyst,* or *Rational ROSE.* CASE tools offer the advantage of consistency and completeness analysis as well as rule-based error checking.

Model-driven analysis approaches are featured in the model-driven methodologies and routes that were introduced in Chapter 3. Let's briefly examine today's three most popular model-driven analysis approaches.

Traditional Approaches   Various traditional approaches to system analysis and design were developed beginning in the 1970s. One of the first formal approaches, which is still widely used today, is *structured analysis.* **Structured analysis** focuses on the flow of data through business and software processes. It is said to be *process-centered.* By process-centered, we mean that the emphasis is on the PROCESS building bocks in your information system framework.

structured analysis a model-driven, PROCESS-centered technique used to either analyze an existing system or define business requirements for a new system, or both. The models are pictures that illustrate the system's component pieces: processes and their associated inputs, outputs, and files.

One of the key tools used to model processes is the *data flow diagram* (Figure 5-2), which depicts the existing and/or proposed processes in a system along with their inputs, outputs, and data. The models show the flow of data between and through processes and show the places where data is stored. Ultimately these process models serve as blueprints for business processes to be implemented and software to be purchased or constructed.

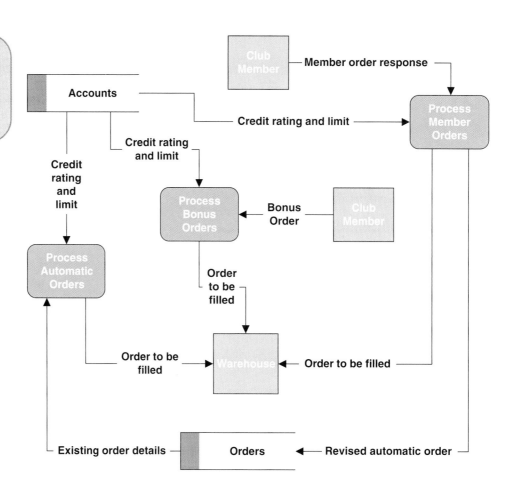

FIGURE 5-2

A Simple Process Model (Also Called a Data Flow Diagram)

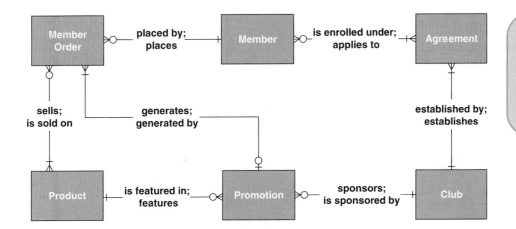

FIGURE 5-3

A Simple Data Model (Also Called an Entity Relationship Diagram)

The practice of structured analysis for software design has greatly diminished in favor of object-oriented methods. However, process modeling is enjoying something of a revival thanks to the renewed emphasis on business process redesign, which is discussed later in this chapter.

Another traditional approach, called **information engineering (IE),** focuses on the structure of stored data in a system rather than on processes. Thus, it was said to be *data-centered,* emphasizing the analysis of KNOWLEDGE (or data) requirements. The key tool to model data requirements is the entity relationship diagram (Figure 5-3). Entity relationship diagrams are still widely used in designing relational databases.

Originally, information engineering was seen as a competing approach to structured analysis. But over time many people made them as complementary: using data flow diagrams to model a system's processes and entity relationship diagrams to model a system's data.

Object-Oriented Approach Traditional approaches deliberately separated the concerns of KNOWLEDGE (data) from those of PROCESSES. Although most systems analysis methods attempted to synchronize data and process models, the attempt did not always work well in practice. **Object** technologies have since emerged to eliminate this artificial separation of data and processes. The **object-oriented approach** views information systems not as data and processes but as a collection of objects that encapsulate data and processes. Objects can contain data attributes. However, the only way to create, read, update, or delete an object's data is through one of its embedded processes (called methods). Object-oriented programming languages, such as *Java, C++*, and the *.NET* languages, are becoming increasingly popular.

The object-oriented approach has a complete suite of modeling tools known as the Unified Modeling Language (UML). One of the UML diagrams, an object class diagram, is shown in Figure 5-4. Some of the UML tools have gained acceptance for systems projects even when the information system will not be implemented with object-oriented technologies.

> Accelerated Systems Analysis Approaches

Discovery prototyping and rapid architected development are examples of accelerated systems analysis approaches that emphasize the construction of prototypes to more rapidly identify business and user requirements for a new system. Most such approaches derive from some variation on the construction of **prototypes,** working but incomplete samples of a desired system. Prototypes cater to the "I'll know what I want when I see it" way of thinking that is characteristic of many users and managers. By "incomplete," we mean that a prototype will not include the error checking, input data validation, security, and processing completeness of a finished application. Nor will it be as polished or offer the user help as in a final system. But because it can be

information engineering (IE) a model-driven and DATA-centered, but PROCESS-sensitive, technique for planning, analyzing, and designing information systems. IE models are pictures that illustrate and synchronize the system's data and processes.

object the encapsulation of the data (called *properties*) that describes a discrete person, object, place, event, or thing, with all of the processes (called *methods*) that are allowed to use or update the data and properties. The only way to access or update the object's data is to use the object's predefined processes.

object-oriented approach a model-driven technique that integrates data and process concerns into constructs called *objects.* Object models are pictures that illustrate the system's objects from various perspectives, such as the structure, behavior, and interactions of the objects.

prototype a small-scale, incomplete, but working sample of a desired system.

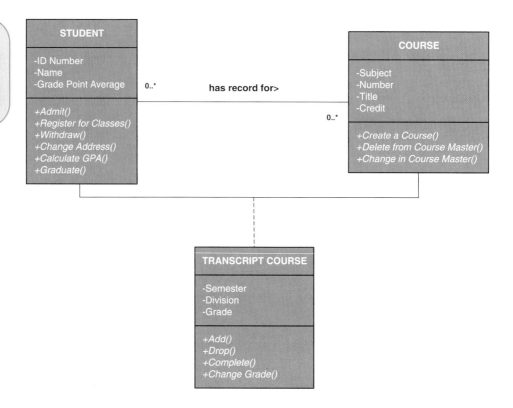

FIGURE 5-4

An Object Model (Using the Unified Modeling Language Standard)

developed quickly, it can quickly identify the most crucial of business-level requirements. Sometimes, prototypes can evolve into the actual, completed information systems and applications.

In a sense, accelerated analysis approaches place much emphasis on the COMMUNICATIONS building blocks in your information system framework by constructing sample forms and reports. At the same time, the software tools used to build prototypes also address the DATA and PROCESS building blocks.

These accelerated approaches are common in the rapid application development (RAD) methodologies and routes that were introduced in Chapter 3. RAD approaches require automated tools. While some repository-based CASE tools include very simple RAD facilities, most analysts use true RAD programming environments such as Sybase *Powerbuilder,* Microsoft *Access,* Microsoft *Visual Basic .NET,* or IBM *Websphere Studio for Application Development (Java*-based).

Let's briefly examine two popular accelerated analysis approaches.

discovery prototyping a technique used to identify the users' business requirements by having them react to a quick-and-dirty implementation of those requirements.

Discovery Prototyping   **Discovery prototyping** uses rapid development technology to help users discover their business requirements. For example, it is very common for systems analysts to use a simple development tool like Microsoft *Access* to rapidly create a simple database, user input forms, and sample reports to solicit user responses as to whether the database, forms, and reports truly represent business requirements. The intent is usually to develop the final new system in a more sophisticated application development tool and language, but the simpler tool allows the analyst to more quickly prototype the user's requirements.

In discovery prototyping, we try to discourage users from becoming preoccupied with the final "look and feel" of the system prototypes—that can be changed during system design! Therein lies the primary criticism of prototyping—software templates exist in prototyping tools to quickly generate some very elegant and visually appealing prototypes. Unfortunately, this can encourage a premature focus on, and commitment to, design represented in the prototype. Users can also be misled to believe (1) that the completed system can be built just as rapidly or (2) that tools like *Access* can be used to build the final system. While tools like *Access* <u>can</u> indeed accelerate systems development, their use in discovery prototyping is fast only because we omit most of the detailed database and application programming required for a complete and secure application. Also, tools like *Access* typically cannot support the database sizes, number of users, and network traffic that are required of most enterprise applications.

Regardless, discovery prototyping is a preferred and recommended approach. Unfortunately, some systems analysts and developers are using discovery prototyping to completely replace model-driven design, only to learn what true engineers have known for years: you cannot prototype without some amount of more formal design . . . enter rapid architected analysis.

Rapid Architected Analysis   **Rapid architected analysis** is an accelerated analysis approach that also builds system models. Rapid architecture analysis is made possible by **reverse-engineering** technology that is included in many automated tools such as CASE and programming languages (as introduced in Chapter 3). Reverse-engineering tools generate system models from existing software applications or system prototypes. The resulting system models can then be edited and improved by systems analysts and users to provide a blueprint for a new and improved system. It should be apparent that rapid architected analysis is a blending of model-driven and accelerated analysis approaches.

There are two different techniques for applying rapid architected analysis:

- Most systems have already been automated to some degree and exist as legacy information systems. Many CASE tools can read the underlying database structures and/or application programs and reverse engineer them into various system models. Those models serve as a point of departure for defining model-driven user requirements analysis.
- If prototypes have been built into tools like Microsoft *Access* or *Visual Basic,* those prototypes can sometimes be reverse engineered into their equivalent system models. The system models usually better lend themselves to analyzing the users' requirements for consistency, completeness, stability, scalability, and flexibility to future change. Also, the system models can frequently be forward engineered by the same CASE tools and ADEs (application development environments) into databases and application templates or skeletons that will use more robust enterprise-level database and programming technology.

Both techniques address the previous issue that engineers rarely prototype in the total absence of a more formal design, and, at the same time, they preserve the advantages of accelerating the systems analysis phases.

> Requirements Discovery Methods

Both model-driven and accelerated systems analysis approaches attempt to express user requirements for a new system, either as models or as prototypes. But both approaches are, in turn, dependent on the more subtle need to actually identify and manage those requirements. Furthermore, the requirements for systems are dependent on the analysts' ability to discover the problems and opportunities that exist in the current system—thus, analysts must become skilled in identifying problems, opportunities, and requirements! Consequently, all approaches to systems analysis require some form of **requirements discovery.** Let's briefly survey a couple of common requirements discovery approaches.

Fact-Finding Techniques   **Fact-finding** is an essential skill for all systems analysts. The fact-finding techniques covered in this book (in fact, in the next chapter) include:

- Sampling of existing documentation, reports, forms, files, databases, and memos.
- Research of relevant literature, benchmarking of others' solutions, and site visits.

---

rapid architected analysis an approach that attempts to derive system models (as described earlier in this section) from existing systems or discovery prototypes.

---

reverse engineering the use of technology that reads the program code for an existing database, application program, and/or user interface and automatically generates the equivalent system model.

---

requirements discovery the process, used by systems analysts, of identifying or extracting system problems and solution requirements from the user community.

---

fact-finding the process of collecting information about system problems, opportunities, solution requirements, and priorities. Also called *information gathering.*

- Observation of the current system in action and the work environment.
- Questionnaires and surveys of the management and user community.
- Interviews of appropriate managers, users, and technical staff.

Joint Requirements Planning The fact-finding techniques listed above are invaluable; however, they can be time-consuming in their classic forms. Alternatively, requirements discovery and management can be significantly accelerated using **joint requirements planning (JRP)** techniques. A JRP-trained or -certified analyst usually plays the role of *facilitator* for a workshop that will typically run from three to five full working days. This workshop can replace weeks or months of classic fact-finding and follow-up meetings.

JRP provides a working environment in which to accelerate all systems analysis tasks and deliverables. It promotes enhanced SYSTEM OWNER and SYSTEM USER participation in systems analysis. But it also requires a facilitator with superior mediation and negotiation skills to ensure that all parties receive appropriate opportunities to contribute to the system's development.

JRP is typically used in conjunction with the model-driven analysis approaches we described earlier, and it is typically incorporated into rapid application development (RAD) methodologies and routes (which were introduced in Chapter 3).

> Business Process Redesign Methods

One of the most interesting contemporary applications of systems analysis methods is **business process redesign (BPR).** The interest in BPR was driven by the discovery that most current information systems and applications have merely automated existing and inefficient business processes. Automated bureaucracy is still bureaucracy; automation does not necessarily contribute value to the business, and it may actually subtract value from the business. Introduced in Chapter 1, BPR is one of many types of projects triggered by the trends we call *total quality management (TQM)* and *continuous process improvement (CPI).*

Some BPR projects focus on all business processes, regardless of their automation. Each business process is thoroughly studied and analyzed for bottlenecks, value returned, and opportunities for elimination or streamlining. Process models, such as data flow diagrams (discussed earlier), help organizations visualize their processes. Once the business processes have been redesigned, most BPR projects conclude by examining how information technology might best be applied to the improved business processes. This may create new information system and application development projects to implement or support the new business processes.

BPR is also applied within the context of information system development projects. It is not uncommon for IS projects to include a study of existing business processes to identify problems, bureaucracy, and inefficiencies that can be addressed in requirements for new and improved information systems and computer applications.

BPR has also become common in IS projects that will be based on the purchase and integration of commercial off-the-shelf (COTS) software. The purchase of COTS software usually requires that a business adapt its business processes to fit the software. An analysis of existing business processes during systems analysis is usually a part of such projects.

> *FAST* Systems Analysis Strategies

Like most commercial methodologies, our hypothetical *FAST* methodology does not impose a single approach on systems analysts. Instead, it integrates all the popular approaches introduced in the preceding paragraphs into a collection of **agile methods.** The SoundStage case study will demonstrate these methods in the context of a typical

---

joint requirements planning (JRP) the use of facilitated workshops to bring together all of the system owners, users, and analysts and some systems designers and builders to jointly perform systems analysis. JRP is generally considered a part of a larger method called *joint application development (JAD),* a more comprehensive application of the JRP techniques to the entire systems development process.

business process redesign (BPR) the application of systems analysis methods to the goal of dramatically changing and improving the fundamental business processes of an organization, independent of information technology.

agile method the integration of various approaches of systems analysis and design for application as deemed appropriate to the problem being solved and the system being developed.

first assignment for a systems analyst. The systems analysis techniques will be applied within the framework of:

- Your information system building blocks (from Chapter 2).
- The *FAST* phases (from Chapter 3).
- *FAST* tasks that implement a phase (described in this chapter).

Given this context for studying systems analysis, we can now explore the systems analysis phases and tasks.

The Scope Definition Phase

Recall from Chapter 3 that the *scope definition phase* is the first phase of the classic systems development process. In other methodologies this might be called the *preliminary investigation phase, initial study phase, survey phase,* or *planning phase.* The scope definition phase answers the question, "Is this project worth looking at?" To answer this question, we must define the scope of the project and the <u>perceived</u> problems, opportunities, and directives that triggered the project. Assuming the project *is* deemed worth looking at, the scope definition phase must also establish the project plan in terms of scale, development strategy, schedule, resource requirements, and budget.[1]

The context for the scope definition phase is shaded in Figure 5-5. Notice that the scope definition phase is concerned primarily with the SYSTEM OWNERS' view of the existing system and the problems or opportunities that triggered the interest. System owners tend to be concerned with the big picture, not details. Furthermore, they determine whether resources can and will be committed to the project.

Figure 5-6 is the first of five task diagrams we will introduce in this chapter to take a closer look at each systems analysis phase. A *task diagram* shows the work (= tasks) that should be performed to complete a phase. Our task diagrams do not mandate any specific methodology, but we will describe in the accompanying paragraphs the approaches, tools, and techniques you might want to consider for each task. Figure 5-6 shows the tasks required for the scope definition phase. It is important to remember that these task diagrams are only templates. The project team and project manager may expand on or alter these templates to reflect the unique needs of any given project.

As shown in Figure 5-6, the final deliverable for the preliminary investigation phase is completion of a PROJECT CHARTER. (Such major deliverables are indicated in each task diagram in all-capital letters.) A project charter defines the project scope, plan, methodology, standards, and so on. Completion of the project charter represents the first milestone in a project.

The scope definition phase is intended to be quick. The entire phase should not exceed two or three days for most projects. The phase typically includes the following tasks:

1.1 Identify baseline problems and opportunities.
1.2 Negotiate baseline scope.
1.3 Assess baseline project worthiness.
1.4 Develop baseline schedule and budget.
1.5 Communicate the project plan.

Let's now examine each of these tasks in greater detail.

---

[1]If your course or reading has already included Chapter 4, you should recognize these planning elements as part of project management. Chapter 4 surveyed and demonstrated the process used by project managers to develop a project plan.

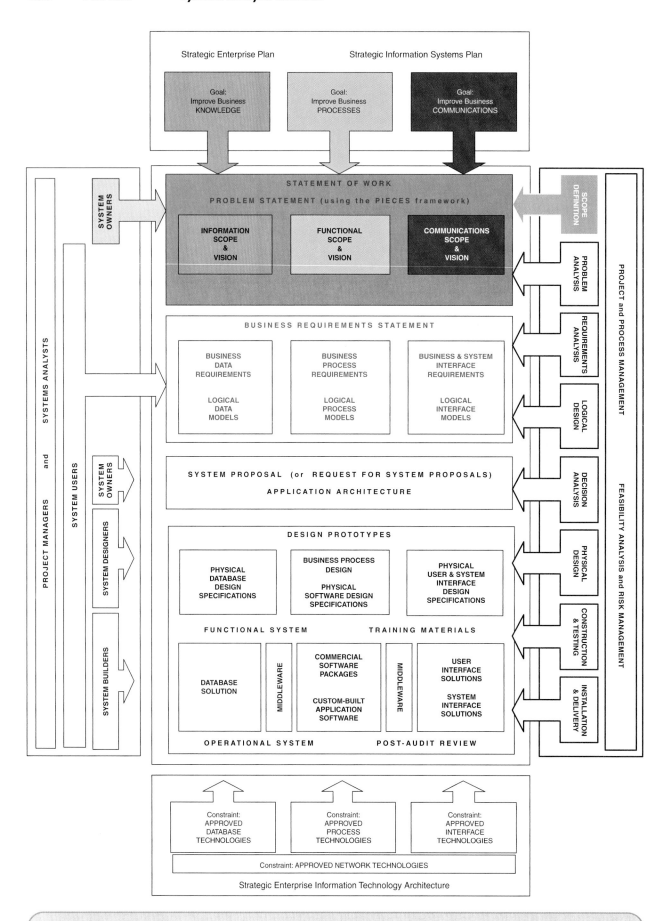

FIGURE 5-5 The Context of the Scope Definition Phase of Systems Analysis

FIGURE 5-6

Tasks for the Scope
Definition Phase of
Systems Analysis

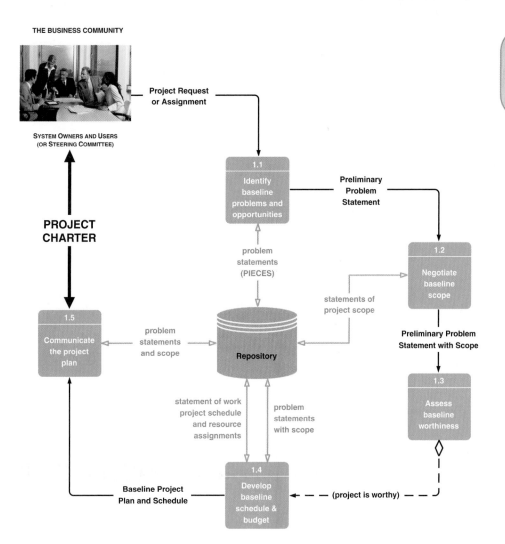

THE BUSINESS COMMUNITY

SYSTEM OWNERS AND USERS
(OR STEERING COMMITTEE)

> Task 1.1—Identify Baseline Problems and Opportunities

One of the most important tasks of the **scope** definition phase is establishing an initial baseline of the problems, opportunities, and/or directives that triggered the project. Each problem, opportunity, and directive is assessed with respect to urgency, visibility, tangible benefits, and priority. Any additional, detailed analysis is not relevant at this stage of the project. It may, however, be useful to list any perceived constraints (limits) on the project, such as deadlines, maximum budget, or general technology.

A senior systems analyst or project manager usually leads this task. Most of the other participants are broadly classified as SYSTEM OWNERS. This includes the executive sponsor(s), the highest-level manager(s) who will pay for and support the project. It also includes managers of all organizational units that may be impacted by the system and possibly includes information systems managers. SYSTEM USERS, SYSTEM DESIGNERS, and SYSTEM BUILDERS are not typically involved in this task.

As shown in Figure 5-6, a PROJECT REQUEST OR ASSIGNMENT triggers the task. This trigger may take one of several alternative forms. It may be as simple as a memorandum of authority from an information systems steering body. Or it may be a memorandum from a business team or unit requesting systems development. Some organizations require that all project requests be submitted on some standard request-for-service form, such as Figure 5-7.

scope the boundaries of a project—the areas of a business that a project may (or may not) address.

FIGURE 5-7

A Request for
Systems Services

SoundStage Entertainment Club *Information System Services* Phone: 494-0666 Fax: 494-0999 Internet: http://www.soundstage.com Intranet: http://www.soundstage.com/iss	**REQUEST FOR INFORMATION SYSTEM SERVICES**

DATE OF REQUEST	SERVICE REQUESTED FOR DEPARTMENT(S)
January 9, 2003	Member Services, Warehouse, Shipping

SUBMITTED BY (key user contact)
Name	Sarah Hartman
Title	Business Analyst, Member Services
Office	B035
Phone	494-0867

EXECUTIVE SPONSOR (funding authority)
Name	Galen Kirkhoff
Title	Vice President, Member Services
Office	G242
Phone	494-1242

TYPE OF SERVICE REQUESTED:

☐ Information Strategy Planning
☒ Business Process Analysis and Redesign
☒ New Application Development
☐ Other (please specify) _____

☐ Existing Application Enhancement
☐ Existing Application Maintenance (problem fix)
☐ Not Sure

BRIEF STATEMENT OF PROBLEM, OPPORTUNITY, OR DIRECTIVE (attach additional documentation as necessary)
The information strategy planning group has targeted member services, marketing, and order fulfillment (inclusive of shipping) for business process redesign and integrated application development. Currently serviced by separate information systems, these areas are not well integrated to maximize efficient order services to our members. The current systems are not adaptable to our rapidly changing products and services. In some cases, separate systems exist for similar products and services. Some of these systems were inherited through mergers that expanded our products and services. There also exist several marketing opportunities to increase our presence to our members. One example includes Internet commerce services. Finally, the automatic identification system being developed for the warehouse must fully interoperate with member services.

BRIEF STATEMENT OF EXPECTED SOLUTION
We envision completely new and streamlined business processes that minimize the response time to member orders for products and services. An order shall not be considered fulfilled until it has been received by the member. The new system should provide for expanded club and member flexibility and adaptability of basic business products and services.
  We envision a system that extends to the desktop computers of both employees and members, with appropriate shared services provided across the network, consistent with the ISS distributed architecture. This is consistent with strategic plans to retire the AS/400 central computer and replace it with servers.

ACTION (ISS Office Use Only)

☐ **Feasibility assessment approved**

☒ **Feasibility assessment waived**

Assigned to Sandra Shepherd

Approved Budget $ 450,000

Start Date ___ASAP___   Deadline ___ASAP___

☐ **Request delayed**

Backlogged until date: _____

☐ **Request rejected**

Reason: _____

Authorized Signatures:

Rebecca J. Todd
Chair, ISS Executive Steering Body

Galen Kirkhoff
Project Executive Sponsor

FORM ISS-100-RFSS (Last revised December, 1999)

The key deliverable of this task, the PRELIMINARY PROBLEM STATEMENT, consists of the problems, opportunities, and directives that were identified. The PROBLEM STATEMENTS are stored in the repository for later use in the project. Figure 5-8 is a sample document that summarizes problems, opportunities, and directives in terms of:

- *Urgency*—In what time frame must/should the problem be solved or the opportunity or directive be realized? A rating scale could be developed to consistently answer this question.
- *Visibility*—To what degree would a solution or new system be visible to customers and/or executive management? Again, a rating scale could be developed for the answers.
- *Benefits*—Approximately how much would a solution or new system increase annual revenues or reduce annual costs? This is often a guess, but if all participants are involved in that guess, it should prove sufficiently conservative.

Problem Statements

Project:	Member services information system	Project manager:	Sandra Shepherd
Created by:	Sandra Shepherd	Last updated by:	Robert Martinez
Date created:	January 9, 2003	Date last updated:	January 15, 2003

Brief Statements of Problem, Opportunity, or Directive	Urgency	Visibility	Annual Benefits	Priority or Rank	Proposed Solution
1. Order response time as measured from time of order receipt to time of customer delivery has increased to an average of 15 days.	ASAP	High	$175,000	2	New development
2. The recent acquisitions of Private Screenings Video Club and Game-Screen will further stress the throughput requirements for the current system.	6 months	Med	75,000	2	New development
3. Currently, three different order entry systems service the audio, video, and game divisions. Each system is designed to interface with a different warehousing system; therefore, the intent to merge inventory into a single warehouse has been delayed.	6 months	Med	515,000	2	New development
4. There is a general lack of access to management and decision-making information. This will become exasperated by the acquisition of two additional order processing systems (from Private Screenings and Game-Screen).	12 months	Low	15,000	3	After new system is developed, provide users with easy-to-learn and -use reporting tools.
5. There currently exist data inconsistencies in the member and order files.	3 months	High	35,000	1	Quick fix; then new development
6. The Private Screenings and GameScreen file systems are incompatible with the SoundStage equivalents. Business data problems include data inconsistencies and lack of input edit controls.	6 months	Med	Unknown	2	New development. Additional quantification of benefit might increase urgency.
7. There is an opportunity to open order systems to the Internet, but security and control are an issue.	12 months	Low	Unknown	4	Future version of newly developed system
8. The current order entry system is incompatible with the forthcoming automatic identification (bar-coding) system being developed for the warehouse.	3 months	High	65,000	1	Quick fix; then new development

FIGURE 5-8 Sample Problem Statements

- *Priority*—Based on the above answers, what are the consensus priorities for each problem, opportunity, or directive. If budget or schedule becomes a problem, these priorities will help to adjust project scope.
- *Possible solutions* (OPT)—At this early stage of the project, possible solutions are best expressed in simple terms such as (a) leave well enough alone, (b) use a quick fix, (c) make a simple to moderate enhancement of the existing system, (d) redesign the existing system, or (e) design a new system. The participants listed for this task are well suited to an appropriately high-level discussion of these options.

The PIECES framework that was introduced in Chapter 3 can be used as a framework for categorizing problems, opportunities, directives, and constraints. For example, Problem 1 in Figure 5-8 could be classified according to PIECES as P.B.—Performance, Response Times. (See Figure 3-4 in Chapter 3). Problem 4 in Figure 5-8 could be classified as I.A.2—Information, Outputs, Lack of necessary information.

The primary techniques used to complete this task include fact-finding and meetings with SYSTEM OWNERS. These techniques are taught in Chapter 6.

> Task 1.2—Negotiate Baseline Scope

Scope defines the boundary of the project—those aspects of the business that will and will not be included in the project. Scope can change during the project; however, the initial project plan must establish the preliminary or baseline scope. Then if the scope changes significantly, all parties involved will have a better appreciation for why the budget and schedule have also changed. This task can occur in parallel with the prior task.

Once again, a senior systems analyst or project manager usually leads this task. Most of the other participants are broadly classified as SYSTEM OWNERS. This includes the executive sponsor, managers of all organizational units that may be impacted by the system, and possibly information systems managers. SYSTEM USERS, SYSTEM DESIGNERS, and SYSTEM BUILDERS are not typically involved in this task.

As shown in Figure 5-6, this task uses the PRELIMINARY PROBLEM STATEMENT produced by the previous task. It should make sense that those problems, opportunities, and directives form the basis for defining scope. The STATEMENTS OF PROJECT SCOPE are added to the repository for later use. These statements are also formally documented as the task deliverable, PRELIMINARY PROBLEM STATEMENT WITH SCOPE.

Scope can be defined easily within the context of your information system building blocks. For example, a project's scope can be described in terms of:

- What types of DATA describe the system being studied? For example, a sales information system may require data about such things as CUSTOMERS, ORDERS, PRODUCTS, and SALES REPRESENTATIVES.
- What business PROCESSES are included in the system being studied? For example, a sales information system may include business processes for CATALOG MANAGEMENT, CUSTOMER MANAGEMENT, ORDER ENTRY, ORDER FULFILLMENT, ORDER MANAGEMENT, and CUSTOMER RELATIONSHIP MANAGEMENT.
- How must the system INTERFACE with users, locations, and other systems? For example, potential interfaces for a sales information system might include CUSTOMERS, SALES REPRESENTATIVES, SALES CLERKS AND MANAGERS, REGIONAL SALES OFFICES, and the ACCOUNTS RECEIVABLE and INVENTORY CONTROL INFORMATION SYSTEMS.

Notice that each statement of scope can be described as a simple list. We don't necessarily "define" the items in the list. Nor are we very concerned with precise requirements analysis. And we definitely are not concerned with any time-consuming steps such as modeling or prototyping.

Once again, the primary techniques used to complete this task are fact-finding and meetings. Many analysts prefer to combine this task with both the previous and the next tasks and accomplish them within a single meeting.

> Task 1.3—Assess Baseline Project Worthiness

This is where we answer the question, "Is this project worth looking at?" At this early stage of the project, the question may actually boil down to a "best guess": Will solving the problems, exploiting the opportunities, or fulfilling the directives return enough value to offset the costs that we will incur to develop this system? It is impossible to do a thorough feasibility analysis based on the limited facts we've collected to date.

Again, a senior systems analyst or project manager usually leads this task. But the SYSTEM OWNERS, inclusive of the executive sponsor, the business unit managers, and the information systems managers, should make the decision.

As shown in Figure 5-6, the completed PRELIMINARY PROBLEM STATEMENT WITH SCOPE triggers the task. This provides the level of information required for this preliminary assessment of worth. There is no physical deliverable other than the GO OR NO-GO DECISION. There are actually several alternative decisions. The project can be approved or canceled, and project scope can be renegotiated (increased or decreased!). Obviously, the remaining tasks in the preliminary investigation phase are necessary only if the project has been deemed worthy and approved to continue.

> Task 1.4—Develop Baseline Schedule and Budget

If the project has been deemed worthy to continue, we can now plan the project in depth. The initial project plan should consist of at least the following:

- A preliminary master plan that includes schedule and resource assignments for the entire project. This plan will be updated at the end of each phase of the project. It is sometimes called a *baseline plan.*
- A detailed plan and schedule for completing the next phase of the project (the problem analysis phase).

The task is the responsibility of the *project manager.* Most project managers find it useful to include as much of the project team, including SYSTEM OWNERS, USERS, DESIGNERS, and BUILDERS, as possible. Chapter 4 coined the term *joint project planning* to describe the team approach to building a project plan.

As shown in Figure 5-6, this task is triggered by the GO OR NO-GO DECISION to continue the project. This decision represents a consensus agreement on the project's scope, problems, opportunities, directives, and worthiness. (This "worthiness" must still be presented and approved.) The PROBLEM STATEMENTS WITH SCOPE are the key input (from the repository). The deliverable of this task is the BASELINE PROJECT PLAN AND SCHEDULE. The STATEMENT OF WORK (see Chapter 4) and PROJECT SCHEDULE AND RESOURCE ASSIGNMENTS are also added to the repository for continuous monitoring and, as appropriate, updating. The schedule and resources are typically maintained in the repository as a project management software file.

The techniques used to create a project plan were covered in depth in Chapter 4. Today, these techniques are supported by project management software such as Microsoft *Project.* Chapter 4 also discussed the detailed steps for completing the plan.

> Task 1.5—Communicate the Project Plan

In most organizations, there are more potential projects than resources to staff and fund those projects. Unless our project has been predetermined to be of the highest priority (by some sort of prior tactical or strategic planning process), then it must be presented and defended to a **steering body** for approval. Most organizations use a steering body to approve and monitor projects and progress. The majority of any steering body should consist of non–information systems professionals or managers. Many organizations designate vice presidents to serve on a steering body. Other

steering body a committee of executive business and system managers that studies and prioritizes competing project proposals to determine which projects will return the most value to the organization and thus should be approved for continued systems development. Also called a *steering committee.*

organizations assign the direct reports of vice presidents to the steering body. And some organizations utilize two steering bodies, one for vice presidents and one for their direct reports. Information systems managers serve on the steering body only to answer questions and to communicate priorities back to developers and project managers.

Regardless of whether or not a project requires steering committee approval, it is equally important to formally launch the project and communicate the project, goals, and schedule to the entire business community. Opening the lines of communication is an important capstone to the preliminary investigation. For this reason, we advocate the "best practices" of conducting a *project kickoff event* and creating an *intranet project Web site.* The project kickoff meeting is open to the entire business community, not just the business units affected and the project team. The intranet project Web site establishes a community portal to all nonsensitive news and documentation concerning the project.

Ideally, the executive sponsor should jointly facilitate the task with the chosen project manager. The visibility of the executive sponsor establishes instant credibility and priority to all who participate in the kickoff meeting. Other kickoff meeting participants should include the entire project team, including assigned SYSTEM OWNERS, USERS, ANALYSTS, DESIGNERS, and BUILDERS. Ideally, the kickoff meeting should be open to any and all interested staff from the business community. This builds community awareness and consensus while reducing both the volume and the consequences of rumor and misinformation. For the intranet component, a Webmaster or Web author should be assigned to the project team.

As shown in Figure 5-6, this task is triggered by the completion of the BASELINE PROJECT PLAN AND SCHEDULE. The PROBLEM STATEMENTS AND SCOPE are available from the repository. The deliverable is the PROJECT CHARTER. The project charter is usually a document. It includes various elements that define the project in terms of participants, problems, opportunities, and directives; scope; methodology; statement of work to be completed; deliverables; quality standards; schedule; and budget. The project charter should be added to the project Web site for all to see. Elements of the project charter may also be reformatted as slides and handouts (using software such as Microsoft *PowerPoint*) for inclusion in the project kickoff event.

Effective interpersonal and communications skills are the keys to this task. These include principles of persuasion, selling change, business writing, and public speaking.

This concludes our discussion of the scope definition phase. The participants in the scope definition phase might decide the project is not worth proposing. It is also possible the steering body may decide that other projects are *more* important. Or the executive sponsor might not endorse the project. In each of these instances, the project is terminated. Little time and effort have been expended. On the other hand, with the blessing of all the system owners and the steering committee, the project can now proceed to the problem analysis phase.

The Problem Analysis Phase

There is an old saying, "Don't try to fix it unless you understand it." That statement aptly describes the *problem analysis phase* of systems analysis. There is always a current or existing system, regardless of the degree to which it is automated with information technology. The problem analysis phase provides the analyst with a more thorough understanding of the problems, opportunities, and/or directives that triggered the project. The problem analysis phase answers the questions, "Are the problems really worth solving?" and "Is a new system really worth building?" In other methodologies, the problem analysis phase may be known as the *study phase, study of the current system, detailed investigation phase,* or *feasibility analysis phase.*

Can you ever skip the problem analysis phase? Rarely! You almost always need <u>some</u> level of understanding of the current system. But there may be reasons to accelerate the problem analysis phase. First, if the project was triggered by a strategic or tactical plan, the worthiness of the project is probably not in doubt—the problem analysis phase would be reduced to understanding the current system, not analyzing it. Second, a project may be initiated by a directive (such as compliance with a governmental directive and deadline). Again, in this case project worthiness is not in doubt. Finally, some methodologies and organizations deliberately consolidate the problem analysis and requirements analysis phases to accelerate systems analysis.

The goal of the problem analysis phase is to study and understand the problem domain well enough to thoroughly analyze its problems, opportunities, and constraints. Some methodologies encourage a very detailed understanding of the current system and document that system in painstaking detail using system models such as data flow diagrams. Today, except when business processes must be redesigned, the effort required and the value added by such detailed modeling is questioned and usually bypassed. Thus, the current version of our hypothetical *FAST* methodology encourages only enough system modeling to refine our understanding of project scope and problem statement, and to define a common vocabulary for the system.

The context for the problem analysis phase is shaded in Figure 5-9. Notice that the problem analysis phase is concerned primarily with both the SYSTEM OWNERS' and the SYSTEM USERS' views of the existing system. Notice that we build on the lists created in the preliminary investigation phase to analyze the KNOWLEDGE, PROCESS, and COMMUNICATIONS building blocks of the existing system. Also notice that we imply minimal system modeling. We may still use the PIECES framework to analyze each building block for problems, causes, and effects.

Figure 5-10 is the task diagram for the problem analysis phase. The final phase deliverable and milestone is producing SYSTEM IMPROVEMENT OBJECTIVES that address problems, opportunities, and directives. Depending on the size of the system, its complexity, and the degree to which project worthiness is already known, the illustrated tasks may consume one to six weeks. Most of these tasks can be accelerated by JRP-like sessions. The problem analysis phase typically includes the following tasks:

2.1 Understand the problem domain.
2.2 Analyze problems and opportunities.
2.3 Analyze business processes.
2.4 Establish system improvement objectives.
2.5 Update or refine the project plan.
2.6 Communicate findings and recommendations.

Let's now examine each of these tasks in greater detail.

> Task 2.1—Understand the Problem Domain

During the problem analysis phase, the *team* initially attempts to learn about the current system. Each SYSTEM OWNER, USER, and ANALYST brings a different level of understanding to the system—different detail, different vocabulary, different perceptions, and different opinions. A well-conducted study can prove revealing to all parties, including the system's own management and users. It is important to study and *understand the problem domain,* that domain in which the business problems, opportunities, directives, and constraints exist.

This task will be led by the project manager but facilitated by the lead systems analyst. It is not uncommon for one individual to play both roles (as Sandra does in the SoundStage case). Other SYSTEMS ANALYSTS may also be involved since they conduct interviews, scribe for meetings, and document findings. A comprehensive study should include representative SYSTEM OWNERS and USERS from all business units that will be supported or impacted by the system and project. It is extraordinarily important that enough users be included to encompass the full scope of the

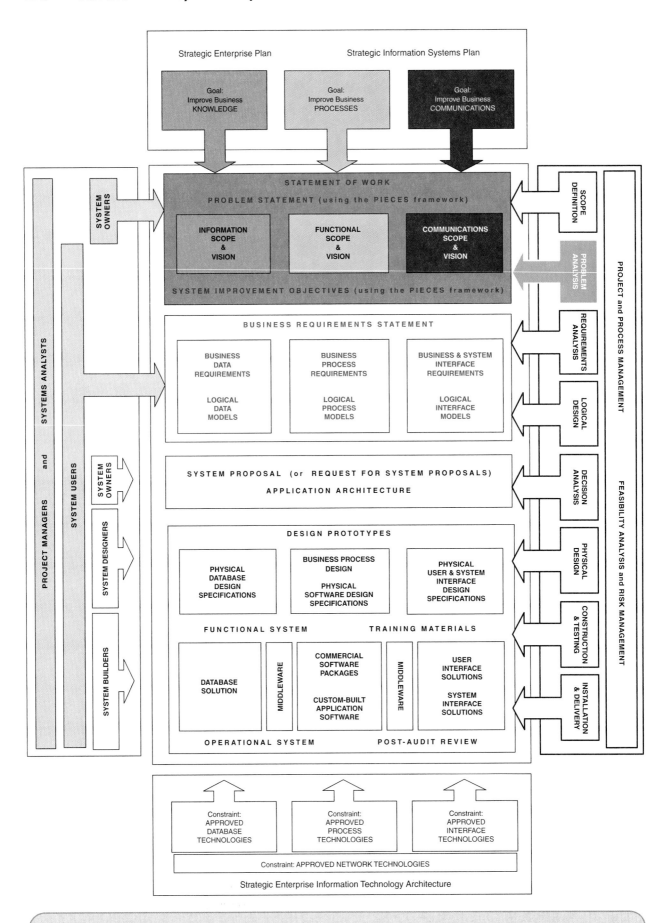

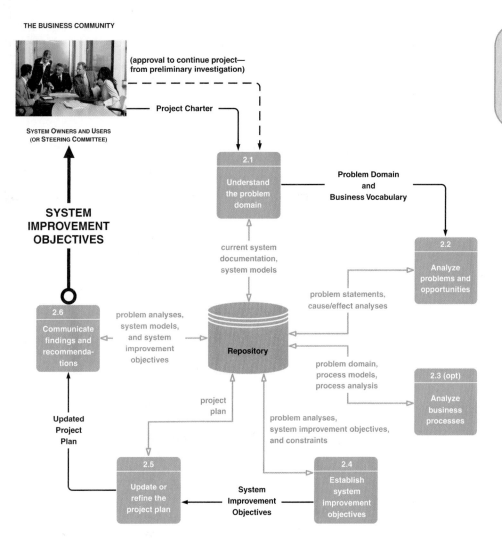

FIGURE 5-10

Tasks for the
Problem Analysis
Phase of Systems
Analysis

system being studied. In some organizations, one or more experienced users are "loaned" to the project <u>full-time</u> as *business analysts;* however, it is rare that any one user can fully represent the interests of all users. Business analysts can, however, serve as facilitators to get the right people involved and sustain effective communication back to the business units and management. SYSTEM DESIGNERS and BUILDERS are rarely involved in this task unless they are interviewed to determine any technical limitations of the current system.

In Figure 5-10, this task is triggered by APPROVAL TO CONTINUE THE PROJECT—from the scope definition phase. (The dashed line indicates this approval is an event or trigger, not a data or information flow.) The approval comes from the SYSTEM OWNERS or steering committee. The key informational input is the PROJECT CHARTER and any CURRENT SYSTEM DOCUMENTATION that may exist in the repository and program libraries for the current system. Current system documentation doesn't always exist. And when it does exist, it must be carefully checked for currency—most such documentation is notoriously out of date because analysts and programmers are not always diligent about updating that documentation as changes occur throughout the lifetime of a system.

The deliverables of this task are an understanding of the PROBLEM DOMAIN AND BUSINESS VOCABULARY. Your understanding of the existing problem domain should be documented so that it can be verified that you truly understand it. There are several ways to document the problem domain. Certainly, drawing SYSTEM MODELS of the current system can help, but they can lead to a phenomenon called "analysis paralysis," in which the desire to produce perfect models becomes counterproductive to the schedule. Another approach might be to use your information system building blocks as a framework for listing and defining the system domain:

- KNOWLEDGE—List all the "things" about which the system currently stores data (in files, databases, forms, etc.). Define each thing in business terms. For example, "An ORDER is a business transaction in which a customer requests to purchase products."

  Additionally, we could list all the reports produced by the current system and describe their purpose or use. For example, "The open orders report describes all orders that have not been filled within one week of their approval to be filled. The report is used to initiate customer relationship management through personal contact."
- PROCESSES—Define each business event for which a business response (process) is currently implemented. For example, "A customer places a new order," or "A customer requests changes to a previously placed order," or "A customer cancels an order."
- COMMUNICATIONS—Define all the locations that the current system serves and all of the users at each of those locations. For example, "The system is currently used at regional sales offices in San Diego, Dallas, St. Louis, Indianapolis, Atlanta, and Manhattan. Each regional sales office has a sales manager, assistant sales manager, administrative assistant, and 5 to 10 sales clerks, all of whom use the current system. Each region is also home to 5 to 30 sales representatives who are on the road most days but who upload orders and other transactions each evening."

  Another facet of interfaces is system interfaces—that is, interfaces that exist between the current information system and other information systems and computer applications. These can be quickly listed and described by the information systems staff.

Ultimately, the organization's systems development methodology and project plan will determine what types and level of documentation are expected.

The business vocabulary deliverable is all too often shortchanged. Understanding the business vocabulary of the system is an excellent way of understanding the system itself. It bridges the communication gap that often exists or develops between business and technology experts.

If you elect to draw SYSTEM MODELS during this task, we suggest that "if you want to learn *anything*, you must not try to learn *everything*—at least not all in this task." To avoid analysis paralysis, we suggest that the following system models may be appropriate:

- KNOWLEDGE—A one-page data model is very useful for establishing business vocabulary and rules. Data modeling is taught in Chapter 8.
- PROCESSES—Today, it is widely accepted that a one- or two-page functional decomposition diagram should prove sufficient to get a feel for the current system processing. Decomposition modeling is taught in Chapter 9.
- COMMUNICATIONS—A one-page context diagram or use-case diagrams are very useful for illustrating the system's inputs and outputs with other organizations, business units, and systems. Context diagrams are discussed below. Use case diagrams are taught in Chapter 7.

Several other techniques and skills are useful for developing an understanding of an existing system. Obviously, fact-finding techniques (taught in the next chapter) are critical to learning about any existing system. Also, joint requirements planning, or JRP, techniques (also taught in the next chapter) can accelerate this task. Finally, the ability to clearly communicate back to users what you've learned about a system is equally crucial.

Context Diagram   The purpose of a context diagram is to analyze how the system interacts with the world around it and to specify in general terms the system inputs and outputs. Context diagrams can be drawn in various ways. Chapter 9 presents the traditional format, which was done as the first step in drawing data flow diagrams.

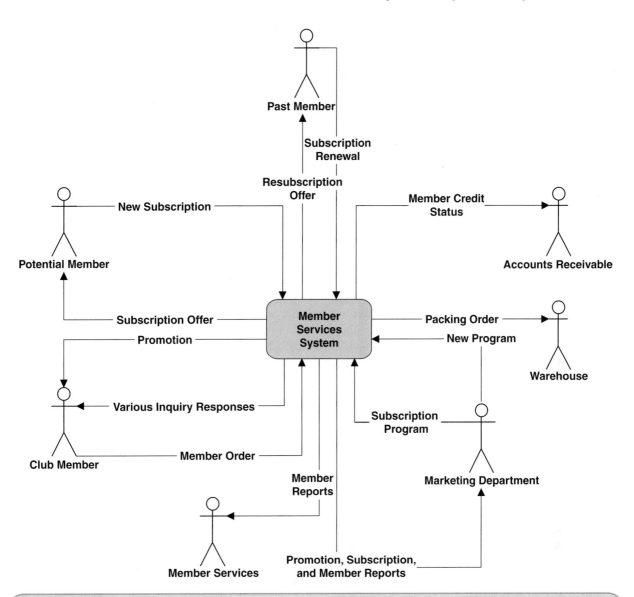

FIGURE 5-11 Context Diagram

Chapter 7 shows a different format for a context diagram. The context diagram shown in Figure 5-11 employs a hybrid approach. It employs use case symbols as use cases are becoming a generally accepted tool of the requirements analysis phase.

The system itself is shown as a "black box" in the middle of the diagram. We are not yet ready to look inside the box. For now we just want to see how everyone will use the box. The stick figures around the outside of the diagram are the persons, organizations, and other information systems that will interact with the system. In use cases, these are called actors, and we can call them that here. In traditional data flow diagrams, they are called external agents. In Chapters 7 and 9 you will learn that once you look inside the system box, other things such as time or devices like sensors can also be actors or external agents. But for a context diagram they are rarely shown.

The lines indicate the inputs (arrows pointing to the system) provided by actors to the system and the outputs (arrows pointing to the actors) created by the system. Each input and output is identified with a noun phrase that describes it.

To build a context diagram ask the users what business transactions the system must respond to; these are the inputs. Also ask the users what reports, notifications, and other outputs must be produced by the system. A system can have many reports that can quickly clutter the diagram; consolidate them as needed to keep the diagram readable. During other phases in the process they will be analyzed separately.

We certainly couldn't build an information system from a context diagram. But it is a solid first step. From this simple diagram we know what inputs the system must respond to and what outputs it must produce. In other words, it helps us understand the problem domain. We will see in Chapter 7 how to detect use cases from a context diagram. That will be the first step in cracking open the "black box." We are following the principles for systems development presented in Chapter 2: "use a problem-solving approach" and "divide and conquer."

> Task 2.2—Analyze Problems and Opportunities

In addition to learning about the current system, the project team must work with system owners and system users to *analyze problems and opportunities.* You might be asking, "Weren't problems and opportunities identified earlier, in the pre-liminary investigation phase?" Yes, they were. But those initial problems may be only symptoms of other problems, perhaps problems not as well known or understood by the users. Besides, we haven't yet really analyzed any of those problems in the classic sense.

cause-and-effect analysis
a technique in which problems are studied to determine their causes and effects.

True problem analysis is a difficult skill to master, especially for inexperienced systems analysts. Experience suggests that most new systems analysts (and many system owners and users) try to solve problems without truly analyzing them. They might state a problem like this: "We need to . . ." or "We want to . . ." In doing so, they are stating the problem in terms of a solution. More effective problem solvers have learned to truly analyze the problem before stating any possible solution. They analyze each per-ceived problem for **causes and effects.** In practice, an effect may actually be a symp-tom of a different, more deeply rooted or basic problem. That problem must also be analyzed for causes and effects, and so on until such a time as the causes and effects do not yield symptoms of other problems. Cause-and-effect analysis leads to true understanding of problems and can lead to not-so-obvious but more creative and valuable solutions.

SYSTEMS ANALYSTS facilitate this task; however, all SYSTEMS OWNERS and USERS should actively participate in the process of cause-and-effect analysis. They are the problem domain experts. SYSTEM DESIGNERS and BUILDERS are not usually involved in this process unless they are called on to analyze technical problems that may exist in the current system.

As shown in Figure 5-10, the team's understanding of the SYSTEM DOMAIN AND BUSINESS VOCABULARY triggers this task. This understanding of the problem domain is crucial because the team members should not attempt to analyze problems unless they understand the domain in which those problems occur. The other informational input to this task is the initial PROBLEM STATEMENTS (from the scope definition phase). The deliverables of this task are the updated PROBLEM STATEMENTS and the CAUSE-EFFECT ANALYSIS for each problem and opportunity. Figure 5-12 illustrates one way to docu-ment a cause-and-effect analysis.

Once again, fact-finding and JRP techniques are crucial to this task. These tech-niques, as well as cause-and-effect analysis, are taught in the next chapter.

> Task 2.3—Analyze Business Processes

This task is appropriate only to *business process redesign (BPR)* projects or system development projects that build on or require significant business process redesign. In such a project, the team is asked to examine its business processes in much greater detail to measure the value added or subtracted by each process as it relates to the total organization. Business process analysis can be politically charged. Sys-tem owners and users alike can become very defensive about their existing business

PROBLEMS, OPPORTUNITIES, OBJECTIVES, AND CONSTRAINTS MATRIX

Project:	Member Services Information System	Project Manager:	Sandra Shepherd
Created by:	Robert Martinez	Last Updated by:	Robert Martinez
Date Created:	January 21, 2003	Date Last Updated:	January 31, 2003

CAUSE-AND-EFFECT ANALYSIS		SYSTEM IMPROVEMENT OBJECTIVES	
Problem or Opportunity	**Causes and Effects**	**System Objective**	**System Constraint**
1. Order response time is unacceptable.	1. Throughput has increased while number of order clerks was downsized. Time to process a single order has remained relatively constant. 2. System is too keyboard-dependent. Many of the same values are keyed for most orders. Net result is (with the current system) each order takes longer to process than is ideal. 3. Data editing is performed by the AS/400. As that computer has approached its capacity, order edit responses have slowed. Because order clerks are trying to work faster to keep up with the volume, the number of errors has increased. 4. Warehouse picking tickets for orders were never designed to maximize the efficiency of order fillers. As warehouse operations grew, order filling delays were inevitable.	1. Decrease the time required to process a single order by 30%. 2. Eliminate keyboard data entry for as much as 50% of all orders. 3. For remaining orders, reduce as many key-strokes as possible by replacing keystrokes with point-and-click objects on the computer display screen. 4. Move data editing from a shared computer to the desktop. 5. Replace existing picking tickets with a paperless communication system between member services and the warehouse.	1. There will be no increase in the order processing workforce. 2. Any system developed must be compatible with the existing Windows 95 desktop standard. 3. New system must be compatible with the already approved automatic identification system (for bar coding).

FIGURE 5-12   A Sample Cause-and-Effect Analysis

processes. The analysts involved must keep the focus on the processes, not the people who perform them, and constantly remind everyone that the goal is to identify opportunities for fundamental business change that will benefit the business and everyone in the business.

One or more systems analysts or business analysts facilitate the task. Ideally, the ANALYSTS should be experienced, trained, or certified in BPR methods. The only other participants should be appropriate SYSTEM OWNERS and USERS. Business process analysis should avoid any temptation to focus on information technology solutions until well after the business processes have been redesigned for maximum efficiency. Some analysts find it useful to assume the existence of "perfect people" and "perfect technology" that can make anything "possible." They ask, "If the world were perfect, would we need this process?"

As depicted in Figure 5-10, a business process analysis task is dependent only on some PROBLEM DOMAIN knowledge (from Task 2.1). The deliverables of this task are business "as is" PROCESS MODELS and PROCESS ANALYSES. The process models can look very much like data flow diagrams (Figure 5-2) except they are significantly annotated to show (1) the volume of data flowing through the processes, (2) the response times of each process, and (3) any delays or bottlenecks that occur in the system. The process analysis data provides additional information such as (a) the cost of each process, (b) the value added by each process, and (c) the consequences of eliminating or streamlining the process. Based on the as-is models and their analysis, the team develops "to be" models that redesign the business processes to eliminate redundancy and bureaucracy and increase efficiency and service.

Several techniques are applicable to this task. Once again, fact-finding techniques and facilitated team meetings (Chapter 6) are invaluable. Also, process modeling techniques (Chapter 9) are critical to BPR success.

> Task 2.4—Establish System Improvement Objectives

objective a measure of success. It is something that you expect to achieve, if given sufficient resources.

constraint something that will limit your flexibility in defining a solution to your objectives. Essentially, constraints cannot be changed.

Given our understanding of the current system's scope, problems, and opportunities, we can now *establish system improvement objectives.* The purpose of this task is to establish the criteria against which any improvements to the system will be measured and to identify any constraints that may limit flexibility in achieving those improvements. The criteria for success should be measured in terms of **objectives.** Objectives represent the first attempt to establish expectations for any new system. In addition to identifying objectives, we must also identify any known constraints. **Constraints** place limitations or delimitations on achieving objectives. Deadlines, budgets, and required technologies are examples of constraints.

The SYSTEMS ANALYSTS facilitate this task. Other participants include the same SYSTEM OWNERS and USERS who have participated in other tasks in this problem analysis phase. Again, we are not yet concerned with technology; therefore, SYSTEM DESIGNERS and BUILDERS are not involved in this task.

This task is triggered by the PROBLEM ANALYSES completed in Tasks 2.2 and 2.3. For each verified and significant problem, the analysts and users should define specific SYSTEM IMPROVEMENT OBJECTIVES. They should also identify any CONSTRAINTS that may limit or prevent them from achieving the system improvement objectives.

System improvement objectives should be precise, measurable statements of business performance that define the expectations for the new system. Some examples are:

> Reduce the number of uncollectible customer accounts by 50 percent within the next year.
> Increase by 25 percent the number of loan applications that can be processed during an eight-hour shift.
> Decrease by 50 percent the time required to reschedule a production lot when a workstation malfunctions.

The following is an example of a poor objective:

> Create a delinquent accounts report.

This is a poor objective because it states only a requirement, not an actual objective. Now, let's reword that objective:

> Reduce credit losses by 20 percent through earlier identification of delinquent accounts.

This gives us more flexibility. Yes, the delinquent accounts report would work. But a customer delinquency inquiry might provide an even better way to achieve the same objective.

System improvement objectives may be tempered by identifiable constraints. Constraints fall into four categories, as listed below (with examples):

- *Schedule:* The new system must be operational by April 15.
- *Cost:* The new system cannot cost more than $350,000.
- *Technology:* The new system must be online, or all new systems must use the DB2 database management system.
- *Policy:* The new system must use double-declining-balance inventory techniques.

The last two columns of Figure 5-12 document typical system improvement objectives and constraints.

> Task 2.5—Update or Refine the Project Plan

Recall that project scope is a moving target. Based on our baseline schedule and budget from the scope definition phase, scope may have grown or diminished in size and complexity. (Growth is much more common!) Now that we're approaching the completion of the problem analysis phase, we should reevaluate project scope and *update or refine the project plan* accordingly.

The project manager, in conjunction with SYSTEM OWNERS and the entire project team, facilitates this task. The SYSTEMS ANALYSTS and SYSTEM OWNERS are the key individuals in this task. The analysts and owners should consider the possibility that not all objectives may be met by the new system. Why? The new system may be larger than expected, and they may have to reduce the scope to meet a deadline. In this case the system owner will rank the objectives in order of importance. Then, if the scope must be reduced, the higher-priority objectives will tell the analyst what's most important.

As shown in Figure 5-10, this task is triggered by completion of the SYSTEM IMPROVEMENT OBJECTIVES. The initial PROJECT PLAN is another key input, and the UPDATED PROJECT PLAN is the key output. The updated plan should now include a detailed plan for the requirements analysis phase that should follow. The techniques and steps for updating the project plan were taught in Chapter 4, "Project Management."

> Task 2.6—Communicate Findings and Recommendations

As with the scope definition phase, the problem analysis phase concludes with a communication task. We must *communicate findings and recommendations* to the business community. The project manager and executive sponsor should jointly facilitate this task. Other meeting participants should include the entire project team, including assigned SYSTEM OWNERS, USERS, ANALYSTS, DESIGNERS, and BUILDERS. And, as usual, the meeting should be open to any and all interested staff from the business community. Also, if an intranet Web site was established for the project, it should have been maintained throughout the problem analysis phase to ensure continuous communication of project progress.

This task is triggered by the completion of the UPDATED PROJECT PLAN. Informational inputs include the PROBLEM ANALYSES, any SYSTEM MODELS, the SYSTEM IMPROVEMENT OBJECTIVES, and any other documentation that was produced during the problem analysis phase. Appropriate elements are combined into the SYSTEM IMPROVEMENT OBJECTIVES, the major deliverable of the problem analysis phase. The format may be a report, a verbal presentation, or an inspection by an auditor or peer group (called a *walkthrough*). An outline for a written report is shown in Figure 5-13.

Interpersonal and communications skills are essential to this task. Systems analysts should be able to write a formal business report and make a business presentation without getting into technical issues or alternatives.

FIGURE 5-13 An Outline for a System Improvement Objectives and Recommendations Report

Analysis of the Current _____ System

I. Executive summary (approximately 2 pages)
   A. Summary of recommendation
   B. Summary of problems, opportunities, and directives
   C. Brief statement of system improvement objectives
   D. Brief explanation of report contents
II. Background information (approximately 2 pages)
   A. List of interviews and facilitated group meetings conducted
   B. List of other sources of information that were exploited
   C. Description of analytical techniques used
III. Overview of the current system (approximately 5 pages)
   A. Strategic implications (if the project is part of or impacts an existing information systems strategic plan)
   B. Models of the current system
      1. Interface model (showing project scope)
      2. Data model (showing project scope)
      3. Geographic models (showing project scope)
      4. Process model (showing functional decomposition only)
IV. Analysis of the current system (approximately 5–10 pages)
   A. Performance problems, opportunities, and cause-effect analysis
   B. Information problems, opportunities, and cause-effect analysis
   C. Economic problems, opportunities, and cause-effect analysis
   D. Control problems, opportunities, and cause-effect analysis
   E. Efficiency problems, opportunities, and cause-effect analysis
   F. Service problems, opportunities, and cause-effect analysis
V. Detailed recommendations (approximately 5–10 pages)
   A. System improvement objectives and priorities
   B. Constraints
   C. Project plan
      1. Scope reassessment and refinement
      2. Revised master plan
      3. Detailed plan for the definition phase
VI. Appendixes
   A. Any detailed system models
   B. Other documents as appropriate

This concludes the problem analysis phase. One of the following decisions must be made after the conclusion of this phase:

- Authorize the project to continue, as is, to the requirements analysis phase.
- Adjust the scope, cost, and/or schedule for the project and then continue to the requirements analysis phase.
- Cancel the project due to (1) lack of resources to further develop the system, (2) realization that the problems and opportunities are simply not as important as anticipated, or (3) realization that the benefits of the new system are not likely to exceed the costs.

With some level of approval from the SYSTEM OWNERS, the project can now proceed to the requirements analysis phase.

The Requirements Analysis Phase

Many inexperienced analysts make a critical mistake after completing the problem analysis phase. The temptation at that point is to begin looking at alternative solutions, particularly technical solutions. One of the most frequently cited errors in new information systems is illustrated in the statement, "Sure the system works, and it is technically impressive, but it just doesn't do what we needed it to do." The *requirements analysis phase* defines the business requirements for a new system.

Did you catch the key word in the quoted sentence? It is "what," not "how"! Analysts are frequently so preoccupied with the *technical* solution that they inadequately define the *business* requirements for that solution. The requirements analysis phase answers the question, "<u>What</u> do the users need and want from a new system?" The requirements analysis phase is critical to the success of any new information system. In different methodologies the requirements analysis phase might be called the *definition phase* or *logical design phase*.

Can you ever skip the requirements analysis phase? Absolutely not! New systems will always be evaluated, first and foremost, on whether or not they fulfill business objectives and requirements, regardless of how impressive or complex the technological solution might be!

It should be acknowledged that some methodologies integrate the problem analysis and requirements analysis phases into a single phase.

Once again, your information systems building blocks (Figure 5-14) can serve as a useful framework for documenting the information systems requirements. Notice that we are still concerned with the SYSTEM USERS' perspectives. Requirements can be defined in terms of the PIECES framework or in terms of the types of data, processes, and interfaces that must be included in the system.

Figure 5-15 illustrates the typical tasks of the requirements analysis phase. The final phase deliverable and milestone is producing a BUSINESS REQUIREMENTS STATEMENT that will fulfill the system improvement objectives identified in the previous phase. One of the first things you may notice in this task diagram is that most of the tasks are not as sequential as those in previous task diagrams. Instead, many of these tasks occur in parallel as the team works toward the goal of completing the requirements statement. The requirements analysis phase typically includes the following tasks:

3.1 Identify and express system requirements.
3.2 Prioritize system requirements.
3.3 Update or refine the project plan.
3.4 Communicate the requirements statement.

Let's now examine each of these tasks in greater detail.

> Task 3.1—Identify and Express System Requirements

The initial task of the requirements analysis phase is to *identify and express requirements*. While this may seem to be an easy or trivial task, it is often the source of many errors, omissions, and conflicts. The foundation for this task was established in the problem analysis phase when we identified system improvement objectives. Minimally, this task translates those objectives into an outline of **functional** and **nonfunctional requirements** that will be needed to meet the objectives. Functional requirements are frequently identified in terms of inputs, outputs, processes, and stored data that are needed to satisfy the system improvement objectives. Examples of nonfunctional requirements include performance (throughput and response time); ease of learning and use; budgets, costs, and cost savings; timetables and deadlines; documentation and training needs; quality management; and security and internal auditing controls.

Rarely will this definition task identify *all* the functional or nonfunctional business requirements. But the outline will frame your thinking as you proceed to later

functional requirement a description of activities and services a system must provide.

nonfunctional requirement a description of other features, characteristics, and constraints that define a satisfactory system.

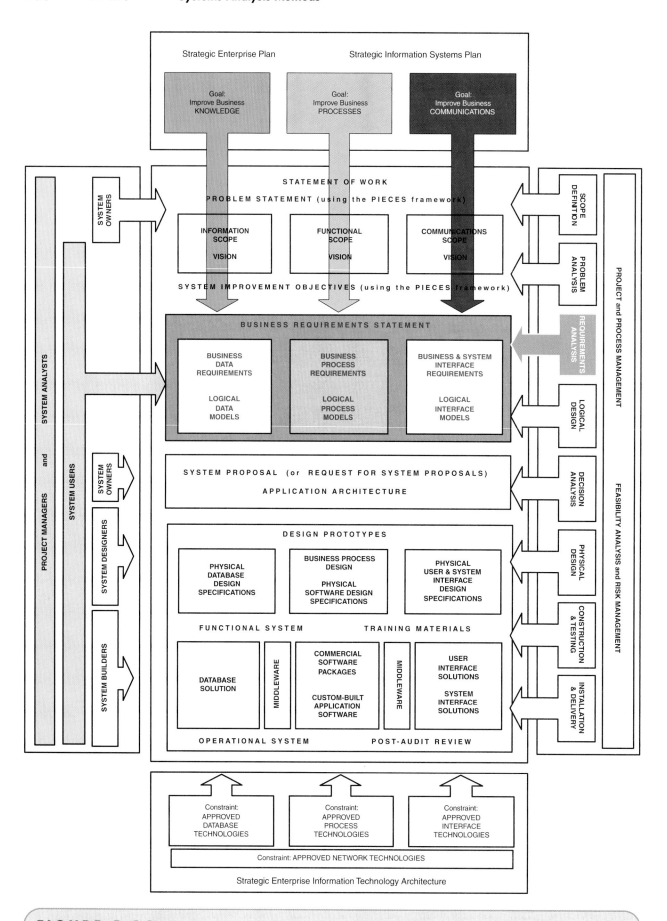

FIGURE 5-14   The Context of the Requirements Analysis Phase of Systems Analysis

THE BUSINESS COMMUNITY

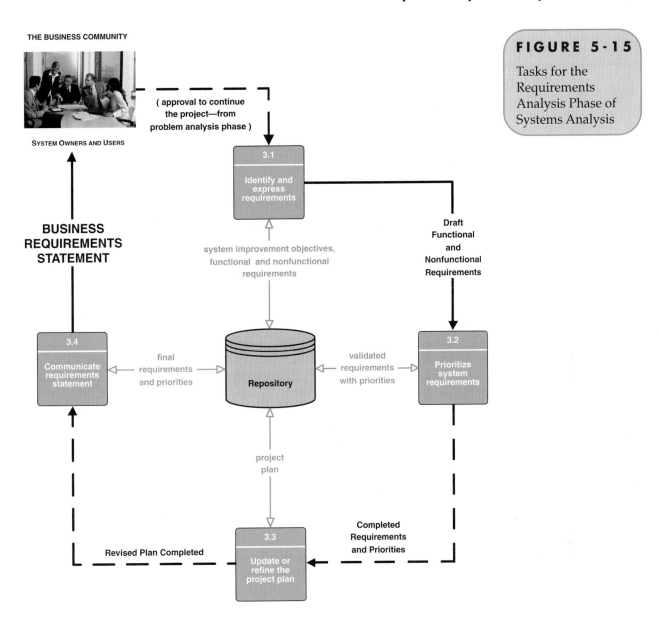

SYSTEM OWNERS AND USERS

(approval to continue
the project—from
problem analysis phase)

3.1
Identify and
express
requirements

Draft
Functional
and
Nonfunctional
Requirements

BUSINESS
REQUIREMENTS
STATEMENT

system improvement objectives,
functional and nonfunctional
requirements

3.4
Communicate
requirements
statement

final
requirements
and priorities

Repository

validated
requirements
with priorities

3.2
Prioritize
system
requirements

project
plan

Revised Plan Completed

3.3
Update or
refine the
project plan

Completed
Requirements
and Priorities

---

tasks that will add new requirements and details to the outline. Thus, neither completeness nor perfection is a goal of this task.

SYSTEMS ANALYSTS facilitate the task. They also document the results. Obviously, SYSTEM USERS are the primary source of business requirements. Some SYSTEM OWNERS may elect to participate in this task since they played a role in framing the system improvement objectives that will guide the task. SYSTEM DESIGNERS and BUILDERS should not be involved because they tend to prematurely redirect the focus to the technology and technical solutions.

As shown in Figure 5-15, this task (and phase) is triggered by the APPROVAL TO CONTINUE THE PROJECT FROM THE PROBLEM ANALYSIS PHASE. The key input is the SYSTEM IMPROVEMENT OBJECTIVES from the problem analysis phase (via the repository). Of course, any and all relevant information from the problem analysis phase is available from the repository for reference as needed.

The only deliverable of this task is the DRAFT FUNCTIONAL AND NONFUNCTIONAL REQUIREMENTS. Various formats can work. In its simplest format, the outline could be divided into four logical sections: the original list of system improvement objectives and, for each objective, a sublist of (a) inputs, (b) processes, (c) outputs, and (d) stored data needed to fulfill the objective. Increasingly, however, system analysts are

use case a business scenario or event for which the system must provide a defined response. Use cases evolved out of object-oriented analysis; however, their use has become common in many other methodologies for systems analysis and design.

expressing functional requirements using a modeling tool called **use cases.** Use cases model business scenarios and events that must be handled by a new system. They are introduced in Chapter 7 and used throughout this book.

The PIECES framework that was used earlier to identify problems, opportunities, and constraints can also be used as a framework for defining draft requirements.

Several techniques are applicable to this task. Joint requirements planning (JRP) is the preferred technique for rapidly outlining business requirements. Alternatively, the analysts could use other fact-finding methods such as surveys and interviews. Both JRP and fact-finding are taught in the next chapter.

> Task 3.2—Prioritize System Requirements

We stated earlier that the success of a systems development project can be measured in terms of the <u>degree</u> to which business requirements are met. But not all requirements are created equal. If a project gets behind schedule or over budget, it may be useful to recognize which requirements are more important than others. Thus, given the validated requirements, system owners and users should *prioritize system requirements.*

Prioritization of requirements can be facilitated using a popular technique called **timeboxing.** Timeboxing attempts to divide requirements into "chunks" that can be implemented within a period of time that does not tax the patience of the user and management community. Timeboxing forces priorities to be clearly defined.

timeboxing a technique that delivers information systems functionality and requirements through versioning. The development team selects the smallest subset of the system that, if fully implemented, will return immediate value to the system owners and users. That subset is developed, ideally with a time frame of six to nine months or less. Subsequently, value-added versions of the system are developed in similar time frames.

SYSTEMS ANALYSTS facilitate the prioritization task. SYSTEM OWNERS and USERS establish the actual priorities. SYSTEM DESIGNERS and BUILDERS are not involved in the task. The task is triggered by the VALIDATED REQUIREMENTS. It should be obvious that you cannot adequately prioritize an incomplete set of requirements. The deliverable of this task is the REQUIREMENTS WITH PRIORITIES. Priorities can be classified according to their relative importance:

- A *mandatory requirement* is one that must be fulfilled by the minimal system, version 1.0. The system is useless without it. Careful! There is a temptation to label too many requirements as mandatory. A mandatory requirement cannot be ranked because it is essential to any solution. In fact, if an alleged mandatory requirement can be ranked, it is actually a desirable requirement.
- A *desirable requirement* is one that is not absolutely essential to version 1.0. It may still be essential to the vision of some future version. Desirable requirements can and should be ranked. Using version numbers as the ranking scheme is an effective way to communicate and categorize desirable requirements.

> Task 3.3—Update or Refine the Project Plan

Here again, recall that project scope is a moving target. Now that we've identified the business system requirements, we should step back and redefine our understanding of the project scope and update our project plan accordingly. The team must consider the possibility that the new system may be larger than originally expected. If so, the team must adjust the schedule, budget, or scope accordingly. We should also secure approval to continue the project into the next phase. (Work may have already started on the design phases; however, the decisions still require review.)

The project manager, in conjunction with SYSTEM OWNERS and the entire project team, facilitates this task. As usual, the project manager and SYSTEM OWNERS are the key individuals in this task. They should consider the possibility that the requirements now exceed the original vision that was established for the project and new system. They may have to reduce the scope to meet a deadline or increase the budget to get the job done.

As shown in Figure 5-15, this task is triggered by completion of the COMPLETED REQUIREMENTS AND PRIORITIES. The up-to-date PROJECT PLAN is the other key input, and it is updated in the repository as appropriate. The tools, techniques, and steps for maintenance of the project plan were covered in Chapter 4, "Project Management."

> Task 3.4—Communicate the Requirements Statement

Communication is an ongoing task of the requirements analysis phase. We must communicate requirements and priorities to the business community throughout the phase. Users and managers will frequently lobby for requirements and priority consideration. Communication is the process through which differences of opinion must be mediated. The project manager and executive sponsor should jointly facilitate this task. Today, a project intranet or portal is frequently used to communicate requirements. Some systems allow users and managers to *subscribe* to requirements documents to ensure they are notified as changes occur. Interpersonal, communications, and negotiation skills are essential to this task.

> Ongoing Requirements Management

The requirements analysis phase is now complete. Or is it? It was once popular to freeze the business requirements before beginning the system design and construction phases. But today's economy has become increasingly fast-paced. Businesses are measured on their ability to quickly adapt to constantly changing requirements and opportunities. Information systems can be no less responsive than the business itself. Thus, requirements analysis really never ends. While we quietly transition to the remaining phases of our project, there remains an ongoing need to continuously manage requirements through the course of the project and the lifetime of the system.

Requirements management defines a process for system owners, users, analysts, designers, and builders to submit proposed changes to requirements for a system. The process specifies how changes are to be requested and documented, how they will be logged and tracked, when and how they will be assessed for priority, and how they will eventually be satisfied (if they are ever satisfied).

The Logical Design Phase

Not all projects embrace model-driven development, but most include some amount of system modeling. A logical design further documents business requirements using system models that illustrate data structures, business processes, data flows, and user interfaces (increasingly using object models, as introduced earlier in the chapter). In a sense, they validate the requirements established in the previous phase.

Once again, your information systems building blocks (Figure 5-16) can serve as a useful framework for documenting the information systems requirements. Notice that we are still concerned with the SYSTEM USERS' perspectives. In this phase, we draw various system models to document the requirements for a new and improved system. The models depict various aspects of our building blocks. Alternatively, prototypes could be built to "discover requirements." Discovery prototypes were introduced earlier in the chapter. Recall that some prototypes can be reverse engineered into system models.

Figure 5-17 illustrates the typical tasks of the logical design phase. The final phase deliverable and milestone is producing a BUSINESS REQUIREMENTS STATEMENT that will fulfill the system improvement objectives identified in the previous phase. One of the first things you may notice in this task diagram is that most of the tasks are not as sequential as in previous task diagrams. Instead, many of these tasks occur in parallel as the team works toward the goal of completing the requirements statement.

The logical design phase typically includes the following tasks:

4.1a Structure functional requirements.
4.1b Prototype functional requirements.
4.2   Validate functional requirements.
4.3   Define acceptance test cases.

Let's now examine each of these tasks in greater detail.

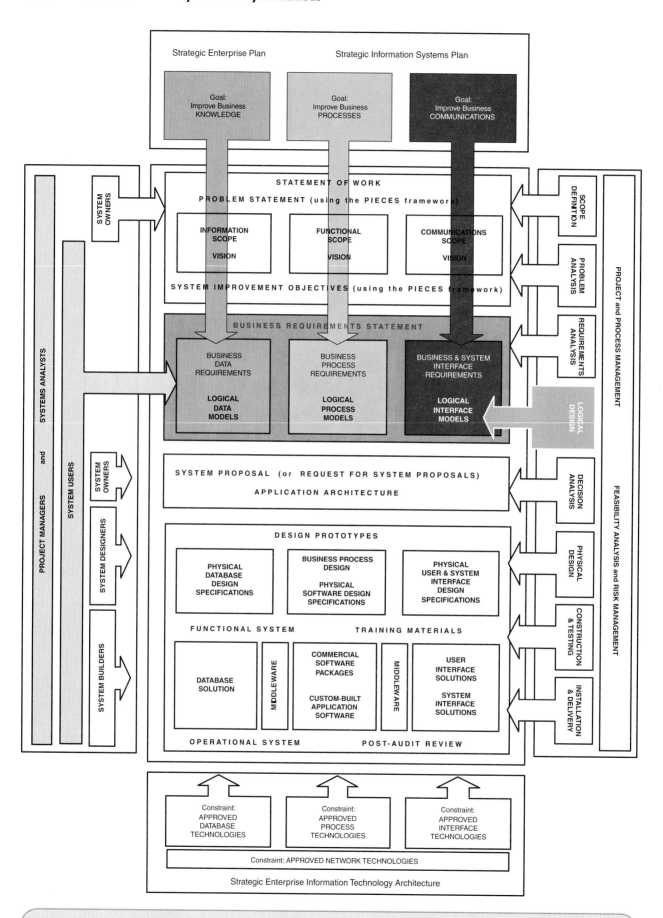

FIGURE 5-16 The Context of the Logical Design Phase of Systems Analysis

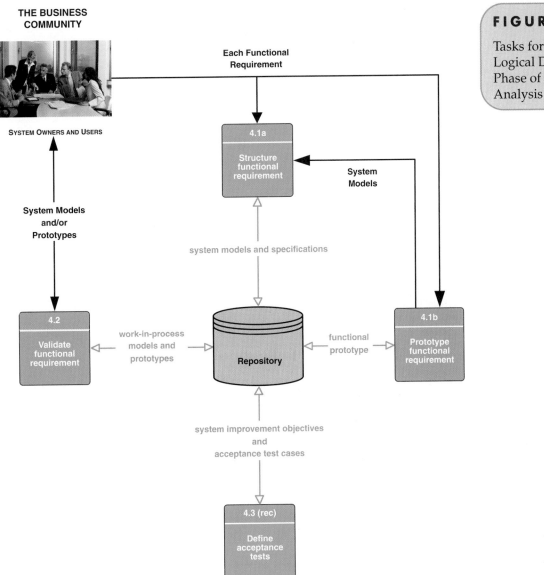

FIGURE 5-17

Tasks for the
Logical Design
Phase of Systems
Analysis

> Task 4.1a—Structure Functional Requirements

One approach to logical design is to *structure the functional requirements.* This means that, using agile methods, you should draw or update one or more system models to illustrate the functional requirement. These may include any combination of data, process, and object models that accurately depict the business and user requirements (but not any technical solution). System models are not complete until all appropriate functional requirements have been modeled. Models are frequently supplemented with detailed logical specifications that describe data attributes, business rules and policies, and the like.

SYSTEMS ANALYSTS facilitate the task. They also document the results. Obviously, SYSTEM USERS are the primary source of factual details needed to draw the models. As shown in Figure 5-17, this task (and phase) is triggered by *each* FUNCTIONAL REQUIREMENT. The outputs are the actual SYSTEM MODELS AND DETAILED SPECIFICATIONS. The level of detail required depends on the methodology being followed. Agile methods usually require "just enough" documentation. How much is enough? That is arguable, but agile methodologists hold that every deliverable should be essential to the forthcoming design and programming phases. This textbook will teach you a variety of different system modeling tools and techniques to apply to logical design.

> Task 4.1b—Prototype Functional Requirements (alternative)

Prototyping is an alternative (and sometimes a prerequisite) to system modeling. Sometimes users have difficulty expressing the facts necessary to draw adequate system models. In such a case, an alternative or complementary approach to system modeling is to build discovery prototypes. Prototyping is typically used in the requirements analysis phase to build sample inputs and outputs. These inputs and outputs help to construct the underlying database and the programs for inputting and outputting the data to and from the database. Although discovery prototyping is optional, it is frequently applied to systems development projects, especially in cases where the users are having difficulty stating or visualizing their business requirements. The philosophy is that the users will recognize their requirements when they see them.

SYSTEMS BUILDERS facilitate this analysis task. SYSTEM ANALYSTS document and analyze the results. As usual, SYSTEM USERS are the primary source of factual input to the task. Figure 5-17 demonstrates that this task is dependent on one or more FUNCTIONAL REQUIREMENTS that have been identified by the users. The system builders and analysts respond by constructing the PROTOTYPES. As described earlier in this chapter, it may be possible to *reverse engineer* some SYSTEM MODELS directly from the prototype databases and program libraries.

> Task 4.2—Validate Functional Requirements

Both SYSTEM MODELS and PROTOTYPES are representations of the users' requirements. They must be validated for completeness and correctness. SYSTEMS ANALYSTS facilitate the prioritization task by interactively engaging system users to identify errors and omissions or make clarifications.

> Task 4.3—Define Acceptance Test Cases

While not a required task, most experts agree that it is not too early to begin planning for system testing. System models and prototypes very effectively define the processing requirements, data rules, and business rules for the new system. Accordingly, these specifications can be used to define TEST CASES that can ultimately be used to test programs for correctness. Either SYSTEM ANALYSTS or SYSTEM BUILDERS can perform this task and validate the test cases with the SYSTEM USERS.

Recall that SYSTEM IMPROVEMENT OBJECTIVES were defined earlier in the project. Test cases can be defined to test these objectives as well.

The Decision Analysis Phase

Given the business requirements for an improved information system, we can finally address how the new system—including computer-based alternatives—*might* be implemented with technology. The purpose of the *decision analysis phase* is to identify candidate solutions, analyze those candidate solutions, and recommend a target system that will be designed, constructed, and implemented. Chances are that someone has already championed a vision for a technical solution. But alternative solutions, perhaps better ones, nearly always exist. During the decision analysis phase, it is imperative that you identify options, analyze those options, and then sell the best solution based on the analysis.

Once again, your information systems building blocks (Figure 5-18) can serve as a useful framework for the decision analysis phase. One of the first things you should notice is that information technology and architecture begin to influence the decisions we must make. In some cases, we must work within standards. In other cases,

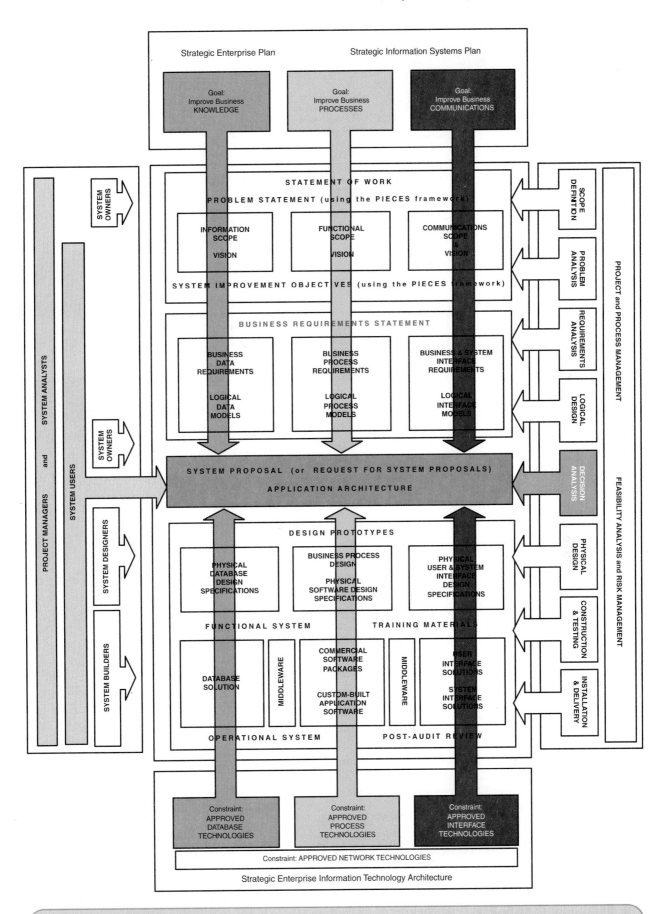

FIGURE 5-18 The Context of the Decision Analysis Phase of Systems Analysis

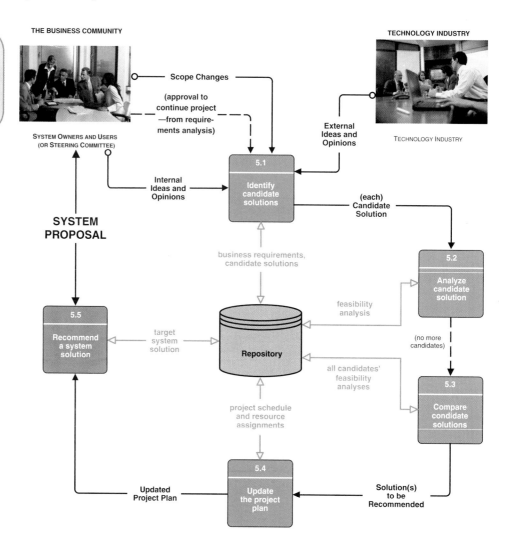

FIGURE 5-19

Tasks for the Decision Analysis Phase of Systems Analysis

we can look to apply different or emerging technology. You should also notice that the perspectives are in transition—from those of the SYSTEM USERS to those of the SYSTEM DESIGNERS. Again, this reflects our transition from pure business concerns or technology. But we are not yet designing. The building blocks indicate our goal as developing a proposal that will fulfill requirements.

Figure 5-19 illustrates the typical tasks of the decision analysis phase. The final phase deliverable and milestone is producing a SYSTEM PROPOSAL that will fulfill the business requirements identified in the previous phase. The decision analysis phase typically includes the following tasks:

5.1 Identify candidate solutions.
5.2 Analyze candidate solutions.
5.3 Compare candidate solutions.
5.4 Update the project plan.
5.5 Recommend a system solution.

Let's now examine each of these tasks in greater detail.

> Task 5.1—Identify Candidate Solutions

Given the business requirements established in the definition phase of systems analysis, we must first identify alternative candidate solutions. Some candidate solutions will be posed by design ideas and opinions from SYSTEM OWNERS and USERS. Others may come from various sources including SYSTEMS ANALYSTS, SYSTEMS DESIGNERS, technical consultants, and other IS professionals. And some technical choices may be limited by a predefined, approved technology architecture. It is the intent of this task not to evaluate the candidates but, rather, simply to define possible candidate solutions to be considered.

The SYSTEMS ANALYSTS facilitate this task. SYSTEM OWNERS and USERS are not normally directly involved in this task, but they may contribute ideas and opinions that start the task. For example, an owner or user may have read an article about, heard about, or learned how some competitor's or acquaintance's similar system was implemented. In any case, it is politically sound to consider the ideas. SYSTEM DESIGNERS and BUILDERS such as database administrators, network administrators, technology architects, and programmers are also a source of ideas and opinions.

As shown in Figure 5-19, this task is formally triggered by the APPROVAL TO CONTINUE THE PROJECT FROM THE REQUIREMENTS ANALYSIS phase. In reality, ideas and opinions have been generated and captured since the preliminary investigation phase—it is human nature to suggest solutions throughout any problem-solving process. Notice that, in addition to coming from the project team itself, IDEAS AND OPINIONS can be generated from both internal and external sources. Each idea generated is considered to be a CANDIDATE SOLUTION to the BUSINESS REQUIREMENTS.

The amount of information describing the characteristics of any one candidate solution may become overwhelming. A candidate matrix, such as Figure 5-20, is a useful tool for effectively capturing, organizing, and comparing the characteristics of different candidate solutions.

As has been the case throughout this chapter, fact-finding and group facilitation techniques like JRP are the principle techniques used to research candidate system solutions. Fact-finding and group facilitation techniques are taught in the next chapter. Also, Chapter 10, "Feasibility Analysis and the System Proposal," will teach you how to actually generate candidate system solutions and document them in the matrix.

> Task 5.2—Analyze Candidate Solutions

Each candidate system solution must be analyzed for feasibility. This can occur as each candidate is identified or after all candidates have been identified. Feasibility analysis should not be limited to costs and benefits. Most analysts evaluate solutions against at least four sets of criteria:

- *Technical feasibility*—Is the solution technically practical? Does our staff have the technical expertise to design and build this solution?
- *Operational feasibility*—Will the solution fulfill the user's requirements? To what degree? How will the solution change the user's work environment? How do users feel about such a solution?
- *Economic feasibility*—Is the solution cost-effective?
- *Schedule feasibility*—Can the solution be designed and implemented within an acceptable time period?

When completing this task, the analysts and users must take care not to make comparisons between the candidates. The feasibility analysis is performed on each individual candidate without regard to the feasibility of other candidates. This approach discourages the analyst and users from prematurely making a decision concerning which candidate is the best.

Again, the SYSTEMS ANALYSTS facilitate the task. Usually SYSTEMS OWNERS and USERS analyze operational, economic, and schedule feasibility. SYSTEMS DESIGNERS and BUILDERS usually contribute to the analyses and play the critical role in analyzing technical feasibility.

Figure 5-19 shows that the task is triggered by the completion of each candidate solution; however, it is acceptable to delay the task until all candidate solutions have been identified. Input to the actual feasibility analyses comes from the various team

FIGURE 5-20 A Candidate Systems Matrix

Characteristics	Candidate 1	Candidate 2	Candidate 3	Candidate . . .
Portion of System Computerized Brief description of that portion of the system that would be computerized in this candidate.	COTS package Platinum Plus from Entertainment Software Solutions would be purchased and customized to satisfy Member Services required functionality.	Member Services and warehouse operations in relation to order fulfillment.	Same as candidate 2.	
Benefits Brief description of the business benefits that would be realized for this candidate.	This solution can be implemented quickly because it's a purchased solution.	Fully supports user's required business processes for SoundStage Inc. Plus more efficient interaction with member accounts.	Same as candidate 2.	
Servers and Workstations A description of the servers and workstations needed to support this candidate.	Technically, architecture dictates Pentium Pro, MS Windows NT class servers and Pentium, MS Windows NT 4.0 workstations (clients).	Same as candidate 1.	Same as candidate 1.	
Software Tools Needed Software tools needed to design and build the candidate (e.g., database management system, emulators, operating systems, languages, etc.). Not generally applicable if applications software packages are to be purchased.	MS Visual C++ and MS Access for customization of package to provide report writing and integration.	MS Visual Basic 5.0 System Architect 3.1 Internet Explorer	MS Visual Basic 5.0 System Architect 3.1 Internet Explorer	
Application Software A description of the software to be purchased, built, accessed, or some combination of these techniques.	Package solution.	Custom solution	Same as candidate 2.	
Method of Data Processing Generally some combination of online, batch, deferred batch, remote batch, and real time.	Client/server.	Same as candidate 1.	Same as candidate 1.	
Output Devices and Implications A description of output devices that would be used, special output requirements (e.g., network, preprinted forms, etc.), and output considerations (e.g., timing constraints).	(2) HP4MV department laser printers. (2) HP5SI LAN laser printers.	(2) HP4MV department laser printers. (2) HP5SI LAN laser printers. (1) PRINTRONIX bar code printer (includes software & drivers). Web pages must be designed to VGA resolution. All internal screens will be designed for SVGA resolution.	Same as candidate 2.	
Input Devices and Implications A description of input methods to be used, input devices (keyboard, mouse, etc.), special input requirements (e.g., new or revised forms from which data would be input), and input considerations (e.g., timing of actual inputs).	Keyboard & mouse.	Apple "Quick Take" digital camera and software. (15) PSC Quickscan laser bar code scanners. (1) HP Scanjet 4C flatbed scanner. Keyboard & mouse.	Same as candidate 2.	
Storage Devices and Implications Brief descriptions of what data would be stored, what data would be accessed from existing stores, what storage media would be used, how much storage capacity would be needed, and how data would be organized.	MS SQL Server DBMS with 100GB arrayed capability.	Same as candidate 1.	Same as candidate 1.	

participants; however, it is not uncommon for external experts (and influences) to also provide data. The feasibility analysis for each candidate is saved in the repository for later comparison to other candidates.

Fact-finding techniques, again, play a role in this systems analysis task. But the ability to perform a feasibility analysis on a candidate system solution is essential. That technique is taught in Chapter 10, "Feasibility Analysis and the System Proposal."

> Task 5.3—Compare Candidate Solutions

Once the feasibility analysis has been completed for each candidate solution, we can compare the candidates and select one or more solutions to recommend to the SYSTEM OWNERS and USERS. At this point, any infeasible candidates are usually eliminated from further consideration. Since we are looking for the most feasible solution of those remaining, we will identify and recommend the candidate that offers the best overall combination of technical, operational, economic, and schedule feasibilities. It should be noted that in selecting such a candidate, it is rare that a given candidate is found to be the most operational, technical, economic, and schedule feasible.

Once again, the SYSTEMS ANALYSTS facilitate the task. SYSTEM DESIGNERS and BUILDERS should be available to answer any technical feasibility questions. But ultimately, the SYSTEMS OWNERS and USERS should be empowered to drive the final analysis and recommendation.

In Figure 5-19, this task is triggered by the completion of the feasibility analysis of all candidate solutions (NO MORE CANDIDATE SOLUTIONS). The input is ALL OF THE CANDIDATES' FEASIBILITY ANALYSES. Once again, a matrix can be used to communicate the large volume of information about candidate solutions. The feasibility matrix in Figure 5-21 allows a side-by-side comparison of the different feasibility analyses for a number of candidates.

The deliverable of this task is the SOLUTION(S) TO BE RECOMMENDED. If more than one solution is recommended, priorities should be established.

Again, feasibility analysis techniques (and the matrix) will be taught in Chapter 10, "Feasibility Analysis and the System Proposal."

> Task 5.4—Update the Project Plan

Hopefully, you noticed a recurring theme throughout this chapter. We are continually updating our project plan as we learn more about a system, its problems, its requirements, and its solutions. We are adjusting scope accordingly. Thus, based on our recommended solution(s), we should once again reevaluate project scope and *update the project plan* accordingly.

The project manager, in conjunction with SYSTEM OWNERS and the entire project team, facilitates this task. The SYSTEMS ANALYSTS and SYSTEM OWNERS are the key individuals in this task. But because we are transitioning into technical system design, we need to begin involving the SYSTEM DESIGNERS and BUILDERS in the project plan updates.

As shown in Figure 5-19, this task is triggered by completion of the SOLUTION(S) TO BE RECOMMENDED. The latest PROJECT SCHEDULE AND RESOURCE ASSIGNMENTS must be reviewed and updated. The UPDATED PROJECT PLAN is the key output. The updated plan should now include a detailed plan for the system design phase that will follow. The techniques and steps for updating the project plan were taught in Chapter 4, "Project Management."

> Task 5.5—Recommend a System Solution

As with the preliminary investigation and problem analysis phases, the decision analysis phase concludes with a communication task. We must *recommend a system solution* to the business community.

FIGURE 5-21   A Feasibility Analysis Matrix

Feasibility Criteria	Weight	Candidate 1	Candidate 2	Candidate 3	Candidate . . .
Operational Feasibility **Functionality.** A description of to what degree the candidate would benefit the organization and how well the system would work. **Political.** A description of how well received this solution would be from user management, user, and organization perspectives.	30%	Only supports Member Services requirements, and current business processes would have to be modified to take advantage of software functionality. **Score: 60**	Fully supports user's required functionality. **Score: 100**	Same as candidate 2. **Score: 100**	
Technical Feasibility **Technology.** An assessment of the maturity, availability (or ability to acquire), and desirability of the computer technology needed to support this candidate. **Expertise.** An assessment of the technical expertise needed to develop, operate, and maintain the candidate system.	30%	Current production release of Platinum Plus package is version 1.0 and has been on the market for only 6 weeks. Maturity of product is a risk, and company charges an additional monthly fee for technical support. Required to hire or train C++ expertise to perform modifications for integration requirements. **Score: 50**	Although current technical staff has only Powerbuilder experience, the senior analysts who saw the MS Visual Basic demonstration and presentation have agreed the transition will be simple and finding experienced VB programmers will be easier than finding Powerbuilder programmers and at a much cheaper cost. MS Visual Basic 5.0 is a mature technology based on version number. **Score: 95**	Although current technical staff is comfortable with Powerbuilder, management is concerned with recent acquisition of Powerbuilder by Sybase Inc. MS SQL Server is a current company standard and competes with SYBASE in the client/server DBMS market. Because of this we have no guarantee future versions of Power-builder will "play well" with our current version SQL Server. **Score: 60**	
Economic Feasibility **Cost to develop:** **Payback period (discounted):** **Net present value:** **Detailed calculations:**	30%	Approximately $350,000. Approximately 4.5 years. Approximately $210,000. See Attachment A. **Score: 60**	Approximately $418,040. Approximately 3.5 years. Approximately $306,748. See Attachment A. **Score: 85**	Approximately $400,000. Approximately 3.3 years. Approximately $325,500. See Attachment A. **Score: 90**	
Schedule Feasibility An assessment of how long the solution will take to design and implement.	10%	Less than 3 months. **Score: 95**	9–12 months. **Score: 80**	9 months. **Score: 85**	
Ranking	100%	60.5	92	83.5	

The project manager and executive sponsor should jointly facilitate this task. Other meeting participants should include the entire project team, including assigned SYSTEM OWNERS, USERS, ANALYSTS, DESIGNERS, and BUILDERS. As usual, the meeting should be open to any and all interested staff from the business community. Also, if an intranet Web site was established for the project, it should have been maintained throughout the problem analysis phases to ensure continuous communication of project progress.

This task is triggered by the completion of the UPDATED PROJECT PLAN. The TARGET SYSTEM SOLUTION (from Task 4.3) is reformatted for presentation as a SYSTEM PROPOSAL. The format may be a report, a verbal presentation, or an inspection by an auditor or peer group (called a *walkthrough*). An outline for a written report is shown in Figure 5-22.

The Next Generations: Systems Analysis

Predicting the future of systems analysis is not easy, but we'll make an attempt. CASE technology will continue to improve, making it easier to model system requirements. First, CASE tools will include object modeling to support emerging object-oriented analysis techniques. While some CASE tools will be purely object-oriented, we believe that the demand for other types of modeling support (e.g., data modeling for databases, process modeling for BPR) will place a premium on comprehensive CASE tools that can support many types of models. Second, the reverse-engineering technology in CASE tools will continue to improve our ability to more quickly generate first-draft system models from existing databases and application programs.

In the meantime, RAD technology will continue to enable accelerated analysis approaches such as prototyping. We also expect the trend for RAD and CASE tools to interoperate through reverse and forward engineering to further simplify both system modeling and discovery prototyping.

Object-oriented analysis will eventually replace structured analysis and information engineering as the best practice for systems analysis. This change may not occur as rapidly as object purists would like, but it will occur all too rapidly for a generation of systems analysts who are skilled in the older methods. There is a grand opportunity for talented young analysts to lead the transition; however, career opportunities will remain strong for analysts who know data modeling that will continue to be used for database design. Also, the process modeling renaissance will continue as BPR projects continue to proliferate.

We also predict our systems proposals will continue to get more interesting. As the Internet, e-commerce, and e-business become increasingly pervasive in our economy, systems analysts will be proposing new alternatives to old problems. There will be a fundamental change in business and information systems to use these new technologies.

One thing will *not* change! We will continue to need systems analysts who understand how to fundamentally investigate and analyze business problems and define the logical business requirements as a preface to system design. But we will all have to do that with increased speed and accuracy to meet the accelerated systems development schedules required in tomorrow's faster-paced economy.

Next Generation

Interpersonal and communications skills are essential to this task. Soft skills such as salesmanship and persuasion become important. (Many schools offer speech and communications courses on these subjects.) Systems analysts should be able to write a formal business report and make a business presentation without getting into technical issues or alternatives.

This concludes the decision analysis phase. And it also concludes our coverage of systems analysis.

FIGURE 5-22 An Outline for a Typical System Proposal

I. Introduction
   A. Purpose of the report
   B. Background of the project leading to this report
   C. Scope of the project
   D. Structure of the report
II. Tools and techniques used
   A. Solution generated
   B. Feasibility analysis (cost-benefit)
III. Information systems requirements
IV. Alternative solutions and feasibility analysis
V. Recommendations
VI. Appendixes

This chapter provided a detailed overview of the systems analysis phases of a project. You are now ready to learn some of the systems skills introduced in this chapter. For most students, this would be the ideal time to study the fact-finding techniques that were identified as critical to almost every phase and task that was described in this chapter. Chapter 6 teaches these skills. It is recommended that you read Chapter 7, "Modeling System Requirements with Use Cases," before proceeding to any of the modeling chapters since use cases are commonly used to facilitate the activity of modeling.

The sequencing of the system modeling chapters is flexible; however, we personally prefer and recommend that Chapter 8, "Data Modeling and Analysis," be studied first. All information systems include databases, and data modeling is an essential skill for database development. Also, it is easier to synchronize early data models with later process models than vice versa. Your instructor may prefer that you first study Chapter 9, "Process Modeling." Advanced courses may elect to jump straight to Chapter 10 to learn about object-oriented analysis and modeling with UML.

If you do jump straight to a system modeling chapter from this chapter, make a commitment to return to Chapter 6 to study the fact-finding techniques. Regardless of how well you master system modeling, that modeling skill is entirely dependent on your ability to discover and collect the business facts that underlie the models.

For those of you who have already completed a systems analysis course, this chapter was probably scheduled only as a review or context for systems design. We suggest that you merely review the system modeling chapters and proceed directly to Chapter 12, "Systems Design." That chapter will pick up where this chapter left off.

Summary

1. Formally, systems analysis is the dissection of a system into its component pieces. As a problem-solving phase, it precedes systems design. With respect to information systems development, systems analysis is the preliminary investigation of a proposed project, the study and problem analysis of the existing system, the requirements analysis of business requirements for the new system, and the decision analysis for alternative solutions to fulfill the requirements.

2. The results of systems analysis are stored in a repository for use in later phases and projects.

3. There are several popular or emerging strategies for systems analysis. These techniques can be used in combination with one another.

   a. Model-driven analysis techniques emphasize the drawing of pictorial system models that represent either a current reality or a target vision of the system.

   i) Structured analysis is a technique that focuses on modeling processes.

   ii) Information engineering is a technique that focuses on modeling data.

   iii) Object-oriented analysis is a technique that focuses on modeling objects that encapsulate the concerns of data and processes that act on that data.

   b. Accelerated analysis approaches emphasize the construction of working models of a system in an effort to accelerate systems analysis.

   i) Discovery prototyping is a technique that focuses on building small-scale, functional subsystems to discover requirements.

   ii) Rapid architected analysis attempts to automatically generate system models from either prototypes or existing systems. The automatic generation of models requires reverse engineering technology.

c. Both model-driven and accelerated system analysis approaches are dependent on requirements discovery techniques to identify or extract problems and requirements from system owners and users.

   i) Fact-finding is the formal process of using research, interviews, questionnaires, sampling, and other techniques to collect information.

   ii) Joint requirements planning (JRP) techniques use facilitated workshops to bring together all interested parties and accelerate the fact-finding process.

d. Business process redesign is a technique that focuses on simplifying and streamlining fundamental business processes before applying information technology to those processes.

4. Each phase of systems analysis (preliminary investigation, problem analysis, requirements analysis, and decision analysis) can be understood in the context of the information system building blocks: KNOWLEDGE, PROCESSES, and COMMUNICATIONS.

5. The purpose of the preliminary investigation phase is to determine the worthiness of the project and to create a plan to complete those projects deemed worthy of a detailed study and analysis. To accomplish the preliminary investigation phase, the systems analyst will work with the system owners and users to: *(a)* list problems, opportunities, and directives; *(b)* negotiate preliminary scope; *(c)* assess project worth; *(d)* plan the project, and *(e)* present the project to the business community. The deliverable for the preliminary investigation phase is a project charter that must be approved by system owners and/or a decision-making body, commonly referred to as the steering committee.

6. The purpose of the problem analysis phase is to answer the questions, Are the problems really worth solving, and is a new system really worth building? To answer these questions, the problem analysis phase thoroughly analyzes the alleged problems and opportunities first identified in the preliminary investigation phase. To complete the problem analysis phase, the analyst will continue to work with the system owner, system users, and other IS management and staff. The systems analyst and appropriate participants will *(a)* study the problem domain; *(b)* thoroughly analyze problems and opportunities; *(c)* optionally, analyze business processes; *(d)* establish system improvement objectives and constraints; *(e)* update the project plan; and *(f)* present the findings and recommendations. The deliverable for the

problem analysis phase is the system improvement objectives.

7. The purpose of the requirements analysis phase is to identify what the new system is to do without the consideration of technology—in other words, to define the business requirements for a new system. As in the preliminary investigation and problem analysis phases, the analyst actively works with system users and owners as well as other IS professionals. To complete the requirements analysis phase, the analyst and appropriate participants will *(a)* define requirements, *(b)* analyze functional requirements using system modeling and/or discovery prototyping, *(c)* trace and complete the requirements statement, *(d)* prioritize the requirements, and *(e)* update the project plan and scope. The deliverable of the requirements analysis phase is the business requirements statement. Because requirements are a moving target with no finalization, requirements analysis also includes the ongoing task of managing changes to the requirements.

8. The purpose of the logical design phase is to document business requirements using system models for the proposed system. These system models can, depending on the methodology, be any combination of process models, data models, and object models. The models depict various aspects of our building blocks. Alternatively, prototypes could be built to "discover requirements." Some discovery prototypes can be reverse engineered into system models. The systems analyst and appropriate participants will *(a)* structure or prototype functional requirements, *(b)* validate functional requirements, and *(c)* define acceptance test cases. These tasks are not necessarily sequential; they can occur in parallel. The deliverable for the logical design phase is the business requirements statement.

9. The purpose of the decision analysis phase is to transition the project from business concerns to technical solutions by identifying, analyzing, and recommending a technical system solution. To complete the decision analysis phase, the analyst and appropriate participants will *(a)* define candidate solutions; *(b)* analyze candidate solutions for feasibility (technical, operational, economic, and schedule feasibility); *(c)* compare feasible candidate solutions to select one or more recommended solutions; *(d)* update the project plan based on the recommended solution; and *(e)* present and defend the target solution. The deliverable of the decision analysis phase is the system proposal.

Review Questions

1. What are the business factors that are driving systems analysis? Based on these factors, what should systems analysis address?
2. What is model-driven analysis? Why is it used? Give several examples.
3. What is the major focus of structured analysis?
4. What is the major focus of information engineering?
5. Why has object-oriented analysis become popular? What problems does it solve?
6. What are the five phases of systems analysis?
7. What is the goal of the scope definition phase?
8. What are the five tasks that you do in the scope definition phase?
9. What is the trigger for communicating the project plan, and who is the audience? Why is communicating the project plan important?
10. Why do many new systems analysts fail to effectively analyze problems? What can they do to become more effective?
11. What is a popular tool used to identify and express the functional requirements of a system?
12. What is a commonly used technique for prioritizing system requirements?
13. When could prototyping be used instead of system modeling for determining functional requirements?
14. Why is the decision analysis phase needed?
15. What are some ways to identify candidate solutions?

Problems and Exercises

1. There are many different approaches to systems analysis. Despite these different approaches, what is the universally accepted definition of systems analysis? What is the general consensus as to when systems analysis begins and when it ends? As a project manager, what is important to know regarding the definition of systems analysis, and what is important to ensure in your organization regarding the definition?
2. As a systems analyst, you will be exposed to and use many different approaches to systems analysis throughout your career. It is important that you understand the conceptual basis of each type of approach, and their essential differences, strengths and weaknesses. Consider the differences in structured analysis, information engineering and data modeling, and object-oriented analysis, all of which represent model-driven analysis, and fill in the matrix shown below.

	CENTRICITY (data, process, etc.)	TYPE OF MODELS USED	ESSENTIAL DIFFERENCES
STRUCTURED ANALYSIS			
INFORMATION ENGINEERING AND DATA MODELING			
OBJECT-ORIENTED ANALYSIS			

3. Accelerated systems analysis approaches are based on the premise that prototypes can help reveal the most important business requirements faster than other methods. Describe the two most commonly used approaches to accelerated analysis. What do they do and how do they do it? What is one of the criticisms of prototyping? Do the accelerated systems analysis approaches completely replace more formal approaches, such as structured analysis?

4. During the scope definition phase, what is one question that you should never lose sight of? And how do you answer this question? What five tasks should occur during the scope definition phase?

5. You are a new systems analyst and eager to prove your abilities on your first project. You are at a problem analysis meeting with the system owners and users and find yourself saying, "We need to do this to solve the problem." Into what common trap are you in danger of falling? What technique could you use to avoid this trap?

6. Your project team has completed the scope definition phase, and is now at the point in the problem analysis phase for establishing system improvement objectives. As the systems analyst on the project team, you are the facilitator of a brainstorming session to define the system improvement objectives. Since several of the project owners and users have never done this before, describe the characteristics of good system improvement objectives and provide some examples. Members of the project team suggest the following objectives:

   a. Reduce the time required to process the order.
   b. The new system must use Oracle to store data.
   c. The data input screens must be redesigned so they are more user-friendly.
   d. The customer satisfaction rate with the online ordering process must be increased by 10 percent.

   Are these examples of good system improvement objectives? Why or why not? If not, how could they be reworded? Also, objectives frequently have constraints that are tied to them; what, if any, do you think the matching constraint might be for each of these objectives?

7. You've made it through the problem analysis phase of the project, and are now beginning the requirements analysis phase. During the first meeting on the business requirements, one of the other analysts on the project team asks the system users, "How should the new system meet your needs?" What common mistake is the analyst making? What are often the consequences of making this mistake?

8. What is the difference between functional and nonfunctional requirements, and what is the purpose of categorizing them into these categories? What are two formats that an analyst can use to document the functional system requirements?

9. Is it important to prioritize system requirements, and if so, when should the requirements be prioritized? What is one technique that can be used, and what is the difference between mandatory and desirable requirements? What is one way to test whether a mandatory requirement is truly a mandatory requirement?

10. Once the system requirements are identified and prioritized, shouldn't everything be frozen to prevent scope or feature creep? Doesn't updating the project plan or allowing stakeholders to continue to request changes just delay system design and construction, and maybe even project completion itself?

11. Why should acceptance test cases be defined during the logical design phase? After all, the technical design hasn't been done yet, let alone building the system. Shouldn't testing activities at least wait until construction is actually underway?

12. How is the logical design phase different from the requirements analysis phase?

13. Let's say you are on the project team of a project that had a great deal of difficulty during the requirements analysis phase, and fell several weeks behind schedule. The project manager wants to try to catch up by either skipping or abbreviating some of the tasks in the logical design phase. After all, the project manager reasons, we really have a clear idea of the requirements now, the designers and builders are really experienced, and they don't really need the logical design in order to do the technical design. Is this a legitimate method to get back on schedule? What are the possible consequences?

14. In identifying and defining possible candidate solutions, what are the typical roles of the various stakeholders who are involved in the project?

15. You are a systems analyst and have been asked to facilitate the analysis and evaluation of several candidate system solutions for their feasibility. What sets of criteria would you typically use? Who do you involve in this task? Should you compare the candidate solutions against each other at this point? Why or why not? What is the typical deliverable coming out of this task?

Projects and Research

1. Select an information system with which you are familiar, and which you feel needs to be improved, based upon your experiences as an employee, customer, other system user, or system owner. Switch roles and perspectives as necessary to perform or answer the following:

    a. Describe the nature of the information system you have selected.
    b. Describe the organization that owns and maintains the information system.
    c. Identify the baseline problems and opportunities, per Task 1.1.
    d. Develop a preliminary problem statement, using the format shown in Figure 5-8.

2. Assume you are the now a systems analyst on the project described in the preceding question. Executive management was extremely impressed by your work on the problem statement. As a result, they have given the project the go-ahead, the baseline schedule and budget have been developed, and the project plan has been approved by the executive steering committee. As the systems analyst, you now have been tasked to do the following:

    a. Develop and document your understanding of the problem domain and business vocabulary, using the textbook's information system building blocks framework as described in Task 2.1.
    b. Analyze problems and opportunities using cause-and-effect analysis (Task 2.2).
    c. Analyze business processes and develop process models (Task 2.3).
    d. Establish system improvement objectives (Task 2.4).
    e. Prepare a Problems, Opportunities, Objectives, and Constraints Matrix, using Figure 5-12 as an example.

3. Communicating findings and recommendations is the final task in the problem analysis phase. As a systems analyst on the project, you have been tasked with preparing the System Improvement Objectives and Recommendations Report. For this exercise, prepare only the Executive Summary portion of the report, using the format shown in Figure 5-13. The executive steering committee will use this summary to make its decisions regarding the recommendations.

4. Your strong work on the project to date has continued to impress executive management. You have received a pay increase and have been tasked with conducting the requirements analysis phase. Specifically:

    a. Identify the system requirements, and prepare an outline of functional and nonfunctional requirements, per Task 3.1. Since your organization uses structured analysis and does not employ use case modeling, list each system improvement objective, and the inputs, processes, outputs, and stored data needed to meet each objective.
    b. Assume that the requirements you identified in the preceding step have been validated. Prioritize the requirements according to their relative importance, using the method described in Task 3.2.

5. Your work has helped keep the project well ahead of schedule, so executive management gives you a couple of weeks of paid vacation. When you return, the project is moving into the decision analysis phase. Your next task is to identify candidate solutions.

    a. Describe the process for identifying candidate solutions. What should you be careful *not* to do at this point?
    b. Develop a candidate systems matrix, using the format in Figure 5-20 as an example, and include three possible solutions.

6. After identifying candidate solutions, the next step is to analyze these solutions.

    a. Describe the process for analyzing candidate solutions. What should the project team *not* do in completing this task?
    b. Develop a Feasibility Analysis Matrix, based upon the candidate solutions identified in the preceding question, and using the format shown in Figure 5-21 as an example. Determine what your weighting factors should be.

Minicases

1. You are the CIO of a major retailer. Recently, you read "Spying on the Sales Floor" in the *Wall Street Journal* on December 21, 2004. You see that your competitors are using video mining to analyze consumer behavior. Should your company also adopt this tool (video mining)? What are the strategic

implications to your company of your competitors' move? What opportunities have been created? Threats?

2. Read "Human Reengineering," by Cooper and Markus, in the *Sloan Management Review*, Summer 1995. In this article, Okuno works on instituting a positive attitude toward change. How does he do this? Discuss the importance of change acceptance by employees to the success of a technology implementation.

3. Refer to Minicase 1. You, as the CIO, believe that the business gains for implementing video mining in your retail stores will outweigh any negative customer perceptions. Your company is Baby's R Us, a child company of Toys R Us. Do an economic feasibility study for this investment. Be sure to include a listing of intangible costs and benefits, as well as an argument for your chosen discount rate. What is the ROI of the video mining? Try to keep your analysis to under 15 pages.

4. Develop a project plan and schedule feasibility study for the video-mining investment into Baby's R Us. Be sure to include a Gantt and PERT/CPM chart, as well as a clear discussion of all tasks that need to be completed.

Suggested Readings

Application Development Trends (monthly periodical). Natick, MA: Software Productivity Group, a ULLO International company. This is our favorite systems development periodical. It follows systems analysis and design strategies, methodologies, CASE, and other relevant trends. Visit its Web site at www.adtmag.com.

Gause, Donald C., and Gerald M. Weinberg. *Are Your Lights On? How to Figure Out What the Problem REALLY Is.* New York: Dorset House Publishing, 1990. Here's a title that should really get you thinking, and the entire book addresses one of the least published aspects of systems analysis: problem solving.

Hammer, Mike. "Reengineering Work: Don't Automate, Obliterate." *Harvard Business Review,* July–August 1990, pp. 104–11. Dr. Hammer is a noted expert on business process redesign. This seminal paper examines some classic cases where the technique dramatically added value to businesses.

Wetherbe, James. *Systems Analysis and Design: Best Practices,* 4th ed. St. Paul, MN: West Publishing, 1994. We are indebted to Dr. Wetherbe for the PIECES framework.

Wood, Jane, and Denise Silver. *Joint Application Design: How to Design Quality Systems in 40% Less Time.* New York: John Wiley & Sons, 1989. This book provides an excellent in-depth presentation of joint application development (JAD).

Yourdon, Edward. *Modern Structured Analysis.* Englewood Cliffs, NJ: Yourdon Press, 1989. This update to the classic DeMarco text on the same subject defines the current state of the practice for the structured analysis approach.

Zachman, John A. "A Framework for Information System Architecture." *IBM Systems Journal* 26, no. 3 (1987). This article presents a popular conceptual framework for information systems surveys and the development of an information architecture.

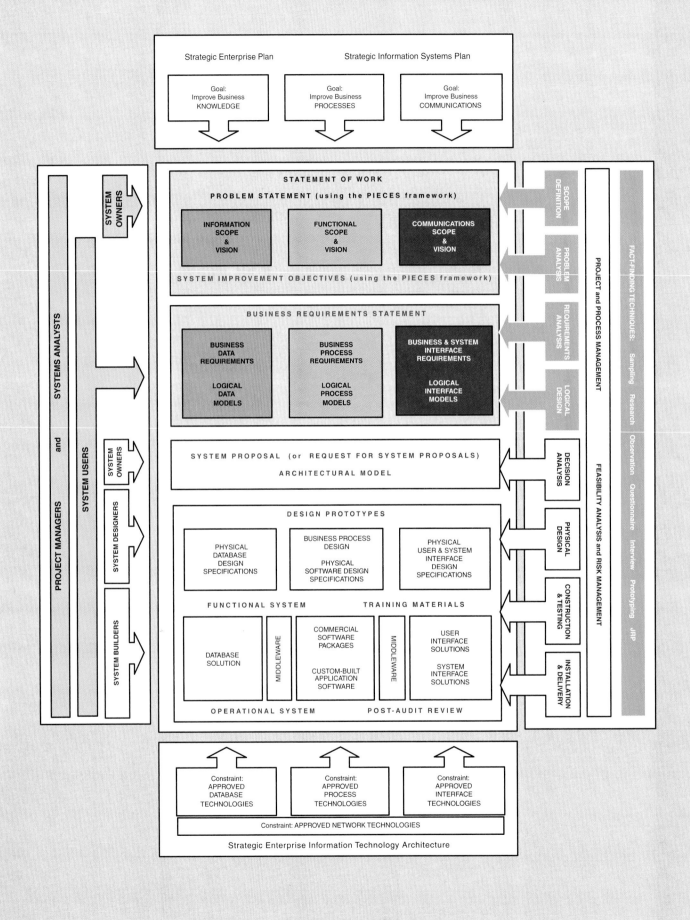

Requirements Gathering

6

Chapter Preview and Objectives

Effective fact-finding techniques are crucial to the development of systems projects. In this chapter you will learn about techniques to discover and analyze information system requirements. You will learn how to use various fact-finding techniques to gather information about the system's problems, opportunities, and directives. You will know that you understand fact-finding techniques and requirements discovery when you can:

- Define system requirements and differentiate between functional and nonfunctional requirements.

- Understand the activity of problem analysis and be able to create an Ishikawa (fishbone) diagram to aid in problem solving.

- Understand the concept of requirements management.

- Identify seven fact-finding techniques and characterize the advantages and disadvantages of each.

- Understand six guidelines for doing effective listening.

- Understand what body language and proxemics are and why a systems analyst should care.

- Characterize the typical participants in a JRP session and describe their roles.

- Complete the planning process for a JRP session, including selecting and equipping the location, selecting the participants, and preparing an agenda to guide the JRP session.

- Describe several benefits of using JRP as a fact-finding technique.

- Describe a fact-finding strategy that will make the most of your time with end users.

Introduction

Bob Martinez has spent most of the week reading. He started with memos related to the proposed Member Services System to better understand the problem. He then reviewed SoundStage's procedures manual for any policies related to member services and promotions. He studied nearly 100 member order forms selected at random, noting the kinds of data recorded in each blank and which blanks were always, sometimes, and never used. He read the documentation for the present member services system. He reviewed data and process diagrams from the prior member service systems development project, noting things that would probably need to be changed in the new system. It was grueling work. But in the end he really felt like he was beginning to understand the system. He produced a report for Sandra, his boss, of the key issues and questions that would need to be answered at the upcoming joint requirements planning meeting.

An Introduction to Requirements Discovery

In Chapter 3 we discussed several phases of systems development. Each phase is important and necessary in order to effectively design, construct, and ultimately implement a system to meet the users' (stakeholders') needs. But to develop such a system, we must first be able to correctly identify, analyze, and understand what the users' requirements are or what the users want the system to do. The process and techniques that a systems analyst uses to identify, analyze, and understand system requirements are referred to as **requirements discovery.** As suggested by the chapter's home page, requirements discovery primarily involves systems analysts working with system users and owners during the earlier system development phases to obtain a detailed understanding of the business requirements of an information system.

What are system requirements? **System requirements** specify what the information system must do or what property or quality the system must have. System requirements that specify what the information system must do are frequently referred to as **functional requirements.** System requirements that specify a property or quality the system must have are frequently referred to as **nonfunctional requirements.**

The PIECES framework (Table 6-1), introduced in Chapter 3, provides an excellent tool for classifying system requirements. The benefit of classifying the various types of requirements is the ability to group requirements for reporting, tracking, and validation purposes. Plus doing so aids in identifying possible overlooked requirements.

Essentially, the purpose of requirements discovery and management is to correctly identify the KNOWLEDGE, PROCESS, and COMMUNICATION requirements for the users of a new system. Failure to correctly identify system requirements may result in one or more of the following:

- The system may cost more than projected.
- The system may be delivered later than promised.
- The system may not meet the users expectations, and that dissatisfaction may cause them not to use it.
- Once in production, the costs of maintaining and enhancing the system may be excessively high.
- The system may be unreliable and prone to errors and downtime.
- The reputation of the IT staff on the team is tarnished because any failure, regardless of who is at fault, will be perceived as a mistake by the team.

The impact in terms of cost can be staggering. Take, for example, Table 6-2, by Barry W. Boehm, a noted expert in information technology economics.[1] He studied several

requirements discovery
the process and techniques used by systems analysts to identify or extract system problems and solution requirements from the user community.

system requirement
something that the information system must do or a property that it must have. Also called a *business requirement.*

functional requirement
something the information system must do.

nonfunctional requirement a property or quality the system must have. Examples include security, ease-of-use, performance, etc.

---

[1]Donald C. Gause and Gerald M. Weinberg, *Exploring Requirements: Quality before Design* (New York: Dorset House Publishing, 1989), pp. 17–18.

TABLE 6-1 PIECES Classification of System Requirements

Nonfunctional Requirement Type	Explanation
Performance	Performance requirements represent the performance the system is required to exhibit to meet the needs of users. • What is the acceptable throughput rate? • What is the acceptable response time?
Information	Information requirements represent the information that is pertinent to the users in terms of content, timeliness, accuracy, and format. • What are the necessary inputs and outputs? When must they happen? • What is the required data to be stored? • How current must the information be? • What are the interfaces to external systems?
Economy	Economy requirements represent the need for the system to reduce costs or increase profits. • What are the areas of the system where costs must be reduced? • How much should costs be reduced or profits be increased? • What are the budgetary limits? • What is the timetable for development?
Control (and security)	Control requirements represent the environment in which the system must operate, as well as the type and degree of security that must be provided. • Must access to the system or information be controlled? • What are the privacy requirements? • Does the criticality of the data necessitate the need for special handling (backups, off-site storage, etc.) of the data?
Efficiency	Efficiency requirements represent the system's ability to produce outputs with minimal waste. • Are there duplicate steps in the process that must be eliminated? • Are there ways to reduce waste in the way the system uses it resources?
Service	Service requirements represent needs in order for the system to be reliable, flexible, and expandable. • Who will use the system, and where are they located? • Will there be different types of users? • What are the appropriate human factors? • What training devices and training materials are to be included in the system? • What training devices and training materials are to be developed and maintained separately from the system, such as stand-alone computer-based training (CBT) programs or databases? • What are the reliability/availability requirements? • How should the system be packaged and distributed? • What documentation is required?

TABLE 6-2 Relative Coasts of Fixing an Error

Phase in which Error Discovered	Cost Ratio
Requirements	1
Design	3–6
Coding	10
Development Testing	15–40
Acceptance Testing	30–70
Operation	40–1,000

software development projects to determine the costs of errors in requirements that weren't discovered until later in the development process.

Based on Boehm's findings, an erroneous requirement that goes undetected and unfixed until the operation phase may cost 1,000 times more than it would if it were detected and fixed in the requirements phase. Therefore, in defining system requirements, it is critical that they meet the following criteria:

- *Consistent*—The requirements are not conflicting or ambiguous.
- *Complete*—The requirements describe all possible system inputs and responses.
- *Feasible*—The requirements can be satisfied based on the available resources and constraints (feasibility analysis is covered in Chapter 11).
- *Required*—The requirements are truly needed and fulfill the purpose of the system.
- *Accurate*—The requirements are stated correctly.
- *Traceable*—The requirements directly map to the functions and features of the system.
- *Verifiable*—The requirements are defined so that they can be demonstrated during testing.

This can be a time-consuming, difficult, and frustrating process that often leads organizations and individuals to take shortcuts to save time and money. But this short-sightedness often leads to the problems mentioned before. Now that we understand our goal, lets look at the process.

The Process of Requirements Discovery

The process of requirements discovery consists of the following activities:

- Problem discovery and analysis.
- Requirements discovery.
- Documenting and analyzing requirements.
- Requirements management.

Let's now examine each one of these activities in detail.

> Problem Discovery and Analysis

As previously stated, requirements solve problems. For systems analysts to be successful, they must be skilled in the activity of problem analysis. To fully understand problem analysis, let's use the following example: A mother takes her young daughter to the doctor because the child is ill. The first thing the doctor tries to do is identify the problem. The child has an earache, a fever, and a runny nose. Are these the problems? The mother has been giving the child pain medicine to ease the pain, but the child has not gotten better. It turns out the mother is treating the symptoms and not the real problem. Fortunately, the doctor is trained to analyze further. After examining the child, the doctor has concluded that the child has an ear infection, which is the root cause of the child's symptoms. Now that the problem has been identified and analyzed, it is time for the doctor to recommend a cure (solution). Normally, an antibiotic is prescribed to cure an ear infection, but in order to do that, the doctor first needs to determine if there are any constraints on the medicine that he can prescribe. How old is the child, and how much does she weigh? Is the child allergic to any medications? Can she swallow pills? Once these constraints are known, a prescription can be generated. Systems analysts use the same problem-solving process as a doctor uses, but instead of diagnosing medical problems they diagnose system problems.

One of the most common mistakes inexperienced systems analysts make when trying to analyze problems is identifying a symptom as a problem. As a result, they may design and implement a solution that more than likely doesn't solve the real problem or that may cause new problems. A popular tool used by development teams to identify, analyze, and solve problems is an **Ishikawa diagram.** The fishbone-shaped diagram is the brainchild of Kaoru Ishikawa, who pioneered quality management processes in the Kawasaki shipyards of Japan and, in the process, became one of the founding fathers of modern management.

Drawing the fishbone diagram begins with the name of the problem of interest entered at the right of the diagram (or the *fish's head*). The possible causes of the problem are then drawn as *bones* off the *main backbone,* each on an arrow pointing to the backbone. Typically, these "bones" are labeled as four basic categories: materials, machines, manpower, and methods (the four Ms). Other names can be used to suit the problem at hand. Alternative or additional categories include places, procedures, policies, and people (the four Ps) or surroundings, suppliers, systems, and skills (the four Ss).

The key is to have three to six main categories that encompass all possible areas of causes. Brainstorming techniques (defined later in the chapter) are commonly performed to add causes to the main bones. When the brainstorming is complete, the fishbone depicts a complete picture of all the possibilities about what could be the root cause for the designated problem. The development team can then use the diagram to decide and agree on what the most likely causes of the problem are and how they should be acted on. Figure 6-1 is an example of a fishbone diagram

Ishikawa diagram a graphical tool used to identify, explore, and depict problems and the causes and effects of those problems. It is often referred to as a *cause-and-effect diagram* or a *fishbone diagram* (because it resembles the skeleton of a fish).

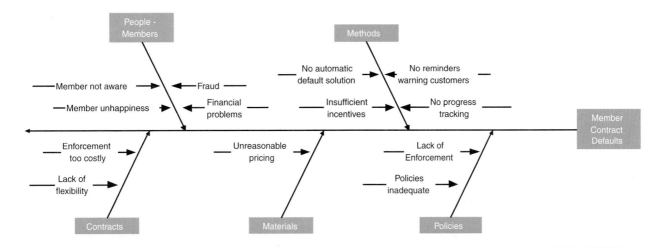

FIGURE 6-1 Sample Fishbone Diagram

depicting the SoundStage problem of members defaulting on contracts. In the diagram, notice that the problem to be solved is in the box at the far right. The five areas that have been identified as categories of causes (People-Members, Methods, Contracts, Materials, and Policies) are listed in boxes above and below the *fish's skeleton* and connected by arrows (bones) pointing to the fish's backbone. The actual causes of the problem for each category are depicted as arrows pointing to the category arrow (bone).

> Requirements Discovery

Given an understanding of problems, the systems analyst can start to define requirements. For today's systems analysts to be successful in defining system requirements, they must be skilled in effective methods for gathering information—fact-finding. **Fact-finding** is a technique that is used across the entire development cycle, but it is extremely critical in the requirements analysis phase. Once fact-finding has been completed, tools such as use cases, data models, process models, and object models will be used to document facts, and conclusions will be drawn from the facts. You will learn about these tools and how to document requirements derived from fact-finding in subsequent chapters of this textbook.

fact-finding the formal process of using research, meetings, interviews, questionnaires, sampling, and other techniques to collect information about system problems, requirements, and preferences. It is also called *information gathering* or *data collection*.

Facts are in the domain of the business application and its end users. Therefore, the analyst must collect those facts in order to effectively apply the documentation tools and techniques. During systems analysis phases, the analyst learns about the vocabulary, problems, opportunities, constraints, requirements, and priorities of a business and a system.

What types of facts must be collected? It would certainly be beneficial if we had a framework to help us determine what facts need to be collected, no matter what project we are working on. Fortunately, we have such a framework. As it turns out, the facts that describe any information system also correspond nicely with the building blocks highlighted on the chapter home page. Notice that fact-finding techniques are used in the early systems development phases to identify information, functional, and communication scopes and visions, as well as to identify business knowledge process, and communication requirements for the system.

> Documenting and Analyzing Requirements

When the systems analyst is performing fact-finding activities, it is important that the analyst assemble or document the gathered information (or *draft requirements*) in an organized, understandable, and meaningful way. These initial documents will provide direction for the modeling techniques the systems analyst will use to analyze the requirements and determine the correct requirements for the project. Once those have been identified, the systems analyst formalizes the requirements by presenting them in a document that will be reviewed and approved by the users.

Documenting the Draft Requirements   Systems analysts use various tools to document their initial findings in draft form. They write *use cases* to describe the system functions from the external users' perspective and in a manner and terminology the users understand. *Decision tables* are used to document an organization's complex business policies and decision-making rules, and *requirements tables* are used to document each specific requirement. Each of these tools is examined in more detail later in the chapter.

Analyzing Requirements   More often than not, fact-finding activities produce requirements that are in conflict with one another. This is because requirements are solicited from many different sources and each person has his or her own opinions and desires for the functionality and features of the new system. The goal of the requirements analysis activity is to discover and resolve the problems with the requirements and reach agreement on any modifications to satisfy the stakeholders. The process is concerned with the "initial" requirements gathered from the stakeholders. These requirements are usually incomplete and documented in an informal way in instruments such as use cases, tables, and reports. The focus at this stage is on reaching agreement on the stakeholder's needs; in other words, the analysis should answer the question, "Do we have the right system requirements for the project?" Inevitably, the draft requirements contain many problems, such as:

- Missing requirements
- Conflicting requirements
- Infeasible requirements
- Overlapping requirements
- Ambiguous requirements

These types of requirements problems are very common in many of the requirement documents written today. If left unresolved, they can be extremely costly to fix later in the development cycle.

It was previously mentioned that stakeholders should agree on the resulting system requirements—thus there is an inevitable negotiation process that exists among stakeholders during analysis. If multiple stakeholders submit requirements that are in conflict with each other or if the proposed requirements are too ambitious, the stakeholders must negotiate, often under the guidance of the systems analyst, to agree on any modifications or simplifications to the system requirements. They also must agree on the criticality and priority of the requirements. This is crucial to ensure the success of the development effort.

The fact-finding and requirements analysis activities are very closely associated with each other and in fact are often interwoven. If requirements discovered during the fact-finding process are found to be problematic, the analyst may go ahead and perform analysis activities on the select items in order to resolve the problems before continuing to elicit additional system needs and desires.

This chapter focuses primarily on the business side of requirements, or, in other words, the logical requirements, but it is important to note that additional technical requirements exist that are physical in nature. Examples of technical requirements include specifying a required software package or hardware platform. These types of requirements will be discussed in depth in Chapter 11.

Formalizing Requirements System requirements are usually documented in a formal way to communicate the requirements to the key stakeholders of the system. This document serves as the contract between the system owners and the development team on what is going to be provided in terms of a new system. Thus, it may go through many revisions and reviews before everyone agrees and authorizes its contents. There is no standard name or format for this document. In fact, many organizations use different names such as requirements statement, requirements specification, requirements definition, functional specification, and the like, and the format is usually tailored to the organization's needs. Companies that provide information systems and software to the U.S. government must use the format and naming conventions specified in the government's published standards document MIL-STD-498.[2] Many organizations have created their own standards adapted from MIL-STD-498 because of its thoroughness and because many people are already familiar with it. In this book we will use the term **requirements definition document,** and Figure 6-2 provides a sample outline of one. Please note that this document will be consolidated with

requirements definition document a formal document that communicates the requirements of a proposed system to key stakeholders and serves as a contract for the systems project. Synonyms include requirements statement, requirements specification, and functional specification.

---

[2]MIL-STD-498 is a standard that merges DOD-STD-2167A and DOD-STD-7935A to define a set of activities and documentation suitable for the development of both weapon systems and automated information systems.

FIGURE 6-2

Sample Requirements Definition Outline

REQUIREMENTS DEFINITION DOCUMENT

1. Introduction
    1.1. Purpose
    1.2. Background
    1.3. Scope
    1.4. Definitions, Acronyms, and Abbreviations
    1.5. References
2. General Project Description
    2.1. Functional Requirements
3. Requirements and Constraints
    3.1. Functional Requirements
    3.2. Nonfunctional Requirements
4. Conclusion
    4.1. Outstanding Issues
Appendix (optional)

other project information to form the requirements statement, which is the final deliverable of the requirements analysis phase. A requirements definition document should consist of the following:

- The functions and services the system should provide.
- Nonfunctional requirements, including the system's features, characteristics, and attributes.
- The constraints, which restrict the development of the system or under which the system must operate.
- Information about other systems with which the system must interface.

Who will read the requirements definition document? This document is probably the most widely read and referenced document of all the project documentation. System owners and users use it to specify their requirements and any changes that may arise. Managers use it to prepare project plans and estimates, and developers use it to understand what is required and to develop tests to validate the system. With this in mind, it is important to note that requirements are read more often than they are written. Therefore, taking the time to write them correctly, concisely, and clearly not only will save time in terms of the schedule but is also more cost-efficient and reduces the risk of costly requirements errors. Performing requirements validation, therefore, is a necessary step toward achieving that goal. Requirements validation is performed on a final draft of the requirements definition document after all input has been solicited from the system owners and users. The purpose of this activity is for the systems analyst to ensure the requirements are written correctly. Examples of errors the systems analyst might find are:

- System models that contain errors.
- Typographical or grammatical errors.
- Conflicting requirements.
- Ambiguous or poorly worded requirements.
- Lack of conformance to quality standards required for the document.

> Requirements Management

Over the lifetime of the project it is very common for new requirements to emerge and existing requirements to change after a requirements definition document has been approved. Some studies have shown that over the life of a project as much as

50 percent or more of the requirements will change before the system is put into production. Obviously, this can be a major headache for the development team. To help alleviate the many problems associated with changing requirements, it is necessary to perform **requirements management.** Requirements management encompasses the policies, procedures, and processes that govern how a change to a requirement is handled. In other words, it specifies how a change request should be submitted, how it is analyzed for impact to scope, schedule, and cost, how it is approved or rejected, and how the change is implemented if approved.

requirements management the process of managing change to the requirements.

Fact-Finding Techniques

In this section we present seven common fact-finding techniques:

- Sampling of existing documentation, forms, and databases.
- Research and site visits.
- Observation of the work environment.
- Questionnaires.
- Interviews.
- Prototyping.
- Joint requirements planning.

An analyst usually applies several of these techniques during a single systems project. To be able to select the most suitable technique for use in any given situation, systems analysts need to learn the advantages and disadvantages of each of the fact-finding techniques.

Fact-Finding Ethics   During fact-finding, systems analysts often come across or analyze information that is sensitive in nature. It could be a file of an aerospace company's pricing structure for a contract bid or even employee profiles, including salaries, performance history, medical history, and career plans. Analysts must take great care to protect the security and privacy of any facts or data with which they have been entrusted. Many people and organizations in this highly competitive atmosphere are looking for an "edge" to get ahead. Careless system analysts who leave sensitive documents in plain view on their desks, or publicly discuss sensitive data could cause great harm to the organization or to individuals. If such data should fall into the wrong people's hands, the systems analyst may lose the respect, credibility, or confidence of users and management. In some cases, the analyst would be responsible for the invasion of a person's privacy and could be liable.

Most corporations make every effort to ensure they conduct business in an ethical manner because the laws may require them to. There have been many cases where corporations have incurred heavy fines for not conducting business properly. To this end, many corporations require that their employees attend annual training seminars on company ethics, and they reinforce the learning by displaying banners or signs that contain the company's code of conduct and ethics statements throughout the workplace in highly visible locations. The company's ethics policies may be in a hard-copy format that is distributed to all employees, or they may be on the company's Web pages, making them easily accessible to employees no matter where they are currently located. Ethics policies document expected and required behavior. Violations of these policies could lead to disciplinary action or even termination. Ethics play a crucial role in fact-finding.

Sampling of Existing Documentation, Forms, and Files

When studying an existing system, systems analysts develop a pretty good feel for the system by studying existing documentation, forms, and files. A good analyst always knows to get facts first from existing documentation rather than from people.

Collecting Facts from Existing Documentation What kind of documents can teach you about a system? The first document the analyst may wish to seek out is the organization chart. An organization chart serves to identify key individual owners and users for a project and their reporting relationships. The analyst may also want to trace the history that led to the project. To accomplish this, the analyst should collect and review documents that describe the problem. These include:

- Interoffice memoranda, studies, minutes, suggestion box notes, customer complaints, and reports that document the problem area.
- Accounting records, performance reviews, work measurement reviews, and other scheduled operating reports.
- Information systems project requests—past and present.

In addition to documents that describe the problem, there are usually documents that describe the business function being studied or designed. These documents may include:

- The company's mission statement and strategic plan.
- Formal objectives for the organization subunits being studied.
- Policy manuals that may place constraints on any proposed system.
- Standard operating procedures (SOPs), job outlines, or task instructions for specific day-to-day operations.
- Completed forms that represent actual transactions at various points in the processing cycle.
- Samples of manual and computerized databases.
- Samples of manual and computerized screens and reports.

Also, analysts often check for documentation of previous system studies and designs performed by former systems analysts and consultants. This documentation may include:

- Various types of flowcharts and diagrams.
- Project dictionaries or repositories
- Design documentation, such as inputs, outputs, and databases.
- Program documentation.
- Computer operations manuals and training manuals.

All documentation collected should be analyzed to determine whether or not the information is current. Outdated documentation should not be discarded; however, analysts should keep in mind that additional fact-finding will be needed to verify or update the facts collected. What is the analyst looking for in all this material? Things that can be gleaned from these documents include:

- The symptoms and (possibly) causes of the problem.
- What persons in the organization have an understanding of the problem.
- The business functions that support the present system.
- The type of data that needs to be collected and reported by the system.
- Things in the documentation that the analyst does not understand and so need to be covered in interviews.

Document and File Sampling Techniques Because it would be impractical to study every occurrence of every form or record in a file or database, system analysts normally use **sampling** techniques to get a large-enough cross section to determine what can happen in the system. The systems analyst should seek to sample enough forms to represent the full nature and complexity of the data. Experienced analysts avoid the pitfalls of sampling blank forms—blank forms tell little about how the form is actually used, when it is not used, or how it is often misused. When studying documents or records from a database table, analysts should study enough samples to identify all the possible processing conditions and exceptions. Statistical sampling techniques can be used to determine if the sample size is large enough to be representative of the total population of records or documents.

sampling the process of collecting a representative sample of documents, forms, and records.

TABLE 6-3 Partial Table of Certainty Factors

Desired Certainty	Certainty Factor
95%	1.960
90	1.645
80	1.281

There are many sampling issues and factors, and this is a good reason for taking an introductory statistics course. One simple and reliable formula for determining sample size is

Sample size = 0.25 × (Certainty factor/Acceptable error)^2

The certainty factor is a value that can simply be looked up in statistical tables based on the desired certainty that the sample selected will be representative of the total population. See Table 6-3 for a partial example.

Suppose an analyst wants to be 90-percent certain that a sample of invoices will contain no unsampled variations. The sample size, SS, is calculated as follows:

$$SS = 0.25(1.645/0.10)^2 = 68$$

The analyst will need to sample 68 invoices to get the desired accuracy. If a higher level of certainty is desired, a larger number of invoices are needed.

If the analyst knows from experience that 1 in every 10 invoices varies from the norm, then he or she can replace the heuristic 0.25 with $p(1 - p)$ where p is the proportion of invoices with variances:

$$SS = p(1 - p)(1.645/0.10)^2$$

By using this formula, the analyst can reduce the number of samples required to get the desired accuracy:

$$SS = 0.10(1 - 0.10)(1.645/0.10)^2 = 25$$

How are the 25 invoices chosen? Two commonly used sampling techniques are randomization and stratification. **Randomization** involves randomly, or without concern, selecting sample data. Therefore, we just randomly choose 25 invoices based on the sample size calculated above. **Stratification** is a thoughtful and systematic approach aimed at reducing the variance of the sample data. For computerized files, stratification sampling can be executed by writing a simple program. For instance, suppose our invoices are in a database that has a volume of approximately 250,000 invoices. Recall that our sample size needs to include 25 invoices. We will simply write a program that prints every 10,000th record (= 250,000/25). For manual files and documents, we could execute a similar scheme.

randomization a sampling technique characterized by having no predetermined pattern or plan for selecting sample data.

stratification a systematic sampling technique that attempts to reduce the variance of estimates by spreading out the sampling—for example, choosing documents or records by formula—and by avoiding very high or very low estimates.

> Research and Site Visits

A second fact-finding technique is thoroughly researching the problem domain. Most problems are not completely unique. Other people have solved them before us. Many times organizations contact or perform site visits with companies they know have previously experienced similar problems. If these companies are "willing to share," valuable information can be obtained that may save tremendous time and cost in the development process.

Computer trade journals and reference books are a good source of information. They can provide information on how others have solved similar problems. With recent advances in cyberspace, analysts rarely have to leave their desks to do research.

Exploring the Internet and intranet via personal computer can provide immeasurable amounts of information.

A similar type of research involves visiting other companies or departments that have addressed similar problems. Memberships in professional societies such as the Association for Information Technology Professionals (AITP) or Association for Information Systems (AIS), among others, can provide a network of useful contacts.

> Observation of the Work Environment

observation a fact-finding technique wherein the systems analyst either participates in or watches a person perform activities to learn about the system.

Observation is one of the most effective data-collection techniques for learning about a system. **Observation** involves the systems analyst becoming an observer of people and activities in order to learn about the system. This technique is often used when the validity of data collected through other methods is in question or when the complexity of certain aspects of the system prevents a clear explanation by the end users.

Collecting Facts by Observing People at Work Even with a well-conceived observation plan, the systems analyst is not assured that fact-finding will be successful. The following story, which appears in a book by Gerald M. Weinberg called *Rethinking Systems Analysis and Design,* gives us an entertaining yet excellent example of some of the pitfalls of observation.[3]

The Railroad Paradox

About thirty miles from Gotham City lay the commuter community of Suburbantown. Each morning, thousands of Surburbanites took the Central Railroad to work in Gotham City. Each evening, Central Railroad returned them to their waiting spouses, children, and dogs.

Suburbantown was a wealthy suburb, and many of the spouses liked to leave the children and dogs and spend an evening in Gotham City with their mates. They preferred to precede their evening of dinner and theater with browsing among Gotham City's lush markets. But there was a problem. To allow time for proper shopping, a Suburbanite would have to depart for Gotham City at 2:30 or 3:00 in the afternoon. At that hour, no Central Railroad train stopped in Suburbantown.

Some Suburbanites noted that a Central train did pass through their station at 2:30, but did not stop. They decided to petition the railroad, asking that the train be scheduled to stop at Suburbantown. They readily found supporters in their door-to-door canvass. When the petition was mailed, it contained 253 signatures. About three weeks later, the petition committee received the following letter from the Central Railroad:

Dear Committee

Thank you for your continuing interest in Central Railroad operations. We take seriously our commitment to providing responsive service to all the people living among our routes, and greatly appreciate feedback on all aspects of our business. In response to your petition, our customer service representative visited the Suburbantown station on three separate days, each time at 2:30 in the afternoon. Although he observed with great care, *on none of the three occasions were there any passengers waiting for a southbound train.*

We can only conclude that there is no real demand for a southbound stop at 2:30, and must therefore regretfully decline your petition.

Yours sincerely,

Customer Service Agent

Central Railroad

---

[3]Gerald M. Weinberg, *Rethinking Systems Analysis and Design,* pp. 23–24. Copyright © 1988, 1982 by Gerald M. Weinberg. Reprinted by permission of Dorset House Publishing, 353 W. 12th St., New York, NY 10014 (212-620-4053/ 800-DH-BOOKS/www.dorsethouse.com). All rights reserved.

What are the lessons learned form the story above? For one, it is necessary to use the appropriate fact-finding technique for the problem at hand. Observation, in this case, was an incorrect choice. Why would anyone be waiting for a 2:30 train when all the town's people knew the train doesn't stop? A second lesson to be learned is to verify fact-finding results with users. Based on the user feedback, you may discover that you need to try other fact-finding techniques to gather additional information. Never jump to conclusions.

Observation Advantages and Disadvantages   Observation can be a very useful and beneficial fact-finding technique provided that you have the ability to observe all aspects of the work being performed by the users and that the work is being performed in the usual manner. You should become aware of the pros and cons of the technique of observation. Advantages and disadvantages include:

Advantages

- Data gathered based on observation can be very reliable. Sometimes, observations are conducted to check the validity of data obtained directly from individuals.
- The systems analyst is able to see exactly what is being done. Complex tasks are sometimes difficult to clearly explain in words. Through observation, the systems analyst can identify tasks that have been missed or inaccurately described by other fact-finding techniques. Also, the analyst can obtain data describing the physical environment of the task (e.g., physical layout, traffic, lighting, noise level).
- Observation is relatively inexpensive compared with other fact-finding techniques. Other techniques usually require substantially more employee release time and copying expenses.
- Observation allows the systems analyst to do work measurements.

Disadvantages

- Because people usually feel uncomfortable when being watched, they may unwittingly perform differently when being observed.
- The work being observed may not involve the level of difficulty or volume normally experienced during that time period.
- Some systems activities may take place at odd times, causing a scheduling inconvenience for the systems analyst.
- The tasks being observed are subject to various types of interruptions.
- Some tasks may not always be performed in the manner in which they are observed by the systems analyst. For example, the systems analyst might have observed how a company filled several customer orders. However, the procedures the systems analyst observed may have been the steps used to fill a number of regular customer orders. If any of those orders had been special orders (e.g., an order for goods not normally kept in stock), the systems analyst would have observed a different set of procedures being executed.
- If people have been performing tasks in a manner that violates standard operating procedures, they may temporarily perform their jobs correctly while you are observing them. In other words, people may let you see what they want you to see.

Guidelines for Observation   How does the systems analyst obtain facts through observation? Does one simply arrive at the observation site and begin recording everything that's viewed? No. Much preparation should take place in advance. The analyst must determine how data will actually be captured. Will it be necessary to have special forms on which to quickly record data? Will the individuals being observed be bothered by having someone watch and record their actions? When are the low, normal, and peak periods of operations for the task to be observed? The systems analyst must identify the ideal time to observe a particular aspect of the system.

An analyst should plan to observe a site when there is a typical workload. Once a typical workload has been observed, observations can be made during peak periods to gather information for measuring the effects caused by the increased volume. As part of the analyst's observation, he or she should obtain samples of documents or forms used by those being observed.

The sampling techniques discussed earlier are also useful for observation. In this case, the technique is called **work sampling,** wherein a large number of observations may be conducted at random intervals. This technique is less threatening to the people being observed because the observation period is not continuous. When using work sampling, an analyst needs to predefine the operations of the job to be observed. A sample size then needs to be calculated as was done for document and file sampling. The analyst should make many random observations, being careful to observe activities at different times of the day. By counting the number of occurrences of each operation during the observations, the analyst will get a feel for how employees spend their days.

The following guidelines are key to honing observation skills:

- Determine the who, what, where, when, why, and how of the observation.
- Obtain permission to observe from appropriate supervisors or managers.
- Inform those who will be observed of the purpose of the observation.
- Keep a low profile.
- Take notes during or immediately following the observation.
- Review observation notes with appropriate individuals.
- Don't interrupt individuals at work.
- Don't focus heavily on trivial activities.
- Don't make assumptions.

Living the System   In this type of observation the systems analyst actively performs the role of the user for a short period of time. This is one of the most effective ways to learn about problems and requirements of the system. By filling the user's "shoes," a systems analyst quickly gains an appreciation for what the user experiences and what she or he has to do to perform the job. This type of role playing gives the systems analyst a firsthand education in the business processes and functions, as well as the problems and challenges associated with them.

> Questionnaires

Another fact-finding technique is conducting surveys through **questionnaires.** The document can be mass-produced and distributed to respondents, who can then complete the questionnaire on their own time. Questionnaires allow the analyst to collect facts from a large number of people while maintaining uniform responses. When dealing with a large audience, no other fact-finding technique can tabulate the same facts as efficiently.

Collecting Facts by Using Questionnaires   Systems analysts have often criticized the use of questionnaires. Many systems analysts claim that the responses lack reliable and useful information. Nevertheless, questionnaires can be an effective means of fact gathering, and many of these criticisms can be attributed to the inappropriate or

---

work sampling a fact-finding technique that involves a large number of observations taken at random intervals.

questionnaire a document that allows the analyst to collect information and opinions from respondents.

ineffective use of the questionnaires. Before using questionnaires, an analyst should understand the pros and cons associated with their use:

Advantages	**Disadvantages**
• Most questionnaires can be answered quickly. People can complete and return questionnaires at their convenience.	• The number of respondents is often low.
• Questionnaires are a relatively inexpensive means of gathering data from a large number of individuals.	• There's no guarantee that an individual will answer or expand on all of the questions.
• Questionnaires allow individuals to maintain anonymity. Therefore, individuals are more likely to provide the real facts, rather than telling you what they think their boss would want them to.	• Questionnaires tend to be inflexible. There's no opportunity for the systems analyst to obtain voluntary information from individuals or to reword questions that may have been misinterpreted.
• Responses can be tabulated and analyzed quickly.	• It's not possible for the systems analyst to observe and analyze the respondent's body language.
	• There is no immediate opportunity to clarify a vague or incomplete answer to any question.
	• Good questionnaires are difficult to prepare.

Types of Questionnaires   There are two formats for questionnaires: free format and fixed format. **Free-format questionnaires** are designed to allow the users to exercise more freedom or latitude in their answers to each question.

Here are two examples of free-format questions:

- What reports do you currently receive and how are they used?
- Are there any problems with these reports (e.g., are they inaccurate, is there insufficient information, or are they difficult to read and/or use)? If so, please explain.

Obviously, responses to such questions may be difficult to tabulate. It is also possible that the respondents' answers may not match the questions asked. In order to ensure useful responses in free-format questionnaires, the analyst should phrase the questions in simple sentences and not use words—such as *good*—that can be interpreted differently by different respondents. The analyst should also ask questions that can be answered with three or fewer sentences. Otherwise, the questionnaire may take up more time than the respondent is willing to sacrifice.

The second type of questionnaire is fixed-format. **Fixed-format questionnaires** are more rigid, requiring that the user select an answer from a predefined set of possible answers. Given any question, the respondent must choose from the available answers. This makes the results much easier to tabulate. On the other hand, the respondent cannot provide additional information that might prove valuable.

There are three types of fixed-format questions:

1. For *multiple-choice questions,* the respondent is given several answers from which to choose. The respondent should be told if more than one answer can be selected. Some multiple-choice questions allow for very brief free-format responses when none of the standard answers apply. Examples of multiple-choice fixed-format questions are:

   Do you feel that backorders occur too frequently?
   ❏ YES   ❏ NO

   Is the current accounts receivable report that you receive useful?
   ❏ YES   ❏ NO

   If no, please explain.

free-format questionnaire a questionnaire designed to offer the respondent greater latitude in the answer. A question is asked, and the respondent records the answer in the space provided after the question.

fixed-format questionnaire a questionnaire containing questions that require selecting an answer from predefined available responses.

2. For *rating questions,* the respondent is given a statement and asked to use supplied responses to state an opinion. To prevent built-in bias, there should be an equal number of positive and negative ratings. The following is an example of a rating fixed-format question:

The implementation of quantity discounts would cause an increase in customer orders.
❑ Strongly agree
❑ Agree
❑ No opinion
❑ Disagree
❑ Strongly disagree

3. For *ranking questions,* the respondent is given several possible answers, which are to be ranked in order of preference or experience. An example of a ranking fixed-format question is:

Rank the following transactions according to the amount of time you spend processing them:
_____ new customer orders
_____ order cancellations
_____ order modifications
_____ payments

Developing a Questionnaire   Good questionnaires can be difficult to develop. The following procedure can prove helpful in developing an effective questionnaire:

1. Determine what facts and opinions must be collected and from whom you should get them. If the number of people is large, consider using a smaller, randomly selected group of respondents.
2. Based on the facts and opinions sought, determine whether free- or fixed-format questions will produce the best answers. A combination format that permits optional free-format clarification of fixed-format responses is often used.
3. Write the questions. Examine them for construction errors and possible misinterpretations. Make sure that the questions don't reveal your personal bias or opinions. Edit the questions.
4. Test the questions on a small sample of respondents. If your respondents had problems with them or if the answers were not useful, edit the questions.
5. Duplicate and distribute the questionnaire.

> Interviews

interview a fact-finding technique whereby the systems analyst collects information from individuals through face-to-face interaction.

The personal interview is generally recognized as the most important and most often used fact-finding technique. Personal **interviews** involve soliciting requirements through direct, face-to-face interaction. Interviewing can be used to achieve any or all of the following goals: find facts, verify facts, clarify facts, generate enthusiasm, get the end user involved, identify requirements, and solicit ideas and opinions. There are two roles assumed in an interview. The systems analyst is the *interviewer,* responsible for organizing and conducting the interview. The system user or system owner is the *interviewee,* who is asked to respond to a series of questions.

There may be one or more interviewers and/or interviewees. In other words, interviews may be conducted one-on-one or many-to-many. Unfortunately, many systems analysts are poor interviewers. In this section you will learn how to conduct proper interviews.

Collecting Facts by Interviewing Users People are the most important element of an information system. More than anything else, people want to be in on things. No other fact-finding technique places as much emphasis on people as interviews. But people have different values, priorities, opinions, motivations, and personalities. Therefore, to use the interviewing technique, a systems analyst must possess good human relations skills for dealing effectively with different types of people. As with other fact-finding techniques, interviewing isn't the best method for all situations. Interviewing has its advantages and disadvantages, which should be weighed against those of other fact-finding techniques for every fact-finding situation:

Advantages

- Interviews give the analyst an opportunity to motivate the interviewee to respond freely and openly to questions. By establishing rapport, the systems analyst is able to give the interviewee a feeling of actively contributing to the systems project.
- Interviews allow the systems analyst to probe for more feedback from the interviewee.
- Interviews permit the systems analyst to adapt or reword questions for each individual.
- Interviews give the analyst an opportunity to observe the interviewee's nonverbal communication. A good systems analyst may be able to obtain information by observing the interviewee's body movements and facial expressions as well as by listening to verbal replies to questions.

Disadvantages

- Interviewing is a very time-consuming, and therefore a costly, fact-finding approach.
- Success of interviews is highly dependent on the systems analyst's human relations skills.
- Interviewing may be impractical due to the location of interviewees.

Interview Types and Techniques There are two types of interviews, unstructured and structured. **Unstructured interviews** are characterized as involving general questions that allow the interviewee to direct the conversation. This type of interview frequently gets off track, and the analyst must be prepared to redirect the interview back to the main goal or subject. For this reason, unstructured interviews don't usually work well for systems analysis and design. **Structured interviews** involve the interviewer asking specific questions designed to elicit specific information from the interviewee. Depending on the interviewee's responses, the interviewer will direct additional questions to obtain clarification or amplification. Some of these questions may be planned and others spontaneous.

Unstructured interviews tend to involve asking **open-ended questions.** Such questions give the interviewees significant latitude in their answers. An example of an open-ended question is "Why are you dissatisfied with the report of uncollectable accounts?" Structured interviews tend to involve asking more **closed-ended questions** that are designed to elicit short, direct responses from the interviewee. Examples of such questions are "Are you receiving the report of uncollectable accounts on time?" and "Does the report of uncollectable accounts contain accurate information?" Realistically, most questions fall between the two extremes.

unstructured interview an interview that is conducted with only a general goal or subject in mind and with few, if any, specific questions. The interviewer counts on the interviewee to provide a framework and direct the conversation.

structured interview an interview in which the interviewer has a specific set of questions to ask of the interviewee.

open-ended question a question that allows the interviewee to respond in any way that seems appropriate.

closed-ended question a question that restricts answers to either specific choices or short, direct responses.

> How to Conduct an Interview

A systems analyst's success is at least partially dependent on the ability to interview. A successful interview will involve selecting appropriate individuals to interview, preparing extensively for the interview, conducting the interview properly, and following up on the interview. Here we examine each of these steps in more detail. Let's assume that the analyst has identified the need for an interview and has determined exactly what kinds of facts and opinions are needed.

Select Interviewees The systems analyst should interview the end users of the information system being studied. A formal organization chart will help identify these individuals and their responsibilities. The analyst should attempt to learn as much as possible about each individual prior to the interview, such as the person's strengths, fears, biases, and motivations. The interview can then be geared to take the characteristics of the individual into account.

The analyst should make an appointment with the interviewee and never just drop in. The appointment should be limited to somewhere between a half hour and an hour. The higher the management level of the interviewee, the less time should be scheduled. If the interviewee is a clerical, service, or blue-collar worker, the analyst must get the supervisor's permission before scheduling the interview. It is also important to ensure that the location for the interview will be available during the time it is scheduled. Interviews should never be conducted in the presence of the analyst's officemates or the interviewee's peers.

Prepare for the Interview Preparation is the key to a successful interview. An interviewee can easily detect when an interviewer is unprepared and may resent the lack of preparation because it wastes valuable time. When the appointment is made, the interviewee should be notified about the subject of the interview. To ensure that all pertinent aspects of the subject are covered, the analyst should prepare an *interview guide.* The interview guide is a checklist of specific questions the interviewer will ask the interviewee. The interview guide may also contain follow-up questions that will be asked only if the answers to other questions warrant the additional answers. A sample interview guide is presented in Figure 6-3. Notice that the agenda is carefully laid out with the specific time allocated to each question. Time should also be reserved for asking follow-up questions and redirecting the interview. Questions should be carefully chosen and phrased. Most questions begin with the standard who, what, when, where, why, and how much type of wording. The following types of questions should be avoided:

- *Loaded questions,* such as "Do we have to have both of these columns on the report?" The question conveys the interviewee's personal opinion on the issue.
- *Leading questions,* such as "You're not going to use this OPERATOR CODE, are you?" The question leads the interviewee to respond, "No, of course not," regardless of actual opinion.
- *Biased questions,* such as "How many codes do we need for FOOD CLASSIFICATION in the INVENTORY FILE? I think 20 ought to cover it." These types of biased questions will influence an interviewee.

Interviewers should always avoid threatening or critical questions. The purpose of the interview is to investigate, not to evaluate or criticize. Additional guidelines for questions include:

- Use clear and concise language.
- Don't include your opinion as part of the question.
- Avoid long or complex questions.
- Avoid threatening questions.
- Don't use "you" when you mean a group of people.

	Interviewee:	Jeff Bentley, Accounts Receivable Manager
	Date:	January 19, 2003
	Time:	1:30 P.M.
	Place:	Room 223, Admin. Bldg.
	Subject:	Current Credit-Checking Policy

Time Allocated	Interviewer Question or Objective	Interviewee Response
1 to 2 min.	**Objective** Open the interview: • Introduce ourselves. • Thank Mr. Bentley for his valuable time. • State the purpose of the interview — to obtain an understanding of the existing credit-checking policies.	
5 min.	**Question 1** What conditions determine whether a customer's order is approved for credit? **Follow-up**	
5 min.	**Question 2** What are the possible decisions or actions that might be taken once these conditions have been evaluated? **Follow-up**	
3 min.	**Question 3** How are customers notified when credit is not approved for their order? **Follow-up**	
1 min.	**Question 4** After a new order is approved for credit and placed in the file containing orders that can be filled, a customer might request that a modification be made to the order. Would the order have to go through credit approval again if the new total order cost exceeds the original cost? **Follow-up**	
1 min.	**Question 5** Who are the individuals who perform the credit checks? **Follow-up**	
1 to 3 min.	**Question 6** May I have permission to talk to those individuals to learn specifically how they carry out the credit-checking process? **Follow-up** If so: When would be an appropriate time to meet with each of them?	
1 min.	**Objective** Conclude the interview: • Thank Mr. Bentley for his cooperation and assure him that he will be receiving a copy of what transpired during the interview.	
21 minutes	Time allotted for questions and objectives	
9 minutes	Time allotted for follow-up questions and redirection	
30 minutes	Time allotted for interview (1:30 p.m. - 2:00 p.m.)	
General Comments and Notes:		

FIGURE 6-3 Sample Interview Guide

Conduct the Interview   Respect your interviewee and his or her time. Dress to match the interviewee. That generally means that you will dress differently to interview managers than you will to interview workers on the loading dock. If the interview will be held in a meeting room other than the interviewee's office, arrive early to make sure it is set up appropriately.

Open the interview by thanking the interviewee in advance. State the purpose and length of the interview and how the gathered data will be used. Then monitor the time so you will keep your promise.

Ask follow-up questions. Probe until you understand the system requirements. Ask especially about exception conditions. As what-if questions, such as "What if the check doesn't clear?" or "What happens if a product is not in stock?"

Listen closely, observe the interviewee, and take notes concerning both verbal and nonverbal responses from the interviewee. It's very important to keep the interview on track; this means anticipating the need to adapt the interview, if necessary. Questions can often be bypassed if they have been answered earlier or they can be deleted if determined to be irrelevant, based on previous answers.

Here is a set of rules that an interviewer should follow:

Do	**Avoid**
• Dress appropriately.	• Assuming an answer is finished or leading nowhere.
• Be courteous.	
• Listen carefully.	• Revealing verbal and nonverbal clues.
• Maintain control of the interview.	
• Probe.	• Using jargon.
• Observe mannerisms and nonverbal communication.	• Revealing your personal biases.
	• Talking instead of listening.
• Be patient.	• Assuming anything about the topic or the interviewee.
• Keep the interviewee at ease.	
• Maintain self-control.	• Tape recording—a sign of poor listening skills.
• Finish on time.	

Conclude the interview by expressing appreciation and providing answers to any questions posed by the interviewee. The conclusion is very important for maintaining rapport and trust with the interviewee.

Follow Up on the Interview   To help maintain good rapport and trust with interviewees, the interviewer should send them a memo that summarizes the interview. This memo should remind the interviewees of their contributions to the systems project and allow them the opportunity to clarify any misinterpretations that the interviewer may have derived during the interview. In addition, the interviewees should be given the opportunity to offer additional information they may have failed to bring out during the interview.

Listening   When most people talk about communication skills, they think of speaking and writing. The skill of listening is rarely mentioned, but it may be the most important skill during the interviewing process. In order to conduct a successful interview, the interviewer must make a distinction between hearing and listening: "To hear is to recognize that someone is speaking, to listen is to understand what the speaker wants to communicate."[4]

We have actually been conditioned most of our lives not to listen. Take, for example, how we can ignore our quarreling brothers and sisters while we enjoy our favorite CD or, as students, how we learn to study by blocking out distractions such as noisy roommates. We have learned not to listen, but we can also learn how to listen effectively and productively.

---

[4]Thomas R. Gildersleeve, *Successful Data Processing Systems Analysis* (Englewood Cliffs, NJ: Prentice Hall, 1978), p. 93.

When working with users trying to solve their problems, analysts may find that getting the users to communicate can be difficult. The following guidelines can help open the lines of communication:

- *Approach the session with a positive attitude.* The interviewer should make the best of any situation, and look at it as a fun, pleasurable experience.
- *Set the other person at ease.* Presenting a nice, cheerful attitude can help the person relax. The interviewer should start by talking about the person's interests or hobbies. Showing an interest in the interviewee's personal life sometimes can serve as an icebreaker and put the person more at ease.
- *Let the other person know you are listening.* The interviewer should always maintain eye contact when listening and use a response such as a head nod or an "uh-huh" to acknowledge what the person is saying. Good posture and leaning forward will tell the speaker that the interviewer is really interested in what the person is saying.
- *Ask questions.* The interviewer should ask questions to make sure he or she clearly understands what the person is saying or to clarify a point. This will show that the interviewer is listening; it will also give the other person the opportunity to expand on the answer.
- *Don't assume anything.* One of the worst things an interviewer can do is to act as if he or she is in a hurry. For example, if an interviewer assumes what the other person is going to say and cuts in and finishes the sentence, he or she will possibly miss what the person intended to say and irritate the speaker. Or if the speaker is interrupted because the interviewer has already heard the information and believes it is not applicable to the topic of the interview, a valuable piece of information may be missed. Don't assume anything! TV host Art Linkletter learned this lesson on his popular television show, *Kids Say the Darnedest Things,* when he asked a child a philosophical question:

  On my show I once had a child tell me he wanted to be an airline pilot. I asked him what he'd do if all the engines stopped out over the Pacific Ocean. He said "First I would tell everyone to fasten their seatbelts, and then I'd find my parachute and jump out."

  While the audience rocked with laughter, I kept my attention on the young man to see if he was being a smart alec. The tears that sprang into his eyes alerted me to his chagrin more than anything he could have said, so I asked him why he'd do such a thing. His answer revealed the sound logic of a child: "I'm going for gas . . . I'm coming back!"[5]

- *Take notes.* The process of taking notes serves two purposes. First, by jotting down brief notes while the other person is speaking, you give the person the impression that what he or she has to say is important enough to be written down. Second, the notes help the interviewer remember the major points of the meeting later.

Body Language and Proxemics

What is body language, and why should a systems analyst care about it during the interviewing process? **Body language** is all the non-verbal information being communicated by an individual. Body language is a form of nonverbal communication that we all use and of which we are usually unaware.

Studies have determined a startling fact: Of a person's total feelings, only 7 percent are communicated verbally (in words), whereas 38 percent are communicated by the tone of voice used and 55 percent are communicated by facial and body expressions. If you only listen to someone's words, you are missing most of what the person has to say.

body language the nonverbal information we communicate.

---

[5]Donald Walton, *Are You Communicating? You Can't Manage without It.* (New York: McGraw-Hill, 1989), p. 31.

For this discussion, we will focus on just three aspects of body language: facial disclosure, eye contact, and posture. *Facial disclosure* means you can sometimes understand how a person feels by watching the expressions on his or her face. Many common emotions have easily recognizable facial expressions associated with them. However, the face is one of the most controlled parts of the body. Some people who are aware that their expressions often reveal what they are thinking are very good at disguising these expressions.

Another form of nonverbal communication is *eye contact.* Eye contact is the least controlled aspect of facial expression. Have you ever spoken to someone who will not look directly at you? How did it make you feel? A continual lack of eye contact may indicate uncertainty. A normal glance is usually from three to five seconds in length; however, direct-eye-contact time should increase with distance. Analysts need to be careful not to use excessive eye contact with users who seem threatened so that they won't further intimidate them. Direct eye contact can cause strong feelings, either positive or negative, in other people.

Posture is the least controlled aspect of the body. As such, body posture holds a wealth of information for the astute analyst. Members of a group who are in agreement tend to display the same posture. A good analyst will watch the audience for changes in posture that could indicate anxiety, disagreement, or boredom. An analyst should normally maintain an "open" body position, signaling approachability, acceptance, and receptiveness. In special circumstances, the analyst may choose to use a confrontation angle of head-on or at a 90-degree angle to another person in order to establish control and dominance.

In addition to the information communicated by body language, individuals also communicate via proxemics. **Proxemics,** the relationship between people and the space around them, is a factor in communications that can be controlled by the knowledgeable analyst.

proxemics the relationship between people and the space around them.

People still tend to be very territorial about their space. Observe where your classmates sit in one of your courses that does not have assigned seats. Or the next time you are involved in a conversation with someone, deliberately move much closer or farther away from the person and see what happens. A good analyst is aware of four spatial zones:

- *Intimate zone*—closer than 1.5 feet.
- *Personal zone*—from 1.5 feet to 4 feet.
- *Social zone*—from 4 feet to 12 feet.
- *Public zone*—beyond 12 feet.

Certain types of communications take place only in some of these zones. For example, an analyst conducts most interviews with system users in the personal zone. But the analyst may need to move back to the social zone if the user displays any signs (body language) of being uncomfortable. Sometimes increasing eye contact can make up for a long distance that can't be changed. Many people use the fringes of the social zone as a "respect" distance.

We have examined some of the informal ways that people communicate their feelings and reactions. A good analyst will use all the information available, not just the written or verbal communications of others.

> Discovery Prototyping

Another type of fact-finding technique is prototyping. Prototyping was introduced in Chapter 3 for use in rapid application development (RAD). As you should recall, the concept behind prototyping is building a small working model of the users' requirements or a proposed design for an information system. This type of prototyping is usually a design technique, but the approach can be applied earlier in the system development life cycle to perform fact-finding and requirements analysis. The process of building a prototype for the purpose of identifying requirements is referred to as **discovery prototyping.**

Discovery prototyping is frequently applied to systems development projects, especially in cases where the development team is having problems defining the system requirements. The philosophy is that the users will recognize their requirements when they see them. It is important that the prototype be developed quickly so that it can be used during the development process. Usually, only the areas where the requirements are not clearly understood are prototyped. This means that a lot of desired functionality may be left out and quality assurance may be ignored. Also, nonfunctional requirements such as performance and reliability may be less stringent than they would be for the final product. Technologies other than the ones used for the final software are frequently used to build the discovery prototypes. In these cases, the prototypes are most likely discarded when the system is finished. This "throwaway" approach is primarily used to gather information and develop ideas for the system concept. Many areas of a proposed system may not be clearly understood, or some features may be a technical challenge for the developers. Creating discovery prototypes enables the developers as well as the users to better understand and refine the issues involved with developing the system. This technique helps minimize the risk of delivering a system that doesn't meet the user's needs or that can't fulfill the technical requirements.

Discovery prototyping has its advantages and disadvantages, which should be weighed against those of other fact-finding techniques for every fact-finding situation:

discovery prototyping the act of building a small-scale representative or working model of the users' requirements in order to discover or verify those requirements.

Advantages	**Disadvantages**
• Allows users and developers to experiment with the software and develop an understanding of how the system might work.	• Developers may need to be trained in the prototyping approach.
• Aids in determining the feasibility and usefulness of the system before high development costs are incurred.	• Users may develop unrealistic expectations based on the performance, reliability, and features of the prototype. Prototypes can only simulate system functionality and are incomplete in nature. Care must be taken to educate the users about this fact and not to mislead them.
• Serves as a training mechanism for users.	
• Aids in building system test plans and scenarios to be used last in the system testing process.	• Doing prototyping may extend the development schedule and increase the development costs.
• May minimize the time spent on fact-finding and help define more stable and reliable requirements.	

> Joint Requirements Planning

Many organizations are using the group work session as a substitute for numerous and separate interviews. One example of the group work session approach is **joint requirements planning (JRP),** wherein highly structured group meetings are conducted for the purpose of identifying and analyzing problems and defining system requirements. This and similar techniques generally require extensive training to work as intended. However, they can significantly decrease the time spent on fact-finding in one or more phases of the life cycle. JRP is becoming increasingly common in systems planning and systems analysis to obtain group consensus on problems, objectives, and requirements. In this section, you will learn about the participants of a JRP session

joint requirements planning (JRP) a process whereby highly structured group meetings are conducted for the purpose of analyzing problems and defining requirements.

and their roles. We will also discuss how to go about planning and conducting a JRP session, the tools and techniques that are used during a JRP session, and the benefits to be achieved through JRP.

JRP Participants   Joint requirements planning sessions include a wide variety of participants and roles. Each participant is expected to attend and actively participate for the entire JRP session. Let's examine the different types of individuals involved in a typical JRP session and their roles:

- *Sponsor*—Any successful JRP session requires a single person, called the *sponsor*, to serve as its champion. This person is normally an individual who is in top management (*not* IT or IS management) and who has authority that spans the different departments and users who are to be involved in the systems project. The sponsor gives full support to the systems project by encouraging designated users to willingly and actively participate in the JRP session. Recalling the "creeping commitment" approach to systems development, it is the sponsor (system owner) who usually makes final decisions regarding the go or no-go direction of the project.

    The sponsor plays a visible role during a JRP session by kicking off the meeting by introducing the participants. Often, the sponsor will also make closing remarks for the session. The sponsor also works closely with the JRP leader to plan the session by helping identify individuals from the user community who should attend and determining the time and location for the JRP session.

- *Facilitator*—JRP sessions involve a single individual who plays the role of the leader or facilitator. The *JRP facilitator* is usually responsible for leading all sessions that are held for a systems project. This individual is someone who has excellent communication skills, possesses the ability to negotiate and resolve group conflicts, has a good knowledge of the business, has strong organizational skills, is impartial to decisions that will be addressed, and does not report to any of the JRP session participants.

    It is sometimes difficult to find an individual within the company who possesses all these traits. Thus, companies often must provide extensive JRP training or hire an expert from outside the organization to fill this role. Many systems analysts are trained to become JRP facilitators.

    The role of the JRP facilitator is to plan the JRP session, conduct the session, and follow through on the results. During the session, the facilitator is responsible for leading the discussion, encouraging the attendees to actively participate, resolving issue conflicts that may arise, and ensuring that the goals and objectives of the meeting are fulfilled. It is the JRP facilitator's responsibility to establish the ground rules that will be followed during the meeting and ensure that the participants abide by these rules.

- *Users and managers*—Joint requirements planning includes a number of participants from the user and management sectors of an organization who are given release time from their day-to-day jobs to devote themselves to active involvement in the JRP sessions. These participants are normally chosen by the project sponsor, who must be careful to ensure that each person has the business knowledge to contribute during the fact-finding sessions. The project sponsor must exercise authority and encouragement to ensure that these individuals will be committed to actively participating.

    A typical JRP session may involve anywhere from a relatively small number of user/management people to a dozen or more. The role of the users during a JRP session is to effectively communicate business rules and requirements, review design prototypes, and make acceptance decisions. The role of the managers during a JRP session is to approve project objectives, establish project priorities, approve schedules and costs, and approve identified training needs and implementation plans. Overall, both users and managers are relied on to ensure that their critical success factors are being addressed.

- *Scribe(s)*—A JRP session also includes one or more *scribes,* who are responsible for keeping records pertaining to everything discussed in the meeting. These records are published and disseminated to the attendees immediately following the meeting in order to maintain the momentum that has been established by the JRP session and its members. The need to quickly publish the records is reflected by the fact that scribes are increasingly using CASE tools to capture many facts (documented using data and process models) that are communicated during a JRP session. Thus, it is advantageous for scribes to possess strong knowledge of systems analysis and design and be skilled with using CASE tools. Systems analysts frequently play this role.

- *IT staff*—A JRP session may also include a number of IT personnel who primarily listen and take notes regarding issues and requirements voiced by the users and managers. Normally, IT personnel do not speak up unless invited to do so. Rather, any questions or concerns they have are usually directed to the JRP facilitator immediately after or before the JRP session. It is the JRP facilitator who initiates and facilitates discussion of issues by users and managers.

  The IT staff in the JRP session usually consists of members of the project team. These members may work closely with the scribe to develop models and other documentation related to facts communicated during the meeting. Specialists may also be called on to gain information regarding special technical issues and concerns that may arise. When the situation warrants, the JRP facilitator may prompt an IT professional to address the technical issue.

How to Plan JRP Sessions   Most JRP sessions span three to five days and occasionally last up to two weeks. The success of any JRP session depends on properly planning and effectively carrying out the plan. Some preparation is necessary well before the JRP session can be performed. Before planning a JRP session, the analyst must work closely with the executive sponsor to determine the scope of the project that is to be addressed through JRP sessions. It is also important that the high-level requirements and expectations of each JRP session be determined. This normally involves interviewing selected individuals who are responsible for departments or functions that are to be addressed by the systems project. Finally, before planning the JRP session, the analyst must ensure that the executive sponsor is willing to commit people, time, and other resources to the session.

Planning for a JRP session involves three steps: selecting a location for the JRP session, selecting JRP participants, and preparing an agenda to be followed during the JRP session. Let's examine each of these planning steps in detail:

1. *Selecting a location for JRP sessions*—When possible, JRP sessions should be conducted away from the company workplace. Most local hotels or universities have facilities designed to host group meetings. By holding the JRP session at an off-site location, the attendees can concentrate on the issues and activities related to the JRP session and avoid interruptions and distractions that would occur at their regular workplace. Regardless of the location of the JRP session, all attendees should be required to attend and be prohibited from returning to their regular workplaces.

   A JRP session typically requires several rooms. A conference room is required in which the entire group can meet to address JRP issues. Also, if the JRP session includes many people, several small breakout rooms may be needed for separate groups of people to meet and focus discussion on specific issues.

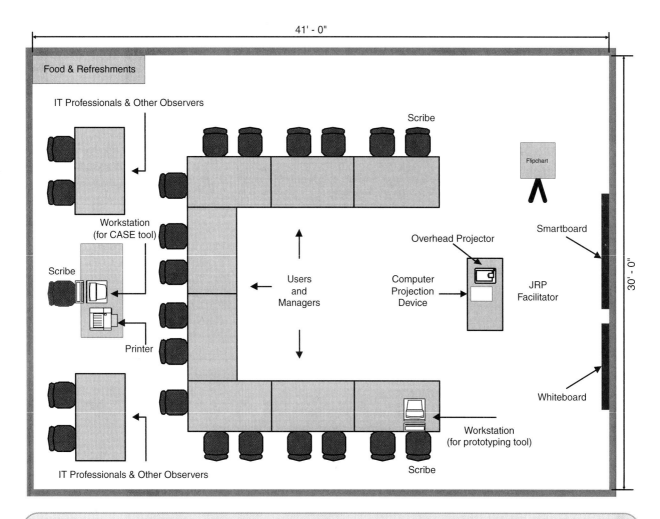

FIGURE 6-4 Typical Room Layout for JRP Session

The conference or main meeting room should comfortably hold all the attendees. The room should be fully equipped with tables, chairs, and other items that meet the needs of all attendees. Figure 6-4 depicts a typical room layout for a JRP session. Typical visual aids for a JRP room should include a whiteboard, smartboard, or blackboard; one or more flipcharts; and one or more projectors.

The room should also include computer equipment needed by scribes to record facts and issues communicated during the session. The computer itself should include software packages to support the various types of records or documentation to be captured and later published by the scribes. Such software may include CASE tool, word processing, spreadsheet, presentation package, prototyping software, printer, copier (or quick access), and computer projection capability. As a guideline, computer equipment (except that used for prototyping) should be located at the rear of the room so that it doesn't interfere or become a distraction for the JRP participants. Personal interaction of the participants, not technology, should be the focus of the session.

Finally, the room should be equipped for teleconferencing so that users at distant locations can participate. The room should include notepads and pencils for users, managers, and other attendees. Attendees should also be provided with nametags, place cards, snacks, and drinks so that they will be as comfortable as possible. Creature comforts are very important since JRP sessions are very intensive and often run the entire day.

2. *Selecting JRP participants*—As mentioned earlier, participants selected include the JRP facilitator, scribe(s), and representatives from the user community. The users should be key individuals who are knowledgeable about their business area. Unfortunately, managers are often very dependent on these individuals to run their business areas and are often hesitant to release them from their duties. Thus, the analyst must ensure that management is committed to the JRP project and willing to not only permit but also require these key individuals to participate.

   Various IT professionals may also be selected to be involved in the JRP session. Usually all IT individuals assigned to the project team are involved in the JRP session. Other IT specialists may also be assigned to address specific technical issues pertaining to the project.

3. *Preparing a JRP session agenda*—Preparation is the key to a successful JRP session. The JRP facilitator must prepare documentation to brief the participants about the scope and objectives of the sessions. In addition, an agenda for each JRP session should be prepared and distributed before each session. The agenda dictates issues to be discussed during the session and the amount of time allotted to each item.

   The agenda should contain three parts: the opening, body, and conclusion. The opening is intended to communicate the expectations of the session, to communicate the ground rules, and to influence or motivate the attendees to participate. The body is intended to detail the topics or issues to be addressed in the JRP session. Finally, the conclusion represents the time set aside to summarize the day's session and to remind the attendees of unresolved issues (to be carried forward).

How to Conduct a JRP Session   The JRP session begins with opening remarks, introductions, and a brief overview of the agenda and objectives for the session. The JRP facilitator will direct the session by following the prepared script. To successfully conduct the session, the facilitator should follow these guidelines:

- Do not unreasonably deviate from the agenda.
- Stay on schedule (agenda topics are allotted specific times).
- Ensure that the scribe is able to take notes (this may mean having the users and managers restate their points more slowly or clearly).
- Avoid the use of technical jargon.
- Apply conflict resolution skills.
- Allow for ample breaks.
- Encourage group consensus.
- Encourage user and management participation without allowing individuals to dominate the session.
- Make sure that attendees abide by the established ground rules for the session.

One of the goals of a JRP session is to generate possible ideas to solve a problem. One approach is brainstorming. **Brainstorming** involves encouraging participants to generate as many ideas as possible, without analyzing the validity of the ideas.

Brainstorming is a formal technique that requires discipline. The following guidelines should be used to ensure effective brainstorming:

brainstorming a technique for generating ideas by encouraging participants to offer as many ideas as possible in a short period of time without any analysis until all the ideas have been exhausted.

1. Isolate the appropriate people in a place that will be free from distractions and interruptions.
2. Make sure that everyone understands the purpose of the meeting (to generate ideas to solve the problem) and focuses on the problem(s).

3. Appoint one person to record ideas. This person should use a flipchart, chalkboard, or overhead projector that can be viewed by the entire group.
4. Remind everyone of the brainstorming rules:
   a. Be spontaneous. Call out ideas as fast as they occur.
   b. Absolutely no criticism, analysis, or evaluation of any kind is permitted while the ideas are being generated. Any idea may be useful, if only to spark another idea.
   c. The goal is quantity of ideas, not necessarily quality.
5. Within a specified time period, team members call out their ideas as quickly as they can think of them.
6. After the group has run out of ideas and all ideas have been recorded, then and only then should the ideas be analyzed and evaluated.
7. Refine, combine, and enhance the ideas that were generated earlier.

With a little practice and attention to these rules, brainstorming can be a very effective technique for generating ideas to solve problems.

As mentioned earlier, the success of a JRP session is highly dependent on planning and the skills of the JRP facilitator and scribes. These skills improve through proper training and experience. Therefore, JRP sessions are usually concluded with an evaluation questionnaire for the participants to complete. The responses will help ensure the likelihood of future JRP successes.

The end product of a JRP session is typically a formal written document. This document is usually created by the JRP facilitator and scribes. It is essential for confirming the specifications agreed on during the session by all participants. The content and organization of the specifications are obviously dependent on the objectives of the JRP session. The analyst may provide a different set of specifications to different participants based on their role—for example, a manager may receive more of a summary version of the document provided to the user participants (especially in cases in which the system owners had minimal actual involvement in the JRP session).

Benefits of JRP   Joint requirements planning offers many benefits as an alternative fact-finding and development approach. More and more companies are beginning to realize its advantages and are incorporating JRP into their existing methodologies. An effectively conducted JRP session offers the following benefits:

- JRP actively involves users and management in the development project (encouraging them to take "ownership" in the project).
- JRP reduces the amount of time required to develop systems. This is achieved by replacing traditional, time-consuming one-on-one interviewing of each user and manager with group meetings. The group meetings allow for more easily obtaining consensus among the users and managers, as well as resolving conflicting information and requirements.
- When JRP incorporates prototyping as a means for confirming requirements and obtaining design approvals, the benefits of prototyping are realized.

Achieving a successful JRP session depends on the JRP facilitator and his or her ability to plan and facilitate the JRP session.

A Fact-Finding Strategy

An analyst needs an organized method for collecting facts. Inexperienced analysts will frequently jump right into interviews. They believe, "Go to the people. That's where the real facts are!" Wrong! This approach fails to recognize an important fact of life: People must complete their day-to-day jobs. You may be thinking, "But

I thought you've been saying that the system is for people and that direct end-user involvement in systems development is essential. Aren't you contradicting yourselves?"

Not at all. Time is money. To waste end users' time is to waste their company's money. To make the most of the time spent with end users, analysts should not jump right into interviews. Analysts should first collect all the facts they can by using other methods. Consider the following step-by-step strategy:

1. Learn from existing documents, forms, reports, and files. Analysts can learn a lot without any people contact.
2. If appropriate, observe the system in action.
3. Given all the facts already collected, design and distribute questionnaires to clear up things that aren't fully understood.
4. Conduct interviews (or group work sessions). Because most of the pertinent facts have already been collected by low-user-contact methods, interviews can be used to verify and clarify the most difficult issues and problems. (Alternatively, consider using JRP techniques to replace or complement interviews.)
5. (Optional). Build discovery prototypes for any functional requirements that are not understood or for requirements that need to be validated.
6. Follow up. Use appropriate fact-finding techniques to verify facts (usually interviews or observation).

The strategy is not sacred. Although a fact-finding strategy should be developed for every pertinent phase of systems development, every project is unique. Sometimes observation and questionnaires may be inappropriate. But the idea should always be to collect as many facts as possible before using interviews.

Learning Roadmap

This chapter introduced you to a wide range of techniques for discovering information system requirements. Most systems development methodologies require some level of documentation and analysis of system requirements. Accordingly, the remaining chapters in this part present a number of systems documentation tools and techniques that can be used during the analysis phase of systems development. Most of you will proceed directly to Chapter 7, "Modeling System Requirements with Use Cases." Use-case models serve as a foundation for the development of subsequent models for modeling additional systems requirements and are presented in Chapters 8 through 11.

Summary

1. The process and techniques that a systems analyst uses to identify, analyze, and understand system requirements are referred to as requirements discovery.

2. System requirements specify what the information system must do, or what property or quality the system must have.

3. The process of requirements discovery consists of the following activities:

   a. Problem discovery and analysis.
   b. Requirements discovery.
   c. Documenting and analyzing requirements.
   d. Requirements management.

4. Fact-finding is a technique that is used across the entire development cycle, but it is extremely critical in the requirements analysis phase.

5. A popular tool used by development teams to identify, analyze, and solve problems is an Ishikawa diagram.

6. Conducting business in an ethical manner is a required practice, and analysts need to be more aware of the implications of not being ethical.

7. There are seven common fact-finding techniques:

   a. The sampling of existing documents and files can provide many facts and details with little or no direct personal communication being necessary. The analyst should collect historical documents, business operations manuals and forms, and information systems documents.
   b. Research is an often-overlooked technique based on the study of other similar applications. It now has become more convenient with the Internet and World Wide Web (WWW). Site visits are a special form of research.
   c. Observation is a fact-finding technique in which the analyst studies people doing their jobs.
   d. Questionnaires are used to collect similar facts from a large number of individuals.
   e. Interviews are the most popular but the most time-consuming fact-finding technique. When interviewing, the analyst meets individually with people to gather information.

      i) When most people talk about communication skills, they think of speaking and writing. The skill of listening hardly gets mentioned at all, but it may be the most important, especially during the interviewing process.
      ii) Research studies have determined a startling fact: Of a person's total feelings, only

7 percent are communicated verbally (in words), whereas 38 percent are communicated by the tone of voice used and 55 percent are communicated by facial and body expressions. If you only listen to someone's words, you are missing most of what the person has to say. Experienced systems analysts pay close attention to body language and proxemics.

   f. Discovery prototyping is frequently applied to systems development projects, especially in cases where the development team is having problems defining the system requirements. The philosophy is that the users will recognize their requirements when they see them. It is important that the prototype be developed quickly so that it can be used during the development process.
   g. Many analysts find flaws with interviewing—separate interviews often lead to conflicting facts, opinions, and priorities. The end result is numerous follow-up interviews and/or group meetings. For this reason, many organizations are using a group work session known as the joint requirements planning session as a substitute for interviews.

      i) Joint requirements planning sessions include a wide variety of participants and roles. Each participant is expected to attend and actively participate for the entire duration of the JRP session.
      ii) An effective JRP session involves extensive planning. Planning for a JRP session involves three steps: selecting a location for the JRP session, selecting JRP participants, and preparing an agenda to be followed during the JRP session.

8. To help alleviate the many problems associated with changing requirements, it is necessary to perform requirements management. Requirements management encompasses the policies, procedures, and processes that govern how a change to a requirement is handled.

9. Because "time is money," it is wise and practical for the systems analyst to use a fact-finding strategy to maximize the value of time spent with the end users.

   a. Learn from existing documents, forms, reports, and files. Analysts can learn a lot without any people contact.

b. If appropriate, observe the system in action.

c. Given all the facts already collected, design and distribute questionnaires to clear up things that aren't fully understood.

d. Conduct interviews (or group work sessions). Because most of the pertinent facts have already been collected by low-user-contact methods, interviews can be used to verify and clarify the most difficult issues and problems. (Alterna-

tively, consider using JRP techniques to replace or complement interviews.)

e. (Optional.) Build discovery prototypes for any functional requirements that are not understood or for requirements that need to be validated.

f. Follow up. Use appropriate fact-finding techniques to verify facts (usually interviews or observation).

Review Questions

1. What is the importance of conducting the requirements discovery process?

2. What are the possible consequences if you fail to identify system requirements correctly and completely?

3. What are some of the criteria deemed to be critical in defining system requirements?

4. The requirements discovery process consists of what activities?

5. Briefly describe the purpose and component parts of an Ishikawa diagram.

6. What technique is commonly used in the requirements discovery phase? Why is it important?

7. Why is analyzing requirements essential?

8. When collecting facts from existing documentation, what kind of documents should system analysts review?

9. What are some of the drawbacks of collecting facts by observing employees in their work environment? How can systems analysts deal with these drawbacks?

10. What are the types of survey questionnaires that systems analysts can use to collect information and opinions?

11. What are some of the ways that you can use to help open the lines of communication in an interview?

12. What is joint requirements planning (JRP)?

13. Why has JRP become popular?

14. Why is the facilitator in JRP so important?

15. What is the main concern in selecting a location for JRP sessions?

Problems and Exercises

1. You are managing a project that was postponed twice because its funding was diverted to higher-priority projects. The system owners do not want that to happen again, so they are very anxious to get the new system started and built as quickly as possible. They are putting a great deal of pressure on you to spend no more than a couple of days on requirements discovery. If anything is missed, they tell you, it can be fixed later on. You really want to make them happy, but a little voice of caution is going off. What are the potential consequences and costs of rushing through the requirements discovery process?

2. You have learned the importance of making sure that requirements are correctly identified. But how do you know when you have a correct requirement—that is, what criteria must each requirement meet in order to be considered correct?

3. What common error does a new systems analyst often make when analyzing a problem? What are the potential consequences of this error? What tool can be used to help avoid this problem?

4. System developers use fact-finding techniques in every project phase. Is fact-finding more important during the requirements analysis phase than for other phases? Why or why not?

5. What ethical issues might arise during the fact-finding process, and how should they be handled?

6. What are some of the common tools and techniques a systems analyst can use to document the initial findings? Should the systems analyst expect the requirements to be complete and correct at this point? If not, what are the common problems? What should be the focus of the project team at this point?

7. What is the deliverable that is created once requirements analysis is completed? Why is this deliverable needed, and what does it include? Who are the audience and/or users of this deliverable, and for what reasons?

8. You are a systems analyst in a software development company that has been hired to do the requirements analysis phase for a large organization. What are three categories of existing documentation that you should collect during requirement discovery? What are some examples of each of these three types of documentation? What should the systems analyst watch out for in collecting documentation?

9. Suppose that you are a systems analyst on a project that involves modifying the sales order process. Since your company receives in the neighborhood of 2,500 sales orders per day, how many do you need to sample if you want 95 percent certainty that you have covered all variations? What if the number of sales orders per day was 25,000 orders?

10. Surveys and questionnaires are frequently used to gather facts. What are some of the advantages and disadvantages of questionnaires? When might you choose free-format questionnaires over fixed-format questionnaires? What is one method of determining the effectiveness of a questionnaire?

11. What are some of the reasons to use joint requirements planning (JRP) as a fact-finding technique? What should be the basis for selecting which users and managers will participate in the JRP session, and who generally selects them? What skills should the facilitator and scribe possess? What is the role of IT staff during JRP sessions? What is the typical duration of the JRP sessions?

12. Provide at least five of the critical success factors for JRP sessions.

13. What one thing should an analyst *not* do when beginning the fact-finding portion of requirements discovery, no matter how tempting?

Projects and Research

1. Systems analysts must have expertise in problem analysis. When systems analysts are starting out, they often find it difficult to differentiate symptoms from problems, and to identify the actual causes of the problem. One tool that can help analysts learn to do this is the Ishikawa, or fishbone, diagram.

   a. Find and select a problem that your organization, school, or other organization is currently attempting to resolve. Describe this problem.
   b. Follow the process described in this chapter and create an Ishikawa diagram.
   c. Which categories did you start with in the diagram, and which categories did you add during the process?
   d. Did this diagram help in finding the actual cause(s) of the problem? Did the cause(s) turn out to be what you originally thought, or something different?

2. Observing the work environment is a technique that predates the information age, but that can still be highly effective. Although not applicable for every situation, observing what people actually do and how they do it can be in some cases much more accurate than asking them! Select a

system—whether hypothetical or real—and do the following:

   a. Provide an overview of the system and what you are trying to learn about the system for a project.
   b. Develop a work observation plan using the guidelines in this chapter. The format is up to you, but it generally should not need more than 1–2 pages.
   c. Develop a work-sampling plan, and describe the sampling procedures you will use.
   d. What are your thoughts about this method compared to other fact-finding methods?

3. You are a systems analyst working on a project to develop an intranet for a large organization with several thousand employees working in offices throughout the United States. This will be the organization's first intranet, and executive management wants it to help increase employee efficiency and commitment to the organization. As part of fact-finding, information needs to be gathered from employees of the organization regarding intranet content and functionality. Due to the size and geographic distribution of the organization, as well as project time constraints, there is insufficient time

and resources for personal interviews, so you have decided that a questionnaire is needed.

   a. What facts and opinions do you need to collect?

   b. Should all employees in the organization be surveyed? Why or why not? If not all employees should be surveyed, how would you select the employees to be surveyed?

   c. What format do you think would work best for this survey questionnaire? If fixed format, what type(s) of fixed-format questions should be used?

   d. How long should the survey questionnaire be in order to get the necessary information without discouraging employees from filling it out?

   e. Create the survey questionnaire, using the question-writing guidelines given in this chapter.

4. Based upon the responses to your intranet survey, you feel that it would be helpful to interview someone in another organization who has had experience in developing and/or maintaining company intranets.

   a. What type of interview do you think would be most appropriate in this situation—unstructured or structured? Why?

   b. Make an appointment with the intranet administrator in your organization or another organization or school to discuss her or his experiences in developing and/or maintaining an intranet. Describe the organization and its intranet.

   c. Prepare an interview guide using the format of Figure 6-3 as an example, ensuring that questions are free of the problems discussed in this chapter.

   d. Conduct the interview, and record the responses.

   e. What do you feel worked well in the interview, and what would you do differently next time?

5. Body language is an extremely important part of communication, as described in the textbook. Analysts need to be aware of not only what is being communicated through the body language of the interviewee but also the impact that their own body language may have upon the interview process. Make an appointment with several co-workers or fellow students to interview regarding the features they would like to see in an intranet; if possible, select interviewees you know well and those that you don't. Prepare for the interviews following the same steps as in the prior question.

   a. Describe the interviewees you selected and the questions you asked.

   b. During each interview, observe the facial expressions of the interviewee. What did you observe? Were the facial expressions always consistent with the responses?

   c. During each interview, observe the eye contact. How long did it last? Observe and describe what happened when you made eye contact for more than three to five seconds with the interviewee.

   d. Try changing your spatial zone during the interview. Did the interviewee show any signs of being uncomfortable? At what point did that occur?

   e. Did you note any differences in body language between those you knew well and those you didn't?

   f. What did you do that was the most successful and the least successful in eliciting information?

6. Analysts typically have access to confidential or sensitive data during the requirements discovery phase of a project, particularly during fact-finding. Analysts need to be aware of situations where there may be a breach of professional ethics, whether by acts of commission or omission, and the possible consequences. Search on the Web and/or in business periodicals in your school library for articles on incidents involving breaches of professional ethics.

   a. What articles did you find?

   b. What was the nature of each of these incidents?

   c. What were the consequences?

   d. What was the analyst's personal responsibility in each incident?

   e. What could have been done at the organizational and/or individual level to prevent the incident or to reduce its severity?

Minicases

1. In Chapter 5, you developed feasibility studies for a project. Economic feasibility assessments are impacted significantly by intangibles, whose value is obtained in part by interviews and questionnaires. Develop interview questions to determine the value to employees of telecommuting.

   a. Begin with unstructured questions posed to one group of employees to determine what matters to the employees and how they view telecommuting.

   b. Once you know what issues surround employee perceptions of telecommuting and why they might like/dislike it, create open-ended, but structured, questions on those issues, and interview a second set of employees. Why are we using two different groups of employees for this process?

2. Develop a questionnaire for mass employee distribution based on your findings from the previous interviews. Why are we completing the analysis with an anonymous survey?

3. You are in charge of developing a new online class registration system for your school. Develop a set of interview questions to determine issues and needs of students, registration staff, and faculty for an online registration system.

4. Discuss the impact that biased or leading questions may have on an analysis. Create one non-biased interview question and one biased or leading question. Pose each of those questions to five people. What kind of responses did you get? Were they what you expected?

Suggested Readings

Andrews, D. C., and N. S. Leventhal. *Fusion Integrating IE, CASE and JAD: A Handbook for Reengineering the Systems Organization.* Englewood Cliffs, NJ: Prentice Hall, 1993.

Berdie, Douglas R., and John F. Anderson. *Questionnaires: Design and Use.* Metuchen, NJ: Scarecrow Press, 1974. A practical guide to the construction of questionnaires. Particularly useful because of its short length and illustrative examples.

Davis, William S. *Systems Analysis and Design.* Reading, MA: Addison-Wesley, 1983. Provides useful pointers for preparing and conducting interviews.

Dejoie, Roy; George Fowler; and David Paradice. *Ethical Issues in Information Systems.* Boston, MA: Boyd and Fraser, 1991. Focuses on the impact of computer technology on ethical decision making in today's business organizations.

Fitzgerald, Jerry; Ardra F. Fitzgerald; and Warren D. Stallings, Jr. *Fundamentals of Systems Analysis,* 2nd ed. New York: John Wiley & Sons, 1981. A useful survey text for the systems analyst. Chapter 6, "Understanding the Existing System," does a good job of presenting fact-finding techniques in the study phase.

Gane, C. *Rapid Systems Development.* New York: Rapid Systems Development, Inc., 1987. This book provides a good discussion on how to lead a group meeting/interview.

Gause, Donald C., and Gerald M. Weinberg. *Exploring Requirements: Quality before Design.* New York: Dorset House Publishing, 1989. An excellent book describing the techniques of requirements discovery.

Gildersleeve, Thomas R. *Successful Data Processing System Analysis.* Englewood Cliffs, NJ: Prentice Hall, 1978. Chapter 4, "Interviewing in Systems Work," provides a compre-

hensive look at interviewing specifically for the systems analyst. A thorough sample interview is scripted and analyzed in this chapter.

London, Keith R. *The People Side of Systems.* New York: McGraw-Hill, 1976. Chapter 5, "Investigation versus Inquisition," provides a very good people-oriented look at fact-finding, with considerable emphasis on interviewing.

Lord, Kenniston W., Jr., and James B. Steiner. *CDP Review Manual: A Data Processing Handbook,* 2nd ed. New York: Van Nostrand Reinhold, 1978. Chapter 8, "Systems Analysis and Design," provides a comprehensive comparison of the merits and demerits of each fact-finding technique. This material is intended to prepare data processors for the Certificate in Data Processing examinations, one of which covers systems analysis and design.

Miller, Irwin, and John F. Freund. *Probability and Statistics for Engineers.* Englewood Cliffs, NJ: Prentice Hall, 1965. Introductory college textbook on probability and statistics.

Mitchell, Ian; Norman Parrington; Peter Dunne; and John Moses. "Practical Prototyping, Part One," *Object Currents,* May 1996. First of a three-part series of articles that explores prototyping and how you can benefit from it. Prototyping is an integral part of JRP.

Robertson, Suzanne, and James Robertson. *Mastering the Requirements Process.* Reading, MA: ACM Press/Addison-Wesley, 1999. This book contains an in-depth coverage of step-by-step procedures for requirements discovery.

Salvendy, G., ed. *Handbook of Industrial Engineering.* New York: John Wiley & Sons, 1974. A comprehensive handbook for industrial engineers; systems analysts are, in a way, a type of industrial engineer. Excellent coverage on sampling and work measurement.

Stewart, Charles J., and William B. Cash, Jr. *Interviewing: Principles and Practices,* 2nd ed. Dubuque, IA: Brown, 1978. Popular college textbook that provides broad exposure to interviewing techniques, many of which are applicable to systems analysis and design.

Walton, Donald. *Are You Communicating? You Can't Manage without It.* New York: McGraw-Hill, 1989. This book is an easy-to-use guidebook on the process of communications and a must for anyone who must work with people and influence them.

Weinberg, Gerald M. *Rethinking Systems Analysis and Design.* Boston: Little, Brown and Company, 1982. A book created to stimulate a new way of thinking.

Wood, Jane, and Denise Silver. *Joint Application Design.* New York: John Wiley & Sons, 1989. This book provides a comprehensive overview of IBM's joint application design technique.

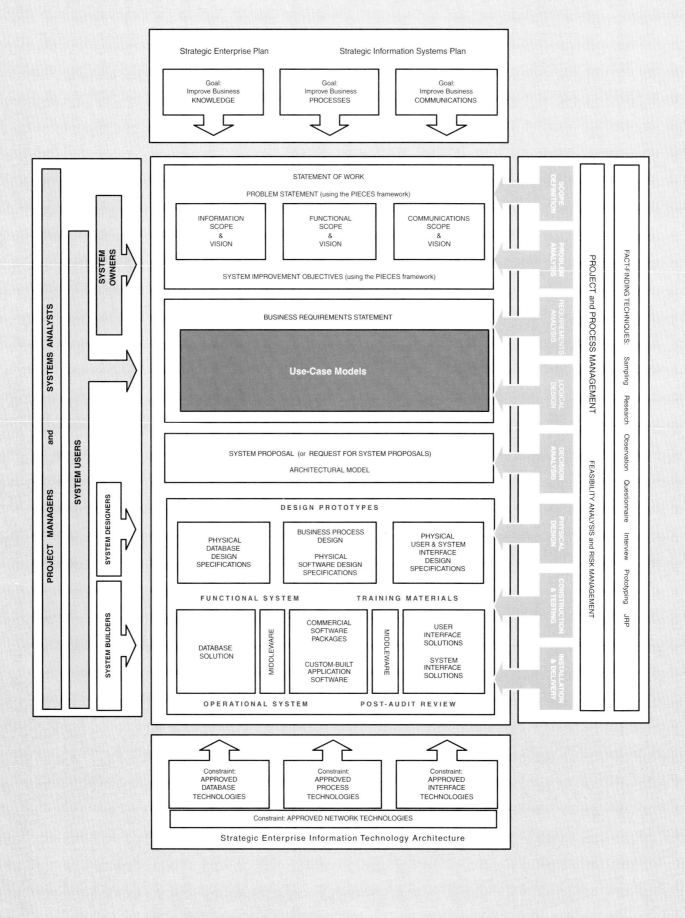

7

Use Cases

Chapter Preview and Objectives

In this chapter you will learn about the tools and techniques necessary to perform use-case modeling to document system requirements. Capturing and documenting system requirements have proved to be critical factors in the outcome of a successful information systems development project. Documenting the requirements from the perspective of the users in a manner that they can understand promotes user involvement, which greatly enhances the probability for the success of the project. You will know that you understand requirements use-case modeling when you can:

▌ Describe the benefits of use-case modeling.

▌ Define actors and use cases and be able to identify them from context diagrams and other sources.

▌ Describe the four types of actors.

▌ Describe the relationships that can appear on a use-case model diagram.

▌ Describe the steps for preparing a use-case model.

▌ Describe how to construct a use-case model diagram.

▌ Describe the various sections of a use-case narrative and be able to prepare one.

▌ Define the purpose of the use-case ranking and priority matrix and the use-case dependency diagram.

Introduction

Following the joint requirements planning (JRP) meeting that was held as one task of the requirements analysis phase, the SoundStage Member Services system project team has built a list of use cases that specify all the required functionality of the system. At first each use case was just a simple verb phrase (such as "Place New Order") that described something one or more users wanted to do with the system. Next, each use case was documented with a narrative describing in detail the desired interaction between the user and the system. Then Bob Martinez and other systems analysts held a series of interviews with users to verify those use-case narratives. Finally they are analyzing which use cases are the highest priority to the system. Bob's boss, Sandra, says that will identify for them what functionality has to be included in the first build cycle of the system. The plan is to take those highest-priority use cases into the logical design and later phases and implement a working version 1.0 of the system on schedule and within budget.

An Introduction to Use-Case Modeling

One of the primary challenges of vital importance to any information systems development team, and especially the systems analyst, is the ability to elicit the correct and necessary system requirements from the stakeholders and specify them in a manner that is understandable to the stakeholders in order for those requirements to be verified and validated. In fact, this has been the case for many years, as the distinguished author Fred Brooks wrote in his famous 1987 article:

> The hardest single part of building a software system is deciding precisely what to build. No other part of the conceptual work is as difficult as establishing the detailed technical requirements, including all the interfaces to people, to machines, and to other software systems. No other work so cripples the resulting system if done wrong. No other part is more difficult to rectify later.

The information technology community has always had problems trying to specify requirements, especially functional requirements, to users. In the past we have used tools such as data models, process models, prototypes, and requirements specifications that we understood and were comfortable with, but they were hard to understand for any user who wasn't educated in software development practices. Because of this, many development projects were and still are plagued with scope creep, cost overruns, and schedule creep problems. Very often systems are developed and deployed that really don't satisfy the user's needs. Some are shelved and not used at all, and a large percentage are canceled even before the development effort is complete. A very well known research firm, the Standish Group, studied 23,000 IT applications in 1994, 1996, and 1998.[1] As shown in Figure 7-1, the 1998 study found that only a little more than a quarter of the projects in 1998 were successful (on budget, on time, and included all features). More than a quarter of them failed (canceled before completion). A little less than half were what Standish considered challenged—the project was complete and operational, but it was completed either over budget, over the time estimate, or without all the features specified by the users. The good news reflected in these studies and others is that the ways and means we are using to develop information systems are improving. The software development industry has learned that in order to successfully plan,

---

[1]The Standish Group International, Inc., "CHAOS: A Recipe for Success" (electronic version), 1999. Retrieved December 5, 2002, from www.pm2go.com/sample_research/chaos1998.pdf. The Standish Group is best known for its independent primary research and analysis of the IT industry.

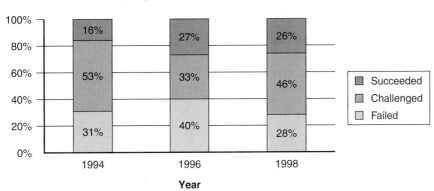

Project Success Rate

FIGURE 7-1

Project Success Rates As Reported by the Standish Group

Source: The Standish Group International, Inc., "Chaos: A Recipe for Success" (electronic version), 1999, www.pm2go.com/sample_research/chaos1998.pdf.

analyze, design, construct, and deploy an information system, the systems analyst first must understand the needs of the stakeholders and the reasons why the system should be developed—a concept called **user-centered development.** By focusing on the users of the system, the analyst can concentrate on how the system will be used and not how it will be constructed. **Use-case modeling** is an approach that facilitates usage-centered development.

Use-case modeling has its roots in object-oriented modeling, and you will learn more about how to apply use-case modeling in the object-oriented analysis and design chapters, but it has gained popularity in nonobject development environments. You will learn throughout the remaining chapters of this book how use-case modeling complements traditional systems analysis and design tools such as data modeling and process modeling as well as provides a basis for architectural decisions and user interface design decisions.

Use-case modeling was originally conceived by Dr. Ivar Jacobson in 1986 and gained popularity after he published his book, *Object-Oriented Software Engineering,* in 1992. Dr. Jacobson used use-case modeling as the framework for his objectory methodology, which he successfully used for developing object-oriented information systems. Use-case modeling has proved to be a valuable aid in meeting the challenges of determining what a system is required to do from a user and stakeholder perspective. It is now widely recognized as a best practice for the defining, documenting, and understanding of an information system's functional requirements.

Using use-case modeling facilitates and encourages user involvement, which is one of the primary critical success factors for ensuring project success. In addition, use-case modeling provides the following benefits:

- Provides a tool for capturing functional requirements.
- Assists in decomposing system scope into more manageable pieces.
- Provides a means of communicating with users and other stakeholders concerning system functionality. Use cases present a common language that is easily understood by various stakeholders.
- Provides a means of identifying, assigning, tracking, controlling, and managing system development activities, especially incremental and iterative development.
- Provides an aid in estimating project scope, effort, and schedule.
- Provides a baseline for testing in terms of defining test plans and test cases.
- Provides a baseline for user help systems and manuals as well as system development documentation.
- Provides a tool for requirements traceability.
- Provides a starting point for the identification of data objects or entities.
- Provides functional specifications for designing user and system interfaces.
- Provides a means of defining database access requirements in terms of adds, changes, deletes, and reads.
- Provides a framework for driving the system development project.

user-centered development a process of systems development based on understanding the needs of the stakeholders and the reasons why the system should be developed.

use-case modeling the process of modeling a system's functions in terms of business events, who initiated the events, and how the system responds to those events.

System Concepts for Use-Case Modeling

use-case diagram a diagram that depicts the interactions between the system and external systems and users. In other words, it graphically describes who will use the system and in what ways the user expects to interact with the system.

functional decomposition the act of breaking a system into subcomponents.

use-case narrative a textual description of the business event and how the user will interact with the system to accomplish the task.

use case a behaviorally related sequence of steps (a scenario), both automated and manual, for the purpose of completing a single business task.

There are two primary artifacts involved when performing use-case modeling. The first is the **use-case diagram,** which graphically depicts the system as a collection of use cases, actors (users), and their relationships. This diagram communicates at a high level the scope of the business events that must be processed by the system. An example of a use-case diagram is shown in Figure 7-2. It shows each system function, or business event (in the ellipses), and the actors, or system users, who interact with those functions. As you can see in Figure 7-2, actors can be placed on either side of the set of use-case figures and can interact with one or more use cases. The use-case diagram is extremely simple. But it begins an important process called **functional decomposition,** the act of breaking a system apart into its subcomponents. It is impossible to understand the entire system at once, but it is possible to understand and specify each part of the system.

The second artifact, called the **use-case narrative,** fills in the details of each business event and specifies how the users interact with the system during that event. The use-case narrative will be discussed in detail later in the chapter.

> Use Cases

Use-case modeling identifies and describes the system functions by using a tool called **use cases.** Use cases describe the system functions from the perspective of external users and in a manner and terminology they understand. To accurately and thoroughly accomplish this demands a high level of user involvement and a subject-matter expert who is knowledgeable about the business process or event.

Use cases are represented graphically by a horizontal ellipse with the name of the use case appearing above, below, or inside the ellipse. A use case represents a single goal of the system and describes a sequence of activities and user interactions in trying to accomplish the goal. The creation of use cases has proved to be an excellent technique to better understand and document system requirements. A use case itself is not considered a functional requirement, but the story (scenario) the use case tells consists of one or more requirements.

Use cases are initially defined during the requirements stages of the life cycle and will be additionally refined throughout the life cycle. During requirements

---

Use Case
Symbol

FIGURE 7-2

Sample Use-Case
Model Diagram

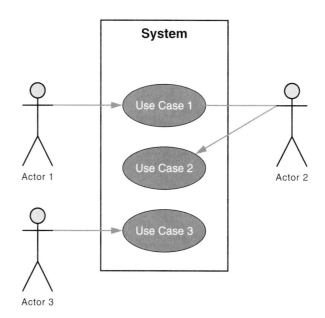

discovery, use cases are used to capture the essence of the business problems and to model (at a high level) the functionality of the proposed system. Additionally, they are the starting point for identifying the data entities (covered in Chapter 8) or objects of the system (covered in Chapter 11). During requirements analysis the use cases are refined to model usage of the system in more detail. In other words, they are updated to specify what the users are trying to accomplish and why. These use cases aid in the definition of any prototypes or user interfaces. During design the use cases are refined to model how the users will actually use the system with regard to any interface and system constraints (covered in Chapter 18). These types of use cases aid in identifying object or system behavior and in designing interface and code specifications, as well as serve as the plan for testing the system. In construction, use cases aid developers in programming and testing. These use cases also serve as the baseline for preparing any user and system documentation, plus they serve as tools for user training. And, because use cases contain an enormous amount of system functionality detail, they will be a constant resource for validating the system.

> Actors

Use cases are initiated or triggered by external users called **actors.** An actor initiates system activity, a use case, for the purpose of completing some business task that produces something of measurable value. Let's use the example of a college student enrolling for the fall semester's courses. The actor would be the *student*, and the business event, or use case, would be *Enrolling in Course*. An actor represents a role fulfilled by a user interacting with the system and is not meant to portray a single individual or job title. In fact, an actor doesn't have to be human. It can be an organization, another information system, an external device such as a heat sensor, or even the concept of time (which will be discussed a little later). An actor is represented graphically as a stick figure labeled with the name of the role the actor plays.

actor anything that needs to interact with the system to exchange information.

Actor Symbol

It is important to note that there are primarily four types of actors:

- *Primary business actor*—the stakeholder that primarily benefits from the execution of the use case by receiving something of measurable or observable value. The primary business actor may or may not initiate the business event. For example, in the business event of an employee receiving a paycheck (something of measurable value) from the payroll system each Friday, the employee does not initiate the event but is the primary recipient of the something of value.
- *Primary system actor*—the stakeholder that directly interfaces with the system to initiate or trigger the business or system event. Primary system actors may interact with primary business actors for the purpose of using the actual system. They facilitate the event through the direct use of the system for the benefit of the primary business actor. Examples include a grocery store clerk who scans the items for the customer buying groceries, a telephone operator who gives directory assistance to a customer, and a bank teller who processes a banking transaction. The primary business actor and primary system actor may be the same person for events where the business actor interfaces with the system directly—for example, a person reserving a rental car via a Web site.
- *External server actor*—the stakeholder that responds to a request from the use case (e.g., a credit bureau authorizing the charging by a credit card).
- *External receiver actor*—the stakeholder that is not the primary actor but receives something of measurable or observable value (output) from the use case (e.g., a warehouse receiving a packing order to prepare a shipment after a customer has placed an order).

In many information systems there are business events triggered by the calendar or the time on a clock. Consider the following examples:

- The billing system for a credit card company automatically generates its bills on the 5th day of the month (billing date).
- A bank reconciles its check transactions every day at 5 P.M.
- On a nightly basis a report is automatically generated listing which courses have been closed to enrollment (no open seats available) and which courses are still open.

temporal event a system event that is triggered by time.

These events are examples of **temporal events.** Who would be the actor? All of the events listed above were performed (or triggered) automatically—when it became a certain date or time. Because of that we say the actor of a temporal event is time.

> Relationships

A relationship is depicted as a line between two symbols on the use-case diagram. The meaning of the relationships may differ depending on how the lines are drawn and what types of symbols they connect. In the following sections we will define the different relationships found on a use-case diagram.

Associations   A relationship between an actor and a use case exists whenever the use case describes an interaction between them. This relationship is referred to as an **association.** As indicated in Figure 7-3, an association is modeled as a solid line connecting the actor and the use case. An association that contains an arrowhead on the end touching the use case (**❶**) indicates the use case was imitated by the actor on the other end of the line. Associations without arrowheads (**❷**) indicate an interaction between the use case and an external server or receiver actor. When any actor is associated with a use case, we say the actor *communicates* with the use case. Associations may be bidirectional or unidirectional.

association a relationship between an actor and a use case in which an interaction occurs between them.

Extends   A use case may contain complex functionality consisting of several steps making the use-case logic difficult to understand. For the purpose of simplifying the use case and making it more easily understood, we can extract the more complex steps into their own use case. The resulting use case is called an **extension use case** in that it extends the functionality of the original use case. The relationship between the extension use case and the use case it is extending is called an *extends* relationship. A use case may have many extends relationships, but an extension use case can be invoked only by the use case it is extending. As depicted in Figure 7-4, the extends relationship is represented as an arrowheaded line (either solid or dashed) beginning at the extension use case and pointing to the use case it is extending. Each extends relationship line is labeled "<<extends>>." Generally extension use cases are not identified in the requirements phase but in the analysis phase.

extension use case a use case consisting of steps extracted from a more complex use case in order to simplify the original case and thus extend its functionality. The extension use case *extends* the functionality of the original use case.

FIGURE 7-3

Example of an Association Relationship

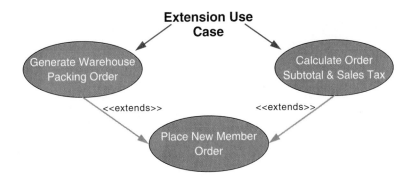

Extension Use Case

FIGURE 7-4

Example of an Extends Relationship

Uses (or Includes)   Very commonly, you may discover two or more use cases that perform steps of identical functionality. It is best to extract these common steps into their own separate use case called an **abstract use case.** An abstract use case represents a form of "reuse" and is an excellent tool for reducing redundancy among use cases. An abstract use case is available for referencing (or use) by any other use case that requires its functionality. The relationship between the abstract use case and the use case that uses it is called a *uses* relationship (some use-case modeling tools refer to it as an *includes* relationship). The uses relationship as presented in Figure 7-5 is depicted as an arrowheaded line (either solid or dashed) beginning at the original use case and pointing to the use case it is using. Each uses relationship line is labeled "<<uses>>." Generally abstract use cases are not identified in the requirements phase but in the analysis phase.

abstract use case a use case that reduces redundancy among two or more other use cases by combining the common steps found in those cases. Another use case *uses* or *includes* the abstract use case.

Depends On   As the project manager or lead developer, it is very helpful to know which use cases have a dependency on other use cases in order to determine the sequence in which use cases need to be developed. Using the banking business as an example, the use case Make a Withdrawal cannot be performed until the use case Make a Deposit has been executed, and that use case cannot execute until the use case Establish Bank Account has occurred. Because of these dependencies the development team will most likely choose to develop the use case Establish Bank Account first, the Make a Deposit use case second, and the Make a Withdrawal use case third for usability and testing purposes. A use-case diagram modeling the system's use-case dependencies using the **depends on** relationship provides a model that is an excellent tool for planning and scheduling purposes. The depends on relationship as

depends on a relationship between use cases indicating that one use case cannot be performed until another use case has been performed.

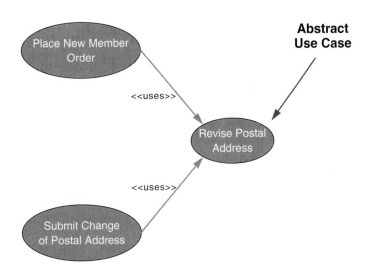

Abstract Use Case

FIGURE 7-5

Example of a Uses Relationship

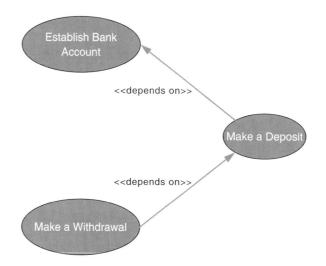

FIGURE 7-6

Example of a
Depends On
Relationship

presented in Figure 7-6 is depicted as an arrowheaded line (either solid or dashed) beginning at one use case and pointing to a use case it is dependent on. The depends on relationship line is labeled "<<depends on>>."

Inheritance When two or more actors share common behavior—in other words, they can initiate the same use case—it is best to extrapolate this common behavior and assign it to a new *abstract* actor in order to reduce redundant communication with the system. For example, a library *patron* is a card-carrying member who is authorized to "Search library inventory" as well as "Check out books" from the library. Since many libraries are public institutions, they welcome *visitors* to use their services onsite such as "Search library inventory," but the visitors are not allowed the extended services (such as "Check out books") that are reserved for the patrons. By creating an abstract actor called *customer,* from which *patron* and *visitor* will inherit, we have to model only once the relationship initiating the use case Search Library Inventory. In the use-case diagram the **inheritance** relationship is depicted by the type of arrow shown in the "After" section of Figure 7-7.

inheritance in use cases, a relationship between actors created to simplify the drawing when an abstract actor inherits the role of multiple real actors.

FIGURE 7-7

Example of an
Inheritance
Relationship

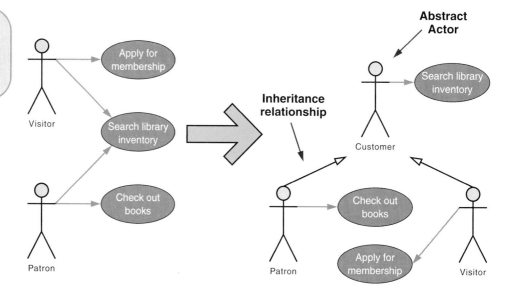

Before **After**

The Process of Requirements Use-Case Modeling

The objective of constructing the requirements use-case model is to elicit and analyze enough requirements information to prepare a model that communicates what is required from a user perspective but is free of specific details about how the system will be built and implemented. Following this approach will later produce a design that is more robust and less likely to be impacted by change. But to effectively estimate and schedule the project, the model may need to include preliminary "system implementation assumptions" to aid in those activities. It is critical that the analyst does not slip into a state of *analysis paralysis* when preparing this model. Speed is the key. Not all of the facts will be learned during this phase of the life cycle, but by utilizing iterative and incremental development, the methodology allows the introduction of new requirements later in the project without seriously impacting the deployment of the final solution. The steps required to produce this model are the following:

1. Identify business actors.
2. Identify business requirements use cases.
3. Construct use-case model diagram.
4. Document business requirements use-case narratives.

> Step 1: Identify Business Actors

Why identify actors first? By focusing on the actors, you can concentrate on how the system will be used and not how it will be built. Focusing on the actors helps to refine and further define the scope and boundaries of the system. Actors also determine the completeness of the system requirements.[2] A benefit of identifying actors first is that doing so identifies candidates we can later interview and observe to complete the use-case modeling effort. Plus, those same individuals can be used to verify and validate the use cases when they are finished.

Where do you look for potential actors? The following references are excellent sources:

- A context diagram that identifies the scope and boundaries of the system.
- Existing system documentation and user manuals.
- Minutes of project meetings and workshops.
- Existing requirements documents, project charter, or statement of work.

When looking for actors, ask the following questions:

- Who or what provides inputs to the system?
- Who or what receives outputs from the system?
- Are interfaces required to other systems?
- Are there any events that are automatically triggered at a predetermined time?
- Who will maintain information in the system?

Actors should be named with a noun or noun phrase.

When you identify an actor, create a textual definition of that actor according to the users' perspective and using their terms. Figure 7-8 is a template of an actor glossary that can be used to document actors. This example contains a partial listing of the SoundStage Member Services System's actors.

---

[2]Frank Armour and Granville Miller, *Advance Use Case Modeling* (Boston: Addison-Wesley, 2001).

FIGURE 7-8

Partial List of
SoundStage
Member Services
System's Actors

Actor Glossary

Term	Synonym	Description
1. Potential member		An individual or corporation that submits a subscription order in order to join the club.
2. Club member	Member	An individual or corporation that has joined the club via an agreement.
3. Past member	Inactive member	A type of member that has fulfilled the agreement obligation but has not placed an order within the last six months but is still in good standing.
4. Marketing		Organization responsible for creating promotion and subscription programs and generating sales for the company.
5. Member services		Organization responsible for providing point of contact for SoundStage Entertainment customers in terms of agreements and orders.
6. Distribution center	Warehouse	Entity that houses and maintains SoundStage Entertainment product inventory and processes customer shipments and returns.
7. Accounts receivable		Organization responsible for processing customer payments and billing as well as maintaining customer account information.
8. Time		Actor concept responsible for triggering temporal events.

> Step 2: Identify Business Requirements Use Cases

A typical information system may consist of dozens of use cases. During requirements analysis we strive to identify and document only the most critical, complex, and important ones, often referred to as *essential* use cases because of time and cost considerations. A **business requirements use case** captures the interactions with the user in a manner that is free of technology and implementation details. Since a use case describes how a real-world actor interacts with the system, an excellent technique for finding business requirements use cases is to examine actors and how they will use the system. When looking for use cases, ask the following questions:

* What are the main tasks of the actor?
* What information does the actor need from the system?
* What information does the actor provide to the system?
* Does the system need to inform the actor of any changes or events that have occurred?
* Does the actor need to inform the system of any changes or events that have occurred?

business requirements use case a use case created during requirements analysis to capture the interactions between a user and the system free of technology and implementation details also called an essential use case.

Again, a context diagram is an excellent source for finding potential use cases. Context diagrams were discussed in Chapter 5. They come from traditional process modeling (Chapter 9) but are useful even for projects that take an object-oriented approach. Let's examine the SoundStage Member Services System's context diagram in Figure 7-9. We can identify potential use cases by looking at the diagram and identifying the primary inputs and outputs of the system and the external parties that submit and receive them. The primary inputs that trigger business events (for example, *Submit Member Order*) within the organization are considered use cases, and the external parties that provide those inputs are considered actors (for example, *Club Member*). It is important to note that inputs that are the result of system requests do

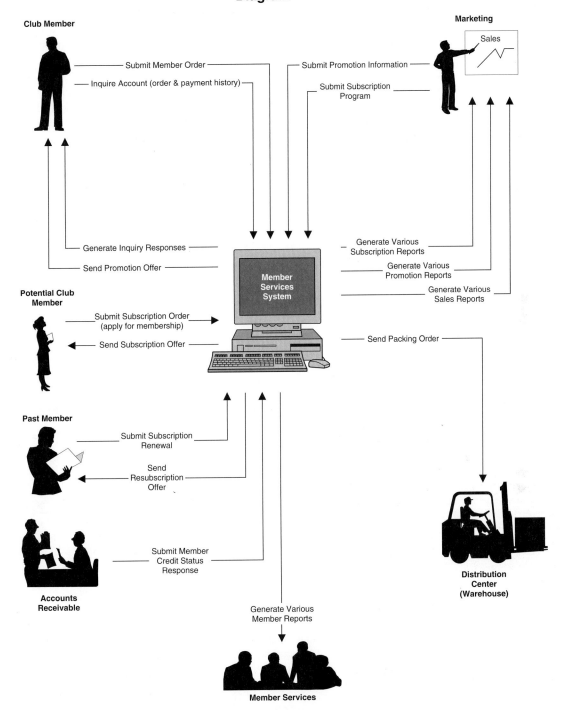

Member Services Context Diagram

FIGURE 7-9 SoundStage Member Services System Context Diagram

not indicate a separate use case—such as a credit card company responding to an authorization request or, as presented in Figure 7-9, the *Accounts Receivable* actor responding with *Member Credit Status Information.*

Use cases are named with a verb phrase specifying the goal of the actor, such as *Submit Subscription Order.* Use cases that are temporal events are usually identified as a result of analyzing the key outputs of the system. For example, any output that is generated on the basis of time or a date, such as monthly or annual reports, is considered a use case, and the actor, as you recall, is time. In Figure 7-9 let's assume that one of the various reports that Member Services receives is a *10-30-60-day default agreement report* that is automatically generated on a daily basis. Since the generation of the report is triggered by time, a use case is required to process the event, and we would name it *Generate Daily 10-30-60-Day Default Agreement Report.* It is important to note that many times individual reports are not listed on a context diagram because they are too numerous and would clutter the diagram and make it hard to read. It is the system analyst's responsibility to research with the appropriate stakeholders the type of outputs they receive and their characteristics, in terms of volume, frequency, and triggering mechanism, in order to identify "hidden use cases."

Figure 7-10 is a template of a use-case glossary that can be used to document use cases. This example contains a partial listing of the SoundStage Member Services System's use cases and actors identified from the context diagram as well as from other sources.

> Step 3: Construct Use-Case Model Diagram

Once the use cases and actors have been identified, a use-case model diagram can be used to graphically depict the system scope and boundaries. The use-case

---

FIGURE 7-10   Partial List of SoundStage Member Services System's Use Cases

Use-Case Glossary

Use-Case Name	Use-Case Description	Participating Actors and Roles
Submit Subscription Order	This use case describes the event of a potential member requesting to join the club by subscribing. ("Take any 12 CDs for one penny and agree to buy 4 more at regular prices within two years.")	• Potential member (primary business) • Distribution Center (external receiver)
Submit Subscription Renewal Order	This use case describes the event of a past member requesting to rejoin the club by subscribing. ("Take any 12 CDs for one penny and agree to buy 4 more at regular prices within two years.")	• Past member (primary business) • Distribution Center (external receiver)
Submit Member Profile Changes	This use case describes the event of a club member submitting changes to his or her profile for such things as postal address, e-mail address, privacy codes, and order preferences.	• Club member (primary business)
Place New Order	This use case describes the event of a club member submitting an order for SoundStage products.	• Club member (primary business) • Distribution Center (external receiver) • Accounts Payable/Receivable (external server)
Revise Order	This use case describes the event of a club member revising an order previously placed. (Order must not have shipped.)	• Club member (primary business) • Distribution Center (external receiver) • Accounts Payable/Receivable (external server)

FIGURE 7-10 Concluded

Cancel Order	This use case describes the event of a club member canceling an order previously placed. (Order must not have shipped.)	• Club member (primary business) • Distribution Center (external receiver) • Accounts Payable/Receivable (external server)
Make Product Inquiry	This use case describes the event of a club member viewing products for possible purchase. (Driven by Web access requirement.)	• Club member (primary business)
Make Purchase History Inquiry	This use case describes the event of a club member viewing her or his purchasing history. (Three-year time limit.)	• Club member (primary business)
Establish New Member Subscription Program	This use case describes the event of the marketing department establishing a new membership subscription plan to entice new members	• Marketing (primary business)
Submit Subscription Program Changes	This use case describes the event of the marketing department changing a subscription plan for club members (e.g., extending the fulfillment period).	• Marketing (primary business)
Establish Past Member Resubscription Program	This use case describes the event of the marketing department establishing a resubscription plan to lure back former members.	• Marketing (primary business)
Submit Member Profile Changes	This use case describes the event of the marketing department establishing a new promotion plan to entice active and inactive members to order the product. (Note: A promotion features specific titles, usually new, that the company is trying to sell at a special price. These promotions are integrated into a catalog sent (or communicated) to all members.)	• Marketing (primary business)
Revise Promotion	This use case describes the event of the marketing department revising a promotion.	• Marketing (primary business)
Generate Daily 10-30-60-Day Default Agreement Report	This use case describes the event of a report that is generated on a daily basis to list the members who have not fulfilled their agreement by purchasing the required number of products outlined when they subscribed. This report is sorted by members who are 10 days past due, 30 days past due, and 60 days past due.	• Time (initiating actor) • Member Services (primary*—external receiver)

*Considered primary because it receives something of measurable value.

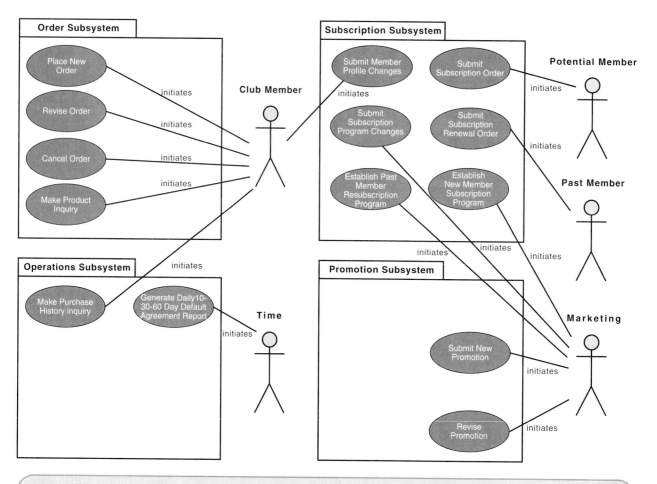

FIGURE 7-11   SoundStage Member Services System's Use-Case Model Diagram

diagram for the use cases listed in Figure 7-10 is shown in Figure 7-11. It was created using Popkin Software's *System Architect* and represents the relationships between the actors and the use cases. In addition, the use cases have been grouped into business subsystems. The subsystems (UML's package symbol) represent logical functional areas of business processes. The portioning of system behavior into subsystems is very important in understanding the system architecture and is a key to defining your development strategy—which use cases will be developed first and by whom. We have labeled the associations between the actors and the use cases "initiates" because the tool did not support lines with arrowheads at the time. We also didn't include the external server and receiver actors because of space limitations. To model all the use cases of a particular system may require the creation of several use-case model diagrams—as you recall, a system may contain dozens of use cases. In that event you may want to create a separate use-case model diagram for each subsystem.

> Step 4: Document Business Requirements Use-Case Narratives

When you are preparing the narratives, it is wise to first document them at a *high level* to quickly obtain an understanding of the events and magnitude of the system. Then return to each use case and expand it to a fully documented business requirement narrative. Figure 7-12 represents a requirements use-case narrative for the

Member Services System

Author (s): _____❶_____   Date: _____❷_____

                                    Version: ❸

Use-Case Name:	Place New Order ❹	**Use-Case Type**
Use-Case ID:	MSS-BUC002.00 ❻	**Business Requirements:** ☑
Priority:	High ❼	❺
Source:	Requirement — MSS-R1.00 ❽	
Primary Business Actor:	Club member ❾	
Other Participating Actors:	• Warehouse (external receiver) • Accounts Receivable (external server) ❿	
Other Interested Stakeholders: ⓫	• Marketing — Interested in sales activity in order to plan new promotions. • Procurement — Interested in sales activity in order to replenish inventory. • Management — Interested in order activity in order to evaluate company performance and customer (member) satisfaction.	
Description: ⓬	This use case describes the event of a club member submitting a new order for SoundStage products. The member's demographic information as well as his or her account standing is validated. Once the products are verified as being in stock, a packing order is sent to the warehouse for it to prepare the shipment. For any product not in stock, a back order is created. On completion, the member will be sent an order confirmation.	

FIGURE 7-12   High-Level Version of Place New Order Use-Case Narrative

Member Services System's Place New Order use case. Notice that it tersely describes the event, which includes the following items:

❶ *Author*—The names of the individuals who contributed to the writing of the use case and who provide a point of contact for anyone requiring additional information about the use case.

❷ *Date*—The date the use case was last modified.

❸ *Version*—The current version of the use case (e.g., 1.0).

❹ *Use-case name*—The use-case name should represent the goal that the use case is trying to accomplish. The name should begin with a verb (e.g., Enter New Member Order).

❺ *Use-case type*—In performing use-case modeling, business requirements use cases, which focus on the strategic vision and goals of the various stakeholders, are constructed first. This type of use case is business-oriented and reflects a high-level view of the desired behavior of the system. It is free from technical details and may include manual activities as well as the activities that will be automated. Business requirements use cases provide a general understanding of the problem domain and scope but don't include the necessary detail to communicate to developers what the system should do.

❻ *Use-case ID*—An identifier that uniquely identifies the use case.

❼ *Priority*—The priority communicates the importance of the use case in terms of high, medium, or low.

❽ *Source*—The source defines the entity that triggered the creation of the use case. This could be a requirement, a specific document, or a stakeholder.

❾ *Primary business actor*—The primary business actor is the stakeholder that primarily benefits from the execution of the use case by receiving something of measurable or observable value.

⑩ *Other participating actors*—Other actors that participate in the use case to accomplish its goal include initiating actors, facilitating actors, server/receiver actors, and secondary actors. Always include the manner in which the actor participates.

⑪ *Interested stakeholder(s)*—A stakeholder is anybody who has a stake in the development and operation of the software system. An interested stakeholder is a person (other than an actor) who has a vested interest in the goal of the use case.

⑫ *Description*—A short summary description that consists of a couple of sentences outlining the purpose of the use case and its activities.

Documenting the Use-Case Course of Events   For each high-level use case identified, we must now expand it to include the use case's typical course of events and its alternate courses. A use case's typical course of events is a step-by-step description starting with the actor initiating the use case and continuing until the end of the business event. In this section we include only the major steps that occur the majority of the time (its typical course). The alternate course documents the exceptions or the conditional branching of the use case. Figure 7-13 represents a requirements use-case narrative for the Member Services System's Place New Order use case. Notice that it includes the following additional items:

❶ *Precondition*—A precondition is a constraint on the state of the system before the use case can be executed. Typically this refers to another use case that must be previously executed.

❷ *Trigger*—The trigger is the event that initiated the execution of the use case. This often is a physical action, such as a customer walking up to a sales counter or a check arriving in the mail. Time can also trigger use cases.

❸ *Typical course of events*—The typical course of events is the normal sequence of activities performed by the actor(s) and the system in order to satisfy the goal of the use case. These include the interactions between the system and the actor and the activities performed by the system in response to the interactions. Note that the actions of the actor are recorded in the left hand column while the actions of the systems are recorded in the right hand column.

❹ *Alternate courses*—Alternate courses document the behaviors of the use case if an exception or variation to the typical course occurs. This can happen when a decision point occurs within the use case or an exception occurs that requires additional steps outside the scope of the typical course.

❺ *Conclusion*—The conclusion specifies when the use case successfully ends—in other words, when the primary actor receives something of measurable value.

❻ *Postcondition*—A postcondition is a constraint on the state of the system after the use case has been successfully executed. This could be data recorded in a database or a receipt delivered to a customer.

❼ *Business rules*—Business rules specify policies and procedures of the business that the new system must abide by. This could include the calculation of shipping charges or rules for granting credit terms.

❽ *Implementation constraints and specifications*—Implementation constraints and specifications specify any nonfunctional requirements that may impact the realization of the use case and may be helpful in any architectural planning and scoping. Items that may be included are security specifications, interface requirements, and so on.

❾ *Assumptions*—Any assumptions that were made by the creator when documenting the use case.

❿ *Open issues*—Any questions or issues that need to be resolved or investigated before the use case can be finalized.

Member Services System

Author (s): _____   Date: _____

Version:

Use-Case Name:	Place New Order	**Use-Case Type**
Use-Case ID:	MSS-BUC002.00	**Business Requirements:** ☑
Priority:	High	
Source:	Requirement — MSS-R1.00	
Primary Business Actor:	Club member	
Other Participating Actors:	• Warehouse (external receiver) • Accounts Receivable (external server)	
Other Interested Stakeholders:	• Marketing — Interested in sales activity in order to plan new promotions. • Procurement — Interested in sales activity in order to replenish inventory. • Management — Interested in order activity in order to evaluate company performance and customer (member) satisfaction.	
Description:	This use case describes the event of a club member submitting a new order for SoundStage products. The member's demographic information as well as his or her account standing is validated. Once the products are verified as being in stock, a packing order is sent to the warehouse for it to prepare the shipment. For any product not in stock, a back order is created. On completion, the member will be sent an order confirmation.	
Precondition: ❶	The party (individual or company) submitting the order must be a member.	
Trigger: ❷	This use case is initiated when a new order is submitted.	

Typical Course of Events: ❸	**Actor Action**	**System Response**
	Step 1: The club member provides his or her demographic information as well as order and payment information.	**Step 2:** The system responds by verifying that all required information has been provided. **Step 3:** The system verifies the club member's demographic information against what has been previously recorded. **Step 4:** For each product ordered, the system validates the product identity. **Step 5:** For each product ordered, the system verifies the product availability. **Step 6:** For each available product, the system determines the price to be charged to the club member. **Step 7:** Once all ordered products are processed, the system determines the total cost of the order. **Step 8:** The system checks the status of the club member's account. **Step 9:** The system validates the club member's payment if provided. **Step 10:** The system records the order information and then releases the order to the appropriate distribution center (warehouse) to be filled. **Step 11:** Once the order is processed, the system generates an order confirmation and sends it to the club member.

FIGURE 7-13 Expanded Version of Place New Order Use-Case Narrative

Business requirements use cases are excellent tools in that they describe the events the organization must process and respond to, but they lack information regarding the interfaces and the activities that are targeted to be automated by information technology. Later, in Chapter 11, you will learn how to evolve the use case to include technical and implementation details.

Alternate Courses: ④	Alt-Step 2: The club member has not provided all the information necessary to process the order. The club member is notified of the discrepancy and prompted to resubmit.
	Alt-Step 3: If the club member information provided is different from what was previously recorded, verify what was recorded is current, then update the club member information accordingly.
	Alt-Step 4: If the product information the club member provided does not match any of SoundStage's products, notify the club member of the discrepancy and request clarification.
	Alt-Step 5: If the quantity ordered of the product is not available, a back order is created.
	Alt-Step 8: If the status of the club member's account is not in good standing, record the order information and place it in hold status. Notify the club member of the account status and the reason the order is being held. Terminate use case.
	Alt-Step 9: If the payment the club member provided (credit card) cannot be validated, notify the club member and request an alternative means of payment. If the club member cannot provide an alternate means, cancel the order and terminate the use case.
Conclusion: ⑤	This use case concludes when the club member receives a confirmation of the order.
Postcondition: ⑥	The order has been recorded and if the ordered products were available, they were released. For any product not available a back order has been created.
Business Rules: ⑦	• The club member responding to a promotion or a member using credits may affect the price of each ordered item.
	• Cash or checks will not be accepted with the orders. If provided, they will be returned to the club member.
	• The club member is billed for products only when they are shipped.
Implementation Constraints and Specifications: ⑧	• GUI to be provided for Member Services associate, and Web screen to be provided for club member.
Assumptions: ⑨	Procurement will be notified of back orders by a daily report (separate use case).
Open Issues: ⑩	1. Need to determine how distribution centers are assigned.

FIGURE 7-13 Concluded

Use Cases and Project Management

As you recall, one of the benefits of use-case modeling is that the use-case model can be used to drive the entire system development effort. Once the business requirements use-case model is complete, the project manager or systems analyst uses the business requirements use cases to plan (estimate and schedule) the build cycles of the project. This is especially crucial when applying the iterative and incremental approach to software development. A build cycle, which consists of the system analysis, design, and construction activities, is scoped on the basis of the importance of the use case and the time it takes to implement the use case. In other words, one or more use cases will be developed in each build cycle. For a use case that is too large or complex to be completed in one build cycle, a simplified version will be implemented initially, followed by the full version in the next build cycle. To determine the importance of the use cases, the project manager or systems analyst will complete a use-case ranking and evaluation matrix and construct a use-case dependency diagram with input from the stakeholders and the development team. You will learn how to use these tools in the following sections.

> Ranking and Evaluating Use Cases

use-case ranking and priority matrix a tool used to evaluate use cases and determine their priority.

In most software development projects the most important use cases are developed first. In order to determine the priority of the use cases, the project manager uses a tool called the **use-case ranking and priority matrix.** This matrix is completed

Use-Case Name	Ranking Criteria, 1 to 5						Total Score	Priority	Build Cycle
	1	2	3	4	5	6			
Submit Subscription Order	5	5	5	4	5	5	29	High	1
Place New Order	4	4	5	4	5	5	27	High	2
Make Product Inquiry	1	1	1	1	1	1	6	Low	3
Establish New Member Subscription Program	4	5	5	3	5	5	27	High	1
Generate Daily 10-30-60-Day Default Agreement Report	1	1	1	1	1	1	6	Low	3
Revise Order	2	2	3	3	4	4	18	Medium	2

FIGURE 7-14 Partial Use-Case Ranking and Priority Matrix

with input from the stakeholders and the development team. This matrix, adapted from Craig Larman's work,[3] evaluates use cases on a scale of 1 to 5 against six criteria. They are as follows:

1. Significant impact on the architectural design.
2. Easy to implement but contains significant functionality.
3. Includes risky, time-critical, or complex functions.
4. Involves significant research or new or risky technology.
5. Includes primary business functions.
6. Will increase revenue or decrease costs.

Once each category has been scored, the individual scores are tallied, resulting in the use case's final score. The use cases with the highest scores are assigned the highest priority and should be developed first.

Figure 7-14 is a partial use-case ranking and priority matrix for the Member Services System. Based on the results of the analysis, the use case Submit Subscription Order should be developed first. But we can't be sure until we analyze the use-case dependencies.

> Identifying Use-Case Dependencies

Some use cases may be dependent on other use cases, with one use case leaving the system in a state that is a precondition for another use case. For example, a precondition of sending a club promotion is that the promotion must first be created. We use a diagram called the **use-case dependency diagram** to model such dependencies. The use-case dependency diagram provides the following benefits:

use-case dependency diagram a graphical depiction of the dependencies among use cases.

- The graphical depiction of the system's events and their states enhances the understanding of system functionality.
- It helps to identify missing use cases. A use case with a precondition that is not satisfied by the execution of any other use case may indicate a missing use case.
- It helps facilitate project management by depicting which use cases are more critical (have the most dependencies) and thus need to have a higher priority.

Figure 7-15 is the use-case dependency diagram for the use cases listed in Figure 7-14. The use cases that are dependent on each other are connected with a dashed

[3]Craig Larman, *Applying UML Patterns* (Upper Saddle River, NJ: Prentice Hall, 1998).

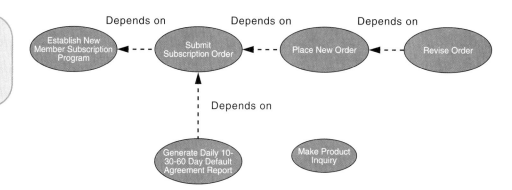

FIGURE 7-15

Sample Use-Case
Dependency
Diagram

line labeled "Depends on." In Figure 7-15, the use case Submit Subscription Order has
a dependency (precondition) on the use case Establish New Member Subscription
Program. Because of this dependency the use case Establish New Member Subscription Program should be developed first even though Submit Subscription Order had
a higher score as reflected in Figure 7-14.

Learning Roadmap

This chapter provided an introduction to use cases and how they can be used to document functional requirements. Also, you have learned that use-case modeling based
on object-oriented concepts is an excellent complementary tool for traditional systems analysis and design tools such as process modeling and data modeling. Many of
you will proceed directly to Chapter 8, "Data Modeling and Analysis." All information
systems include databases, and data modeling is an essential skill for database development. Also, it is easier to synchronize early data models with later process models
than vice versa. Your instructor may prefer that you first study Chapter 9, "Process
Modeling." Process modeling is an effective way to analyze and document functional
system requirements. Courses that want to follow an object-oriented approach may
elect to jump straight to Chapter 10, "Object-Oriented Analysis and Modeling Using
the UML," which teaches emerging object-modeling techniques using the unified
modeling language.

Summary

1. There are two primary artifacts involved when performing use-case modeling. The first is the use-case
   diagram, which graphically depicts the system as a
   collection of use cases, actors (users), and their relationships. Details of each business event and
   how the users interact with the system are described in the second artifact, called the use-case
   narrative, which is the textual description of the

   business event and how the user will interact with
   the system to accomplish the task.

2. Use-case modeling utilizes two constructs: actors
   and use cases. An actor represents anything that
   needs to interact with the system to exchange information. An actor is a user, a role, which could
   be an external system as well as a person. A use
   case is a behaviorally related sequence of steps (a

scenario), both automated and manual, for the purpose of completing a single business task.

3. There are primarily four types of actors:

    a. *Primary business actor*—The stakeholder that primarily benefits from the execution of the use case by receiving something of measurable or observable value.

    b. *Primary system actor*—The stakeholder that directly interfaces with the system to initiate or trigger the business or system event.

    c. *External server actor*—The stakeholder that responds to a request from the use case.

    d. *External receiver actor*—The stakeholder that is not the primary actor but receives something of measurable or observable value (output) from the use case.

4. Temporal events are business events that are performed (or triggered) automatically—when it becomes a certain date or time. Because of that, we say the actor of a temporal event is time.

5. A relationship is depicted as a line between two symbols on the use-case diagram.

    a. An association is a relationship between an actor and a use case.

    b. The relationship between the extension use case and the use case it is extending is called an extends relationship.

    c. The relationship between the abstract use case and the use case that uses it is called a uses relationship.

    d. The inheritance relationship occurs when an actor inherits the ability to initiate a use case from another.

    e. The depends-on relationship indicates a dependency between use cases. In other words, the precondition of one use case is dependent on the postcondition of another use case.

6. The steps required to produce a requirements use-case model are the following:

    a. Identify business actors.
    b. Identify business requirements use cases.
    c. Construct use-case model diagram.
    d. Document business requirements use-case narratives.

7. The use-case ranking and priority matrix and the use-case dependency diagram are tools used by project managers for prioritizing and scheduling use-case development.

Review Questions

1. What is user-centered development and why is it critical to the success of the system development process?
2. How is use-case modeling related to user-centered development?
3. In addition to encouraging user involvement, use-case modeling provides numerous other benefits. List the benefits that use-case modeling provides.
4. Use-case modeling uses two primary artifacts— the use-case diagram and the use-case narrative. How are these two artifacts used and what are their differences?
5. Use case diagrams consist of three components. What are these three components, and what is their purpose?
6. How are use cases used throughout the entire system development life cycle?
7. Of the four primary categories of actors, who is the primary system actor?
8. What are the different types of relationships employed in a use-case diagram, and what is their purpose?
9. What is the objective of constructing the requirements use-case model and what steps are to be followed?
10. Why is identifying the actors the first step in use-case modeling?
11. What should we be aware of when we are looking for business requirements use cases?
12. What is a use case's typical course of events?
13. Why is ranking and evaluating of use cases essential?
14. What are the six criteria in the use-case ranking and priority matrix?
15. What is the use-case dependency diagram, and why do we use it?

Problems and Exercises

1. According to author Fred Brooks, what is the single most difficult thing to do in systems development? How does use-case modeling help in this area?

2. In use case modeling, what two main artifacts does the systems analyst use? Describe each of these artifacts and explain their purpose.

3. What should a systems analyst always keep in mind in identifying and developing use cases regarding their purpose? Since requirements fact-finding has been completed previously, is it really necessary to spend much time with users at this point? Just what should a use case represent? Is a use case the same as a functional requirement?

4. During what part of the development life cycle are use cases first defined? When are they used during the development life cycle, and for what purpose?

5. Match the following stakeholders and external users with the correct actor. What is a temporal event? Who or what is considered to be the actor in a temporal event, and why?

Stakeholders and external users

- United States Postal Service
- Computerized door lock with key pad
- Rental car agent
- Sales manager generating regional sales report
- Sales manager receiving regional sales report
- Automatic lawn sprinkler system
- Driver purchasing gasoline with ATM card
- Bank loan authorization service

Actor

Primary business actor

Primary system actor

External server actor

External receiver actor

Time

6. What is the type of relationship for each of the following examples?

- The relationship between the use case "Print Form" and several other use cases that involve printing different types of forms.
- The relationship between a motorcycle officer and a handheld citation writing device.
- The relationship between a customer and a sales clerk who can each query the inventory system to see if an item is in stock, and an actor created specifically to minimize duplicative system communication.
- The relationship between the use case "Calculate GPA" and the lengthy use case "Create Transcript."
- The relationship between the use case "Ship Order" and the use case "Submit Order."

7. Y&J Cookbooks is a fictional small business owned and operated by a retired couple. Up to this time, Y&J Cookbooks has sold its books by mail order only. The owners now want to develop an online system to sell rare and out-of-print cookbooks over the Internet. Visitors will be able to browse a variety of cookbooks, but they will have to create a customer account before being able to make a purchase. Payments will be accepted only online with a major credit card, and the credit card will be verified before the order can be approved. Based upon this information, identify the main business actors.

8. In use-case modeling, once you identify the business actors, what perspective and language should you use in defining them? Use that perspective and language to construct an actor glossary using Figure 7-8 as an example.

9. A context diagram can help tremendously in identifying different use cases. Prepare a high-level context diagram for the Y&J Cookbooks Web site, using Figure 7-9 as an example.

10. The next step in requirements use-case modeling is to identify the business requirements use cases. What should each use case capture? What effective technique can you use to identify use cases? What questions might you ask in order to identify use cases? What is the difference between a use case and an essential use case?

11. The third step in use-case modeling is to construct the use-case model diagrams. Based upon the Y&J Cookbooks actor glossary and context diagram, create a high-level use-case model diagram, showing the interactions between the shopper/customer actor and the system, including searching and browsing for books, purchasing, and managing the customer account. Make sure to show the relationships between the actor and each of these use cases.

12. The next step is creating use-case narratives to document the business requirements. Why is preparation of the narratives generally done in a two-step process, and what are these two steps? Based upon the preceding high-level use-case

model diagram, create an expanded narrative, using Figure 7-13 as an example.

13. What is the relationship of use-case modeling to project management? Why is this important? Why are use cases ranked, and what tool is used to rank them? Who provides the input for ranking them? What criteria are used for ranking? Explain why use-case dependencies need to be identified, and provide an example. What tool is used to identify dependencies?

Projects and Research

1. At the beginning of Chapter 7, there is a quote taken from an article by Frederick P. Brooks Jr., who is generally considered to be one of the leading authors and contributors to the field of project management and software development. Search the Web for this article and for other articles by and/or about Fred Brooks.

    a. In conducting your search, how many references to the author did you find?

    b. Based upon the information presented in the previous chapters, explain Brooks's statement that "the hardest single part of building a software system is deciding precisely what to build."

    c. What was the name of the article in which Brooks made the preceding statement, and what was the article's main theme?

    d. What is Brooks's best known book that is still in print and widely read decades after its original publication? What was the main theme of this book?

    e. What do you consider to be Brooks's greatest contribution to date? Why?

2. The Standish Group, which was mentioned in Chapter 7, is an independent research group that studies changes and trends in information technology. In 1994, the Standish Group published its groundbreaking CHAOS Report, which "expose[d] the overwhelming failure of IT application development projects in today's MIS environment." Since that time, the Standish Group has published periodic updates to their original report. Go to their Web site at *www.standishgroup.com,* and follow the links to their public access area, where you can find a summary of their latest CHAOS research report, as well as the original 1994 report itself.

    a. What criteria does the Standish Group use to determine whether a project succeeded, was challenged, or failed?

    b. Based upon the latest research report, what percentage of projects succeeded, were challenged, or failed?

    c. How do these latest rates compare to the project success, challenge, and failure rates shown in Figure 7-1 of the textbook? How would you describe the overall trends, if any?

    d. Based upon your reading and experience, what do you believe to be the reason(s) for the changes in project success, challenge, and failure rates?

    e. Do you think that current project management and system development methodologies will remain essentially the same but continue to be refined, or do you foresee the emergence of dramatically different methodologies for managing projects and developing systems over the next decade?

3. Select an information system used in your organization or in your school. Interview a systems analyst or designer who is familiar with the system. Based upon the information provided, do the following:

    a. Describe the information system and organization you selected.

    b. Create a context diagram of the system.

    c. Identify the business actors.

    d. Create an actor glossary.

    e. Identify the business requirements essential use cases.

    f. Create a use case glossary.

4. Based upon the information provided regarding the system you selected in the preceding question:

    a. Construct a use-case model diagram that includes all major subsystems.

    b. Prepare expanded use-case narratives for three of the essential use cases.

    c. Prepare a use-case ranking and priority matrix, then use it to rank and evaluate the use cases.

    d. Identify use case dependencies.

    e. Prepare a use-case dependency diagram.

5. Search the Web or professional journals in your school library for research articles on new and emerging developments in use-case modeling. Select two articles, then do the following:

    a. Provide a bibliography for each article. (Use the format used by your school.)
    b. Create an abstract in your own words for each article.
    c. Compare and contrast the methodologies described in each article. Describe which one you feel is more viable and/or significant, and explain why.

6. Conduct interviews with several developers regarding their views on use-case modeling. If possible, try to find developers from different organizations and/or with different lengths of experience, as well as different types of experience (i.e., a developer who has experience mostly as a systems analyst, one mostly as a designer, and one as a builder).

    a. Describe the developers that you interviewed in terms of their experience. For example, how long have they worked in IT, what is their area of expertise, and how familiar are they are with use-case modeling?

    b. What is the nature of their organization(s)?
    c. What questions did you ask?
    d. What are the viewpoints of each developer regarding the importance and value of use-case modeling?
    e. Do these developers actually employ use-case modeling in their current organization? Why or why not?
    f. If they were CIOs of their organization for a day, would they change their organization's IT architecture regarding use-case modeling? If so, how?
    g. Using the capability maturity model, how would you rate the maturity level of their organization? Why?
    h. What conclusions, if any, can you draw from the interviews regarding the practical application of use-case modeling?
    i. What were your views regarding the importance and value of use-ease modeling before the interviews? Did your views change any as a result of the interviews? If so, how did they change and why?

Minicases

1. In a mincase for Chapter 6, you interviewed stakeholders for an online class registration system. In that exercise, you were to develop an understanding of any issues and needs those stakeholders had in regards to the system. Review your findings from those interviews.

    a. Visit other school registration systems. Is there any functionality or ease of use differences? Are there any features that you think the stakeholders would particularly like/dislike? Make notes and create screen dump examples of other systems.

    b. Create a follow-up interview with those stakeholders you previously spoke to and determine specific functionality and ease-of-use requirements for your school.

2. Create a use-case description for at least one of the functionality requirements you found in the previous problem. Follow the example shown in Figure 7-10.
3. Identify all of the actors for the school registration system. Which uses cases will each initiate?
4. Using your answer to the previous problem, draw a use-case diagram of the school registration system.

Suggested Readings

Ambler, Scott W. *The Object Primer.* New York: Cambridge University Press, 2001. Very good information about documenting use cases and their use.

Armour, Frank, and Granville Miller. *Advance Use Case Modeling.* Boston: Addison-Wesley, 2001. This book presents excellent coverage of the use-case modeling process.

Brooks, Jr., F.P., 1987. "No SilverBullet—Essence and Accidents of Software Engineering." *Computer* 20(4), April, 10–19. *proc. IFIP Congress,* Dublin, Ireland, 1986

Jacobson, Ivar; Magnus Christerson; Patrik Jonsson; and Gunnar Overgaard. *Object-Oriented Software Engineering— A Use Case Driven Approach.* Workingham, England: Addison-Wesley, 1992. This book presents detailed coverage of how to identify and document use cases.

Larman, Craig. *Applying UML and Patterns.* Upper Saddle River, NJ: Prentice Hall, 1998. This book provides a comprehensive overview of a use-case modeling process.

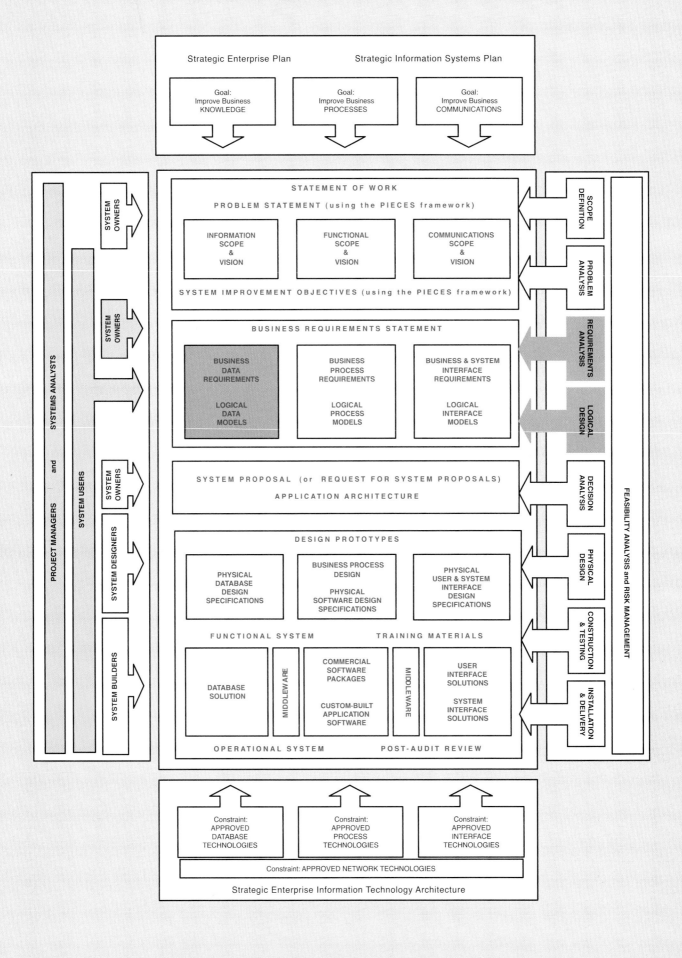

Data Modeling and Analysis

Chapter Preview and Objectives

In this chapter you will learn how to use a popular data-modeling tool, *entity relationship diagrams,* to document the data that must be captured and stored by a system, independently of showing how that data is or will be used—that is, independently of specific inputs, outputs, and processing. You will also learn about a data analysis technique called *normalization* that is used to ensure that a data model is a "good" data model. You will know data modeling and data analysis as systems analysis tools and techniques when you can:

■ Define systems modeling and differentiate between logical and physical system models.

■ Define data modeling and explain its benefits.

■ Recognize and understand the basic concepts and constructs of a data model.

■ Read and interpret an entity relationship data model.

■ Explain when data models are constructed during a project and where the models are stored.

■ Discover entities and relationships.

■ Construct an entity relationship context diagram.

■ Discover or invent keys for entities and construct a key-based diagram.

■ Construct a fully attributed entity relationship diagram and describe all data structures and attributes to the repository or encyclopedia.

■ Normalize a logical data model to remove impurities that can make a database unstable, inflexible, and nonscalable.

■ Describe a useful tool for mapping data requirements to business operating locations.

Introduction

As the SoundStage Member Services system project moves from requirements analysis into logical design, the first task according to their methodology is to analyze the data requirements for the new system. Bob Martinez remembers a favorite professor in college who always said, "Get the data right and the system will be able to elegantly support all your present requirements and even requirements users don't yet envision; get the data wrong and it will be a pain in the neck to meet requirements today, tomorrow, and forever."

Bob enjoyed his database classes in college and always did well in them. Of course, the member services system is larger and more detailed than any data project he did in school. Fortunately, he has the database from the previous version of the system to start with, plus forms and reports from the previous system, plus notes from user interviews, plus use-case narratives created during the requirements analysis phase. Sandra has asked Bob to take the first shot at pulling it all together into a logical data model. He's determined to impress her.

What Is Data Modeling?

data modeling a technique for organizing and documenting a system's data. Sometimes called *database modeling*.

Systems models play an important role in systems development. This chapter will present **data modeling** as a technique for defining business requirements for a database. Data modeling is sometimes called *database modeling* because a data model is eventually implemented as a database.

Figure 8-1 is an example of a simple data model called an *entity relationship diagram*, or *ERD*. This diagram makes the following business assertions:

- We need to store data about CUSTOMERS, ORDERS, and INVENTORY PRODUCTS.
- The value of CUSTOMER NUMBER <u>uniquely</u> identifies one and only one CUSTOMER. The value of ORDER NUMBER <u>uniquely</u> identifies one and only one ORDER. The value of PRODUCT NUMBER <u>uniquely</u> identifies one and only one INVENTORY PRODUCT.
- For a CUSTOMER we need to know the CUSTOMER NAME, SHIPPING ADDRESS, BILLING ADDRESS, and BALANCE DUE. For an ORDER we need to know ORDER DATE and ORDER TOTAL COST. For an INVENTORY PRODUCT we need to know PRODUCT NAME, PRODUCT UNIT OF MEASURE, and PRODUCT UNIT PRICE.

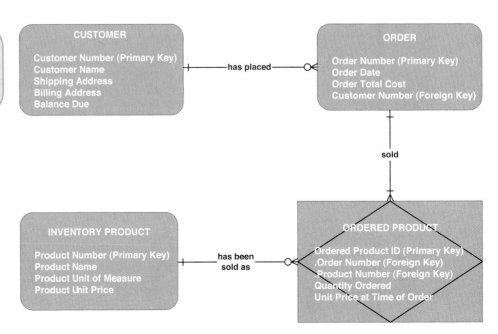

FIGURE 8-1

An Entity Relationship Data Model

- A CUSTOMER has placed zero, one, or more current or recent ORDERS.
- An ORDER is placed by exactly one CUSTOMER. The value of CUSTOMER NUMBER (as recorded in ORDER) identifies that CUSTOMER.
- An ORDER sold one or more ORDERED PRODUCTS. Thus, an ORDER must contain at least one ORDERED PRODUCT.
- An INVENTORY PRODUCT may have been sold as zero, one, or more ORDERED PRODUCTS.
- An ORDERED PRODUCT identifies a single INVENTORY PRODUCT on a single ORDER. The ORDER NUMBER (for an ORDERED PRODUCT) identifies the ORDER, and the PRODUCT NUMBER (for an ORDERED PRODUCT) identifies the INVENTORY PRODUCT. Together, they identify one and only one ORDERED PRODUCT.
- For each ORDERED PRODUCT we need to know QUANTITY ORDERED and UNIT PRICE AT TIME OF ORDER.

After you study this chapter, you will be able to read data models and construct them.

System Concepts for Data Modeling

There are several notations for data modeling. The actual model is frequently called an **entity relationship diagram (ERD)** because it depicts data in terms of the entities and relationships described by the data. There are several notations for ERDs. Most are named after their inventor (e.g., Chen, Martin, Bachman, Merise) or after a published standard (e.g., IDEF1X). These data modeling "languages" generally support the same fundamental concepts and constructs. We have adopted the Martin (information engineering) notation because of its widespread use and CASE tool support.

Let's explore some basic concepts that underlie all data models.

> **entity relationship diagram (ERD)** a data model utilizing several notations to depict data in terms of the entities and relationships described by that data.

> Entities

All systems contain data—usually lots of data! Data describes "things." Consider a school system. A school system includes data that describes things such as STUDENTS, TEACHERS, COURSES, and CLASSROOMS. For any of these things, it is not difficult to imagine some of the data that describes any given instance of the thing. For example, the data that describes a particular student might include NAME, ADDRESS, PHONE NUMBER, DATE OF BIRTH, GENDER, RACE, MAJOR, and GRADE POINT AVERAGE, to name a few.

We need a concept to abstractly represent all instances of a group of similar things. We call this concept an entity. An **entity** is something about which the business needs to store data. In system modeling, we find it useful to assign each abstract concept to a shape. In this book, an entity will be drawn as a rectangle with rounded corners (see margin). This shape represents all instances of the named entity. For example, the entity STUDENT represents all students in the system. Thus, an entity identifies specific classes of entities and is distinguishable from the other entities.

Categories of entities (and examples) include:

> **entity** a class of persons, places, objects, events, or concepts about which we need to capture and store data.

An Entity

Persons: AGENCY, CONTRACTOR, CUSTOMER, DEPARTMENT, DIVISION, EMPLOYEE, INSTRUCTOR, STUDENT, SUPPLIER. Notice that a person entity class can represent individuals, groups, or organizations.

Places: SALES REGION, BUILDING, ROOM, BRANCH OFFICE, CAMPUS.

Objects: BOOK, MACHINE, PART, PRODUCT, RAW MATERIAL, SOFTWARE LICENSE, SOFTWARE PACKAGE, TOOL, VEHICLE MODEL, VEHICLE. An object entity can represent actual objects (such as a specific software license) or specifications for a type of object (such as specifications for different software packages).

Events: APPLICATION, AWARD, CANCELLATION, CLASS, FLIGHT, INVOICE, ORDER, REGISTRATION, RENEWAL, REQUISITION, RESERVATION, SALE, TRIP.

Concepts: ACCOUNT, BLOCK OF TIME, BOND, COURSE, FUND, QUALIFICATION, STOCK.

It is important to distinguish between an entity and its instances. An **entity instance** is a single occurrence of an entity. For example, the entity STUDENT may have multiple instances: Mary, Joe, Mark, Susan, Cheryl, and so forth. In data modeling, we do not concern ourselves with individual students because we recognize that each student is described by similar pieces of data.

entity instance a single occurrence of an entity.

> Attributes

If an entity is something about which we want to store data, then we need to identify what specific pieces of data we want to store about each instance of a given entity. We call these pieces of data **attributes.** As noted at the beginning of this section, each instance of the entity STUDENT might be described by the following attributes: NAME, ADDRESS, PHONE NUMBER, DATE OF BIRTH, GENDER, RACE, MAJOR, GRADE POINT AVERAGE, and others.

attribute a descriptive property or characteristic of an entity. Synonyms include *element, property,* and *field.*

We can now extend our graphical abstraction of the entity to include attributes by recording those attributes inside the entity shape along with the name (see margin).

Some attributes can be logically grouped into superattributes called **compound attributes.** For example, a student's NAME is actually a compound attribute that consists of LAST NAME, FIRST NAME, and MIDDLE INITIAL. In the margin, we demonstrate one <u>possible</u> notation for compound attributes. Notice that a period is placed at the beginning of each primitive attribute that is included in the composite attribute.

Domains   When analyzing a system, we should define those values for an attribute that are legitimate or that make business sense. The values for each attribute are defined in terms of three properties: data type, domain, and default.

The **data type** for an attribute defines what type of data can be stored in that attribute. Data typing should be familiar to those of you who have written computer programs; declaring types for variables is common to most programming languages. For purposes of systems analysis and business requirements definition, it is useful to declare logical (nontechnical) data types for business attributes. For the sake of argument, we will use the logical data types shown in Table 8-1.

An attribute's data type constrains its domain. The **domain** of an attribute defines what values the attribute can legitimately take on. Eventually, system designers must use technology to enforce the business domains of all attributes. Table 8-2 demonstrates how logical domains might be expressed for each data type.

STUDENT
Name
.Last Name
.First Name
.Middle Initial
Address
.Street Address
.City
.State or Province
.Country
.Postal Code
Phone Number
.Area Code
.Exchange Number
.Number Within Exchange
Date of Birth
Gender
Race
Major
Grade Point Average

**Attributes and
Compound Attributes**

compound attribute an attribute that consists of other attributes. Synonyms in different data modeling languages are numerous: *concatenated attribute, composite attribute,* and *data structure.*

data type a property of an attribute that identifies what type of data can be stored in the attribute.

domain a property of an attribute that defines what values the attribute can legitimately take on.

TABLE 8-1 Representative Logical Data Types for Attributes

Logical Data Type	Logical Business Meaning
NUMBER	Any number, real or integer.
TEXT	A string of characters, inclusive of numbers. When numbers are included in a TEXT attribute, it means we do not expect to perform arithmetic or comparisons with those numbers.
MEMO	Same as TEXT but of an indeterminate size. Some business systems require the ability to attach potentially lengthy notes to a given database record.
DATE	Any date in any format.
TIME	Any time in any format.
YES/NO	An attribute that can assume only one of these two values.
VALUE SET	A finite set of values. In most cases, a coding scheme would be established (e.g., FR = freshman, SO = sophomore, JR = junior, SR = senior, etc.).
IMAGE	Any picture or image.

TABLE 8-2 Representative Logical Domains for Logical Data Types

Data Type	Domain	Examples
NUMBER	For integers, specify the range: {minimum–maximum}	{10–99}
	For real numbers, specify the range and precision: {minimum.precision–maximum.precision}	{1.000–799.999}
TEXT	TEXT (maximum size of attribute) *Actual values are usually infinite; however, users may specify certain narrative restrictions.*	TEXT (30)
MEMO	*Not applicable. There are no logical restrictions on size or content.*	*Not applicable.*
DATE	Variation on the MMDDYYYY format. To accommodate the year 2000, do not abbreviate year to YY.	MMDDYYYY
	Formatting characters are rarely stored; therefore, do not include hyphens or slashes.	MMYYYY YYYY
TIME	For AM/PM times: HHMMT or For military times: HHMM	HHMMT HHMM
YES/NO	{YES, NO}	{YES, NO} {ON, OFF}
VALUE SET	{value#1, value#2, . . . value#n} or {table of codes and meanings}	{FRESHMAN, SOPHOMORE, JUNIOR, SENIOR} {FR = FRESHMAN SO = SOPHOMORE JR = JUNIOR SR = SENIOR}
IMAGE	*Not applicable; however, any known characteristics of the images will eventually prove useful to designers.*	*Not applicable.*

Finally, every attribute should have a logical **default value** that represents the value of an attribute if its value is not specified by the user. Table 8-3 shows possible default values for an attribute. Notice that NOT NULL is a way to specify that each instance of the attribute must <u>have</u> a value, while NULL is a way to specify that some instances of the attribute may be optional, or not have a value.

Identification   An entity has many instances, perhaps thousands or millions. There exists a need to uniquely identify each instance based on the data value of one or more attributes. Thus, every entity must have a **key.** For example, each instance of the entity STUDENT might be uniquely identified by the key STUDENT NUMBER attribute. No two students can have the same STUDENT NUMBER.

Sometimes more than one attribute is required to uniquely identify an instance of an entity. A key consisting of a group of attributes is called a **concatenated key.** For example, each DVD entity instance in a video store might be uniquely identified by the

default value the value that will be recorded if a value is not specified by the user.

key an attribute, or a group of attributes, that assumes a unique value for each entity instance. It is sometimes called an *identifier.*

concatenated key a group of attributes that uniquely identifies an instance of an entity. Synonyms include *composite key* and *compound key.*

> **TABLE 8-3 Permissible Default Values for Attributes**

Default Value	Interpretation	Examples
A legal value from the domain (as described above)	For an instance of the attribute, if the user does not specify a value, then use this value.	0 1.00 FR
NONE or NULL	For an instance of the attribute, if the user does not specify a value, then leave it blank.	NONE NULL
REQUIRED or NOT NULL	For an instance of the attribute, requires that the user enter a legal value from the domain. (This is used when no value in the domain is common enough to be a default but some value must be entered.)	REQUIRED NOT NULL

concatenation of TITLE NUMBER plus COPY NUMBER. TITLE NUMBER by itself would be inadequate because the store may own many copies of a single title. COPY NUMBER by itself would also be inadequate since we presumably have a *copy #1* for every title we own. We need both pieces of data to identify a specific tape (e.g., *copy #7* of *Star Wars: Revenge of the Sith*). In this book, we will give a name to the group as well as the individual attributes. For example, the concatenated key for DVD would be recorded as follows:

DVD ID
.TITLE NUMBER
.COPY NUMBER

Frequently, an entity may have more than one key. For example, the entity EMPLOYEE may be uniquely identified by SOCIAL SECURITY NUMBER, or company-assigned EMPLOYEE NUMBER, or E-MAIL ADDRESS. Each of these attributes is called a *candidate key*. A **candidate key** is a "candidate to become the primary key" of instances of an entity. It is sometimes called a *candidate identifier*. (A candidate key may be a single attribute or a concatenated key.) A **primary key** is that candidate key that will most commonly be used to uniquely identify a single entity instance. The default for a primary key is always NOT NULL because if the key has no value, it cannot serve its purpose to identify an instance of an entity. Any candidate key that is not selected to become the primary key is called an **alternate key.** A common synonym is *secondary key.* In the margin, we demonstrate our notation for primary and alternate keys. All candidate keys must be either primary or alternate; therefore, we do not use a separate notation for candidate keys. All attributes that are not part of the primary key are called nonkey attributes.

Sometimes, it is also necessary to identify a subset of an entity's instances as opposed to a single instance. For example, we may require a simple way to identify all male students and all female students. A **subsetting criteria** is an attribute (or concatenated attribute) whose finite values divide all entity instances into useful subsets. This is sometimes referred to as an *inversion entry.* In our STUDENT entity, the attribute GENDER divides the instances of STUDENT into two subsets: male students and female students. In general, subsetting criteria are useful only when an attribute has a finite (meaning "limited") number of legitimate values. For example, GRADE POINT AVERAGE would not be a good subsetting criteria because there are 999 possible values between 0.00 and 4.00 for that attribute. The margin art demonstrates a notation for subsetting criteria.

> Relationships

Conceptually, entities and attributes do not exist in isolation. The things they represent interact with and impact one another to support the business mission. Thus, we

---

candidate key one of a number of keys that may serve as the primary key of an entity. Also called *candidate identifier.*

primary key a candidate key that will most commonly be used to uniquely identify a single entity instance.

alternate key a candidate key that is not selected to become the primary key. A synonym is *secondary key.*

subsetting criteria an attribute(s) whose finite values divide entity instances into subsets. Sometimes called *inversion entry.*

introduce the concept of a relationship. A **relationship** is a natural business association that exists between one or more entities. The relationship may represent an event that links the entities or merely a logical affinity that exists between the entities. Consider, for example, the entities STUDENT and CURRICULUM. We can make the following business assertions that link students and courses:

- A current STUDENT IS ENROLLED IN one or more CURRICULA.
- A CURRICULUM IS BEING STUDIED BY zero, one, or more STUDENTS.

The underlined verb phrases define business relationships that exist between the two entities.

We can graphically illustrate this association between STUDENT and CURRICULUM as shown in Figure 8-2. The connecting line represents a relationship. Verb phrases describe the relationship. Notice that all relationships are implicitly bidirectional, meaning they can be interpreted in both directions (as suggested by the above business assertions). Data modeling methods may differ in their naming of relationships—some include both verb phrases, and others include a single verb phrase.

Cardinality   Figure 8-2 also shows the complexity or *degree* of each relationship. For example, if we know how to read it, Figure 8-2 can answer the following questions:

- Must there exist an instance of STUDENT for each instance of CURRICULUM? No!
- Must there exist an instance of CURRICULUM for each instance of STUDENT? Yes!
- How many instances of CURRICULUM can exist for each instance of STUDENT? Many!
- How many instances of STUDENT can exist for each instance of CURRICULUM? Many!

We call this concept *cardinality*. **Cardinality** defines the minimum and maximum number of occurrences of one entity that may be related to a single occurrence of the other entity. Because all relationships are bidirectional, cardinality must be defined in both directions for every relationship. A popular graphical notation for cardinality is shown in Figure 8-3. Sample cardinality symbols were demonstrated in Figure 8-2.

Conceptually, cardinality tells us the following rules about the data entities shown in Figure 8-2:

- When we insert a STUDENT instance in the database, we <u>must</u> link (associate) that STUDENT to at least one instance of CURRICULUM. In business terms, "a student cannot be admitted without declaring a major." (Most schools would include an instance of CURRICULUM called "undecided" or "undeclared.")
- A STUDENT <u>can</u> study more than one CURRICULUM, and we must be able to store data that indicates all CURRICULA for a given STUDENT.
- We must insert a CURRICULUM before we can link (associate) STUDENTS to that CURRICULUM. That is why a CURRICULUM can have zero students—no students have yet been admitted to that CURRICULUM.
- Once a CURRICULUM has been inserted into the database, we can link (associate) many STUDENTS with that CURRICULUM.

Degree   Another measure of the complexity of a data relationship is its degree. The **degree** of a relationship is the number of entities that participate in the relationship. All the relationships we've explored so far are *binary* (degree = 2). In other words, two different entities participated in the relationship.

STUDENT

Student Number
  (Primary Key)
Social Security Number
  (Alternate Key)
Name
.Last Name
.First Name
.Middle Initial
Address
.Street Address
.City
.State or Province
.Country
.Postal Code
Phone Number
.Area Code
.Exchange Number
.Number Within Exchange
Date of Birth
Gender (Subsetting Criteria 1)
Race (Subsetting Criteria 2)
Major (Subsetting Criteria 3)
Grade Point Average

Keys and Subsetting Criteria

---

relationship a natural business association between one or more entities.

---

cardinality the minimum and maximum number of occurrences of one entity that may be related to a single occurrence of the other entity.

---

degree the number of entities that participate in a relationship.

STUDENT — is enrolled in ———— is being studied by — CURRICULUM

FIGURE 8-2

A Relationship (Many-to-Many)

FIGURE 8-3

Cardinality
Notations

CARDINALITY INTERPRETATION	MINIMUM INSTANCES	MAXIMUM INSTANCES	GRAPHIC NOTATION
Exactly one (one and only one)	1	1	– or –
Zero or one	0	1	
One or more	1	many (>1)	
Zero, one, or more	0	many (>1)	
More than one	>1	>1	

recursive relationship a relationship that exists between instances of the same entity.

Relationships may also exist between different instances of the same entity. We call this a **recursive relationship** (degree = 1). For example, in your school a course may be a prerequisite for other courses. Similarly, a course may have several other courses as its prerequisite. Figure 8-4 demonstrates this many-to-many recursive relationship.

Relationships can also exist between more than two different entities. These are sometimes called *N*-ary relationships. An example of a *3-ary,* or *ternary, relationship* is shown in Figure 8-5. An *N*-ary relationship is illustrated with a new entity construct called an *associative entity*. An **associative entity** is an entity that inherits its primary key from more than one other entity (called *parents*). Each part of that concatenated key points to one and only one instance of each of the connecting entities.

associative entity an entity that inherits its primary key from more than one other entity.

In Figure 8-5 the associative entity ASSIGNMENT (notice the unique shape) matches an EMPLOYEE, a LOCATION, and a PROJECT. For each instance of ASSIGNMENT, the key

FIGURE 8-4

A Recursive
Relationship

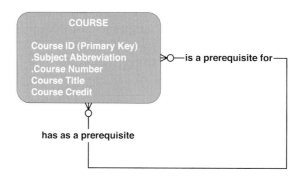

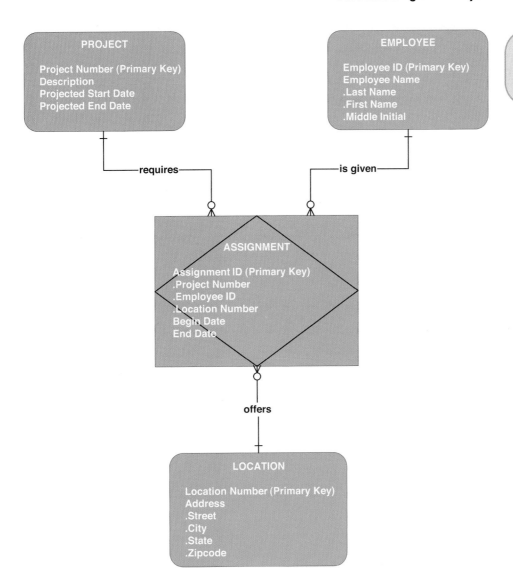

FIGURE 8-5

A Ternary
Relationship

indicates which EMPLOYEE ID, which LOCATION NUMBER, and which PROJECT NUMBER are combined to form that assignment.

Also as shown in Figure 8-5, an associative entity can be described by its own nonkey attributes. In addition to the primary key, an ASSIGNMENT is described by the attributes BEGIN DATE and END DATE. If you think about it, none of these attributes describes an EMPLOYEE, LOCATION, or PROJECT—they describe a single instance of the relationship between an instance of each of those entities.

Foreign Keys   A relationship implies that instances of one entity are related to instances of another entity. We should be able to identify those instances for any given entity. The ability to identify specific related entity instances involves establishing foreign keys. A **foreign key** is a primary key of one entity that is contributed to (duplicated in) another entity to identify instances of a relationship. A foreign key (always in a **child entity**) always matches the primary key (in a **parent entity**). In Figure 8-6(a), we demonstrate the concept of foreign keys with our simple data model. Notice that the <u>maximum</u> cardinality for DEPARTMENT is "one," whereas the maximum cardinality for CURRICULUM is "many." In this case, DEPARTMENT is called the *parent* entity and CURRICULUM is the *child* entity. The primary key is always contributed by the parent to the child as a foreign key. Thus, an instance of CURRICULUM now has a foreign key DEPARTMENT NAME whose value points to the instance of DEPARTMENT that offers that curriculum. (Foreign keys are never contributed from child to parent.)

foreign key a primary key of an entity that is used in another entity to identify instances of a relationship.

child entity a data entity that derives one or more attributes from another entity, called the parent. In a one-to-many relationship the child is the entity on the "many" side.

parent entity a data entity that contributes one or more attributes to another entity, called the child. In a one-to-many relationship the parent is the entity on the "one" side.

(a)

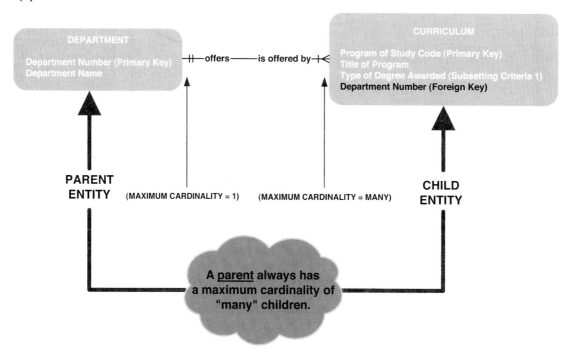

(b)

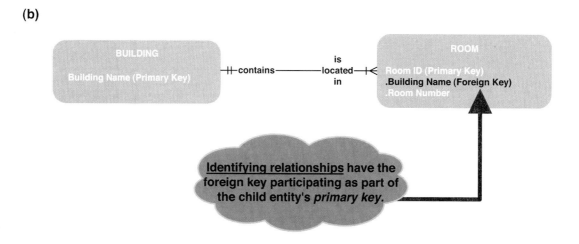

FIGURE 8-6 Foreign Keys

nonidentifying relationship a relationship in which each participating entity has its own independent primary key.

In our example, the relationship between CURRICULUM and DEPARTMENT is referred to as a *nonidentifying relationship.* **Nonidentifying relationships** are those in which each of the participating entities has its own independent primary key. In other words, none of the primary-key attributes is shared. The entities CURRICULUM and DEPARTMENT are also referred to as *strong* or independent entities because neither depends on any other entity for its identification. Sometimes, however, a foreign key may participate as part of the primary key of the child entity. For example, in Figure 8-6(b) the parent entity BUILDING contributes its key to the entity ROOM. Thus, BUILDING NAME serves as a foreign key to relate a ROOM and BUILDING and in conjunction with ROOM ID

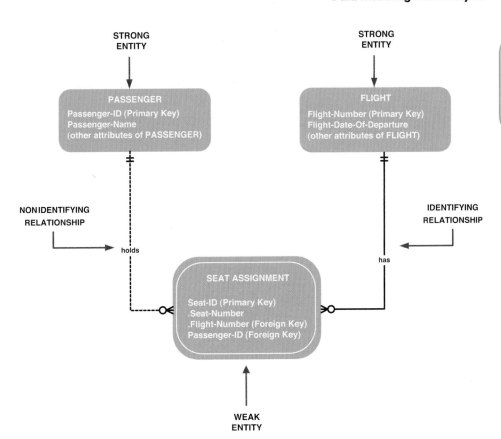

FIGURE 8-7

Notations for Weak Entity and Nonidentifying Relationship

to uniquely identify a given instance of ROOM. In those situations the relationship between the parent entity and the child entity is referred to as an *identifying relationship*. **Identifying relationships** are those in which the parent entity contributes its primary key to become part of the primary key of the child entity. The child entity of any identifying relationship is frequently referred to as a *weak entity* because its identification is dependent on the parent entity's existence.

Most popular CASE tools and data modeling methods use different notations to distinguish between identifying and nonidentifying relationships and between strong and weak entities. In Figure 8-7, we use a *dashed line* notation to represent the nonidentifying relationship between PASSENGER and SEAT ASSIGNMENT. Because part of the primary key of SEAT ASSIGNMENT is the foreign key FLIGHT NUMBER from the parent entity FLIGHT, the relationship is an identifying relationship and is represented using a *solid line*. Finally, seat assignment is a weak entity because it receives the primary key of flight to compose its own primary key. A weak entity is represented using a symbol composed of a *rounded rectangle within a rounded rectangle*.

NOTE: To reinforce the above concepts of identifying and nonidentifying relationships and strong versus weak entities and to be consistent with most popular data modeling methods and most widely used CASE tools, the authors use the above modeling notations on all subsequent data modeling examples presented in the book.

What if you cannot differentiate between parent and child? For example, in Figure 8-8(a) on page 280 we see that a CURRICULUM enrolls zero, one, or more STUDENTS. At the same time, we see that a STUDENT is enrolled in one or more CURRICULA. The maximum cardinality on both sides is "many." So, which is the parent and which is the child? You can't tell! This is called a *nonspecific relationship*. A **nonspecific relationship** (or *many-to-many relationship*) is one in which many instances of one entity are associated with many instances of another entity. Such relationships are suitable only for preliminary data models and should be resolved as quickly as possible.

identifying relationship
a relationship in which the parent entity's key is also part of the primary key of the child entity.

nonspecific relationship
a relationship where many instances of an entity are associated with many instances of another entity. Also called *many-to-many relationship*.

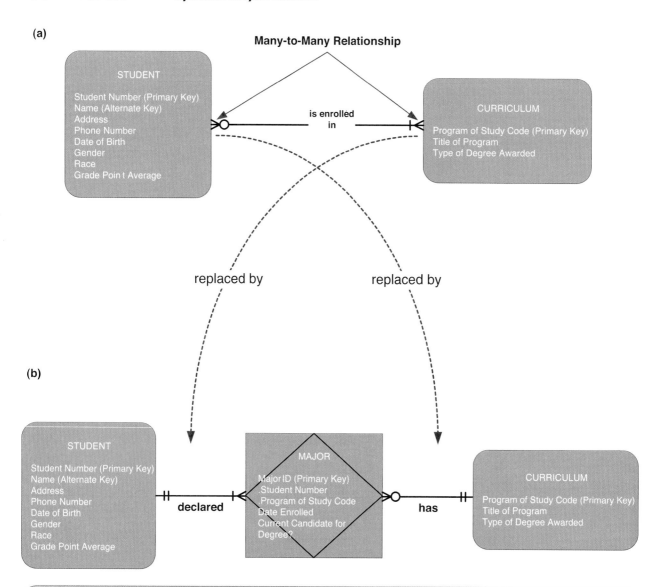

FIGURE 8-8 Resolving Nonspecific Relationships with an Associative Entity

Many nonspecific relationships can be resolved into a pair of one-to-many relationships. As illustrated in Figure 8-8(b), each entity becomes a parent. A new, *associative entity* is introduced as the child of each parent. In Figure 8-8(b), each instance of MAJOR represents <u>one</u> STUDENT's enrollment in <u>one</u> CURRICULUM. If a student is pursuing two majors, that student will have two instances of the entity MAJOR.

Study Figure 8-8(b) carefully. For associative entities, the cardinality from child to parent is always <u>one and only one</u>. That makes sense because an instance of MAJOR must correspond to one and only one STUDENT and one and only one CURRICULUM. The cardinality from parent to child depends on the business rule. In our example, a STUDENT must declare <u>one or more</u> MAJORS. Conversely, a CURRICULUM is being studied by zero, one, or more MAJORS—perhaps it is new and no one has been admitted to it yet. An associative entity can also be described by its own nonkey attributes (such as DATE ENROLLED and CURRENT CANDIDATE FOR DEGREE?). Finally, associative entities inherit the primary keys of the parents; thus, all associative entities are weak entities.

Not all nonspecific relationships can and should be automatically resolved as described above. Occasionally nonspecific relationships result from the failure of the analyst to identify the existence of other entities. For example, examine the relationship between CUSTOMER and PRODUCT in Figure 8-9(a). Recognize that the relationship

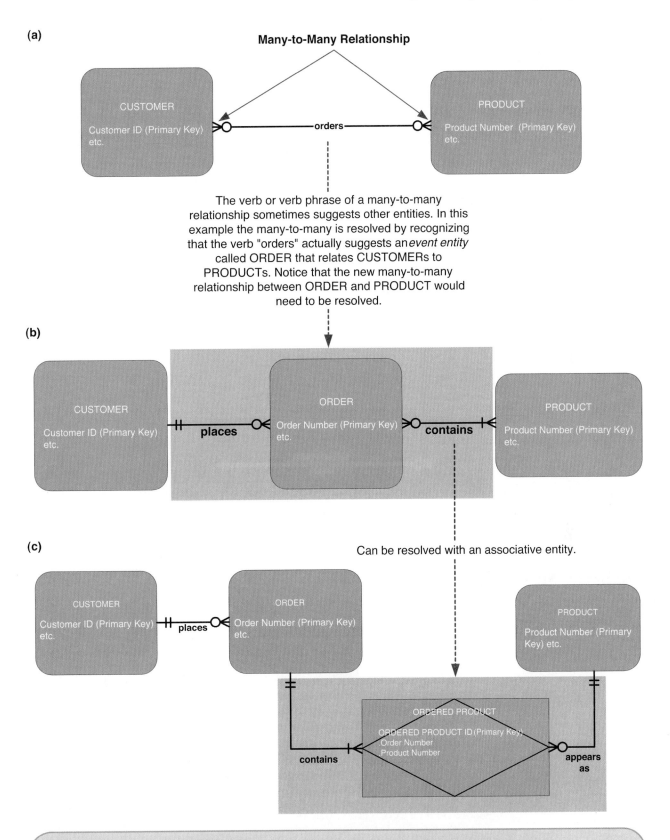

(a)

Many-to-Many Relationship

CUSTOMER

Customer ID (Primary Key)
etc.

—orders—

PRODUCT

Product Number (Primary Key)
etc.

The verb or verb phrase of a many-to-many relationship sometimes suggests other entities. In this example the many-to-many is resolved by recognizing that the verb "orders" actually suggests an *event entity* called ORDER that relates CUSTOMERs to PRODUCTs. Notice that the new many-to-many relationship between ORDER and PRODUCT would need to be resolved.

(b)

CUSTOMER

Customer ID (Primary Key)
etc.

places

ORDER

Order Number (Primary Key)
etc.

contains

PRODUCT

Product Number (Primary Key)
etc.

(c)

Can be resolved with an associative entity.

CUSTOMER

Customer ID (Primary Key)
etc.

places

ORDER

Order Number (Primary Key)
etc.

PRODUCT

Product Number (Primary Key) etc.

contains

ORDERED PRODUCT

ORDERED PRODUCT ID (Primary Key)
.Order Number
.Product Number

appears as

FIGURE 8-9   Resolving Nonspecific Relationships by Recognizing a Fundamental Business Entity

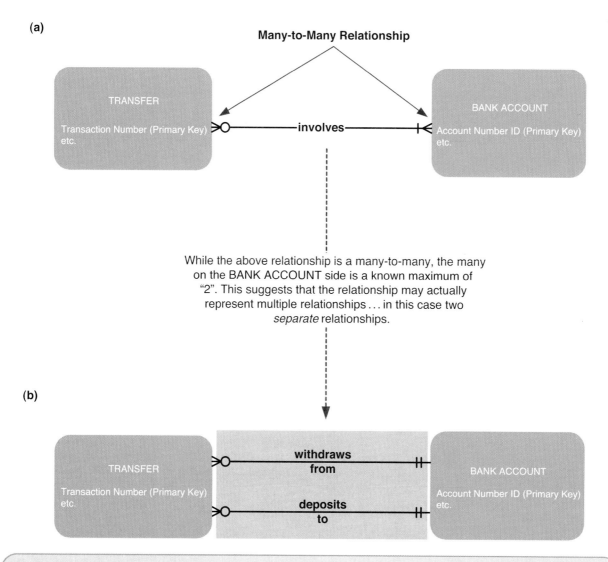

FIGURE 8-10 Resolving Nonspecific Relationships by Recognizing Separate Relationships

"orders" between CUSTOMER and PRODUCT suggests an event about which a user might want to store data. That event represents an event entity called ORDER depicted in Figure 8-9(b). In reality, CUSTOMER and PRODUCT do not have a natural and direct relationship as was depicted in Figure 8-9(a). Rather, they are related indirectly, by way of an ORDER. Thus, our many-to-many relationship was replaced by separate relationships between CUSTOMER, ORDER, and PRODUCT. Notice that the relationship between ORDER and PRODUCT is a many-to-many relationship. That relationship would need to be resolved by replacing it with an associative entity and two one-to-many relationships, as is illustrated in Figure 8-9(c).

Finally, some nonspecific relationships can be resolved by introducing separate relationships. Notice the many-to-many relationship between TRANSFER and BANK ACCOUNT shown in Figure 8-10(a). While it is true that a TRANSFER transaction involves many BANK ACCOUNTS and a BANK ACCOUNT may be involved in many TRANSFER transactions, we must be careful! Data modeling notations can sometimes mislead us. Technically, a single TRANSFER transaction involves two BANK ACCOUNTS. When we know the specific maximum number of occurrences of a many-to-many relationship, it often suggests that our original relationship is weak or too general. Notice in Figure 8-10(b) that our relationship "involves" was replaced by two separate one-to-many relationships that more accurately describe the business relationships between a TRANSFER and BANK ACCOUNTS.

Generalization   Most people associate the concept of generalization with modern object-oriented techniques. In reality, the concepts have been applied by data modelers for many years. Generalization is an approach that seeks to discover and exploit the commonalities between entities. **Generalization** is a technique wherein the attributes that are common to several types of an entity are grouped into their own entity. Consider, for example, a typical school. A school enrolls STUDENTS and employs EMPLOYEES (in a university, a person could be both). There are several attributes that are common to both entities; for example, NAME, GENDER, RACE, MARITAL STATUS, and possibly even a key based on SOCIAL SECURITY NUMBER. We could consolidate these common attributes into an entity supertype called PERSON. An entity **supertype** is an entity whose instances store attributes that are common to one or more entity subtypes.

The entity supertype will have one or more *one-to-one* relationships to entity *subtypes*. These relationships are sometimes called "is a" relationships (or "was a," or "could be a") because each instance of the supertype "is <u>also</u> an" instance of one or more subtypes. An entity **subtype** is an entity whose instances inherit some common attributes from an entity supertype and then add other attributes that are unique to an instance of the subtype. In our example, "a PERSON <u>is an</u> employee, or a student, or both." The top half of Figure 8-11 illustrates this generalization ❶ as a hierarchy. Notice that the subtypes STUDENT and EMPLOYEE have inherited attributes from PERSON, as well as adding their own.

Extending the metaphor, we see that an entity can be both a supertype and a subtype. Returning to Figure 8-11, we see that a STUDENT (which was a subtype of PERSON) has its own subtypes. In the diagram, we see that a STUDENT is ❷ either a PROSPECT, <u>or</u> a CURRENT STUDENT, <u>or</u> a FORMER STUDENT (having left for any reason other than graduation), <u>and</u> ❸ a STUDENT could be an ALUMNUS. These additional subtypes inherit all the attributes from STUDENT as well as those from PERSON. Finally, notice that all subtypes are weak entities.

Through inheritance, the concept of generalization in data models permits us to reduce the number of attributes through the careful sharing of common attributes. The subtypes not only inherit the attributes but also the data types, domains, and defaults of those attributes. This can greatly enhance the consistency with which we treat attributes that apply to many different entities (e.g., dates, names, addresses, currency, etc.).

In addition to inheriting attributes, subtypes also inherit relationships to other entities. For instance, all EMPLOYEES and STUDENTS inherit the relationship ❹ between PERSON and ADDRESS. But only EMPLOYEES inherit the relationship ❺ with CONTRACTS. And only an ALUMNUS can be related to ❻ an AWARDED DEGREE.

> **generalization** a concept wherein the attributes that are common to several types of an entity are grouped into their own entity.

> **supertype** an entity whose instances store attributes that are common to one or more entity subtypes.

> **subtype** an entity whose instances may inherit common attributes from its entity supertype.

The Process of Logical Data Modeling

Now that you understand the basic concepts of data models, we can examine the process of data modeling. When do you do it? How many data models may be drawn? What technology exists to support the process?

Data modeling may be performed during various types of projects and in multiple phases of projects. Data models are progressive; there is no such thing as the "final" data model for a business or application. Instead, a data model should be considered a living document that will change in response to a changing business. Data models should ideally be stored in a repository so that they can be retrieved, expanded, and edited over time. Let's examine how data modeling may come into play during systems planning and analysis.

> Strategic Data Modeling

Many organizations select application development projects based on strategic information systems plans. Strategic planning is a separate project. This project produces

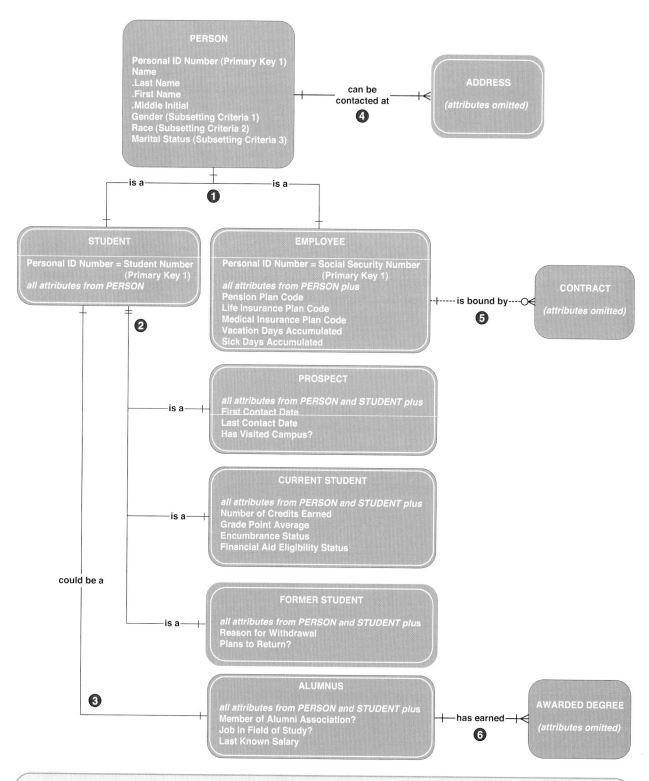

FIGURE 8-11 A Generalization Hierarchy

an information systems strategy plan that defines an overall vision and architecture for information systems. Almost always, this architecture includes an *enterprise data model.* Information engineering is a methodology that embraces this approach.

An enterprise data model typically identifies only the most fundamental of entities. The entities are typically defined (as in a dictionary), but they are not described in terms of keys or attributes. The enterprise data model may or may not include relationships (depending on the planning methodology's standards and the level of detail desired by executive management). If relationships are included, many of them will be nonspecific (a concept introduced earlier in the chapter).

How does an enterprise data model affect subsequent applications development? Part of the information strategy plan identifies application development projects and prioritizes them according to whatever criteria management deems appropriate. As those projects are started, the appropriate subsets of the information systems architecture, including a subset of the enterprise data model, are provided to the applications development team as a point of departure.

The enterprise data model is usually stored in a corporate repository. When the application development project is started, the subset of the enterprise data model (as well as the other models) is exported from the corporate repository into a project repository. Once the project team completes systems analysis and design, the expanded and refined data models are imported back into the corporate repository.

> Data Modeling during Systems Analysis

In systems analysis and in this chapter, we will focus on *logical* data modeling as a part of systems analysis. The data model for a single information system is usually called an **application data model.**

> **application data model**
> a data model for a complete, single information system.

Data models are rarely constructed during the scope definition phase of systems analysis. The short duration of that phase makes them impractical. If an enterprise data model exists, the subset of that model that is applicable to the project might be retrieved and reviewed as part of the phase requirement to establish context. Alternatively, the project team could identify a simple list of entities, the things about which team members think the system will have to capture and store data.

Unfortunately, data modeling is rarely associated with the problem analysis phase of systems analysis. Some analysts prefer to draw process models (Chapter 9) to document the current system, but many analysts report that data models are far superior for the following reasons:

- Data models help analysts to quickly identify business vocabulary more completely than process models.
- Data models are almost always built more quickly than process models.
- A complete data model can fit on a single sheet of paper. Process models often require dozens of sheets of paper.
- Process modelers frequently and too easily get hung up on unnecessary detail.
- Data models for existing and proposed systems are far more similar than process models for existing and proposed systems. Consequently, there is less work to throw away as you move into later phases.

We agree! A problem analysis phase model includes only entities and relationships, but no attributes—it is called a **context data model.** The intent is to refine our understanding of scope, not to get into details about the entities and business rules. Many relationships may be nonspecific.

> **context data model** a data model that includes entities and relationships but no attributes.

Many automated tools provide the ability to read existing system files and databases and translate them into "physical" data models. These physical data models can then be transformed into their equivalent "logical" data model. This translation capability benefits both the problem analysis and the requirements analysis phases.

The requirements analysis results in a logical data model that is developed in stages as follows:

1. We begin by constructing the *context data model* to establish the project scope. If a context data model was already developed during problem analysis, that model may be revised to reflect new requirements and project scope.

key-based data model a data model that includes entities and relationships with precise cardinalities resolving non-specific relationships into associative entities, and also including primary and alternate keys.

fully attributed data model a data model that includes all entities, attributes, relationships, subsetting criteria, and precise cardinalities.

2. Next, a **key-based data model** will be drawn. This model will eliminate non-specific relationships, add associative entities, and include primary and alternate keys. The key-based model will also include precise cardinalities and any generalization hierarchies.

3. Next, a **fully attributed data model** will be constructed. The fully attributed model includes all remaining descriptive attributes and subsetting criteria. Each attribute is defined in the repository with data types, domains, and defaults (in what is sometimes called a *fully described data model*).

4. The completed data model is analyzed for adaptability and flexibility through a process called *normalization*. The final analyzed model is referred to as a *normalized data model*.

This data requirements model requires a team effort that includes systems analysts, users and managers, and data analysts. A data administrator often sets standards for and approves all data models.

Ultimately, during the decision analysis phase, the data model will be used to make implementation decisions—the best way to implement the requirements with database technology. In practice, this decision may have already been standardized as part of a database architecture. For example, SoundStage has already standardized on two database management systems: Microsoft *Access* for personal and work-group databases, and Microsoft *SQL Server* for enterprise databases.

Finally, data models cannot be constructed without appropriate facts and information as supplied by the user community. These facts can be collected through a number of techniques such as sampling of existing forms and files, research of similar systems, surveys of users and management, and interviews of users and management. The fastest method of collecting facts and information and simultaneously constructing and verifying the data models is joint requirements planning. JRP uses a carefully facilitated group meeting to collect the facts, build the models, and verify the models—usually in one or two full-day sessions. Fact-finding and information-gathering techniques were fully explored in Chapter 6. Table 8-4 summarizes some questions that may be useful for fact-finding and information gathering as it pertains to data modeling.

> Looking Ahead to Systems Design

During system design, the logical data model will be transformed into a physical data model (called a *database schema*) for the chosen database management system. This model will reflect the technical capabilities and limitations of that database technology, as well as the performance tuning requirements suggested by the database administrator. Any further discussion of database design is deferred until Chapter 14.

> Automated Tools for Data Modeling

metadata data about data.

Data models are stored in a repository. In a sense, the data model is **metadata**—that is, data about the business's data. Computer-aided systems engineering (CASE) technology, introduced in Chapter 3, provides the repository for storing the data model and its detailed descriptions. Most CASE products support computer-assisted data modeling and database design. Some CASE products (such as Logic Works' *ERwin*) only support data modeling and database design. CASE takes the drudgery out of drawing and maintaining these models and their underlying details.

Using a CASE product, you can easily create professional, readable data models without the use of paper, pencil, erasers, and templates. The models can be easily modified to reflect corrections and changes suggested by end users—you don't have to start over! Also, most CASE products provide powerful analytical tools that can check your models for mechanical errors, completeness, and consistency. Some CASE

TABLE 8-4 JRP and Interview Questions for Data Modeling

Purpose	Candidate Questions
Discover the system entities	What are the subjects of the business? In other words, what types of persons, organizations, organizational units, places, things, materials, or events are used in or interact with this system about which data must be captured or maintained? How many instances of each subject exist?
Discover the entity keys	What unique characteristic (or characteristics) distinguishes an instance of each subject from other instances of the same subject? Are there any plans to change this identification scheme in the future?
Discover entity subsetting criteria	Are there any characteristics of a subject that divide all instances of the subject into useful subsets? Are there any subsets of the above subjects for which you have no convenient way to group instances?
Discover attributes and domains	What characteristics describe each subject? For each of these characteristics, (1) what type of data is stored? (2) who is responsible for defining legitimate values for the data? (3) what are the legitimate values for the data? (4) is a value required? and (5) is there any default value that should be assigned if you don't specify otherwise?
Discover security and control needs	Are there any restrictions on who can see or use the data? Who is allowed to create the data? Who is allowed to update the data? Who is allowed to delete the data?
Discover data timing needs	How often does the data change? Over what period of time is the data of value to the business? How long should we keep the data? Do you need historical data or trends? If a characteristic changes, must you know the former values?
Discover generalization hierarchies	Are all instances of each subject the same? That is, are there special types of each subject that are described or handled differently? Can any of the data be consolidated for sharing?
Discover relationships and degrees	What events occur that imply associations between subjects? What business activities or transactions involve handling or changing data about several different subjects of the same or a different type?
Discover cardinalities	Is each business activity or event handled the same way, or are there special circumstances? Can an event occur with only some of the associated subjects, or must all the subjects be involved?

products can even help you analyze the data model for consistency, completeness, and flexibility. The potential time and quality savings are substantial.

As mentioned earlier, some CASE tools support reverse engineering of existing file and database structures into data models. The resulting data models represent "physical" data models that can be revised and reengineered into a new file or database, or they may be translated into their equivalent "logical" model. The logical data model could then be edited and forward engineered into a revised physical data model, and subsequently a file or database implementation.

CASE tools do have their limitations. Not all data modeling conventions are supported by all CASE products. And different CASE tools adopt slightly different

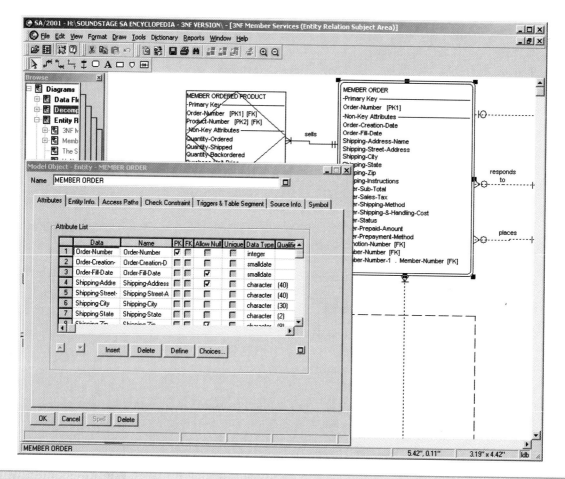

FIGURE 8-12 Screen Capture of *System Architect* CASE Tool

notations for the same data-modeling methods. Therefore, it is very likely that any given CASE product may force a company to adapt its methodology's data-modeling symbols or approach so that it is workable within the limitations of the CASE tool.

All the SoundStage data models in the next section of this chapter were created with Popkin Systems and Software's CASE tool, *System Architect 2001.* For the case study, we provide you with the printouts exactly as they came off our printers. We did not add color. The only modifications by the artist were the bullets that call your attention to specific items of interest on the printouts. All of the entities, attributes, and relationships on the SoundStage data models were automatically cataloged into *System Architect's* project repository (which it calls an encyclopedia). Figure 8-12 illustrates some of *System Architect's* screens as used for data modeling.

How to Construct Data Models

You now know enough about data models to read and interpret them. But as a systems analyst or knowledgeable end user, you must learn how to construct them. We will use the SoundStage Entertainment Club project to teach you how to construct data models.

NOTE: This example teaches you to draw the data model from scratch. In reality, you should always look for an existing data model. If such models exist, the data management or data administration group usually maintains them. Alternatively, you could reverse engineer a data model from an existing database.

> Entity Discovery

The first task in data modeling is relatively easy. You need to discover the fundamental entities in the system that are or might be described by data. You should not restrict your thinking to entities about which the end users know they want to store data. There are several techniques that may be used to identify entities:

- During interviews or JRP sessions with system owners and users, pay attention to key words in their discussion. For example, during an interview with an individual discussing SoundStage's business environment and activities, a user may state, "We have to keep track of all our <u>members</u> and their bound <u>agreements</u>." Notice that the key words in this statement are MEMBERS and AGREEMENTS. Both are entities!
- During interviews or JRP sessions, specifically ask system owners and users to identify things about which they would like to capture, store, and produce information. Those things often represent entities that should be depicted on the data model.
- Another technique for identifying entities is to study existing forms, files, and reports. Some forms identify event entities. Examples include ORDERS, REQUISITIONS, PAYMENTS, DEPOSITS, and so forth. But most of these same forms also contain data that describes other entities. Consider a registration form used in your school's course enrollment system. A REGISTRATION is itself an event entity. But the average registration form also contains data that describes other entities, such as STUDENT (a person), COURSES (which are concepts), INSTRUCTORS (other persons), ADVISOR (yet another person), DIVISIONS (another concept), and so forth. Studying the computerized registration system's computer files, databases, or outputs could also discover these same entities.
- If use-case narratives have been written during the requirements analysis phase, they can be a source of data attributes and entities. Scan each use-case narrative for nouns. Every noun is a potential data attribute or entity. You will have to massage the resulting list of nouns because not all of them will be attributes or entities. Some will be references to users or other information systems. Some will be references to things that are part of the user interface, not data. Some will be synonyms for other attributes or entities elsewhere on the list, and you would not want to duplicate them. Chapter 10 explains how to do this, taking an object-oriented approach to build a list of objects and their attributes. You can use a very similar approach to discover data entities and their attributes.
- Technology may also help you identify entities. Some CASE tools can reverse engineer existing files and databases into <u>physical</u> data models. The analyst must usually clean up the resulting model by replacing physical names, codes, and comments with their logical, business-friendly equivalents.

While these techniques may prove useful in identifying entities, they occasionally play tricks on you. A simple, quick quality check can eliminate false entities. Ask your user to specify the number of instances of each entity. A true entity has multiple instances—dozens, hundreds, thousands, or more! If not, the entity does not exist.

As entities are discovered, give them simple, meaningful, business-oriented names. Entities should be named with nouns that describe the person, event, place, object, or thing about which we want to store data. Try not to abbreviate or use acronyms. Names should be singular so as to distinguish the logical concept of the entity from the actual instances of the entity. Names may include appropriate adjectives or clauses to better describe the entity—for instance, an externally generated CUSTOMER ORDER must be distinguished from an internally generated STOCK ORDER.

For each entity, define it in business terms. Don't define the entity in technical terms, and don't define it as "data about . . . " Try this: Use an English dictionary to create a draft definition, and then customize it for the business at hand. Your entity names and definitions should establish an initial glossary of business terminology that will serve both you and future analysts and users for years.

TABLE 8-5 Fundamental Entities for the SoundStage Project

Entity Name	Business Definition
AGREEMENT	A contract whereby a member agrees to purchase a certain number of products within a certain time. After fulfilling that agreement, the member becomes eligible for bonus credits that are redeemable for free or discounted products.
MEMBER	An active member of one or more clubs. Note: A target system objective is to reenroll inactive members as opposed to deleting them.
MEMBER ORDER	An order generated for a member as part of a monthly promotion, or an order initiated by a member. Note: The current system only supports orders generated from promotions; however, customer-initiated orders have been given a high priority as an added option in the proposed system.
TRANSACTION	A business event to which the Member Services System must respond.
PRODUCT	An inventoried product available for promotion and sale to members. Note: System improvement objectives include (1) compatibility with new bar code system being developed for the warehouse, and (2) adaptability to a rapidly changing mix of products.
PROMOTION	A monthly or quarterly event whereby special product offerings are made available to members.

Our SoundStage management and users initially identified the entities listed in Table 8-5. Notice how the definitions contribute to establishing the vocabulary of the system.

> The Context Data Model

The next task in data modeling is to construct the context data model. The context data model should include the fundamental business entities that were previously discovered as well as their natural relationships.

Relationships should be named with verb phrases that, when combined with the entity names, form simple business sentences or assertions. Some CASE tools, such as *System Architect,* let you name the relationships in both directions. Otherwise, always name the relationship from parent to child.

We have completed this task in Figure 8-13. This figure represents a data model created in *System Architect.* Once we begin mapping attributes, new entities and relationships may surface. The numbers below reference the same numbers in Figure 8-13. The ERD communicates the following:

❶ An AGREEMENT <u>binds</u> one or more MEMBERS. While relationships may be named in only one direction (parent to child), the other direction is implicit. For example, it is implicit that a MEMBER <u>is bound to</u> one and only one AGREEMENT.

❷ A MEMBER <u>has conducted</u> zero, one, or more TRANSACTIONS. Implicitly, a given TRANSACTION was <u>conducted by</u> one and only one MEMBER.

❸ A MEMBER ORDER <u>is a</u> TRANSACTION. In fact, a given MEMBER ORDER may correspond to many TRANSACTIONS (for example, a new member order, a canceled member order, a changed member order, etc.). But a given TRANSACTION may or may not <u>represent</u> a MEMBER ORDER.

❹ A PROMOTION <u>features</u> one or more PRODUCTS. Implicitly, a PRODUCT <u>is featured in</u> zero, one, or more PROMOTIONS. For example, a CD that appeals to both

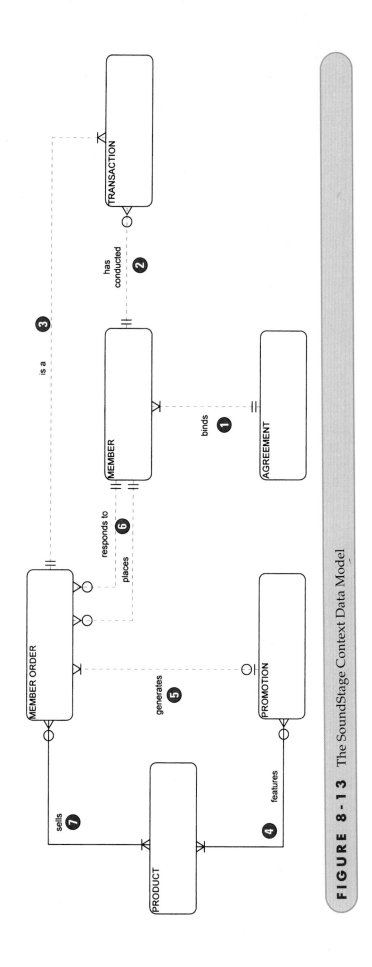

FIGURE 8-13 The SoundStage Context Data Model

country/western and light-rock audiences might be featured in the promotion for both. Since products greatly outnumber promotions, most products are never featured in a promotion.

❺ A PROMOTION <u>generates</u> many MEMBER ORDERS. Implicitly, a MEMBER ORDER <u>is generated for</u> zero or one PROMOTION. Why zero? In the new system, a member will be able to initiate his or her own order.

❻ It is permissible for more than one relationship to exist between the same two entities if the separate relationships communicate different business events or associations. Thus, a MEMBER <u>responds to</u> zero, one, or more MEMBER ORDERS. This relationship supports the promotion-generated orders. A MEMBER <u>places</u> zero, one, or more MEMBER ORDERS. This relationship supports member-initiated orders. In both cases, a MEMBER ORDER <u>is placed by</u> (<u>is responded to by</u>) exactly one MEMBER.

Although we didn't need it for this double relationship, some CASE tools (including *System Architect*) provide a symbol for recording Boolean relationships (such as AND, OR). Thus, for any two relationships, a Boolean symbol could be used to establish that instances of the relationships must be mutually exclusive (= OR) or mutually contingent (= AND).

❼ A MEMBER ORDER <u>sells</u> one or more PRODUCTS. Implicitly, a PRODUCT is <u>sold on</u> zero, one, or more MEMBER ORDERS. Note that this is a nonspecific relationship, which will later be resolved.

If you read each of the preceding items carefully, you probably learned a great deal about the SoundStage system. Data models have become increasingly popular as a tool for describing the business context for system projects.

> The Key-Based Data Model

The next task is to identify the keys of each entity. The following guidelines are suggested for keys:[1]

1. The value of a key should not change over the lifetime of each entity instance. For example, NAME would be a poor key since a person's last name could change by marriage or divorce.
2. The value of a key cannot be null.
3. Controls must be installed to ensure that the value of a key is valid. This can be accomplished by precisely defining the domain and using the database management system's validation controls to enforce that domain.
4. Some experts (Bruce) suggest you avoid **intelligent keys.** An intelligent key is a business code whose structure communicates data about an entity instance (such as its classification, size, or other properties). A code is a group of characters and/or digits that identifies and describes something in the business system. Some experts argue that because those characteristics can change, it violates rule 1 above.

   We disagree. Business codes can return value to the organization because they can be quickly processed by humans without the assistance of a computer.

   a. There are several types of codes. They can be combined to form effective means for entity instance identification.

   (1) *Serial codes* assign sequentially generated numbers to entity instances. Many database management systems can generate and constrain serial codes to a business's requirements.

intelligent key a business code whose structure communicates data about an entity instance.

---

[1]Adapted from Thomas A. Bruce, *Designing Quality Databases with IDEF1X Information Models.* Copyright © 1992 by Thomas A. Bruce. Reprinted by permission of Dorset House Publishing, 353 W. 12th St., New York, NY 10014 (212-620-4053/1-800-DH-BOOKS/www.dorsethouse.com). All rights reserved.

(2) *Block codes* are similar to serial codes except that block numbers are divided into groups that have some business meaning. For instance, a satellite television provider might assign 100–199 as PAY PER VIEW channels, 200–299 as CABLE channels, 300–399 as SPORT channels, 400–499 as ADULT PROGRAMMING channels, 500–599 as MUSIC-ONLY channels, 600–699 as INTERACTIVE GAMING channels, 700–799 as INTERNET channels, 800–899 as PREMIUM CABLE channels, and 900–999 as PREMIUM MOVIE AND EVENT channels.

(3) *Alphabetic codes* use finite combinations of letters (and possibly numbers) to describe entity instances. For example, each STATE has a unique two-character alphabetic code. Alphabetic codes must usually be combined with serial or block codes to uniquely identify instances of most entities.

(4) In *significant position codes,* each digit or group of digits describes a measurable or identifiable characteristic of the entity instance. Significant digit codes are frequently used to code inventory items. The codes you see on tires and lightbulbs are examples of significant position codes. They tell us about characteristics such as tire size and wattage, respectively.

(5) *Hierarchical codes* provide a top-down interpretation for an entity instance. Every item coded is factored into groups, subgroups, and so forth. For instance, we could code employee positions as follows:
    — First digit identifies classification (clerical, faculty, etc.).
    — Second and third digits indicate level within classification.
    — Fourth and fifth digits indicate calendar of employment.

  b. The following guidelines are suggested when creating a business coding scheme:
    (1) Codes should be expandable to accommodate growth.
    (2) The full code must result in a unique value for each entity instance.
    (3) Codes should be large enough to describe the distinguishing characteristics but small enough to be interpreted by people *without a computer.*
    (4) Codes should be convenient. A new instance should be easy to create.

5. Consider inventing a surrogate key instead to substitute for large concatenated keys of independent entities. This suggestion is not practical for associative entities because each part of the concatenated key is a foreign key that must precisely match its parent entity's primary key.

   Figure 8-14 is the key-based data model for the SoundStage project. Notice that the primary key is specified for each entity.

❶ Many entities have a simple, single-attribute primary key.

❷ We resolved the nonspecific relationship between MEMBER ORDER and PRODUCT by introducing the associative entity MEMBER ORDERED PRODUCT. Each associative entity instance represents one product on one member order. The parent entities contributed their own primary keys to comprise the associative entity's concatenated key. *System Architect* places a "PK1" next to ORDER NUMBER to indicate that it is "part one" of the concatenated primary key and a "PK2" beside PRODUCT NUMBER to indicate that it is "part two" of the concatenated key. Also notice that each attribute in that concatenated key, by itself, is a foreign key that points back to the correct parent entity instance.

   Likewise, the nonspecific relationship between PRODUCT and PROMOTION was resolved using an associative entity, TITLE PROMOTION, that also inherits the keys of the parent entities.

When developing this model, look out for a couple of things. If you cannot define keys for an entity, it may be that the entity doesn't really exist—that is, multiple occurrences of the so-called entity do not exist. Thus, assigning keys is a good quality check before fully attributing the data model. Also, if two or more entities have identical keys, they are in all likelihood the same entity.

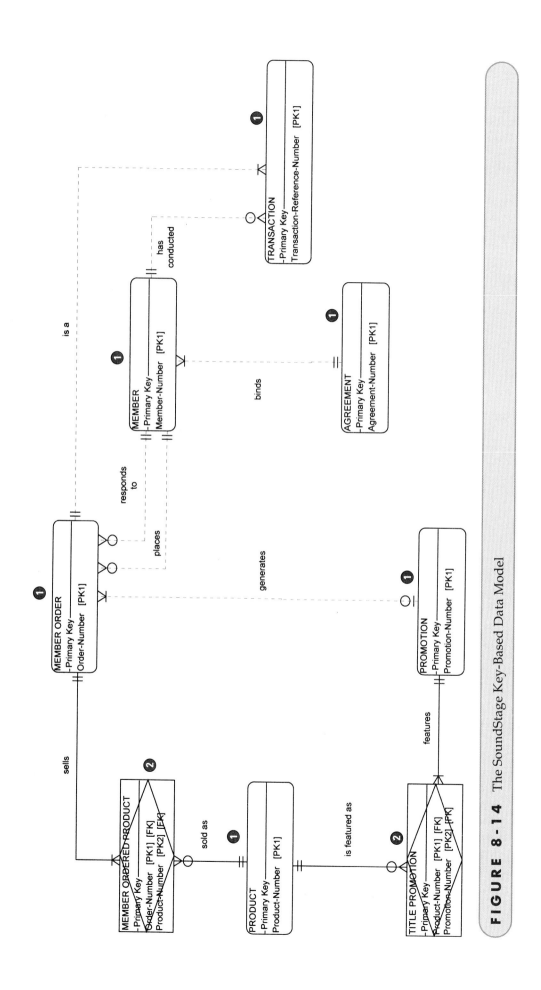

FIGURE 8-14 The SoundStage Key-Based Data Model

> Generalized Hierarchies

At this time, it would be useful to identify any generalization hierarchies in the business domain. The SoundStage project at the beginning of this chapter identified at least one supertype/subtype structure. Subsequent discussions did uncover a generalization hierarchy. Thus, our key-based model was revised as shown in Figure 8-15. We had to lay out the model somewhat differently because of the hierarchy; however, the relationships and keys that were previously defined have been retained. We call your attention to the following:

❶ The SoundStage CASE tool automatically draws a dashed box around a generalization hierarchy.

❷ The subtypes inherit the keys of the supertypes.

❸ We disconnected PROMOTION from PRODUCT as it was shown earlier and reconnected it to the subtype TITLE. This was done to accurately assert the business rule that MERCHANDISE is never featured on a PROMOTION—only TITLES.

> The Fully Attributed Data Model

It may seem like a trivial task to identify the remaining data attributes; however, analysts not familiar with data modeling frequently encounter problems. To accomplish this task, you must have a thorough understanding of the data attributes for the system. These facts can be discovered using top-down approaches (such as brainstorming) or bottom-up approaches (such as form and file sampling). If an enterprise data model exists, some (perhaps many) of the attributes may have already been identified and recorded in a repository.

The following guidelines are offered for attribution:

• Many organizations have naming standards and approved abbreviations. The data administrator usually maintains such standards.

• Choose attribute names carefully. Many attributes share common base names such as NAME, ADDRESS, DATE. Unless the attributes can be generalized into a supertype, it is best to give each variation a unique name such as:

CUSTOMER NAME	CUSTOMER ADDRESS	ORDER DATE
SUPPLIER NAME	SUPPLIER ADDRESS	INVOICE DATE
EMPLOYEE NAME	EMPLOYEE ADDRESS	FLIGHT DATE

Also, remember that a project does not live in isolation from other projects, past or future. Names must be distinguishable across projects.

Some organizations maintain reusable, global templates for these common base attributes. This promotes consistent data types, domains, and defaults across all applications.

• Physical attribute names on existing forms and reports are frequently abbreviated to save space. Logical attribute names should be clearer—for example, translate the order form's attribute COD into its logical equivalent, AMOUNT TO COLLECT ON DELIVERY; translate QTY into QUANTITY ORDERED; and so forth.

• Many attributes take on only YES or NO values. Try naming these attributes as questions. For example, the attribute name CANDIDATE FOR A DEGREE? suggests the values are YES and NO.

Each attribute should be mapped to only one entity. If an attribute truly describes different entities, it is probably several different attributes. Give each a unique name.

• Foreign keys are the exception to the nonredundancy rule—they identify associated instances of related entities.

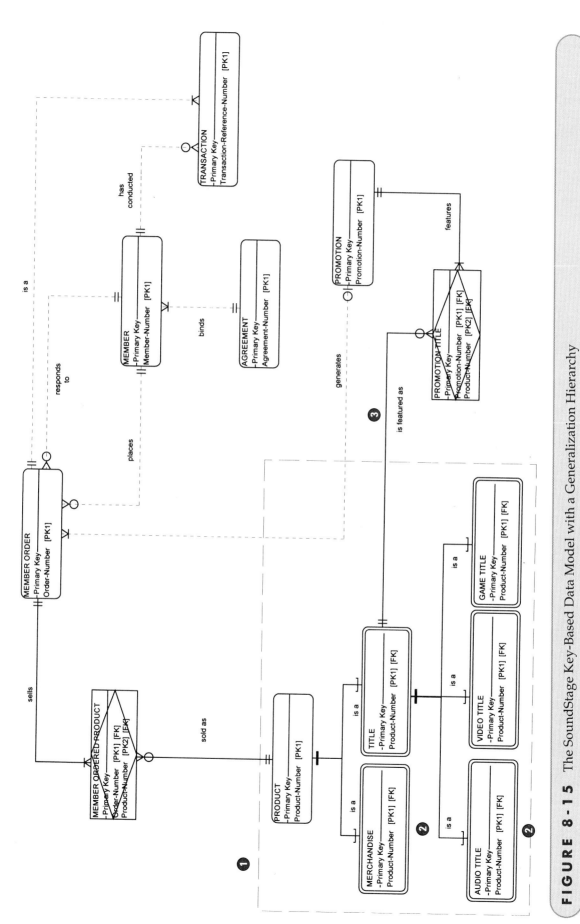

FIGURE 8-15 The SoundStage Key-Based Data Model with a Generalization Hierarchy

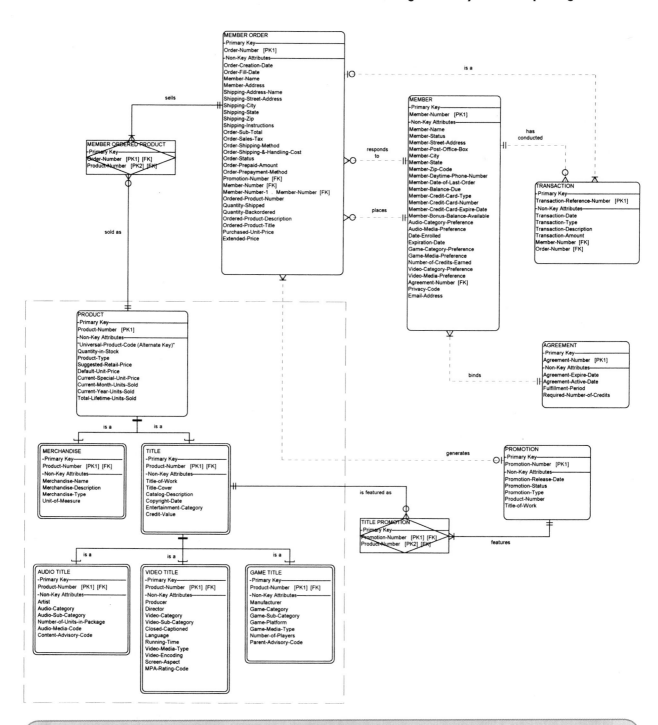

FIGURE 8-16 The SoundStage Fully Attributed Data Model

- An attribute's domain should not be based on logic. For example, in the SoundStage case we learned that the values of MEDIA were dependent on the type of product. If the product type is a video, the media could be VHS tape, 8mm tape, laserdisc, or DVD. If the product type is audio, the media could be cassette tape, CD, or MD. The best solution would be to assign separate attributes to each domain: AUDIO MEDIA and VIDEO MEDIA.

Figure 8-16 provides the mapping of data attributes to entities for the definition phase of our SoundStage systems project. While the fully attributed model identifies all the attributes to be captured and stored in our future database, the descriptions for those attributes are incomplete; they require domains. Most CASE tools provide extensive facilities for describing the data types, domains, and defaults for all attributes to the repository. Additionally, each attribute should be defined for future reference.

Analyzing the Data Model

While a data model effectively communicates database requirements, it does not necessarily represent a *good* database design. It may contain structural characteristics that reduce flexibility and expansion or create unnecessary redundancy. Therefore, we must prepare our fully attributed data model for database design and implementation.

This section will discuss the characteristics of a *quality* data model—one that will allow us to develop an ideal database structure. We'll also present the process used to analyze data model quality and make necessary modifications before database design.

> What Is a Good Data Model?

What makes a data model good? We suggest the following criteria:

- *A good data model is simple.* As a general rule, the data attributes that describe any given entity should describe only that entity. Consider, for example, the following entity definition:

COURSE REGISTRATION =   COURSE REGISTRATION NUMBER (PRIMARY KEY) +
COURSE REGISTRATION DATE +
STUDENT ID NUMBER (A FOREIGN KEY) +
STUDENT NAME +
STUDENT MAJOR +
One or more COURSE NUMBERS

Do STUDENT NAME and STUDENT MAJOR really describe an instance of course registration? Or do they describe a different entity, say, STUDENT? The same argument could be applied to STUDENT ID NUMBER, but on further inspection, that attribute is needed to "point" to the corresponding instance of the STUDENT entity. Another aspect of simplicity is stated as follows: Each attribute of an entity instance can have only one value. Looking again at the previous example, we see that COURSE NUMBER can have as many values for one COURSE REGISTRATION as the student elects.

- *A good data model is essentially nonredundant.* This means that each data attribute, <u>other than foreign keys</u>, describes at most one entity. In the prior example, it is not difficult to imagine that STUDENT NAME and STUDENT MAJOR might also describe a STUDENT entity. We should choose. Based on the previous bullet, the logical choice would be the STUDENT entity. There may also exist subtle redundancies in a data model. For example, the same attribute might be recorded more than once under different names (synonyms).

- *A good data model should be flexible and adaptable to future needs.* In the absence of this criterion, we would tend to design databases to fulfill only today's business requirements. Then, when a new requirement becomes known, we can't easily change the databases without rewriting many or all of the programs that used those databases. While we can't change the reality that most projects are application-driven, we can make our data models as application-independent as possible to encourage database structures that can be extended or modified without impact to current programs.

So how do we achieve the above goals? How can you design a database that can adapt to future requirements that you cannot predict? The answer lies in data analysis.

> Data Analysis

The technique used to improve a data model in preparation for database design is called *data analysis.* **Data analysis** is a process that prepares a data model for implementation as a simple, nonredundant, flexible, and adaptable database. The specific technique is called *normalization.* **Normalization** is a data analysis technique that organizes data attributes such that they are grouped to form nonredundant, stable, flexible, and adaptive entities. Normalization is a three-step technique that places the data model into first normal form, second normal form, and third normal form.[2] Don't get hung up on the terminology—it's easier than it sounds. For now, let's establish an initial understanding of these three formats:

- Simply stated, an entity is in **first normal form (1NF)** if there are no attributes that can have more than one value for a single instance of the entity. Any attributes that can have multiple values actually describe a separate entity, possibly an entity and relationship.
- An entity is in **second normal form (2NF)** <u>if it is already in 1NF and</u> if the values of all non-primary-key attributes are dependent on the full primary key—not just part of it. Any nonkey attributes that are dependent on only part of the primary key should be moved to any entity where that partial key is actually the full key. This may require creating a new entity and relationship on the model.
- An entity is in **third normal form (3NF)** <u>if it is already in 2NF and</u> if the values of its non-primary-key attributes are not dependent on any other non-primary-key attributes. Any nonkey attributes that are dependent on other nonkey attributes must be moved or deleted. Again, new entities and relationships may have to be added to the data model.

> Normalization Example

There are numerous approaches to normalization. We have chosen to present a nontheoretical and nonmathematical approach. We'll leave the theory, relational algebra, and detailed implications to the database courses and textbooks.

As usual, we'll use the SoundStage case study to demonstrate the steps. Let's begin by referring to the fully attributed data model that was developed earlier (see Figure 8-16). Is it a normalized data model? No. Let's identify the problems and walk through the steps of normalizing our data model.

First Normal Form   The first step in data analysis is to place each entity into 1NF. In Figure 8-16, which entities are not in 1NF?

You should find two—MEMBER ORDER and PROMOTION. Each contains a *repeating group,* that is, a group of attributes that can have multiple values for a single instance of the entity *(denoted by the brackets).* These attributes repeat many times "as a group." Consider, for example, the entity MEMBER ORDER. A single MEMBER ORDER may contain many products; therefore, the attributes ORDERED PRODUCT NUMBER, ORDERED PRODUCT DESCRIPTION, ORDERED PRODUCT TITLE, QUANTITY ORDERED, QUANTITY SHIPPED, QUANTITY BACKORDERED, PURCHASED UNIT PRICE, and EXTENDED PRICE may (and probably do) repeat for each instance of MEMBER ORDER.

Similarly, since a PROMOTION may feature more than one PRODUCT TITLE, the PRODUCT NUMBER and TITLE OF WORK attributes may repeat. How do we fix these anomalies in our model?

Figures 8-17 and 8-18 demonstrate how to place these two entities into 1NF. The original entity is depicted on the left side of the page. The 1NF entities are on the right

data analysis a technique used to improve a data model for implementation as a database.

normalization a data analysis technique that organizes data into groups to form nonredundant, stable, flexible, and adaptive entities.

first normal form (1NF) an entity whose attributes have no more than one value for a single instance of that entity.

second normal form (2NF) an entity whose non-primary-key attributes are dependent on the full primary key.

third normal form (3NF) an entity whose non-primary-key attributes are not dependent on any other non-primary-key attributes.

---

[2]Database experts have identified additional normal forms. Third normal form removes most data anomalies. We leave a discussion of advanced normal forms to database textbooks and courses.

MEMBER ORDER (1NF)

Order-Number (Primary Key)
Order-Creation-Date
Order-Automatic-Fill-Date
Member-Number (Foreign Key)
Member-Number-2 (Foreign Key)
Member-Name
Member-Address
Shipping-Address
Shipping Instructions
Promotion-Number (Foreign Key)
Order-Sub-Total-Cost
Order-Sales-Tax
Ship-Via-Method
Shipping-Charge
Order-Status
Prepaid-Amount
Prepaid-Method

MEMBER ORDER (unnormalized)

Order-Number (Primary Key)
Order-Creation-Date
Order-Fill-Date
Member-Number (Foreign Key)
Member-Number-2 (Foreign Key)
Member-Name
Member-Address
Shipping-Address
Shipping Instructions
Promotion-Number (Foreign Key)
1 {ORDERED-PRODUCT-NUMBER} N
0 {ORDERED-PRODUCT-DESCRIPTION} N
0 {ORDERED-PRODUCT-TITLE} N
1 {QUANTITY-ORDERED} N
1 {QUANTITY-SHIPPED} N
1 {QUANTITY-BACKORDERED} N
1 {PURCHASED-UNIT-PRICE} N
1 {EXTENDED-PRICE} N
Order-Sub-Total-Cost
Order-Sales-Tax
Ship-Via-Method
Shipping-Charge
Order-Status
Prepaid-Amount
Prepaid-Method

CORRECTION

sells

MEMBER ORDERED PRODUCT (1NF)

Order-Number (Primary Key 1 and Foreign Key)
PRODUCT-NUMBER (PRIMARY KEY 2 AND FOREIGN KEY)
ORDERED-PRODUCT-DESCRIPTION
ORDERED-PRODUCT-TITLE
QUANTITY-ORDERED
QUANTITY-SHIPPED
QUANTITY-BACKORDERED
PURCHASED-UNIT-PRICE
EXTENDED-PRICE

sold
as

PRODUCT (1NF)

Product-Number (Primary Key)
Universal-Product-Code (Alternate Key)
Quantity-in-Stock
Product-Type
Suggested-Retail-Price
Default-Unit-Price
Current-Special-Unit-Price
Current-Month-Units-Sold
Current-Year-Units-Sold
Total-Lifetime-Units-Sold

FIGURE 8-17 First Normal Form (1NF)

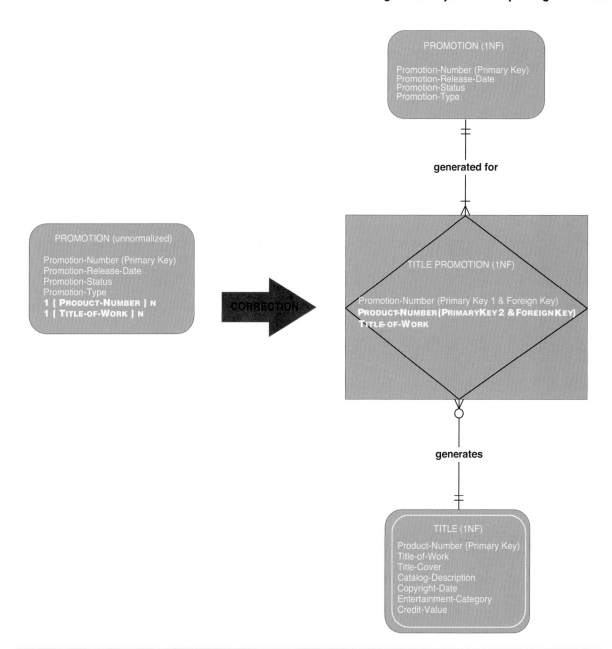

FIGURE 8-18 First Normal Form (1NF)

side of the page. Each figure shows how normalization changed the data model and attribute assignments. For your convenience, the attributes that are affected are in boldface type and in small capital letters.

In Figure 8-17, we first removed the attributes that can have more than one value for an instance of the MEMBER ORDER entity. That alone places MEMBER ORDER in 1NF. But what do we do with the removed attributes? We can't remove them entirely from the model—they are part of the business requirements! Therefore, we moved the entire group of attributes to a new entity, MEMBER ORDERED PRODUCT. Each instance of these attributes describes one PRODUCT on a single MEMBER ORDER. Thus, if a specific order contains five PRODUCTS, there will be five instances of the MEMBER ORDERED PRODUCT entity. Each entity instance has only one value for each attribute; therefore, the new entity is also in first normal form.

Another example of 1NF is shown in Figure 8-18 for the PROMOTION entity. As before, we moved the repeating attributes to a different entity, TITLE PROMOTION.

All other entities are already in 1NF because they do not contain any repeating groups.

Second Normal Form The next step of data analysis is to place the entities into 2NF. Recall that it is required that you have already placed all entities into 1NF. Also recall that 2NF looks for an attribute whose value is determined by only part of the primary key—not the entire concatenated key. Accordingly, entities that have a single-attribute primary key are already in 2NF. That takes care of PRODUCT (and its subtypes), MEMBER ORDER, MEMBER, PROMOTION, AGREEMENT, and TRANSACTION. Thus, we need to check only those entities that have a concatenated key—MEMBER ORDERED PRODUCT and TITLE PROMOTION.

First, let's check the MEMBER ORDERED PRODUCT entity. Most of the attributes are dependent on the full primary key. For example, QUANTITY ORDERED makes no sense unless you have *both* an ORDER NUMBER and a PRODUCT NUMBER. Think about it! By it-self, ORDER NUMBER is inadequate since the order could have as many quantities ordered as there are products on the order. Similarly, by itself, PRODUCT NUMBER is inadequate since the same product could appear on many orders. Thus, QUANTITY ORDERED requires both parts of the key and is dependent on the full key. The same could be said of QUANTITY SHIPPED, QUANTITY BACKORDERED, PURCHASE UNIT PRICE, and EXTENDED PRICE.

But what about ORDERED PRODUCT DESCRIPTION and ORDERED PRODUCT TITLE? Do we really need ORDER NUMBER to determine a value for either? No! Instead, the values of these attributes are dependent only on the value of PRODUCT NUMBER. Thus, the attributes are *not* dependent on the full key; we have uncovered a *partial dependency* anomaly that must be fixed. How do we fix this type of normalization error?

Refer to Figure 8-19 on the next page. To fix the problem, we simply move the nonkey attributes, ORDERED PRODUCT DESCRIPTION and ORDERED PRODUCT TITLE, to an entity that has only PRODUCT NUMBER as its key. If necessary, we would have to create this entity, but the PRODUCT entity with that key already exists. But we have to be careful because PRODUCT is a supertype. Upon inspection of the subtypes, we discover that the attributes are already in the MERCHANDISE and TITLE entities, albeit under a synonym. Thus, we didn't actually have to move the attributes from the MEMBER ORDERED PRODUCT entity; we just deleted them as redundant attributes.

Next, let's examine the TITLE PROMOTION entity. The concatenated key is the combination of PROMOTION NUMBER and PRODUCT NUMBER. TITLE OF WORK is dependent on the PRODUCT NUMBER portion of the concatenated key. Thus, TITLE OF WORK is removed from TITLE PROMOTION (see Figure 8-20). Notice that TITLE OF WORK already existed in the entity TITLE, which has a product number as its primary key.

Third Normal Form We can further simplify our entities by placing them into 3NF. <u>Entities are required to be in 2NF before beginning 3NF analysis.</u> Third normal form analysis looks for two types of problems, *derived data* and *transitive dependencies*. In both cases, the fundamental error is that nonkey attributes are dependent on other nonkey attributes.

The first type of 3NF analysis is easy—examine each entity for derived attributes. **Derived attributes** are those whose values can be either calculated from other attributes or derived through logic from the values of other attributes. If you think about it, storing a derived attribute makes little sense. First, it wastes disk storage space. Second, it complicates what should be simple updates. Why? Every time you change the base attributes, you must remember to reperform the calculation and also change its result.

For example, look at the MEMBER ORDERED PRODUCT entity in Figure 8-21. The attribute EXTENDED PRICE is calculated by multiplying QUANTITY ORDERED by PURCHASED UNIT PRICE. Thus, EXTENDED PRICE (a nonkey attribute) is not dependent on the primary key

derived attribute an attribute whose value can be calculated from other attributes or derived from the values of other attributes.

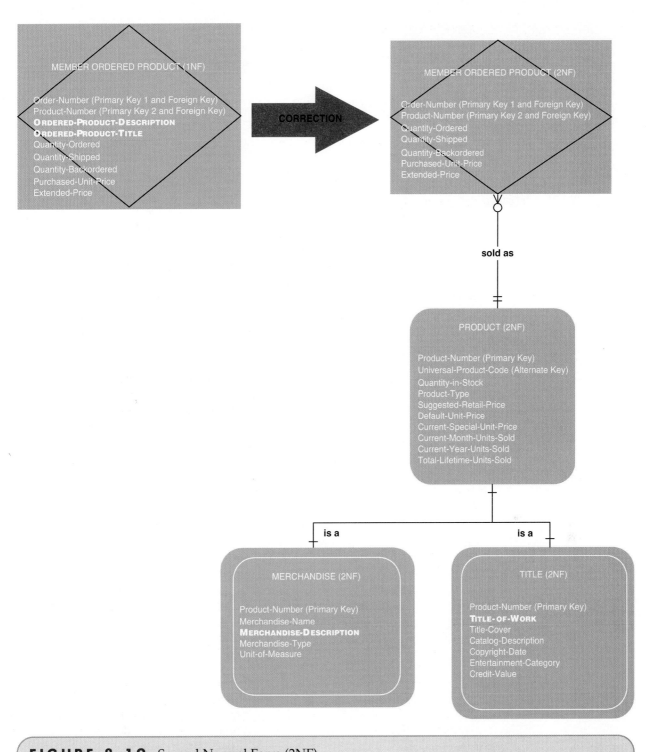

FIGURE 8-19 Second Normal Form (2NF)

as much as it is dependent on the nonkey attributes, QUANTITY ORDERED and PURCHASED
UNIT PRICE. Thus, we correct the entity by deleting EXTENDED PRICE.

Sounds simple, right? Well, not always! There is disagreement on how far you take
this rule. Some experts argue that the rule should be applied only within a single entity. Thus, these experts would not delete a derived attribute if the attributes required
for the derivation were assigned to *different* entities. We agree based on the argument
that a derived attribute that involves multiple entities presents a greater danger for

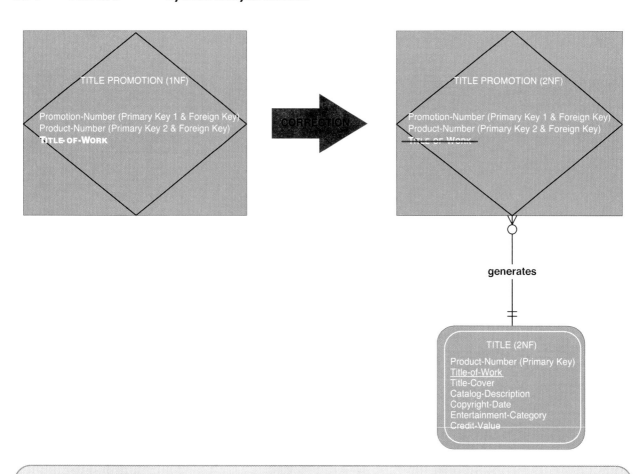

FIGURE 8-20 Second Normal Form (2NF)

data inconsistency caused by updating an attribute in one entity and forgetting to subsequently update the derived attribute in another entity.

Another form of 3NF analysis checks for *transitive dependencies.* A **transitive dependency** exists when a nonkey attribute is dependent on another nonkey attribute (other than by derivation). This error usually indicates that an undiscovered entity is still embedded within the problem entity. Such a condition, if not corrected, can cause future flexibility and adaptability problems if a new requirement eventually requires that we implement that undiscovered entity as a separate database table.

transitive dependency
when the value of a nonkey attribute is dependent on the value of another nonkey attribute other than by derivation.

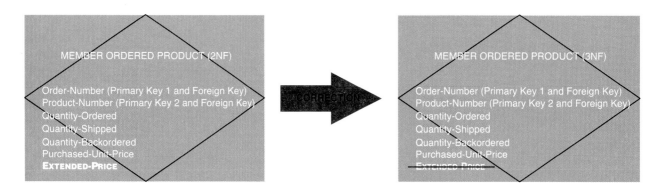

FIGURE 8-21 Third Normal Form (3NF)

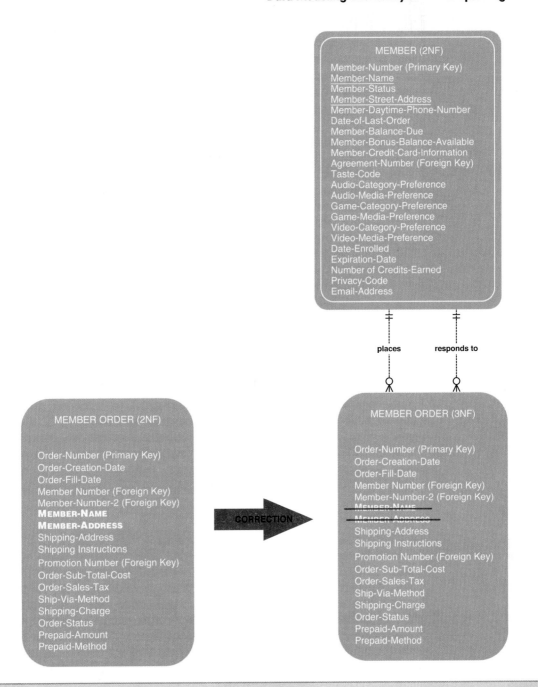

FIGURE 8-22 Third Normal Form (3NF)

Transitive analysis is performed only on entities that do not have a concatenated key. In our example, this includes PRODUCT, MEMBER ORDER, PROMOTION, AGREEMENT, MEMBER, and TRANSACTION. For the entity PRODUCT, all the nonkey attributes are dependent on the primary key and only the primary key. Thus, PRODUCT is already in third normal form. A similar analysis of PROMOTION, AGREEMENT, and TRANSACTION reveals that they are also in third normal form.

But look at the entity MEMBER ORDER in Figure 8-22. In particular, examine the attributes MEMBER NAME and MEMBER ADDRESS. Are these attributes dependent on the primary key, ORDER NUMBER? No! The primary key ORDER NUMBER in no way determines the value of MEMBER NAME and MEMBER ADDRESS. On the other hand, the values of MEMBER NAME and MEMBER ADDRESS are dependent on the value of another non-*primary* key in the entity, MEMBER NUMBER.

How do we fix this problem? MEMBER NAME and MEMBER ADDRESS need to be moved from the MEMBER ORDER entity to an entity whose primary key is just MEMBER NUMBER. If necessary, we would create that entity, but in our case we already have a MEMBER entity with the required primary key. As it turns out, we don't need to really move the problem attributes because they are already assigned to the MEMBER entity. We did, however, notice that MEMBER ADDRESS was a synonym for MEMBER STREET ADDRESS. We elected to keep the latter term in MEMBER.

Several normal forms beyond 3NF exist. Each successive normal form makes the data model simpler, less redundant, and more flexible. However, systems analysts (and most database experts) rarely take data models beyond 3NF. Consequently, we will leave further discussion of normalization to database textbooks.

The first few times you normalize a data model, the process will appear slow and tedious. However, with time and practice, it becomes quick and routine. Many experienced modelers significantly reduce the modeling time and effort by doing normalization during attribution (they are able to do normalization at the time they are developing the fully attributed data model). It may help to always remember the following ditty (source unknown), which nicely summarizes first, second, and third normal forms:

> An entity is said to be in third normal form if every non-primary-key attribute is dependent on the primary key, the whole primary key, and nothing but the primary key.

Simplification by Inspection Normalization is a fairly mechanical process. But it is dependent on naming consistencies in the original data model (before normalization). When several analysts work on a common application, it is not unusual to create problems that won't be taken care of by normalization. These problems are best solved through simplification by inspection, a process wherein a data entity in 3NF is further simplified by such efforts as addressing subtle data redundancy.

The final, normalized data model is presented in Figure 8-23 on the next page.

CASE Support for Normalization Many CASE tools claim to support normalization concepts. They read the data model and attempt to isolate possible normalization errors. On close examination, most CASE tools can normalize only to first normal form. They accomplish this in one of two ways. They look for many-to-many relationships and resolve those relationships into associative entities. Or they look for attributes specifically described as having multiple values for a single entity instance. (One could argue that the analyst should have recognized that as a 1NF error and not described the attributes as such.)

It is exceedingly difficult for a CASE tool to identify second and third normal form errors. That would require that the CASE tool have the intelligence to recognize partial and transitive dependencies. In reality, such dependencies can be discovered only through less-than-routine examination by the systems analysts or database experts.

Mapping Data Requirements to Locations

While a logical data model is effective for describing what data is to be stored for a new system, it does not communicate the requirements on a business operating location basis. We need to identify what data and access rights are needed at which locations. Specifically, we might ask the following business questions:

- Which subsets of the entities and attributes are needed to perform the work at each location?
- What level of access is required?

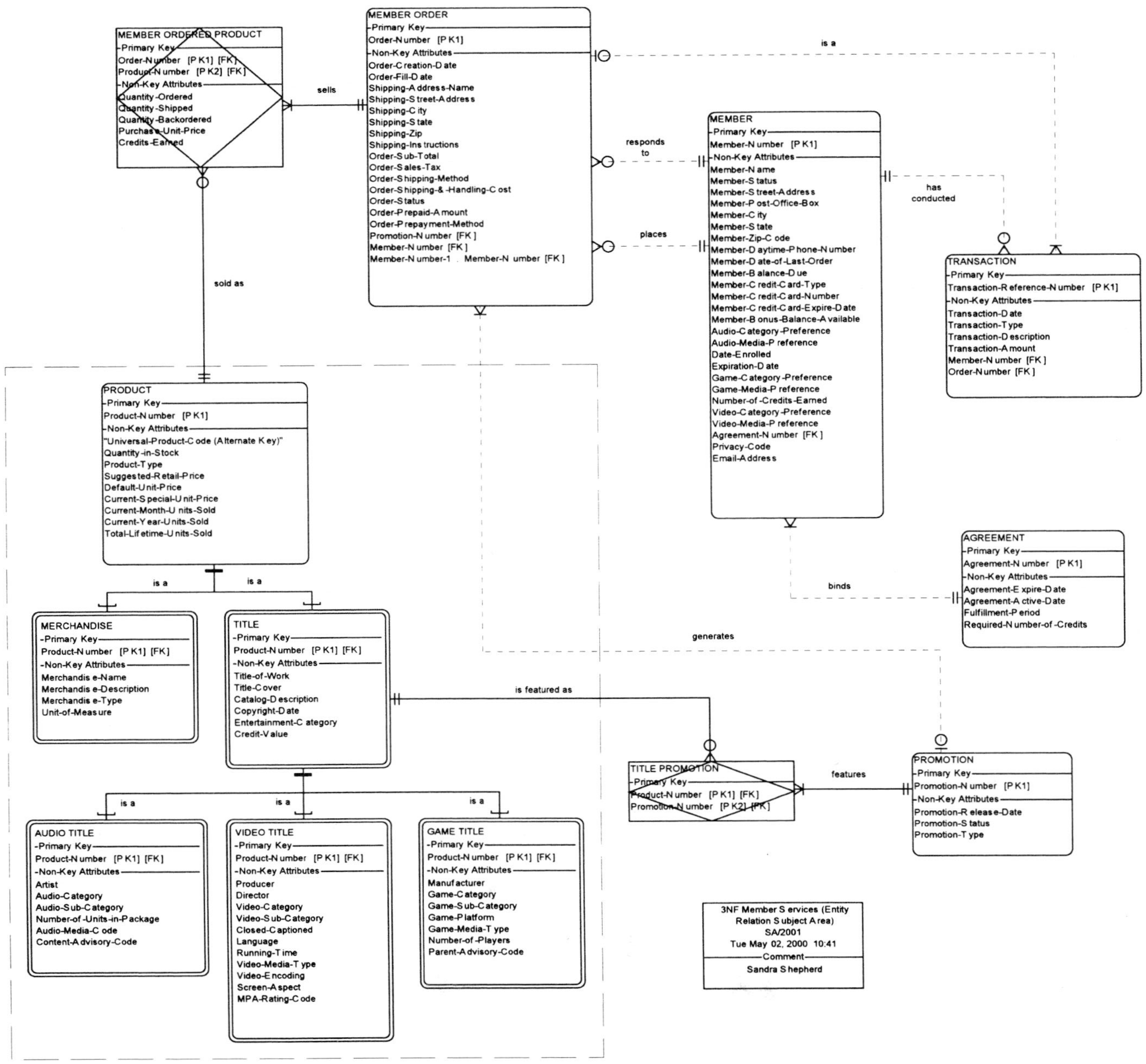

FIGURE 8-23 SoundStage Logical Data Model in Third Normal Form (3NF)

Entity . Attribute \ Location	Customers	Kansas City	.Marketing	.Advertising	.Warehouse	.Sales	.A/R	Boston	.Sales	.Warehouse	San Francisco	.Sales	San Diego	.Warehouse
Customer	INDV					ALL	ALL		SS	SS		SS		SS
Customer Number	R				R	CRUD	R		CRUD	R		CRUD		R
Customer Name	RU				R	CRUD	R		CRUD	R		CRUD		R
Customer Address	RU				R	CRUD	R		CRUD	R		CRUD		R
Customer Credit Rating	X					R	RU		R			R		
Customer Balance Due	R					R	RU		R			R		
Order	INDV	ALL		SS	ALL				SS	SS		SS		SS
Order Number	SRD	R	CRUD	R	CRUD	R			CRUD	R		CRUD		R
Order Date	SRD	R	CRUD	R	CRUD	R			CRUD	R		CRUD		R
Order Amount	SRD	R	CRUD		CRUD	R			CRUD	R		CRUD		R
Ordered Product	INDV	ALL		SS	ALL				SS	SS		SS		SS
Quantity Ordered	SUD	R	CRUD	R	CRUD	R			CRUD			CRUD		
Ordered Item Unit Price	SUD	R	CRUD		CRUD	R			CRUD			CRUD		
Product	ALL	ALL	ALL	ALL	ALL				ALL	ALL		ALL		ALL
Product Number	R	CRUD	R	R	R				R	R		R		R
Product Name	R	CRUD	R	R	R				R	R		R		R
Product Description	R	CRUD	RU	R	R				R	R		R		R
Product Unit of Measure	R	CRUD	R	R	R				R	R		R		R
Product Current Unit Price	R	CRUD	R		R				R	R		R		R
Product Quantity on Hand	X				RU	R			R	RU		R		RU

INDV = individual   **ALL** = ALL   **SS** = subset   X = no access

S = submit   C = create   R — read   U = update   D = delete

FIGURE 8-24 Data-to-Location-CRUD Matrix

- Can the location *create* instances of the entity?
- Can the location *read* instances of the entity?
- Can the location *delete* instances of the entity?
- Can the location *update* existing instances of the entity?

data-to-location-CRUD matrix a matrix that is used to map data requirements to locations.

Systems analysts have found it useful to define these logical requirements in the form of a *data-to-location-CRUD matrix*. A **data-to-location-CRUD matrix** is a table in which the rows indicate entities (and possible attributes); the columns indicate locations; and the cells (the intersection of rows and columns) document level of access where C = create, R = read, U = update or modify, and D = delete or deactivate. Figure 8-24 illustrates a typical data-to-location-CRUD matrix. The decision to include or not include attributes is based on whether locations need to be restricted as to which attributes they can access. Figure 8-24 also demonstrates the ability to document that a location requires access only to a subset (designated SS) of entity instances. For example, each sales office might need access only to those customers in its region.

In some methodologies and CASE tools, you can define *views* of the data model for each location. A view includes only the entities and attributes to be accessible for a single location. If views are defined, they must also be kept in sync with the master data model. (Most CASE tools do this automatically.)

Most of you will proceed directly to Chapter 9, "Process Modeling." Whereas data modeling was concerned with data independently from how that data is captured and used (data at rest), process modeling shows how the data will be captured and used (data in motion). To take an object-oriented approach, you may jump to Chapter 10, "Object-Oriented Analysis and Modeling with UML." Object modeling has many parallels with data modeling. An object includes attributes, but it also includes all the processes that can act on and use those methods.

If you want to immediately learn how to implement data models as databases, you should skim or read Chapter 14, "Designing Databases." In that chapter, the logical data models are transformed into physical database schemas. With CASE tools, the code for creating the database can be generated automatically.

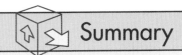 Summary

1. Data modeling is a technique for organizing and documenting a system's DATA. Data modeling is sometimes called database modeling because a data model is usually implemented as a database.

2. There are several notations for data modeling. The actual model is frequently called an entity relationship diagram (ERD) because it depicts data in terms of the entities and relationships described by the data.

3. An entity is something about which the business needs to store data. Classes of entities include persons, places, objects, events, and concepts.

4. An entity instance is a single occurrence of an entity class.

5. Pieces of data we want to store about each instance of a given entity are called attributes. An attribute is a descriptive property or characteristic of the entity. Some attributes can be logically grouped into superattributes called compound attributes.

6. When analyzing a system, we should define those values for an attribute that are legitimate or that make business sense. The values for each attribute are defined in terms of three properties—data type, domain, and default:

   a. The data type defines what class of data can be stored in the attribute.

   b. The domain of an attribute defines what values an attribute can legitimately take on.

   c. The default value for an attribute is the value that will be recorded if not specified by the user.

7. Every entity must have an identifier or key. A key is an attribute, or a group of attributes, that assumes a unique value for each entity instance.

   a. A group of attributes that uniquely identifies an instance of an entity is called a concatenated key.

   b. A candidate key is a "candidate to become the primary identifier" of instances of an entity.

    c. A primary key is the candidate key that will most commonly be used to uniquely identify a single entity instance.

    d. Any candidate key that is not selected to become the primary key is called an alternate key.

    e. Sometimes, it is also necessary to identify a subset of entity instances as opposed to a single instance. A subsetting criteria is an attribute (or concatenated attribute) whose finite values divide all entity instances into useful subsets.

8. A relationship is a natural business association that exists between one or more entities. The relationship may represent an event that links the entities or merely a logical affinity that exists between the entities. All relationships are implicitly bidirectional, meaning they can be interpreted in both directions.

9. Cardinality defines the minimum and maximum number of occurrences of one entity for a single occurrence of the related entity. Because all relationships are bidirectional, cardinality must be defined in both directions for every relationship.

10. The degree of a relationship is the number of entity classes that participate in the relationship. Not all relationships are binary. Some relationships may be recursive relationships, wherein the relationship exists between different instances of the same entity. Relationships can also exist between more than two different entities, as in the case of a 3-ary, or ternary, relationship.

11. An associative entity is an entity that inherits its primary key from more than one other entity (parents). Each part of that concatenated key points to one and only one instance of each of the connecting entities.

12. A foreign key is a primary key of one entity that is contributed to (duplicated in) another entity to identify instances of a relationship. A foreign key (always in a child entity) always matches the primary key (in a parent entity).

13. Nonidentifying relationships are those in which each of the participating entities has its own independent primary key, of which none of the primary-key attributes is shared. The entities in a nonidentifying relationship are referred to as *strong* or independent entities because neither depends on any other entity for its identification. Identifying relationships are those in which the parent entity contributes its primary key to become part of the primary key of the child

entity. The child entity of any identifying relationship is referred to as a *weak* entity because its identification is dependent on the existence of the parent entity's existence.

14. A nonspecific relationship (or many-to-many relationship) is one in which many instances of one entity are associated with many instances of another entity. Such relationships are suitable only for preliminary data models and should be resolved as quickly as possible.

15. Generalization is an approach that seeks to discover and exploit the commonalities between entities. It is a technique wherein the attributes are grouped to form entity supertypes and subtypes.

    a. An entity supertype is an entity whose instances store attributes that are common to one or more entity subtypes.

    b. An entity subtype is an entity whose instances inherit some common attributes from an entity supertype and then add other attributes that are unique to an instance of the subtype.

16. A logical data model is developed in the following stages:

    a. Entities are discovered and defined.

    b. A context data model is built. A context data model contains only business entities and relationships identified by the system owners and users.

    c. A key-based data model is built. The key-based model eliminates nonspecific relationships and adds associative entities. All entities in the model are given keys.

    d. A fully attributed model is built. This model shows all the attributes to be stored in the system.

    e. A fully described model is built. Each attribute is defined in the dictionary and described in terms of properties such as domain and security.

    f. The completed data model is then analyzed for adaptability and flexibility through a process called normalization. The final analyzed model is referred to as a third normal form data model.

17. A logical data model does not communicate data requirements on a business operating location basis. Systems analysts have found it useful to define these requirements in the form of a data-to-location-CRUD matrix.

Review Questions

1. What is the difference between logical and physical models?
2. Why is it important to create an implementation-independent model of a system?
3. Why is it necessary to create an implementation-dependent model of a system?
4. What is an entity? What are entity instances?
5. A *relationship* is a natural business association between entities. What is the relationship between student and teacher? Does it depend on how many classes a student can take, or how many classes a teacher can teach?
6. What is cardinality? Give an example.

Problems and Exercises

1. What is a reasonable domain for the data attribute for a student's last name?
2. What default value would you choose for a student's last name?
3. What default value would you choose for gender?
4. The student table you are working with contains the attributes: STUDENT ID, NAME, PHONE NUMBER, and MAJOR. Normalize to 3NF.
5. What attributes would you have in a table to describe a movie?
6. A many-to-many relationship (also called a non-specific relationship) can and generally should be resolved into a pair of one-to-many relationships with an associative entity. When is this not the case?
7. Give an example of a many-to-many relationship. Resolve using an entity or an associative entity. Which did you use? Why?
8. Describe each of the first three normal forms. Give an example of each.
9. A customer goes to a shoe store and purchases several pairs of shoes. Diagram this relationship.
10. Give an example each of ternary, identifying, and nonidentifying relationships.
11. On the surface, data modeling appears not to require much creativity. Why is this incorrect?
12. Can a well-designed database give a business a strategic advantage? How?

Projects and Research

1. Go to the school library. Ask the librarian at the circulation desk to print out a copy of the information they keep on you. What types of data are being stored? Is there anything that surprises you or seems irrelevant to checking out books? If so, ask why the information is collected.
2. Create a list of attributes for the student entity from the information you found in the previous problem. Normalize to third normal form.
3. Go to a grocery store and make a purchase. What type of data would a good information system maintain on a transaction? What does a good information system do for a business?
4. Compare your answer from the above question (grocery store) to that of at least one other student. How were your answers different?
5. What legal and privacy issues are related to databases used by grocery stores? Go to *http://www.findlaw.com* and research some recent court cases on the topic. Present your work in a short (five-page) paper to your class.
6. How can databases, and the information kept in databases, be used by businesses for a strategic advantage? In a small group, go to a business of your choosing. Brainstorm to create a database that offers that company either a solution to an existing problem, or exploits an existing opportunity in business. Remember to be creative.

Minicases

1. Consider a fictitious online grocery store called Wow Munchies. This is a national franchise, complete with marketing, accounting, shipping, and customer service departments. The CIO has decided to update the database corresponding to the Web store so that it collects pertinent transaction and shopping information for the different departments. She is undecided what software or hardware will support this database.

   Step 1. How will you determine what data is important to collect and maintain in the new database? Be specific.

   Step 2. Create surveys, questionnaires, and the like, and administer them to appropriate personnel in the affected departments. Review the forms that each department uses.

   Step 3. What kind of responses did you get? Go to an online grocery store, and see what data is being collected by "rival" companies. Did you miss anything? Did any of the responses you get require a secondary meeting with department personnel? If so, revise your surveys and questions and reinterview.

   Step 4. Utilize the methods you outlined in the above question, and ascertain the entities and attributes you will need to include in your data.

   Step 5. Draft the relationships and cardinality between the entities. What kind of data model are you using? Implementation specific or nonspecific? Why?

   Step 6. Revise your data model so that there are no many-to-many relationships, and the model is normalized to third normal form.

   Step 7. Submit all questionnaires, surveys, forms, and responses to your professor, along with your final data model draft. Include a short explanation as to how you derived your entities and attributes and the relationships and any assumptions or limitations your data model may have.

2. Research a car rental agency and create a data model for its database for car rentals. What departments are affected by the rental of a car? Review any forms publicly available and create surveys and interviews as necessary to help you determine what your database should contain. Be sure to normalize to third normal form and to resolve any many-to-many relationships. Submit your data model and all supporting documents to your professor.

3. Consider the Mafia. Assuming that organized crime groups maintain databases to evade capture and to run their businesses, what information would they keep? Why?

4. What legal, ethical, and privacy issues are associated with databases used in crime fighting? Research and present a short paper (ten pages or less) to your class.

Suggested Readings

Bruce, Thomas A. *Designing Quality Databases with IDEF1X Information Models.* New York: Dorset House Publishing, 1992. We actually use this book as a textbook in our database analysis and design course. IDEF1X is a rich, standardized syntax for data modeling (which Bruce calls information modeling). The graphical language looks different, but it communicates the same system concepts presented in our book. The language is supported by at least two CASE tools: Logic Works' *ERwin* and Popkin's *System Architect.* The book includes two case studies.

Hay, David C. *Data Model Patterns: Conventions of Thought.* New York: Dorset House Publishing, 1996. What a novel book! This book starts with the premise that most business data models are derivatives of some basic, repeatable patterns. CASE vendors, how about including these patterns in CASE tools as reusable templates?

Martin, James, and Clive Finkelstein. *Information Engineering,* 3 vols. New York: Savant Institute, 1981. Information engineering is a formal, database, and fourth-generation language-oriented methodology. The graphical data modeling language of information engineering is virtually identical to ours. Data modeling is covered in Volumes I and II.

Reingruber, Michael, and William Gregory. *The Data Modeling Handbook.* New York: John Wiley & Sons, 1994. This is an excellent book on data modeling and is particularly helpful in ensuring the quality and accuracy of data models.

Schlaer, Sally, and Stephen J. Mellor. *Object-Oriented Systems Analysis: Modeling the World in Data.* Englewood Cliffs, NJ: Yourdon Press, 1988. Forget the title! "Object-oriented" means something different today than it did in 1988, but the book is still one of the easiest-to-read books on the subject of data modeling.

Teorey, Toby J. *Database Modeling & Design: The Fundamental Principles,* 2nd ed. San Francisco: Morgan Kaufman Publishers, 1994. This book is somewhat more conceptual than the others in the list, but it provides useful insights into the practice of data modeling.

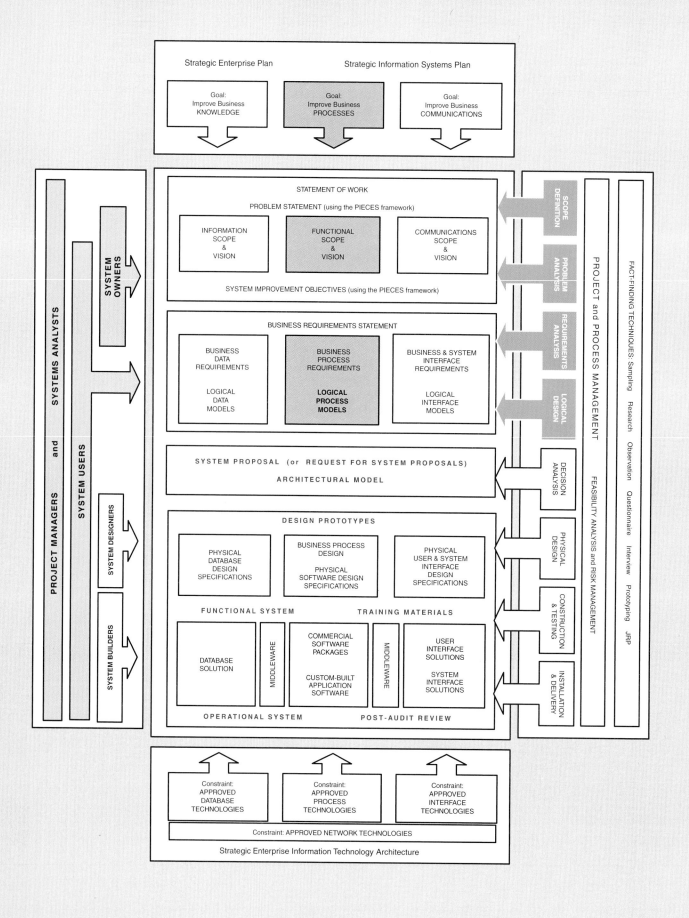

Strategic Enterprise Plan   Strategic Information Systems Plan

Goal:
Improve Business
KNOWLEDGE

Goal:
Improve Business
PROCESSES

Goal:
Improve Business
COMMUNICATIONS

SYSTEMS ANALYSTS

PROJECT MANAGERS   and

SYSTEM USERS

SYSTEM OWNERS

SYSTEM DESIGNERS

SYSTEM BUILDERS

STATEMENT OF WORK

PROBLEM STATEMENT (using the PIECES framework)

INFORMATION
SCOPE
&
VISION

FUNCTIONAL
SCOPE
&
VISION

COMMUNICATIONS
SCOPE
&
VISION

SYSTEM IMPROVEMENT OBJECTIVES (using the PIECES framework)

SCOPE DEFINITION

PROBLEM ANALYSIS

BUSINESS REQUIREMENTS STATEMENT

BUSINESS
DATA
REQUIREMENTS

LOGICAL
DATA
MODELS

BUSINESS
PROCESS
REQUIREMENTS

LOGICAL
PROCESS
MODELS

BUSINESS & SYSTEM
INTERFACE
REQUIREMENTS

LOGICAL
INTERFACE
MODELS

REQUIREMENTS ANALYSIS

LOGICAL DESIGN

SYSTEM PROPOSAL (or REQUEST FOR SYSTEM PROPOSALS)

ARCHITECTURAL MODEL

DECISION ANALYSIS

DESIGN PROTOTYPES

PHYSICAL
DATABASE
DESIGN
SPECIFICATIONS

BUSINESS PROCESS
DESIGN

PHYSICAL
SOFTWARE DESIGN
SPECIFICATIONS

PHYSICAL
USER & SYSTEM
INTERFACE
DESIGN
SPECIFICATIONS

PHYSICAL DESIGN

FUNCTIONAL SYSTEM   TRAINING MATERIALS

DATABASE
SOLUTION

MIDDLEWARE

COMMERCIAL
SOFTWARE
PACKAGES

CUSTOM-BUILT
APPLICATION
SOFTWARE

MIDDLEWARE

USER
INTERFACE
SOLUTIONS

SYSTEM
INTERFACE
SOLUTIONS

CONSTRUCTION & TESTING

INSTALLATION & DELIVERY

OPERATIONAL SYSTEM   POST-AUDIT REVIEW

PROJECT and PROCESS MANAGEMENT

FEASIBILITY ANALYSIS and RISK MANAGEMENT

FACT-FINDING TECHNIQUES: Sampling   Research   Observation   Questionnaire   Interview   Prototyping   JRP

Constraint:
APPROVED
DATABASE
TECHNOLOGIES

Constraint:
APPROVED
PROCESS
TECHNOLOGIES

Constraint:
APPROVED
INTERFACE
TECHNOLOGIES

Constraint: APPROVED NETWORK TECHNOLOGIES

Strategic Enterprise Information Technology Architecture

Process Modeling

Chapter Preview and Objectives

In this chapter you will learn how to draw data flow diagrams, a popular process model that documents a system's processes and their data flows. You will know process modeling as a systems analysis tool when you can:

▌ Define systems modeling and differentiate between logical and physical system models.

▌ Define process modeling and explain its benefits.

▌ Recognize and understand the basic concepts and constructs of a process model.

▌ Read and interpret a data flow diagram.

▌ Explain when to construct process models and where to store them.

▌ Construct a context diagram to illustrate a system's interfaces with its environment.

▌ Identify use cases and external and temporal business events for a system.

▌ Perform event partitioning and organize events in a functional decomposition diagram.

▌ Draw event diagrams and then merge those event diagrams into system diagrams.

▌ Draw primitive data flow diagrams and describe the elementary data flows and processes in terms of data structures and procedural logic (Structured English and decision tables), respectively.

▌ Document the distribution of processes to locations.

▌ Synchronize data and process models using a CRUD matrix.

Introduction

Bob Martinez has been working on the SoundStage Member Services system for several weeks. He understands the system pretty well, but it is still easy to get confused and to forget details.

"The problem," Bob said to his boss, Sandra Shepherd, "is that the system is too big to keep it all in your head at one time."

"I'm glad you said that," Sandra answered, "because your next assignment is to break the system down into parts that you can get your head around. Each part is called a process, and you'll need to define the inputs and outputs to that process plus who or what each input and output comes from or goes to. And, by the way, you'll need to specify the logic for the process."

"I thought the use cases did all that," Bob replied.

"They aren't specific enough," Sandra stated. "Would you like to turn a project over to a programmer located across the country with no more specifics than what a use case has? Who knows what you'd end up with? Welcome to the world of process modeling, Bob. Get to work."

An Introduction to Process Modeling

In Chapter 5 you were introduced to systems analysis activities that called for drawing system models. System models play an important role in system development. As a systems analyst or user, you will constantly deal with unstructured problems. One way to structure such problems is to draw models. A **model** is a representation of reality. Just as a picture is worth a thousand words, most system models are pictorial representations of reality. Models can be built for existing systems as a way to better understand those systems or for proposed systems as a way to document business requirements or technical designs. An important concept is the distinction between logical and physical models.

model a pictorial representation of reality.

Logical models show *what* a system is or does. They are implementation-*in*dependent; that is, they depict the system independent of any technical implementation. As such, logical models illustrate the *essence* of the system. Popular synonyms include *essential model, conceptual model,* and *business model.* **Physical models** show not only *what* a system is or does but also *how* the system is physically and technically implemented. They are implementation-*de*pendent because they reflect technology choices and the limitations of those technology choices. Synonyms include *implementation model* and *technical model.*

logical model a nontechnical pictorial representation that depicts what a system is or does. Synonyms are *essential model, conceptual model,* and *business model.*

Systems analysts have long recognized the value of separating business and technical concerns. That is why they use logical system models to depict business requirements and physical system models to depict technical designs. Systems analysis activities tend to focus on the logical system models for the following reasons:

physical model a technical pictorial representation that depicts what a system is or does and how the system is implemented. Synonyms are *implementation model* and *technical model.*

- Logical models remove biases that are the result of the way the current system is implemented or the way that any one person thinks the system might be implemented. Thus, we overcome the "we've always done it that way" syndrome. Consequently, logical models encourage creativity.
- Logical models reduce the risk of missing business requirements because we are too preoccupied with technical details. Such errors can be costly to correct after the system is implemented. By separating what the system must do from how the system will do it, we can better analyze the requirements for completeness, accuracy, and consistency.
- Logical models allow us to communicate with end users in nontechnical or less technical languages. Thus, we don't lose requirements in the technical jargon of the computing discipline.

In this chapter we will focus exclusively on *logical* process modeling during systems analysis. **Process modeling** is a technique for organizing and documenting the structure and flow of data through a system's PROCESSES and/or the logic, policies, and procedures to be implemented by a system's PROCESSES. In the context of information system building blocks (see the home page at the beginning of the chapter), *logical* process models are used to document an information system's PROCESS focus from the system owners' and users' perspective (the intersection of the PROCESS column with the system owner and system user rows). Also notice that one special type of process model, called a *context diagram,* illustrates the COMMUNICATION focus from the system owners' and users' perspective.

Process modeling originated in classical software engineering methods; therefore, you may have encountered various types of process models such as program structure charts, logic flowcharts, or decision tables in an application programming course. In this chapter, we'll focus on a systems analysis process model, *data flow diagrams.*

A **data flow diagram (DFD)** is a tool that depicts the flow of data through a system and the work or processing performed by that system. Synonyms include *bubble chart, transformation graph,* and *process model.* We'll also introduce a DFD planning tool called *decomposition diagrams.* Finally, we'll also study *context diagrams,* a processlike model that actually illustrates a system's interfaces to the business and outside world, including other information systems.[1]

A simple data flow diagram is illustrated in Figure 9-1. In the design phase, some of these business processes might be implemented as computer software (either built in-house or purchased from a software vendor). If you examine this data flow diagram, you should find it easy to read, even before you complete this chapter—that has always been the advantage of DFDs. There are only three symbols and one connection:

- The rounded rectangles represent *processes* or work to be done. Notice that they are illustrated in the PROCESS color from your information system framework.
- The squares represent *external agents*—the boundary of the system. Notice that they are illustrated in the INTERFACE color from your information system framework.
- The open-ended boxes represent *data stores,* sometimes called files or databases. If you have already read Chapter 8, these data stores correspond to all instances of a single entity in a data model. Accordingly, they have been illustrated with the DATA color from your information systems framework.
- The arrows represent *data flows,* or inputs and outputs, to and from the processes.

There is sometimes a tendency to confuse data flow diagrams with flowcharts because program design frequently involves the use of flowcharts. However, data flow diagrams are very different. Let's summarize the differences.

- Processes on a data flow diagram can operate in parallel. Thus, several processes might be executing or working simultaneously. This is consistent with the way businesses work. On the other hand, processes on flowcharts can execute only one at a time.
- Data flow diagrams show the flow of data through the system. Their arrows represent paths down which data can flow. Looping and branching are not typically shown. On the other hand, flowcharts show the sequence of processes or operations in an algorithm or program. Their arrows represent pointers to the next process or operation. This may include looping and branching.

process modeling a technique used to organize and document a system's processes.

data flow diagram (DFD) a process model used to depict the flow of data through a system and the work or processing performed by the system. Synonyms are *bubble chart, transformation graph,* and *process model.*

Author's Note: there are several competing symbol sets for DFDs. Throughout this chapter, the authors have chosen to use the Gane and Sarson notation because of its wide popularity.

---

[1]In classic structured analysis, context diagrams are considered to be another type of process model. But in object-oriented analysis, they illustrate scope and interfaces. In this edition, we have chosen the latter definition.

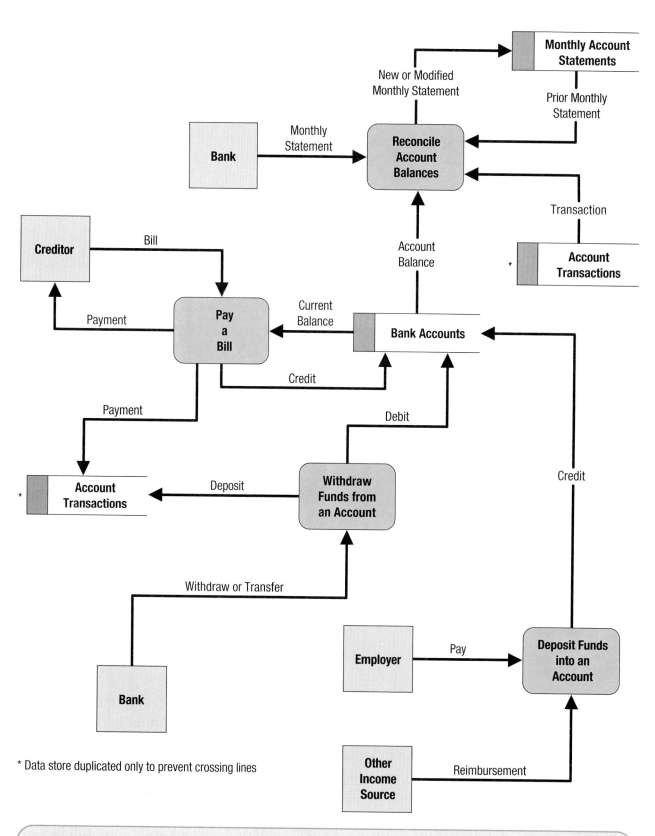

* Data store duplicated only to prevent crossing lines

FIGURE 9-1 A Simple Data Flow Diagram

- Data flow diagrams can show processes that have dramatically different timing. For example, a single DFD might include processes that happen hourly, daily, weekly, yearly, and on demand. This doesn't happen in flowcharts.

Data flow diagrams have been popular for more than 20 years, but the interest in DFDs has been recently renewed because of their applicability in *business process redesign (BPR)*. As businesses have come to realize that most data processing systems have merely automated outdated, inefficient, and bureaucratic business processes, there is renewed interest in streamlining the business processes. This is accomplished by first modeling the business processes for the purpose of analyzing, redesigning, and/or improving them. Subsequently, information technology can be applied to the improved business processes in creative ways that maximize the value returned to the business. We'll revisit this trend at the end of the chapter.

System Concepts for Process Modeling

> External Agents

All information systems respond to events and conditions in the system's environment. The environment of an information system includes *external agents* that form the boundary of the system and define places where the system interfaces with its environment. An **external agent** defines a person, an organization unit, another system, or another organization that lies outside the scope of the project but that interacts with the system being studied. External agents provide the net inputs into a system and receive net outputs from a system. A common synonym is *external entity* (not to be confused with *data entity* as introduced in Chapter 8).

The term *external* means "external to the system being analyzed or designed." In practice, an external agent may actually be outside of the business (such as government agencies, customers, suppliers, and contractors), or it may be inside the business but outside the project and system scope (such as other departments, other business functions, and other internal information systems). An external agent is represented by a square on the data flow diagram. The DeMarco/Yourdon equivalent is a rectangle (see margin).

It is important to recognize that work and activities are occurring inside the external agent, but that work and those activities are said to be "out of scope" and not subject to change. Thus, the data flows between your system and these boundaries should not cause substantive change to the work or activities performed by the external agents.

The external agents of an information system are rarely fixed. As project scope and goals change, the scope of an information system can either grow or shrink. If the system scope grows, it can consume some of the original external agents—in other words, what was once considered outside the system is now considered inside the system (*as new processes*).

Similarly, if the system scope shrinks (because of budget or schedule constraints), processes that were once considered to be inside the system may become external agents.

External agents on a logical data flow diagram may include people, business units, other internal systems with which a system must interact, and external organizations. Their inclusion on the logical DFD means that your system interacts with these agents. They are almost always one of the following:

- An office, department, division, or individual within your company that provides net inputs to that system, receives net outputs from that system, or both.
- An organization, agency, or individual that is outside your company but that provides net inputs to, or receives net outputs from, your system. Examples include CUSTOMERS, SUPPLIERS, CONTRACTORS, BANKS, and GOVERNMENT AGENCIES.

external agent an outside person, organization unit, system, or organization that interacts with a system. Also called *external entity*.

External Agent

Gane and Sarson shape

External Agent

DeMarco/Yourdon shape

External Agent Symbols

- Another business or information system—possibly, though not necessarily, computer-based—that is separate from your system but with which your system must interface. It is becoming common to interface information systems with those of other businesses.
- A system's end users or managers. In this case, the user or manager is either a net source of data to be input to a system and/or a net destination of outputs to be produced by a system.

External agents should be named with descriptive, singular nouns, such as REGISTRAR, SUPPLIER, MANUFACTURING SYSTEM, or FINANCIAL INFORMATION SYSTEM. External agents represent fixed, *physical* systems; therefore, they can have physical names or acronyms—even on a <u>logical</u> DFD. For example, an external agent representing our school's financial management information system would be called FMIS. If an external agent describes an individual, we recommend job titles or role names instead of proper names (for example, use ACCOUNT CLERK, not *Mary Jacobs*).

To avoid crossing data flow lines on a DFD, it is permissible to duplicate external agents on DFDs. But as a general rule, external agents should be located on the perimeters of the page, consistent with their definition as a system boundary.

> Data Stores

Most information systems capture data for later use. The data is kept in a data store, the last symbol on a data flow diagram. It is represented by the open-end box (see margin). A **data store** is an "inventory" of data. Synonyms include *file* and *database* (although those terms are too implementation-oriented for essential process modeling). If data flows are *data in motion*, think of data stores as *data at rest*.

Ideally, essential data stores should describe "things" about which the business wants to store data. These things include:

Persons: AGENCY, CONTRACTOR, CUSTOMER, DEPARTMENT, DIVISION, EMPLOYEE, INSTRUCTOR, OFFICE, STUDENT, SUPPLIER. Notice that a person entity can represent either individuals, groups, or organizations.

Places: SALES REGION, BUILDING, ROOM, BRANCH OFFICE, CAMPUS.

Objects: BOOK, MACHINE, PART, PRODUCT, RAW MATERIAL, SOFTWARE LICENSE, SOFTWARE PACKAGE, TOOL, VEHICLE MODEL, VEHICLE. An object entity can represent actual objects (such as SOFTWARE LICENSE) or specifications for a type of object (such as SOFTWARE PACKAGE).

Events: APPLICATION, AWARD, CANCELLATION, CLASS, FLIGHT, INVOICE, ORDER, REGISTRATION, RENEWAL, REQUISITION, RESERVATION, SALE, TRIP.

Concepts: ACCOUNT, BLOCK OF TIME, BOND, COURSE, FUND, QUALIFICATION, STOCK.

NOTE: If the above list looks familiar, it should: A data store represents *all occurrences* of a data entity—defined in Chapter 8 as something about which we want to store data. As such, the data store represents the synchronization of a system's process model with its data model.

If data modeling is done before process modeling, identification of most data stores is simplified by the following rule:

There should be one data store for each data entity on an entity relationship diagram. (We even include associative and weak entity data stores on our models.)

If, on the other hand, process modeling is done before data modeling, data store discovery tends to be more arbitrary. In that case, our best recommendation is to identify existing implementations of files or data stores (e.g., computer files and databases, file cabinets, record books, catalogs) and then rename them to reflect the logical "things" about which they store data. Consistent with information engineering strategies, we recommend that data models precede the process models.

Generally, data stores should be named as the plural of the corresponding data model entity. Thus, if the data model includes an entity named CUSTOMER, the process

data store stored data intended for later use. Synonyms are *file* and *database*.

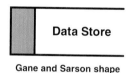

Data Store

Gane and Sarson shape

Data Store

DeMarco/Yourdon shape

Data Store Symbols

models will include a data store named CUSTOMERS. This makes sense because the data store, by definition, stores all instances of the entity. Avoid physical terms such as *file, database, file cabinet, file folder,* and the like.

As was the case with boundaries, it is permissible to duplicate data stores on a DFD to avoid crossing data flow lines. Duplication should be minimized.

> Process Concepts

Recall from Chapter 2 that a fundamental building block of information systems is PROCESSES. All information systems include processes—usually many of them. Information system processes respond to business events and conditions and transform DATA (another building block) into useful information. Modeling processes helps us to understand interactions with the system's environment, other systems, and other processes.

A System *Is* a Process   We have used the word *system* throughout this book to describe almost any orderly arrangement of ideas or constructs. People speak of educational systems, computer systems, management systems, business systems, and, of course, information systems. In the oldest and simplest of all system models, a system *is* a process.

In systems analysis, models are used to view or present a system. As shown in Figure 9-2, the simplest process model of a system is based on inputs, outputs, and the system itself—viewed as a process. The process symbol defines the boundary of the system. The system is inside the boundary; the environment is outside that boundary. The system exchanges inputs and outputs with its environment. Because the environment is always changing, well-designed systems have a feedback and control loop to allow the system to adapt itself to changing conditions.

Consider a business as a system. It operates within an environment that includes customers, suppliers, competitors, other industries, and the government. Its inputs include materials, services, new employees, new equipment, facilities, money, and orders (to name but a few). Its outputs include products and/or services, waste

> **Process name**
>
> Gane & Sarson shape;
> used throughout
> this book

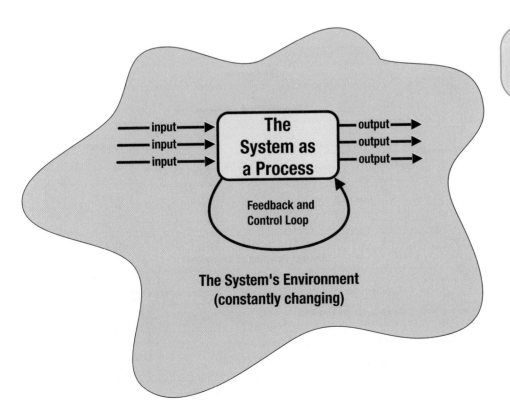

FIGURE 9-2

The Classic Process Model of a System

materials, retired equipment, former employees, and money (payments). It monitors its environment to make necessary changes to its product line, services, operating procedures, competitors, and the economy.

A rounded rectangle (the Gane and Sarson notation) is used throughout this chapter to represent a process (see margin). **A process** is work performed on, or in response to, incoming data flows or conditions. A synonym is *transform*. Different process modeling notations use a circle (the DeMarco/Yourdon notation) or a rectangle (the SSADM/IDEF0 notation). The choice is often dependent on your methodology and CASE tool features.

Although processes can be performed by people, departments, robots, machines, or computers, we once again want to focus on *what* work or action is being performed (the *logical* process), not on who or what is doing that work or activity (the *physical* process). For instance, in Figure 9-1 we included the logical process WITHDRAW FUNDS FROM AN ACCOUNT. We did not indicate how this would be done. Intuitively, we can think of several physical implementations such as using an ATM, using a bank's drive-through service, or actually going inside the bank.

Process Decomposition   A complex system is usually too difficult to fully understand when viewed as a whole (meaning *as a single process*). Therefore, in systems analysis we separate a system into its component subsystems, which are decomposed into smaller subsystems, until we have identified manageable subsets of the overall system (see Figure 9-3). We call this technique *decomposition*. **Decomposition** is the act of breaking a system into its component subsystems, processes, and subprocesses. Each level of *abstraction* reveals more or less detail (as desired) about the overall system or a subset of that system. You have already applied decomposition in various ways. Most of you have *outlined* a term paper—this is a form of decomposition. Many of you have partitioned a medium- to large-size computer program into subprograms that could be developed and tested independently before they are integrated. This is also decomposition.

In systems analysis, decomposition allows you to partition a system into logical subsystems of processes for improved communication, analysis, and design. A diagram similar to Figure 9-3 can be a little difficult to construct when dealing with all

process work performed by a system in response to incoming data flows or conditions. A synonym is *transform*.

decomposition the act of breaking a system into subcomponents.

FIGURE 9-3

A System Consists of Many Subsystems and Processes

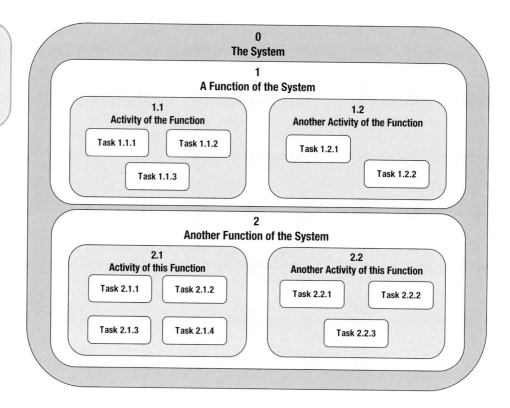

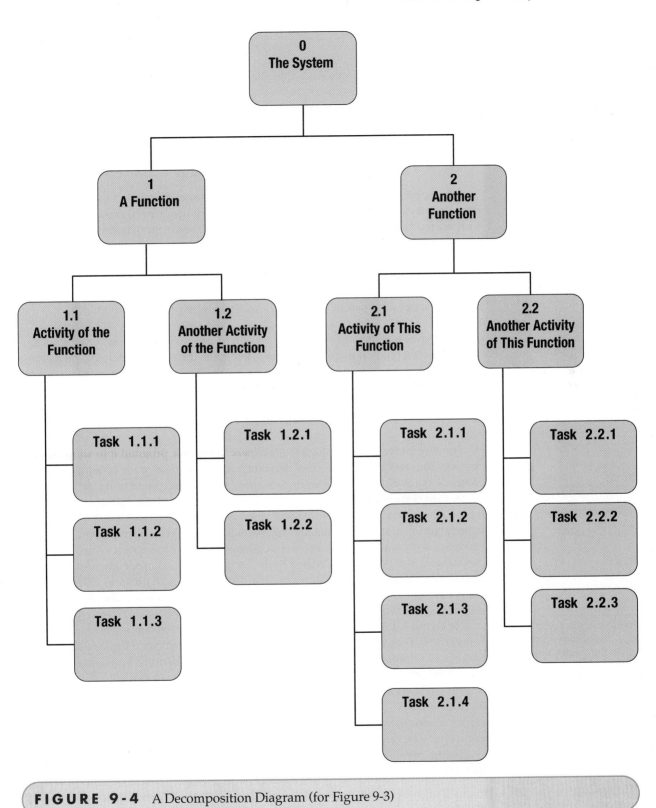

FIGURE 9-4   A Decomposition Diagram (for Figure 9-3)

but the smallest of systems. Figure 9-4 demonstrates an alternative layout that is supported by many CASE tools and development methodologies. It is called a *decomposition diagram*. We'll use it extensively in this chapter. A **decomposition diagram,** also called a *hierarchy chart,* shows the top-down functional decomposition and structure of a system. A decomposition diagram is essentially a planning

decomposition diagram
a tool used to depict the decomposition of a system. Also called *hierarchy chart.*

tool for more detailed process models, namely, data flow diagrams. The following rules apply:

- Each process in a decomposition diagram is either a *parent process,* a *child process* (of a parent), or both.
- A parent *must* have two or more children—a single child does not make sense because that would not reveal any additional detail about the system.
- In most decomposition diagramming standards, a child may have only one parent.
- Finally, a child of one parent may be the parent of its own children.

The upper and lower halves of the decomposition diagram in Figure 9-4 demonstrate two styles for laying out the processes and connections. You may use either or both as necessary to present an uncluttered model. Some models may require multiple pages for maximum clarity.

The connections on a decomposition diagram do not contain arrowheads because the diagram is meant to show *structure,* not *flow.* Also, the connections are not named. Implicitly they all have the same name—CONSISTS OF—since the sum of the child processes for a parent process *equals* the parent process.

Logical Processes and Conventions   *Logical* processes are work or actions that <u>must</u> be performed no matter <u>how</u> you implement the system. Each logical process is (or will be) implemented as one or more physical processes that may include work performed by people, work performed by robots or machines, or work performed by computer software. It doesn't matter which implementation is used, however, because logical processes should only indicate <u>that</u> there is work that must be done.

Naming conventions for logical processes depend on where the process is in the decomposition diagram/data flow diagram and the type of process depicted. There are three types of logical processes: *functions, events,* and *elementary processes.*

function a set of related and ongoing activities of a business.

A **function** is a set of related and <u>ongoing</u> activities of the business. A function has no start or end; it just continuously performs its work as needed. For example, a manufacturing system may include the following functions (subsystems): PRODUCTION PLANNING, PRODUCTION SCHEDULING, MATERIALS MANAGEMENT, PRODUCTION CONTROL, QUALITY MANAGEMENT, and INVENTORY CONTROL. Each of these functions may consist of dozens or hundreds of more discrete processes to support specific activities and tasks. Function names are nouns that describe the entire function. Additional examples are ORDER ENTRY, ORDER MANAGEMENT, SALES REPORTING, CUSTOMER RELATIONS, and RETURNS AND REFUNDS.

event a logical unit of work that must by completed as a whole. Sometimes called a *transaction.*

An **event** is a logical unit of work that must be completed as a whole. An event is triggered by a discrete input and is completed when the process has responded with appropriate outputs. Events are sometimes called *transactions.* Functions consist of processes that respond to events. For example, the MATERIALS MANAGEMENT function may respond to the following events: TEST MATERIAL QUALITY, STOCK NEW MATERIALS, DISPOSE OF DAMAGED MATERIALS, DISPOSE OF SPOILED MATERIALS, REQUISITION MATERIALS FOR PRODUCTION, RETURN UNUSED MATERIALS FROM PRODUCTION, ORDER NEW MATERIALS, and so on. Each of these events has a trigger and response that can be defined by its inputs and outputs.

Using *modern* structured analysis techniques such as those advocated by McMenamin, Palmer, Yourdon, and the Robertsons (see the Suggested Readings at the end of the chapter), analysts decompose system functions into business events. Each business event is represented by a single process that will respond to that event. Event process names tend to be very general. We will adopt the convention of naming event processes as follows: PROCESS _____, RESPOND TO _____, or GENERATE _____, where the blank would be the name of the event (or its corresponding input). Sample event process names are PROCESS CUSTOMER ORDER, PROCESS CUSTOMER ORDER CHANGE, PROCESS CUSTOMER CHANGE OF ADDRESS, RESPOND TO CUSTOMER COMPLAINT, RESPOND TO ORDER INQUIRY, RESPOND TO PRODUCT PRICE CHECK, GENERATE BACK-ORDER REPORT, GENERATE CUSTOMER ACCOUNT STATEMENT, and GENERATE INVOICE.

An event process can be further decomposed into elementary processes that illustrate in detail how the system must respond to an event. **Elementary processes** are discrete, detailed activities or tasks required to complete the response to an event. In other words, they are the lowest level of detail depicted in a process model. A common synonym is *primitive process.* Elementary processes should be named with a strong action verb followed by an object clause that describes what the work is performed on (or for). Examples of elementary process names are VALIDATE CUSTOMER IDENTIFICATION, VALIDATE ORDERED PRODUCT NUMBER, CHECK PRODUCT AVAILABILITY, CALCULATE ORDER COST, CHECK CUSTOMER CREDIT, SORT BACK ORDERS, GET CUSTOMER ADDRESS, UPDATE CUSTOMER ADDRESS, ADD NEW CUSTOMER, and DELETE CUSTOMER.

Logical process models omit any processes that do nothing more than move or route data, thus leaving the data unchanged. Thus, you should omit any process that corresponds to a secretary or clerk receiving and simply forwarding a variety of documents to the next processing location. In the end, you should be left only with logical processes that:

- *Perform computations* (calculate grade point average).
- *Make decisions* (determine availability of ordered products).
- *Sort, filter, or otherwise summarize data* (identify overdue invoices).
- *Organize data into useful information* (generate a report or answer a question).
- *Trigger other processes* (turn on the furnace or instruct a robot).
- *Use stored data* (create, read, update, or delete a record).

Be careful to avoid three common mechanical errors with processes (illustrated in Figure 9-5):

- Process 3.1.2 has inputs but no outputs. We call this a *black hole* because data enters the process and then disappears. In most cases, the modeler simply forgot the output.
- Process 3.1.3 has outputs but no input. In this case, the input flows were likely forgotten.
- In Process 3.1.1 the inputs are insufficient to produce the output. We call this a *gray hole.* There are several possible causes, including (1) a misnamed process, (2) misnamed inputs and/or outputs, or (3) incomplete facts. Gray holes are the most common errors—and the most embarrassing. Once handed to a programmer, the input data flows to a process (to be implemented as a program) must be sufficient to produce the output data flows.

> Data Flows

Processes respond to inputs and generate outputs. Thus, at a minimum, all processes have at least one input and one output *data flow.* Data flows are the communications between processes and the system's environment. Let's examine some of the basic concepts and conventions of data flows.

Data in Motion   A data flow is *data in motion.* The flow of data between a system and its environment or between two processes inside a system is *communication.* Let's study this form of communication.

A **data flow** represents an input of data to a process or the output of data (or information) from a process. A data flow is also used to represent the creation, reading, deletion, or updating of data in a file or database (called a *data store* on the DFD). Think of a data flow as a highway down which packets of known composition travel. The name implies what type of data may travel down that highway. This highway is depicted as a solid line with arrow (see margin).

The *packet* concept is critical. Data that should travel together should be shown as a single data flow, no matter how many *physical* documents might be included. The

elementary process
discrete, detailed activity or task required to complete the response to an event. Also called *primitive process.*

data flow data that is input or output to or from a process.

Data flow name ⟶

Data Flow Symbol

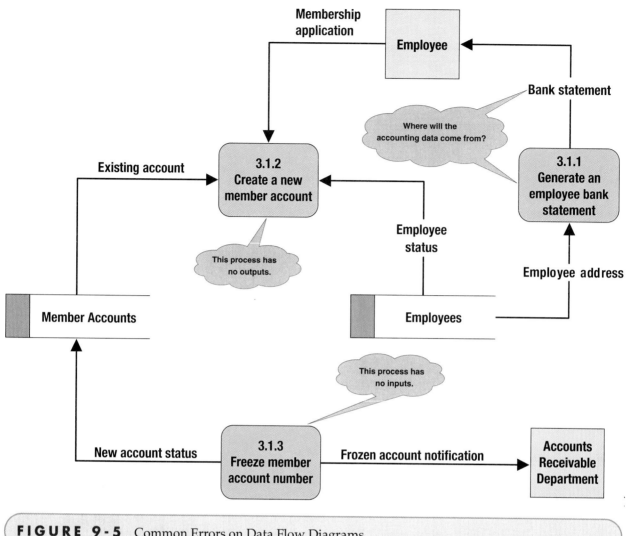

FIGURE 9-5   Common Errors on Data Flow Diagrams

packet concept is illustrated in Figure 9-6, which shows the correct and incorrect ways to show a logical data flow packet.

The *known composition* concept is equally important. A data flow is composed of either actual data attributes (also called *data structures*—more about them later) or other data flows. A **composite data flow** is a data flow that consists of other data flows. They are used to combine similar data flows on high-level data flow diagrams

composite data flow a data flow that consists of other data flows.

FIGURE 9-6

The Data Flow Packet Concept

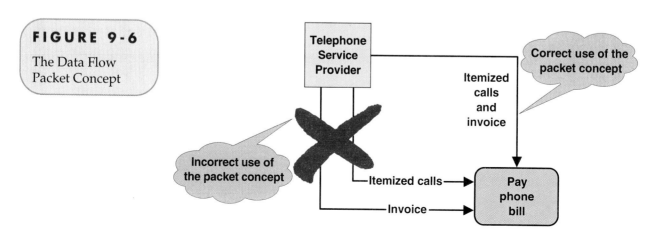

(a) **High-Level DFD**

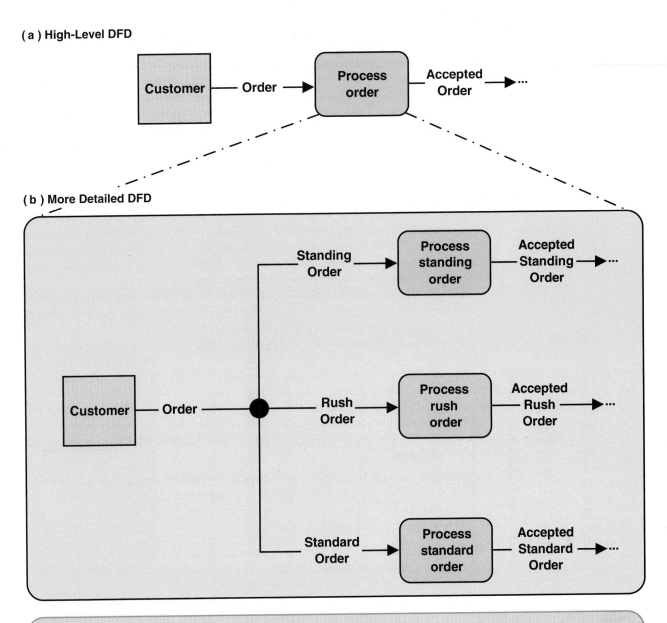

(b) **More Detailed DFD**

FIGURE 9-7   Composite and Elementary Data Flows

to make those diagrams easier to read. For example, in Figure 9-7(a), a high-level DFD consolidates all types of orders into a composite data flow called ORDER. In Figure 9-7(b), a more detailed data flow diagram shows specific types of orders: STANDING ORDER, RUSH ORDER, and STANDARD ORDER. These different orders require somewhat different processing. (The small, black circle is called a *junction.* It indicates that any given ORDER is an instance of only one of the order types.)

Another common use of composite data flows is to consolidate all reports and inquiry responses into one or two composite flows. There are two reasons for this. First, these outputs can be quite numerous. Second, many modern systems provide extensive user-defined reports and inquiries that cannot be predicted before the system's implementation and use.

Some data flow diagramming methods also recognize *nondata* flows called *control flows.* A **control flow** represents a condition or nondata event that triggers a process. Think of it as a condition to be monitored while the system works. When the system realizes that the condition meets some predetermined state, the process to

control flow a condition or nondata event that triggers a process.

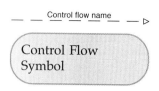

Control flow name

Control Flow
Symbol

which it is input is started. The classic information system example is *time*. For example, a report generation process may be triggered by the temporal event END-OF-MONTH. In real-time systems, control flows often represent real-time conditions such as TEMPERATURE and ALTITUDE. In most methodologies that distinguish between data and control flows, the control flow is depicted as a dashed line with arrow (see margin).

Typically, information systems analysts have dealt mostly with data flows; however, as information systems become more integrated with real-time systems (such as manufacturing processes and computer-integrated manufacturing), the need to distinguish the concept of control flows becomes necessary.

Logical **Data Flows and Conventions**   While we recognize that data flows can be implemented a number of ways (e.g., telephone calls, business forms, bar codes, memos, reports, computer screens, and computer-to-computer communications), we are interested only in *logical* data flows. Thus, we are only interested <u>that</u> the flow is needed (not <u>how</u> we will implement that flow). Data flow names should discourage premature commitment to any possible implementation.

Data flow names should be descriptive nouns and noun phrases that are singular, as opposed to plural (ORDER—not ORDERS). We do not want to imply that occurrences of the flow <u>must</u> be implemented as a *physical* batch.

Data flow names also should be unique. Use adjectives and adverbs to help to describe how processing has changed a data flow. For example, if an input to a process is named ORDER, the output should not be named ORDER. It might be named VALID ORDER, APPROVED ORDER, ORDER WITH VALID PRODUCTS, ORDER WITH APPROVED CREDIT, or any other more descriptive name that reflects what the process <u>did</u> to the original order.

Logical data flows to and from data stores require special naming considerations (see Figure 9-8). (Data store names are plural, and the numbered bullets match the note to the figure.)

- Only the *net* data flow is shown. Intuitively, you may realize that you have to get a record to update it or delete it. But unless data is needed for some other purpose (e.g., a calculation or decision), the "read" action is not shown. This keeps the diagram uncluttered.

❶ A data flow from a data store to a process indicates that data is to be "read" for some specific purpose. The data flow name should clearly indicate what data is to be read. This is shown in Figure 9-8.

❷ A data flow from a process to a data store indicates that data is to be created, deleted, or updated in/from that data store. Again, as shown in Figure 9-8, these data flows should be clearly named to reflect the specific action performed (such as NEW CUSTOMER, CUSTOMER TO BE DELETED, or UPDATED ORDER ADDRESS).

Notice that the names suggest the classic actions that can be performed on a file, namely, CREATE, READ, UPDATE, and DELETE (CRUD). In a real DFD, we would not actually record these action names on the diagram.

No data flow should never go unnamed. Unnamed data flows are frequently the result of flowchart thinking (step 1, step 2, etc.). If you can't give the data flow a reasonable name, it probably does not exist. Consistent with our goal of *logical* modeling, data flow names should describe the data flow without describing how the flow is or could be implemented. Suppose, for example, that end users explain their system as follows: *"We fill out Form 23 in triplicate and send it to . . ."* The logical name for the "Form 23" data flow might be COURSE REQUEST. This logical name eliminates physical, implementation biases—the idea that we must use a *paper form* and the notion that we must use carbon copies. Ultimately, this will free us to consider other physical alternatives such as Touch-Tone phone responses, online registration screens, or even e-business Internet pages.

Finally, all data flows must begin and/or end at a process because data flows are the inputs and outputs of a process. Consequently, all the data flows on the left side of Figure 9-9 are illegal. The corrected diagrams are shown on the right side.

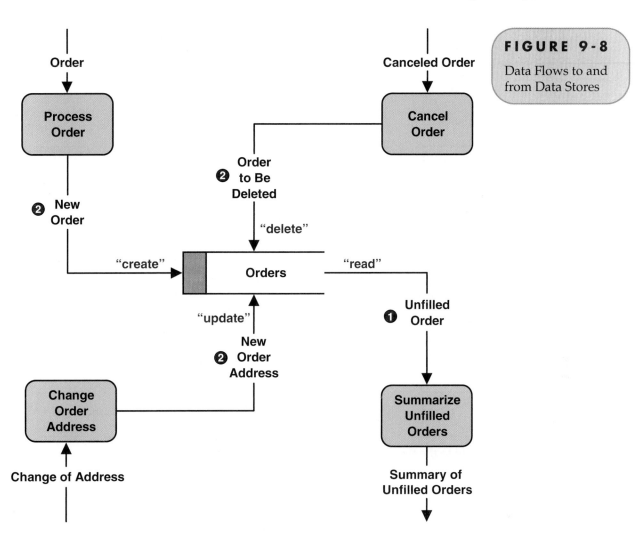

FIGURE 9-8
Data Flows to and from Data Stores

Data Flow Conservation For many years we have tried to improve business processes by automating them. It hasn't always worked or worked well because the business processes were designed to process data flows in a precomputing era. Consider the average business form. It is common to see the form divided into sections that are designed for different audiences. The first recipient completes his part of the form, the next recipient completes her part, and so forth. At certain points in this processing sequence, a copy of the form might even be detached and sent to another recipient who initializes a new multiple-part form that requires transcribing much of the same data from the initial form. In our own university, we've seen examples where poor form design requires that the same data be typed a dozen times.

Now, if the flow of current data is computerized based on the current business forms and processes, the resulting computer programs will merely automate these inefficiencies. This is precisely what has happened in most businesses! Today, a new emphasis on *business process redesign* encourages management, users, and systems analysts to identify and eliminate these inefficiencies <u>before</u> designing any new information system. We can support this trend in *logical* data flow diagrams by practicing data conservation. **Data conservation,** sometimes called "starving the processes," requires that a data flow contain only the data that is truly needed by the receiving process. By ensuring that processes receive only as much data as they really need, we simplify the interface between those processes. To practice data conservation, we must precisely define the data composition of each (noncomposite) data flow. Data composition is expressed in the form of *data structures.*

data conservation the practice of ensuring that a data flow contains only data needed by the receiving process.

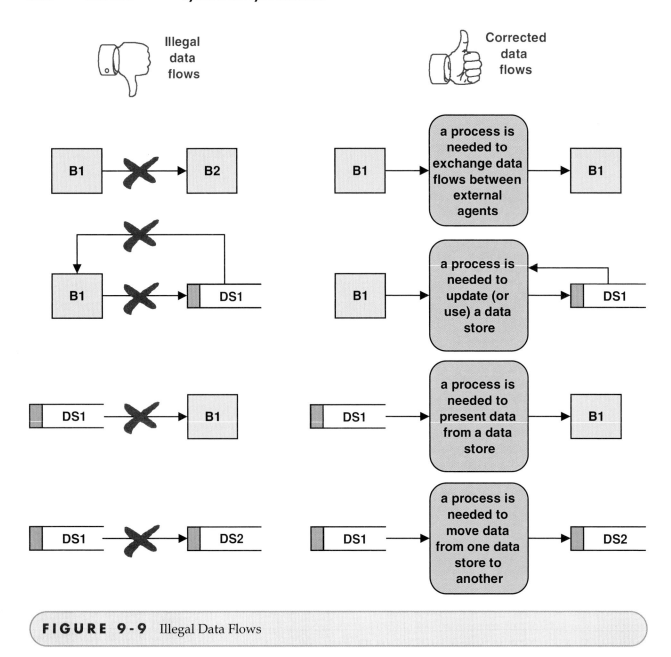

FIGURE 9-9 Illegal Data Flows

Data Structures Ultimately, a data flow contains data items called *attributes*. A **data attribute** is the smallest piece of data that has meaning to the end users and the business. (This definition also applies to *attributes* as they were presented in Chapter 8.) Sample attributes for the data flow ORDER might include ORDER NUMBER, ORDER DATE, CUSTOMER NUMBER, SHIPPING ADDRESS (which consists of attributes such as STREET ADDRESS, CITY, and ZIP CODE), ORDERED PRODUCT NUMBERS, QUANTITY(ies) ORDERED, and so on. Notice that some attributes occur once for each instance of ORDER, while others may occur several times for a single instance of ORDER.

The data attributes that comprise a data flow are organized into **data structures.** Data flows can be described in terms of the following types of data structures:

- A *sequence* or group of data attributes that occur one after another.
- The *selection* of one or more attributes from a set of attributes.
- The *repetition* of one or more attributes.

The most common data structure notation is a Boolean algebraic notation that is required by many CASE tools. Other CASE tools and methodologies support proprietary,

data attribute the smallest piece of data that has meaning to the users and the business.

data structure a specific arrangement of data attributes that define a single instance of a data flow.

FIGURE 9-10 A Data Structure for a Data Flow

Data Structure	English Interpretation

ORDER =
    ORDER NUMBER +
    ORDER DATE +
    [PERSONAL CUSTOMER NUMBER,
      CORPORATE ACCOUNT NUMBER]
    SHIPPING ADDRESS = ADDRESS +
    (BILLING ADDRESS = ADDRESS) +
    1 {PRODUCT NUMBER +
     PRODUCT DESCRIPTION +
     QUANTITY ORDERED +
     PRODUCT PRICE +
     PRODUCT PRICE SOURCE +
     EXTENDED PRICE} N +
    SUM OF EXTENDED PRICES +
    PREPAID AMOUNT
    (CREDIT CARD NUMBER + EXPIRATION DATE)
    (QUOTE NUMBER)
 	
 ADDRESS =
    (POST OFFICE BOX NUMBER) +
    STREET ADDRESS +
    CITY +
    [STATE MUNICIPALITY] +
    (COUNTRY) +
    POSTAL CODE
	An instance of ORDER consists of:
	    ORDER NUMBER and
	    ORDER DATE and
	    Either PERSONAL CUSTOMER NUMBER
	     or CORPORATE ACCOUNT NUMBER
	    and SHIPPING ADDRESS (which is equivalent
	     to ADDRESS)
	    and optionally: BILLING ADDRESS (which is
	     equivalent to ADDRESS)
	    and one or more instances of:
	     PRODUCT NUMBER and
	     PRODUCT DESCRIPTION and
	     QUANTITY ORDERED and
	     PRODUCT PRICE and
	     PRODUCT PRICE SOURCE and
	     EXTENDED PRICE
	    and SUM OF EXTENDED PRICES and
	    PREPAID AMOUNT and
	    optionally: both CREDIT CARD NUMBER and
	     EXPIRATION DATE
	    and optionally: QUOTE NUMBER
	
	 An instance of ADDRESS consists of:
	    optionally: POST OFFICE BOX NUMBER and
	    STREET ADDRESS and
	    CITY and
	    Either STATE OR MUNICIPALITY
	    and optionally: COUNTRY
	     and POSTAL CODE

but essentially equivalent, notations. A sample data structure for the data flow ORDER is presented in Figure 9-10. This algebraic notation uses the following symbols:

=   Means "consists of" or "is composed of."

+   Means "and" and designates *sequence.*

[...]   Means "only one of the attributes within the brackets may be present"— designates *selection.* The attributes in the brackets are separated by commas.

{...}   Means that the attributes in the braces may occur many times for one instance of the data flow—designates *repetition.* The attributes inside the braces are separated by commas.

(...)   Means the attribute(s) in the parentheses are optional—no value—for some instances of the data flow.

In our experience, <u>all</u> data flows can be described in terms of these fundamental constructs. Figure 9-11 demonstrates each of the fundamental constructs using examples. Returning to Figure 9-10, notice that the constructs are combined to describe the data content of the data flow.

FIGURE 9-11 Data Structure Constructs

Data Structure	Format by Example (relevant portion is **boldfaced**)	English Interpretation (relevant portion is **boldfaced**)
Sequence of Attributes—The sequence data structure indicates one or more attributes that may (or must) be included in a data flow.	WAGE AND TAX STATEMENT = **TAXPAYER IDENTIFICATION NUMBER + TAXPAYER NAME + TAXPAYER ADDRESS + WAGES, TIPS, AND COMPENSATION + FEDERAL TAX WITHHELD + . . .**	An instance of WAGE AND TAX STATEMENT consists of: **TAXPAYER IDENTIFICATION NUMBER and TAXPAYER NAME and TAXPAYER ADDRESS and WAGES, TIPS, AND COMPENSATION and FEDERAL TAX WITHHELD and . . .**
Selection of Attributes—The selection data structure allows you to show situations where different sets of attributes describe different instances of the data flow.	ORDER = **(PERSONAL CUSTOMER NUMBER, CORPORATE ACCOUNT NUMBER) +** ORDER DATE + . . .	An instance of ORDER consists of: **Either PERSONAL CUSTOMER NUMBER or CORPORATE ACCOUNT NUMBER;** and ORDER DATE and . . .
Repetition of Attributes—The repetition data structure is used to set off a data attribute or group of data attributes that may (or must) repeat themselves a specified number of times for a single instance of the data flow. The minimum number of repetitions is usually zero or one. The maximum number of repetitions may be specified as "n" meaning "many" where the actual number of instances varies for each instance of the data flow.	CLAIM = POLICY NUMBER + POLICYHOLDER NAME + POLICYHOLDER ADDRESS + **0 {DEPENDENT NAME + DEPENDENT'S RELATIONSHIP} N + 1 {EXPENSE DESCRIPTION + SERVICE PROVIDER + EXPENSE AMOUNT} N**	An instance of CLAIM consists of: POLICY NUMBER and POLICYHOLDER NAME and POLICYHOLDER ADDRESS and **zero or more instances of: DEPENDENT NAME and DEPENDENT'S RELATIONSHIP and one or more instances of: EXPENSE DESCRIPTION and SERVICE PROVIDER and EXPENSE ACCOUNT**
Optional Attributes—The optional notation indicates that an attribute, or group of attributes in a sequence or selection data structure may not be included in all instances of a data flow. Note: For the repetition data structure, a minimum of "zero" is the same as making the entire repeating group "optional."	CLAIM = POLICY NUMBER + POLICYHOLDER NAME + POLICYHOLDER ADDRESS + **(SPOUSE NAME + DATE OF BIRTH) + . . .**	An instance of CLAIM consists of: POLICY NUMBER and POLICYHOLDER NAME and POLICYHOLDER ADDRESS and **optionally, SPOUSE NAME and DATE OF BIRTH and . . .**
Reusable Attributes—For groups of attributes that are contained in many data flows, it is desirable to create a separate data structure that can be reused in other data structures.	DATE = MONTH + DAY + YEAR	Then, the reusable structures can be included in other data flow structures as follows: ORDER = ORDER NUMBER . . . + DATE INVOICE = INVOICE NUMBER . . . + DATE PAYMENT = CUSTOMER NUMBER . . . + DATE

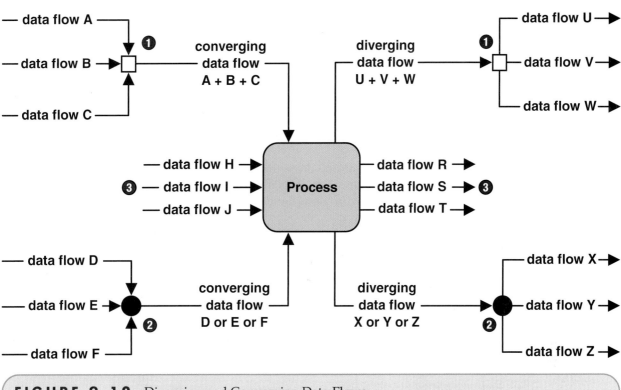

FIGURE 9-12   Diverging and Converging Data Flows

The importance of defining the data structures for every data flow should be apparent—you are defining the business data requirements for each input and output! These requirements must be determined before any process could be implemented as a computer program. This standard notation provides a simple but effective means for communicating between end users and programmers.

Domains   An attribute is a piece of data. In analyzing a system, it makes sense that we should define those values for an attribute that are legitimate, or that make sense. The values for each attribute are defined in terms of two properties: data type and domain. The **data type** for an attribute defines what class of data can be stored in that attribute, whereas the **domain** of an attribute defines what values an attribute can legitimately take on. The concepts of data type and domain were introduced in Chapter 8. See that discussion and Tables 8-1 and 8-2 for a more complete description of data type and domain.

data type a class of data that can be stored in an attribute.

domain the legitimate values for an attribute.

Divergent and Convergent Flows   It is sometimes useful to depict diverging or converging data flows on a data flow diagram. A **diverging data flow** is one that splits into multiple data flows. Diverging data flows indicate that all or parts of a single data flow are routed to different destinations.[2] A **converging data flow** is the merger of multiple data flows into a single data flow. Converging data flows indicate that data flows from different sources can (must) come together as a single packet for subsequent processing.

Diverging and converging data flows are depicted as shown in Figure 9-12. Notice that we do not include a process to "route" the flows. The flows simply diverge from

diverging data flow a data flow that splits into multiple data flows.

converging data flow the merger of multiple data flows into a single data flow.

---

[2]Some experts suggest that diverging data flows should be used only when all data in the flow is routed to all destinations. We prefer the classic DeMarco definition that allows all *or* parts of the flow to be routed to different processes.

or converge to a common flow. The following notations, not supported by all CASE tools, are used in this book:

❶ The small square *junction* means "and." This means that each time the process is performed, it must input (or output) all the diverging or converging data flows. *(Some DFD notations simply place a + between the data flows.)*

❷ The small black *junction* means "exclusive or." This means that each time the process is performed, it must input (or output) only one of the diverging or converging data flows. *(Some DFD notations simply place an * between the data flows.)*

❸ In the absence of one diverging or converging data flow, the reader should assume an "inclusive or." This means that each time the process is performed, it may input any <u>or</u> all of the depicted data flows.

With the above rules, the most complex of business process and data flow combinations can be depicted.

The Process of Logical Process Modeling

Now that you understand the basic concepts of process models, we can examine building a process model. When do you do it? How many process models may be drawn? What technology exists to support the development of process models?

> Strategic Systems Planning

Many organizations select application development projects based on strategic information system plans. Strategic planning is a separate project that produces an information systems strategy plan that defines an overall vision and architecture for information systems. This architecture frequently includes an *enterprise process model.* (The plan usually has other architectural components that are not important to this discussion.)

An enterprise process model typically identifies only business areas and functions. Events and detailed processes are rarely examined. Business areas and functions are identified and mapped to other enterprise models such as the enterprise data model (Chapter 8). Business areas and functions are subsequently prioritized into application development projects. Priorities are usually based on which business areas, functions, and supporting applications will return the most value to the business as a whole.

An enterprise process model is stored in a corporate repository. Subsequently, as application development projects are started, subsets of the enterprise process model are exported to the project teams to serve as a starting point for building more detailed process models (including data flow diagrams). Once the project team completes systems analysis and design, the expanded and refined process models are returned to the corporate repository.

> Process Modeling for Business Process Redesign

Business process redesign (BPR) has been discussed several times in this book and chapter. Recall that BPR projects analyze business processes and then redesign them to eliminate inefficiencies and bureaucracies before any (re)application of information technology. To redesign business processes, we must first study the existing processes. Process models play an integral role in BPR.

Each BPR methodology recommends its own process model notations and documentation. Most of the models are a cross between data flow diagrams and flowcharts. The diagrams tend to be very *physical* because the BPR team is trying to isolate the implementation idiosyncrasies that cause inefficiency and reduce value. BPR data flow diagrams/flowcharts may include new symbols and information to illustrate timing, throughput, delays, costs, and value. Given this additional data, the BPR team then attempts to simplify the processes and data flows in an effort to maximize efficiency and return the most value to the organization.

Opportunities for the efficient use of information technology may also be recorded on the physical diagrams. If so, the BPR diagram becomes an input to systems analysis (described next).

> Process Modeling during Systems Analysis

In systems analysis and in this chapter, we focus exclusively on *logical* process modeling as a part of business requirements analysis. In your information system framework, logical process models have a process focus and a SYSTEM OWNER and/or SYSTEM USER perspective. They are typically constructed as deliverables of the requirements analysis phase of a project. While logical process models are not concerned with implementation details or technology, they may be constructed (through *reverse engineering*) from existing application software, but this technology is much less mature and reliable than the corresponding reverse *data* engineering technology.

In the heyday of the original structured analysis methodologies, process modeling was also performed in the problem analysis phase of systems analysis. Analysts would build a *physical process model of the current system,* a *logical model of the current system,* and a *logical model of the target system.* Each model would be built top-down—from very general models to very detailed models. While conceptually sound, this approach led to modeling overkill and significant project delays, so much so that even structured techniques guru Ed Yourdon called it "analysis paralysis."

Today, most modern structured analysis strategies focus exclusively on the *logical model of the target system* being developed. Instead of being built either top-down or bottom-up, they are organized according to a commonsense strategy called *event partitioning.* **Event partitioning** factors a system into subsystems based on business events and responses to those events. This strategy for event-driven process modeling is illustrated in Figure 9-13 and described as follows:

1 A system **context data flow diagram** is constructed to establish *initial* project scope. This simple, one-page data flow diagram shows only the system's main interfaces with its environment.

2 A **functional decomposition diagram** is drawn to partition the system into logical subsystems and/or functions. (This step is omitted for very small systems.)

3 An **event-response** or **use-case list** is compiled to identify and confirm the business events to which the system must provide a response. The list will also describe the required or possible responses to each event.

4 One process, called an **event handler,** is added to the decomposition diagram for each event. The decomposition diagram now serves as the outline for the system.

5 Optionally, an **event diagram** is constructed and validated for each event. This simple data flow diagram shows only the event handler and the inputs and outputs for each event.

6 One or more **system diagrams** are constructed by merging the event diagrams. These data flow diagrams show the "big picture" of the system.

7 **Primitive diagrams** are constructed for event processes that require additional processing details. These data flow diagrams show all the elementary processes, data stores, and data flows for single events. **8** The logic of each elementary process and **9** the data structure of each elementary data flow are described using the tools described earlier in the chapter.

event partitioning a structured analysis strategy in which a system is factored into subsystems based on business events and responses to those events.

context data flow diagram a diagram that shows the system as a "black box" and its main interfaces with its environment.

functional decomposition diagram a diagram that partitions the system into logical subsystems and/or functions.

event-response list a list of the business events to which the system must provide a response similar to a use-case list.

event handler a process that handles a given event in the event-response list.

event diagram a data flow diagram for a single event handler and the agents and data stores that provide inputs or receive outputs.

system diagram a data flow diagram that merges event diagrams for the entire system or part of the system.

primitive diagram a data flow diagram that depicts the elementary processes, data stores, and data flows for a single event.

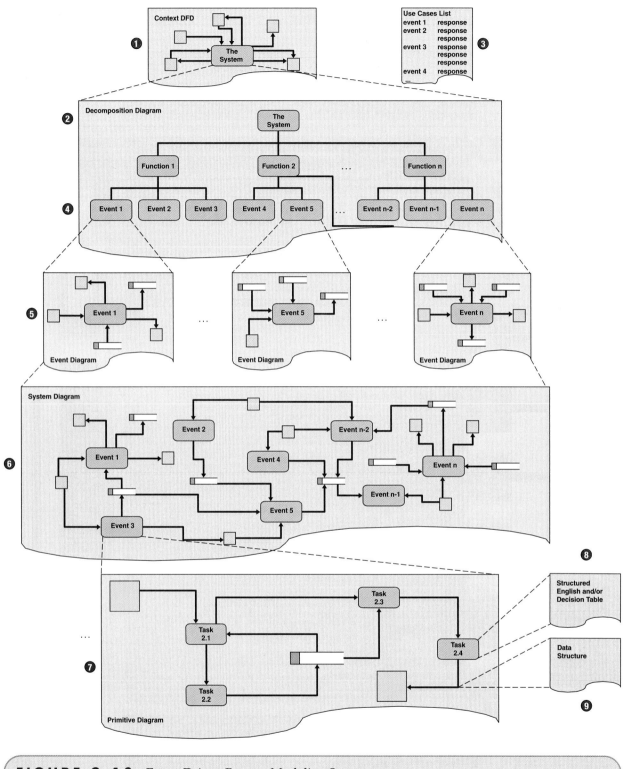

FIGURE 9-13   Event-Driven Process Modeling Strategy

The above process models collectively document all the business processing requirements for a system. We'll demonstrate the technique in our SoundStage case study.

The logical process model from systems analysis describes business processing requirements of the system, not technical solutions. Recall from Chapter 5 that the purpose of the decision analysis phase is to determine the best way to implement those requirements with technology. In practice, this decision may have already been standardized as part of an application architecture. For example, the SoundStage application architecture requires that the development team first determine if an acceptable system can be purchased. If not, the current application architecture specifies that software built in-house be written in either Microsoft's *Visual Basic .NET* or *C#.*

> Looking Ahead to Systems Design

During system design, the logical process model will be transformed into a physical process model (called an *application schema*) for the chosen technical architecture. This model will reflect the technical capabilities and limitations of the chosen technology. Any further discussion of physical process/application design is deferred until Chapter 13.

> Fact-Finding and Information Gathering for Process Modeling

Process models cannot be constructed without appropriate facts and information as supplied by the user community. These facts can be collected through a number of techniques such as sampling of existing forms and files, research of similar systems, surveys of users and management, and interviews of users and management. The fastest method of collecting facts and information and simultaneously constructing and verifying the process models is *joint requirements planning (JRP)*. JRP uses a carefully facilitated group meeting to collect the facts, build the models, and verify the models—usually in one or two full-day sessions.

Fact-finding, information gathering, and JRP techniques were explored in Chapter 6.

> Computer-Aided Systems Engineering (CASE) for Process Modeling

Like all system models, process models are stored in the repository. Computer-aided systems engineering (CASE) technology, introduced in Chapter 3, provides the repository for storing the process model and its detailed descriptions. Most CASE products support computer-assisted process modeling. Most support decomposition diagrams and data flow diagrams. Some support extensions for business process analysis and redesign.

Using a CASE product, you can easily create professional, readable process models without using paper, pencil, eraser, and templates. The models can be easily modified to reflect corrections and changes suggested by end users. Also, most CASE products provide powerful analytical tools that can check your models for mechanical errors, completeness, and consistency. Some CASE products can even help you analyze the data model for consistency, completeness, and flexibility. The potential time savings and quality are substantial.

CASE tools do have their limitations. Not all process model conventions are supported by all CASE products. Therefore, any given CASE product may force the company to adapt its methodology's process modeling symbols or approach so that it is workable within the limitations of its CASE tool.

All the SoundStage process models in the next section of this chapter were created with Popkin's CASE tool, *System Architect 2001.* For the case study, we provide you with the printouts exactly as they came off our printers. We did not add color. The only modifications by the artist were the occasional bullets that call your attention to specific items of interest on the printouts. All the processes, data flows, data stores, and boundaries on the SoundStage process models were automatically cataloged into *System Architect's* project repository (which it calls an encyclopedia). Figure 9-14 illustrates some of *System Architect's* screens as used for data modeling.

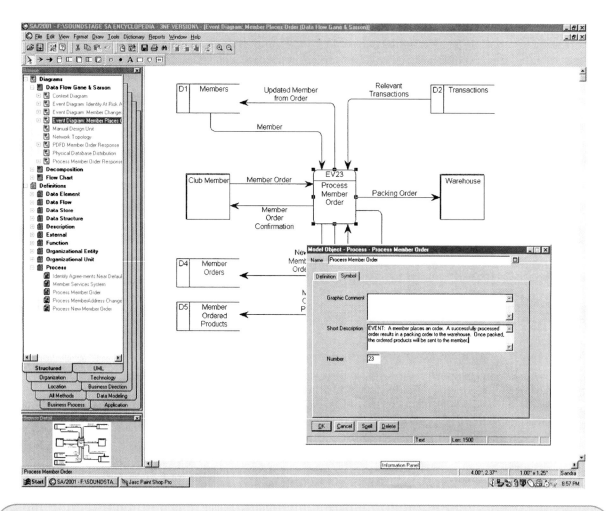

FIGURE 9-14   CASE for Process Modeling (using *System Architect 2001* by Popkin Software & Systems)

How to Construct Process Models

As a systems analyst or knowledgeable end user, you must learn how to draw decomposition and data flow diagrams to model business process requirements. We will use the SoundStage Entertainment Club project to teach you how to draw these process models.

Let's assume the preliminary investigation and problem analysis phases of the project have been completed and the project team understands the current system's strengths, weaknesses, limitations, problems, opportunities, and constraints. The team has also already built the data model (in Chapter 8) to document business data requirements for the new system. Team members will now build the corresponding process models.

> The Context Data Flow Diagram

First, we need to document the initial project scope. All projects have scope. A project's scope defines what aspect of the business a system or application is supposed to support and how the system being modeled must interact with other systems and the business as a whole. In your information system framework, scope is defined as the COMMUNICATION focus from the SYSTEM OWNERS' perspective. It is documented with a **context data flow diagram.** Because the scope of any project is always subject to change, the context diagram is also subject to constant change. A synonym is *environmental model* [Yourdon, 1990].

context data flow diagram a process model used to document the scope for a system. Also called *environmental model.*

We suggest the following strategy for documenting the system's boundary and scope:

1. Think of the system as a container in order to distinguish the inside from the outside. Ignore the inner workings of the container. This is sometimes called "black box" thinking.
2. Ask your end users what business transactions a system must respond to. These are the *net inputs* to the system. For each net input, determine its source. Sources will become *external agents* on the context data flow diagram.
3. Ask your end users what responses must be produced by the system. These are the net outputs to the system. For each net output, determine its destination. Destinations will also become *external agents.* Requirements for reports and queries can quickly clutter the diagram. Consider consolidating them into composite data flows.
4. Identify any *external* data stores. Many systems require access to the files or databases of other systems. They may use the data in those files or databases. Sometimes they may update certain data in those files and databases. But generally, they are not permitted to change the structure of those files and databases—therefore, they are outside the project scope.
5. Draw a context diagram from all of the preceding information.

If you try to include all the inputs and outputs between a system and the rest of the business and outside world, a typical context data flow diagram might show as many as 50 or more data flows. Such a diagram would have little, if any, communication value. Therefore, we suggest you show only those data flows that represent the main objective or most important inputs and outputs of the system. Defer less common data flows to more detailed DFDs to be drawn later.

The context data flow diagram contains one and only one process (see Figure 9-15). Sometimes, this process is identified by the number "0"; however, our CASE tool did not allow this. External agents are drawn around the perimeter. Data flows define the interactions of your system with the boundaries and with the external data stores.

As shown in the context data flow diagram, the main purpose of the system is to process NEW SUBSCRIPTIONS in response to SUBSCRIPTION OFFERS, create NEW PROMOTIONS for products, and respond to MEMBER ORDERS by sending PACKING ORDERS to the warehouse to be filled. (Notice that we made all data flow names singular.) Management has also emphasized the need for VARIOUS REPORTS. Finally, the Web extensions to this system require that the system provide members with VARIOUS INQUIRY RESPONSES regarding orders and accounts.

> The Functional Decomposition Diagram

Recall that a decomposition diagram shows the top-down functional decomposition or structure of a system. It also provides us with the beginnings of an outline for drawing our data flow diagrams.

Figure 9-16 on page 341 is the functional decomposition diagram for the Sound-Stage project. Let's study this diagram. First, notice that the processes are depicted as rectangles, not rounded rectangles. This is merely a limitation of our CASE tool's implementation of decomposition diagrams—you also may have to adapt to your CASE tool.

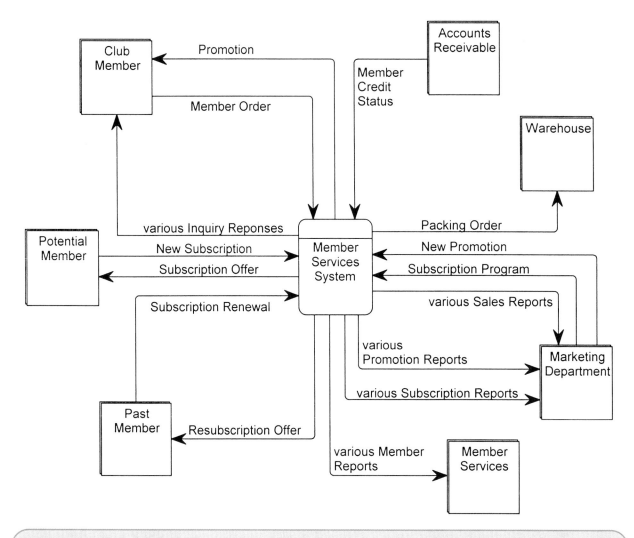

FIGURE 9-15 The Context Data Flow Diagram (created with *System Architect 2001*)

The following is an item-by-item discussion of the decomposition diagram. The circled numbers correspond to specific points of interest on the diagram.

❶ The root process corresponds to the entire system.
❷ The system is initially factored into subsystems and/or functions. These subsystems and functions do not necessarily correspond to organization units on an organization chart. Increasingly, analysts and users are being asked to ignore organizational boundaries and to build cross-functional systems that streamline processing and data sharing.
❸ We like to separate the operational and reporting aspects of a system. Thus, we factored each subsystem accordingly. Later, if this structure doesn't make sense, we can change it.

Larger systems might have first been factored into subsystems *and* functions. There is no limit to the number of child processes for a parent process. Many authors used to recommend a maximum of five to nine processes per parent, but any such limit is too artificial. Instead, structure the system such that it makes sense for the business!

Factoring a parent process into a single child process doesn't make sense. It would provide no additional detail. Therefore, if a process is to be factored, it should be factored into at least two child processes.

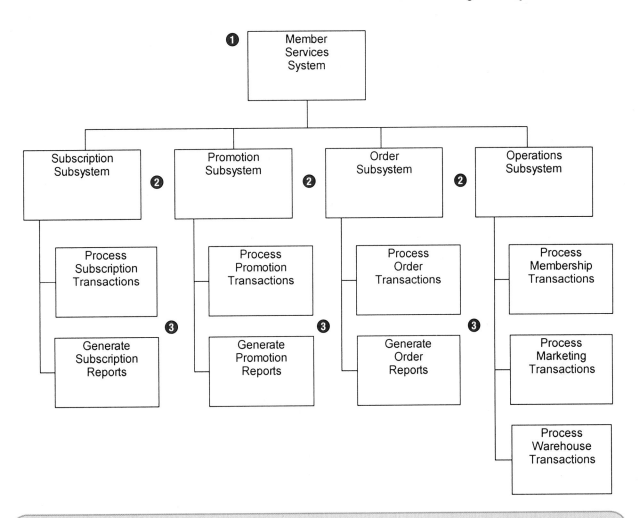

FIGURE 9-16 A Functional Decomposition Diagram (created with *System Architect 2001*)

> The Event-Response or Use-Case List

After constructing the decomposition diagram, we next determine what business events the system must respond to and what responses are appropriate. Events are not hard to find. Some of the inputs on the context diagram are associated with events, but the context diagram rarely shows all the events. Essentially, there are three types of events:

- *External events* are so named because they are initiated by external agents. When these events happen, an input data flow occurs for the system. For example, the event CUSTOMER PLACES A NEW ORDER is recognized in the form of the input data flow ORDER from the external agent CUSTOMER.
- *Temporal events* trigger processes on the basis of time, or something that merely happens. When these events happen, an input *control flow* occurs. Examples of temporal events include TIME TO REMIND CUSTOMERS TO PAY PAST INVOICES or END OF MONTH.
- *State events* trigger processes based on a system's change from one state or condition to another. Like temporal events, state events will be illustrated as an input *control flow*.

Information systems usually respond mostly to external and temporal events. State events are usually associated with real-time systems such as elevator or robot control.

One of the more popular and successful approaches for finding and identifying events and responses is a technique called *use cases* (Chapter 7) developed by Dr. Ivar Jacobson. This technique is rooted in object-oriented analysis but is easily adapted to structured analysis and data flow diagramming. **Use-case** analysis is the process of identifying and modeling business events, who initiated them, and how the system responds to them.

use case an analysis tool for finding and identifying business events and responses.

Use cases identify and describe necessary system processes from the perspective of users. Each use case is initiated by users or external systems called *actors.* An actor is anything that needs to interact with the system to exchange information and so is analogous to external agents in DFDs.

The context data flow diagram identifies the key actors as *external agents.* It also identifies *some* of the use cases. The key word is "some." Recall that the context diagram shows only the main inputs and outputs of a system. There are almost always more inputs and outputs than are depicted—usually many more. Some of the inputs and outputs depicted are really composites of many types of and variations on those inputs and outputs (e.g., the "various reports" on our context diagram). Also, the context diagram may not illustrate the many exception inputs and outputs such as errors, inquiries, and follow-ups.

One way to expand the use cases is to interview the external agents (actors) depicted on the diagram. The agents can (1) identify the events (use cases) for which they believe the system may have to provide a response and (2) identify other actors (new external agents) that were not originally shown on the context diagram.

Another way to identify use cases (events) is to study the data model, assuming a data model was developed before drawing data flow diagrams, and study the life history of each entity on that data model. Instances of these entities must be created, updated, and eventually deleted. Events or use cases trigger these actions on the entity. It is not difficult to get users talking about the events that could create, update, and delete entity instances. After all, they live these events daily. This approach was used to build the use-case list for the SoundStage project.

A partial table of use cases is illustrated in Figure 9-17 (pages 343–344). For each use case, you will find:

- The actor that initiates the event (which will become an external agent on our DFDs).
- The event (which will be handled by a process on our DFDs).
- The input or trigger (which will become a data or control flow on our DFDs).
- All outputs and responses (which will also become data flows on our DFDs). Notice that we used parentheses to denote temporal events.
- Outputs (but be careful not to imply implementation). When we used the term *report* we were not necessarily implying a paper-based document. Notice that our responses include changes to stored data about entities from the data model. These include create new instances of the entity, update existing instances of the entity, and delete instances of the entity.

The number of use cases for a system is usually quite large. This is necessary to ensure that the system designers build a complete system that will respond to all the business events. As a final step, consider assigning each event to one of the subsystems and functions identified in the decomposition diagram (drawn in the previous step).

> Event Decomposition Diagrams

To further partition our functions in the decomposition diagram, we simply add event handling processes (one per use case) to the decomposition (see Figure 9-18 on page 345). If the entire decomposition diagram will not fit on a single page, add separate pages for subsystems or functions. The root process on a subsequent page

FIGURE 9-17 A Partial Use-Case Table

Actor/External Agent	Event (or Use Case)	Trigger	Responses
Marketing	Establishes a new membership subscription plan to entice new members.	NEW MEMBER SUBSCRIPTION PROGRAM	Generate SUBSCRIPTION PLAN CONFIRMATION. Create AGREEMENT in the database.
Marketing	Establishes a new membership resubscription plan to lure back former members.	PAST MEMBER RESUBSCRIPTION PROGRAM	Generate SUBSCRIPTION PLAN CONFIRMATION. Create AGREEMENT in the database.
Marketing	Changes a subscription plan for current members (e.g., extending the fulfillment period).	SUBSCRIPTION PLAN CHANGE	Generate AGREEMENT CHANGE CONFIRMATION. Update AGREEMENT in the database.
(time)	A subscription plan expires.	(current date)	Generate AGREEMENT CHANGE CONFIRMATION. Logically Delete (void) AGREEMENT in the database.
Marketing	Cancels a subscription plan before its planned expiration date.	SUBSCRIPTION PLAN CANCELLATION	Generate AGREEMENT CHANGE CONFIRMATION. Logically Delete (void) AGREEMENT in the database.
Member	Joins the club by subscribing. ("Take any 12 CDs for one penny and agree to buy 4 more at regular prices within two years.")	NEW SUBSCRIPTION	Generate MEMBER DIRECTORY UPDATE CONFIRMATION. Create MEMBER in the database. Create first MEMBER ORDER and MEMBER ORDERED PRODUCTS in the database.
Member	Changes address (including e-mail and privacy code).	CHANGE OF ADDRESS	Generate MEMBER DIRECTORY UPDATE CONFIRMATION. Update MEMBER in the database.
Accounts Receivable	Changes member's credit status.	CHANGE OF CREDIT STATUS	Generate CREDIT DIRECTORY UPDATE CONFIRMATION. Update MEMBER in the database.

FIGURE 9-17 (Concluded)

Actor/External Agent	Event (or Use Case)	Trigger	Responses
(time)	90 days after Marketing decides to no longer sell a product.	(current date)	Generate CATALOG CHANGE CONFIRMATION. Logically Delete (deactivate) PRODUCT in the database.
Member	Wants to pick products for possible purchase. (Logical requirement is driven by vision of Web-based access to information.)	PRODUCT INQUIRY	Generate CATALOG DESCRIPTION.
Member	Places order.	NEW MEMBER ORDER	Generate MEMBER ORDER CONFIRMATION. Create MEMBER ORDER and MEMBER ORDERED PRODUCT in the database.
Member	Revises order.	MEMBER ORDER CHANGE REQUEST	Generate MEMBER ORDER CONFIRMATION. Update MEMBER ORDER and/or MEMBER ORDERED PRODUCTS in the database.
Member	Cancels order.	MEMBER ORDER CANCELLATION	Generate MEMBER ORDER CONFIRMATION. Logically Delete MEMBER ORDER and MEMBER ORDERED PRODUCTS in the database.
(time)	90 days after the order.	(current date)	Physically Delete MEMBER ORDER and MEMBER ORDERED PRODUCTS in the database.
Member	Inquires about his or her purchase history (three-year time limit).	MEMBER PURCHASE INQUIRY	Generate MEMBER PURCHASE HISTORY.
(each) Club	(end of month).	(current date)	Generate MONTHLY SALES ANALYSIS. Generate MONTHLY MEMBER AGREEMENT EXCEPTION ANALYSIS. Generate MEMBERSHIP ANALYSIS REPORT.

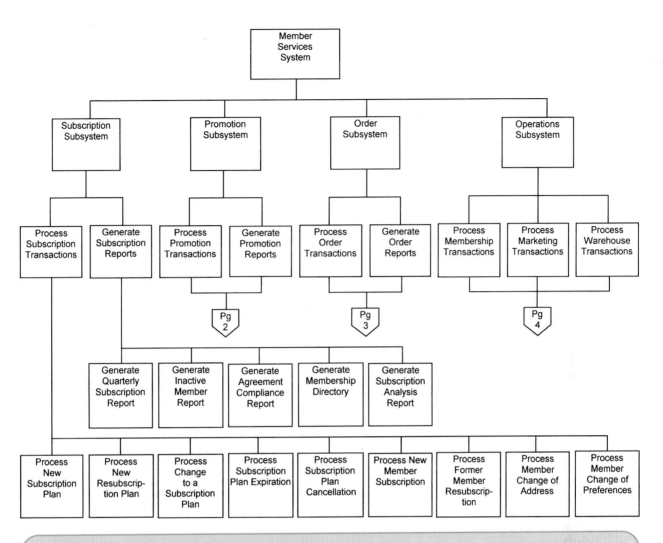

FIGURE 9-18   A Partial Event Decomposition Diagram (created with *System Architect 2001*)

should be duplicated from an earlier page to provide a cross-reference. Figure 9-18 shows only the event processes for the MEMBERSHIPS subsystem. Events for the PROMOTIONS and ORDERS functions would be on separate pages.

There is no need to factor the decomposition diagram beyond the events and reports. That would be like outlining down to the final paragraphs or sentences in a paper. The decomposition diagram, as constructed, will serve as a good outline for the later data flow diagrams.

> Event Diagrams

Using our decomposition diagram as an outline, we can draw one event diagram for each event process. This is an optional, but useful, step. An **event diagram** is a context diagram for a single event. It shows the inputs, outputs, and data store interactions for the event. By drawing an event diagram for each process, users do not become overwhelmed by the overall size of the system. They can examine each use case as its own context diagram.

Before drawing any event diagrams, you may find it helpful to have a list of all the data stores available. Because SoundStage already completed the data model for this project, team members simply created a list of each entity name on that data model

event diagram a data flow diagram that depicts the context for a single event.

**DATA STORES
(ENTITIES)**

AGREEMENTS

MEMBERS

MEMBER ORDERS

MEMBER ORDERED PRODUCTS

PRODUCTS

PROMOTIONS

TITLE PROMOTIONS

(see margin). It is useful to review the definition and attributes for each entity/data store on the list.

Most event diagrams contain a single process—the same process that was named to handle the event on the decomposition diagram. For each event, illustrate the following:

- The inputs and their sources. Sources are depicted as external agents. The data structure for each input should be recorded in the repository.
- The outputs and their destinations. Destinations are depicted as external agents. The data structure for each output should be recorded in the repository.
- Any data stores from which records must be "read" should be added to the event diagram. Data flows should be added and named to reflect what data is read by the process.
- Any data stores in which records must be created, deleted, or updated should be included in the event diagram. Data flows to the data stores should be named to reflect the nature of the update.

The simplicity of event diagramming makes the technique a powerful communication tool between users and technical professionals.

A complete set of event diagrams for the SoundStage case study would double the length of this chapter without adding substantive educational value. Thus, we will demonstrate the model with three simple examples.

Figure 9-19 illustrates a simple event diagram for an external event. Most systems have many such simple event diagrams because all systems must provide for routine maintenance of data stores.

Figure 9-20 depicts a somewhat more complex external event, one for the business transaction MEMBER ORDER. Notice that business transactions tend to use and update more data stores and have more interactions with external agents.

Can an event diagram have more than one process on it? The answer is maybe. Some event processes may trigger other event processes. In this case, the combination of events should be shown on a single event diagram. In our experience, most event diagrams have one process. An occasional event diagram may have two or perhaps three processes. If the number of processes exceeds three, you are probably drawing what is called an *activity diagram* (prematurely), not an event diagram—in other words, you're getting too involved with details. Most event processes do not directly communicate with one another. Instead, they communicate across shared data stores. This allows each event process to do its job without worrying about other processes keeping up.

Figure 9-21 on page 348 shows an event diagram for a temporal event. We added an external entity CALENDAR or TIME to serve as a source for this control flow.

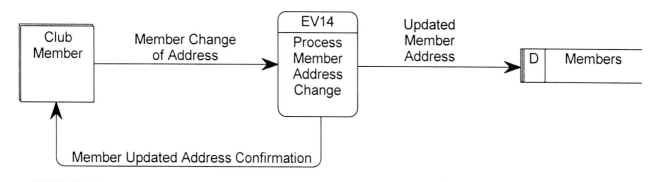

FIGURE 9-19   A Simple External Event Diagram (created with *System Architect 2001*)

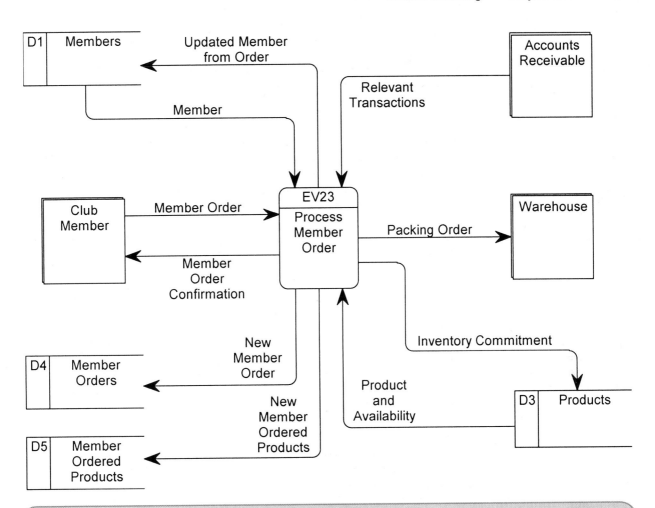

FIGURE 9-20 A More Complex External Event Diagram (created with *System Architect 2001*)

Each event process should be described to the CASE repository with the following properties:

- Event sentence—for business perspective.
- Throughput requirements—the volume of inputs per some time period.
- Response time requirements—how fast the typical event must be handled.
- Security, audit, and control requirements.
- Archival requirements (from a business perspective).

For example, consider the event diagram in Figure 9-19: "A member submits a change of address."

- Occurs 25 times per month.
- Should be processed within 15 days.
- Must protect privacy of addresses unless the member authorizes release.
- Should retain a semipermanent record of some type.

All the above properties can be added to the descriptions associated with the appropriate processes, data flows, and data stores on the model.

> The System Diagram(s)

The event diagrams serve as a meaningful context to help users validate the accuracy of each event to which the system must provide a response. But these events do not

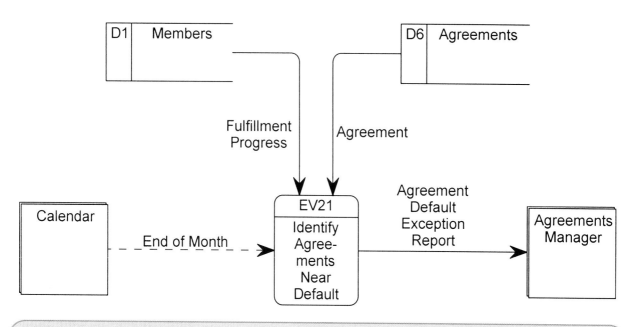

FIGURE 9-21 A Temporal Event Diagram (created with *System Architect 2001*)

exist in isolation. They collectively define systems and subsystems. It is, therefore, useful to construct one or more system diagrams that show all the events in the system or a subsystem.

The system diagram is said to be "exploded" from the single process that we created on the original context diagram (Figure 9-15). The system diagram shows either (1) all the events for the system on a single diagram or (2) all the events for a single subsystem on a single diagram. Depending on the size of the system, a single diagram may be too large.

While the SoundStage project is moderate in size, it still responds to too many events to squeeze all those processes onto a single diagram. Instead, Bob Martinez elected to draw a subsystem diagram for each of the major subsystems. Figure 9-22 (pages 350–351) shows the subsystem diagram for the ORDERS SUBSYSTEM. It consolidates all the transaction and report-writing events for that subsystem onto a single diagram. (The reporting events may be omitted or consolidated into composites if the diagram is too cluttered.) Notice that the system diagram demonstrates how event processes communicate using shared data stores.

If necessary, and after drawing the four subsystem diagrams for this project, Bob could have drawn a system diagram that illustrates only the interactions between those four subsystems. This is a relic of the original top-down data flow diagramming strategy of the original structured analysis methodology. In practice, this higher-level diagram requires so much consolidation of data flows and data stores that its communication value is questionable. To Bob, this was busywork, and his time was better spent on the next set of data flow diagrams.

We now have a set of event diagrams (one per business event) and one or more system/subsystem diagrams. The event diagram processes are merged into the system diagrams. It is very important that each of the data flows, data stores, and external agents that were illustrated on the event diagrams be represented on the system diagrams. Notice that we duplicate data stores and external agents to minimize crossing of lines. Most CASE tools include facilities to check for balancing errors.

Before we leave this topic, we should introduce the concept of balancing. **Balancing** is the synchronizing of data flow diagrams at different levels of detail to preserve consistency and completeness of the models. Balancing is a quality assurance technique. Balancing requires that, if you explode a process to another DFD to reveal

balancing a concept that requires that data flow diagrams at different levels of detail reflect consistency and completeness.

more detail, you must include the same data flows and data stores on the child diagram that you included in the parent diagram's original process (or their logical equivalents).

> Primitive Diagrams

Some event processes on the system diagram may be exploded into a primitive data flow diagram to reveal more detail. This is especially true of the more complex business transaction processes (e.g., order processing). Other events, such as generation of reports, are simple enough that they do not require further explosion.

Event processes with more complex event diagrams should be exploded into a more detailed, primitive data flow diagram such as that illustrated in Figure 9-23 on page 352. This primitive DFD shows detailed processing requirements for the event. This DFD shows several elementary processes *for* the event process. Each elementary process is cohesive—that is, it does only one thing. On a primitive diagramming it is permissable to have flows connecting the elementary processes.

When Bob drew this primitive data flow diagram, he had to add new data flows between the processes. In doing so, he tried to practice good data conservation, making sure each process has only the data it truly needs. The data structure for each data flow had to be described in his CASE tool's repository. Also notice that he used data flow junctions to split and merge appropriate data flows on the diagram.

Note that the primitive DFD contains some new exception data flows that were *not* introduced in Figure 9-22. It should not be hard to imagine a computer program structure when examining this DFD.

The combination of the context diagram, system diagram, event diagrams, and primitive diagrams completes our process models. Collectively, this *is* the process model. A well-crafted and complete process model can effectively communicate business requirements between end users and computer programmers, eliminating much of the confusion that often occurs in system design, programming, and implementation.

> Completing the Specification

The data flow diagrams are complete. Where do you go from here? That depends on your choice of methodology. If you are practicing the pure structured analysis methodology (from which data flow diagramming was derived), you must complete the specification. To do so, each data flow, data store, and *elementary* process (meaning one that is not further exploded into a more detailed DFD) must be described to the encyclopedia or data dictionary. CASE tools provide facilities for such descriptions.

Data flows are described by data structures, as explained earlier in this chapter. Figure 9-24 (page 353) demonstrates how *System Architect 2001,* the SoundStage CASE tool, can be used to describe a data flow. Notice that this CASE tool uses an algebraic notation for data structures, as described in this chapter. Ultimately, each data element or attribute should also be described in the data dictionary to specify data type, domain, and default value (as was described in Chapter 8). Data stores correspond to all instances of a data entity from our data model. Thus, they are best described in the data dictionary that corresponds to each entity and its attributes, as was taught in Chapter 8. Some analysts like to translate each data store's content into a relational data structure similar to that used in Figure 9-24 to describe data flows. We consider this to be busywork—let the entity descriptions from the data model describe the contents of a data store. Besides, defining data structures for data stores could lead to synchronization errors between the data and process models—if you would make any changes to an entity in the data model, you would be forced to remember to make those same changes in the corresponding data store's data structure. This requires too much effort (unless you have a CASE tool capable of doing it automatically for you).

Process Logic   Decomposition diagrams and data flow diagrams will prove very effective tools for identifying processes, but they are not good at showing the logic inside

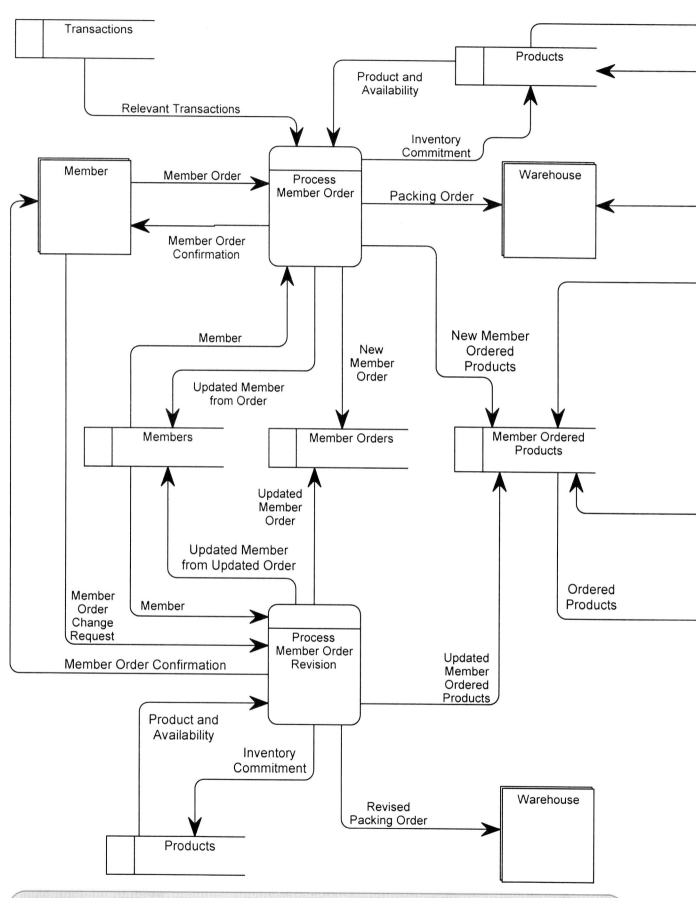

FIGURE 9-22 A System Diagram (created with *System Architect 2001*)

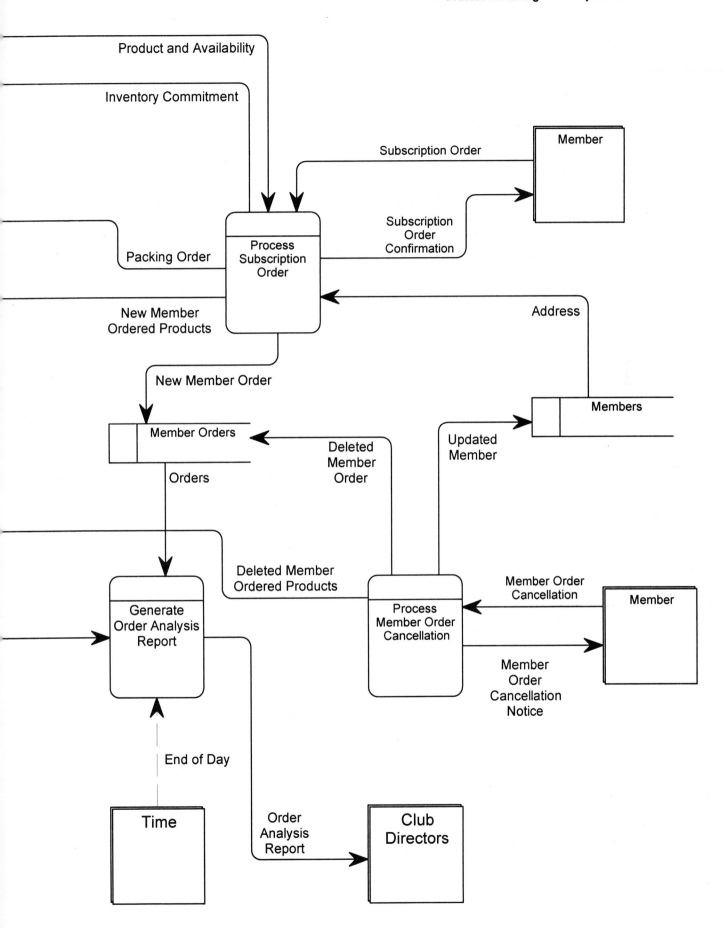

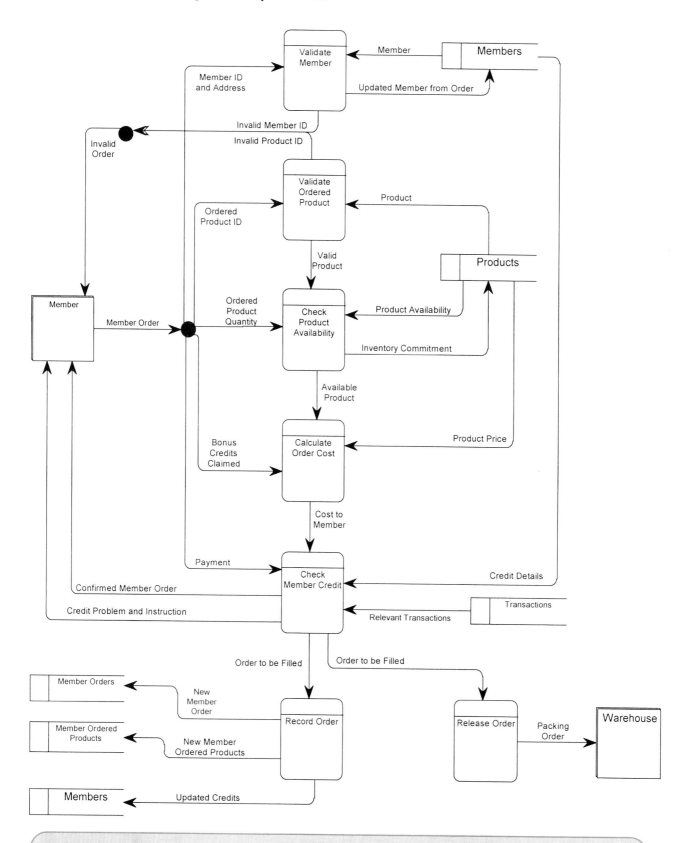

FIGURE 9-23 A Primitive Diagram (created with *System Architect 2001*)

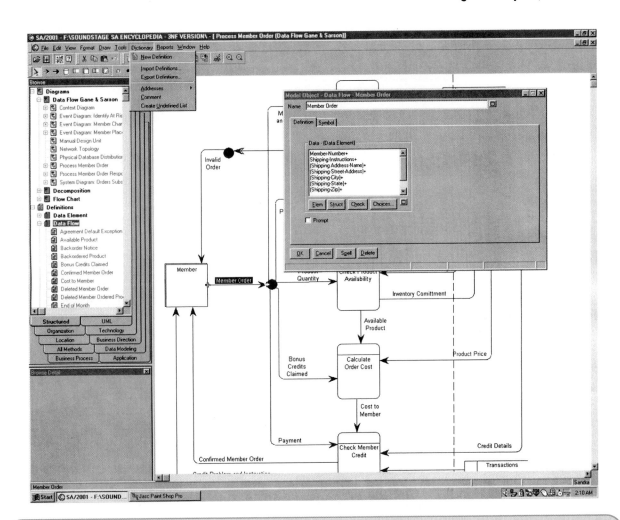

FIGURE 9-24 A Data Flow (created with *System Architect 2001*)

those processes. Eventually, we will need to specify detailed *instructions* for the elementary processes on a data flow diagram. Consider, for example, an elementary process named CHECK CUSTOMER CREDIT. By itself, the named process is insufficient to explain the logic needed to CHECK CUSTOMER CREDIT. We need an effective way to model the logic of an elementary process. Ideally, our logic model should be equally effective for communicating with users (who must verify the business accuracy of the logic) and programmers (who may have to implement the business logic in a programming language).

We can rule out flowcharts. While they do model process logic, most end users tend to be extremely intimidated by them. The same would be true of pseudocode and other popular programming logic tools. We can also rule out natural English. It is too often imprecise and frequently subject to interpretation (and misinterpretation). Figure 9-25 summarizes some common problems encountered by those who attempt to use natural English as a procedural language.

To address this problem, we require a tool that marries some of the advantages of natural English with some of the rigor of programming logic tools. **Structured English** is a language and syntax, based on the relative strengths of structured programming and natural English, for specifying the underlying logic of elementary processes on process models (such as *data flow diagrams*). An example of Structured English is shown in Figure 9-26. (The numbers and letters at the beginning of each statement are optional. Some end users like them because they further remove the programming "look and feel" from the specification.)

Structured English a language syntax for specifying the logic of a process.

FIGURE 9-25 Problems with Natural English as a Procedure Specification Language

- Many of us do not write well, and we also tend not to question our writing abilities.
- Many of us are too educated! It's often difficult for a highly educated person to communicate with an audience that may not have had the same educational opportunities. For example, the average college graduate (including most analysts) has a working vocabulary of 10,000 to 20,000 words; on the other hand, the average noncollege graduate has a working vocabulary of around 5,000 words.
- Some of us write everything like it was a program. If business procedures required such precision, we'd write everything in a programming language.
- Too often, we allow the jargon and acronyms of computing to dominate our language.
- English statements frequently have an excessive or confusing scope. How would you carry out this procedure: "If customers walk in the door and they do not want to withdraw money from their account or deposit money to their account or make a loan payment, send them to the trust department." Does this mean that the only time you should not send the customer to the trust department is when he or she wishes to do all three of the transactions? Or does it mean that if a customer does not wish to perform at least one of the three transactions, that customer should not be sent to the trust department?
- We overuse compound sentences. Consider the following procedure: "Remove the screws that hold the outlet cover to the wall. Remove the outlet cover. Disconnect each wire from the plug, but first make sure the power to the outlet has been turned off." An unwary person might try to disconnect the wires before turning off the power!
- Too many words have multiple definitions.
- Too many statements use imprecise adjectives. For example, a loan officer asks a teacher to certify that a student is in good academic standing. What is "good"?
- Conditional instructions can be imprecise. For example, if we state that "all applicants under the age of 19 must secure parental permission," do we mean less than 19, or less than or equal to 19?
- Compound conditions tend to show up in natural English. For example, if credit approval is a function of several conditions—credit rating, credit ceiling, annual dollar sales for the customer in question—then different combinations of these factors can result in different decisions. As the number of conditions and possible combinations increases, the procedure becomes more and more tedious and difficult to write.

Source: Adapted from Leslie Matthies, *The New Playscript Procedure* (Stamford, CT: Office Publications, Inc., 1977)

Structured English is *not* pseudocode. It does not concern itself with declarations, initialization, linking, and such technical issues. It does, however, borrow some of the logical constructs of *structured programming* to overcome the lack of structure and precision in the English language. Think of it as the marriage of natural English language with the syntax of structured programming.

FIGURE 9-26

Using Structured English to Document an Elementary Process

1. For each CUSTOMER NUMBER in the data store CUSTOMERS:
   a. For each LOAN in the data store LOANS that matches the above CUSTOMER NUMBER:
      1) Keep a running total of NUMBER OF LOANS for the CUSTOMER NUMBER.
      2) Keep a running total of ORIGINAL LOAN PRINCIPAL for the CUSTOMER NUMBER.
      3) Keep a running total of CURRENT LOAN BALANCE for the CUSTOMER NUMBER.
      4) Keep a running total of AMOUNTS PAST DUE for the CUSTOMER NUMBER.
   b. If the TOTAL AMOUNTS PAST DUE for the CUSTOMER NUMBER is greater than 100.00 then
         1) Write the CUSTOMER NUMBER and data in the data flow LOANS AT RISK.
      Else
         1) Exclude the CUSTOMER NUMBER and data from the data flow LOANS AT RISK.

The overall structure of a Structured English specification is built using the fundamental constructs that have governed structured programming for nearly three decades. These constructs (summarized in Figure 9-27) are:

- A *sequence* of simple, declarative sentences—one after another. Compound sentences are discouraged because they frequently create ambiguity. Each sentence uses strong, action verbs such as GET, FIND, RECORD, CREATE, READ, UPDATE, DELETE, CALCULATE, WRITE, SORT, MERGE, or anything else recognizable or understandable to users. A formula may be included as part of a sentence (e.g., CALCULATE GROSS PAY = HOURS WORKED × HOURLY WAGE).
- A *conditional* or *decision structure* indicates that a process must perform different steps under well-specified conditions. There are two variations (and a departure) on this construct.
    — The IF-THEN-ELSE construct specifies that one set of steps should be taken if a specified condition is true but that a different set of steps should be specified if the specified condition is false. The steps to be taken are typically a *sequence* of one or more sentences as described above.
    — The CASE construct is used when there are more than two sets of steps to choose from. Once again, these steps usually consist of the aforementioned *sequential* statements. The case construct is an elegant substitute for an *IF-THEN-ELSE IF-THEN-ELSE IF-THEN* . . . construct (which is very convoluted to the average user).
    — For logic based on multiple conditions and combinations of conditions (which programmers call a *nested IF*), *decision tables* are a far more elegant logic modeling tool. Decision tables will be introduced shortly.
- An *iteration,* or *repetition,* structure specifies that a set of steps should be repeated based on some stated condition. There are two variations on this construct:
    — The DO-WHILE construct indicates that certain steps are repeated zero, one, or more times based on the value of the stated condition. Note that these steps may not execute at all if the condition is not true when the condition is first tested.
    — The REPEAT-UNTIL construct indicates that certain steps are repeated one or more times based on the value of the stated condition. Note that a REPEAT-UNTIL set of steps must execute at least once, unlike the DO-WHILE set of actions.

Additionally, Structured English places the following restrictions on process logic:

- Only strong, imperative verbs may be used.
- Only names that have been defined in the project dictionary may be used. These names may include those of data flows, data stores, entities (from data models; see Chapter 8), attributes (the specified data fields or properties contained in a data flow, data store, or entity), and domains (the specified legal values for attributes).
- Formulas should be stated clearly using appropriate mathematical notations. In short, you can use whatever notation is recognizable to the users. Make sure each operand in a formula is either input to the process in a data flow or a defined constant.
- Undefined adjectives and adverbs (the word *good,* for instance) are not permitted unless clearly defined in the project dictionary as legal values for data attributes.
- Blocking and indentation are used to set off the beginning and ending of constructs and to enhance readability. (Some authors and models encourage the use of special verbs such as ENDIF, ENDCASE, ENDDO, and ENDREPEAT to terminate constructs. We dislike this practice because it gives the Structured English too much of a pseudocode or programming look and feel.)
- When in doubt, user readability should always take priority over programmer preferences.

Structured English Procedural Structures

Construct	Sample Template
Sequence of steps – Unconditionally perform a sequence of steps.	[Step 1] [Step 2] … [Step n]

Simple condition steps – If the specified condition is true, then perform the first set of steps. Otherwise, perform the second set of steps.
 	
Use this construct if the condition has only two possible values.
 	
(Note: The second set of conditions is optional.)
	If [truth condition]
	   then
	       [sequence of steps or other conditional steps]
	   else
	       [sequence of steps or other conditional steps]
	~~**End If**~~

Complex condition steps – Test the value of the condition and perform the appropriate set of steps.
 	
Use this construct if the condition has more than two values.
	Do the following based on [condition]:
	   Case 1: If [condition] = [value] then
	      [sequence of steps or other conditional steps]
	   Case 2: If [condition] = [value] then
	      [sequence of steps or other conditional steps]
	  …
	   Case n: If [condition] = [value] then
	      [sequence of steps or other conditional steps]
	~~**End Case**~~

Multiple conditions – Test the value of multiple conditions to determine the correct set of steps. Use a decision table instead of nested if-then-else Structured English constructs to simplify the presentation of complex logic that involves combinations of conditions. *A **decision table** is a tabular presentation of complex logic in which rows represent conditions and possible actions and columns indicate which combinations of conditions result in specific actions.*	<table><tr><td>DECISION TABLE</td><td>Rule</td><td>Rule</td><td>Rule</td><td>Rule</td></tr><tr><td>[Condition]</td><td>value</td><td>value</td><td>value</td><td>value</td></tr><tr><td>[Condition]</td><td>value</td><td>value</td><td>value</td><td>value</td></tr><tr><td>[Condition]</td><td>value</td><td>value</td><td>value</td><td>value</td></tr><tr><td>[Sequence of steps or conditional steps]</td><td>X</td><td></td><td></td><td></td></tr><tr><td>[Sequence of steps or conditional steps]</td><td></td><td>X</td><td>X</td><td></td></tr><tr><td>[Sequence of steps or conditional steps]</td><td></td><td></td><td></td><td>X</td></tr></table> Although it isn't a Structured English construct, a decision table can be named, and referenced within a Structured English procedure.
One-to-many iteration – Repeat the set of steps until the condition is false. Use this construct if the set of steps must be performed <u>at least</u> once, regardless of the condition's initial value.	**Repeat the following until** [truth condition]: [sequence of steps or conditional steps] ~~**End Repeat**~~

Zero-to-many iteration – Repeat the set of steps until the condition is false.
 	
Use this construct if the set of steps is conditional based on the condition's initial value.
	Do while [truth condition]:
	   [sequence of steps or conditional steps]
	~~**End Do**~~
	         - OR -
	For [truth condition]:
	   [sequence of steps or conditional steps]
	~~**End For**~~

FIGURE 9-27 Structured English Constructs

Structured English should be precise enough to clearly specify the required business procedure to a programmer or user. Yet it should not be so inflexible that a programmer or user spends hours arguing over syntax.

Many processes are governed by complex combinations of conditions that are not easily expressed with Structured English. This is most commonly encountered in business policies. A **policy** is a set of rules that governs some process in the business.

In most firms, policies are the basis for decision making. For instance, a credit card company must bill cardholders according to various policies that adhere to restrictions imposed by state and federal governments (maximum interest rates and minimum payments, for instance). Policies consist of *rules* that can often be translated into computer programs if the users and systems analysts can accurately convey those rules to the computer programmer.

There are ways to formalize the specification of policies and other complex combinations of conditions. One such logic modeling tool is a **decision table.** While people who are unfamiliar with them tend to avoid them, decision tables are very useful for specifying complex policies and decision-making rules. Figure 9-28 illustrates the three components of a simple decision table:

- *Condition stubs* (the upper rows) describe the conditions or factors that will affect the decision or policy.

policy a set of rules that governs how a process is to be completed.

decision table a tabular form of presentation that specifies a set of conditions and their corresponding actions.

A SIMPLE POLICY STATEMENT

CHECK CASHING IDENTIFICATION CARD

A customer with check cashing privileges is entitled to cash personal checks of up to $75.00 and payroll checks from companies pre-approved by *LMART*. This card is issued in accordance with the terms and conditions of the application and is subject to change without notice. This card is the property of *LMART* and shall be forfeited upon request of *LMART*.

SIGNATURE *Charles C. Parker, Jr.*
EXPIRES **May 31, 2003**

THE EQUIVALENT POLICY DECISION TABLE

Conditions and Actions	Rule 1	Rule 2	Rule 3	Rule 4
C1: Type of check	personal	payroll	personal	payroll
C2: Check amount less than or equal to $75.00	yes	doesn't matter	no	doesn't matter
C3: Company accredited by *LMART*	doesn't matter	yes	doesn't matter	no
A1: Cash the check	X	X		
A2: Don't cash the check			X	X

Condition Stubs { C1, C2, C3

Action Stubs { A1, A2

Rules

FIGURE 9-28 A Sample Decision Table

- *Action stubs* (the lower rows) describe, in the form of statements, the possible policy actions or decisions.
- *Rules* (the columns) describe which actions are to be taken under a specific combination of conditions.

The figure depicts a check-cashing policy that appears on the back of a check-cashing card for a grocery store. This same policy has been documented with a decision table. Three conditions affect the check-cashing decision: the type of check, whether the amount of the check exceeds the maximum limit, and whether the company that issued the check is accredited by the store. The actions (decisions) are either to cash the check or to refuse to cash the check. Notice that each combination of conditions defines a rule that results in an action, denoted by an X.

Both decision tables and Structured English can describe a single elementary process. For example, a legitimate statement in a Structured English specification might read DETERMINE WHETHER OR NOT TO CASH THE CHECK USING THE DECISION TABLE, LMART CHECK CASHING POLICY.

Elementary processes can be described by Structured English and/or decision tables. Because they are "elementary," they should be described in one page or less of either tool. Figure 9-29, demonstrates how *System Architect 2001* can be used to describe an elementary process. Like many CASE tools, *System Architect* does not support decision table construction. Fortunately, decision tables are easily constructed using the table features in most word processors and spreadsheets.

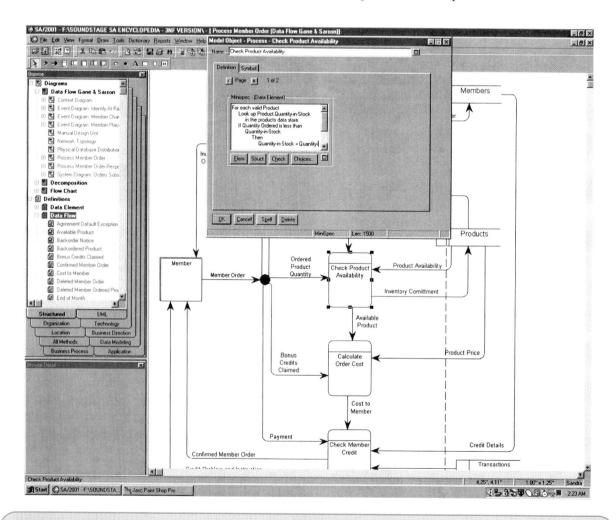

FIGURE 9-29 An Elementary Process (created with *System Architect 2001*)

Synchronizing of System Models

Data and process models represent different views of the same system, but these views are interrelated. Modelers need to synchronize the different views to ensure consistency and completeness of the total system specification. In this section, we'll review the basic synchronization concepts for data and process models.

> Data and Process Model Synchronization

The linkage between data and process models is almost universally accepted by all major methodologies. In short, there should be one data store in the process models for each entity in the data model. Some methodologies exempt associative entities from this requirement, but we believe it is simpler (and more consistent) to apply the rule to all entities on the data model.

Figure 9-30 illustrates a typical *data-to-process-CRUD matrix*. The decision to include or not include attributes is based on whether processes need to be restricted as to which attributes they can access.

Data-to-Process-CRUD Matrix

Entity . Attribute	Process Customer Application	Process Customer Credit Application	Process Customer Change of Address	Process Internal Customer Credit Change	Process New Customer Order	Process Customer Order Cancellation	Process Customer Change to Outstanding Order	Process Internal Change to Customer Order	Process New Product Addition	Process Product Withdrawal from Market	Process Product Price Change	Process Change to Product Specification	Process Product Inventory Adjustment
Customer	C	C			R	R	R	R					
.Customer Number	C	C			R	R	R	R					
.Customer Name	C	C	U		R		R	R					
.Customer Address	C	C	U		RU		RU	RU					
.Customer Credit Rating		C		U	R		R	R					
.Customer Balance Due					RU	U	R	R					
Order					C	D	RU	RU					
.Order Number					C		R	R					
.Order Date					C		U	U					
.Order Amount					C		U	U					
Ordered Product					C	D	CRUD	CRUD		RU			
.Quantity Ordered					C		CRUD	CRUD					
.Ordered Item Unit Price					C		CRUD	CRUD					
Product					R	R	R	R	C	D	RU	RU	RU
.Product Number					R	R	R	R	C			R	
.Product Name					R		R	R	C			RU	
.Product Description					R		R	R	C			RU	
.Product Unit of Measure					R		R	R	C		RU	RU	
.Product Current Unit Price					R		R	R			U		
.Product Quantity on Hand					RU	U	RU	RU					RU

C = create   R = read   U = update   D = delete

FIGURE 9-30 Sample Data-to-Process-CRUD Matrix

The synchronization quality check is stated as follows:

Every entity should have *at least* one C, one R, one U, and one D entry for system completeness. If not, one or more event processes were probably omitted from the process models. More importantly, users and management should validate that all possible creates, reads, updates, and deletes have been included.

The matrix provides a simple quality check that is simpler to read than either the data or process models. Of course, any errors and omissions should be recorded both on the matrix and in the corresponding data and process models to ensure proper synchronization.

> Process Distribution

Process models illustrate the essential work to be performed by the system as a whole. However, processes must be distributed to locations where work is to be performed. Some work may be unique to one location. Other work may be performed at multiple locations. Before we design the information system, we should identify and document what processes must be performed at which locations. This can be accomplished through a *process-to-location-association matrix*. A **process-to-location-association matrix** is a table in which the rows indicate processes (event or elementary processes), the columns indicate locations, and the cells (the intersection of rows and columns) document which processes must be performed at which locations. Figure 9-31 illustrates a typical process-to-location-association matrix. Once it is validated for accuracy, the system designer will use this matrix to determine which processes should be implemented centrally or locally.

Some methodologies and CASE tools may support views of the process model that are appropriate to a location. If so, these views (subsets of the process models) must be kept in sync with the master process models of the system as a whole.

process-to-location-association matrix a table used to document processes and the locations at which they must be performed.

Process-to-Location-Association Matrix

Process	Customers	Kansas City — Marketing	Advertising	Warehouse	Sales	Accounts Receivable	Boston — Sales	Warehouse	San Francisco — Sales	San Diego — Warehouse
Process Customer Application	X				X		X		X	
Process Customer Credit Application	X					X				
Process Customer Change of Address	X				X		X		X	
Process Internal Customer Credit Change						X				
Process New Customer Order	X				X		X		X	
Process Customer Order Cancellation	X				X		X		X	
Process Customer Change to Outstanding Order	X				X		X		X	
Process Internal Change to Customer Order					X		X		X	
Process New Product Addition			X							
Process Product Withdrawal from Market			X							
Process Product Price Change			X							
Process Change to Product Specification			X	X						
Process Product Inventory Adjustment					X			X		X

FIGURE 9-31 Sample Process-to-Location-Association Matrix

If you are taking a traditional approach you will proceed directly to Chapter 11, "Feasibility Analysis and the System Proposal." Given data and process models that describe *logical* system requirements, Chapter 11 will examine the methods and techniques for identifying candidate *physical* solutions that will fulfill the logical requirements, techniques for analyzing the feasibility of each of those solutions, and approaches for presenting the solution that you deem most feasible. This system proposal culminates systems analysis in our *FAST* methodology.

At your instructor's discretion, some of you may jump to Chapter 10, "Object-Oriented Modeling and Analysis with UML." This alternative to data and process modeling is rapidly gaining acceptance in industry and may eventually make both data and process modeling obsolete. Of course, object modeling presents many parallels with data and process modeling. You'll see some similar constructs because an object is defined as the encapsulation of related data and the processes that will be allowed to create, read, update, and use that data.

If you are interested in how we will use the logical DFDs in this chapter during systems design, you might want to preview Chapter 13, "Process Design and Modeling." In that chapter, we will teach you how to transform the logical data flow diagrams into physical data flow diagrams that model the technology architecture of a system to be designed and implemented.

 ## Chapter Review

1. We construct logical models to better understand business problem domains and business requirements.

2. Process modeling is a technique for organizing and documenting the process requirements and design for a system. This chapter focused on a process model called a data flow diagram, which depicts the flow of data through a system's processes.

3. External agents are entities that are outside the scope of a system and project but that provide net inputs to or net outputs from a system. As such, they form the boundary of the system.

4. Data stores present files of data to be used and maintained by the system. A data store on a process model corresponds to all instances of an entity on a data model.

5. A system *is* a process. A process is work performed on, or in response to, inputs and conditions.

6. Just as systems can be recursively decomposed into subsystems, processes can be recursively decomposed into subprocesses. A decomposition diagram shows the functional decomposition of a system into processes and subprocesses. It is a planning tool for subsequent data flow diagrams.

7. Logical processes show essential work to be performed by a system without showing how the processes will be implemented. There are three types of logical processes: functions (very high level), events (middle level of detail), and elementary processes (very detailed).

8. Elementary processes are further described by procedural logic. Structured English is a tool for expressing this procedural logic. Structured English is a derivative of structured programming logic constructs married to natural English.

9. Complex elementary processes may be described by policies that are expressed in decision tables, which show complex combinations of conditions that result in specific actions.

10. Data flows are the inputs to and the outputs from processes. They also illustrate data store accesses and updates.

11. All data flows consist of either other data flows or discrete data structures that include descriptive attributes. A data flow should contain only the amount of data needed by a process; this is called data conservation.

12. Process modeling may be used in different types of projects, including business process redesign and application development. For application

development projects, this chapter taught an event-driven data flow diagramming strategy as follows:

a. Draw a context data flow diagram that shows how the system interfaces to other systems, the business, and external organizations.

b. Draw a functional decomposition diagram that shows the key subsystems and/or functions that comprise the system.

c. Create an event list that identifies the external and temporal events to which the system must provide a response. External events are triggered by the external agents of a system. Temporal events are triggered by the passing of time.

d. Update the decomposition diagram to include processes to handle the events (one process per event).

e. For each event, draw an event diagram that shows its interactions with external entities, data stores, and, on occasion, other triggers to other events.

f. Combine the event diagrams into one or more system diagrams.

g. For each event on the system diagram, either describe it as an elementary process using Structured English or explode it into a primitive data flow diagram that includes elementary processes that must be subsequently described by either Structured English or decision tables, or by both. When processes are exploded on data flow diagrams to reveal greater detail, it is important to maintain consistency between the different types of diagrams; this is called synchronization.

13. Most computer-aided software engineering tools support both decomposition diagramming and data flow diagramming.

Review Questions

1. What is a logical model, and what are its common synonyms?
2. Why are logical models valuable tools for systems analysts?
3. What is a data flow diagram, and what are its common synonyms?
4. How is a data flow diagram different from a flowchart?
5. Why is a system considered to a process?
6. What is decomposition, and why is it needed? What is the tool used to depict the decomposition of a system?
7. What are the three types of logical processes?
8. What are the common mechanical errors when depicting processes on a data flow diagram and other process models?
9. What is Structured English, and why is it used when constructing process logic?
10. What are the naming conventions of logical data flows?
11. What is data conservation and why is it needed?
12. What are external agents and why can the external agents of an information system change?
13. What are some examples of the event-driven modeling used in systems analysis?
14. What process model is used to document the scope for an information system, and what is depicted in this process model?
15. Why is it important to synchronize data and process models?

Problems and Exercises

1. You are working as a student assistant for an engineering firm and are paid by the hour. Every two weeks, you turn in a time sheet to your supervisor, and three workdays later, your paycheck is direct deposited into your checking account. List the different entities or objects, logical processes, data flows, and data stores that are involved, starting from the time you submit your time sheet.

2. Match the terms in the first column with the definitions or examples in the second column:

1. Structured English	A. Disassembling a system into its components
2. Process	B. Logical unit of work that must be completed as a whole
3. Logical model	C. Tool for logic modeling
4. Primitive process	D. Set of business activities that are related and ongoing

5. Policy   E. Technique for organizing and documenting a system's processes

6. DFD   F. Procedure specification language

7. Decision table   G. Depiction of what a system is or does

8. Decomposition   H. Depiction of system data flow

9. Event   I. Depiction of system decomposition

10. Physical model   J. Work performed by system in response to incoming data flows

11. Function   K. Detailed, separate activity/task needed to complete event response

12. Hierarchy chart   L. Process completion rules

13. Process modeling   M. Depiction of what a system is or does, and how it is implemented

3. In a decomposition diagram, how do you show one child for a parent, and how do you show more than one parent for a child? Why don't the connections on a decomposition diagram show arrowheads, like most other diagrams? Why aren't the connections named?

4. Consider carpool lanes in Sacramento, California. Between the hours of 6:00 a.m. and 10:00 a.m. and 3:00 p.m. and 7:00 p.m., Monday through Friday, they are restricted to passenger vehicles with two or more people of any age, motorcycles, and hybrid (gas/electric) vehicles with one or more persons. For all other vehicles or conditions, the driver is subject to a traffic citation. Outside those time periods, there are no restrictions as to their usage. Based upon this information, write a policy decision table for use by highway patrol officers.

5. You work in the headquarters office of the investigation division of a law enforcement agency and are developing an automated case-tracking system for your headquarters office to replace the current manual system. Cases are opened when a request-for-investigation form is received from other divisions in your agency; no cases are initiated internally. A new case folder is created, containing any criminal record information based upon checking various criminal justice databases, then sent to the appropriate field investigation office. When the case is completed, headquarters receives an investigation report from the field office, the case is closed, and a copy of the completed investigation report is sent to the originating office. Every week, a listing showing cases opened, completed, and in progress is sent to each field office. What are some of the strategies you might use for setting the scope and boundaries of the system?

6. Based upon the preceding question, what did you determine are the system's:

   a. Net inputs
   b. Net outputs
   c. External agents
   d. External stores

7. Now that you have your net inputs, net outputs, external agents, and external stores, draw a context data flow diagram.

8. You are now modeling the logic of each elementary process for the case-tracking system and have decided to write it in Structured English in order to communicate effectively with both users and programmers. For the Open/Closed/In Progress Case Listing Report to Field Office, write a Structured English statement to document the process of keeping a running total of cases opened, closed, and in progress. Additionally, you want to add another column to the report showing the number of cases still in progress that are over six months old.

9. You are now ready to create the functional decomposition diagram for the case-tracking system. What is the root process for the function decomposition diagram? What subsystems would you typically include? What processes would you show, and to what subsystem would they belong? Use this information to create a functional decomposition diagram.

10. Your next step, after drawing the decomposition diagram, is creating the event-response or use-case list. There should be a use case for each event initiated by an external agent. Temporal events should also be shown. For each external agent, there should be at least one use case. Start a partial use-case table that includes the events OTHER DIVISION SENDS REQUEST FOR INVESTIGATION and FIELD OFFICE SENDS COMPLETED INVESTIGATION REPORT. Also include the event GENERATE OPEN/CLOSED/IN PROGRESS CASE LISTING REPORT.

11. An event diagram is equivalent to a context diagram for one event. The event diagram includes the inputs, outputs, and data store interactions related to that specific event. Its purpose is to help users focus on a single event without becoming overwhelmed or confused by a picture of the entire system. Select one of the events from the case-tracking system, and draw an event diagram.

12. Match the definitions or examples in the first column with the terms in the second column:

A. Smallest meaningful data segment	1. Domain
B. Combination of data flows that are similar	2. Junction
C. Expressed in the form of data structures	3. External agent
D. Condition to be monitored	4. Event partitioning
E. An attribute's legitimate values	5. Composite data flow
F. "Starving the processes"	6. Control flow
G. Output of data from a process or input to the process	7. Data composition
H. Data class that can be stored in an attribute	8. Data attribute
I. Arrangement of data attributes that comprise a data flow	9. Data conservation
J. Outside entity that interacts with a system	10. Data flow
K. Symbol that given data flow is instance of only one type	11. Data structure
L. Data at rest	12. Data type
M. System broken into subsystems based on business events	13. Data store

13. Although data and process models depict the same system with different views, system designers must synchronize these different views to make sure that their models are consistent and complete. One way to ensure this is through a data-to-process-CRUD matrix. Select a system with which you are familiar. Identify at least three of the entities used in that system and their attributes. Next, identify the processes associated with those entities. Then build a data-to-process-CRUD matrix, using Figure 9-30 as your template. As a quality check, show your matrix to someone else who is familiar with the system, and have her or him review it for completeness and correctness.

Projects and Research

1. Suppose you are starting work on a project for an organization that has never used any modeling techniques or tools in designing a system. (Yes, it is hard to imagine, but it does exist.) Your manager is reluctant to change from the way they have always done things. Write a one- to two-page issue paper (or a PowerPoint presentation as an alternative) on why systems modeling is worth the time and resources involved.

2. The textbook uses the Gane and Sarson process modeling notations and compares them to the notations used by the DeMarco/Yourdon and SSADM/IDEF0 process modeling methodologies. Research at least two of these or other process modeling methodologies, then compare and contrast them.

   a. What other process modeling methodologies did you find?
   b. What, if any, are the significant differences in their process modeling methodologies, other than in their notation methods?
   c. What are their similarities?

   d. Which notation method does your organization use?
   e. If you were asked to recommend one of these methodologies for your organization, which one would you choose? Why?

3. Until fairly recently, magazines and periodicals were available in printed versions only. Publishers are now offering an increasing number of periodicals in either the traditional printed version or in a digital format that can be downloaded over the Internet. Consider the processes involved in these two methods and:

   a. Create a high-level data flow diagram describing the typical traditional methods of renewing a subscription via mail to the print version of a magazine.
   b. Create another high-level data flow diagram describing the processes for renewing a subscription via the Internet to the digital format version of a magazine.

c. What, if any, are the essential differences between the two diagrams?

d. Based upon the data flow diagrams, which format is more efficient in renewing a magazine subscription? Do you think the same holds true from the perspective of the subscriber? What about from the perspective of the publisher?

e. What about receiving and reading the magazine? From your own perspective, what are the advantages and disadvantages of a magazine published in digital format versus the traditional print version?

4. In 1978, Tom DeMarco wrote what is considered to be the classic text on structured systems analysis methodology—*Structured Analysis and System Specification*. James Wetherbe is considered by the authors of the textbook to be "one of the strongest advocates of system concepts and system thinking as part of the discipline of systems analysis and design," and has written numerous articles and books on "systems think." Edward Yourdon is another widely acknowledged leader in systems design and is noted for formalizing the event-driven approach methodology in his 1989 book, *Modern Structured Analysis*. Search the Internet for their Web sites, if any, and for recent articles and/or books by these three leaders in systems analysis and design.

   a. What articles and/or Web sites did you find?
   b. Describe some of their more recent work.
   c. Compare and contrast each of these authors in terms of their perspective on systems analysis and design; specifically, do you see their approaches as complementary or opposed?

   d. What emerging changes or trends in systems analysis and design do they foresee?
   e. Do you feel that they still represent the leading thinkers in systems analysis and design? Why or why not?

5. Look at the different information systems used in your school or organization. Find a system with incomplete and/or outdated documentation (this should not be hard to do in most organizations!). Update and complete the documentation for this system using data flow diagrams, data structures, and other process models described in this chapter. Use the Gane and Sarson notation method, unless your organization supports a different notation method.

6. At the conclusion of this chapter, the textbook mentions that data and process modeling may eventually become obsolete due to the increasing popularity and usage of object-oriented modeling and analysis with UML. Research articles discussing this topic in your school library and/or the Internet.

   a. What articles did you find?
   b. What were the authors' positions on this topic?
   c. Compare and contrast the articles you found; which one made the more convincing case?

As you gaze into your crystal ball, what modeling methodology do you think will be the most widely used one 10 years from now? Do you predict it will be either of these two methodologies or a different one entirely? Why?

Minicases

1. Go to a small company of your choice. What does the business do? Write a one- to two-page paper describing the business and its existing system. Then draw a context-level diagram and a system-level diagram for the existing system. Do you see any inefficiencies or weaknesses in the current system? Describe.

2. In the previous case, you documented (at a high level) an existing information system at a business of your choice. Now describe the system *you* think is appropriate for this business. Consider efficiency, flow of information from one department to another, and so forth. Is there pertinent information that the previous system did not utilize? How can your new system offer a business advantage? Document this advantage in a two-page paper, as well as in a context- and system-level diagram.

3. In Chapter 2 you addressed the following problem: Government service departments are deeply burdened by the amount of data that they hold and process. Interview someone from a service department and draft a short essay. Example service departments that must sift through vast amounts of data are those that deal with, for example, missing persons, child protective services, DMV, and tracking of persons on probation following a crime. You should include, but are not limited to topics such as:

   • What is the department (or person's) job?
   • What kind of data do they collect and analyze?
   • What kind of analyses do they do on the data?
   • How much information do they collect and from whom, and what programs do they use?

In this exercise, utilize the information you gathered from your interview and draft a complete DFD of the *existing* information system for that department. You may need to go back and reinterview your contact or get forms, reports, and soon, so that you have a complete picture of the flow of data.

4. In the previous case, you documented the existing information system of a government department.

Now think about what the flow of information *should* be like. Discuss in a short paper the existing flow of information, and what type of information you think should be in the system. Is there a big difference between what you think they should have and what they do have?

Suggested Readings

Copi, I. R. *Introduction to Logic.* New York: Macmillan, 1972. Copi provides a number of problem-solving illustrations and exercises that aid in the study of logic. The poker chip problem in our exercises was adapted from one of Copi's reasoning exercises.

DeMarco, Tom. *Structured Analysis and System Specification.* Englewood Cliffs, NJ: Prentice Hall, 1978. This is the classic book on the structured systems analysis methodology, which is built heavily around the use of data flow diagrams. The progression through (1) *current physical system DFDs,* (2) *current logical system DFDs,* (3) *target logical system DFDs,* and (4) *target physical system DFDs* is rarely practiced anymore, but the essence of DeMarco's pioneering work lives on in event-driven structured analysis. DeMarco created the data structure and logic notations used in this book.

Gildersleeve, T. R. *Successful Data Processing Systems Analysis.* Englewood Cliffs, NJ: Prentice Hall, 1978. The first edition of this book includes an entire chapter on the construction of decision tables. Gildersleeve does an excellent job of demonstrating how narrative process descriptions can be translated into condition and action entries in decision tables. Unfortunately, the chapter was deleted from the second edition.

Harmon, Paul, and Mark Watson. *Understanding UML: The Developers Guide.* San Francisco: Morgan Kaufman Publishers, 1998. This book does an excellent job of introducing use cases.

Martin, James, and Carma McClure. *Action Diagrams: Towards Clearly Specified Programs.* Englewood Cliffs, NJ: Prentice Hall, 1986. This book describes a formal grammar of Structured English that encourages the natural progression of a process (program) from Structured English to code. Action diagrams are supported directly in some CASE tools.

Matthies, Leslie H. *The New Playscript Procedure.* Stamford, CT: Office Publications, 1977. This book provides a thorough explanation and examples of the weaknesses of

the English language as a tool for specifying business procedures.

McMenamin, Stephen M., and John F. Palmer. *Essential Systems Analysis.* New York: Yourdon Press, 1984. This was the first book to suggest event partitioning as a formal strategy to improve structured analysis. The book also strengthened the distinction between logical and physical process models and the increased importance of the logical models (which they called *essential* models).

Robertson, James, and Suzanne Robertson. *Complete Systems Analysis* (Vols. 1 and 2). New York: Dorset House Publishing, 1994. This is the most up-to-date and comprehensive book on the event-driven approach to structured analysis, even though we feel it still overemphasizes the current system and physical models more than the Yourdon book described below.

Seminar notes for *Process Modeling Techniques.* Atlanta: Structured Solutions, Inc., 1991. You probably can't get a copy of these notes, but we wanted to acknowledge the instructors of the *AD/Method* methodology course that stimulated our thinking and motivated our departure from classical structured analysis techniques to the event-driven structured analysis techniques taught in this chapter. Structured Solutions was acquired by Protelicess, Inc.

Wetherbe, James, and Nicholas P. Vatarli. *Systems Analysis and Design: Best Practices,* 4th ed. St. Paul, MN: West Publishing, 1994. Jim Wetherbe has always been one of the strongest advocates of system concepts and system thinking as part of the discipline of systems analysis and design. Jim has shaped many minds, including our own. The authors provide a nice chapter on system concepts in this book—and the rest of the book is must reading for those of you who truly want to learn to "systems think."

Yourdon, Edward. *Modern Structured Analysis.* Englewood Cliffs, NJ: Yourdon Press, 1989. This was the first mainstream book to abandon classic structured analysis's overemphasis on the current physical system models and to formalize McMenamin and Palmer's event-driven approach.

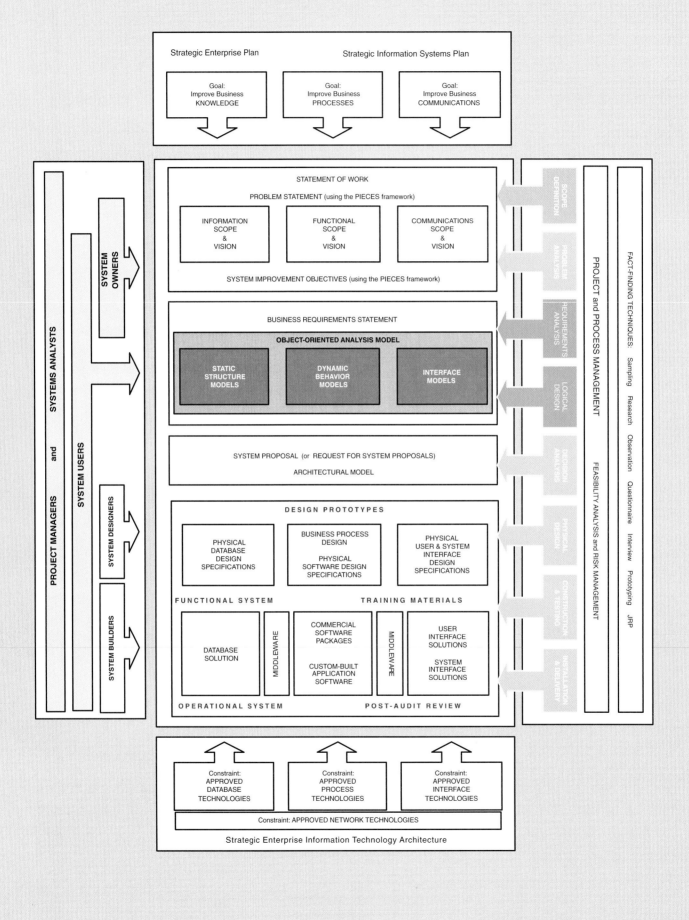

Strategic Enterprise Plan

Strategic Information Systems Plan

Goal:
Improve Business
KNOWLEDGE

Goal:
Improve Business
PROCESSES

Goal:
Improve Business
COMMUNICATIONS

STATEMENT OF WORK

PROBLEM STATEMENT (using the PIECES framework)

INFORMATION
SCOPE
&
VISION

FUNCTIONAL
SCOPE
&
VISION

COMMUNICATIONS
SCOPE
&
VISION

SYSTEM IMPROVEMENT OBJECTIVES (using the PIECES framework)

BUSINESS REQUIREMENTS STATEMENT

OBJECT-ORIENTED ANALYSIS MODEL

STATIC
STRUCTURE
MODELS

DYNAMIC
BEHAVIOR
MODELS

INTERFACE
MODELS

SYSTEM PROPOSAL (or REQUEST FOR SYSTEM PROPOSALS)

ARCHITECTURAL MODEL

DESIGN PROTOTYPES

PHYSICAL
DATABASE
DESIGN
SPECIFICATIONS

BUSINESS PROCESS
DESIGN

PHYSICAL
SOFTWARE DESIGN
SPECIFICATIONS

PHYSICAL
USER & SYSTEM
INTERFACE
DESIGN
SPECIFICATIONS

FUNCTIONAL SYSTEM

TRAINING MATERIALS

DATABASE
SOLUTION

MIDDLEWARE

COMMERCIAL
SOFTWARE
PACKAGES

CUSTOM-BUILT
APPLICATION
SOFTWARE

MIDDLEWARE

USER
INTERFACE
SOLUTIONS

SYSTEM
INTERFACE
SOLUTIONS

OPERATIONAL SYSTEM

POST-AUDIT REVIEW

Constraint:
APPROVED
DATABASE
TECHNOLOGIES

Constraint:
APPROVED
PROCESS
TECHNOLOGIES

Constraint:
APPROVED
INTERFACE
TECHNOLOGIES

Constraint: APPROVED NETWORK TECHNOLOGIES

Strategic Enterprise Information Technology Architecture

SYSTEM OWNERS

SYSTEM DESIGNERS

SYSTEM BUILDERS

SYSTEM USERS

SYSTEMS ANALYSTS

PROJECT MANAGERS   and

SCOPE DEFINITION

PROBLEM ANALYSIS

REQUIREMENTS ANALYSIS

LOGICAL DESIGN

DECISION ANALYSIS

PHYSICAL DESIGN

CONSTRUCTION & TESTING

INSTALLATION & DELIVERY

PROJECT and PROCESS MANAGEMENT

FEASIBILITY ANALYSIS and RISK MANAGEMENT

FACT-FINDING TECHNIQUES: Sampling   Research   Observation   Questionnaire   Interview   Prototyping   JRP

10

Object-Oriented Analysis and Modeling Using the UML

Chapter Preview and Objectives

This is the first of two chapters on object-oriented tools and techniques for systems development. This chapter focuses on object modeling during systems analysis. You will know object modeling as a systems analysis technique when you can:

▌ Define object modeling and explain its benefits.

▌ Recognize and understand the basic concepts and constructs of object modeling.

▌ Define the UML and its various types of diagrams.

▌ Evolve a business requirements use-case model into a system analysis use-case model.

▌ Construct an activity diagram.

▌ Discover objects and classes and their relationships.

▌ Construct a class diagram.

An Introduction to Object-Oriented Analysis

Let's suppose that SoundStage had a policy that all new information systems would be developed using object-oriented technologies. After all, object-oriented programming languages, such as *Java* and the *.NET* languages, are growing in popularity. The reason is because object-oriented programming can promote better code reuse to hold down programming costs. Also, an object-oriented approach is more appropriate for projects where geographically separated groups of programmers have to collaborate to produce an integrated system. Each team can be responsible for developing independent pieces of programming code to implement one or more objects with a defined interface. We'll learn more about objects later.

object-oriented analysis (OOA) an approach used to (1) study existing objects to see if they can be reused or adapted for new uses and (2) define new or modified objects that will be combined with existing objects into a useful business computing application.

An object-oriented (OO) approach to programming requires techniques for **object-oriented analysis (OOA)** and object-oriented design (OOD). Some of the object-oriented diagrams, such as class diagrams (taught in this chapter) and sequence diagrams (taught in Chapter 18) would be inappropriate except when the system will be implemented in an object-oriented environment. Other diagrams developed for object-oriented analysis and design can be used in any kind of environment. Use cases, for example, are now used in both object-oriented and traditional, structured analysis. Activity diagrams (taught in this chapter) and deployment diagrams (Chapter 18), though developed for object-oriented analysis and design, can be used in any kind of methodology.

So if the SoundStage Member Services system project took an OO approach, how would Bob's path be different? Would Bob have done traditional process modeling (Chapter 9)? Probably not. Information systems developed with OO technologies have processes like all information systems. But those processes (called behaviors in OOA) would be designed as part of the object classes rather than separately and often not until the systems design phase (Chapter 18). Would Bob have done traditional data modeling (Chapter 8)? Perhaps, but not in the same way. During the systems analysis phase, Bob would have analyzed and documented the data attributes of the system using a class diagram (as taught in this chapter) instead of an ERD. If the system data was to be stored in a relational database, then during the systems design phase, Bob would translate the class diagram into an ERD and follow the steps for data design taught in Chapter 14. But with either approach, Bob would have still followed the same phases of requirements analysis, systems analysis, and so on. Regardless of the tools and techniques, systems analysis and design is still systems analysis and design.

History of Object Modeling

object modeling a technique for identifying objects within the systems environment and identifying the relationships between those objects.

The object-oriented approach is centered around a technique referred to as **object modeling.** The object modeling technique prescribes the use of methodologies and diagramming notations that are completely different from the ones used for data modeling and process modeling. In the late 80s and early 90s many different object-oriented methods were being used throughout industry. The most notable of these were Grady Booch's *Booch Method,* James Rumbaugh's *Object Modeling Technique (OMT),* and Ivar Jacobson's *Object-Oriented Software Engineering (OOSE).* The existence of so many methods and associated modeling techniques was a major problem for the object-oriented system development industry. It was not uncommon for a developer to have to learn several object modeling techniques depending on what was being used on the project at the time. Because so many were being used, this was limiting the ability to share models across projects (reduced reusability) and development teams. Consequently, it hampered communication between team members and users, which led to many errors being introduced into the project. These problems and others led to the effort to design a standard modeling language.

In 1994 Grady Booch and James Rumbaugh joined forces, merging their respective object-oriented development methods with the goal of creating a single, standard process for developing object-oriented systems. Ivar Jacobson joined them in 1995, and the three altered their focus to create a standard object modeling language instead of a standard object-oriented approach or method.[1] Referencing their own work as well as that of countless others in the OO industry, the **Unified Modeling Language (UML)** version 1.0 was released in 1997. The current version is 2.0.

The UML does not prescribe a method for developing systems—only a notation that is now widely accepted as a standard for object modeling. The Object Management Group (OMG), the industry's standards body, adopted the UML in November 1997 and has continually worked to improve it based on industry needs. In this chapter and Chapter 18, "Object-Oriented Design and Modeling Using the UML," we will present an introduction to the UML and some of its diagrams.[2]

There are many underlying concepts for object modeling. In the next section you will learn about those concepts and how to apply them while developing object models during systems analysis.

> **Unified Modeling Language** a set of modeling conventions that is used to specify or describe a software system in terms of objects.

System Concepts for Object Modeling

Object-oriented analysis is based on several concepts. Some of these concepts require a new way of thinking about systems and the development process. As depicted on the home page at the beginning of the chapter, object-oriented analysis is concerned with defining the static structure and dynamic behavior models of the information system instead of defining data and process models, which is the goal of traditional development approaches. These OOA concepts have presented a formidable challenge to veteran developers, who must relearn how they have traditionally viewed systems. As you will soon see, these concepts are not foreign to how you have already come to view your own environment.

> Objects, Attributes, Methods, and Encapsulation

The object-oriented approach to system development is based on the concept of objects that exist within a system's environment. Objects are everywhere. Let's consider your environment. Look around. What are some of the objects present within your environment? Perhaps you see a door, a window, or the room itself. What about this book—it's an object, as is the very page you are reading. Perhaps you also have a student workbook, which is also an object. If there are other individuals in the room, they are objects too. You may also see a phone, a chair, and perhaps a table. All these are objects that may be clearly visible within your immediate environment.

Consider the *Webster's Dictionary* definition of *object:* "something that is or is capable of being seen, touched, or otherwise sensed."

The objects mentioned above are those that one would be able to see or touch. What about objects that you might sense? Perhaps you are waiting for a phone call. That phone call is something that you are sensing. You may be waiting for a meeting. Once again, that meeting is something that you can identify, relate to, and anticipate even though you can't actually see the meeting. Thus, according to *Webster's Dictionary,* an anticipated phone call or meeting may be considered an object.

The previous examples pertain to objects that may exist within your immediate environment. Similarly, in the object-oriented approach to systems development, it is important to identify the objects that exist within a system's environment. In

---

[1] Booch, Rumbaugh, and Jacobson created an object modeling methodology called the *Rational Unified Process,* marketed by IBM.

[2] There are excellent books dedicated to the detailed use of the UML, and many are listed at the end of this chapter.

object something that is or is capable of being seen, touched, or otherwise sensed and about which users store data and associate behavior.

object-oriented approaches to systems development, the definition of an **object** is as presented in the margin.

Three aspects of this definition need to be examined closely. First, let's consider the term *something,* which can be characterized as a type of object much like the objects that we identified within your current environment. The types of objects may include a *person, place, thing,* or *event.* An employee, customer, instructor, and student are examples of person objects. A particular warehouse, regional office, building, and room are examples of place objects. Examples of thing objects include a product, a vehicle, a computer, a videotape, or a window appearing on a user's display monitor. Finally, examples of event objects include an order, payment, invoice, application, registration, and reservation.

Now let's consider the *data* aspect of our definition. In object-oriented circles, this part of our definition refers to what are called **attributes.**

attribute the data that represents characteristics of interest about an object.

For example, we might be interested in the following attributes for an object called "customer": CUSTOMER NUMBER, FIRST NAME, LAST NAME, HOME ADDRESS, WORK ADDRESS, TYPE OF CUSTOMER, HOME PHONE, WORK PHONE, CREDIT LIMIT, AVAILABLE CREDIT, ACCOUNT BALANCE, and ACCOUNT STATUS. In reality, there may be many customer objects for which we would be interested in these attributes. Each individual customer is referred to as an **object instance.** For example, for each customer the attributes would assume values specific to that customer—such as 412209, Lonnie, Bentley, 2625 Darwin Drive, West Lafayette, Indiana, 47906, and so forth. Let's consider your current environment. Perhaps there's another person in the room. Each of you represents an instance of a person object. Each of you can be described according to some common attributes such as LAST NAME, SOCIAL SECURITY NUMBER, PHONE NUMBER, and ADDRESS.

object instance each specific person, place, thing, or event, as well as the values for the attributes of that object. Sometimes referred to simply as an object.

Thus, object-oriented approaches to systems development are concerned with identifying attributes that are of interest regarding an object. With advances in technology, attributes have evolved to include more than simple data characteristics as those represented in the previous example. Today, objects may include newer attribute types, such as a picture, sound, or even video.

behavior the set of things that an object can do and that correspond to functions that act on the object's data (or attributes). In object-oriented circles, an object's behavior is commonly referred to as a *method, operation,* or *service* (we may use the terms interchangeably throughout our discussion).

Let's now consider the last aspect of our definition for an object—the **behavior** of an object. This represents a substantially different way of viewing objects. When you look at the door object within your environment, you may simply see a motionless object that is incapable of thinking—much less carrying out some action. In object-oriented approaches to systems development, that door can be associated with behavior that it is assumed can be performed. For example, the door can *open,* it can *shut,* it can *lock,* or it can *unlock.* All of these behaviors are associated with the door and are accomplished by the door and no other object.

Consider another object—a telephone. What behaviors can be associated with a phone? With advances in technology we actually have phones that are voice-activated and can *answer, dial, hang up,* and carry out other behaviors. Thus, object-oriented approaches to systems development simply require an adjustment to how we commonly perceive objects.

Another important object-oriented principle is that an object is solely responsible for carrying out any functions or behaviors that act on its own data (or attributes). For example, only YOU (an object) may CHANGE (behavior) your LAST NAME and HOME ADDRESS (attributes about you). This leads us to an important concept in understanding objects: **encapsulation.** Applied to an object, both attributes and behavior of the object are packaged together. They are considered part of that object. The only way to access or change an object's attributes is through that object's specific behaviors.

encapsulation the packaging of several items together into one unit.

In object-oriented development, models depicting objects are often drawn. Let's examine the modeling notation (signs and symbols) used to represent an object in these object models. Figure 10-1(a) shows two object instances, each drawn using a rectangle with the name of the object instance. The name consists of the value of the attribute that uniquely identifies it, followed by a colon, and then the name of the class in which the object has been categorized. The entire name phrase is centered in the rectangle and is also underlined. In Figure 10-1(a) the attribute CUSTOMER NUMBER, whose value is

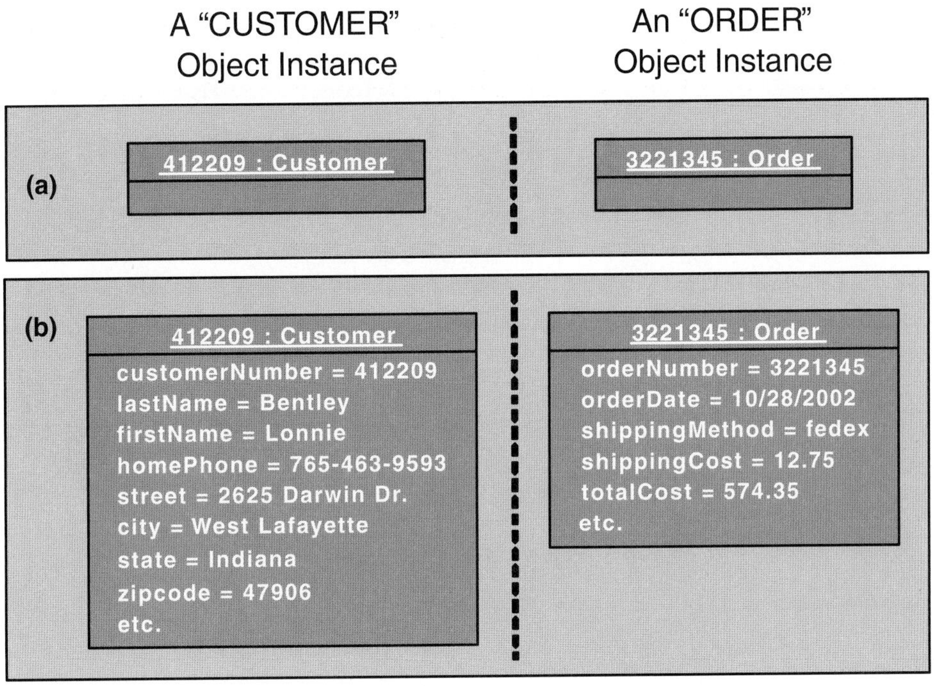

A "CUSTOMER"
Object Instance

An "ORDER"
Object Instance

FIGURE 10-1

Object Instances

(a)

412209 : Customer

3221345 : Order

(b)

412209 : Customer

customerNumber = 412209
lastName = Bentley
firstName = Lonnie
homePhone = 765-463-9593
street = 2625 Darwin Dr.
city = West Lafayette
state = Indiana
zipcode = 47906
etc.

3221345 : Order

orderNumber = 3221345
orderDate = 10/28/2002
shippingMethod = fedex
shippingCost = 12.75
totalCost = 574.35
etc.

412209, uniquely identifies that instance of CUSTOMER. Thus, 412209 is the name of the object instance and CUSTOMER is its classification. Optionally, the object instance can also be drawn as shown in Figure 10-1(b). The attribute values for the object instance are recorded within the symbol and are separated from the object name by a line.

> Classes, Generalization, and Specialization

Another important concept of object modeling is the concept of categorizing objects into **object classes.** Let's consider some of the objects within your current environment. It would be natural for you to classify both your *Systems Analysis and Design Methods* textbook and another textbook, such as *Introduction to Programming,* as BOOKS [see Figure 10-2(a)]. Both these object instances have some similar attributes and behavior. For example, similar attributes might be ISBN NUMBER, TITLE, COPYRIGHT DATE, EDITION, and so on. Likewise, they have similar behavior, such as being able to OPEN and CLOSE. There may be several other objects within your environment that could be classified because of their similarities. For example, you and other individuals in the room might be classified as PERSON.

How are object classes represented in object modeling using the UML notation? As depicted in Figure 10-2(b), they are drawn very similar to an object instance, except that the values of the attributes are omitted and the name of the class is not underlined. In addition, the class symbol may include a list of behaviors. Also, as shown in Figure 10-2(b), to simplify the appearance of diagrams containing numerous object class symbols, sometimes the object classes are drawn without the list of behaviors and attributes. Most object modeling tools allow you to do this in order to customize the model to your liking.

We can also recognize subclasses of objects [see Figure 10-3(a)]. For example, some of the individuals in the room might be classified as STUDENTS and others as TEACHERS. Thus, STUDENT and TEACHER object classes are members of the object class PERSON. When levels of object classes are identified, the concept of **inheritance** is applied.

The approach that seeks to discover and exploit the commonalities between object classes is referred to as **generalization/specialization.** In examining Figure 10-3(b), you notice that the object classes STUDENT and TEACHER contain attributes and behaviors

object class a set of object instances that share the same attributes and behaviors. Often referred to simply as a class.

inheritance the concept wherein methods and/or attributes defined in an object class can be inherited or reused by another object class.

generalization/ specialization a technique wherein the attributes and behaviors that are common to several types of object classes are grouped (or abstracted) into their own class, called a *supertype*. The attributes and methods of the supertype object class are then inherited by those object classes (*subtypes*). Sometimes abbreviated as gen/spec.

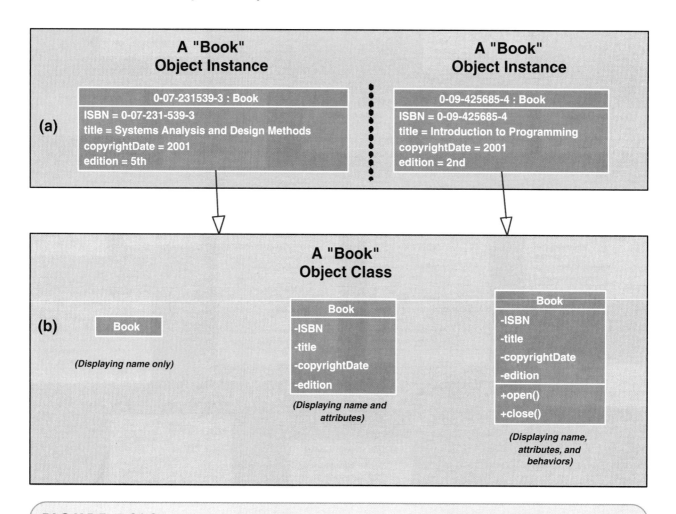

FIGURE 10-2   Representing Object Classes in the UML

---

which are unique to them (making them more *specialized*) but that they also have access to the *generalized* attributes and behaviors of the PERSON object class via inheritance.

In our example, the object class PERSON is referred to as a **supertype** (or generalization class) whereas STUDENT and TEACHER are referred to as **subtypes** (or specialization classes). The object class supertype will have one or more *one-to-one* relationships to object class *subtypes* because, in our example, any one person (object instance) will be at most one teacher or one student or possibly both. Also, any one teacher will be only one person. These relationships are sometimes called "*is a*" *relationships* because of how you express the relationship in a sentence. For example, "A STUDENT *is a* type of PERSON" or "A TEACHER *is a* type of PERSON."

In object-oriented systems development, objects are categorized according to classes and subclasses. Identifying classes realizes numerous benefits. For example, consider the fact that a new attribute of interest, called GENDER, needs to be added to both the teacher and the student object classes. Because the attribute is common to both, the attribute could be added once, to the class PERSON—implying that both the teacher and the student object classes will inherit that attribute. Looking down the road toward program maintenance, we note that the implication is substantial. Program maintenance is enhanced by the need to simply make modifications in one place. For example, let's assume the attribute LAST NAME currently had a field size of 15 characters. Let's also assume that through analysis of our data we found many last

supertype an entity that contains attributes and behaviors that are common to one or more class subtypes. Also referred to as *abstract* or *parent* class.

subtype an object class that inherits attributes and behaviors from a supertype class and then may contain other attributes and behaviors that are unique to it. Also referred to as *child* class and, if it exists at the lowest level of the inheritance hierarchy, as *concrete* class.

(a)

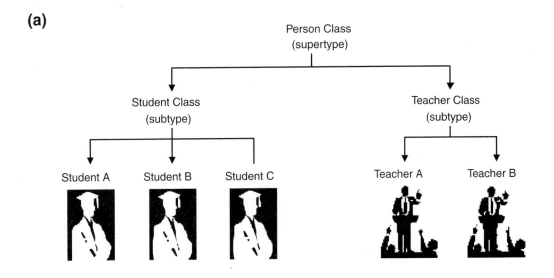

(b)

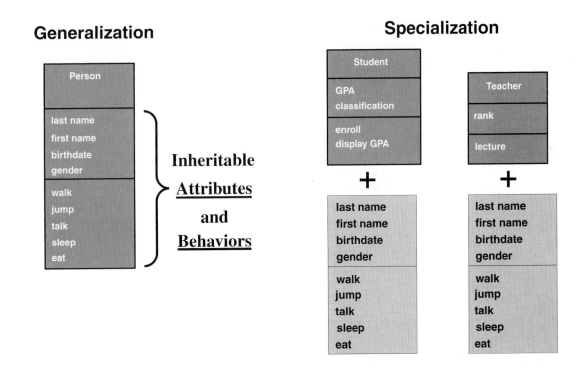

FIGURE 10-3   Supertype and Subtype Relationships between Object Classes

names that were more than 15 characters. Because of this we need to change the LAST NAME attribute field size to 25 to be able to hold the entire values of all last names. By taking advantage of inheritance, we have to make that change only once in the PERSON class. Without inheritance, we would have had to make a change to both the STUDENT and the TEACHER classes. The preceding example is fairly simple, but considering that a large application may contain dozens of classes with hundreds of attributes and

FIGURE 10-4

Representing a
Generalization/
Specialization
Relationship Using
the UML

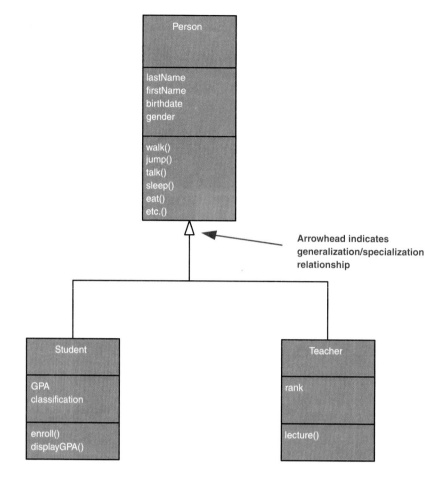

behaviors, the time and money saved by having to make modifications in only one place is considerable.

How is generalization/specialization (supertype, subtype classes) depicted using the UML notation? Figure 10-4 illustrates how to depict the supertype-subtype relationship between the PERSON, STUDENT, and TEACHER object classes. All the attributes and behaviors of the PERSON object class are inherited by the STUDENT and TEACHER object classes. The attributes and behaviors that uniquely apply to a STUDENT or TEACHER are recorded directly in the subtype class symbol.

> Object Class Relationships

object class relationship
a natural business association
that exists between one or
more objects and classes.

Conceptually, objects do not exist in isolation. The things that they represent interact with and impact one another to support the business mission. Thus an **object class relationship** is inevitable. You, for example, interact with this textbook by reading it, a telephone by using it, and perhaps other individuals in the room by communicating with them. Similarly, objects interact with other objects within a systems environment. Consider, for example, the object classes CUSTOMER and ORDER that may exist in a typical information system. We can make the following business assertions about how customers and orders are associated (or interact):

- A CUSTOMER PLACES zero or more ORDERS.
- An ORDER IS PLACED BY one and only one CUSTOMER.

We can graphically illustrate this association between CUSTOMER and ORDER as shown in Figure 10-5(a). The connecting line represents a relationship between the

(a)

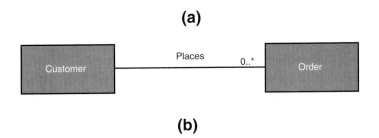

(b)

Multiplicity	UML Multiplicity Notation	Association with Multiplicity	Association Meaning
Exactly 1	1 — or — *leave blank*	Employee — Works for — 1 — Department Employee — Works for — Department	An employee works for one and only one department.
Zero or 1	0..1	Employee — Has — 0..1 — Spouse	An employee has either one or no spouse.
Zero or more	0..* — or — *	Customer — Makes — 0..* — Payment Customer — Makes — * — Payment	A customer can make no payment up to many payments.
1 or more	1..*	University — Offers — 1..* — Course	A university offers at least 1 course up to many courses.
Specific range	7..9	Team — Has scheduled — 7..9 — Game	A team has either 7, 8, or 9 games scheduled.

FIGURE 10-5 Object/Class Associations and Multiplicity Notations

classes. UML refers to this line as an *association,* and we will use this term through the remaining parts of this chapter. The verb phrase describes the association. All relationships are implicitly bidirectional, meaning that they can be interpreted in both directions (as suggested by the above business assertions).

Figure 10-5(a) also shows the complexity or degree of each association. For example, for the above business assertions, we must also answer the following questions:

- Must there exist an instance of CUSTOMER for each instance of ORDER? (Yes)
- Must there exist an instance of ORDER for each instance of CUSTOMER? (No)
- How many instances of ORDER can exist for each instance of CUSTOMER? (Many)
- How many instances of CUSTOMER can exist for each instance of ORDER? (One)

multiplicity the minimum and maximum number of occurrences of one object class for a single occurrence of the related object class.

We call this concept **multiplicity.** Because all associations are by default bidirectional, meaning the CUSTOMER class "knows about" the ORDER class and the ORDER class "knows about" the CUSTOMER class, multiplicity must be defined in both directions for every association. The possible UML graphical notation for multiplicity between classes is shown in Figure 10-5(b). If you have learned data modeling in Chapter 8, you will realize that multiplicity is essentially the same concept as cardinality. The notations are different, but the relationships are nearly the same.

Some objects are made up of other objects. For example if you buy something over the Internet, your one order could be composed of multiple items (a CD, a DVD, a book, etc.). Other examples include a club, which is made up of several club members, and a computer contains a case, CPU, motherboard, power supply, and so on. This kind of relationship is called **aggregation.** This relationship is characterized by the phrases "whole-part" and "is part of."

aggregation a relationship in which one larger "whole" class contains one or more smaller "parts" classes. Conversely, a smaller "part" class is part of a "whole" larger class.

Composition is a stronger form of aggregation. Think of the word *component* for composition. In composition the "whole" is completely responsible for the creation and destruction of its parts, and each "part" is associated to only one "whole" object. The relationship between club and club member would not be composition, because members have a life outside the club and can, in fact, belong to multiple clubs. But the Internet order and order items would be composition. If you cancel the order, then all the items on that order will get canceled with it. A behavior performed on the whole will also be performed on all its parts. For example, if we printed the order, each order item would be automatically printed also.

composition an aggregation relationship in which the "whole" is responsible for the creation and destruction of its "parts." If the "whole" were to die, the "part" would die with it.

In earlier versions of UML, aggregation was drawn with a hollow diamond, with the diamond connected to the "whole" object class, as shown in Figure 10-6(a). Notice that multiplicity must be specified for both sides of the relationship.

Composition is drawn with a filled diamond, as shown in Figure 10-6(b). Because each "part" can belong to only one "whole," multiplicity needs to be specified only for the "part." Figure 10-6(b) also illustrated multilevel composition. A book is composed of chapters, which are each composed of pages, and so forth.

In UML 2.0 the notation for aggregation has been dropped. Why? While the composition relationship has definite distinctions that play out in programming, aggregation has always been more indistinct. For example, couldn't the relationship between club and club member simply be a one-or-more association between independent object classes? Because of this, some practitioners consider aggregation (the weaker form) to be essentially meaningless in any practical sense.

> Messages and Message Sending

message communication that occurs when one object invokes another object's method (behavior) to request information or some action.

Object classes interact or "communicate" with one another by passing **messages.** Recall the concept of encapsulation, wherein an object is a package of attributes and behavior. Only that object can perform its behavior and act on its data.

Let's consider the CUSTOMER and ORDER objects mentioned earlier. A CUSTOMER object checking the current status of an ORDER sends a message to an ORDER object by

(a)

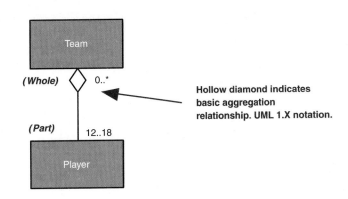

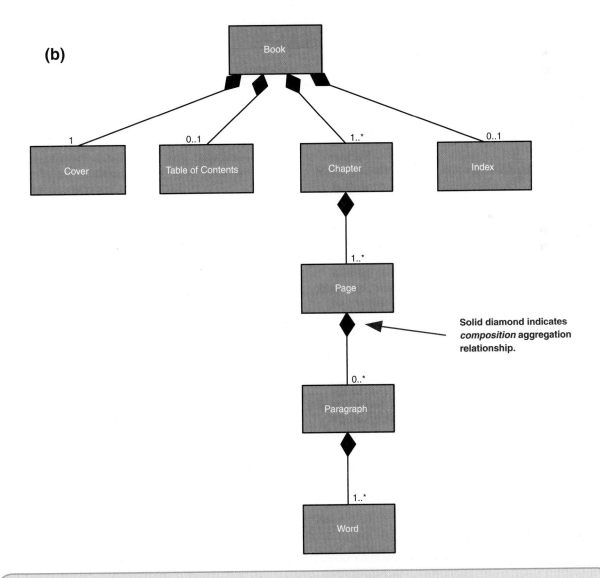

FIGURE 10-6 Aggregation Relationships

FIGURE 10-7

Messaging

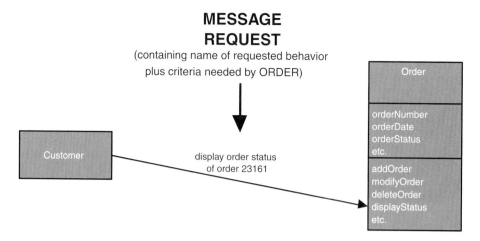

invoking the ORDER object's *display status* behavior (a behavior that accesses and displays the ORDER STATUS attribute).

The object sending a message does not need to know how the receiving object is organized internally or how the behavior is to be accomplished, only that it responds to the request in a predefined way. This concept of messaging is illustrated in Figure 10-7. A message can be sent only between two objects that have an association. Chapter 18 presents a discussion on how to document and specify messages.

> Polymorphism

polymorphism literally meaning "many forms," the concept that different objects can respond to the same message in different ways.

An important concept that is closely related to messaging is **polymorphism.** Let's consider the WINDOW and DOOR objects within your environment. Both objects have a common behavior that they may perform; they may both close. How a DOOR object carries out that behavior may differ substantially from the way in which a WINDOW carries out that behavior. A DOOR "swings shut"; a WINDOW "slides downward." Thus, the behavior close may take on two different forms. Once again, let's consider the WINDOW object. Not all WINDOWS would actually accomplish the close behavior in the same way. Some WINDOW objects, like DOOR objects, swing shut! Thus, the close behavior takes on different forms for a given object class.

override a technique whereby a subclass (subtype) uses an attribute or behavior of its own instead of an attribute or behavior inherited from the class (supertype).

Polymorphism is applied in object-oriented applications when a behavior in the supertype needs to be **overridden** by a behavior in the subtype. Examine the generalization/specialization relationship in Figure 10-8. The EMPLOYEE class contains a behavior called "compute pay" to calculate how much each EMPLOYEE will be paid. Because FULL-TIME EMPLOYEES and PART-TIME EMPLOYEES get paid differently (full-time employees receive an annual salary in 52-week increments, and part-time employees get paid only for the hours they work), two behaviors that perform different calculations are required. But because of polymorphism, the behaviors can be named the same to simplify message sending. The subtype that requires the unique behavior will contain in its behavior list the same behavior that is listed for its parent (supertype). When the PART-TIME EMPLOYEE object receives a message to "compute pay," it will automatically use the compute-pay behavior in its own behavior list because it overrides what it inherits from its parent. Polymorphism is very useful when making enhancements to an existing system, because adding new classes to an existing generalization/specialization relationship in order to satisfy new business rules or requirements may not be possible or practical.

So how is polymorphism related to message sending? Once again, the requesting object knows what service (or behavior) to request and from which object. However, the requesting object does not need to worry about how a behavior is accomplished.

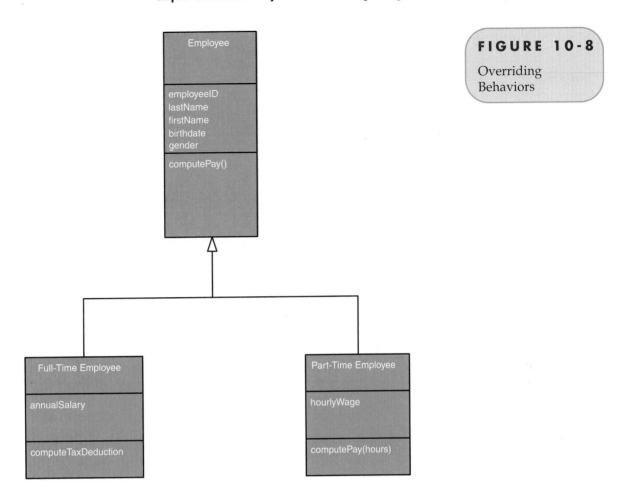

FIGURE 10-8
Overriding Behaviors

The UML Diagrams

Think of the UML diagrams as if they were blueprints for constructing a house. Where a set of blueprints typically provides the builder with perspectives for plumbing, electricity, heating, air conditioning, and the like, each UML diagram provides the development team with a different perspective on the information system.

Figure 10-9 describes the 13 diagrams of UML 2.0. This list is organized not alphabetically but in an order that allows the description of each diagram to build on the descriptions of those above it. It would take an entire college course to cover each diagram in depth. As we study an overview of the systems analysis life cycle, three chapters will delve into the core UML diagrams:

Chapter 7—*FAST* requirements analysis phase.
- Use-case diagrams.

Chapter 10—*FAST* logical design phase.
- Activity diagrams.
- System sequence diagrams (a high-level kind of sequence diagram).
- Class diagrams.

Chapter 18—*FAST* physical design phase.
- Sequence diagrams.
- Class diagrams (with more detail).
- State machine diagrams.
- Communication diagrams.
- Component diagrams.
- Deployment diagrams.

FIGURE 10-9 UML 2.0 Diagrams

Diagram	Description
Use case	Depicts the interactions between the system and external systems and users. In other words, it graphically describes who will use the system and in what ways the user expects to interact with the system. The use-case narrative is used, in addition, to textually describe the sequence of steps of each interaction.
Activity	Depicts the sequential flow of activities of a use-case or business process. It can also be used to model logic with the system.
Class	Depicts the system's object structure. It shows object classes that the system is composed of as well as the relationships between those object classes.
Object	Similar to a class diagram, but instead of depicting object classes, it models actual object instances with current attribute values. The object diagram provides the developer with a snapshot of the system's object at one point in time.
State machine	Models how events can change the state of an object over its lifetime, showing both the various states that an object can assume and the transitions between those states.
Composite structure	Decomposes the internal structure of a class, component, or use case.
Sequence	Graphically depicts how objects interact with each other via messages in the execution of a use case or operation. It illustrates how messages are sent and received between objects and in what *sequence*.
Communication	Called a collaboration diagram in UML 1.X, it depicts the interaction of objects via messages. Thus, it is similar to a sequence diagram. But while a sequence diagram focuses on the timing or sequence of messages, a communication diagram focuses on the structural organization of objects in a network format.
Interaction overview	Combines features of sequence and activity diagrams to show how objects interact within each activity of a use case.
Timing	Another interaction diagram that focuses on timing constraints in the changing state of a single object or group of objects. A timing diagram is especially useful when designing embedded software for devices.
Component	Depicts the organization of programming code divided into components and how the components interact.
Deployment	Depicts the configuration of software components within the physical architecture of the system's hardware "nodes."
Package	Depicts how classes or other UML constructs are organized into packages (corresponding to *Java* packages or *C++* and *.NET* namespaces) and the dependencies of those packages.

The Process of Object Modeling

As mentioned earlier, in performing object-oriented analysis (OOA), as with any other systems analysis method, the purpose is to gain a better understanding of the system and its functional requirements. In other words, OOA requires that we identify the required system functionality from the user's perspective and identify the objects, along with their data attributes, associated behavior, and relationships, which support the required system functionality. In Chapter 7 you were introduced to use-case modeling, which is used to identify required system functionality. In this chapter you will learn to refine the use-case model created in Chapter 7, learn to document complex use cases with activity diagrams, and learn to perform object modeling to document the identified objects and the data and behavior they encapsulate, plus their relationships with other objects.

There are three general activities in performing object-oriented analysis:

1. Modeling the functions of the system.
2. Finding and identifying the business objects.
3. Organizing the objects and identifying their relationships.

> Modeling the Functional Description of the System

Recall that in Chapter 7 you were taught the process of use-case modeling to document functional system requirements using business requirements use cases. During this activity the use cases were documented to contain only general information about the business event. The goal was to quickly document all of the business events (use cases) in order to define and validate requirements. In performing object-oriented analysis, each previously defined use case will be refined to include more and more detail based on the facts we learned throughout the development process, such as user interface requirements. To prepare to perform object modeling, we need to evolve the business requirements use-case model into the analysis use-case model.

> Constructing the Analysis Use-Case Model

In object-oriented analysis we evolve the requirements use-case model into the analysis use-case model by performing the following steps:

1. Identify, define, and document new actors.
2. Identify, define, and document new use cases.
3. Identify any reuse possibilities.
4. Refine the use-case model diagram (if necessary).
5. Document system analysis use-case narratives.

Step 1: Identify, Define, and Document New Actors   Between the time the business requirements use-case model was created and the time it is subsequently approved by the system owners, the systems analyst and the rest of the development team, through talking with stakeholders and researching project artifacts, continue to learn more about what is required in order for the system to be successful. During these efforts it is possible that additional actors may be discovered and thus need to be defined and documented. For example, when analyzing the *Place New Order* use case (see Figure 7-13) initiated by the CLUB MEMBER, we identified the need for the CLUB MEMBER to be able to enter the order information via the Internet, but the member could also submit orders by mail. For the order information to be input into the system, someone else would have to interact with the system to accomplish this, thus the need for another actor. The newly identified actor named *Member Services Associate*, along with any other new actors, would need to be defined in the actor glossary previously prepared (Figure 7-8).

Step 2: Identify, Define, and Document New Use Cases   The new actor MEMBER SERVICES ASSOCIATE, discovered in step 1, leads to a new interaction with the system—thus a new use case. As a general rule of thumb, each type of user interface used to process a business event will require its own use case. Using the banking industry as an example, the use case of making a deposit at an ATM machine will be different from the use case of making a deposit using a bank teller. The goal of the process is the same and many of the steps will be the same, but the actual system user may be different or how the user interacts with the system using a specific technology (ATM machine versus a workstation with a GUI designed for a bank teller) may be different. The newly identified use cases would need to be defined in the use-case actor glossary previously prepared.

Step 3: Identify Any Reuse Possibilities   As stated in step 2 above, when you have two use cases that have the same business goal but the interface technology or the actual system user may be different, both use cases may share common steps. As you recall from Chapter 7, to eliminate redundant steps, we can extract these common steps into their own separate use case called an *abstract use case*. In addition, when we analyze the use cases and find a use case that contains complex functionality consisting of several steps, making it difficult to understand, we can extract the more complex steps into their own use case called an *extension use case*. These new use cases would also be defined in the use-case glossary previously prepared.

Step 4: Refine the Use-Case Model Diagram (if Necessary)   With the discovery of new actors and/or use cases, we now would update the use-case model diagram previously constructed (see Figure 7-10) to include these items. Figure 10-10

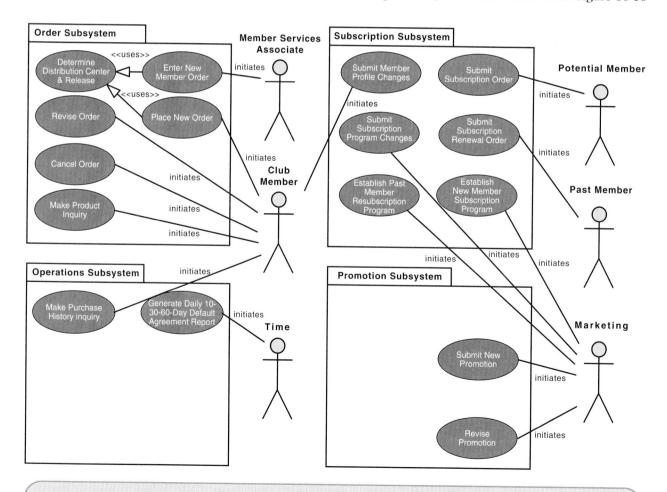

FIGURE 10-10   Revised Member Services System Use-Case Model Diagram

is the revised use-case model diagram, which includes the newly identified actor MEMBER SERVICES ASSOCIATE and the newly identified use cases *Enter New Member Order* and *Determine Appropriate Distribution Center and Release Order to Be Filled.*

Step 5: Document System Analysis Use-Case Narratives   Once all business requirements use cases have been reviewed and approved by the users, each use case will be refined to include more information in order to specify the system functionality in detail. The resulting use cases are called **system analysis use cases** and still should be free of most implementation details except high-level information describing the means (Windows GUI, Internet browser, telephony, etc.) the system user will use to interact with the system. System analysis use cases include a narrative from the perspective of the system user and are more conversational (with the system) in nature than business requirements use cases. Figures 10-11 and 10-12 are evolutions of the business requirements use case *Place New Order.* Figure 10-11 depicts the CLUB MEMBER as the primary system actor, using the system to enter the order, and Figure 10-12 depicts the MEMBER SERVICES ASSOCIATE using the system to enter the order from the information received from the club member.

> **system analysis use case**
> a use case that documents the interaction between the system user and the system. It is highly detailed in describing what is required but is free of most implementation details and constraints.

System analysis use cases will be further refined during the design phases of the life cycle to specify the how or implementation specifics. It is important that all open issues and *to be determines* (TBDs) be resolved before going forward into design because such decisions may impact the nature of the design. Please note the additional elements found in system analysis use cases.

❶ *Use-case type*—In performing use-case modeling, the first cases to be constructed are business requirements use cases, which focus on the strategic vision and goals of the various stakeholders. This *type* of use case is business-oriented and reflects a high-level view of the desired behavior of the system. It is free from technical details and may include both manual activities and the activities that will be automated. Business requirements use cases provide a general understanding of the problem domain and scope, but they don't include the necessary detail to communicate to developers what the system should do.

To reflect implementation details such as user interface constraints, tactical use cases, called *system use cases,* are derived from the business use cases. One or more system analysis use cases may evolve from a single business use case. Developers use this type of use case to specify detailed requirements, assist in estimating and planning, communicate programming requirements, and form the basis for user documentation. Each system use case corresponds to a test case that will be executed to verify that the system satisfies the customer's requirements.

System use cases continue to be refactored throughout the systems development life cycle. In following an iterative approach to development, it is wise to track where each use case is in terms of its evolution from the requirements level through analysis and on to design.

❷ *Primary system actor*—The primary system actor is the stakeholder that actually uses and interfaces with the system. It is for this stakeholder that the interface must be designed.

❸ *Abstract use case*—Example of calling an abstract use case.

Documenting Abstract and Extension Use-Case Narratives   Documenting the narratives of extension and abstract use cases is very similar to documenting regular use cases, but there are a few major differences. Abstract and extension use cases are not initiated by actors; they are invoked by other use cases. Also, abstract and extension use cases tend to be much shorter and don't require as many data fields.

Member Services System

Author(s): **K. Dittman** _____   Date: ___**11/01/02**_____

Version: _**1.00**_____

Use Case Name:	Place New Order	Use Case Type
Use Case ID:	MSS-SUC002.00	Business Requirements: ☐
Priority:	High	System Analysis: ☑
Source:	Requirement — MSS-R1.00 Requirements Use Case — MSS-BUC002.00	**①**

Primary Business Actor:	Club Member (Alias — Active Member, Member)
Primary System Actor:	Club Member (Alias — Active Member, Member) **②**
Other Participating Actors:	• Warehouse (Alias — Distribution Center) (external receiver) • Accounts Receivable (external server)
Other Interested Stakeholders:	• Marketing — Interested in sales activity in order to plan new promotions. • Procurement — Interested in sales activity in order to replenish inventory. • Management — Interested in order activity in order to evaluate company performance and customer (member) satisfaction.
Description:	This use case describes the event of a club member submitting a new order for SoundStage products via the World Wide Web. The member selects the items he or she wishes to purchase. Once the member has completed shopping, the member's demographic information as well as account standing will be validated. Once the products are verified as being in stock, a packing order is sent to the warehouse for it to prepare the shipment. For any product not in stock, a back order is created. On completion, the member will be sent an order confirmation.
Precondition:	The individual submitting the order must be an active club member. The member must log in to the system (provide identification) to enter an order.
Trigger:	This use case is initiated when the member selects the option to enter a new order.

Typical Course of Events:	**Actor Action**	**System Response**
	Step 1: The member requests the option to enter a new order.	**Step 2:** The system responds by displaying the catalog of the SoundStage products.
	Step 3: The member browses the available items and selects the ones he or she wishes to purchase, along with the quantity.	**Step 4:** Once the member has completed the selections, the system retrieves from file and presents the member's demographic information (shipping and billing addresses).
	Step 5: The member verifies demographic information (shipping and billing addresses). If no changes are necessary, the member responds accordingly (to continue).	**Step 6:** For each product ordered, the system verifies the product availability and determines an expected ship date, determines the price to be charged to the member, and determines the cost of the total order. If an item is not immediately available, it indicates the product is back-ordered or that it has not been released for shipping (for preorders). If an item is no longer available, that is indicated also. The system then displays a summary of the order to the member for verification.
	Step 7: The member verifies the order. If no changes are necessary, the member responds accordingly (to continue).	**Step 8:** The system checks the status of the member's account. If satisfactory, the system prompts the member to select the desired payment option (to be billed later or pay immediately with a credit card).
	Step 9: The member responds by selecting the desired payment option.	**Step 10:** The system displays a summary of the order, including the desired payment option, to the member for verification.
	Step 11: The member verifies the order. If no changes are necessary, the member responds accordingly (to continue).	**Step 12:** The system records the order information (including back orders if necessary). **Step 13:** Invoke abstract use case _MSS-AUC001.00, Determine Appropriate Distribution Center and Release Order to Be Filled._ **③** **Step 14:** Once the order is processed, the system generates an order confirmation and displays it to the member as well as sending it to the member via e-mail.

Alternate Courses:	**Alt-Step 3:** The member enters search criteria to retrieve a specific item or to display a reduced list of items to browse and order from.
	Alt-Step 5: If changes are required, the member updates the appropriate shipping, billing, or e-mail addresses and tells the system to store them accordingly. The system will validate the changes and, if successful, will store the new information to file.
	Alt-Step 7: If the order requires changes, the member can delete any item no longer wanted or change the order quantity. Once the member has completed the order changes, the system reprocesses the order (**go to step 6**). If the member requests to do additional shopping, **go to step 3**. If the member needs to change the demographic information, **go to step 5**.
	Alt-Step 8: If the member's account is not in good standing, display to the member the account status, the reason the order is being held, and what actions are necessary to resolve the problem. In addition, an e-mail is sent to the member with the same information. The system prompts the member to hold the order for later processing or cancel the order. If the member wishes to hold the order, the system records the order information and places it in hold status and then displays the SoundStage main page. If the member chooses to cancel the order, the system clears the inputted information and then displays the SoundStage main page. Terminate the use case.
	Alt-Step 10: If the member selects the option to pay by credit card, the system prompts the member to enter the credit card information (number and expiration date) and reminds the member that the billing address on file must match the billing address of the credit card provided. The member enters the required information and requests that the system continue. The system validates the credit card account provided. If the account cannot be validated, the system notifies the member and requests an alternative means of payment. If the member cannot provide an alternative means at this time, he or she can choose either to hold or to cancel the order. If the member wishes to hold the order, the system records the order information and places it in hold status and then displays the SoundStage main page. If the member chooses to cancel the order, the system clears the inputted information and then displays the SoundStage main page. Terminate the use case.
	Alt-Step 11: If the order requires changes, the member can delete any item no longer wanted or change the order quantity. Once the member has completed the order changes, the system reprocesses the order (**go to step 6**). If the member requests to do additional shopping, **go to step 3**. If the member needs to change the demographic information, **go to step 5**.
	Alt-Step 12: If all items ordered are on back order, the order is not released to the distribution center.
Conclusion:	This use case concludes when the member receives a confirmation of the order.
Postcondition:	The order has been recorded and, if the ordered products were available, released to the distribution center. For any product not available a back order has been created.
Business Rules:	• Member must have a valid e-mail address to submit online orders. • Member is billed for products only when they are shipped.
Implementation Constraints and Specifications:	• Use case must be available to the member 24 × 7. • Frequency — It is estimated that this use case will be executed 3,500 times per day. It should support up to 50 concurrent members.
Assumptions:	• Product can be transferred among distribution centers to fill orders. • Procurement will be notified of back orders by a daily report (separate use case). • The member responding to a promotion or using credits may affect the price of each ordered item. • The member can cancel the order at any time.
Open Issues:	None

FIGURE 10-11 *Place New Order* Use Case

Member Services System

Author(s): **K. Dittman**   Date: **11/01/02**
                                                                                     Version: **1.00**

Use Case Name:	Enter New Member Order	Use Case Type	
Use Case ID:	MSS-SUC003.00	**Business Requirements:**	☐
Priority:	High	**System Analysis:**	☑
Source:	Requirement — MSS-R1.00 Requirements Use Case — MSS-BUC002.00		
Primary Business Actor:	Club Member (Alias — Active Member, Member)		
Primary System Actor:	Member Services Associate (Alias — User)		
Other Participating Actors:	• Warehouse (Alias — Distribution Center) (external receiver) • Accounts Receivable (external server)		
Other Interested Stakeholders:	• Marketing — Interested in sales activity in order to plan new promotions. • Procurement — Interested in sales activity in order to replenish inventory. • Management — Interested in order activity in order to evaluate company performance and customer (member) satisfaction.		
Description:	This use case describes the event of a Member Services Associate entering a new order for SoundStage products that either has been submitted by mail by a member or is being telephoned in by a member. The member's demographic information as well as account standing will be validated. Once the products are verified as being in stock, a packing order is sent to the distribution center for it to prepare the shipment. For any product not in stock, a back order is created. On completion, the member will be sent an order confirmation.		
Precondition:	The individual submitting the order must be a member. The Member Services Associate must be logged in to the system.		
Trigger:	This use case is initiated when the Member Services Associate selects the option to enter a new order.		

Typical Course of Events:	Actor Action	System Response	
	Step 1: The Member Services Associate request the option to enter a new order.	**Step 2:** The system responds by prompting the user to enter the ID or name of the member submitting the order.	
	Step 3: The Member Services Associate provides the member name or ID.	**Step 4:** The system retrieves the member's demographic information on file and displays it to the user. If there are multiple members who match the criteria provided by the user, the system displays a list and prompts the user to select the correct one.	
	Step 5: The Member Services Associate verifies demographic information (shipping and billing addresses). If no changes are necessary, the associate responds accordingly (to continue).	**Step 6:** The system responds by prompting the user to enter the ID and quantity of each item to be ordered.	
	Step 7: The Member Services Associate enters the ID and quantity of each item provided.	**Step 8:** For each product ordered, the system validates the product identity. **Step 9:** For each product ordered, the system verifies the product availability and determines an expected ship date, determines the price to be charged to the member, and determines the cost of the total order. If an item is not immediately available, it indicates that the product is back-ordered or that it has not been released for shipping (for preorders). If an item is no longer available, that is indicated also. The system then displays a summary of the order to the user for verification.	
	Step 10: The Member Services Associate verifies the order with the information provided by the member. If no changes are necessary, the associate responds accordingly (to continue).	**Step 11:** The system checks the status of the member's account. If satisfactory, the system prompts the user to select the desired payment option (to be billed later or pay immediately with a credit card).	

	Step 12: The Member Services Associate responds by selecting the desired payment option indicated by the member.	**Step 13:** The system displays a final summary of the order, including the desired payment option, to the user for verification.
	Step 14: The Member Services Associate verifies the order. If no changes are necessary they respond accordingly (to continue).	**Step 15:** The system records the order information (including back orders if necessary).
		Step 16: Invoke abstract use case *MSS-AUC001.00, Determine Appropriate Distribution Center and Release Order to Be Filled.*
		Step 17: Once the order is processed, the system generates an order confirmation and displays it to the Member Services Associate as well as sending it to the member via e-mail or U.S. mail.
Alternate Courses:	**Alt-Step 4:** If the member cannot be found on file, notify user of discrepancy. **Alt-Step 5:** If changes are required, the member updates the appropriate shipping, billing, or e-mail addresses and tells the system to store them accordingly. The system will validate the changes and, if successful, will store the new information to file. **Alt-Step 8:** If the product information the member provided does not match any of SoundStage"s products, the system displays the discrepancy to the user and prompts the user for clarification. (Member Services Associate may have to contact member to resolve at a later time; if so, order may have to be placed in hold status.) **Alt-Step 4:** If the order requires changes, the user can delete any item no longer wanted or change the order quantity. Once the member has completed the order changes, the system reprocesses the order (**go to step 8**) **Alt-Step 11:** If the member's account is not in good standing, display to the user the member account status, the reason the order is being held, and what actions are necessary to resolve the problem. (The Member Services Associate may have to contact the member to resolve at a later time. In addition, an e-mail is sent to the member with the same information if the member has a valid e-mail account on file.) The system prompts the user to hold the order for later processing or cancel the order. If the user wishes to hold the order, the system records the order information and places it in hold status. If the user chooses to cancel the order, the system clears the inputted information. Terminate the use case. **Alt-Step 13:** If the user selects the option to pay by credit card, the system prompts the member to enter the credit card information (number and expiration date) provided by the member and reminds the user that the billing address on file must match the billing address of the credit card provided. The user enters the required information and requests that the system continue. The system validates the credit card account provided. If the account cannot be validated, the system notifies the user and requests an alternative means of payment. If the user cannot provide an alternative means at this time, the user can choose either to hold or to cancel the order. If the user wishes to hold the order, the system records the order information and places it in hold status. If the user chooses to cancel the order, the system clears the inputted information. Terminate the use case. **Alt-Step 14:** If the order requires changes, the user can delete any item no longer wanted or change the order quantity. Once the member has completed the order changes, the system reprocesses the order (**go to step 8**). **Alt-Step 15:** If all items ordered are on back order, the order is not released to the distribution center.	
Conclusion:	This use case concludes when the member receives a confirmation of the order.	
Postcondition:	The order has been recorded and, if the ordered products were available, released. For any product not available a back order has been created.	
Business Rules:	• Member must have a valid e-mail address to submit online orders. • Member is billed for products only when they are shipped.	
Implementation Constraints and Specifications:	• Use case must be available to the Member Services Associate from 7:00 A.M. to 9:00 P.M. EST. • Frequency — It is estimated that this use case will be executed 4,500 times per day. It should support up to 25 concurrent users.	
Assumptions:	• Product can be transferred among distribution centers to fill orders. • Procurement will be notified of back orders by a daily report (separate use case). • The member responding to a promotion or using credits may affect the price of each ordered item. • The member can cancel the order at any time.	
Open Issues:	None	

FIGURE 10-12 *Enter New Member Order Use Case*

Member Services System

Author(s): <u>K. Dittman</u>   Date: <u>11/01/02</u>
                                                               Version: <u>1.00</u>

Use Case Name:	Determine Appropriate Distribution Center and Release Order to Be Filled.	**Use Case Type**
		Abstract: ☑
Use Case ID:	MSS-AUC001.00	Extension: ☐
Priority:	High	**①**
Source:	MSS-SUC002.00 **②** MSS-SUC003.00	
Participating Actors:	• Warehouse (Alias — Distribution Center) (external receiver)	
Description:	This use case describes the event of selecting the distribution center that services the shipping address provided by the club member for a particular order. The order information (packing order) is then sent (released) to that distribution center to be filled.	
Precondition:	The order is ready to be released to the appropriate distribution center.	
Typical Course of Events:	**Step 1:** The system selects the appropriate distribution center based on the state and zip code of the shipping address. **Step 2:** Once the distribution center has been selected, a packing order containing the items to ship is formatted. **Step 3:** The packing order is transmitted to the distribution center (shipping and receiving system) to be used to prepare the shipment.	
Alternate Courses:	**Alt-Step 1:** If the shipping address is an international address, route the packing order to the Indianapolis, IN, location.	
Postcondition:	The packing slip has been transmitted (released) to the appropriate distribution center.	

FIGURE 10-13   Example of an Abstract Use-Case Narrative

Figure 10-13 is an example of an abstract use case. Please note the differences in elements of the narrative.

① *Use-case type*—An abstract use case is used when it's invoked by two or more use cases. An extension use case is used when it extends the functionality of a single use case.

② *Invoked by*—The IDs or names of the use cases that invoke this particular use case.

Please be aware that an abstract use case can invoke other abstract and/or extension use cases and that an extension use case can invoke other abstract and/or extension use cases—thus providing many avenues for use-case reusability.

After the system analysis use cases have been defined, they contain a level of detail that is adequate for the objects involved in the use cases to be realistically identified. These objects represent things or entities in the business domain—things of interest about which we would like to capture information. At this point, we will concentrate on describing these objects with a sentence or two. Later, we will expand our definitions to contain more detailed facts that we learn about each object.

> Modeling the Use-Case Activities

activity diagram a diagram that can be used to graphically depict the flow of a business process, the steps of a use case, or the logic of an object behavior (method).

The UML offers an additional diagram called an **activity diagram** to model the process steps or activities of the system. They are similar to flowcharts in that they graphically depict the sequential flow of *activities* of either a business process or a use case. They are different from flowcharts in that they provide a mechanism to depict activities that occur in parallel. Because of this they are very useful to model actions that will be performed when an operation is executing as well as the results of those actions—such as modeling the events that cause windows to be displayed or closed. Activity diagrams are flexible in that they can be used during both analysis and design. Figure 10-14 is an example of an activity diagram constructed on the use case *Enter New Member Order.* At least one activity diagram can be constructed for each use case. More than one can be constructed if the use case is long or contains complex logic. System analysts use activity diagrams to better understand the flow and sequencing of the use-case steps.

Figure 10-14 illustrates the following activity diagram notations:

❶ *Initial node*—the solid circle representing the start of the process.

❷ *Actions*—the rounded rectangles representing individual steps. The sequence of actions makes up the total activity shown by the diagram.

❸ *Flow*—the arrows on the diagram indicating the progression through the actions. Most flows do not need words to identify them unless coming out of decisions.

❹ *Decision*—the diamond shapes with one flow coming in and two or more flows going out. The flows coming out are marked to indicate the conditions.

❺ *Merge*—the diamond shapes with two or more flows coming in and one flow going out. This combines flows that were previously separated by decisions. Processing continues with any one flow coming into the merge.

❻ *Fork*—a black bar with one flow coming in and two or more flows going out. Actions on parallel flows beneath the fork can occur in any order or concurrently.

❼ *Join*—a black bar with two or more flows coming in and one flow going out, noting the end of concurrent processing. All actions coming into the join must be completed before processing continues.

❽ *Activity final*—the solid circle inside the hollow circle representing the end of the process.

The activity diagram shown in Figure 10-14 graphically illustrates the steps of the use case, but it does not specify who is doing those steps. That may not be a problem. Often you draw an activity diagram just to get a handle on the logic. But if you want to specify who does what, you can divide the activity diagram into *partitions* showing the actions performed by a specific class or actor. Figure 10-15 is an activity diagram for the *Place New Order* use case (Figure 10-11) with a simple one-dimensional partitioning of actions by member and system. The partitions are often called *swim lanes* because they resemble the lanes used by competition swimmers. An activity diagram might have three or more swim lanes showing receiver actors. You could also partition an activity diagram into a two-dimensional grid.

Figure 10-15 illustrates two additional features of activity diagrams:

❾ *Subactivity indicator*—the rake symbol in an action indicates that this action is broken out in another separate activity diagram. This helps you keep the activity diagram from becoming overly complex.

❿ *Connector*—A letter inside a circle gives you another tool for managing complexity. A flow coming into a connector jumps to the flow coming out of a connector with a matching letter.

These two examples do not exhaust all the functionality of activity diagrams. Actions can be invoked by signals based on time or an outside process. Actions can also send signals as well as receive them. You can even indicate the passing of parameters and other special kinds of information. But we have covered enough to get you started in drawing activity diagrams.

So would you draw an activity diagram for every use case? No, save them for the use cases (or even just sections of use cases) that have complex logic. Activity diagrams can help you think through system logic. They also are useful for communicating that

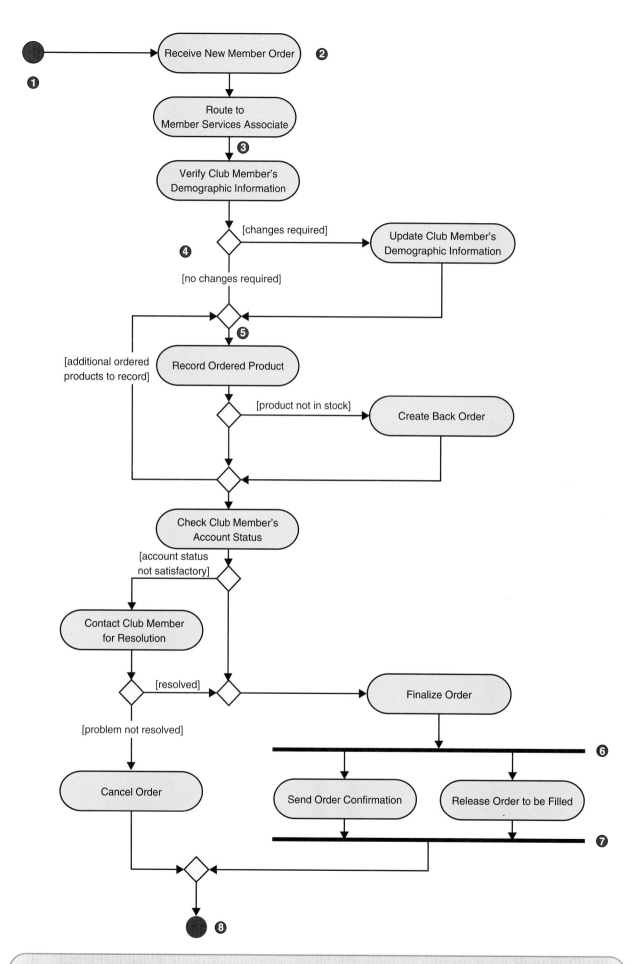

FIGURE 10-14 Activity Diagram of the *Enter New Member Order* Use Case

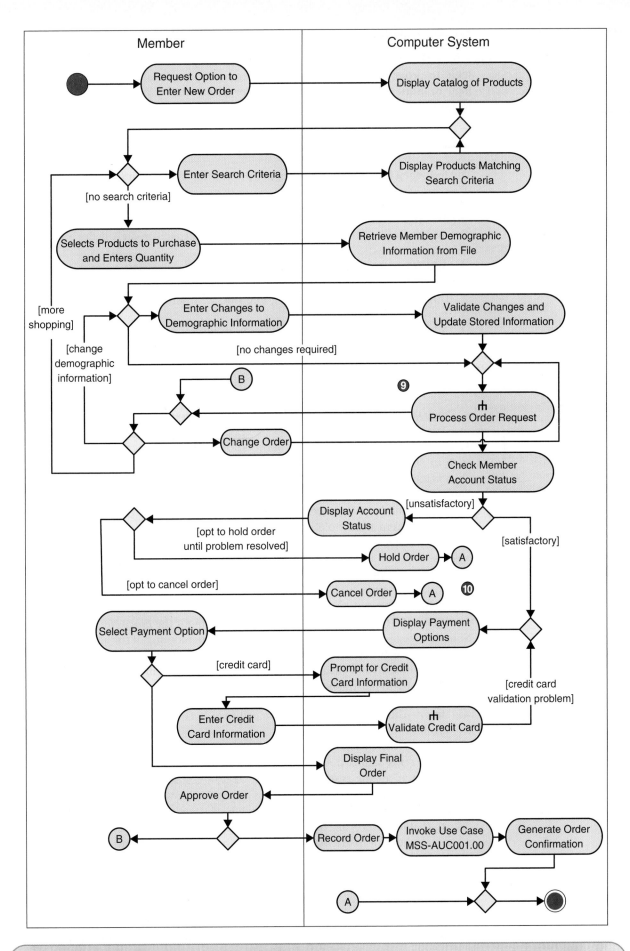

logic to programmers. Be careful, however, of trying to use them to communicate logic to users. Nontechnical users may have trouble following them. You are better off to employ use-case narratives with users.

> Guidelines for Constructing Activity Diagrams

The following list presents an excellent process for constructing activity diagrams:

- Start with one initial node as a starting point.
- Add partitions if they are relevant to your analysis.
- Add an action for each major step of the use case (or each major step an actor initiates).
- Add flows from each action to another action, a decision point, or an end point. For maximum precision of meaning, each action should have only one flow coming in and one flow going out, with all forks, joins, decisions, and merges shown explicitly.
- Add decisions where flows diverge with alternating routes. Be sure to bring them back together with a merge.
- Add forks and joins where activities are performed in parallel.
- End with a single notation for activity final.

> Drawing System Sequence Diagrams

system sequence diagram a diagram that depicts the interaction between an actor and the system for a use case scenario.

Another tool used by some OO methodologists in the logical design phase is the **system sequence diagram.** As discussed earlier, a sequence diagram depicts how objects interact with each other via messages in the execution of a use case or operation. We have not yet started analyzing the individual object classes; that will come next as we build our first version of the class diagram. For now we are still thinking about the system as a whole.

As we have said, the object-oriented world is driven by messages sent between objects. A system sequence diagram helps us begin to identify the high-level messages that enter and exit the system. Later these messages will become the responsibility of individual objects, which will fulfill those responsibilities by communicating with other objects. We will save that for Chapter 18.

Figure 10-16 shows a system sequence diagram for the *Place New Order* use case. Note that the system sequence diagram does not include any of the alternative courses of the use case. It depicts a single scenario, a single path through the use case. So a full set of system sequence diagrams might have several diagrams for a single use case.

Figure 10-16 illustrates the following system sequence diagram notations:

❶ *Actor*—the initiating actor of the use case is shown with the use case actor symbol.

❷ *System*—the box indicates the system as a "black box" or as a whole. The colon (:) is standard sequence diagram notation to indicate a running "instance" of the system.

❸ *Lifelines*—the dashed vertical lines extending downward from the actor and system symbols, which indicate the life of the sequence.

❹ *Activation bars*—the bars that are set over the lifelines indicate the period of time when the participant is active in the interaction. Some methodologists leave them off the system sequence diagram, but we have included them to be consistent with the full sequence diagram.

❺ *Input messages*—horizontal arrows from the actor to the system indicate the message inputs. The UML convention for messages is to begin the first word with a lowercase letter and append additional words with an initial uppercase letter and no space. In parentheses include any parameters that you know at this point, following the same naming convention and separating individual

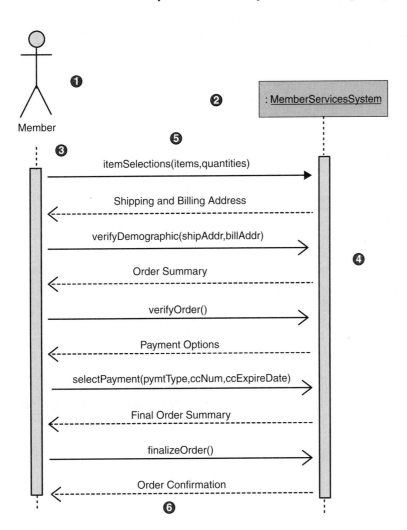

FIGURE 10-16

System Sequence Diagram for *Place New Order* Use Case

parameters with commas. You might wonder how the user will pass these messages. The answer is that the user will interact with the user interface, which will pass the messages for the user in the appropriate format. We'll say more about that in Chapter 18.

❻ *Output messages*—horizontal arrows from the system to the actor are shown as dashed lines. Since these take the form of Web forms, reports, e-mails, etc. these messages do not need to use the standard notation, though you can if you want.

Figure 10-17 is a system sequence diagram for a login validation illustrating the following additional notations:

❼ *Receiver Actor*—other actors or external systems that receive messages from the system can be included.

❽ *Frame*—a box can enclose one or more messages to divide off a fragment of the sequence. These can show loops, alternate fragments, or optional (opt) steps. For an optional fragment, the condition shown in square brackets indicates the conditions under which the steps will be performed.

> Guidelines for Constructing System Sequence Diagrams

- Identify which scenario of the use case you will depict. The purpose of the diagram is to discover messages, not to model logic. So though you can include optional and alternate messages for an entire use case, it is more important to clearly communicate a single scenario.

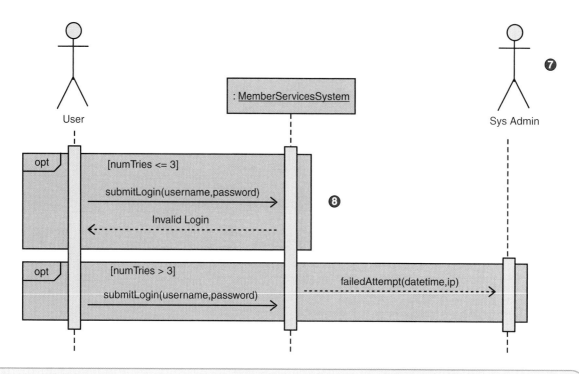

FIGURE 10-17 System Sequence Diagram for Login Validation

- Draw a rectangle representing the system as a whole and extend a lifeline under it.
- Identify each actor who directly provides an input to the system or directly receives an output from the system. Extend lifelines under the actor(s).
- Examine the use-case narrative to identify system inputs and outputs. Ignore messages inside the system. Draw each external message as a horizontal arrow from the actor's lifeline to the system or from the system to the actor. Label inputs according to UML convention, which will help identify behaviors and attributes in business objects.
- Add frames to indicate optional messages with conditions. Frames can also indicate loops and alternate fragments (these will be discussed in Chapter 18).
- Confirm that the messages are shown in the proper sequence from top to bottom.

> Finding and Identifying the Business Objects

In trying to identify objects, many methodology experts recommend the technique of searching the requirements document or other associated documentation and underlining the nouns that may represent potential objects. This could be a monumental task. There are just too many nouns. Use-case modeling provides a solution to this problem by breaking down the entire scope of a system into use cases. This abridged format simplifies the technique and makes underlining the nouns more efficient. Let's now examine the steps involved to identify and find business objects for object modeling during systems analysis.

Step 1: Find the Potential Objects This step is accomplished by reviewing each use case to find nouns that correspond to business entities or events. For example, Figure 10-18 depicts the use case *Place New Order* with all the nouns highlighted. Each noun that is found in reviewing the use case is added to a list of potential objects that will be analyzed further (see Figure 10-19).

Member Services System

Author(s): _____

Date: _____

Version: _____

Use Case Name:	Place New Order	Use Case Type	
Use Case ID:	MSS-SUC002.00	Business Requirements:	☐
Priority:	High	System Analysis:	☑
Source:	Requirement — MSS-R1.00 Requirements Use Case — MSS-BUC002.00		
Primary Business Actor:	Club Member (Alias — Active Member, Member)		
Primary System Actor:	Club Member (Alias — Active Member, Member)		
Other Participating Actors:	• Warehouse (Alias — Distribution Center) (external receiver) • Accounts Receivable (external server)		
Other Interested Stakeholders:	• Marketing — interested in sales activity in order to plan new promotions. • Procurement — interested in sales activity in order to replenish inventory. • Management — interested in order activity in order to evaluate company performance and customer (member) satisfaction.		
Description:	This use case describes the event of a member submitting a new order for SoundStage products via the World Wide Web. The member selects the items he or she wishes to purchase. Once the member has completed shopping, the member's demographic information as well as account standing will be validated. Once the products are verified as being in stock, a packing order is sent to the distribution center for it to prepare the shipment. For any product not in stock, a back order is created. On completion, the member will be sent an order confirmation.		
Precondition:	The individual submitting the order must be an active club member. The member must log in to the system (provide identification) to enter an order.		
Trigger:	This use case is initiated when the member selects the option to enter a new order.		

Typical Course of Events:	Actor Action	System Response
	Step 1: The member requests the option to enter a new order.	**Step 2:** The system responds by displaying the catalog of the SoundStage products.
	Step 3: The member browses the available items and selects the ones he or she wishes to purchase, along with the quantity.	**Step 4:** Once the member has completed the selections, the system retrieves from file and presents the member's demographic information (shipping and billing addresses).
	Step 5: The member verifies demographic information (shipping and billing addresses). If no changes are necessary, the member responds accordingly (to continue).	**Step 6:** For each product ordered, the system verifies the product availability and determines an expected ship date, determines the price to be charged to the member, and determines the cost of the total order. If an item is not immediately available, it indicates the product is back-ordered or that it has not been released for shipping (for preorders). If an item is no longer available, that is indicated also. The system then displays a summary of the order to the member for verification.
	Step 7: The member verifies the order. If no changes are necessary, the member responds accordingly (to continue).	**Step 8:** The system checks the status of the member's account. If satisfactory, the system prompts the member to select the desired payment option (to be billed later or pay immediately with a credit card).
	Step 9: The member responds by selecting the desired payment option.	**Step 10:** The system displays a summary of the order, including the desired payment option, to the member for verification.
	Step 11: The member verifies the order. If no changes are necessary, the member responds accordingly (to continue).	**Step 12:** The system records the order information (including back orders if necessary).

FIGURE 10-18 Sample Use-Case Narrative with Nouns Highlighted

		Step 13: Invoke abstract use case *MSS-AUC001.00, Determine Appropriate Distribution Center and Release Order to Be Filled.*
		Step 14: Once the order is processed, the system generates an order confirmation and displays it to the member as well as sending it to the member via e-mail.
Alternate Courses:		**Alt-Step 3:** The member enters search criteria to retrieve a specific item or to display a reduced list of items to browse and order from.
		Alt-Step 5: If changes are required, the member updates the appropriate shipping, billing, or e-mail addresses and tells the system to store them accordingly. The system will validate the changes and, if successful, will store the new information to file.
		Alt-Step 7: If the order requires changes, the member can delete any item no longer wanted or change the order quantity. Once the member has completed the order changes, the system reprocesses the order (**go to step 6**). If the member requests to do additional shopping, **go to step 3**. If the member needs to change the demographic information, **go to step 5**.
		Alt-Step 8: If the member's account is not in good standing, display to the member the account status, the reason the order is being held, and what actions are necessary to resolve the problem. In addition, an e-mail is sent to the member with the same information. The system prompts the member to hold the order for later processing or cancel the order. If the member wishes to hold the order, the system records the order information and places it in hold status and then displays the SoundStage main page. If the member chooses to cancel the order, the system clears the inputted information and then displays the SoundStage main page. Terminate the use case.
		Alt-Step 10: If the member selects the option to pay by credit card, the system prompts the member to enter the credit card information (number and expiration date) and reminds the member that the billing address on file must match the billing address of the credit card provided. The member enters the required information and requests that the system continue. The system validates the credit card account provided. If the account cannot be validated, the system notifies the member and requests an alternative means of payment. If the member cannot provide an alternative means at this time, he or she can choose either to hold or to cancel the order. If the member wishes to hold the order, the system records the order information and places it in hold status and then displays the SoundStage main page. If the member chooses to cancel the order, the system clears the inputted information and then displays the SoundStage main page. Terminate the use case.
		Alt-Step 11: If the order requires changes, the member can delete any item no longer wanted or change the order quantity. Once the member has completed the order changes, the system reprocesses the order (**go to step 6**). If the member requests to do additional shopping, **go to step 3**. If the member needs to change the demographic information, **go to step 5**.
		Alt-Step 12: If all items ordered are on back order, the order is not released to the distribution center.
Conclusion:		This use case concludes when the member receives a confirmation of the order.
Postcondition:		The order has been recorded and, if the ordered products were available, released to the distribution center. For any product not available a back order has been created.
Business Rules:		• Member must have a valid e-mail address to submit online orders. • Member is billed for products only when they are shipped.
Implementation Constraints and Specifications:		• Use case must be available to the member 24 × 7. • Frequency — It is estimated that this use case will be executed 3,500 times per day. It should support up to 50 concurrent members.
Assumptions:		• Product can be transferred among distribution centers to fill orders. • Procurement will be notified of back orders by a daily report (separate use case). • The member responding to a promotion or using credits may affect the price of each ordered item. • The member can cancel the order at any time.
Open Issues:		None

FIGURE 10-18 (Concluded)

FIGURE 10-19 Member Services System Potential Object List

Accounts receivable	Member Account Standing
Actions	Member's Account Status
Active Member	New Order
Available Items	New Promotions
Back Order	Option
Back-Ordered	Order Activity
Billing Addresses	Order Confirmation
Catalog	Order Total Cost
Club Member	Ordered Products
Company Performance	Packing Order
Credit Card	Payment Option
Credit Card Expiration Date	Preorders
Credit Card Number	Price to Be Charged
Credits	Problem
Customer Satisfaction	Procurement
Daily Report	Product Availability
Demographic Information	Product Ordered
Distribution Center	Promotion
E-Mail	Purchase
E-Mail Addresses	Quantity
Event	Reason
Expected Ship Date	Requirement
External Receiver	Requirements Use Case
External Server	Sales Activity
File	Search Criteria
Hold Status	Selections
Identification	Shipment
In Stock	Shipping Address
Individual	Shopping
Inventory	SoundStage Products
Items	Summary of the Order
Main Page	System
Management	Warehouse
Marketing	World Wide Web
Member	

Step 2: Select the Proposed Objects   Not all of the candidates (nouns) on our list represent useful business objects that are within the scope of our problem domain. By analyzing each candidate and asking the following questions, we can determine whether the candidate should stay or be removed from the list:

- Is the candidate a synonym of another object? In other words, is it really the same object with a different name?
- Is the candidate outside the scope of the system?
- Is the candidate a role without unique behavior, or is it an external role?
- Is the candidate unclear and in need of focus?
- Is the candidate an action or an attribute that describes another object?

If you answer yes to any of the questions above during the analysis of a candidate, the candidate should be removed from the list. If you find that any of the candidates are attributes, make sure you record them on a separate list so that they won't be forgotten. They are used later in the process to construct the class diagram. If you are unsure about a particular candidate, it is better to leave the candidate on the list; it is much easier to remove candidates later if we determine they are not objects than it would be to add them back after the class diagram has been constructed.

Figure 10-20 shows the process of cleaning up our list of candidate objects. An "**x**" marks the candidates we are discarding, and a "✓" marks the candidates we are keeping as objects. Also listed is the explanation of why we are keeping or discarding each candidate. Finally, Figure 10-21 presents the results of our cleaning-up process, as well as other objects discovered from the other use cases.

> Organizing the Objects and Identifying Their Relationships

class diagram a graphical depiction of a system's static object structure, showing object classes that the system is composed of as well as the relationships between those object classes.

Now that we have identified the business objects of the system, it is time to organize those objects and document any major conceptual relationships between the objects. A **class diagram** is used to graphically depict the objects and their associations. On this diagram we will also include multiplicity, generalization/specialization relationships, and aggregation relationships.

Step 1: Identifying Associations and Multiplicity   In this step, we need to identify associations that exist between object classes. Recall that an association between two object classes is what one object "needs to know" about the other. This allows for one object class to cross-reference another and to be able to send it messages. Once the associations have been identified, the multiplicity that governs the association must be defined.

It is very important that the analyst identify not only associations that are obvious or recognized by the users. One way to help ensure that possible relationships are identified is to use an object class matrix. This matrix lists the object classes as column headings as well as row headings. The matrix can then be used as a checklist to ensure that each object class appearing on a row is checked against *each* object class appearing in a column for possible associations. The name of the association and the multiplicity can be recorded directly in the intersection cell of the matrix. Figure 10-22 is a matrix that includes a sample of the proposed objects of the Member Services System. To interpret the contents of the cells, start with the object on the left (heading of row), read the contents of the cell, and then finish with the object at the top of the column. For example:

- A CLUB MEMBER places zero to many MEMBER ORDERS.
- A CLUB MEMBER has purchased zero to many MEMBER ORDERED PRODUCTS.
- A CLUB MEMBER and PRODUCT have no association between them.
- A MEMBER ORDER is placed by one and only one CLUB MEMBER.
- And so on . . .

Step 2: Identifying Generalization/Specialization Relationships   Once we have identified the basic associations and their multiplicity, we must determine if any generalization/specialization relationships exist. Recall that generalization/specialization relationships, also known as classification hierarchies or "is a" relationships, consist of supertype (*abstract* or *parent*) classes and subtype (*concrete* or *child*) classes. The supertype class is general in that it contains the common attributes and behaviors of the hierarchy. The subtype class is specialized in that it contains attributes and behaviors unique to the object but it inherits the supertype class's attributes and behaviors.

FIGURE 10-20 Analyzing the Potential Object List

Potential Object		Reason
Accounts Receivable	X	Not relevant for current project
Actions	X	Needs better focus—probably will be a comments attribute in MEMBER ORDER
Active Member	✓	Type of MEMBER
Available Items	X	Synonym of PRODUCT
Back Order	X	Responsibility of Procurement system - Not relevant for current project
Back-Ordered	X	Responsibility of Procurement system - Not relevant for current project
Billing Addresses	✓	Type of ADDRESS
Catalog	X	Same as PRODUCT. Potential Interface item to be addressed in object-oriented design
Club Member	✓	Type of MEMBER
Company Performance	X	Not relevant for current project
Credit Card	✓	CREDIT CARD ACCOUNT
Credit Card Expiration Date	X	Attribute of CREDIT CARD ACCOUNT
Credit Card Number	X	Attribute of CREDIT CARD ACCOUNT
Credits	X	Attribute of MEMBER
Customer Satisfaction	X	Not relevant for current project
Daily report	X	Potential Interface item to be addressed in object-oriented design
Demographic Information	X	Attribute of MEMBER
Distribution Center	✓	DISTRIBUTION CENTER
E-Mail	X	Potential Interface item to be addressed in object-oriented design
E-Mail Addresses	✓	Type of ADDRESS
Event	X	Not relevant for current project
Expected Ship Date	X	Attribute of MEMBER ORDERED PRODUCT
External Receiver	X	Not relevant for current project
External Server	X	Not relevant for current project
File	X	Not relevant for current project
Hold Status	X	Attribute of MEMBER ORDER
Identification	X	Attribute of MEMBER
In Stock	X	Attribute of PRODUCT
Individual	X	Synonym of MEMBER
Inventory	X	Attribute of PRODUCT
Items	X	Synonym of PRODUCT
List	X	Potential Interface item to be addressed in object-oriented design
Main Page	X	Potential Interface item to be addressed in object-oriented design
Management	X	Not relevant for current project
Marketing	X	Not relevant for current project
Member	✓	MEMBER
Member Account Standing	X	Attribute of MEMBER
Member's Account Status	X	Attribute of MEMBER
New Order	✓	MEMBER ORDER
New Promotions	✓	PROMOTION

FIGURE 10-20 (Concluded)

Option	X	Potential Interface item to be addressed in object-oriented design
Order Activity	X	Potential Interface item to be addressed in object-oriented design (report)
Order Confirmation	X	Potential interface item to be addressed in object-oriented design
Order Total Cost	X	Attribute of MEMBER ORDER
Ordered Products	✓	MEMBER ORDERED PRODUCT
Packing Order	X	Potential Interface item to be addressed in object-oriented design
Payment Option	X	Attribute of MEMBER ORDER
Preorders	✓	Type of MEMBER ORDER
Price to Be Charged	X	Attribute of MEMBER ORDERED PRODUCT
Problem	X	Needs better focus—probably will be a comments attribute in MEMBER ORDER
Procurement	X	Not relevant for current project
Product Availability	X	Attribute of PRODUCT
Product Ordered	X	Synonym of MEMBER ORDERED PRODUCT
Promotion	✓	PROMOTION
Purchase	X	Synonym of MEMBER ORDER
Quantity	X	Attribute of MEMBER ORDERED PRODUCT
Reason	X	Needs better focus—probably will be a comments attribute in MEMBER ORDER
Requirement	X	Not relevant for current project
Requirements Use Case	X	Not relevant for current project
Sales Activity	X	Potential Interface item to be addressed in object-oriented design (report)
Search Criteria	X	Potential Interface item to be addressed in object-oriented design
Selections	X	Synonym of MEMBER ORDERED PRODUCT
Shipment	X	Not relevant for current project—responsibility of shipping and receiving
Shipping Address	✓	Type of ADDRESS
Shopping	X	Potential Interface item to be addressed in object-oriented design
SoundStage Products	X	Synonym of PRODUCT
Summary of the Order	X	Potential interface item to be addressed in object-oriented design
System	X	Not relevant for current project
Warehouse	X	Synonym of DISTRIBUTION CENTER
World Wide Web	X	Potential Interface item to be addressed in object-oriented design

Generalization/specialization relationships may be discovered by looking at the class diagram. Do any associations exist between two classes that have a one-to-one multiplicity? If so, can you say the sentence "Object X *is a* type of object Y" and it be true? If it is true, you may have a generalization/specialization relationship. Also look for classes that have common attributes and behaviors. It may be possible to combine the common attributes and behaviors into a new supertype class. Why do we want

FIGURE 10-21 Member Services System Proposed Object List

Proposed Object List

ACTIVE MEMBER
BILLING ADDRESS
CLUB MEMBER
CREDIT CARD ACCOUNT
DISTRIBUTION CENTER
E-MAIL ADDRESS
MEMBER
MEMBER ORDER
PROMOTION
MEMBER ORDERED PRODUCT
PREORDER
PRODUCT
SHIPPING ADDRESS

PLUS

AGREEMENT
AUDIO TITLE
FORMER MEMBER
GAME TITLE
INACTIVE MEMBER
MERCHANDISE
RETURN
TITLE
TRANSACTION
VIDEO TITLE

	CLUB MEMBER	**MEMBER ORDER**	**MEMBER ORDERED PRODUCT**	**PRODUCT**
CLUB MEMBER		Places zero to many	Has purchased zero to many	XX
MEMBER ORDER	Is placed by one and only one		Contains one to many	XX
MEMBER ORDERED PRODUCT	Was purchased by one and only one	Is part of one and only one		Relates to one and only one
PRODUCT	XX	XX	Sold as zero to many	

FIGURE 10-22 Sample Object Association Matrix

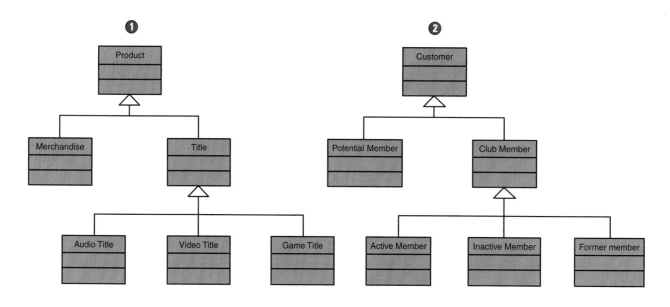

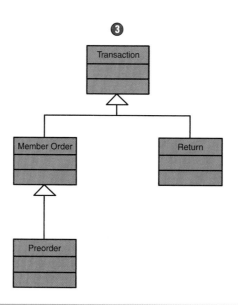

FIGURE 10-23 Generalization/Specialization Hierarchies in the Member Services System

generalization/specialization relationships? They allow us to take advantage of inheritance, which facilitates the reuse of objects and programming code.

Please draw your attention to Figure 10-23. When analyzing the class diagram, we identified three generalization/specialization hierarchies:

❶ A PRODUCT hierarchy that allows us to keep track of all SoundStage products that can be purchased and enables us to add different types of products in the future, such as BOOK TITLES.

❷ A CUSTOMER hierarchy that allows us to keep track of all MEMBERS (past, present, and potential). It allows us to send special promotions to inactive members to encourage them to start ordering products again. It also allows us to identify former members who terminated their membership or whose membership was terminated because their account was in bad standing, as in the case of members who defaulted on their agreement. It also enables us to add different types of customers in the future, such as corporate customers.

❸ A TRANSACTION hierarchy that allows us to keep track of the various transactions CUSTOMERS conduct. Currently, MEMBER ORDERS, PREORDERS, and RETURNS are recorded, but the hierarchy could be modified to include reservations of TITLES to be released in the future.

Step 3: Identifying Aggregation/Composition Relationships

In this step, we must determine if any basic aggregation or composition relationships exist. Recall that aggregation is a unique type of relationship in which one object "is part of" another object. It is often referred to as a *whole/part relationship* and can be read as "Object A *contains* object B and object B *is part of* object A." Aggregation relationships are asymmetric, in that object B is part of object A but object A is not part of object B. These relationships do not imply inheritance, in that object B does not inherit behavior or attributes from object A. However, behavior applied to the whole is automatically applied to the parts. For example, if I want to send object A to a customer, object B would be sent also.

When analyzing the class diagram, we identified one composition relationship, the relationship between a MEMBER ORDER and the ORDERED PRODUCTS it contains.

Step 4: Prepare the Class Diagram

Figure 10-24 is a partial UML class diagram for the Member Services System.[3] Notice that the model depicts business object classes within the domain of the SoundStage Member Services system. The object/class notation on the model does not depict behaviors (methods). These will be identified and defined in Chapter 18.

The model also reflects the associations and multiplicity that were identified in step 1, three generalization/specialization relationships that were discovered in step 2, and one aggregation/composition relationship discovered in step 3. Notice at the bottom of each class the word **persistent** appears. Typically, this means that the objects the class describes will be stored permanently in a database. All business domain classes tend to be persistent. Objects that are created temporarily by a software program are called **transient objects.** In an object-oriented programming language, all code exists inside an object class. So there is class code for the user interface and for controlling the system. These transient objects are created while the program is running and discarded later when no longer needed. Transient objects are usually modeled during object-oriented design, which we will cover in Chapter 18.

> **persistent class** a class that describes an object that outlives the execution of the program that created it.

> **transient object class** a class that describes an object that is created temporarily by the program and lives only during that program's execution.

Finally, if you have learned data modeling from Chapter 8 or a previous course, you should note the following differences between object class attributes and data entity attributes

- There is no need in a class diagram to include a primary key attribute unless it is a real business attribute. For example, the Product class has attributes for productNumber and UPC. While either of them could be a primary key if this class's data was stored in a relational database, neither of them are included for that purpose. They are included because that information is relevant to one or more use cases. An artificial primary key, such as an auto-incrementing ID value, would not be included in a class diagram.

- There is no need for foreign keys in a class diagram. If the data attributes in the class will eventually be stored in a relational database, then the database will have foreign keys. But in object-oriented programming languages, there will be transient objects between your business class and the database that will handle that, so there is no need to include them in your business class. We will learn more about those transient objects in Chapter 18.

---

[3] The diagram was constructed using Popkin Software's *System Architect.*

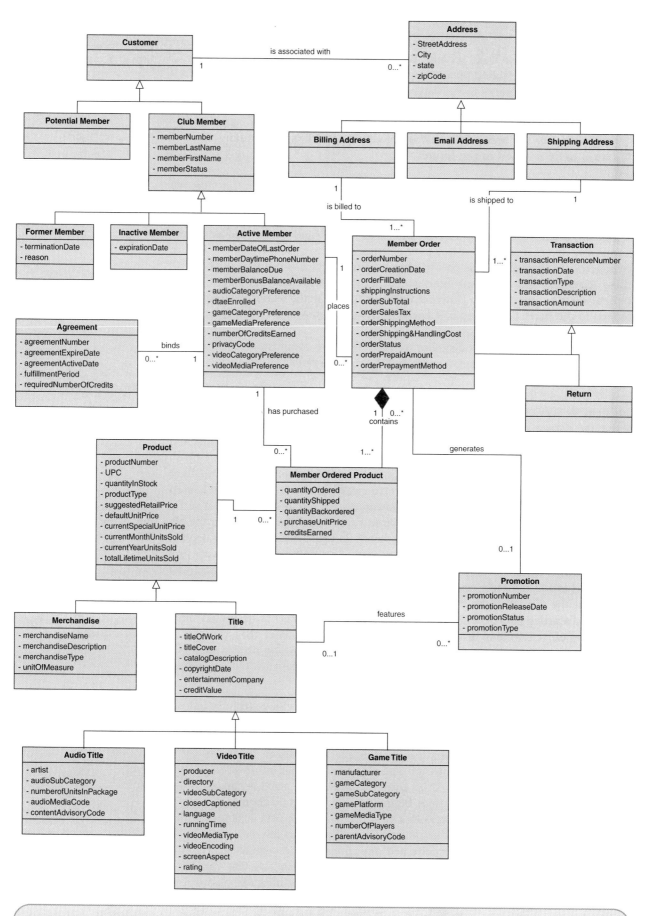

FIGURE 10-24 Member Services System Class Diagram

This chapter introduced the newer object-oriented approach to systems development. Specifically, this chapter focused on object modeling tools and techniques for systems analysis. You are now ready to learn about the object-oriented approach as it applies to systems design. Object-oriented design is covered in Chapter 18. In that chapter, you will learn how the object models developed in this chapter are expanded to include design decisions for a new system.

 ## Chapter Review

1. The approach of using object modeling during systems analysis and design is called object-oriented analysis (OOA). Object-oriented analysis techniques are used (1) to study existing objects to see if they can be reused or adapted for new uses and (2) to define new or modified objects that will be combined with existing objects into a useful business computing application.

2. The object-oriented approach is centered around a technique referred to as object modeling. Object modeling is a technique for identifying objects within the systems environment and identifying the relationships between those objects.

3. There are many underlying concepts for object modeling, including:

   a. Systems consist of objects, where an object is something that is or is capable of being seen, touched, or otherwise sensed and about which users store data and associate behavior. The data, or attributes, represent characteristics of interest about an object. The behavior of an object refers to those things that the object can do and that correspond to functions that act on the object's data (or attributes). Each object encapsulates the attributes and behavior together as a single unit.

   b. Objects can be categorized into classes. A class is a set of objects that share common attributes and behavior. Objects may be grouped into multiple levels of classes. The most general class in the grouping is the supertype (or generalization of the class). The more refined class is referred

   to as the subtype class (or specialization class). All subtype classes "inherit" the attributes and behavior of the supertype class.

   c. Objects and classes have relationships. A relationship is a natural business association that exists between one or more objects and classes. The degree, or multiplicity, of a relationship specifies the business rules governing the relationship. Some relationships are more "structural," meaning that a class may be related to another class in that one class may represent an assembly of one or more other class types. This type of relationship is referred to as an aggregation structure.

   d. Objects communicate by passing messages. A message is passed when one object invokes another object's behavior to request information or some action.

   e. A type of behavior may be completed differently for different objects/classes. This concept is referred to as polymorphism.

4. One of the most critical aspects of performing object-oriented development is correctly identifying the objects and their relationships early in the development process. Use-case modeling is a popular approach that assists in object identification.

5. In trying to identify objects, many methodology experts recommend the technique of searching the requirements document or other associated documentation and underlining the nouns that may represent potential objects. This could be a monumental task! There are just too many nouns.

Use-case modeling provides a solution to this problem by breaking down the entire scope of system functionality into many smaller statements of system functionality called use cases. This smaller format simplifies and makes more efficient the technique of underlining the nouns.

6. Activity diagrams are used to better understand the use-case logic in terms of the flow of steps and their sequencing.
7. A class diagram is used to organize the objects found as a result of use-case modeling and to document the relationships between the objects.

Review Questions

1. What is the most commonly accepted notation standard for object modeling?
2. *Object* is defined as "something that is or is capable of being seen, touched or otherwise sensed and about which users store data and associate behavior." Please explain what it means by *something, data,* and *behavior* in this definition.
3. What is encapsulation?
4. Consider that textbooks and cookbooks are both objects belonging to the class of *Book.* Please give an example of a class and its objects.
5. What is the relationship between inheritance and supertype/subtype?
6. In object-oriented analysis and modeling, objects and classes do not exist in isolation. Why is this?
7. How should analysts show the object or class relationship using UML?

8. In terms of aggregation relationships, what is the difference between the use of a hollow diamond and the use of a solid diamond?
9. What is polymorphism, and when is it applied?
10. What are the five groups of UML diagrams?
11. What are the differences between sequence diagrams and collaboration diagrams?
12. What are the three major activities in performing object-oriented analysis?
13. What is an activity diagram? When is the diagram used?
14. What are some ways to determine if a candidate object is useful and should be kept, or whether it should be discarded?
15. What are the steps in organizing the objects and identifying their relationships?

Problems and Exercises

1. Since its inception in 1997, the Unified Modeling Language (UML) has quickly gained wide acceptance and usage throughout the world.

   a. In terms of object modeling, what *does* UML provide to designers? What *doesn't* UML provide?
   b. What was the reason that UML was developed?
   c. What might object modeling look like today if UML had *not* been developed?

2. Object-oriented analysis (OOA) and object modeling have become familiar terms in many organizations, but their underlying concepts are not always intuitive and can be difficult to understand, especially by nontechnical users who are involved in a systems development project.

   a. In nontechnical terms, explain what an object is and what the object-oriented analysis approach is.
   b. Also in nontechnical terms, explain the technique of object modeling.
   c. What are the main differences between object-oriented analysis and traditional systems analysis in how they approach system development?

   d. Do you think it would be easier to learn object-oriented analysis methods if you were a systems designer experienced in traditional development methods, or if this was the first analysis method you were learning? Explain your answer.

3. Consider a movie DVD as an example of an object.

   a. Using the textbook's terminology, what type of object is a movie DVD?
   b. What are some of the attributes of a movie DVD?
   c. What is an object instance of a movie DVD?
   d. Represent the class of *Movie DVD* in an object model using UML notation, as shown in Figure 11-2. Include the class name, attributes, and behaviors.
   e. Would the object class of *Movie DVD* be a considered a supertype or subtype? Give examples.

4. For this exercise, consider a different example of an object—a dog.

   a. What type of object is a dog?
   b. What are some of the attributes of a dog?

c. Show an object instance of *Dog,* using Figure 10-1 as an example.

d. What are some of the behaviors of the class of *Dog?*

e. Represent the class of *Dog* in the UML, using Figure 10-2(b) as an example.

5. Again, consider the class of *Dog.*

a. Provide five or six examples of the association between the class of *Dog* and the class of *Person.*

b. Show the object class associations and multiplicity notations for the class of *Dog.*

c. What type of aggregation relationships might exist for the class of *Dog?*

6. Objects and classes can send messages to each other in order to interact.

a. Give an example of a message request from the object classes of *Dog* to *Person,* and the return behavior of *Person.*

b. In message sending, what *doesn't* the sending object need to know about the receiving object?

c. What needs to exist in order to be able to send a message between the two objects?

7. Polymorphism is a concept that is important to understand in object-oriented analysis. You need to explain the concept to the system users who are on the project team. In nontechnical terms:

a. Define the concept of polymorphism.

b. Explain how polymorphism is related to messaging.

c. Explain what overriding behaviors are.

8. You are teaching an introductory class in object-oriented analysis and design. Explain:

a. The different groups of UML diagrams, and what each group of diagrams depicts and/or models.

b. What use-case modeling identifies.

c. What the three major tasks are in conducting object-oriented analysis.

d. How and why business requirements use-case models are refined and changed into analysis use-case models.

9. During the design phase, abstract and extension use-case narratives are also developed.

a. Why are different narratives used for abstract and extension use cases?

b. What are some of the differences in documenting abstract and extension use-case narratives compared to documenting regular use cases?

c. Can an abstract use case be invoked by a single use case?

d. Is an extension use case reusable?

e. Can an extension use case be invoked by a single use case?

f. Can an abstract use case invoke a regular use case?

10. UML activity diagrams are used to model system process activities and to help system analysts visualize the flow and sequencing of use cases.

a. How are they different from flowcharts, and how is this difference useful?

b. How are they similar?

c. What does the solid black bar in an activity diagram represent?

11. In object-oriented analysis and modeling, it is extremely important to identify all potential objects. This can be accomplished, as suggested by a number of experts, by going through the requirements documentation to find all the nouns, since each one can represent a possible object.

a. What is the problem with this method?

b. How does use-case modeling help identify potential objects?

c. Once potential (candidate) objects are identified, should each one become an actual object?

d. How should candidate objects be selected?

e. What if a candidate object turns out to actually be an attribute?

12. The last step in conducting the object-oriented analysis is organizing the objects and identifying their relationships.

a. Does a class diagram show the structure of a system as dynamic or static?

b. Why are associations between objects identified before defining multiplicity?

c. What is the purpose of an object class matrix?

d. If you have 72 objects and classes, how many empty (null) cells will there be in the matrix?

13. After identifying object class associations and multiplicity, you must perform several other steps before organization of the objects can be considered to be complete.

a. What are these other steps?

b. Why are generalization/specialization relationships important to identify during the design phase?

c. What are two techniques for identifying possible generalization/specialization relationships?

d. What is the essential difference between generalization/specialization relationships and aggregation relationships.

e. Can a business domain class contain a transient object?

Projects and Research

1. Since its introduction in 1997, the Unified Modeling Language (UML) has quickly become a commonly accepted standard and a widely used tool for object modeling. Go to *www.omg.org,* which is the Web site of the Object Management Group (OMG), the standards body for UML, and take a look at its UML Resource Page, as well as its links to other sites such as IBM and Popkin Software.

   a. What is the most current version of UML?
   b. What is the Object Management Group?
   c. In reviewing some of the historical articles on the Web site, why do you think that UML became a leading tool for object modeling so quickly?
   d. Review the specifications for UML 2.0 (they are available for download or viewing free of charge)—what did you find most interesting and/or valuable about the new version?
   e. Many languages used in information technology have come and gone, but certain ones, such as COBOL, are still in wide use decades after they were introduced. Based upon the articles available on the Web site, what do you think will be the life span for UML? Why?
   f. Does UML have any new or emerging competitors at this time?

2. Find and talk to a system designer who is experienced in modeling with UML.

   a. What types of systems does the designer work on with UML?
   b. Does the designer use UML via a CASE tool? If so, which one?
   c. What does the designer like best about UML?
   d. What does the designer like least?
   e. If the designer could choose a modeling language, would it be UML? Why or why not?
   f. What features would the designer like to see added to UML?

3. You are currently working as a freelance systems designer and have been asked to do some of the design work for a case-tracking information system that is being developed for a local law firm which specializes in civil cases. The business objective is to implement a system that tracks a civil case from the time that the law firm begins working on the lawsuit through its final adjudication.

   What are some of the main objects and classes you would expect to find in a law firm that specializes in civil filings?

   a. Describe each these classes, including their name and attributes, using Figure 10-2 as an example.

   b. Describe any generalization/specialization relationships, using Figure 10-4 as an example.
   c. Identify the object/class associations, and create an object/class association table, using Figure 10-5 as an example. Make sure to include the multiplicity for each association.
   d. Identify the aggregation relationships, and prepare an aggregation relationship table, using Figure 10-6 as an example.

4. The law firm likes your work and wants to extend your contract so you can continue designing its case-tracking system. Assuming you negotiate an acceptable rate, for your next tasks:

   a. Create at least two detailed use-case narratives, using Figure 10-11 as an example.
   b. Create an abstract use-case narrative, using Figure 10-12 as an example.
   c. Create an activity diagram for each of these use case narratives, using Figure 10-13 as an example.

5. At this point, you want to make sure that you have included everything and haven't left anything out. So for your next task:

   a. Find potential objects, using the techniques described in the textbook.
   b. Create the potential (candidate) object list, using Figure 10-16 as an example, and determine whether each candidate should be kept or discarded.
   c. Create a proposed object list, using Figure 10-17 as an example.
   d. Create an object association matrix, using Figure 10-18 as an example, and identify the associations and multiplicity that exist between objects and classes.
   e. Did you find objects and classes that you had not previously identified?

6. You are almost done with your object-oriented analysis and modeling of the case-tracking system for the law firm. Based upon your design work in the preceding questions, your final tasks are to:

   a. Create a generalization/specialization hierarchies diagram, using Figure 10-19 as an example.
   b. Create a class diagram, using Figure 10-20 as an example.
   c. At this point, if you could choose between using object-oriented analysis and modeling with UML, or a traditional structured design method, which would you choose? Why?

Minicases

1. Take the information you have gathered and your assessment of needs *so far* and suggest a system to meet the department's current needs, as well as future needs and opportunities. Prepare a paper that includes a situation background, an overview of the system you are suggesting, and the specific technological requirements of said system. Your paper should be no more than 12 pages long (1.5 spacing).

2. Prepare a full feasibility analysis, including Economic, Operational, Schedule, Legal, and Technical analyses for the system you are suggesting in problem 1. Your analysis should be no more than 30 pages long (1.5 spaced).

3. Create the use-case descriptions and diagram(s) for the system in problems 1 and 2. Be sure to create your use cases so that they are complete and clear. Remember, in real life, the system analyst/designer is often not the person who develops the system, and, in fact, the teams rarely meet at all. Clarity and completeness are essential.

4. Prepare a presentation on the material from problems 1–3 and present it to your class. Utilize interesting presentation media (such as video, sound, etc.).

Suggested Readings

Ambler, Scott W. *The Object Primer.* New York: Cambridge University Press, 2001. Very good information about documenting use cases and their use.

Armour, Frank, and Granville Miller. *Advance Use Case Modeling.* Boston: Addison-Wesley, 2001. This book presents excellent coverage of the use-case modeling process.

Booch, G. *Object-Oriented Design with Applications.* Menlo Park, CA: Benjamin Cummings, 1994. Many Booch concepts were integrated into the UML.

Coad, P., and E. Yourdon. *Object-Oriented Analysis,* 2nd ed. Englewood Cliffs, NJ: Prentice Hall, 1991. This book provides a very good overview of object-oriented concepts. However, the object model techniques are somewhat limited in comparison to UML and other object-oriented modeling approaches.

Eriksson, Hans-Erik, and Magnus Penker. *UML Toolkit.* New York: John Wiley & Sons, 1998. This book provides detailed coverage of the UML.

Fowler, Martin, and Kendall Scott. *UML Distilled—Applying the Standard Object Modeling Language.* Reading, MA: Addison-Wesley, 1997. A good short guide introducing the concepts and notation of the UML.

Harman, Paul, and Mark Watson. *Understanding UML—The Developer's Guide.* San Francisco: Morgan Kaufmann Publishers, 1997. This is an excellent reference book. The examples were prepared using Popkin's *System Architect.*

Jacobson, Ivar; Magnus Christerson; Patrik Jonsson; and Gunnar Overgaard. *Object-Oriented Software Engineering—A Use Case Driven Approach.* Wokingham, England: Addison-Wesley, 1992. This book presents detailed coverage of how to identify and document use cases.

Larman, Craig. *Applying UML and Patterns—An Introduction to Object-Oriented Analysis and Design.* Englewood Cliffs, NJ: Prentice Hall, 1997. This is an excellent reference book explaining the concepts of OO development utilizing the UML.

Martin, J., and J. Odell. *Object-Oriented Analysis and Design.* Englewood Cliffs, NJ: Prentice Hall, 1992.

Rumbaugh, James; Michael Blaha; William Premerlani; Frederick Eddy; and William Lorensen. *Object-Oriented Modeling and Design.* Englewood Cliffs, NJ: Prentice Hall, 1991. This book presents detailed coverage of the object modeling technique and its application throughout the entire systems development life cycle. Many OMT constructs are now in the UML.

Rumbaugh, James; Ivar Jacobson; and Grady Booch. *The Unified Modeling Language Reference Manual.* Reading, MA: Addison-Wesley, 1999. This book presents detailed coverage of the UML by the primary authors who created it.

Rumbaugh, James; Ivar Jacobson; and Grady Booch. *The Unified Modeling Language Users Guide.* Reading, MA: Addison-Wesley, 1999. This book presents detailed coverage of the UML by the primary authors who created it.

Taylor, David A. *Object-Oriented Information Systems—Planning and Implementation.* New York: John Wiley & Sons, 1992. This book is a very good entry-level resource for learning the concepts of object-oriented technology and techniques.

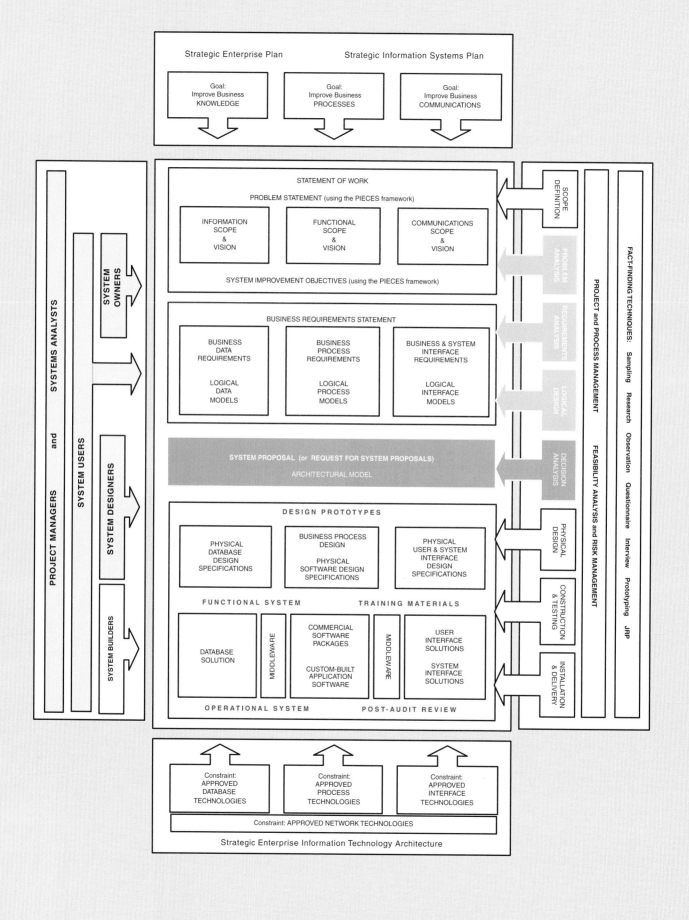

11

Feasibility Analysis and the System Proposal

Chapter Preview and Objectives

Good systems analysts thoroughly evaluate alternative solutions before proposing change. In this chapter you will learn how to analyze and document those alternatives on the basis of four feasibility criteria: operational, technical, schedule, and economic. You will also learn how to make a system proposal in the form of a written report and a formal presentation. You will know that you understand the feasibility analysis and recommendation skills needed by the systems analyst when you can:

▌ Identify feasibility checkpoints in the system's life cycle.

▌ Identify alternative system solutions.

▌ Define and describe six types of feasibility and their respective criteria.

▌ Perform various cost-benefit analyses using time-adjusted costs and benefits.

▌ Write suitable system proposal reports for different audiences.

▌ Plan for a formal presentation to system owners and users.

Introduction

As all the analysis has been going on for the SoundStage Member Services system project, Bob Martinez has been getting more and more excited about it. The programmer in Bob would like to jump in and start coding the information system. But Sandra, his boss, had him research packaged solutions on the market. They were pricey. But then Sandra ran the numbers for the labor costs of in-house programming. Bob realized that the packaged solutions weren't that expensive relatively and could be put in place a whole lot faster. There would still be programming to do, because the packaged solutions would need to be customized to meet all their requirements.

The final decision on which solution to select would be made by the steering committee that was overseeing the project. Sandra said the executives on the steering committee were currently very budget conscious. They would be scrutinizing the numbers and would approve the project to continue only if it showed a solid return on investment. Bob would have a small part in the system proposal presentation. He rehearsed and studied up on the facts to make sure he was ready for any question. He didn't want to blow it. He was surprised to realize he was now glad that some of his college courses required him to dress in a business suit and make a formal presentation.

Feasibilty Analysis and the System Proposal

In today's business world, it is becoming increasingly apparent that analysts must learn to think like business managers. Computer applications are expanding at a record pace. Now more than ever, management expects information systems to pay for themselves. Information is a major capital investment that must be justified, just as marketing must justify a new product and manufacturing must justify a new plant or equipment. Systems analysts are called on more than ever to help answer the following questions: Will the investment pay for itself? Are there other investments that will return even more on their expenditure?

This chapter deals with feasibility analysis issues of interest to the systems analyst and users of information systems. It also emphasizes the importance of making recommendations to management in the form of a system proposal that is a formal written report and/or oral presentation. As is illustrated in the chapter home page, feasibility analysis is appropriate to the systems analysis phases but particularly important to the decision analysis phase. The system proposal represents the deliverable and presents the technical KNOWLEDGE, PROCESS, and COMMUNICATION solution.

> Feasibility Analysis—A Creeping Commitment Approach

feasibility the measure of how beneficial or practical an information system will be to an organization.

feasibility analysis the process by which feasibility is measured.

Let's begin with a formal definition of feasibility and feasibility analysis. **Feasibility** is the measure of how beneficial or practical the development of an information system will be to an organization. **Feasibility analysis** is the process by which feasibility is measured.

Feasibility should be measured throughout the life cycle. In earlier chapters we called this a *creeping commitment* approach to feasibility. The scope and complexity of an apparently feasible project can change after the initial problems and opportunities are fully analyzed or after the system has been designed. Thus, a project that is feasible at one point may become infeasible later.

Figure 11-1 shows feasibility checkpoints during the systems analysis phases of our life cycle. The checkpoints are represented by red diamonds. The diamonds indicate that a feasibility reassessment and management review should be conducted at the end of the prior phase (before the next phase). A project may be canceled or revised at any checkpoint, despite whatever resources have been spent.

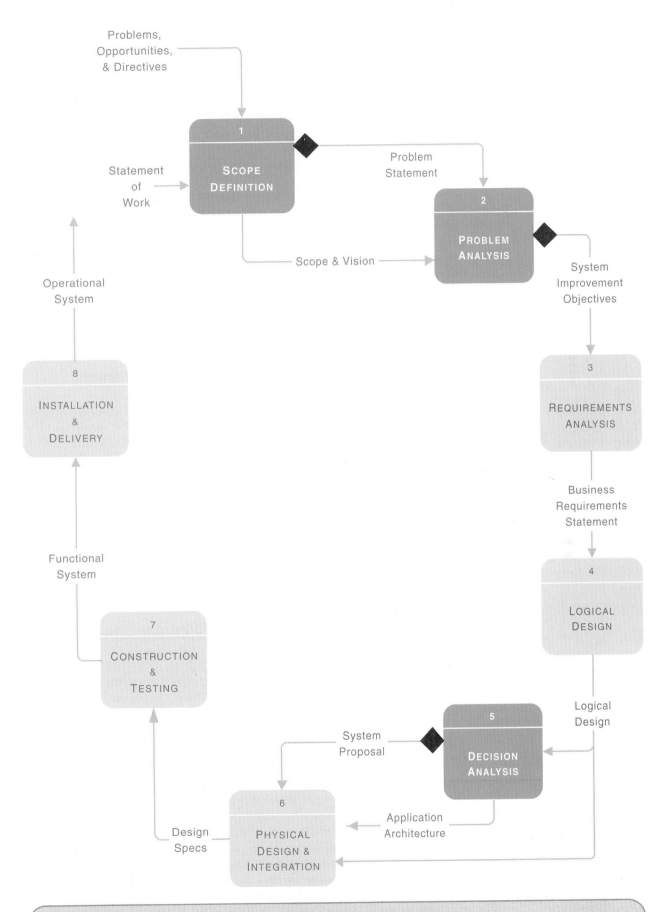

FIGURE 11-1 Feasibility Checkpoints during Systems Analysis

The idea of canceling a project is often difficult to face. A natural inclination may be to justify continuing a project based on the time and money that has already been spent. However, a fundamental principle of management is never to throw good money after bad—cut your losses and move on to a more feasible project. Deciding to cancel doesn't mean the costs already spent are not important. Costs must eventually be recovered if the investment is ever to be considered a success. Let's briefly examine the checkpoints in Figure 11-1.

> Systems Analysis—Scope Definition Checkpoint

The first feasibility analysis is conducted during the scope definition phase. At this early stage of the project, feasibility is rarely more than a measure of the urgency of the problem and the first-cut estimate of development costs. It answers the question, Do the problems (or opportunities) warrant the cost of a detailed study and analysis of the current system? Realistically, feasibility can't be accurately measured until the problems (and opportunities) and requirements are better understood.

After estimating the benefits of solving the problems and opportunities, analysts estimate the costs of developing the expected system. Experienced analysts routinely increase these costs by 50 to 100 percent (or more) because experience tells them the problems are rarely well defined and user requirements are typically understated.

> Systems Analysis—Problem Analysis Checkpoint

The next checkpoint occurs after a more detailed study and problem analysis of the current system. Because the problems are better understood, the analysts can make better estimates of development costs and of the benefits to be obtained from a new system. The minimum value of solving a problem is equal to the cost of that problem. For example, if inventory carrying costs are $35,000 over acceptable limits, then the minimum value of an acceptable information system would be $35,000. It is hoped an improved system will be able to do better than that; however, it must return this minimum value.

Development costs, at this point, are still just guesstimates. Analysts have yet to fully define user requirements or to specify a design solution to those requirements.

If the cost estimates significantly increase from the preliminary investigation phase to the problem analysis phase, the likely culprit is scope. Scope has a tendency to increase in many projects. If increased scope threatens feasibility, then scope might be reduced.

> Systems Design—Decision Analysis Checkpoint

The decision analysis phase represents a major feasibility analysis activity since it charts one of many possible implementations as the target for systems design.

Problems and requirements should be known by now. During the decision analysis phase, alternative solutions are defined in terms of their input/output methods, data storage methods, computer hardware and software requirements, processing methods, and people implications. The following list presents the typical range of options that can be evaluated by the analyst:

- Do nothing. Leave the current system alone. Regardless of management's opinion or your own opinion, this option should be considered and analyzed as a baseline option against which all others can and should be evaluated.
- Reengineer the (manual) business processes, not the computer-based processes. This may involve streamlining activities, reducing duplication and unnecessary tasks, reorganizing office layouts, and eliminating redundant and unnecessary forms and processes, among others.
- Enhance existing computer processes.
- Purchase a packaged application.
- Design and construct a new computer-based system.

After defining these options, each is analyzed for operational, technical, schedule, and economic feasibility. One alternative is recommended to system owners for approval and the basis for general and detailed design.

Six Tests for Feasibility

So far, we've defined feasibility and feasibility analysis, and we've identified feasibility checkpoints during systems analysis. Feasibility can be viewed from multiple perspectives. Below we present six categories of feasibility tests.

- *Operational feasibility* is a measure of how well a solution meets the identified system requirements to solve the problems and take advantage of the opportunities envisioned for the system.
- *Cultural (or political) feasibility* is a measure of how people feel about a solution and how well it will be accepted in a given organizational climate.
- *Technical feasibility* is a measure of the practicality of a specific technical solution and the availability of technical resources and expertise to implement and maintain it.
- *Schedule feasibility* is a measure of how reasonable the project timetable is.
- *Economic feasibility* is a measure of the cost-effectiveness of a project or solution.
- *Legal feasibility* is a measure of how well a solution can be implemented within existing legal and contractual obligations.

Actually, few systems are infeasible. Instead, different solution options tend to be more or less feasible than others. Let's take a closer look at the four feasibility criteria.

> Operational Feasibility

Operational feasibility is the measure of how well a proposed system solves the problems and takes advantage of the opportunities identified during the scope definition and problem analysis phases and how well it satisfies the system requirements identified in the requirements analysis phase. Operational feasibility also asks if, given what is now known about the problem and the cost of the solution, the problem is still worth solving. The PIECES framework (Chapter 3) can be used as the basis for analyzing the urgency of a problem or the effectiveness of a solution.

operational feasibility a measure of how well a solution meets the identified system requirements to solve the problems and take advantage of the opportunities envisioned for the system.

> Cultural (or Political) Feasibility

This is related to operational feasibility. But where operational feasibility deals more with how well the solution will meet system requirements, **cultural/political feasibility** deals with how the end users feel about the proposed system. You could say that operational feasibility evaluates whether a system *can* work, and cultural/political feasibility asks whether a system *will* work in a given organizational climate.

In an information age, knowledge is power. It is common for an information system to change the structure of how information is routed and controlled, changing to some extent the power structure of the organization. Some users and managers may feel threatened and fight implementation of the system.

Recognize that increasingly the culture of an organization is multicultural. Employees and divisions may have been merged in from different companies with widely varying perspectives on how work should be structured and what information systems should do and not do. With international organizations, the information system must also be accepted by multiple national cultures. The following questions address this concern:

cultural (or political) feasibility a measure of how well the solution will be accepted in a given organizational climate.

- Does management support the system?
- How do the end users feel about their role in the new system?

- What end users or managers may resist or not use the system? Can this problem be overcome? If so, how?
- How will the working environment of the end users change? Can or will end users and management adapt to the change?

> Technical Feasibility

technical feasibility a measure of the practicality of a technical solution and the availability of technical resources and expertise.

Today, very little is technically impossible. Consequently, technical feasibility looks at what is practical and reasonable. Technical feasibility addresses three major issues:

1. Is the proposed technology or solution practical?
2. Do we currently possess the necessary technology?
3. Do we possess the necessary technical expertise?

Is the Proposed Technology or Solution Practical? The technology for any defined solution is normally available. The question is whether that technology is mature enough to be easily applied to our problems. Some firms like to use state-of-the-art technology, but most firms prefer to use mature and proven technology. A mature technology has a larger customer base for obtaining advice concerning problems and improvements.

Do We Currently Possess the Necessary Technology? Assuming the solution's required technology is practical, we must next ask ourselves, Is the technology available in our information systems shop? If the technology is available, we must ask if we have the capacity. For instance, will our current printer be able to handle the new reports and forms required of a new system?

If the answer to either of these questions is no, then we must ask ourselves, Can we get this technology? The technology may be practical and available, and, yes, we need it. But we simply may not be able to afford it at this time. Although this argument borders on economic feasibility, it is truly technical feasibility. If we can't afford the technology, then the alternative that requires the technology is not practical and is technically infeasible!

Do We Possess the Necessary Technical Expertise? This consideration of technical feasibility is often forgotten during feasibility analysis. Even if a company has the technology, that doesn't mean it has the skills required to properly apply that technology. For instance, a company may have a database management system (DBMS). However, the analysts and programmers available for the project may not know that DBMS well enough to properly apply it. True, all information systems professionals can learn new technologies; however, that learning curve will impact the technical feasibility of the project—specifically, it will impact the schedule.

> Schedule Feasibility

schedule feasibility a measure of how reasonable a project timetable is.

Given the available technical expertise, are the project deadlines reasonable—that is, what is the **schedule feasibility** of the project? Some projects are initiated with specific deadlines. It is necessary to determine whether the deadlines are mandatory or desirable. For instance, a project to develop a system to meet new government reporting regulations may have a deadline that coincides with when the new reports must be initiated. Penalties associated with missing such a deadline may make meeting it mandatory. If the deadlines are desirable rather than mandatory, the analyst can propose alternative schedules.

It is preferable (unless the deadline is absolutely mandatory) to deliver a properly functioning information system two months late than to deliver an error-prone, useless information system on time! While missing deadlines can be problematic, developing inadequate systems can be disastrous. It's a choice between the lesser of two evils.

> Economic Feasibility

The bottom line in many projects is **economic feasibility.** During the early phases of the project, economic feasibility analysis amounts to little more than judging whether the possible benefits of solving the problem are worthwhile. Costs are practically impossible to estimate at that stage because the end user's requirements and alternative technical solutions have not been identified. However, as soon as specific requirements and solutions have been identified, the analyst can weigh the costs and benefits of each alternative. This is called a cost-benefit analysis. Cost-benefit analysis is discussed later in this chapter.

economic feasibility a measure of the cost-effectiveness of a project or solution.

> Legal Feasibility

Information systems have a legal impact. First of all, there are copyright restrictions. For any system that includes purchased components, one has to make sure that the license agreements are not violated. For one thing this means installing only licensed copies. But license agreements and copy protection can also restrict how you integrate the data and processes with other parts of the system. If you are working with contract programmers, the ownership of the program source code and nondisclosure agreements have to be worked out in advance.

legal feasibility is a measure of how well a solution can be implemented within existing legal and contractual obligations.

Union contracts can add constraints to the information system on how workers are paid and how their work is monitored. Legal requirements for financial reporting must be met. System requirements for sharing data with partners could even run up against antitrust laws. Finally, many information systems today are international in scope. Some countries mandate where data on local employees and local transactions must be stored and processed. Countries differ on the number of hours that make up a workweek or how long employees break for lunch.

> The Bottom Line

We have now discussed the fact that any alternative solution can be evaluated according to six criteria: operational, cultural/political, technical, schedule, economic, and legal feasibility. How does an analyst pick the best solution? It's not easy. Operational and economic issues often conflict. For example, the solution that provides the best operational impact for end users may also be the most expensive and, therefore, the least economically feasible. The final decision can be made only by sitting down with end users, reviewing the data, and choosing the best overall alternative.

Cost-Benefit Analysis Techniques

Economic feasibility has been defined as a cost-benefit analysis. How can costs and benefits be estimated? How can those costs and benefits be compared to determine economic feasibility? Most schools offer complete courses on these subjects—courses on financial management, financial decision analysis, and engineering economics and analysis. This section presents an overview of the techniques.

> How Much Will the System Cost?

Costs fall into two categories. There are costs associated with developing the system, and there are costs associated with operating a system. The former can be estimated from the outset of a project and should be refined at the end of each phase of the project. The latter can be estimated only after specific computer-based solutions have been defined. Let's take a closer look at the costs of information systems.

The costs of developing an information system can be classified according to the phase in which they occur. Systems development costs are usually onetime costs that will not recur after the project has been completed. Many organizations have standard cost categories that must be evaluated. In the absence of such categories, the following list should help:

- *Personnel costs*—The salaries of systems analysts, programmers, consultants, data entry personnel, computer operators, secretaries, and the like, who work on the project make up the personnel costs. Because many of these individuals spend time on many projects, their salaries should be prorated to reflect the time spent on the projects being estimated.
- *Computer usage*—Computer time will be used for one or more of the following activities: programming, testing, conversion, word processing, maintaining a project dictionary, prototyping, loading new data files, and the like. If a computing center charges for usage of computer resources such as disk storage or report printing, the cost should be estimated.
- *Training*—If computer personnel or end users have to be trained, the training courses may incur expenses. Packaged training courses may be charged out on a flat fee per site, a student fee (such as $395 per student), or an hourly fee (such as $75 per class hour).
- *Supply, duplication, and equipment costs.*
- *Cost of any new computer equipment and software.*

Sample development costs for a typical solution are displayed in Figure 11-2. When analysts are estimating development costs, it is important that money be set aside for the possibility that a system will incur costs after it is operating. The lifetime benefits must recover both the developmental and the operating costs. Unlike system development costs, operating costs tend to recur throughout the lifetime of the system. The costs of operating a system over its useful lifetime can be classified as fixed or variable.

Fixed costs occur at regular intervals but at relatively fixed rates. Examples of fixed operating costs include:

fixed cost a cost that occurs at a regular interval and at a relatively fixed rate.

- Lease payments and software license payments.
- Prorated salaries of information systems operators and support personnel (although salaries tend to rise, the rise is gradual and tends not to change dramatically from month to month).

Variable costs occur in proportion to some usage factor. Examples include:

variable cost a cost that occurs in proportion to some usage factor.

- Costs of computer usage (e.g., CPU time used, terminal connect time used, storage used), which vary with the workload.
- Supplies (e.g., preprinted forms, printer paper used, punched cards, floppy disks, magnetic tapes, and other expendables), which vary with the workload.
- Prorated overhead costs (e.g., utilities, maintenance, and telephone service), which can be allocated throughout the lifetime of the system using standard techniques of cost accounting.

Sample operating cost estimates for a solution are also displayed in Figure 11-2.

> What Benefits Will the System Provide?

Benefits normally increase profits or decrease costs, both highly desirable characteristics of a new information system. As much as possible, benefits should be quantified in dollars and cents; they should also be classified as tangible or intangible.

tangible benefit a benefit that can be easily quantified.

Tangible benefits are those that can be easily quantified. Tangible benefits are usually measured in terms of monthly or annual savings or of profit to the firm. For example, consider the following scenario:

Estimated Costs for Client-Server System Alternative

DEVELOPMENT COSTS

Personnel:

2	Systems Analysts (400 hours/ea $50.00/hr)	$40,000
4	Programmer/Analysts (250 hours/ea $35.00/hr)	$35,000
1	GUI Designer (200 hours/ea $40.00/hr)	$8,000
1	Telecommunications Specialist (50 hours/ea $50.00/hr)	$2,500
1	System Architect (100 hours/ea $50.00/hr)	$5,000
1	Database Specialist (15 hours/ea $45.00/hr)	$675
1	System Librarian (250 hours/ea $15.00/hr)	$3,750

Expenses:

4	Smalltalk training registration ($3,500.00/student)	$14,000

New Hardware & Software:

1	Development Server	$18,700
1	Server software (operating system, misc.)	$1,500
1	DBMS server software	$7,500
7	DBMS client software ($950.00 per client)	$6,650

Total Development Costs: $143,275

PROJECTED ANNUAL OPERATING COSTS

Personnel:

2	Programmer/Analysts (125 hours/ea $35.00/hr)	$8,750
1	System Librarian (20 hours/ea $15.00/hr)	$300

Expenses:

1	Maintenance Agreement for server	$995
1	Maintenance Agreement for server DBMS software	$525
	Preprinted forms (15,000/year @ .22/form)	$3,300

Total Projected Annual Costs: **$13,870**

FIGURE 11-2   Costs for a Proposed Systems Solution

While processing student housing applications, we discover that considerable data is being redundantly typed and filed. An analysis reveals that the same data is typed seven times, requiring an average of 44 additional minutes of clerical work per application. The office processes 1,500 applications per year. That means a total of 66,000 minutes or 1,100 hours of redundant work per year. If the average salary of a secretary is $15 per hour, the cost of this problem and the benefit of solving the problem is $16,500 per year.

Alternatively, tangible benefits might be measured in terms of unit cost savings or profit. For instance, an alternative inventory valuation scheme may reduce inventory carrying cost by $0.32 per unit of inventory. Some examples of tangible benefits are listed in the margin.

Other benefits are intangible. **Intangible benefits** are those that are believed to be difficult or impossible to quantify. Unless these benefits are at least identified, it is entirely possible that many projects would not be feasible. Examples of intangible benefits are listed in the margin on the next page.

TANGIBLE BENEFITS

Fewer Processing Errors
Increased Throughput
Decreased Response Time
Elimination of Job Steps
Increased Sales
Reduced Credit Losses
Reduced Expenses

---

intangible benefit a benefit that is believed to be difficult or impossible to quantify.

Unfortunately, if a benefit cannot be quantified, it is difficult to accept the validity of an associated cost-benefit analysis that is based on incomplete data. Some analysts dispute the existence of intangible benefits. They argue that all benefits are quantifiable; some are just more difficult to quantify than others. Suppose, for example, that improved customer goodwill is listed as a possible intangible benefit. Can we quantify goodwill? You might try the following analysis:

1. What is the result of customer ill will? The customer will submit fewer (or no) orders.
2. To what degree will a customer reduce orders? A user may find it difficult to specifically quantify this impact, but you could try to have the end user estimate the possibilities (or invent an estimate to which the end user can react). For instance:
   a. There is a 50 percent (.50) chance that the regular customer would send a few orders—fewer than 10 percent of all its orders—to competitors to test their performance.
   b. There is a 20 percent (.20) chance that the regular customer would send as many as half its orders (.50) to competitors, particularly those orders we are historically slow to fulfill.
   c. There is a 10 percent (.10) chance that a regular customer would send us an order only as a last resort. That would reduce that customer's normal business with us to 10 percent of its current volume (90 percent, or .90, loss).
   d. There is a 5 percent (.05) chance that a regular customer would choose not to do business with us at all (100 percent or 1.00 loss).
3. We can calculate an estimated business loss as follows:

$$
\begin{aligned}
\text{Loss} = \ &.50 \times (.10 \text{ loss of business}) \\
&+ .20 \times (.50 \text{ loss of business}) \\
&+ .10 \times (.90 \text{ loss of business}) \\
&+ .50 \times (1.00 \text{ loss of business}) \\
= \ &.29 \\
= \ &29\% \text{ statistically estimated loss of business}
\end{aligned}
$$

4. If the average customer does $40,000 per year of business, then we can expect to lose 29 percent, or $11,600, of that business. If we have 500 customers, this can be expected to amount to a total of $5,800,000.
5. Present this analysis to management, and use it as a starting point for quantifying the benefit.

> Is the Proposed System Cost-Effective?

There are three popular techniques for assessing economic feasibility, also called *cost-effectiveness:* payback analysis, return on investment, and net present value.

The choice of techniques should consider the audiences that will use them. Virtually all managers who have come through business schools are familiar with all three techniques. One concept that should be applied to each technique is the adjustment of cost and benefits to reflect the time value of money.

The Time Value of Money   A concept shared by all three techniques is the **time value of money**—a dollar today is worth more than a dollar one year from now. You could invest that dollar today and, through accrued interest, have more than one dollar a year from now. Thus, you'd rather have that dollar today than in one year. That's why your creditors want you to pay your bills promptly—they can't invest what they don't have. The same principle can be applied to costs and benefits *before* a cost-benefit analysis is performed.

Some of the costs of a system will be accrued after implementation. Additionally, all benefits of the new system will be accrued in the future. Before cost-benefit analysis, these costs should be brought back to current dollars. An example should clarify the concept.

Suppose we are going to realize a benefit of $20,000 two years from now. What is the current dollar value of that $20,000 benefit? If the current return on investments is running about 10 percent, an investment of $16,528 today would give us our $20,000 in two years (we'll show you how to calculate this later). Therefore, the current value of the estimated benefit is $16,528—that is, we'd rather have $16,528 today than the promise of $20,000 two years from now.

Because projects are often compared against other projects that have different lifetimes, time-value analysis techniques have become the preferred cost-benefit methods for most managers. By time-adjusting costs and benefits, you can improve the following cost-benefit techniques.

Payback Analysis   The **payback analysis** technique is a simple and popular method for determining if and when an investment will pay for itself. Because system development costs are incurred long before benefits begin to accrue, it will take some time for the benefits to overtake the costs. After implementation, you will incur additional operating expenses that must be recovered. Payback analysis determines how much time will elapse before accrued benefits overtake accrued and continuing costs. This period of time is called the **payback period.**

In Figure 11-3 we see an information system that will be developed at a cost of $418,040. The estimated net operating costs for each of the next six years are also recorded in the table. The estimated net benefits over the same six operating years are also shown. What is the payback period?

First, we need to adjust the costs and benefits for the time value of money (that is, adjust them to current dollar values). Here's how: The present value of a dollar in year *n* depends on something typically called a **discount rate.** The discount rate is a percentage similar to interest rates that you earn on your savings account. In most cases the discount rate for a business is the **opportunity cost** of being able to invest money in other projects, including the possibility of investing in the stock market, money market funds, bonds, and the like. Alternatively, a discount rate could represent

payback analysis a technique for determining if and when an investment will pay for itself.

payback period the period of time that will elapse before accrued benefits overtake accrued costs.

	A	B	C	D	E	F	G	H	I
4	**Cash flow description**	Year 0	Year 1	Year 2	Year 3	Year 4	Year 5	Year 6	
5	Development cost:	($418,040)							
6	Operation & maintenance cost:		($15,045)	($16,000)	($17,000)	($18,000)	($19,000)	($20,000)	
7	Discount factors for 12%:	1.000	0.893	0.797	0.712	0.636	0.567	0.507	
8	Time-adjusted costs (adjusted to present value):	($418,040)	($13,435)	($12,752)	($12,104)	($11,448)	($10,773)	($10,140)	
9	Cumulative time-adjusted costs over lifetime:	($418,040)	($431,475)	($444,227)	($456,331)	($467,779)	($478,552)	($488,692)	
10									
11	Benefits derived from operation of new system:	$0	$150,000	$170,000	$190,000	$210,000	$230,000	$250,000	
12	Discount factors for 12%:	1.000	$0.893	$0.797	$0.712	$0.636	$0.567	$0.507	
13	Time-adjusted benefits (current of present value):	$0	$133,950	$135,490	$135,280	$133,560	$130,410	$126,750	
14	Cumulative time-adjusted benefits over lifetime:	$0	$133,950	$269,440	$404,720	$538,280	$668,690	$795,440	
15		0	1	2	3	4	5	6	
16	Cumulative lifetime time-adjusted costs + benefits:	($418,040)	($297,525)	($174,787)	($51,611)	$70,501	$190,138	$306,748	
17									

FIGURE 11-3

Payback Analysis for a Project

FIGURE 11-4 Partial Table For Present Value Of A Dollar

Periods	8%	9%	10%	11%	12%	13%	14%
1	0.926	0.917	0.909	0.901	0.893	0.885	0.877
2	0.857	0.842	0.826	0.812	0.797	0.783	0.769
3	0.794	0.772	0.751	0.731	0.712	0.693	0.675
4	0.735	0.708	0.683	0.659	0.636	0.613	0.592
5	0.681	0.650	0.621	0.593	0.567	0.543	0.519
6	0.630	0.596	0.564	0.535	0.507	0.480	0.456
7	0.583	0.547	0.513	0.482	0.452	0.425	0.400
8	0.540	0.502	0.467	0.434	0.404	0.376	0.351

present value the current value of a dollar at any time in the future.

what the company considers an acceptable return on its investments. This number can be learned by asking any financial manager, officer, or comptroller.

Let's say the discount rate for our sample company is 12 percent. The current value, actually called the **present value,** of a dollar at any time in the future can be calculated using the following formula:

$$PV_n = 1/(1 + i)^n$$

where PV_n is the present value of $1.00 n years from now and i is the discount rate. Therefore, the present value of a dollar two years from now is

$$PV_2 = 1/(1 + .12)^2 = 0.797$$

Earlier we stated that a dollar today is worth more than a dollar a year from now. But it looks as if it is worth less. This is an illusion. The present value is interpreted as follows. If you have 79.7 cents today, it is better than having 79.7 cents two years from now. How much better? Exactly 20.3 cents better since that 79.7 cents would grow into one dollar in two years (assuming our 12 percent discount rate).

To determine the present value of any cost or benefit in year 2, you simply multiply 0.797 times the estimated cost or benefit. For example, the estimated operating expense in year 2 is $16,000. The present value of this expense is $16,000 × 0.797, or $12,752 (rounded up). Fortunately, you don't have to calculate discount factors. There are tables similar to the partial one shown in Figure 11-4 that show the present value of a dollar for different time periods and discount rates. Simply multiply this number times the estimated cost or benefit to get the present value of that cost or benefit. More detailed versions of this table can be found in many accounting and finance books as well as in spreadsheet functions.

Better still, most spreadsheets include built-in functions for calculating the present value of any cash flow, be it cost or benefit. All the examples in this module were done with Microsoft *Excel.* The same tables can be prepared with *Lotus 1-2-3.* The beauty of a spreadsheet is that once the rows, columns, and functions have been set up, you simply enter the costs and benefits and let the spreadsheet discount the numbers to present value. (In fact, you can also program the spreadsheet to perform the cost-benefit analysis.)

In Figure 11-3, notice that we have brought all costs and benefits for our example back to present value. Also notice that the discount rate for year 0 is 1.000. Why? The present value of a dollar in year 0 is exactly $1. In other words, if you hold a dollar today, it is worth exactly $1.

Now that we've discounted the costs and benefits, we can complete our payback analysis. Look at the cumulative lifetime costs and benefits. The lifetime costs are gradually increasing over the six-year period because operating costs are being incurred. But also notice that the lifetime benefits are accruing at a much faster pace.

Lifetime benefits will overtake the lifetime costs between years 3 and 4. By charting the cumulative lifetime time-adjusted costs and benefits, we can estimate that the break-even point (when Costs + Benefits = 0) will occur approximately 3.5 years after the system begins operating.

Is this information system a good or bad investment? It depends. Many companies have a payback period guideline for all investments. In the absence of such a guideline, you need to determine a reasonable guideline before you determine the payback period. Suppose that the guideline states that all investments must have a payback period less than or equal to four years. Because our example has a payback period of 3.5 years, it is a good investment. If the payback period for the system were greater than four years, the information system would be a bad investment.

It should be noted that you can perform payback analysis without time-adjusting the costs and benefits. The result, however, would show a 2.8-year payback that looks more attractive than the 3.5-year payback that we calculated. Thus, non-time-adjusted paybacks tend to be overly optimistic and misleading.

Return-on-Investment Analysis   The **return-on-investment (ROI) analysis** technique compares the lifetime profitability of alternative solutions or projects. The ROI for a solution or project is a percentage rate that measures the relationship between the amount the business gets back from an investment and the amount invested. The lifetime ROI for a potential solution or project is calculated as follows:

> **return-on-investment (ROI) analysis** a technique that compares the lifetime profitability of alternative solutions.

$$\text{Lifetime ROI} = (\text{Estimated lifetime benefits} - \text{Estimated lifetime costs}) / \text{Estimated lifetime costs}$$

Let's calculate the lifetime ROI for the same systems solution we used in our discussion of payback analysis. Once again, all costs and benefits should be time-adjusted over a period of six years. The time-adjusted costs and benefits were presented in rows 9 and 16 of Figure 11-3. The estimated lifetime benefits minus estimated lifetime costs equal

$$\$795,440 - \$488,692 = \$306,748$$

Therefore, the lifetime ROI is

$$\text{Lifetime ROI} = \$306,748/\$488,692 = .628 = 63\%$$

This is a lifetime ROI, *not* an annual ROI. Simple division by the lifetime of the system (63 ÷ 6) yields an average ROI of 10.5 percent per year. This solution can be compared with alternative solutions. The solution offering the highest ROI is the best alternative. However, as was the case with payback analysis, the business may set a minimum acceptable ROI for all investments. If none of the alternative solutions meets or exceeds that minimum standard, then none of the alternatives is economically feasible. Once again, spreadsheets can greatly simplify ROI analysis through their built-in financial analysis functions.

As with payback analysis, we could have calculated the ROI without time-adjusting the costs and benefits. This would, however, result in a misleading 129.4 percent lifetime or a 21.6 percent annual ROI. Consequently, we recommend time-adjusting all costs and benefits to current dollars.

Net Present Value   The **net present value** of an investment alternative is considered the preferred cost-benefit technique by many managers, especially those who have substantial business schooling. Once again, you initially determine the costs and benefits for each year of the system's lifetime. And once again, we need to adjust all the costs and benefits back to present dollar values.

> **net present value** an analysis technique that compares the annual discounted costs and benefits of alternative solutions.

Figure 11-5 illustrates the net present value technique. Costs are represented by negative cash flows, while benefits are represented by positive cash flows. We have brought all costs and benefits for our example back to present value. Notice again that the discount rate for year 0 (used to accumulate all development costs) is 1.000 because the present value of a dollar in year 0 is exactly $1.

FIGURE 11-5

Net Present Value
Analysis for a
Project

	A	B	C	D	E	F	G	H	I	J
1	Net Present Value Analysis for Client-Server System Alternative									
2		(Numbers rounded to nearest $1)								
3										
4	Cash flow description	Year 0	Year 1	Year 2	Year 3	Year 4	Year 5	Year 6	Total	
5	Development cost:	($418,040)								
6	Operation & maintenance cost:		($15,045)	($16,000)	($17,000)	($18,000)	($19,000)	($20,000)		
7	Discount factors for 12%:	1.000	0.893	0.797	0.712	0.636	0.567	0.507		
8	Present value of annual costs:	($418,040)	($13,435)	($12,752)	($12,104)	($11,448)	($10,773)	($10,140)		
9	Total present value of lifetime costs:								($488,692)	
10										
11	Benefits derived from operation of new	$0	$150,000	$170,000	$190,000	$210,000	$230,000	$250,000		
12	Discount factors for 12%:	1.000	$0.893	$0.797	$0.712	$0.636	$0.567	$0.507		
13	Present value of annual benefits:	$0	$133,950	$135,490	$135,280	$133,560	$130,410	$126,750		
14	Total present value of lifetime benefits:								$795,440	
15										
16	NET PRESENT VALUE OF THIS ALTERNATIVE:								$306,748	
17										

After discounting all costs and benefits, subtract the sum of the discounted costs from the sum of the discounted benefits to determine the net present value. If it is positive, the investment is good. If negative, the investment is bad. When comparing multiple solutions or projects, the one with the highest positive net present value is the best investment. (This works even if the alternatives have different lifetimes!) In our example the solution being evaluated yields a net present value of $306,748. This means that if we invest $306,748 at 12 percent for six years, we will make the same profit that we'd make by implementing this information systems solution. This is a good investment provided no other alternative has a net present value greater than $306,748.

Once again, spreadsheets can greatly simplify net present value analysis through their built-in financial analysis functions.

Feasibility Analysis of Candidate Systems

During the decision analysis phase of system analysis, the systems analyst identifies candidate system solutions and then analyzes those solutions for feasibility. We discussed the criteria and techniques for analysis in this chapter. In this section, we evaluate a pair of documentation techniques that can greatly enhance the comparison and contrast of candidate system solutions. Both use a matrix format. We have found these matrices useful for presenting candidates and recommendations to management.

> Candidate Systems Matrix

candidate systems matrix a tool used to document similarities and differences between candidate systems.

The first matrix allows us to compare candidate systems on the basis of several characteristics. The **candidate systems matrix** documents similarities and differences between candidate systems; however, it offers no analysis.

The columns of the matrix represent candidate solutions. Experienced analysts always consider multiple implementation options. At least one of those options should be the existing system because it serves as a baseline for comparing alternatives.

The rows of the matrix represent characteristics that differentiate the candidates. For purposes of this book, we based some of the characteristics on the information system building blocks. The breakdown is as follows:

- *Stakeholders*—Identify how the system will interact with people and other systems.
- *Knowledge*—Identify how data stores will be implemented (e.g., conventional files, relational databases, other database structures), how inputs will be

FIGURE 11-6 Candidate Systems Matrix Template

	Candidate 1 Name	**Candidate 2 Name**	**Candidate 3 Name**
Stakeholders			
Knowledge			
Processes			
Communications			

captured (e.g., online, batch, etc.), how outputs will be generated (e.g., on a schedule, on demand, printed, on screen, etc.).

- *Processes*—Identify how (manual) business processes will be modified, how computer processes will be implemented. For the latter, we have numerous options, including online versus batch processes and packaged versus built-in-house software.
- *Communications*—Identify how processes and data will be distributed. Once again, we might consider several alternatives—for example, centralized versus decentralized versus distributed (or duplicated) versus cooperative (client/server) solutions. Network distribution types and strategies will be discussed in Chapter 13.

The cells of the matrix document whatever characteristics help the reader understand the differences between options. Figure 11-6 illustrates the basic structure of the matrix.

Before considering any solutions, we must consider any constraints on solutions. Solution constraints take the form of architectural decisions intended to bring order and consistency to applications. For example, a technology architecture may restrict solutions to relational databases or client/server networks.

There are several approaches for identifying candidate solutions, including:

- *Recognizing users' ideas and opinions*—Throughout a systems project, users may suggest manual or technology-related solutions. They should be given consideration.
- *Consulting methodology and architecture standards*—Many organizations' development methodology and architecture standards may dictate how technology solutions are to be selected and what technology(ies) may be represented.
- *Brainstorming possible solutions*—Brainstorming is an effective technique for identifying possible solutions. It is particularly effective when done using an organized approach or framework, such as the IS building blocks or other IS characteristics. Brainstorming should encompass solutions that represent buy, build, and a combination of buy and build options.
- *Seeking references*—The analyst should solicit ideas and opinions from other persons and organizations that have implemented similar systems.
- *Browsing appropriate journals and periodicals*—Such literature may feature advertisements and articles concerning automation strategies, successes, failures, and technologies.

A combination of the above approaches could be used independently by the development team members to derive a number of possible alternative system solutions.

A sample, partially completed candidate systems matrix listing three of the five candidates is shown in Figure 11-7. In the figure, the matrix is used to provide

FIGURE 11-7 Sample Candidate Systems Matrix

Characteristics	Candidate 1	Candidate 2	Candidate 3	Candidate . . .
Portion of System Computerized Brief description of that portion of the system that would be computerized in this candidate.	COTS package Platinum Plus from Entertainment Software Solutions would be purchased and customized to satisfy Member Services required functionality.	Member Services and warehouse operations in relation to order fulfillment.	Same as candidate 2.	
Benefits Brief description of the business benefits that would be realized for this candidate.	This solution can be implemented quickly because it's a purchased solution.	Fully supports user-required business processes for SoundStage Inc. Plus more efficient interaction with member accounts.	Same as candidate 2.	
Servers and Workstations A description of the servers and workstations needed to support this candidate.	Technically, architecture dictates Pentium III, MS Windows 2000 class servers and workstations (clients).	Same as candidate 1.	Same as candidate 1.	
Software Tools Needed Software tools needed to design and build the candidate (e.g., database management system, emulators, operating systems, languages). Not generally applicable if applications software packages are to be purchased.	MS Visual C++ and MS Access for customization of package to provide report writing and integration.	MS Visual Basic 5.0 System Architect 2001 Internet Explorer	MS Visual Basic 5.0 System Architect 2001 Internet Explorer	
Application Software A description of the software to be purchased, built, accessed, or some combination of these techniques.	Package solution	Custom solution	Same as candidate 2.	
Method of Data Processing Generally some combination of online, batch, deferred batch, remote batch, and real time.	Client/Server	Same as candidate 1.	Same as candidate 1.	
Output Devices and Implications A description of output devices that would be used, special output requirements (e.g., network, preprinted forms, etc.), and output considerations (e.g., timing constraints).	(2) HP4MV department laser printers (2) HP5SI LAN laser printers	(2) HP4MV department laser printers (2) HP5SI LAN laser printers (1) PRINTRONIX bar code printer (includes software & drivers) Web pages must be designed to VGA resolution. All internal screens will be designed for SVGA resolution.	Same as candidate 2.	
Input Devices and Implications A description of input methods to be used, input devices (e.g., keyboard, mouse, etc.), special input requirements (e.g., new or revised forms from which data would be input), and input considerations (e.g., timing of actual inputs).	Keyboard & mouse	Apple "Quick Take" digital camera and software (15) PSC Quickscan laser bar code scanners (1) HP Scanjet 4C flatbed scanner Keyboard & mouse	Same as candidate 2.	
Storage Devices and Implications Brief descriptions of what data would be stored, what data would be accessed from existing stores, what storage media would be used, how much storage capacity would be needed, and how data would be organized.	MS SQL Server DBMS with 100GB arrayed capability.	Same as candidate 1.	Same as candidate 1.	

FIGURE 11-8   Feasibility Analysis Matrix Template

	Weighting	**Candidate 1**	**Candidate 2**	**Candidate 3**
Description				
Operational feasibility				
Cultural feasibility				
Technical feasibility				
Economic feasibility				
Schedule feasibility				
Legal feasibility				
Weighted score				

overview characteristics concerning the portion of the system to be computerized, the business benefits, and the software tools and/or applications needed. Subsequent pages would provide additional details concerning other characteristics such as those mentioned previously. Two columns can be similar except for their entries in one or two cells. Multiple pages would be used if we were considering more than three candidates. A simple word processing "table" template can be duplicated to create a candidate systems matrix.

> Feasibility Analysis Matrix

The second matrix complements the candidate systems matrix with an analysis and ranking of the candidate systems. It is called a **feasibility analysis matrix.**

The columns of the matrix correspond to the same candidate solutions as shown in the candidate systems matrix. Some rows correspond to the feasibility criteria presented in this chapter. Rows are added to describe the general solution and a ranking of the candidates. The general format is shown in Figure 11-8.

The cells contain the feasibility assessment notes for each candidate. Each row can be assigned a rank or score for each criterion (for operational feasibility, candidates can be ranked 1, 2, 3, etc.). After ranking or scoring all candidates on each criterion, a final ranking or score is recorded in the last row. Not all feasibility criteria are necessarily equal in importance; consequently, before assigning final rankings, candidates for which any criterion is deemed infeasible can be eliminated. In reality, this doesn't happen very often.

A completed feasibility analysis matrix is presented in Figure 11-9. In the figure, the feasibility assessment is provided for each candidate solution. In this example, a score is recorded directly in the cell for each candidate's feasibility criteria assessment. The weightings allow you to quantify the analysis. But be aware that any solution that is completely infeasible on any criteria should be eliminated. For instance, a solution that could be implemented only by violating contracts with suppliers could not be considered.

feasibility analysis matrix a tool used to rank candidate systems.

FIGURE 11-9 Sample Feasibility Analysis Matrix

	Wt	Candidate 1	Candidate 2	Candidate 3
Description		Purchase commercial off-the-shelf package for member services.	Write new application in-house using new company standard VB. NET and SQL Server database	Rewrite current in-house application using Powerbuilder.
Operational feasibility	15%	Supports only Member Services requirements. Current business process would have to be modified to take advantage of software functionality. Also, there is concern about security in the system. **Score: 60**	Fully supports user-required functionality. **Score: 100**	Fully supports user-required functionality. **Score: 100**
Cultural feasibility	15%	Possible user resistance to nonstandard user interface of proposed purchased package. **Score: 70**	No foreseeable problems **Score: 100**	No foreseeable problems **Score: 100**
Technical feasibility	20%	Current production release of Platinum Plus package is version 1.0 and has been on the market for only 6 weeks. Maturity of product is a risk, and company charges and additional monthly fee for technical support. Required to hire or train Java J2EE expertise to perform modifications for integration requirements. **Score: 50**	Solution requires writing application in VB. NET. Although current technical staff has only Powerbuilder experience, it should be relatively easy to find programmers with VB. NET experience. **Score: 95**	Although current technical staff is comfortable with Powerbuilder, management is concerned about acquisition of Powerbuilder by Sybase Inc. MS SQL Server is the current company standard for database, which competes with Sybase DBMS. We have no guarantee that future versions of Powerbuilder will "play well" with our current version of SQL Server. **Score: 60**
Economic feasibility Cost to develop: Payback (discounted): Net present value: Detailed calculations:	30%	Approx. $350,000 Approx. 4.5 years Approx. $210,000 See Attachment A **Score: 60**	Approx. $418,000 Approx. 3.5 years Approx. $307,000 See Attachment A **Score: 85**	Approx. $400,000 Approx. 3.3 years Approx. $325,000 See Attachment A **Score: 90**
Schedule feasibility	10%	Less than 3 months **Score: 95**	9–12 months **Score: 80**	9 months **Score: 85**
Legal feasibility	10%	No foreseeable problems **Score: 100**	No foreseeable problems **Score: 100**	No foreseeable problems **Score: 100**
Weighted score	100%	67	92.5	87.5

The System Proposal

Recall from Chapter 5 that the decision analysis phase involves identifying candidate solutions, analyzing those solutions, comparing and then selecting the best overall solution, and then recommending a solution. We've just learned how to do the first three tasks. Let's now learn about recommending a solution.

Recommending a solution involves producing a **system proposal.** This deliverable is usually a formal written report or oral presentation intended for system owners and users. Therefore, the systems analysts should be able to write a formal business report and make a business presentation without getting into technical issues or alternatives. Let's survey some important concepts of written reports and presentations.

system proposal a report or presentation of a recommended solution.

> Written Report

The written report is the most abused method used by analysts to communicate with system users. There is a tendency to generate large, voluminous reports that look impressive. Sometimes such reports are necessary, but often they are not. If a manager receives a 300-page technical report, the manager may skim it but not read it—and you can be certain it won't be studied carefully.

Length of the Written Report   Trial and error has taught us about report size. The following are general guidelines on limiting report size:

- To executive-level managers—one or two pages.
- To middle-level managers—three to five pages.
- To supervisory-level managers—less than 10 pages.
- To clerk-level personnel—less than 50 pages.

It is possible to organize a larger report to include subreports for managers who are at different levels. These subreports are usually included as early sections in the report and summarize the report, focusing on the bottom line.

Organization of the Written Report   There is a general pattern to organizing any report. Every report consists of both primary and secondary elements. *Primary elements* present the actual information that the report is intended to convey. Examples include the introduction and the conclusion.

While the primary elements present the actual information, all reports also contain secondary elements. *Secondary elements* package the report so that the reader can easily identify the report and its primary elements. Secondary elements also add a professional polish to the report.

As indicated in Figure 11-10, the primary elements can be organized in one of two formats: factual and administrative. The *factual format* is traditional and best suited to

---

FIGURE 11-10   Formats For Written Reports

Factual Format	Administrative Format
I. Introduction	I. Introduction
II. Methods and procedures	II. Conclusions and recommendations
III. Facts and details	III. Summary and discussion of facts and details
IV. Discussion and analysis of facts and details	IV. Methods and procedures
V. Recommendations	V. Final conclusion
VI. Conclusion	VI. Appendixes with facts and details

FIGURE 11-11 Secondary Elements for a Written Report

Letter of transmittal
Title page
Table of contents
List of figures, illustrations, and tables
Abstract or executive summary
  (*The primary elements—the body of the report in either the factual or administrative format—are presented in this portion of the report.*)
Appendixes

readers who are interested in facts and details as well as conclusions. This is the format we would use to specify detailed requirements and design specifications to system users. But the factual format is not appropriate for most managers and executives.

The *administrative format* is a modern, results-oriented format preferred by many managers and executives. This format is designed for readers who are interested in results, not facts. It presents conclusions or recommendations first. Any reader can read the report straight through, until the point at which the level of detail exceeds the reader's interest.

Both formats include some common elements. The *introduction* should include four components: purpose of the report, statement of the problem, scope of the project, and a narrative explanation of the contents of the report. The *methods and procedures section* should briefly explain how the information contained in the report was developed—for example, how the study was performed or how the new system will be designed. The bulk of the report will be in the *facts section.* This section should be named to describe the type of factual data presented (e.g., "Existing Systems Description," "Analysis of Alternative Solutions," or "Design Specifications"). The *conclusion* should briefly summarize the report, verifying the problem statement, findings, and recommendations.

Figure 11-11 shows the secondary, or packaging, elements of the report and their relationship to the primary elements. Many of these elements are self-explanatory. We briefly discuss here those that may not be. No report should be distributed without a *letter of transmittal* to the recipient. This letter should be clearly visible, not inside the cover of the report. A letter of transmittal states what type of action is needed on the report. It can also call attention to any features of the project or report that deserve special attention. In addition, it is an appropriate place to acknowledge the help you've received from various people.

The *abstract* or *executive summary* is a one- or two-page summary of the entire report. It helps readers decide if the report contains information they need to know. It can also serve as the highest-level summary report. Virtually every manager reads these summaries. Most managers will read on, possibly skipping the detailed facts and appendixes.

Writing the Report   Figure 11-12 illustrates the proper procedure for writing a formal report. Here are some guidelines to follow:

- *Paragraphs should convey a single idea.* They should flow nicely, one to the next. Poor paragraph structure can almost always be traced to outlining deficiencies.
- *Sentences should not be too complex.* The average sentence length should not exceed 20 words. Studies suggest that sentences longer than 20 words are difficult to read and understand.
- *Write in the active voice.* The passive voice becomes wordy and boring when used consistently.

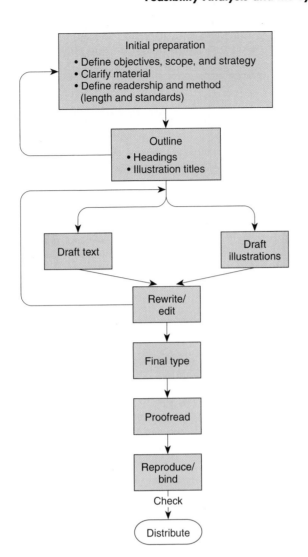

- *Eliminate jargon, big words, and deadwood.* For example, replace "DBMS" with "database management system," substitute "so" for "accordingly," try "useful" instead of "advantageous," and use "clearly" instead of "it is clear that."

Every businessperson should have a copy of *The Elements of Style* by William Strunk, Jr., and E. B. White. This classic paperback may set a record in value-to-cost ratio. Barely bigger than a pocket-size book, it is a gold mine of information.

> Formal Presentation

To communicate information to the many different people involved in a systems development project, a systems analyst is frequently required to make formal presentations. **Formal presentations** are special meetings used to sell new ideas and gain approval for new systems. They may also be used for any of these purposes: sell a new system, sell new ideas, sell change, head off criticism, address concerns, verify conclusions, clarify facts, and report progress. In many cases, a formal presentation may set up or supplement a more detailed written report.

formal presentation a special meeting used to sell new ideas and gain approval for new systems.

Effective and successful presentations require significant preparation. The time allotted to presentations is frequently brief; therefore, organization and format are critical issues. You cannot improvise and expect acceptance.

Presentations offer the advantage of impact through immediate feedback and spontaneous responses. The audience can respond to the presenter, who can use

FIGURE 11-13 Typical Outline and Time Allocation for an Oral Presentation

I. Introduction (one-sixth of total time available)
   A. Problem statement
   B. Work completed to date
II. Part of the presentation (two-thirds of total time available)
   A. Summary of existing problems and limitations
   B. Summary description of the proposed system
   C. Feasibility analysis
   D. Proposed schedule to complete project
III. Questions and concerns from the audience (Time here is not to be included in the time allotted for presentation and conclusion; it is determined by those asking the questions and voicing their concerns.)
IV. Conclusion (one-sixth of total time available)
   A. Summary of proposal
   B. Call to action (request for whatever authority you require to continue systems development)

emphasis, timed pauses, and body language to convey messages not possible with the written word. The disadvantage to presentations is that the material presented is easily forgotten because the words are spoken and the visual aids are transient. That's why presentations are often followed by a written report, either summarized or detailed.

Preparing for the Formal Presentation   Presenters must know their audience. This is especially crucial when your presentation is trying to sell new ideas and a new system. The systems analyst is frequently thought of as the dreaded agent of change in an organization. As Machiavelli wrote in his classic book *The Prince,*

> There is nothing more difficult to carry out, nor more dangerous to handle, than to initiate a new order of things. For the reformer has enemies in all who profit by the old order, and only lukewarm defenders in all those who would profit from the new order, this lukewarmness arising partly from fear of their adversaries— and partly from the incredulity of mankind, who do not believe in anything new until they have had actual experience of it.[1]

People tend to be opposed to change. There is comfort in the familiar way things are today. Yet a substantial amount of the analyst's job is to bring about change—in methods, procedures, technology, and the like. A successful analyst must be an effective salesperson. It is entirely appropriate (and strongly recommended) for an analyst to formally study salesmanship. To effectively present and sell change, you must be confident in your ideas and have the facts to back them up. Again, preparation is the key!

First, define expectations of the presentation—for instance, that the goal is to seek approval to continue the project, that another goal is to confirm facts, and so forth. A presentation is a summary of ideas and proposals that is directed toward the presenter's expectations.

Executives are usually put off by excessive detail. To avoid this, a presentation should be carefully organized around the allotted time (usually 30 to 60 minutes). Although each presentation differs, the organization and time allocation suggested in Figure 11-13 provide an idea of how this works. This figure illustrates some typical

---

[1]Niccolo Machiavelli, *The Prince and Discourses,* trans. Luigi Ricci (New York: Random House, 1940, 1950). Reprinted by permission of Oxford University Press.

topics of an oral presentation and the amount of time to allow for each. Note that this particular outline is for a systems analysis presentation. Other types of presentations might be slightly different.

What else can you do to prepare for the presentation? Because of the limited time, use visual aids—predrawn flipcharts, overhead slides, Microsoft PowerPoint slides, and the like—to support your position. Just like a written paragraph, each visual aid should convey a single idea. When preparing pictures or words, use the guidelines shown in Figure 11-14.

Microsoft PowerPoint contains software guides called *wizards* to assist the most novice users with creating professional-looking presentations. The wizard steps the user through the development process by asking a series of questions and tailoring the presentation based on responses. To hold your audience's attention, consider distributing photocopies of the visual aids at the start of the presentation. This way, the audience doesn't have to take as many notes.

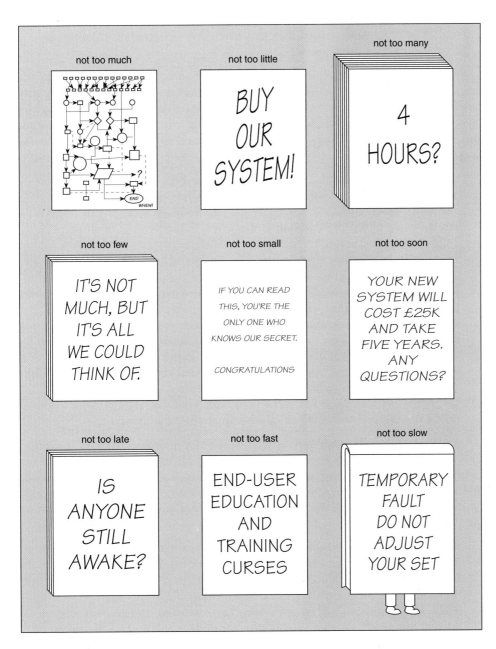

FIGURE 11-14

Guidelines for Visual Aids

Source: Copyright Keith London.

Finally, practice the presentation in front of the most critical audience you can assemble. Play your own devil's advocate, or, better yet, get somebody else to raise criticisms and objections. Practice your responses to these issues.

Conducting the Formal Presentation   A few additional guidelines may improve the actual presentation:

- *Dress professionally.* The way you dress influences people. John T. Malloy's books, *Dress for Success* and *The Woman's Dress for Success Book,* are excellent reading for both wardrobe advice and the results of studies regarding the effects of clothing on management.
- *Avoid using the word "I" when making the presentation.* Use "you" and "we" to assign ownership of the proposed system to management.
- *Maintain eye contact with the group and keep an air of confidence.* If you don't show management that you believe in your proposal, why should management believe in it?
- *Be aware of your own mannerisms.* Some of the most common mannerisms include using too many hand gestures, pacing, and repeatedly saying "you know" or "OK." Although mannerisms alone don't contradict the message, they can distract the audience.

Sometimes while you are making a presentation, some members of the audience may not be listening. This lack of attention may take several forms. Some people may be engaged in competing conversations, some may be daydreaming, some may be busy glancing at their watches, some who are listening may have puzzled expressions, and some may show no expression. The following suggestions may prove useful for keeping people listening:

- *Stop talking.* The silence can be deafening. The best public speakers know how to use dramatic pauses for special emphasis.
- *Ask a question, and let someone in the audience answer it.* This involves the audience in the presentation and is a very effective way of stopping a competing conversation.
- *Try a little humor.* You don't have to be a talented comedian. But everybody likes to laugh. Tell a joke on yourself.
- *Use props.* Use some type of visual aid to make your point clearer. Draw on the chalkboard, illustrate on the back of your notes, or create a physical model to make the message easier to understand.
- *Change your voice level.* By making your voice louder or softer, you force the audience to listen more closely or make it easier for the audience to hear. Either way, you've made a change from what the audience was used to, and that is the best way to get and hold attention.
- *Do something unexpected.* Drop a book; toss your notes; jingle your keys. Doing the unexpected is almost always an attention grabber.

A formal presentation will usually include time for questions from the audience. This time is very important because it allows you to clarify any points that were unclear and draw additional emphasis to important ideas. It also allows the audience to interact with you. However, sometimes answering questions after a presentation may be difficult and frustrating. We suggest the following guidelines when answering questions:

- *Always answer a question seriously, even if you think it is a silly question.* Remember, if you make someone feel stupid for asking a "dumb" question, that person will be offended. Also, other members of the audience won't ask their questions for fear of the same treatment.
- *Answer both the individual who asked the question and the entire audience.* If you direct all your attention to the person who asked the question,

the rest of the audience will be bored. If you don't direct enough attention to the person who asked the question, that person won't be satisfied. Try to achieve a balance. If the question is not of general interest to the audience, answer it later with that specific person.

- *Summarize your answers.* Be specific enough to answer the question, but don't get bogged down in details.
- *Limit the amount of time you spend answering any one question.* If additional time is needed, wait until after the presentation is over.
- *Be honest.* If you don't know the answer to a question, admit it. Never try to bluff your way out of a question. The audience will eventually find out, and you will destroy your credibility. Instead, promise to find out and report back. Or ask someone in the audience to do some research and present the findings later.

Following Up the Formal Presentation   As mentioned earlier, it is extremely important to follow up a formal presentation because the spoken word and impressive visual aids used in a presentation often do not leave a lasting impression. For this reason, most presentations are followed by written reports that provide the audience with a more permanent copy of the information that was communicated.

Chapter Review

1. Feasibility is a measure of how beneficial the development of an information system would be to an organization. Feasibility analysis is the process by which we measure feasibility. It is an ongoing evaluation of feasibility at various checkpoints in the life cycle. At any of these checkpoints, the project may be canceled, revised, or continued. This is called a creeping commitment approach to feasibility.

2. There are six feasibility tests: operational, cultural/political, technical, schedule, economic, and legal.

   a. Operational feasibility is a measure of problem urgency or solution acceptability. It includes a measure of how the end users and managers feel about the problems or solutions.

   b. Cultural (or political) feasibility is a measure of how people feel about a solution and how well it will be accepted.

   c. Technical feasibility is a measure of how practical solutions are and whether the technology is already available within the organization. If the technology is not available to the firm, technical feasibility also looks at whether it can be acquired.

   d. Schedule feasibility is a measure of how reasonable the project schedule or deadline is.

   e. Economic feasibility is a measure of whether a solution will pay for itself or how profitable a solution will be. For management, economic feasibility is the most important of our four measures.

   f. Legal feasibility is a measure of how well a solution can be implemented within existing legal and contractual obligations.

3. To analyze economic feasibility, you itemize benefits and costs. Benefits are either tangible (easy to measure) or intangible (hard to measure). To properly analyze economic feasibility, try to estimate the value of all benefits. Costs fall into two categories: development and operating.

   a. Development costs are onetime costs associated with analysis, design, and implementation of the system.

   b. Operating costs may be fixed over time or variable with respect to system usage.

4. Given the costs and benefits, economic feasibility is evaluated by the techniques of cost-benefit analysis. Cost-benefit analysis determines if a project or solution will be cost-effective—if lifetime benefits will exceed lifetime costs. There are three popular ways to measure cost-effectiveness: payback analysis, return-on-investment analysis, and net present value analysis.

   a. Payback analysis defines how long it will take for a system to pay for itself.

   b. Return-on-investment and net present value analyses determine the profitability of a system.

c. Net present value analysis is preferred because it can compare alternatives with different lifetimes.

5. A candidate systems matrix is a useful tool for documenting the similarities and differences between candidate systems being considered.

6. A feasibility analysis matrix is used to evaluate and rank candidate systems. Both the candidate systems matrix and the feasibility analysis matrix are useful for presenting the results of a feasibility analysis as part of a system proposal.

7. Written reports are the most common communications vehicle used by analysts. Reports consist of both primary and secondary elements. Primary elements contain factual information. Secondary elements package the report for ease of use. Reports may be organized in either the factual or administrative format. The factual format presents the details before conclusions; the administrative format reverses that order. Managers like the administrative format because it is results-oriented and gets right to the bottom-line question.

8. Formal presentations are a special type of meeting at which a person presents conclusions, ideas, or proposals to an interested audience. Preparation is the key to effective presentations.

9. The system proposal may be a formal written report or an oral presentation.

Review Questions

1. What does a creeping commitment approach to feasibility analysis mean?
2. What are the feasibility analysis checkpoints in the development cycle? What should be done at each checkpoint?
3. What are the objectives of the operational feasibility test?
4. Why is it important to find out how the end users and managers feel about the problem solution that the system analyst has identified?
5. When is usability analysis performed? What is the objective of the usability analysis?
6. What is the objective of the technical feasibility test?
7. What are the characteristics of development costs and operating costs? List three examples of each kind of cost.

8. List five examples of tangible benefits.
9. Why is the time-value-of-money concept an essential consideration when accessing economic feasibility?
10. What are the most commonly used techniques to determine the cost-effectiveness of a project?
11. For what are the candidate systems matrix and feasibility analysis matrix used?
12. For written reports, what is the difference between the factual format and the administrative format?
13. What are the steps in writing a report?
14. What are the advantages and disadvantages of presentations?
15. What should be done to follow up the formal presentation?

Problems and Exercises

1. The textbook describes a creeping commitment approach to feasibility.

   a. Explain this approach and why the textbook recommends it.
   b. What are the some of the changes or events that might occur which make this approach advisable?
   c. Should an organization cancel a project if it becomes infeasible?

2. The textbook describes three checkpoints for measuring feasibility.

   a. What are these checkpoints?
   b. Typically, how accurately can feasibility be determined at each checkpoint?

   c. Which checkpoint, if any, is the most critical one?

3. What are the four categories of feasibility tests, and what is the criteria each of them uses to measure feasibility?

4. You are a systems designer on a project which is getting close to finishing the systems design phase. A working prototype has been developed, and you've been tasked with doing a usability analysis. Draft a one- or two-page plan detailing your approach to conducting the usability analysis.

5. You are a systems analyst working in the IT shop of a medium-size organization with about 300 employees. The organization is in the system design phase of a project to develop an

electronic activity reporting system for all employees, replacing the current hard copy method. All of the work is being done in-house except for several consultants, who are providing ancillary services, such as IV&V. The application will use employees' existing desktops, although several dedicated servers will need to be acquired. The user interface is very intuitive, but the project calls for about a half day of training for all employees on policies and procedures for using the new application. The system is not using any new technology, and the IT technical staff have a great deal of expertise. Create a worksheet, detailing the estimated one-time development costs and ongoing operating costs. By the way, in your organization, salary and benefits for systems analysts average $40 per hour; you can use this as a basis for estimating salary and benefits for other classifications involved in the project.

6. In the project described above, it was noted that the electronic activity reporting system will be replacing the current manual system. Describe the tangible benefits that might be expected. Take a "best guess" approach, and calculate the annual savings to the organization. Show your assumptions in the calculations.

7. You are designing a Web-based system where your regional offices can submit their sales reports online instead of filling them out by hand and mailing them in. Three candidate solutions have been identified. Their estimated lifetime benefits and estimated lifetime costs are shown below. All have been time-adjusted over the projected five-year lifetime of each alternative.

	Estimated Lifetime Benefits	Estimated Lifetime Costs
Candidate Solution 1:	$640,000	$172,000
Candidate Solution 2:	$640,000	$160,000
Candidate Solution 3:	$640,000	$185,000

According to return-on-investment analysis, which candidate solution offers the highest ROI? If the organization sets a minimum lifetime ROI of 80 percent, which of these solutions is economically feasible?

8. What are the different techniques or methods for identifying candidate solutions? If you had to choose just one of these methods, which would it be and why?

9. You are working as a system designer for a company that manufactures heavy-duty power tools used by contractors. Every month, your regional sales and service centers batch together the hard copy repair orders for work performed under warranty. They are sent to headquarters, where they are run through a legacy mainframe batch process. A report is then generated, which the engineers analyze for signs of any problem trends in the new models. The company's CEO has decided that this process is far too slow in today's highly competitive business environment and wants to replace the legacy system as soon as possible with something more contemporary. Identify at least three candidate solutions, and describe them in a candidate systems matrix, using Figure 11-7 as an example.

10. Prepare a feasibility analysis matrix, using the candidate solutions you identified and described in the preceding question. Use Figure 11-9 as your template, but choose the weighting factors that you feel would be most appropriate in this situation. For purposes of this exercise, you may provide an estimate of the economic feasibility.

11. Once the feasibility analysis matrix has been completed, it is time to write the feasibility report. For this exercise, prepare a feasibility report to executive-level managers, using the appropriate format shown in Figure 11-10.

12. You have been asked to present the feasibility analysis and recommendation to the executive managers of every department in your organization at their weekly meeting. Prepare a set of PowerPoint slides to be used as a visual aid during your presentation.

13. Name at least 10 things you should *not* do if you want your presentation to be informative, persuasive, and well-received.

Projects and Research

1. Steve McConnell is an author who has written numerous books on software engineering and development. In his book *Rapid Development,* McConnell points out that in engineering, design is usually a much smaller part of the total project than the actual construction. He compares bridge building projects, where design is about 10 percent of the total effort and construction about 90 percent, to software development projects, where design is generally at least 50 percent of the total

project effort. Explore and expand on this theme of the unique differences in software engineering compared to other types of engineering, and summarize your analysis and findings in a one- to two-page paper. What do other software engineering leaders have to say on this topic?

2. You work as a system analyst in the headquarters of your state's highway patrol, which has field offices throughout the state. Currently, traffic accident reports are handwritten in the field by the highway patrol officers, reviewed by their sergeant, stored temporarily, then batched and sent monthly to headquarters. Each one is entered into a legacy mainframe system by key data operators, then after the reports from the patrol offices in each county have been input, a computer operator runs the edit program using JCL.

   Reports with major errors or omissions are rejected and returned to the county highway patrol office of origin for correction. After the edit program is completed for all the counties, an update program is run adding the monthly batch of traffic accident reports to the master file of reports. Statistical reports are generated quarterly and yearly. The entire process from the time the batches of reports are received to the point the master file is updated generally takes about three months. Executive management is interested in replacing the system with something that is more modern, less labor-intensive, more accurate, and easier for users to access and that will reduce turnaround time for preparing statistical reports. Your assignment, as a member of the project team, is to prepare the feasibility study report (FSR).

   a. What are some of the options or alternatives that you think should be considered? (Identify at least three in addition to "do nothing" or "maintain the status quo").
   b. Prepare a candidate systems matrix describing the characteristics of each of these alternatives, using the candidate systems matrix template shown in Figure 11-6.
   c. Expand the candidate systems matrix, using the template shown in Figure 11-7.
   d. Evaluate each of these alternatives for operational, technical, and schedule feasibility, using the techniques described in the textbook and using the template shown in Figure 11-9.

3. Based upon the scenario described in the preceding question:

   a. Prepare an estimated-costs worksheet for each alternative, using the format shown in Figure 11-2.
   b. Assess the economic feasibility of each alternative, using one of the three techniques described in the textbook. Which technique did you use and why?
   c. Add the economic feasibility analysis to the feasibility analysis matrix from the preceding question.
   d. Compare and score each of these alternatives. Use different weighting factors for each of the feasibility criteria than those used in the textbook.
   e. What weighting factors did you choose for the different criteria in your feasibility analysis matrix? Why?

4. Management was impressed by your excellent work on the feasibility analysis matrix and has asked you to prepare the system proposal report. Write a system proposal whose primary audience will be the midlevel business and IT managers, but which also will be read by the executive sponsor and chief information officer. Use the appropriate format shown in Figure 11-10.

5. Midlevel management was extremely impressed by your system proposal. They now want you to prepare and present a formal presentation to the top management of the department.

   a. Describe the steps you should go through to prepare for the formal presentation.
   b. Prepare a PowerPoint slide presentation, using the guidelines suggested in the textbook, or in other books and articles on the do's and don'ts of PowerPoint presentations.
   c. What do you consider to be the most critical thing to know in preparing for the formal presentation? Why

6. A wide variety of formats, templates, and methods exist for preparing system proposals and feasibility study reports. Search the Web to see what other tools and techniques you can find.

   a. Describe the formats that you found and their sources.
   b. Compare the different formats that you found to each other and to the one in the textbook. What are some of the differences?
   c. Do you think there is one format with clear-cut advantages over the others? If so, describe which one, and why you feel it is better.
   d. Create what you believe to be the ideal FSR template for your organization.

Minicases

The grocery store, Wow Munchies, from an earlier chapter is considering developing an online site for customers to purchase food. The owner of the store believes that this capability will enable the store to grab market share from nearby Fast Food Co., which has a Web site and delivers food to the customer. This site will allow customers to purchase any item that is currently in stock in the store. The store will *not* deliver the food, but will have the food bagged and ready for pickup at the time designated by the customer. Wow Munchies has a single storefront.

1. Conduct an operational feasibility study. Do you think the Web site will enable Wow Munchies to gain market share, as is its purpose? What factors will affect the operational success of this site? Submit your paper, supporting documents, charts, and any interviews you conducted.

2. Conduct a technical feasibility study. What would you recommend the company use in the creation and maintenance of their site (e.g. languages, specific host, encryption)? Why does your choice affect the feasibility of said site? Submit your paper, supporting documents, charts, and any interviews you conducted.

3. Conduct an economic feasibility study for the investment into an e-commerce site. What discount rate are you using? Why? Submit your paper, supporting documents, charts, and any interviews you conducted.

4. Develop a timeline and schedule feasibility study for completion of this Web site. Do you see any mitigating factors that might cause a delay in the timeline or deadline overrun? Submit both a short paper and a Gantt chart.

Suggested Readings

Bovee, Courtland L., and John V. Thill. *Business Communications Today,* 2nd ed. New York: Random House, 1989.

Gildersleeve, Thomas R. *Successful Data Processing Systems Analysis,* 2nd ed. Englewood Cliffs, NJ: Prentice Hall, 1985. This book provides an excellent chapter on cost-benefit analysis techniques. Chapter 5 discusses presentations. We are indebted to Gildersleeve for the creeping commitment concept.

Gore, Marvin, and John Stubbe. *Elements of Systems Analysis,* 4th ed. Dubuque, IA: Brown, 1988. The feasibility analysis chapter suggests an interesting matrix approach to identifying, cataloging, and analyzing the feasibility of alternative solutions for a system.

Smith, Randi Sigmund. *Written Communications for Data Processing.* New York: Van Nostrand Publishing, 1976.

Stuart, Ann. *Writing and Analyzing Effective Computer System Documentation.* New York: Holt, Rinehart and Winston, 1984.

Uris, Auren. *The Executive Deskbook,* 3rd ed. New York: Van Nostrand Reinhold, 1988.

Walton, Donald. *Are You Communicating? You Can't Manage without It.* New York: McGraw-Hill, 1989.

Wetherbe, James. *Systems Analysis and Design: Traditional, Structured, and Advanced Concepts and Techniques,* 2nd ed. St. Paul, MN: West, 1984. Wetherbe pioneered the PIECES framework for problem classification. In this chapter we extended that framework to analyze operational feasibility of solutions.

Part Three

Systems Design Methods

The chapters in Part Three introduce you to systems design methods. Chapter 12, "Systems Design," provides the context for all the subsequent chapters by introducing the activities of systems design. Systems design includes the preparation of detailed computer-based specifications that will fulfill the requirements specified during systems analysis and construction of system prototypes. With respect to information systems development, systems design consists of the configuration, procurement, and design and integration phases.

Chapter 13, "Application Architecture and Modeling," introduces physical process and data design. It specifically addresses design decisions regarding distribution issues for shared data and processes. This results in an application architecture that consists of design units that can be assigned to different team members for detailed design, construction, and unit testing.

Chapter 14, "Database Design," introduces the design of physical data stores from the data model developed in Chapter 8.

Chapter 15, "Output Design and Prototyping," teaches output design and prototyping. Different types, formats, and media for outputs are presented. The use of the most common types of graphs is discussed. The chapter demonstrates how to design and prototype printed and display outputs.

Chapter 16, "Input Design and Prototyping," teaches input design and prototyping. Formats, methods, media, human factors, and internal controls for inputs are stressed. The proper usage of screen-based controls for data input on graphical user interface (GUI) screen designs is discussed. The chapter also emphasizes prototyping as a way of finding, documenting, and communicating input design requirements.

Chapter 17, "User Interface Design," teaches user interface design and prototyping. You will learn how to develop a friendly and effective interface for an application. The design of the user interface is crucial because user acceptance of the system is frequently dependent on a friendly, easy-to-use interface. A GUI-based interface for obtaining the inputs and outputs designed in Chapters 15 and 16 is demonstrated.

Finally, Chapter 18, "Object-Oriented Design and Modeling Using the UML," introduces you to tools and techniques used to perform systems design using an object-oriented approach to systems development.

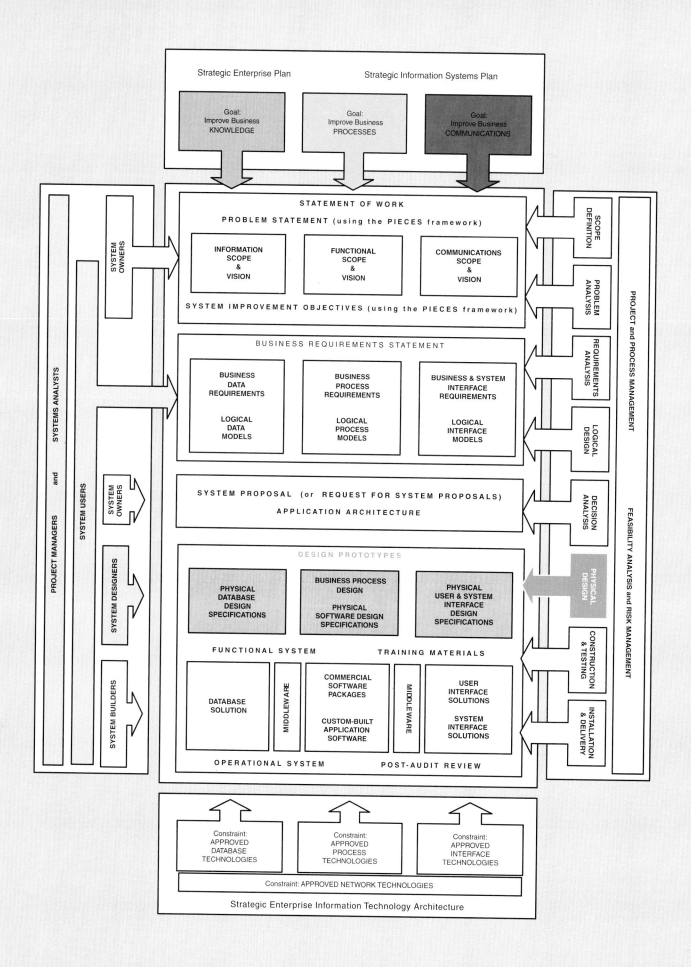

Systems Design

Chapter Preview and Objectives

In this chapter you will learn more about the design phase of systems development. You will know that you understand the process of systems design when you can:

▮ Describe the design phase in terms of your information building blocks.

▮ Identify and differentiate between several systems design strategies.

▮ Describe the design phase tasks in terms of a computer-based solution for an in-house development project.

▮ Describe the design phase in terms of a computer-based solution involving procurement of a commercial systems software solution.

Although some techniques of systems design are introduced in this chapter, it is not the intent of this chapter to teach the techniques of systems design. This chapter teaches only the process of systems design and introduces you to some techniques that will be taught in later chapters.

Introduction

The system proposal for the SoundStage Member Services system has been approved. Now the systems project team is converting from logical design (specifying "what" the system must do) to physical design (specifying "how" the system will work). There are many tasks to do, including designing the database, specifying how the system will work, and prototyping the user interface. Parts of the member services system are being assembled from purchased components. That will save programming time, but add steps to the process to design and test how the components will interface with components they write. Bob Martinez will be given several tasks during the process. He is anxious to get started. The system is finally starting to take shape, if only on paper and in prototypes.

What Is Systems Design?

system design the specification of a detailed computer-based solution.

In Chapter 3 you learned about the systems development process. In that chapter we purposefully limited our discussion to only briefly examining each phase. In this chapter, we take a much closer look at the systems design phase that follows systems analysis. Information **systems design** is defined as those tasks that focus on the specification of a detailed computer-based solution. It is also called *physical design*. Thus, whereas systems analysis emphasized the <u>business</u> problem, systems design focuses on the <u>technical</u> or <u>implementation</u> concerns of the system.

As was illustrated in the chapter home page at the start of this chapter, systems design is driven by the technical concerns of SYSTEM DESIGNERS. Hence, it addresses the IS building blocks from the SYSTEM DESIGNERS' perspective. The SYSTEMS ANALYSTS serve as facilitators of systems design.

Most of us define the process of design too restrictively. We envision ourselves drawing blueprints of the computer-based systems to be programmed and developed by ourselves or our own programmers. Thus, we design inputs, outputs, files, databases, and other computer components. Recruiters of computer-educated graduates refer to this restrictive definition as the "not-invented-here syndrome." In reality, many companies purchase more software than they write in-house. That shouldn't surprise you. Why reinvent the wheel? Many systems are sufficiently generic that computer vendors have written adequate—but rarely, if ever, perfect—software packages that can be bought and possibly modified to fulfill end-user requirements.

This chapter examines systems design from the perspectives of both in-house development, or "build," projects and software procurement, or "buy," projects. Let's begin our study by first examining some overall strategies for systems design.

Systems Design Approaches

There are many strategies or techniques for performing systems design. They include *modern structured design, information engineering, prototyping, JAD, RAD,* and *object-oriented design.* These strategies are often viewed as competing alternative approaches to systems design, but in reality certain combinations complement one another. Let's briefly examine these strategies and the scope or goals of the projects to which they are suited. The intent is to develop a high-level understanding only. The subsequent chapters will teach you the actual techniques.

NOTE: Recall from Chapter 3 that methodology "routes" are sometimes defined for these approaches.

> Model-Driven Approaches

Structured design, information engineering, and object-oriented design are examples of model-driven approaches. **Model-driven design** emphasizes the drawing of pictorial system models to document the technical or implementation aspects of a new system.

The design models are often derived from logical models that were developed earlier, in model-driven analysis (discussed in Chapter 5). Ultimately, the system design models become the blueprints for constructing and implementing the new system.

Today, model-driven approaches are almost always enhanced by the use of automated tools. Some designers draw system models with general-purpose graphics software such as *Visio Professional* or *Corel Flow*. Other designers and organizations require the use of repository-based CASE or modeling tools such as *System Architect, Microsoft Visio, Visible Analyst,* or *IBM's Rational.* CASE tools offer consistency and completeness as well as rule-based error checking.

Let's briefly examine the most commonly encountered model-driven design approaches. Model-driven design approaches are featured in the model-driven methodologies and routes (introduced in Chapter 3).

Modern Structured Design   Structured design techniques help developers deal with the size and complexity of programs. **Modern structured design** is a process-oriented technique for breaking up a large program into a hierarchy of modules, which results in a computer program that is easier to implement and maintain (change). Synonyms (although technically inaccurate) are *top-down program design* and *structured programming*.

The concept is simple. Design a program as a top-down hierarchy of modules. A module is a group of instructions—a paragraph, block, subprogram, or subroutine. The top-down structure of these modules is developed according to various design rules and guidelines. (Thus, merely drawing a hierarchy or structure chart for a program is *not* structured design.)

Structured design is considered a process-oriented technique because its emphasis is on the PROCESS building blocks in our information system—specifically, software processes. Structured design seeks to factor a program into the top-down hierarchy of modules that have the following properties:

- Modules should be highly *cohesive;* that is, each module should accomplish one and only one function. This makes the modules reusable in future programs.
- Modules should be loosely *coupled;* in other words, modules should be minimally dependent on one another. This minimizes the effect that future changes in one module will have on other modules.

As will be discussed in Chapter 18, cohesion and coupling are important concepts also in the objected-oriented world. The software model derived from structured design is called a *structure chart* (Figure 12-1). The structure chart is derived by studying the flow of data through the program. Structured design is performed during systems design. It does not address all aspects of design—for instance, structured design will not help you design inputs, outputs, or databases.

Structured design has lost some of its popularity with many of today's applications that call for newer techniques that focus on *event-driven* and *object-oriented programming* techniques. However, it is still a popular technique involving the design of mainframe-based application software and is used to address coupling and cohesion issues at the "system" level.

Information Engineering   In Chapter 5 you learned that *information engineering (IE)* is a model-driven and DATA-centered, but PROCESS-sensitive, technique for planing, analyzing, and designing information systems. The primary tool of IE is a data model

---

model-driven design a system design approach that emphasizes drawing system models to document technical and implementation aspects of a system.

modern structured design a system design technique that decomposes the system's processes into manageable components.

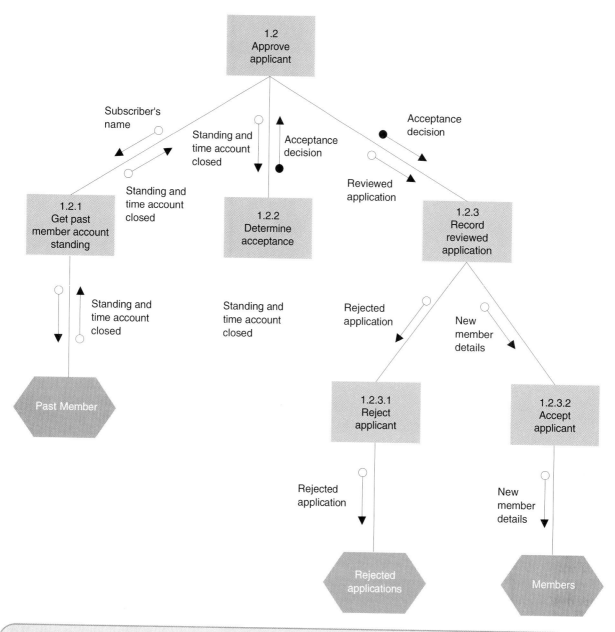

FIGURE 12-1 The End Product of Structured Design

diagram (see Figure 12-2). IE involves conducting a business area requirements analysis from which information system applications are carved out and prioritized. The applications identified in IE become projects to which other systems analysis *and* design methods are intended to be applied in order to develop the production systems. These methods may include some combination of modern structured analysis (discussed in Chapter 5), modern structured design, prototyping, and object-oriented analysis and design.

Prototyping Traditionally, physical design has been a paper-and-pencil process. Analysts drew pictures that depicted the layout or structure of outputs, inputs, and databases and the flow of dialogue and procedures. This is a time-consuming process that is prone to considerable errors and omissions. Frequently, the resulting paper specifications were inadequate, incomplete, or inaccurate.

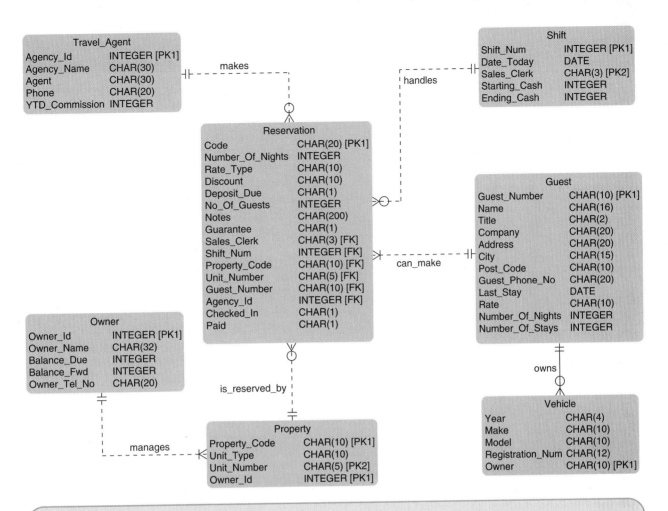

FIGURE 12-2 Sample Information Engineering Physical Entity Relationship Diagram

Today many analysts and designers prefer prototyping, a modern engineering-based approach to design. The prototyping approach is an iterative process involving a close working relationship between the designer and the users. This approach has several advantages:

- Prototyping encourages and requires active end-user participation. This increases end-user morale and support for the project. End users' morale is enhanced because the system appears real to them.
- Iteration and change are a natural consequence of systems development—that is, end users tend to change their minds. Prototyping better fits this natural situation because it assumes that a prototype evolves, through iteration, into the required system.
- It has often been said that end users don't fully know their requirements until they see them implemented. If so, prototyping endorses this philosophy.
- Prototypes are an active, not passive, model that end users can see, touch, feel, and experience.
- An approved prototype is a working equivalent to a paper design specification, with one exception—errors can be detected much earlier.
- Prototyping can increase creativity because it allows for quicker user feedback, which can lead to better solutions.
- Prototyping accelerates several phases of the life cycle, possibly bypassing the programmer. In fact, prototyping consolidates parts of phases that normally occur one after the other.

There are also disadvantages or pitfalls to using the prototyping approach. Most of these can be summed up in one statement: Prototyping encourages ill-advised shortcuts through the life cycle. Fortunately, the following pitfalls can all be avoided through proper discipline:

- Prototyping encourages a return to the "code, implement, and repair" life cycle that used to dominate information systems. As many companies have learned, systems developed in prototyping languages can present the same maintenance problems that have plagued legacy systems developed in languages such as *COBOL.*
- Prototyping does not negate the need for the systems analysis phases. A prototype can solve the wrong problems and opportunities just as easily as a conventionally developed system can.
- You cannot completely substitute any prototype for a paper specification. No engineer would prototype an engine without some paper design. Yet many information systems professionals try to prototype without a specification. Prototyping should be used to complement, not replace, other methodologies. The level of detail required of the paper design may be reduced, but it is not eliminated.
- Numerous design issues are not addressed by prototyping. These issues can be inadvertently forgotten if you are not careful.
- Prototyping often leads to premature commitment to a design (usually the first design that is developed).
- During prototyping, the scope and complexity of the system can quickly expand beyond original plans. This can easily get out of control.
- Prototyping can reduce creativity in designs. The very nature of any implementation—for instance, a prototype of a report—can prevent analysts, designers, and end users from looking for better solutions.
- Prototypes often suffer from slower performance than their third-generation-language counterparts (albeit this difference is rapidly becoming a nonissue).

Prototypes can be quickly developed using many of the 4GLs and object-oriented programming languages available today. Figure 12-3 depicts a prototype screen for a system. Prototypes can be built for simple outputs, computer dialogues, key functions, entire subsystems, or even the entire system. Each prototype system is reviewed by end users and management, who make recommendations about requirements, methods, and formats. The prototype is then corrected, enhanced, or refined to reflect the new requirements. Prototyping technology makes such revisions in a relatively straightforward manner. The revision and review process continues until the prototype is accepted. At that point, the end users are accepting both the requirements and the design that fulfills those requirements.

Design by prototyping doesn't necessarily fulfill all design requirements. For instance, prototypes don't always address important performance issues and storage constraints. Prototypes rarely incorporate internal controls. The analyst or designer must still specify these.

Object-Oriented Design Object-oriented design (OOD) is the newest design strategy. The concepts behind this strategy (and technology) are covered extensively in Chapter 18, "Object-Oriented Design and Modeling Using the UML," but a simplified introduction is appropriate here. This technique is an extension of the object-oriented analysis strategy presented in Chapter 10. Figure 12-4 shows one of the many diagrams used in object-oriented design.

Object technologies and techniques are an attempt to eliminate the separation of concerns about DATA and PROCESS. OOD techniques are used to refine the object

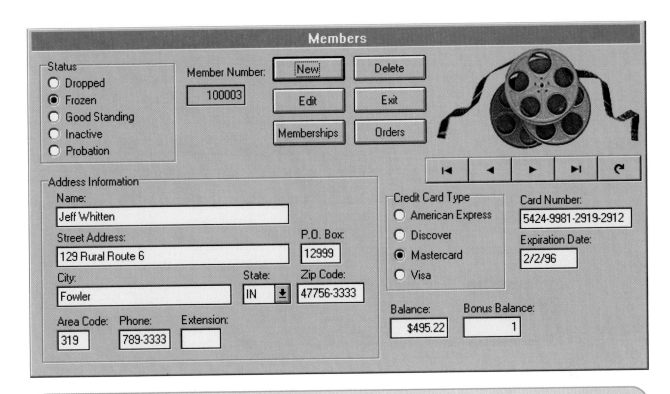

FIGURE 12-3 Sample Prototype Screen

requirements definitions identified earlier during analysis and to define design-specific objects.

For example, based on a design implementation decision, during OOD the designer may need to revise the data or process characteristics for an object that was defined during systems analysis. Likewise, a design implementation decision may necessitate that the designer define a new set of objects that will make up an interface screen that the user(s) may interact with in the new system.

> Rapid Application Development

Another popular design strategy used today is rapid application development. **Rapid application development (RAD)** is the merger of various structured techniques (especially the data-driven information engineering) with prototyping techniques and joint application development techniques to accelerate systems development.

RAD calls for the interactive use of structured techniques and prototyping to define the users' requirements and design the final system. Using structured techniques, the developer first builds preliminary data and process models of the business requirements. Prototypes then help the analyst and users to verify those requirements and to formally refine the data and process models. The cycle of models, then prototypes, then models, then prototypes, and so forth, ultimately results in a combined business requirements and technical design statement to be used for constructing the new system.

The expedition of the design effort is enhanced through the emphasis on user participation in joint application development sessions. Recall that *joint application development (JAD),* introduced in Chapter 5 and discussed in more detail in Chapter 6, is a technique that complements other systems analysis and design techniques by emphasizing *participative development* among SYSTEM OWNERS, USERS, DESIGNERS, and

rapid application development (RAD) a systems design approach that utilizes structured, prototyping, and JAD techniques to quickly develop systems.

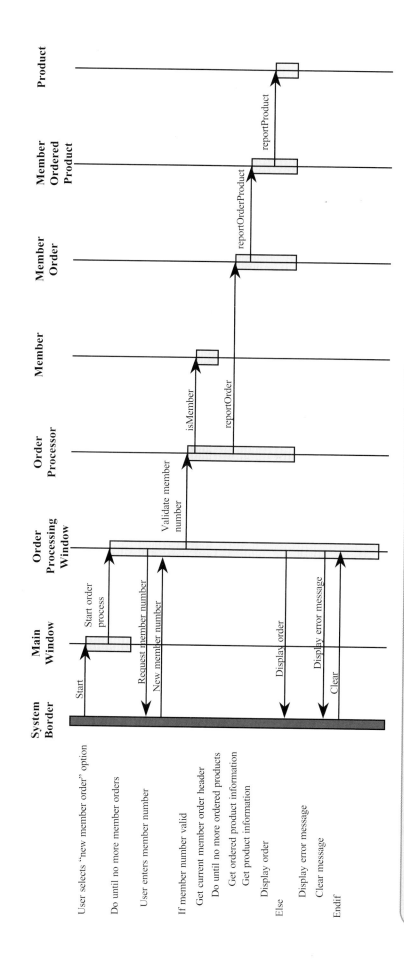

FIGURE 12-4 Sample Object-Oriented Design Model

BUILDERS. During the JAD sessions for systems design, the systems designer will take on the role of facilitator for possibly several full-day workshops intended to address different design issues and deliverables. JAD is an essential element contributing greatly to the acceleration emphasis of RAD.

> *FAST* Systems Design Strategies

Like most commercial methodologies, our hypothetical *FAST* methodology does not impose a single approach on systems design. Instead, it integrates all the popular approaches introduced in the preceding paragraphs. The SoundStage case study will demonstrate these methods in the context of a typical first assignment for a systems analyst. The systems analysis techniques will be applied within the framework of:

- Your information system building blocks (from Chapter 2).
- The systems development phases (from Chapter 3).
- The tasks that implement a phase (described in this chapter).

Given this context, we can now study systems design. We will begin by studying systems design as it relates to an in-house development, or "build," project. Afterward, we will examine how the systems design phases are affected when a decision has been made to acquire, or "buy," a commercial software package as a solution.

Systems Design for In-House Development— The "Build" Solution

Let's begin by placing systems design for in-house development projects into context relative to the system life cycle. As is illustrated in Figure 12-5, an approved system proposal from the decision analysis phase triggers the design phase. The goal of the design phase is twofold. First, the analyst seeks to design a system that both fulfills requirements and will be friendly to its end users. Human engineering will play a pivotal role during design. Second, and still very important, the analyst seeks to present clear and complete specifications to the computer programmers and technicians. As is shown is Figure 12-5, the approved physical design specifications will trigger the construction phase of our in-house development project.

Figure 12-6 is a task diagram depicting the work (= tasks) that should be performed to complete the design phase. This task diagram does not mandate any specific methodology, but we will describe in the accompanying paragraphs the approaches, tools, and techniques you might want to consider for each design task. This task diagram is only a template. The project team and project manager may expand on or alter the template to reflect the unique needs of any given project.

Let's now examine each systems design task in detail.

> Task 5.1—Design the Application Architecture

The purpose of this first design task is to specify an application architecture. An **application architecture** defines the technologies to be used by (and used to build) one, more, or all information systems in terms of their data, processes, interfaces, and network components. Thus, designing the application architecture involves considering network technologies and making decisions on how the systems' DATA, PROCESSES, and INTERFACES are to be distributed among the business locations.

This task is accomplished by analyzing the data models and process models that were initially created during requirements analysis. Given the data models, process models, and target solution, distribution decisions will need to be made. As decisions on data, processes, and interfaces are made, they are documented. An example is the

application architecture
a specification of the technologies to be used to implement information systems.

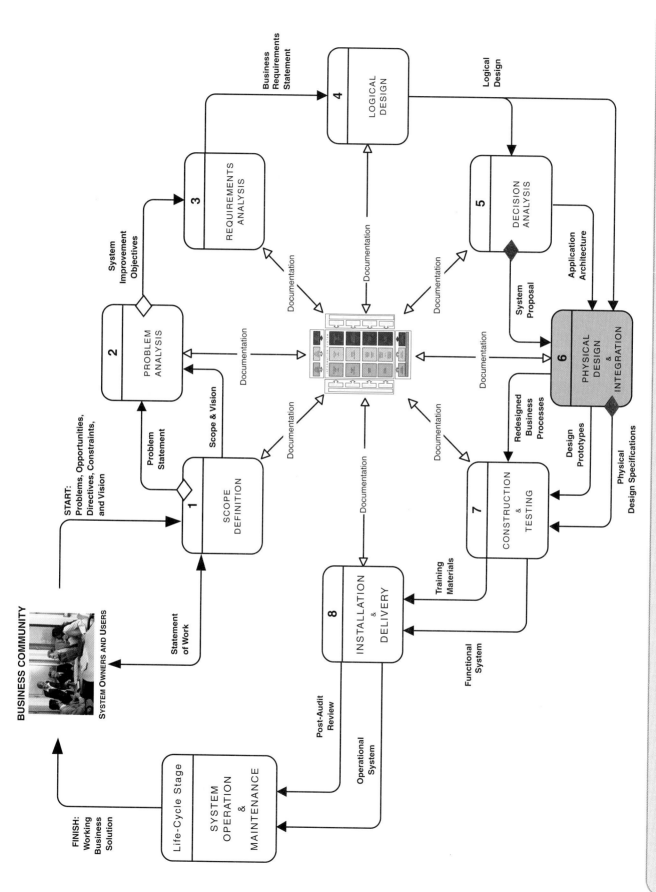

BUSINESS COMMUNITY

SYSTEM OWNERS AND USERS

START:
Problems, Opportunities,
Directives, Constraints,
and Vision

FINISH: Working Business Solution

Life-Cycle Stage

1 SCOPE DEFINITION

2 PROBLEM ANALYSIS

3 REQUIREMENTS ANALYSIS

4 LOGICAL DESIGN

5 DECISION ANALYSIS

6 PHYSICAL DESIGN & INTEGRATION

7 CONSTRUCTION & TESTING

8 INSTALLATION & DELIVERY

SYSTEM OPERATION & MAINTENANCE

Statement of Work

Problem Statement

Scope & Vision

System Improvement Objectives

Business Requirements Statement

Logical Design

Application Architecture

System Proposal

Redesigned Business Processes

Design Prototypes

Physical Design Specifications

Training Materials

Functional System

Operational System

Post-Audit Review

Documentation

FIGURE 12-5 The Context of Systems Design for In-House Development

454

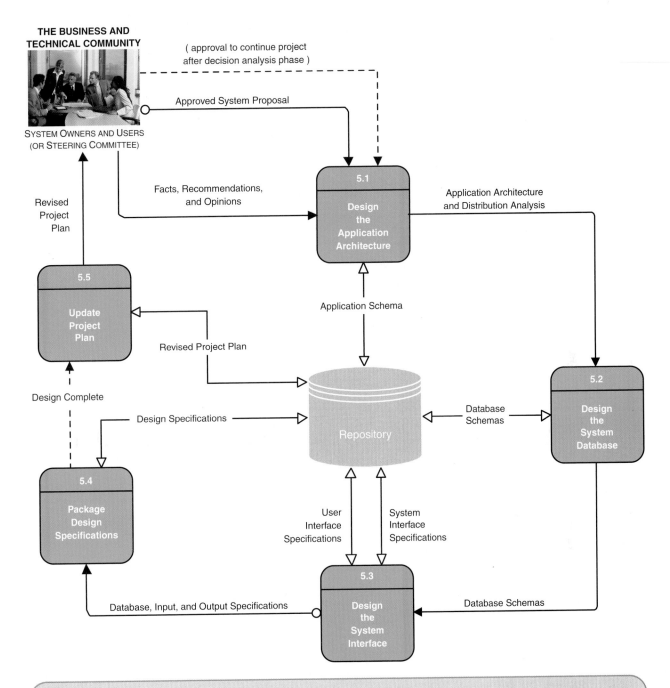

THE BUSINESS AND TECHNICAL COMMUNITY

SYSTEM OWNERS AND USERS (OR STEERING COMMITTEE)

(approval to continue project after decision analysis phase)

Approved System Proposal

Facts, Recommendations, and Opinions

Revised Project Plan

5.1 Design the Application Architecture

Application Architecture and Distribution Analysis

5.5 Update Project Plan

Revised Project Plan

Application Schema

Design Complete

Repository

Database Schemas

5.2 Design the System Database

5.4 Package Design Specifications

Design Specifications

User Interface Specifications

System Interface Specifications

Database, Input, and Output Specifications

5.3 Design the System Interface

Database Schemas

FIGURE 12-6 The Systems Design Tasks for In-House Development

physical data flow diagram (PDFD) that is used to establish physical processes and data stores (databases) across a network (see Figure 12-7). You will learn about PDFDs to document application architecture in Chapter 13.

To complete this activity, the analyst may involve a number of SYSTEM DESIGNERS and SYSTEM USERS. System users may be involved in this activity to help address business data, process, and location issues. Several different SYSTEM DESIGNER specialists may be instrumental in the completion of this activity, including a *data and database administrator, network administrator and engineers, applications administrator,* and various other experts, as needed (e.g., an expert on automatic data capture for addressing bar-coding technology and issues).

physical data flow diagram a process model used to communicate the technical implementation characteristics of an information system.

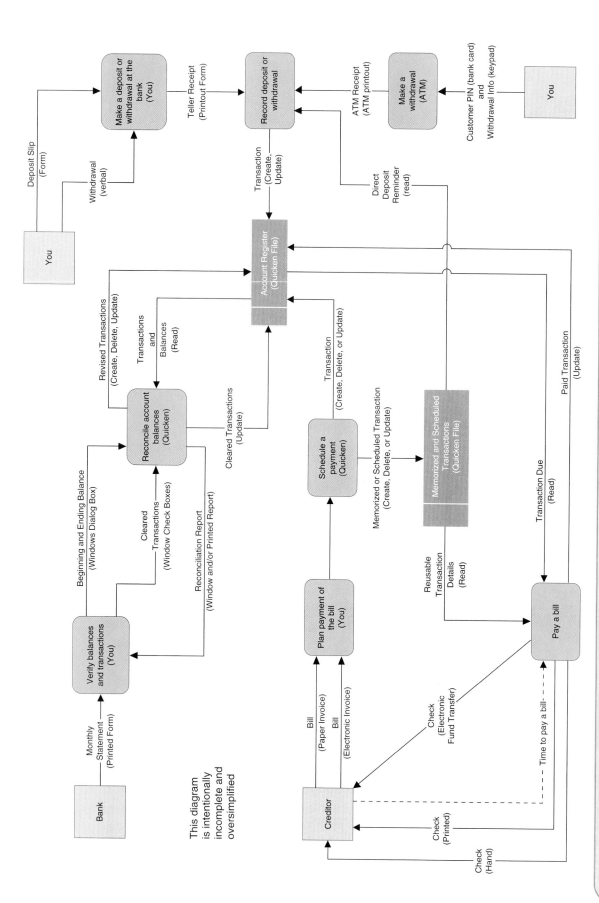

This diagram is intentionally incomplete and oversimplified

FIGURE 12-7 A Sample Physical Data Flow Diagram

The key inputs to this task are the facts, recommendations, and opinions that are solicited from various sources and the approved system proposal from the decision analysis phase. The principal deliverable of the task is the application architecture and distribution analysis that serves as a blueprint for subsequent detailed design phase activities.

> Task 5.2—Design the System Database(s)

Typically the next system design task is to develop the corresponding database design specifications. The design of data goes far beyond the simple layout of records. Databases are a shared resource. Many programs will typically use them. Future programs may use databases in ways not originally envisioned. Consequently, the designer must be especially attentive to designing databases that are adaptable to future requirements and expansion.

The designer must also analyze how programs will access the data in order to improve performance. You may already be somewhat familiar with various programming data structures and their impact on performance and flexibility. These issues affect database organization decisions. Other issues to be addressed during database design include record size and storage volume requirements. Finally, because databases are shared resources, the designer must also design internal controls to ensure proper security and disaster recovery techniques, in case data is lost or destroyed.

The purpose of this task is to prepare technical design specifications for a database that will be adaptable to future requirements and expansion. While the SYSTEMS ANALYSTS who may participate in database modeling facilitate this task, the SYSTEM DESIGNERS are responsible for the completion of this activity. The *data administrator* may participate (or complete) the database design. Recognize that the new system most likely uses some portion of an existing database. This is where the knowledge of the database administrator is crucial. Finally, SYSTEM BUILDERS may also participate when asked to build a prototype database for the project.

As is illustrated in Figure 12-6, a key input to this activity is the application architecture and distribution analysis decisions from the prior design task. The deliverable of the task includes the resulting database schemas. An example of a database schema was presented earlier, in Figure 12-2. A *database schema* is the structural model for a database. It is a picture or map of the records and relationships to be implemented by the database. You will learn how to develop database schemas in Chapter 14.

> Task 5.3—Design the System Interface

Once the database has been designed and possibly a prototype built, the systems designer can work closely with system users to develop input, output, and dialogue specifications. Because end users and managers will have to work with inputs and outputs, the designers must be careful to solicit their ideas and suggestions, especially regarding format. Their ideas and opinions must also be sought regarding an easy-to-learn and easy-to-use dialogue for the new system.

Transaction outputs will frequently be designed as preprinted forms onto which transaction details will be printed. Reports and other outputs are usually printed directly onto paper or displayed on a terminal screen. The precise format and layout of the outputs must be specified. Finally, internal controls must be specified to ensure that the outputs are not lost, misrouted, misused, or incomplete. Figure 12-8 is a sample output design. You will learn how to design outputs in Chapter 15.

For inputs, it is crucial to design the data capture method to be used. For instance, you may design a form on which data to be input will be initially recorded. You want

FIGURE 12-8

A Sample Output
Prototype Screen

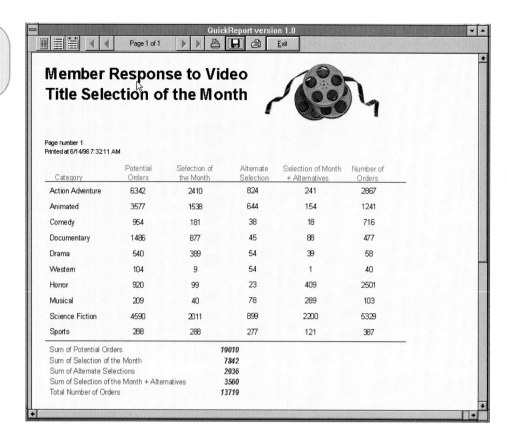

to make it easy for the data to be recorded on the form, but you also want to simplify the entry of the data from the form into the computer or onto a computer-readable medium. This is particularly true if the data is to be input by people who are not familiar with the business application. Also, any time you input data to the system, you can make mistakes. We need to define editing controls to ensure the accuracy of input data. A sample input prototype screen was depicted earlier, in Figure 12-3. You will learn how to design inputs in Chapter 16.

For interface or dialogue design, the design must consider such factors as terminal familiarity, possible errors and misunderstandings that the end user may have or may encounter, the need for additional instructions or help at certain points, and screen content and layout. You are trying to anticipate every little error or keystroke that an end user might make—no matter how improbable. Furthermore, you are trying to make it easy for the end user to understand what the screen is displaying at any given time. Figure 12-9 is a sample interface design. You will learn how to do interface design in Chapter 17.

SYSTEM USERS should be involved in this activity! The inputs, outputs, and interface dialogues are what they will see and work with. The degree to which they are involved is emphasized in design efforts that involve prototyping. They will be asked to provide feedback regarding each input/output prototype. SYSTEM DESIGNERS are responsible for the completion of this activity. They may draw on the expertise of systems designers that specialize in *graphical user interface* design. In addition, SYSTEM BUILDERS may construct the various screen designs for the users to review during design by prototyping.

As was illustrated in Figure 12-6, the key input to this activity is the database schema(s) from the previous task and the user and system interface specifications that are available from the project's repository. The deliverable of the design task is the completed database, input, and output specifications.

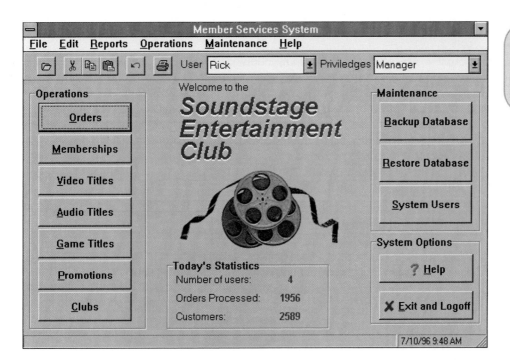

FIGURE 12-9

A Sample Dialogue Interface Prototype Screen

> Task 5.4—Package Design Specifications

This final design task involves packaging all the specifications from the previous design tasks into a set of specifications that will guide the computer programmer's activities during the construction phase of the systems development methodology.

There is more to this task than packaging, however. How much more depends on two things: (1) where you draw the line between the system designer's and computer programmer's responsibilities, and (2) whether the methodology and solution calls for the design of the overall program structure. Most organizations have adopted accelerated systems development approaches that do not require the latter. Program structure dealt with quality issues that were of concern to developers of systems that used older programming languages and tended to be mainframe-based applications.

The SYSTEMS ANALYST, who may be aided by the SYSTEM DESIGNERS, usually completes this task. Before proceeding with the packaging of the design specifications and the construction phase, the systems design should be reviewed with all appropriate audiences. While SYSTEM USERS have already seen and approved the outputs, inputs, and dialogue for the new system, the overall work and data flow for the new system should get a final walkthrough and approval. SYSTEM OWNERS should get a final chance to question the project's feasibility and determine whether the project should be adjusted, terminated, or approved to proceed to construction. At this stage of a project the company's *audit staff* may become heavily involved. The staff will pass judgment on the internal controls in a new system.

As was illustrated in Figure 12-6, the inputs to this task are the various database, input, and output specifications that were created earlier. Once these specifications have been reviewed, approved, and organized as design specifications that are suitable for constructing the new system, they are made available to the team of system builders via the project repository. It is more common for a project manager to make design specifications available via a shared repository than to provide each individual developer with a copy of a printed set of organized specifications.

> Task 5.5—Update the Project Plan

Now that we're approaching the completion of the design phase, we should reevaluate project feasibility and *update the project plan* accordingly. The *project manager,* in conjunction with SYSTEM OWNERS and the entire project team, facilitates this task. The SYSTEMS ANALYSTS and SYSTEM OWNERS are the key individuals in this task. The analysts and owners should consider the possibility that, based on the completed design work, the overall project schedule, cost estimates, and other estimates may need to be adjusted.

As shown in Figure 12-6, this task is triggered when the project manager determines that the design is complete. The key deliverable of the task is the updated project plan. The updated plan should now include a detailed plan for the construction phase that should follow. Recall that the techniques and steps for updating the project plan were taught in Chapter 4, Project Management.

Systems Design for Integrating Commercial Software— The "Buy" Solution

Let's now examine systems design for solutions that involve acquiring a commercial off-the-shelf (COTS) software product. The life cycle for projects that involve purchase, or "buy," solutions is illustrated in Figure 12-10. Notice that the business requirements statement (for software) and its integration as a business solution trigger a series of phases absent from the in-house development process we just learned about. The most notable differences between the buy and the in-house development projects is the inclusion of a new *procurement phase* and a special decision analysis phase (process labeled "5A") to address software and services.

When new software is needed, the selection of appropriate products is often difficult. Decisions are complicated by technical, economic, and political considerations. A poor decision can ruin an otherwise successful analysis and design. The systems analyst is becoming increasingly involved in the procurement of software packages (as well as peripherals and computers to support specific applications being developed by that analyst). The purpose of the procurement and decision analysis phases is to do the following:

1. Identify and research specific products that could support our recommended solution for the target information system.
2. Solicit, evaluate, and rank vendor proposals.
3. Select and recommend the best vendor proposal.
4. Contract with the awarded vendor to obtain the product.

In this section we will examine the tasks involved in completing the procurement and decision analysis phases for a buy solution. As is depicted in Figure 12-10, a buy solution affects how other phases in the life cycle are also completed (phases that are impacted are shaded in light blue). After examining the procurement and decision analysis phases, we will explore the impacts that a buy solution would have on how those phases would be completed.

Figure 12-11 is a task diagram depicting the work (= tasks) that should be performed to complete the procurement and decision analysis phases for a buy project solution. This task diagram does not mandate any specific methodology, but we will describe in the accompanying paragraphs the approaches, tools, and techniques you might want to consider for each design task. This task diagram is only a template. The project team and project manager may expand on or alter the template to reflect the unique needs of any given project.

The first two tasks (4.1 and 4.2) are procurement phase tasks, and the remaining tasks (5A.1, 5A.2, and 5A.3) are decision analysis–related tasks. Let's now examine each task in detail.

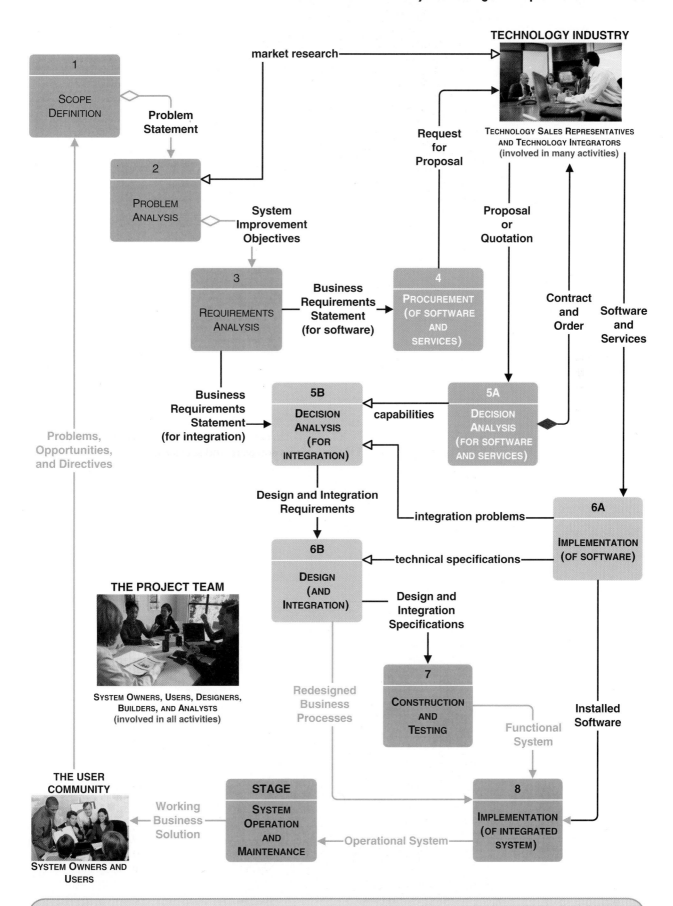

FIGURE 12-10 The Context of Systems Design for Commercial Off-the-Shelf Software Solution

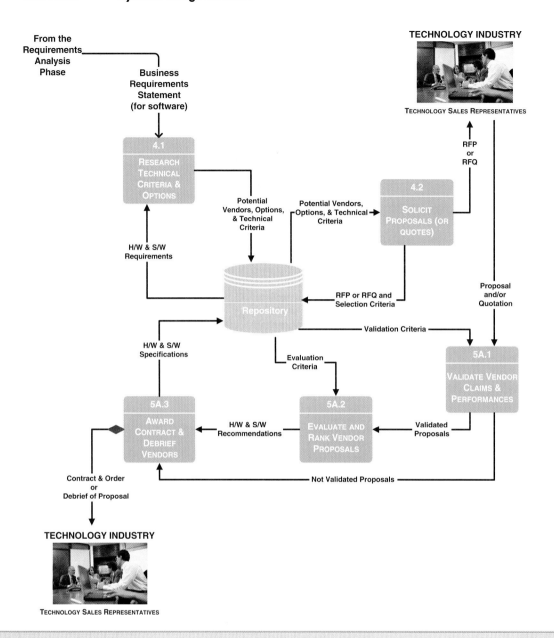

FIGURE 12-11 Tasks for the Procurement Phase

> Task 4.1—Research Technical Criteria and Options

The first task is to research technical alternatives. This task identifies specifications that are important to the software and/or hardware that is to be selected. The task involves focusing on the software and/or hardware requirements established in the requirements analysis phase. These requirements specify the functionality, features, and critical performance parameters for our new software/hardware.

Most analysts read appropriate magazines and journals and search the Internet to help them identify the technical and business issues and specifications that will become important to the selection decision. Other sources of information for conducting research include the following:

- *Internal standards* may exist for hardware and software selection. Some companies insist that certain technology will be bought from specific vendors if those vendors offer it. For instance, some companies have standardized on specific brands of microcomputers, terminals, printers, database management systems, network managers, data communications software, spreadsheets, and programming languages. A little homework here can save you a lot of unnecessary research.

- *Information services* are primarily intended to constantly survey the marketplace for new products and advise prospective buyers on what specifications to consider. They also provide information such as the number of installations and general customer satisfaction with the products.
- *Trade newspapers and periodicals* offer articles and experiences on various types of hardware and software that you may be considering. Many can be found in school and company libraries. Subscriptions (sometimes free) are also available.

The research should also identify potential vendors that supply the products to be considered. After the analysts have completed their homework, they will initiate contact with these vendors. Thus, the analysts will be better equipped to deal with vendor sales pitches after doing their research!

The purpose of this task is to research technical alternatives to specify important criteria and options that will be important for the new hardware and/or software that is to be selected. This task is facilitated by the *project manager.* SYSTEM DESIGNERS are responsible for the completion of this task. The designer may seek input from various technical experts, including data and database administrators, network administrators, and applications administrators.

As is illustrated in Figure 12-11, a key input to this task is the business requirements statement (for software) established in the requirements analysis phase. The designer will also obtain additional product and vendor facts from various sources. Designers are careful not to get their information solely from a salesperson—not that sales representatives are dishonest, but the number-one rule of salesmanship is to emphasize the product's strengths and deemphasize its weaknesses. The principal deliverable of this task includes a list of potential vendors, product options, and technical criteria.

To complete this task, designers must conduct extensive research to gain important facts concerning the hardware/software product and vendor. They must be careful to screen their various sources. The sources are used to identify potential vendors from which the products might be obtained. This step may be optional if your company has a commitment or contract to acquire certain products from a particular source. Finally, the designer must review the product, vendor, and supplier findings.

> Task 4.2—Solicit Proposals or Quotes from Vendors

The next task is to solicit proposals or quotes from vendors. If your company is committed to buying from a single source (IBM, for example), the task is quite informal. You simply contact the supplier and request price quotations and terms. But most decisions offer numerous alternatives. In this situation, good business sense dictates that you use the competitive marketplace to your advantage.

The solicitation task requires the preparation of one of two documents: a **request for quotations (RFQ)** or a **request for proposals (RFP)**. The request for quotations is used when you have already decided on the specific product but that product can be acquired from several distributors. Its primary intent is to solicit specific configurations, prices, maintenance agreements, conditions regarding changes made by buyers, and servicing. The request for proposals is used when several different vendors and/or products are candidates and you want to solicit competitive proposals and quotes. RFPs can be thought of as a superset of RFQs. Both define selection criteria that will be used in a later validation.

request for quotation (RFQ) a formal document that communicates business, technical, and support requirements for an application software package to a single vendor that has been determined as being able to supply that application package and services.

request for proposal (RFP) a formal document that communicates business, technical, and support requirements for an application software package to vendors that may wish to compete for the sale of that application package and services.

The primary purpose of the RFP is to communicate requirements and desired features to prospective vendors. Requirements and desired features must be categorized as mandatory (must be provided by the vendor), extremely important (desired from the vendor but can be obtained in-house or from a third-party vendor), or desirable (can be done without). Requirements might also be classified by two alternate criteria: those that satisfy the needs of the systems and those that satisfy our needs from the vendor (for example, service).

This task is facilitated by the *project manager.* The SYSTEM DESIGNER is also responsible for completing this activity and may seek the input from *data* and *database administrators, network administrators,* and *applications administrators* when writing the RFP or RFQ.

The key input to this task is the potential vendors, options, and technical criteria that resulted from previous research. The principal deliverable of this task is the RFP or RFQ that is to be received by candidate vendors. The quality of an RFP has a significant impact on the quality and completeness of the resulting proposals. A suggested outline for an RFP is presented in Figure 12-12, since an actual RFP is too lengthy to include in this book.

FIGURE 12-12 Request for Proposals

Request for Proposals (RFP)

I. Introduction
   A. Background
   B. Brief summary of needs
   C. Explanation of RFP document
   D. Call for action on part of vendor
II. Standards and instructions
   A. Schedule of events leading to contract
   B. Ground rules that will govern selection decision
      1. Who may talk with whom and when
      2. Who pays for what
      3. Required format for a proposal
      4. Demonstration expectations
      5. Contractual expectations
      6. References expected
      7. Documentation expectations
III. Requirements and features
   A. Hardware
      1. Mandatory requirements, features, and criteria
      2. Essential requirements, features, and criteria
      3. Desirable requirements, features, and criteria
   B. Software
      1. Mandatory requirements, features, and criteria
      2. Essential requirements, features, and criteria
      3. Desirable requirements, features, and criteria
   C. Service
      1. Mandatory requirements
      2. Essential requirements
      3. Desirable requirements
IV. Technical questionnaires
V. Conclusion

Many of the skills you developed in Part Two, such as process and data modeling, can be very useful for communicating requirements in the RFP. Vendors are very receptive to these tools because they find it easier to match products and options and package a proposal that is directed toward your needs. Other important skills include report writing (discussed in Chapter 11) and questionnaires (covered in Chapter 6).

> Task 5A.1—Validate Vendor Claims and Performances

Soon after the RFPs or RFQs are sent to prospective vendors, you will begin receiving proposal(s) and/or quotation(s). Because proposals cannot and should not be taken at face value, claims and performance must be validated. This task is performed independently for each proposal; proposals are not compared with one another.

The purpose of this task is to validate requests for proposals and/or quotations received from vendors. SYSTEM DESIGNERS are responsible for the completion of this activity. Once again, the designer may involve the following individuals in validating the proposals: *data* and *database administrators, network administrators,* and *applications administrators.*

This task is triggered by the receipt of proposal(s) and/or quotation(s) from prospective vendors. The key outputs of this task are those vendor proposals that proved to be validated proposals or claims and others whose claims were not validated.

To complete this task, the designer must collect and review all facts pertaining to the product requirements and features. The designer must review the vendor proposals and should eliminate any proposal that does not meet all the mandatory requirements. If the requirements were clearly specified, no vendor should have submitted such a proposal. For proposals that cannot meet one or more extremely important requirements, verify that the requirements or features can be fulfilled by some other means. For each vendor proposal not eliminated, the designer must validate the vendor claims and promises against validation criteria. Claims about mandatory, extremely important, and desirable requirements and features can be validated by completed questionnaires and checklists (included in the RFP) with appropriate vendor-supplied references to user and technical manuals. Promises can be validated only by ensuring that they are written into the contract. Finally, performance is best validated by a demonstration, which is particularly important when you are evaluating software packages. Demonstrations allow you to obtain test results and findings that confirm capabilities, features, and ease of use.

> Task 5A.2—Evaluate and Rank Vendor Proposals

The validated proposals can now be evaluated and ranked. The evaluation and ranking is, in reality, another cost-benefit analysis performed during systems development. The evaluation criteria and scoring system should be established before the actual evaluation occurs so as not to bias the criteria and scoring to subconsciously favor any one proposal.

The executive sponsor, ideally, should facilitate this task. SYSTEM DESIGNERS are responsible for the completion of this activity. The designer may involve several experts in evaluating and ranking the proposals, including *data* and *database administrators, network administrators,* and *applications administrators.*

The inputs to this task include validated proposals and the evaluation criteria to be used to rank the proposals. The key deliverable of this task is the hardware and/or software recommendations.

The ability to perform a feasibility assessment is an extremely important skill requirement for completing this task. Feasibility assessment techniques and skills were covered in Chapter 11. To complete this task, designers must first collect and review all details concerning the validated proposals. They must then establish an evaluation criteria and scoring system. There are many ways to go about this. Some methods suggest that requirements be weighted on a point scale. Better approaches use dollars and cents! Monetary systems are easier to defend to management than points. One such technique is to evaluate the proposals on the basis of "hard" and "soft" dollars. Hard-dollar costs are the costs you will have to pay to the selected vendor for the equipment or software. Soft-dollar costs are additional costs you will incur if you select a particular vendor (for instance, if you select vendor A, you may incur an additional expense to vendor B to overcome a shortcoming of vendor A's proposed system). This approach awards the contract to the vendor who fulfills all essential requirements while offering the lowest total hard-dollar cost plus soft-dollar penalties for desired features not provided (for a detailed explanation of this method, see Isshiki, 1982, or Joslin, 1977, in the Suggested Readings). Once the evaluation criteria and scoring system have been established, the last step toward completing our task is to do the actual evaluation and ranking of the vendor proposals.

> Task 5A.3—Award (or Let) Contract and Debrief Vendors

Having ranked the vendor proposals, the next activity usually includes presenting a recommendation to management for final approval. Once again, communication skills, especially salesmanship, are important if the analyst is to persuade management to follow the recommendations. Given management's approval of the recommendation, a contract must then be drawn up and awarded to the winning vendor. This activity often also includes debriefing losing vendors, being careful not to burn bridges.

The purpose of this activity is to negotiate a contract with the vendor who supplied the winning proposal and to debrief the vendors that submitted losing proposals. Ideally, the executive sponsor who must approve recommendations and project continuation should facilitate the activity. But it is the SYSTEM DESIGNER who must make and defend the recommendation and award the contract. In doing so, the system designer may involve a company lawyer in drafting the contract. Report writing and presentation skills are important for completing this task.

The key inputs include the hardware and software recommendation and the non-validated proposals from the previous evaluation tasks. Pending the approval of the executive sponsor, a contract order would subsequently be produced for the winning vendor. A debriefing of proposals would be provided for the losing vendors.

To complete this task, the designer must first present a hardware and software recommendation for final approval. Once the final hardware and software approval decision is made, a contract must then be negotiated with the winning vendor. Certain special conditions and terms may have to be written into the standard contract and order. Ideally, no computer contract should be signed without the advice of a lawyer. The analyst must be careful to read and clarify all licensing agreements. No final decision should be approved without the consent of a qualified accountant or management. Purchasing, leasing, and leasing with a purchase option involve complex tax considerations. Finally, out of common courtesy and to maintain good relationships, provide a debriefing of proposals for losing vendors. The purpose of this meeting is not to allow the vendors a second chance to be awarded the contract; rather, the briefing is intended to inform the losing vendors of precise weaknesses in their proposals and/or products.

> Impact of Buy Decision on Remaining Life-Cycle Phases

It is not enough merely to purchase or build systems that fulfill the target system requirements. The analyst must integrate or interface the new system to the myriad of other existing systems that are essential to the business. Many of these systems may use dramatically different technology, techniques, and file structures.

The analyst must consider how the target system fits into the federation of systems of which it is a part. The integration requirements that are specified are vital to ensuring that the target system will work in harmony with those systems.

As was depicted in Figure 12-10, the decision to buy a commercial software package solution can impact additional phases (denoted in light blue) of the life cycle. Upon completion of the decision analysis (for software and services) phase and its intensive evaluation of the commercial product, we have become knowledgeable about the product's capabilities (or shortcomings). During decision analysis for integration we will need to make revisions to reflect this new knowledge in our data and process models that comprised the business requirements statement. When software and services are received from the vendor(s), the software must be implemented. During implementation we may encounter integration problems that must also be reflected in our business requirements statement. These capabilities and integration problems are reflected in the design and integration requirements.

Finally, given the design and integration requirements we must now complete the design phase. Completion of the design phase involves many of the same tasks that were discussed earlier in the chapter. The primary difference is that we are not simply "developing" an entire system. Rather, we may be designing technical specifications for developing a small subset of programs, software utilities, and other components necessary for the business processes and the commercial software product to be integrated and work together properly. Let's consider an example. Our existing business system may use bar-coding technology to capture data. Yet our software product may require that data be entered via the keyboard. We may need to customize the software product to allow data to be entered via the keyboard or from a batch file containing scanned data.

This chapter provided a detailed overview of systems design for a project. You are now ready to learn some of the systems design skills introduced in this chapter. Because systems design is dependent on requirements specified during systems analysis, we recommend that you first complete Chapters 5 through 11. Chapter 5 gives you an overview of systems analysis. Chapters 6 through 11 teach different system analysis tools and techniques that provide for basic inputs to the systems design activities presented in Part Three.

The order of the system design chapters that follow is flexible.

Learning Roadmap

Chapter Review

1. Formally, information systems design is defined as those tasks that focus on the specifications of a detailed computer-based solution. Whereas systems analysis emphasizes the business problem, systems design focuses on the technical or implementation concerns of the system.

2. Systems design is driven by the technical concerns of system designers. Therefore, with respect to the information systems building blocks, systems design addresses the information system building blocks from the system designer's perspective.

3. Systems design differs for in-house development, or "build" projects, versus "buy" projects, where a systems software package is bought.

4. There are many popular strategies or techniques for performing systems design. These techniques can be used in combination with one another.

   a. Modern structured design, a technique that focuses on processes.
   b. Information engineering (IE), a technique that focuses on data and strategic planning to produce application projects.
   c. Prototyping, a technique that is an iterative process involving a close working relationship between designers and users to produce a model of the new system.
   d. Joint application development (JAD), a technique that emphasizes participative development among system owners, users, designers, and builders. During JAD sessions for systems design, the system designer takes on the role of the facilitator.
   e. Rapid application development (RAD), a technique that represents a merger of various structured techniques with prototyping and JAD to accelerate systems development.

   f. Object-oriented design (OOD), a new design strategy that follows up object-oriented analysis to refine object requirement definitions and to define new design-specific objects.

5. For in-house development (build) projects, the systems design involves developing technical design specifications that will guide the construction and implementation of the new system. To complete the design phase, the system designer must complete the following tasks:

   a. Design the application architecture.
   b. Design the system database(s).
   c. Design the system interface.
   d. Package the design specifications.
   e. Update the project plan.

6. Systems design for solutions that involve acquiring a commercial off-the-shelf (COTS) software product include a procurement and decision analysis phase that addresses software and services. Completion of these phases involves the following tasks:

   a. Research technical criteria and options.
   b. Solicit proposals (or quotes) from vendors.
   c. Validate vendor claims and performances.
   d. Evaluate and rank vendor proposals.
   e. Award (or let) contract and debrief vendors.

7. It is not enough merely to purchase or build systems that fulfill the target system requirements. The analyst must integrate or interface the new system to the myriad of other existing systems that are essential to the business. Many of these systems may use dramatically different technology, techniques, and file structures.

Review Questions

1. What is the essential difference between systems analysis and systems design?
2. What are some of the different model-driven methodologies?
3. What are some of the benefits of prototyping?
4. What are the five high-level tasks involved in conducting system design for a development project to be built in-house?
5. Why is it necessary to design the application architecture?
6. In designing the system database(s), what should designers always keep in mind?

7. What is a database schema?
8. What is the goal when designing the system interface?
9. What specific factors should system designers focus on when designing the system interface?
10. What is the phase needed in systems design if the software is being purchased instead of being developed in-house? What is the purpose of this additional phase?
11. What is a request for quotations (RFQ)?

Problems and Exercises

1. What is the primary target of systems design and what phases are included in systems design? If the systems analysis was poorly done or incomplete, can a good systems design effort overcome that?

2. Match the terms in the first column with the definitions or examples in the second column.

1.	Information engineering	A. Information engineering
2.	JAD	B. Structured design module properties
3.	Modern structured design	C. Participative development emphasis
4.	Prototyping	D. IBM's Rational
5.	System design	E. Derived model from structured design
6.	Physical Entity Relationship	F. Combined data and process
7.	Coupling and cohesion	G. Model-driven, data-centered, process-sensitive technique
8.	RAD	H. Pictorial system models emphasis
9.	Model-driven design	I. Functional incomplete model built using RAD
10.	Code, implement, and repair	J. Process decomposition technique
11.	Repository-based CASE tool	K. Computer-based solution specification tasks
12.	OOD	L. Merger of JAD, prototyping, and structured techniques
13.	Structure chart	M. Potential prototyping pitfall

3. Prototyping has many strengths, but it also has a number of weaknesses and hazards. Discuss some of these weaknesses and hazards. What strategies could be implemented to reduce the risk of their occurring?

4. Consider the issues raised by the preceding question and write a one- or two-page policy and procedures memorandum to all systems analysts and designers in your organization regarding prototyping.

5. You are a systems designer in an organization. One of the other designers on your team has recently retired. Your manager comes to you and asks you to sit in on the interviews to find a replacement. What qualities should you look for when you do the interviews?

6. You are designing a data interface screen for a new system that is under development. The purpose of this data interface screen is to enter changes of address submitted by drivers to their state's DMV. Each key data operator will enter on this screen from these hand-printed forms about a thousand changes of address per day. What is one very important principle to keep in mind?

7. In your organization, it is traditional to give everyone involved in the project a printed copy of the design specifications after they have been approved. It costs more, but management feels this is one way to acknowledge everyone's effort on the project and to keep people committed. If the organization doesn't mind the cost, is there anything wrong with this?

8. Complete the following sentences:

A critical part of designing the _____ is deciding how to distribute the system's data, _____ and _____ to different _____.

Databases are a resource typically _____ by many _____ and they may be used by future _____ not yet known for purposes _____.

In designing _____, the key is to make it _____ for the _____ to understand what to do next, and to anticipate every type of _____ that a user could make.

9. You are a systems designer who is responsible for reviewing vendor proposals. A vendor who has done satisfactory work for your company in the past has submitted a proposal that does not meet several critical requirements. What should you do?

10. Match the definitions or examples in the first column with the terms in the second column.

A.	Procurement phase	1. DBA
B.	Shows physical processes and databases across network	2. Distribution analysis
C.	Replaced text-based display	3. Application architecture
D.	Vendor evaluation criteria and scoring system	4. Database schema
E.	Competitive proposal solicitation document	5. RFQ
F.	Part of a blueprint for detailed design phase activities	6. Auditor

G. Specialist responsible
for database architecture

7. COTS

H. Solicitation document
for specific product

8. PDFD

I. Specialist responsible
for internal controls

9. Hard dollar
costs

J. Technologies used to
build information system

10. "Buy" solution

K. Structural model for
database

11. GUI

L. Commercial software
product

12. RFP

11. The life cycle for a project that you are working on involves a "buy" solution to purchase a commercial off-the-shelf product for the company's marketing specialists. Your company wants to solicit competitive proposals. Use the format shown in Figure 12–12 to prepare a request for proposal (REP). (Note: For purposes of this exercise, it is not necessary to develop a fully detailed RFP, but your RFP should contain at least the high-level details and information called for in each section.)

12. You work for a consulting company that has been hired to do the systems design portion of the project. The systems analysis portion was done by another consulting company. During systems design, you find what you are definitely sure is a mistake in the requirements. You are not sure just how serious it is, but you know this for sure: if this mistake, made by another company, is pointed out to the project manager, systems design work will have to be halted until the mistake is fixed. This will put your company behind schedule and either it will have to pay you and its other consultants for sitting around while the mistake is being fixed, or it will lay all of you off. What is your ethical obligation in this situation?

13. Data security and privacy are increasingly important issues. What are some examples of security and privacy issues that systems analysts, system designers, and database administrators need to be aware of in developing and maintaining a relational database system?

Projects and Research

1. You are a systems analyst who has been working for several years in the IT shop of a cabinet manufacturing company. The company is known for the quality of its products and for being an industry leader. The former chief information officer (CIO), who was from the "old school," recently retired and has just been replaced by a new, more dynamic and progressive CIO. The new CIO, in an effort to raise the maturity level of the organization, is conducting a series of brainstorming sessions to develop its first IT architecture plan. You and the other systems analysts and designers have each been asked to provide input on which systems design approach or approaches the organization should adopt as its approved standard. Use the information in the textbook, your own experience, and any supplemental research you conduct to write a memo to your CIO that:

a. Provides relevant background regarding your organization—for example, its vision, mission, strategic goals objectives, level of maturity, organizational structure, and culture.

b. Describes the different systems design approaches.

c. Compares and contrasts their methods, strengths, and weaknesses.

d. Recommends a specific approach or combination of approaches for adoption as the standard for your organization. The recommendation should also include a justification for the basis of your recommendation.

2. You are one of a large team of systems analysts and designers on an enterprise-level project that touches every part of your organization, both in its headquarters office and in regional offices throughout the country. Following the recommendation of its staff, executive management has decided to do the systems design in-house. Due to the scale of this project and the size of your organization, this project involves the participation of hundreds of system owners and users who are located both in headquarters and in the regional offices. For many of these system owners and users, this is the first time they have been involved in a project of this nature. One of your responsibilities is to make sure that they understand their respective roles in this phase, and its importance relative to the overall success of the project.

a. Write an e-mail (or e-mails) to the system owners and users in your organization regarding the design phase and their roles in it, using

Tasks 5.1–5.5 as a guide. Your objective is to ensure that they understand their roles and are committed to the success of the project.

b. After you compose the e-mail(s), explain the scenario to several people in your organization. Have them read and critique your e-mail(s) for clarity, completeness, and persuasiveness. What were the results? If this were a real situation, would they have understood their roles and would your explanation have had a positive impact upon their commitment?

3. In the first question, you were asked to look at a variety of approaches to systems design, including prototyping. In the past several years, the number of prototyping methodologies and application tools appears to have increased exponentially. Research on the Internet some of the different prototype technologies that are available. In addition, talk with several systems designers who use prototyping, and ask them for their thoughts regarding the different prototyping application tools on the market. Prepare a written analysis describing your research and reporting your findings.

4. You work in the IT shop of a sales organization with a half dozen satellite offices located in your state. The organization wants to develop and implement a Web-based information system so that its satellite offices can submit their sales reports on a real-time basis. But your IT shop is small, everyone is already fully committed to maintenance and support activities or to other projects, and besides, no one has any experience in developing a Web-based information system. So your management has decided to outsource the design and development. Your job is to do the following:

a. Interview IT vendors in your local community regarding their experiences with requests for proposals (RFPs); that is, find out what common deficiencies they see in RFPs, and what key things need to be included in order to prepare an appropriate proposal. Prepare a short memorandum to your management describing the interviews.

b. Research some of the different RFP templates that are available, including the template used in Figure 12-12. Select one, and explain why you chose the one that you did.

c. Using your selected template, write a request for proposal.

5. Now that you have completed the request for proposal in the preceding question, your next assignment is to plan the systems design phase of this project. Using what you have read to date regarding systems design, prepare a high-level project plan showing the major tasks, resources and estimated hours required, time frames, and dependencies (refer back to Chapter 4).

6. Numerous evaluation criteria and scoring systems are used by organizations in the public and private sectors to evaluate and rank vendor proposals. For instance, California has an optional "best value" approach that state agencies can use for vendor ranking and selection. In addition, the textbook references books on other methods in its list of Suggested Readings. Research on the Internet these and/or other approaches or methods used by private and public sector agencies. Also, try to interview the staff members in these organizations who are responsible for evaluating and ranking vendor proposals, as well as vendors that prepare the proposals.

Minicases

1. In the previous chapter, you worked on designing a system for a government department. Pick a specific task in the creation of that system (e.g. develop a Web site), and gather at least two proposals from different vendors for it. Validate claims and performances that the vendors submit to you. Analyze your findings and submit your results and recommendations to your professor.

2. You are developing a complex system for a large company. The code will be complex, and the language is fairly new to the programming team. Your boss has requested that you use an evolutionary prototype in the development. Is this the

appropriate prototype model to use? If it is not, how should you handle your boss's request?

3. In Chapter 8, you researched a car rental agency, and created a data model for the rental of cars. Utilize the work you previously did, as well as preliminary interviews, to create a prototype for a system that rents cars.

   Note (to student): Why do you need to do more interviews? This is so that you can develop an interface as well as functionality that the *user* wants and needs. Remember, aside from process functionality, the *user* will determine the success or failure of a system. What do *they* want the interface to be and act like? What format do *they* want to handle data in?

4. In the previous problem, you created a prototype for a car rental system based on a previous chapter's work and preliminary interviews. Now that you have created your prototype, return to your client and present your prototype.

   a. What is the client's reaction? Does the client like the functionality? What about the interface design? Document the responses, as well as the body language.

   b. Rework your system to incorporate the client's suggestions and wishes. Is there anything the client wanted that was unreasonable, or not feasible at this time? If there is, document it.

   c. Submit your initial prototype, your revised system prototype, and a paper discussing client needs, their response to your prototype, how you addressed their suggestions into your revised system, and any additional background information on your system.

Suggested Readings

Application Development Strategies (monthly periodical). Arlington, MA: Cutter Information Corporation. This is our favorite theme-oriented periodical that follows system development strategies, methodologies, CASE, and other relevant trends. Each issue focuses on a single theme.

Boar, Benard. *Application Prototyping: A Requirements Definition Strategy for the 80s.* New York: John Wiley & Sons, 1984. This is one of the first books to appear on the subject of systems prototyping. It provides a good discussion of when and how to do prototyping, as well as thorough coverage of the benefits that may be realized through this approach.

Coad, Peter, and Yourdon, Edward. *Object-Oriented Design,* 2nd ed. Englewood Cliffs, NJ: Yourdon Press, 1991. Chapter 1 is a great way to expose yourself to objects and the relationship of object methods to everything that preceded them.

Connor, Denis. *Information System Specification and Design Road Map.* Englewood Cliffs, NJ: Prentice Hall, 1985. This book compares prototyping with other popular analysis and design methodologies. It makes a good case for not prototyping without a specification.

Gane, Chris. *Rapid Systems Development.* Englewood Cliffs: NJ: Prentice Hall, 1989. This book presents a nice overview of RAD that combines model-driven development and prototyping in the correct balance.

Isshiki, Koichiro R. *Small Business Computers: A Guide to Evaluation and Selection.* Englewood Cliffs, NJ: Prentice Hall, 1982. Although it is oriented toward small computers, this book surveys most of the better-known strategies for evaluating vendor proposals. It also surveys most of the steps of the selection process, although they are not put in the perspective of the entire systems development life cycle.

Joslin, Edward O. *Computer Selection,* rev. ed. Fairfax Station, VA: Technology Press, 1977. Although somewhat dated, the concepts and selection methodology originally suggested in this classic book are still applicable. The book provides keen insights into vendor, customer, and end-user relations.

Lantz, Kenneth E. *The Prototyping Methodology.* Englewood Cliffs, NJ: Prentice Hall, 1986. This book provides excellent coverage of the prototyping methodology.

Wood, Jane, and Denise Silver. *Joint Application Design: How to Design Quality Systems in 40% Less Time.* New York: John Wiley & Sons, 1989. This book provides an excellent in-depth presentation of joint application development (JAD).

Yourdon, Edward. *Modern Structured Analysis.* Englewood Cliffs, NJ: Yourdon Press, 1989. Chapter 4, "Moving into Design," shows how modern structured design picks up from modern structured analysis.

Zachman, John A. "A Framework for Information System Architecture," *IBM Systems Journal* 26, no. 3 (1987). This article presents a popular conceptual framework for information systems design.

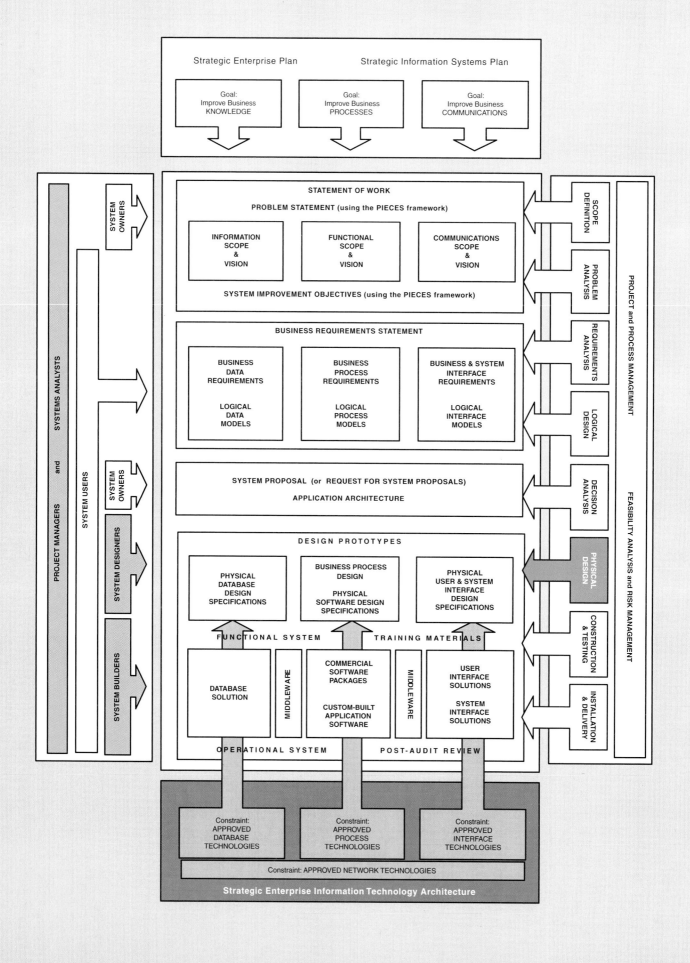

13

Application Architecture and Modeling

Chapter Preview and Objectives

This chapter teaches you techniques for designing the overall information system application architecture with a focus on physical process models. Information application architecture and physical process modeling include techniques for distributing knowledge, processes, and communications to network locations in a distributed computing environment. Physical data flow diagrams are used to document the architecture and design in terms of design units—cohesive collections of data and processes at specific locations—that can be designed, prototyped, or constructed in greater detail and subsequently implemented as stand-alone subsystems. You will know that you understand application architecture and process design when you can:

▌ Define an information system's architecture in terms of KNOWLEDGE, PROCESSES, and COMMUNICATIONS—the building blocks of all information systems. Consistent with modern trends, these building blocks will be distributed across a NETWORK.

▌ Differentiate between logical and physical data flow diagrams and explain how physical data flow diagrams are used to model an information system's architecture.

▌ Describe both centralized and distributed computing alternatives for information system design, including various client/server and Internet-based computing options.

▌ Describe database and data distribution alternatives for information system design.

▌ Describe user and system interface alternatives for information system design.

▌ Describe various software development environments for information system design.

▌ Describe strategies for developing or determining the architecture of an information system.

▌ Draw physical data flow diagrams for an information system's architecture and processes.

Introduction

It had been decided that part of the SoundStage Member Services system would be purchased, and part of it would be programmed in house. The question remained of what the architecture would be for the programmed part of the system. Bob Martinez had learned *C# .NET* in college and was a big fan. He pointed out that using the same language and the same *.NET* framework, they could program both a client/server desktop application for the in-house part of the application and a Web application for the *e*-commerce part of the application. The two applications could even share some components. And since the *C#* syntax was essentially the same as JavaScript, the server-side and client-side Web application code would be very similar.

Bob convinced his boss, Sandra. However, as she pointed out, SoundStage's approved application architecture was written before *.NET* was released and did not include it as a development option. Sandra had Bob research and write up a variance request for the technology committee. It took a couple of drafts, but he finally got it written in a way she thought the committee would approve. She was right. Now they could start designing "how" to implement the system, integrating the purchased components with *.NET*.

Application Architecture

Chapter 12 presented a high-level overview of the entire systems design process. You learned that early during system design you develop an architectural blueprint that will serve as an outline for subsequent internal and external design. This chapter focuses exclusively on that blueprint and current alternatives for application architecture. (Subsequent chapters focus on the detailed internal and external design of each architectural component.) The architectural blueprint will communicate the following design decisions:

- The degree to which the information system will be centralized or distributed—Most contemporary systems are distributed across networks, including both intranets and the Internet.
- The distribution of stored data across a network—Most modern databases are either distributed or duplicated across networks, either in a client/server or network computing pattern.
- The implementation technology for all software to be developed in-house—Which programming language and tools will be used?
- The integration of any commercial off-the-shelf software—And the need for customization of that software.
- The technology to be used to implement the user interface—Including inputs and outputs.
- The technology to be used to interface with other systems.

application architecture
a specification of the technologies to be used to implement information systems.

These considerations define the application architecture for the information system. An **application architecture** specifies the technologies to be used to implement one or more (possibly all) information systems. It serves as an outline for detailed design, construction, and implementation.

In most chapters, we have initially taught concepts and principles before introducing tools and techniques. For this chapter, we are going to first introduce the primary tool, physical data flow diagrams. This will work for two reasons. First, you already know the system concepts and basic constructs of data flow diagrams from Chapter 9. Second, the tool is an elegant and relatively simple way to introduce the different types of application architecture that we want you to learn.

Although you will learn a new technique in this chapter, physical data flow diagrams, this is not as important as the application architecture concepts used to partition an information system across a computer network.

Physical Data Flow Diagrams

Data flow diagrams (DFDs) were introduced in Chapter 9 as a systems analysis tool for modeling the *logical* (meaning "nontechnical") business requirements of an information system. With just a few extensions of the graphical language, DFDs can also be used as a systems design tool for modeling the *physical* (meaning "technical") architecture and design of an information system. **Physical data flow diagrams** model the technical and human design decisions to be implemented as part of an information system. They communicate technical choices and other design decisions to those who will actually construct and implement the system. In other words, physical DFDs serve as a technical blueprint for system construction and implementation.

physical data flow diagram a process model used to communicate the technical implementation characteristics of an information system.

Physical data flow diagrams were conceived by Gane, Sarson, and DeMarco as part of a formal software engineering methodology called *structured analysis and design*. This methodology was especially well suited to mainframe *COBOL* transaction-based information systems and software. The methodology required rigorous and detailed specification of both logical and physical representations of an information system. In sequence, systems analysts or software engineers would develop the following system models and associated detailed specifications:

1. *Physical DFDs of the current system*—These physical DFDs were intended to help analysts identify and analyze physical problems in the existing system during the problem analysis phase of systems analysis.
2. *Logical DFDs of the current system*—These logical DFDs were merely a transformation of the above physical DFDs that remove all physical detail. They were used as a point of departure for the requirements analysis phase of systems analysis.
3. *Logical DFDs of the target system*—These logical DFDs and their accompanying specifications (data structures and Structured English) were intended to represent the detailed nontechnical requirements for the new system.
4. *Physical DFDs of the target system*—These physical DFDs were intended to propose and model the technology choices and design decisions for all logical processes, data flows, and data stores. These diagrams (the focus of this chapter) are developed during the systems design stage of the project.
5. *Structure charts of the software elements of the target system*—The above physical DFDs would be transformed into structure charts that illustrate a top-down hierarchy of software modules that would conform to accepted principles of good software design.

The above methodology was labor-intensive and required significant precision and rigor to accomplish its intended result. Today, the complete structured analysis and design methodology as described above is rarely practiced—it is not as well suited to today's object-oriented and component-based software technologies—but data flow diagramming (both logical and physical) remains a useful and much practiced legacy of the structured analysis and design era of systems development.

In this chapter, we will examine the graphical conventions for *physical* DFDs. Physical DFDs use the same basic shapes and connections as logical DFDs (Chapter 9), namely: (a) *processes,* (b) *external agents,* (c) *data stores,* and (d) *data flows.* A sample physical DFD is shown in Figure 13-1. For now, just notice that the physical DFD primarily shows more technical and implementation detail than its logical DFD equivalent.

> Physical Processes

Recall that processes are the key shapes on any DFD. That's why they are called *process models.* Physical DFDs depict the planned, physical implementation of each process. A *physical process* is either a *processor,* such as a computer or person, or the

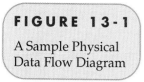

FIGURE 13-1

A Sample Physical Data Flow Diagram

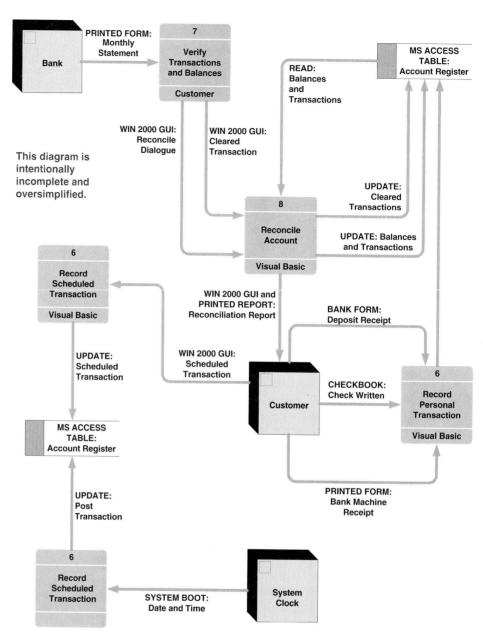

technical implementation of specific work to be performed, such as a computer program or manual process.

Earlier in the project, during requirements analysis, we specified *logical* processes needed to fulfill essential business requirements. These logical processes were modeled in the logical data flow diagrams in Chapter 9. Now, during system design, we must specify how these logical processes will be physically implemented. As implied in the above definition for physical processes, there are two characteristics of physical data flow diagrams:

- Logical processes are frequently <u>assigned to</u> specific physical processors such as PCs, servers, mainframes, people, or other devices in a computer network. To this end, we might draw a physical DFD to model the network's structure.
- Each logical process must be <u>implemented as</u> one *or more* physical processes as some logical processes must be split into multiple physical processes for one or more of the following reasons:
  — To split the process into a portion to be performed by people and a portion to be performed by the computer.

— To split the process into a portion to be implemented with one technology and a portion to be implemented with a different technology.
— To show multiple but different implementations of the same logical process (such as one process for paper orders and a different process for Internet orders).
— To add processes that are necessary to handle exceptions or to implement security requirements and audit trails.

In all cases, if you split a logical process into multiple physical processes, or add additional physical processes, you have to add all necessary data flows to preserve the essence of the original logical process. In other words, the physical processes must still meet the logical process requirements.

IDs are optional, but they can be useful for matching physical processes with their logical counterparts (especially if the logical process is to be implemented with multiple physical processes). Process names use the same action verb + noun/object clause convention as the one we introduced in Chapter 9. This name is recorded in the center of the shape (see margin). In the bottom of the shape, the implementation is recorded. This convention may have to be adjusted depending on the capabilities of your CASE or automated diagramming tool. The following names demonstrate various possible implementations of the same logical process:

	ID (optional)
	Action Verb
	+
	Noun or Object Phrase
	Implementation

Logical Process	Sample Physical Process Implementations			
4.3 Check Customer Credit	4.3 Check Customer Credit Acct Clerk	4.3 Check Customer Credit COBOL/CICS	4.3 Check Customer Credit Visual Basic	4.3 Check Customer Credit Quickbooks

If your CASE tool limits the size of names, you may have to develop and use a set of abbreviations for the technology (and possibly abbreviate your action verbs and object clauses).

If a logical process is to be implemented partially by people and partially by software, it must be split into separate physical processes, and appropriate data flows must be added between the physical processes. The name of a physical process to be performed by people, not software, should indicate who would perform that process. We recommend you use titles or roles, not proper names. The following is an example:

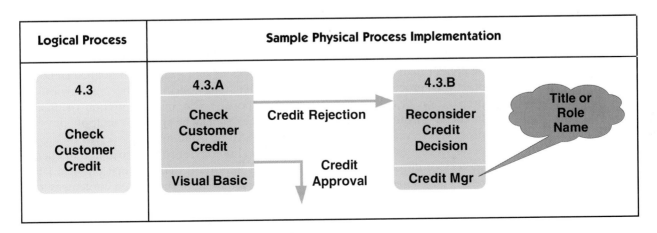

We didn't just change the manual process, RECONSIDER CREDIT DECISION, to an external agent, CREDIT MANAGER, because the entire logical process, CHECK CUSTOMER CREDIT is in the project scope. For that reason, both aspects of the physical implementation are also in the scope. The design is not complete until we specify the process for both the automated and the manual aspects of the business requirement.

For computerized processes, the implementation method is, in part, chosen from one of the following possibilities:

- A purchased application software package (e.g., *Sap,* an enterprise software application, or *Ariba,* an Internet-based procurement/purchasing software application).
- A system or utility program (e.g., Microsoft's *Exchange Server,* an e-mail/messaging system, or IBM's *WebSphere Commerce Business,* an electronic commerce framework).
- An existing application program from a program library, indicated simply as LIBRARY or NAME of library.
- A program to be written. Typically, the implementation method specifies the language or tool to be used to construct the program. Example implementation methods include VB, .NET, C++, JAVA, MS ACCESS, PERL, or ORACLE DEVELOPER.

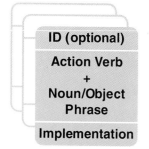

One final physical process construct should be introduced, the *multiprocess* (see margin). The multiprocess indicates multiple implementations of the same physical processor or process. For example, we can use this symbol to indicate multiple PCs, the implementation of a named program on multiple PCs, or the implementation of work to be performed by multiple people. Some CASE tools do not support this construct. If they do not, you may need to resort to plural names to imply multiplicity of a process or processor.

Many designers prefer a more physical naming convention for computer processes. Instead of a noun + verb phrase, they would substitute the file name of the computer program's physical source code. Consider the following examples:

Logical Process	Sample Physical Process Implementations	
4.3 **Check Customer Credit**	**4.3.A** **CHK_CREDIT.COB** **COBOL + CICS**	**4.3.B** **appCheckCredit.vbx** **Visual Basic**

Many organizations have naming conventions and standards for program names.

Again, the number of physical processes on a physical DFD will almost always be greater than the number of logical processes on its equivalent logical DFD. For one thing, processes may be added to reflect data collection, filtering, forwarding, preparation, or quality checks—all in response to the implementation vision that has been selected. Also, some logical processes may be split into multiple physical processes to reflect portions of a process to be done manually versus by a computer, to be implemented with different technology, or to be distributed to clients, servers, or different host computers. It is important that the final physical DFDs reflect all manual and computer processes required for the chosen implementation strategy.

> Physical Data Flows

Recall that all processes on any DFD must have at least one input and one output data flow. A *physical data flow* represents any of the following: (1) the planned implementation of an input to or output from a physical process; (2) a database command or actions such as create, read, update, or delete; (3) the import of data from or the export of data to another information system across a network; or (4) the flow of data between two modules or subroutines within the same program.

Physical data flows are named as indicated by the templates in the margin. Figure 13-2 demonstrates the application of one of these naming conventions as applied to several types of physical data flows.

Physical DFDs must also indicate any data flows to be implemented as business forms. For instance, FORM 23: COURSE REQUEST might be a one-part business form used by students to register for classes. Business forms frequently use a multiple-copy (carbon or carbonless) implementation. At some point in processing, the different copies are split and travel to different manual processes. This is shown on a physical DFD as a diverging data flow (introduced in Chapter 9). Each copy should be uniquely named. For example, at a restaurant, the customer receives FORM: CREDIT CARD VOUCHER (CUSTOMER COPY) and the merchant retains FORM: CREDIT CARD VOUCHER (MERCHANT COPY).

Most logical data flows are carried forward to the physical DFDs. Some may be consolidated into single physical data flows that represent business forms. Others may be split into multiple flows as a result of having split logical processes into multiple physical processes. Still others may be duplicated as multiple flows with different technical implementations. For example, the logical data flow ORDER might be implemented as all of the following: FORM: ORDER, PHONE: ORDER (verbal order taken over the phone), HTML: ORDER (order submitted over the Internet), FAX: ORDER (order received by fax), and MESSAGE: ORDER (an order submitted via e-mail).

> Physical External Agents

External agents are carried over from the logical DFD to the physical DFD unchanged. Why? By definition, external agents were classified during systems analysis as outside the scope of the systems and therefore not subject to change. Only a change in requirements can initiate a change in external agents.

> Physical Data Stores

From Chapter 9 you know that each data store on the logical DFD now represents all instances of a named entity on an entity relationship diagram (from Chapter 9). Physical data stores implement the logical data stores. A *physical data store* represents the implementation of one of the following: (1) a database, (2) a table in a database, (3) a computer file, (4) a tape or media backup of anything important, (5) any temporary file or batch as needed by a program (e.g., TAX TABLES), or (6) any type of noncomputerized file.

When most people think of data stores, they think of computer files and databases. But many data stores are not computerized. File cabinets of paper records immediately come to mind; however, most businesses are replete with more subtle forms of manual data stores such as address cards, paper catalogs, cheat sheets of various important and reusable information, standards manuals, standard operating procedures manuals, directories, and the like. Despite predictions about the demise of paper files, they will remain a part of many systems well into the foreseeable future—if for no other reasons than (1) there is psychological comfort in paper and (2) the government frequently requires it.

The name of a physical data store uses the format indicated in the margin. Some examples of physical data stores are shown in Figure 13-3 (see page 483).

Some designs require that temporary files be created to act as a queue or buffer between physical processes that have different timing. Such files are documented in the same manner, except their names indicate their temporary status.

Margin templates:

Implementation method:
data flow name

OR

Data flow name
(implementation method)

| ID (opt) | Implementation Method: Data Store Name |

| ID (opt) | Data Store Name (Implementation Method) |

FIGURE 13-2

Physical Data
Flows

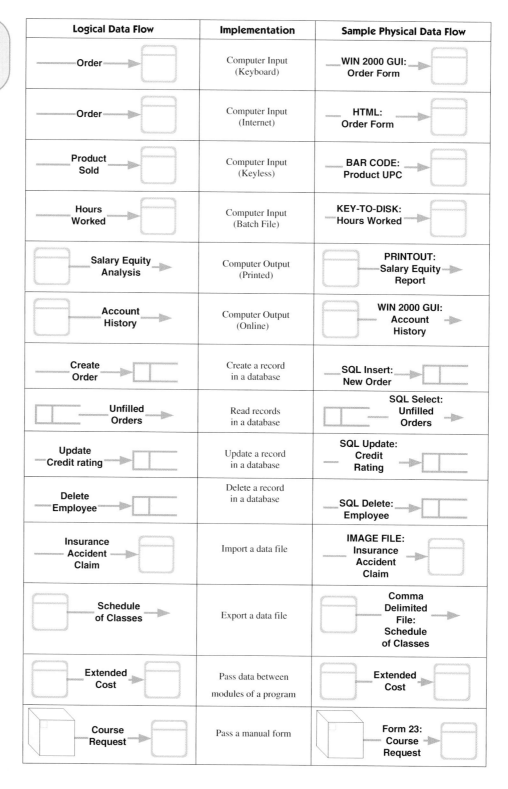

Logical Data Flow	Implementation	Sample Physical Data Flow
Order	Computer Input (Keyboard)	WIN 2000 GUI: Order Form
Order	Computer Input (Internet)	HTML: Order Form
Product Sold	Computer Input (Keyless)	BAR CODE: Product UPC
Hours Worked	Computer Input (Batch File)	KEY-TO-DISK: Hours Worked
Salary Equity Analysis	Computer Output (Printed)	PRINTOUT: Salary Equity Report
Account History	Computer Output (Online)	WIN 2000 GUI: Account History
Create Order	Create a record in a database	SQL Insert: New Order
Unfilled Orders	Read records in a database	SQL Select: Unfilled Orders
Update Credit rating	Update a record in a database	SQL Update: Credit Rating
Delete Employee	Delete a record in a database	SQL Delete: Employee
Insurance Accident Claim	Import a data file	IMAGE FILE: Insurance Accident Claim
Schedule of Classes	Export a data file	Comma Delimited File: Schedule of Classes
Extended Cost	Pass data between modules of a program	Extended Cost
Course Request	Pass a manual form	Form 23: Course Request

Physical processes, data flows, external agents, and data stores make up the physical data flow diagrams. And these physical DFDs model the proposed or planned architecture of an information system application. We can subsequently use that physical model to design the internal and external details for each data store (Chapter 14) and data flow (Chapters 15–17). Now that you understand the basic components of physical DFDs, let's use them to introduce some of today's architectural choices for information system design.

Logical Data Store	Implementation	Physical Data Store
Human Resources	A database (multiple tables)	Oracle : Human Resources DB
Marketing	A database view (subset of a database)	SQL Server: Northeast Marketing DB
Purchase Orders	A table in a database	MS Access: Purchase Orders
Accounts Receivable	A legacy file	VSAM File: Accounts Receivable
Tax Rates	Static data	ARRAY: Tax Table
Orders	An off-line archive	TAPE Backup: Closed Orders
Employees	A file of paper records	File Cabinet: Personnel Records
Faculty/Staff Contact Data	A directory	Handbook: Faculty/Staff Directory
Course Enrollments By Date	Archived reports (for reuse and recall)	REPORT MGR: Course Enrollment Reports

FIGURE 13-3

Physical Data Stores

Information Technology Architecture

Information technology (IT) architecture can be a complex subject worthy of its own course and textbook. (See the Suggested Readings at the end of this chapter.) In this section, we will attempt to summarize contemporary IT alternatives and trends that are influencing design decisions as we go to press. It should be noted that new alternatives are continuously evolving. The best systems analysts will not only learn more about these technologies but will also understand how they work and their limitations. Such a level of detail is beyond the scope of this book. Systems analysts must continuously read popular trade journals to stay abreast of the latest technologies and techniques that will keep their customers and their information systems competitive.

The information system framework provides one suitable framework for understanding IT architecture. Accordingly, our building blocks are being distributed or duplicated across networks. We call the approach *distributed systems architecture:*

- Architectural standards and/or technology constraints are represented in the bottom row of the framework of this chapter's home page. Notice that these standards or decisions are determined either as part of a separate architecture project (preferred and increasingly common) or as part of each system development project.
- The upward-pointing arrows indicate the technology standards that will influence or constrain the design models.

> Distributed Systems

Today's information systems are no longer monolithic, mainframe computer-based systems. Instead, they are built on some combination of networks to form distributed systems. A **distributed system** is one in which the components of an information system are distributed to multiple locations in a computer network. Accordingly, the processing workload required to support these components is also distributed across multiple computers on the network.

The opposite of distributed systems are centralized systems. In **centralized systems,** a central, multiuser computer (usually a mainframe) hosts all components of an information system. The users interact with this host computer via terminals (or, today, a PC emulating a terminal), but virtually <u>all</u> of the actual processing and work is done on the host computer.

Distributed systems are inherently more complicated and more difficult to implement than centralized solutions. So why is the trend toward distributed systems?

- Modern businesses are already distributed, and, thus, they need distributed system solutions.
- Distributed computing moves information and services closer to the customers that need them.
- Distributed computing consolidates the incredible power resulting from the proliferation of personal computers across an enterprise (and society in general). Many of these personal computers are only used to a fraction of their processing potential when used as stand-alone PCs.
- In general, distributed system solutions are more user-friendly because they use the PC as the user interface processor.
- Personal computers and network servers are much less expensive than mainframes. (But admittedly, the total cost of ownership is at least as expensive once the networking complexities are added in.)

There is a price to be paid for distributed systems. Network data traffic can cause congestion that actually slows performance. Data security and integrity can also be more easily compromised in a distributed solution. Still, there is no arguing the trend toward distributed systems architecture. While many centralized, legacy applications still exist, they are gradually being transformed into distributed information systems.

Figure 13-4 compares various distributed systems architectures. Conceptually, any information system application can be mapped to five layers:

- The *presentation layer* is the actual user interface—the presentation of inputs and outputs to the user.
- The *presentation logic layer* is any processing that must be done to generate the presentation. Examples include editing input data and formatting output data.
- The *application logic layer* includes all the logic and processing required to support the actual business application and rules. Examples include credit checking, calculations, data analysis, and the like.
- The *data manipulation layer* includes all the commands and logic required to store and retrieve data to and from the database.
- The *data layer* is the actual stored data in a database.

Figure 13-4 shows these conceptual layers as rows. The columns in the figure illustrate how the layers can be implemented in different distributed information system architectures. There are three types of distributed systems architecture:

- *File server architecture.*
- *Client/server architecture.*
- *Internet-based architecture.*

Let's discuss each in greater detail.

---

distributed system a system in which components are distributed across multiple locations and computer networks.

centralized system a system in which all components are hosted by a central, multiuser computer.

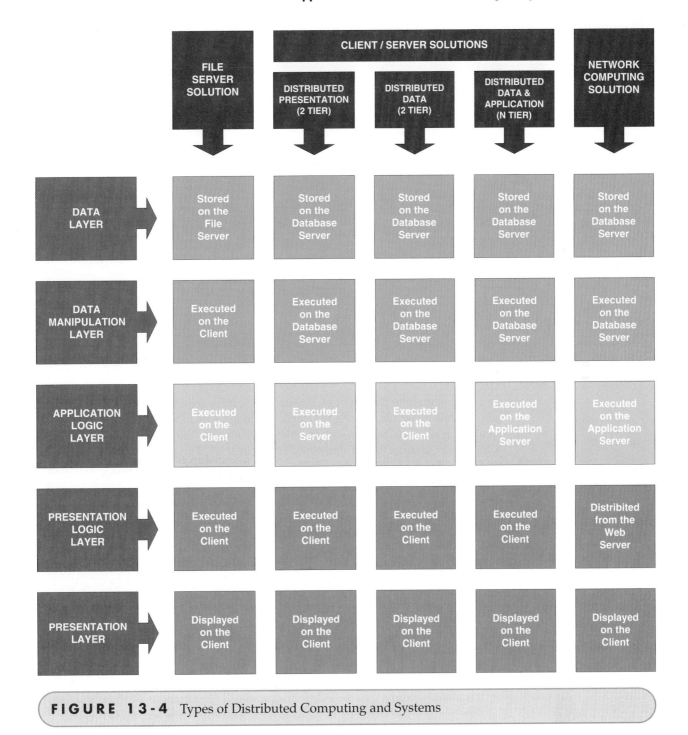

	FILE SERVER SOLUTION	CLIENT / SERVER SOLUTIONS			NETWORK COMPUTING SOLUTION
		DISTRIBUTED PRESENTATION (2 TIER)	DISTRIBUTED DATA (2 TIER)	DISTRIBUTED DATA & APPLICATION (N TIER)	
DATA LAYER	Stored on the File Server	Stored on the Database Server	Stored on the Database Server	Stored on the Database Server	Stored on the Database Server
DATA MANIPULATION LAYER	Executed on the Client	Executed on the Database Server	Executed on the Database Server	Executed on the Database Server	Executed on the Database Server
APPLICATION LOGIC LAYER	Executed on the Client	Executed on the Server	Executed on the Client	Executed on the Application Server	Executed on the Application Server
PRESENTATION LOGIC LAYER	Executed on the Client	Executed on the Client	Executed on the Client	Executed on the Client	Distributed from the Web Server
PRESENTATION LAYER	Displayed on the Client	Displayed on the Client	Displayed on the Client	Displayed on the Client	Displayed on the Client

FIGURE 13-4 Types of Distributed Computing and Systems

File Server Architecture   Today very few personal computers and workstations are used to support stand-alone information systems. Organizations need to share data and services. Local area networks allow many PCs and workstations to be connected to share resources and communicate with one another. A **local area network (LAN)** is a set of client computers (usually PCs) connected to one or more servers (usually a more powerful PC or larger computer) through either cable or wireless connections over relatively short distances—for instance, in a single department or in a single building.

In the simplest LAN environments, a file server architecture is used to implement information systems. A **file server system** is a LAN-based solution in which a server computer hosts only the data layer. All other layers of the information system application

local area network (LAN) a set of client computers connected over a relatively short distance to one or more servers.

file server system a LAN in which a server hosts the data of an information system.

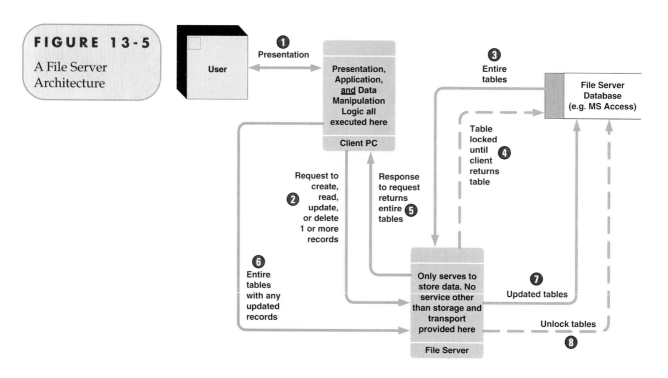

FIGURE 13-5

A File Server
Architecture

are implemented on the client PC. (Note: File servers are also typically used to share other nondatabase files across networks—examples include word processing documents, spreadsheets, images and graphics, engineering drawings, presentations, etc.) A file server architecture is illustrated in Figure 13-5. This architecture is typical of those used for many PC database engines such as Microsoft *Access* and *FoxPro*. While your *Access* database may be stored on a network server, the actual *Access* program must be installed or executed from each PC that uses the database.

File server architectures are practical only for small database applications shared by relatively few users because the entire file or table of records must be first downloaded to the client PC, where the data manipulation logic will be executed to read a single desired record. There are several disadvantages to this approach:

- Large amounts of unnecessary data must be moved between the client and the server. This data traffic can significantly reduce network and application performance.
- The client PC must be robust (a so-called *fat client*). It is doing virtually all of the actual work, including data manipulation, application logic, presentation logic, and presentation. It must also have enough disk capacity to store the downloaded tables.
- Database integrity can be easily compromised. Think about it. If any record has been downloaded to be updated, the entire file has been downloaded. Other users (clients) must be prevented (or locked) from making changes to any other record in that file. The greater the number of simultaneous users, the more this locking requirement slows response time.

Very few mission critical information systems can be implemented with file server technology. So why are file server database management systems such as MS *Access* so popular? First, file server tools such as *Access* can be used to develop fairly robust applications for individuals and small work groups. Second, and perhaps more significantly, file server databases such as *Access* can be used to rapidly construct prototypes for more robust client/server architectures.

Client/Server Architectures   The prevailing distributed computing model of the current era is called *client/server computing* (although it is rapidly giving way to

Internet-based models). A **client/server system** is a solution in which the presentation, presentation logic, application logic, data manipulation, and data layers are distributed between client PCs and one or more servers.

The client computers may be any combination of personal computers or workstations, "sometimes connected" notebook computers, handheld computers (e.g., Palm or Windows Mobile Platforms), Web TVs, or any devices with embedded processors that could connect to the network (e.g., robots or controllers on a manufacturing shop floor). Clients may be thin or fat. A **thin client** is a personal computer that does not have to be very powerful (or expensive) in terms of processor speed and memory because it only presents the interface (screens) to the user—in other words, it acts only as a terminal. Examples include Remote Desktop and *X/Windows.* In thin-client computing, the actual application logic executes on a remote application server. A **fat client** is a personal computer, notebook computer, or workstation that is typically more powerful (and expensive) in terms of processor speed, memory, and storage capacity. Almost all PCs are considered fat clients.

A server in the client/server model must be more powerful and capable than a server in the file server model. In fact, a mainframe computer can play the role of server in a client/server solution. More typical, however, are network servers running client/server-capable operating systems such as UNIX, Windows Server 2003, or Linux. Several types of servers may be used in a client/server solution. These may reside on separate physical servers or be consolidated into fewer servers:

- A **database server** hosts one or more shared databases (like a file server) but also executes all database commands and services for information systems (unlike a file server). Most database servers host an SQL database engine such as *Oracle,* Microsoft *SQL Server,* or IBM *DB2 Universal Database.*
- A **transaction server** hosts services that ultimately ensure that all database updates for a single business transaction succeed or fail as a whole. Examples include IBM *CICS* and BEA.
- An **application server** hosts application logic and services for an information system. It must communicate on the front end with the clients (for presentation) and on the back end with database servers for data access and update. An application server is often integrated with the transaction server. Most application servers are based on either the *CORBA* object-sharing standard or the Microsoft *COM+* standard.
- A **messaging** or **groupware server** hosts services for e-mail, calendaring, and other work group functionality. This type of functionality can actually be integrated into information system applications. Examples include Lotus *Notes* and Microsoft *Exchange Server.*
- A **Web server** hosts Internet or intranet Web sites. It communicates with fat and thin clients by returning to them documents (in formats such as *HTML*) and data (in formats such as *XML*). Some Web servers are specifically designed to host e-commerce applications (e.g., IBM's *WebSphere Commerce Business*).

Client/server architecture itself comes in several types, each of which deserves its own explanation. Each of these C/S types is also compared to the others in Figure 13-4.

Client/Server—Distributed Presentation

Most centralized (or mainframe) computing applications use an older character user interface (CUI) that is cumbersome and awkward when compared to today's graphical user interfaces (GUIs) such as Microsoft *Windows* and UNIX *X/Windows* (not to mention Web browsers such as Mozilla *Firefox* and Microsoft *Internet Explorer*). As personal computers rapidly replaced dumb terminals, users became increasingly comfortable with this newer technology. And as they developed familiarity and experience with PC productivity tools such as word processors and spreadsheets, they wanted their centralized, legacy computing applications to have a similar look and feel using the GUI model.

client/server system a distributed computing solution in which the presentation, presentation logic, application logic, data manipulation, and data layers are distributed between client PCs and one or more servers.

thin client a personal computer that does not have to be very powerful.

fat client a personal computer, notebook computer, or work station that is typically powerful.

database server a server that hosts one or more databases.

transaction server a server that hosts services which ensure that all database updates for a transaction succeed or fail as a whole.

application server a server that hosts application logic and services for an information system.

messaging or **groupware server** a server that hosts services for groupware.

Web server a server that hosts Internet or intranet Web sites.

distributed presentation
a client/server system in which presentation and presentation logic are shifted from the server to reside on the client.

Enter distributed presentation. A **distributed presentation** client/server system is a solution in which the presentation and presentation logic layers are shifted from the server of a legacy system to reside on the client. The application logic, data manipulation, and data layers remain on the server (usually a mainframe). Sometimes called the "poor person's client/server," this alternative builds on and enhances centralized computing applications. Essentially, the old CUIs are stripped from the legacy applications and regenerated as GUIs that will run on the PC. In other words, only the user interface (or presentation layer) is distributed to the client.

Distributed presentation offers several advantages. First, it can be implemented relatively quickly because most aspects of the legacy application remain unchanged. Second, users get a fast, friendly, and familiar user interface to legacy systems—one that looks at least somewhat familiar to their PC productivity tools. Finally, the useful lifetime of legacy applications can be extended until resources warrant a wholesale redevelopment of the application. The disadvantages are that the application's functionality cannot be significantly improved and the solution does not maximize the potential of the client's desktop computer by dealing only with the user interface.

A class of CASE tools, sometimes called *screen scrapers,* automatically read the CUI and generate a first-cut GUI that can be modified by a GUI editor. Figure 13-6 demonstrates this technology. Figure 13-7 shows a physical DFD for a distributed presentation solution.

FIGURE 13-6

Building a GUI from a CUI

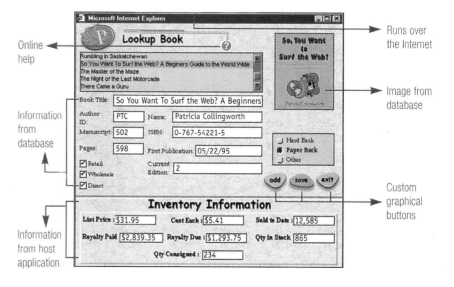

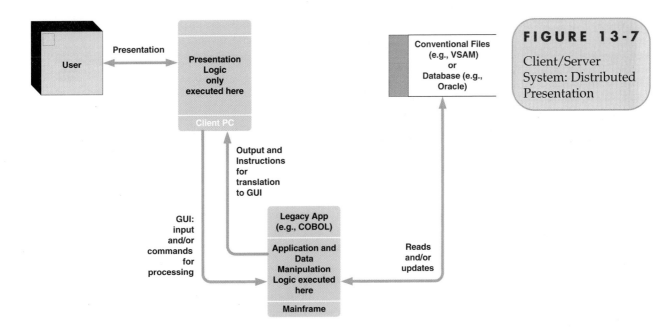

FIGURE 13-7

Client/Server
System: Distributed
Presentation

Client/Server—Distributed Data   This is the simplest form of true client/server computing. A local area network usually connects the clients to the server. A **distributed data** client/server system is a solution in which the data and data manipulation layers are placed on the server(s), and the application logic, presentation logic, and presentation are placed on the clients. This is also called *two-tiered client/server computing*. A two-tiered, distributed data client/server system is illustrated as a physical DFD in Figure 13-8.

It is important to understand the difference between file server systems and distributed data client/server systems. Both store their actual database on a server. But only client/server systems execute all data manipulation commands (e.g., SQL instructions to create, read, update, and delete records) on a server. Recall that in file server

distributed data a
client/server system in which
the data and data manipulation layers are placed on
servers and other layers are
placed on clients. Also called
*two-tiered client/server
computing.*

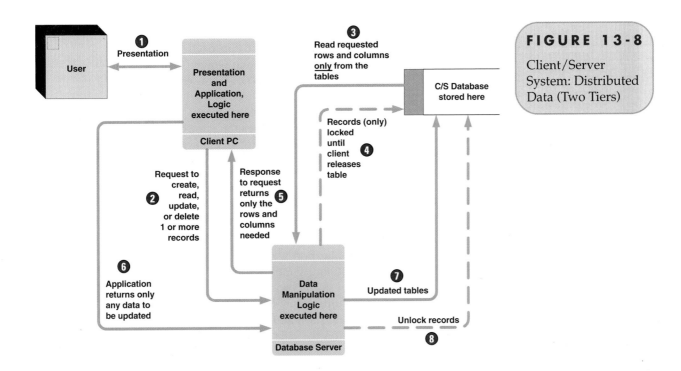

FIGURE 13-8

Client/Server
System: Distributed
Data (Two Tiers)

systems, those data manipulation commands must be implemented on the client. Distributed data client/server solutions offer several advantages over file server solutions:

- There is much less network traffic because only the database requests and the database records that are needed are actually transported to and from the client workstations.
- Database integrity is easier to maintain. Only the records in use by a client must typically be locked. Other clients can simultaneously work on other records in the same table or database.

The client workstation must still be fairly robust ("fat") to provide the processing for the application logic layer. This logic is usually written in a client/server programming language such as Sybase Corporation's *PowerBuilder,* Microsoft's *Visual Basic .NET* or *C#.* Those programs must be compiled for and execute on the client. To improve application efficiency and reduce network traffic, some business logic may be distributed to the database server in the form of stored procedures (discussed in the next chapter).

The *database server* is fundamental to this architecture. Database servers store the database, but they also execute the database instructions directly on those servers. The clients merely send their database instructions to the server. The server returns only the result of the database command processing—not entire databases or tables. All high-end database engines such as *Oracle* and Microsoft *SQL Server* use this approach. A distributed data architecture may involve more than one database server. Data may be distributed across several database servers or duplicated on several database servers.

The key potential disadvantage to the two-tiered client/server is that the application logic must be duplicated and thus maintained on all the clients, possibly hundreds or thousands. The designer must plan for version upgrades and provide controls to ensure that each client is running the most current release of the business logic, as well as ensure that other software on the PC (purchased or developed in-house) does not interfere with the business logic.

Client/Server—Distributed Data and Application

When the number of clients grows, two-tiered systems frequently suffer performance problems associated with the inefficiency of executing all the application logic on the clients. Also, in multiple-user transaction processing systems (also called *online application processing,* or *OLAP*), transactions must be managed by software to ensure that all the data associated with the transaction is processed as a single unit. This generally requires a distribution that uses a multitiered client/server approach. A **distributed data and application** client/server system is a solution in which (1) the data and data manipulation layers are placed on their own server(s), (2) the application logic is placed on its own server, and (3) only the presentation logic and presentation are placed on the clients. This is also called *three-tiered,* or *n-tiered, client/server computing.*

The three-tiered client/server solution uses the same database servers as those in the two-tiered approach. Additionally, the three-tiered system introduces an application and/or transaction server. By moving the application logic to its own server, that logic now only needs to be maintained on the server. The three-tiered solution is depicted as a physical data flow diagram in Figure 13-9.

Three-tiered client/server logic can be written and partitioned across multiple servers using languages such as Microsoft *Visual Basic .NET* and *C#* in combination with a transaction monitor. High-end tools such as *Forté* provide an even greater opportunity to distribute application logic and data across a complex network. As with the database server solution, some business logic could be distributed to the database server in the form of stored procedures.

In a three-tiered system, the clients execute a minimum of the overall system's components. Only the user interface and some relatively stable or personal application logic need be executed on the clients. This simplifies client configuration and management.

distributed data and application a client/server system in which the data and manipulation layers are placed on their own server(s), the application logic is placed on its own server, and the presentation logic and presentation are placed on the clients. Also called *three-tiered,* or *n-tiered, client/server computing.*

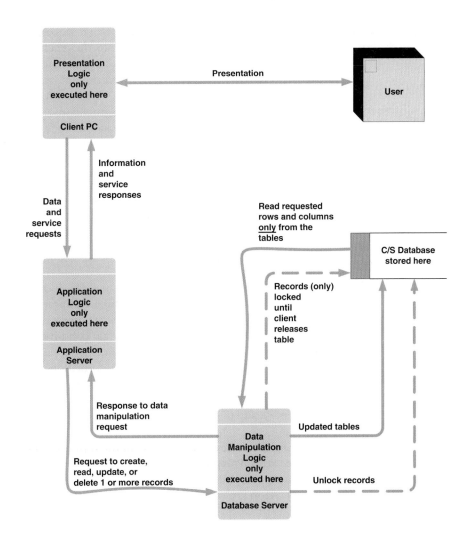

FIGURE 13-9

Client/Server System: Distributed Data and Application (Three Tiers)

The biggest drawback of the three-tiered client/server is its complexity in design and development. The most difficult aspect of a three-tiered client/server application design is partitioning. **Partitioning** is the act of determining how to best distribute or duplicate application components across the network. Fortunately, CASE tools are constantly improving to provide greater assistance with partitioning.

partitioning the act of determining how to best distribute or duplicate application components across a network.

Internet-Based Computing Architectures   Some consider Internet-based system architectures to be the latest evolution of client/server. We present Internet-based computing alternatives in this section as a fundamentally different form of distributed architecture that is rapidly reshaping the design thought processes of systems analysts and information technologists.

A **network computing system** is a multitiered solution in which the presentation and presentation logic layers are implemented in client-side Web browsers using content downloaded from a Web server. The presentation logic layer then connects to the application logic layer that runs on an application server, which subsequently connects to the database server(s) on the backside. Think about it! All information systems running in browsers—financials, human resources, operations—all of them! E-commerce is part of this formula, and as we go to press, e-commerce applications are getting most of the attention. But the same Internet technologies being used to build e-commerce solutions are being used to reshape the internal information systems of most businesses—we call it *e-business* (although that term is also subject to multiple interpretations). Network computing is, in our view, a fundamental shift away from what we just described as client/server.

network computing system a multitiered solution in which the presentation and presentation logic layers are implemented in client-side Web browsers using content downloaded from a Web server.

Very few new technologies have witnessed as explosive a growth in business and society as the Internet or the World Wide Web. The Internet extends the reach of our information and transaction processing systems to include potential customers, customers, partners, remotely located employees, suppliers, the government, and even competitors. During the late 1990s the Internet was largely being used to establish a company's presence in a virtual marketplace and to disseminate public information about products and services and provide a new foundation for customer-focused service. Today, however, most businesses are focused on developing e-commerce solutions that will allow customers to directly interact with and conduct business on the Web (such as direct-to-consumer shopping). We've even seen the invention of the virtual business, a business that "does business" entirely on the Web, such as Amazon.com (books and media), ETrade (stocks and bonds), eBay (auctions), and Buy.com (electronics and appliances). One of the most intriguing debates is whether these "click-and-mortar" virtual companies can turn a profit and actually compete with more traditional "brick-and-mortar" companies—many of which are diversifying rapidly to enter cybermarkets.

intranet a server network that uses Internet technology to integrate desktop, work group, and enterprise computing.

But the greatest potential of this Internet technology may actually be its application to traditional information systems applications and development on intranets. An **intranet** is a secure network, usually corporate, that uses Internet technology to integrate desktop, work group, and enterprise computing into a single cohesive framework. Everything runs in (or at least from) a browser—your productivity applications such as word processing and spreadsheets; any and all traditional information systems applications you need for your job (financials, procurement, human resources, etc.); all e-mail, calendaring, and work group services (allowing, for example, virtual meetings and group editing of documents); and of course all of the external Internet links that are relevant to your job.

The appeal of this concept should not be hard to grasp. Each employee's "start page" is a *portal* into all computer information systems and services he or she needs to do his or her entire job. Because everything runs in a Web browser, there is no longer a need to worry about, or develop for, multiple different computer architectures (Intel versus Motorola versus RISC) or worry about different desktop operating systems. A physical data flow diagram for network computing is shown in Figure 13-10. Notice that a Web server is added to the prior three-tiered model. The DFD also shows both e-commerce (business-to-consumer) and e-business (business-to-business) dimensions of network computing.

Does this all sound too good to be true? Something of a cyber-Camelot? By the time you read this, our own institution will have likely implemented its first mission critical e-business information system. Purdue, like all enterprises, has a procurement (or purchasing) function. We buy everything from pencils to furniture to computers to radioactive isotopes—literally tens of thousands of different supplies, materials, and other products. As we go to press, Purdue has redesigned its procurement system to be a combination intranet/Internet/extranet application. Here's how it works:

> The entire application runs in a Web browser. Any employee of the university, once authenticated, can initiate a purchase requisition via his or her Web browser (the intranet dimension). Employees can even "shop" for items from a Web-mall of approved suppliers with which the university has standing contracts (the Internet dimension). When a requisition is submitted, it will be smart enough to know who must approve it (at every level) based on cost and type of items ordered. The system will be able to automatically check for available funds to pay for the purchase. Employees will be able to audit the electronic flow of the requisition through the approval process and into purchase order status. Managers will be able to revise the approval flow to get additional input as needed. Most final orders will be transmitted electronically over a secure business-to-businesses extranet between the university and its suppliers.

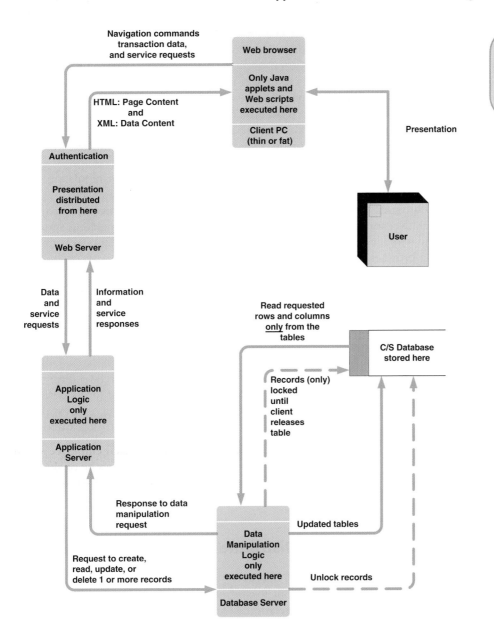

FIGURE 13-10

Network
Computing System:
Internet/Intranet

When ordered goods are received, the recipient will indicate receipt via the same Web-based system, and payment, via electronic funds in most cases, will be made. At any time, managers will be able to generate useful procurement information for employees, departments, suppliers, whatever! A paperless office? Very close!

Such intriguing system development possibilities are being fueled by some fundamental emerging technologies that you should make a part of your curriculum plan of study if possible:

- The programming language of choice for the application logic in network computing architectures is likely to be *Java*. Essentially, *Java* is a reasonably platform-independent programming language designed specifically to exploit both Internet and object-oriented programming standards. *Java* is designed to execute in your Internet browser, making it less susceptible to differences in computing platforms and operating systems (but it's not yet perfect).

- The interface language of choice for the presentation and presentation logic layers in network computing architectures is currently *HTML,* or hypertext markup language. *HTML* is used to create the pages that run in your browser. Soon another player in this layer will dominate—*XML,* the extensible markup language. This widely embraced standard allows developers to also define the structure of the data to be passed to Web pages, a critical requirement for Web-based e-commerce and intranet-based information systems. *XML* may eventually replace *HTML,* or the two standards may merge into one very powerful language.
- As with traditional information systems, the data and data manipulation layers will likely continue to be implemented with SQL database engines.
- Web browsers will continue to be important. In fact, Web browsers may ultimately be more important than your choice of a desktop operating system. Is it any wonder that *Windows,* with each new version, looks and feels more like a browser.

All of these Internet and intranet solutions involve leading-edge technologies and standards that have no doubt changed even since these words were written. Technology vendors will undoubtedly play a significant role in the evolution of the technology. However the specific technologies play out, we expect the Internet and intranets to become the most common architectural models for tomorrow's information systems.

So where are we in our study of information technology architecture? You've learned that several distributed systems and network options exist for modern information systems. Most can be broadly classified as either client/server or network computing architectures. To be sure, there is much more for you to learn about the underlying communications technology, but that is the subject of at least one additional book. The AIS/ACM/AITP Model Curriculum for Undergraduate Information Systems recommends that all information systems graduates complete at least one course in information technology architecture and one course in fundamental data communications.

> Data Architectures—Distributed Relational Databases

The underlying technology of client/server and network computing has made it possible to distribute data without loss of control. This control is accomplished through advances in distributed relational database technology. A *relational database* stores data in a tabular form. Each file is implemented as a table. Each field is a column in the table. Each record in the file is a row in the table. Related records between two tables (e.g., CUSTOMERS and ORDERS) are implemented by intentionally duplicating columns in the two tables (in this example, CUSTOMER NUMBER is stored in both the CUSTOMERS and ORDERS tables). A *distributed relational database* distributes or duplicates tables to multiple database servers located in geographically important locations (such as different sales regions). The software required to implement distributed relational databases is called a *distributed relational database management system.* A **distributed relational database management system** (or *distributed RDBMS*) is a software program that controls access to and maintenance of the stored data in the relational format. It also provides for backup, recovery, and security. It is sometimes called a *client/server database management system.*

In a distributed RDBMS, the underlying database engine that processes all database commands executes on the database server. This arrangement reduces the data traffic on the network. This is a significant advantage for all but the smallest systems (as measured in number of users). A distributed relational DBMS also provides more sophisticated backup, recovery, security, integrity, and processing (although the differences seem to erode with each new PC RDBMS release).

distributed relational database management system software that implements distributed relational databases.

Examples of distributed RDBMSs include Oracle Corporation's *Oracle,* IBM's *DB2 Universal Database* family, Microsoft's *SQL Server,* and Sybase Corporation's *Sybase.* Most RDBMSs support two types of distributed data:

- *Data partitioning* truly distributes rows and columns to specific database servers with little or no duplication between servers. Different columns can be assigned to different database servers (*vertical partitioning*) or different rows in a table can be allocated to different database servers (*horizontal partitioning*).
- *Data replication* duplicates some or all tables (rows and columns) on more than one database server. Entire tables can be duplicated on some database servers, while subsets of rows in a table can be duplicated to other database servers. The RDBMS with replication technology not only controls access to and management of each database server database but also propagates updates on one database server to any other database server where the data is duplicated.

For a given information system application, the data architecture must specify the RDBMS technology and the degree to which data will be partitioned or replicated. One way to document these decisions is to record them in the physical data stores as shown below. Notice how we used the ID area to indicate codes for partitioning (P) and replication (M for the master copy and R for the replicated copy). In the case of the former, we should specify which rows and/or columns are to be partitioned to the physical database.

Logical Data Store	Physical Data Stores Using Partitioning	Physical Data Stores Using Replication
1 CUSTOMERS	1P.# Oracle 7: REGION 1 CUSTOMERS 1P.# Oracle 7: REGION 2 CUSTOMERS	Not applicable. Branch offices do not need access to data about customers outside of their own sales region.
2 PRODUCTS	Not applicable. All branch offices need access to data for all products, regardless of sales region.	2M Oracle 8i: PRODUCTS (Master) 2R Oracle 8i: PRODUCTS (Replicated Copy)

An application's DATA architecture is selected based on the desired client/server or network computing model and the database technology needed to support that model. Many organizations have standardized on both their PC RDBMS of choice and their preferred distributed, enterprise RDBMS of choice. For example, SoundStage has standardized on Microsoft *Access* and *SQL Server.* Generally, a qualified database administrator should be included in any discussions about the database technology to be used and the design implications for any databases that will use that technology.

> Interface Architectures—Inputs, Outputs, and Middleware

Another fundamental information technology decision must be made regarding inputs, outputs, and intersystem connectivity. The decision used to be simple—batch inputs versus online inputs. Today we must consider modern alternatives such as automatic identification, pen data entry, various graphical user interfaces, electronic data interchange, imaging, and voice recognition, among others. Let's briefly examine these alternatives and their physical DFD constructs.

Batch Inputs or Outputs   In batch processing, transactions are accumulated into batches for periodic processing. The batch inputs are processed to update databases and produce appropriate outputs. Most outputs tend to be generated to paper or microfiche on a scheduled basis. Others might be produced on demand or within a specified time period (e.g., 24 hours).

Contrary to popular belief, batch input technologies are not quite obsolete. You rarely see punched cards and tape batches today, but some application requirements lend themselves to batch processing. Perhaps the inputs arrive in natural batches (e.g., mail), or perhaps outputs are generated in natural batches (e.g., invoices). Many organizations still collect and process time cards in batches. There is, however, a definite trend away from batch input to online approaches. In the meantime, key-to-disk file is the most common, and its physical data flow construct would look as shown below. First, notice that the logical name is singular, but the batch name is plural. Also notice that the batch goes into a temporary data store, which is read by a payroll process triggered by date.

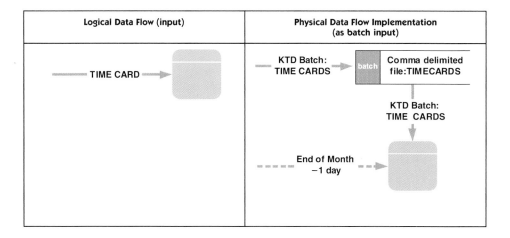

Batch output is quite another story. Many applications lend themselves to batch output. Examples include generation of invoices, account statements, grade reports, paychecks, W-2 tax forms, and many others. Batch outputs often share one common physical characteristic, the use of a preprinted form. It should not be difficult for you to envision a preprinted form to be loaded in the printer to produce any of the aforementioned output examples. A physical data flow construct would look something like the following. Again, note the plural name reflective of batch processing.

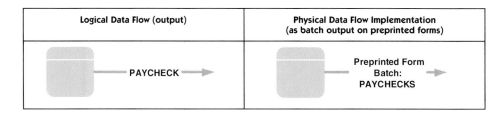

As older batch-based systems become candidates for replacement, other physical implementation alternatives should be explored.

Online Inputs and Outputs   The majority of systems have slowly evolved from batch processing to online or real-time processing. Online inputs and outputs provide for a more conversational dialogue between the user and computer applications. They also provide nearly immediate feedback in response to transactions, problems, and inquiries. In today's fast-paced economy, most business transactions and inquiries are best processed as soon as possible. Errors are identified and corrected more quickly because there is no time lapse between data entry and input (as was the case in batch processing). Furthermore, online methods permit greater human interaction in decision making.

Today most systems are being designed for online processing, even if the data arrives in natural batches. Technically, all GUI and Web applications are online or real-time, and since we've already learned that those architectures are preferred in client/server and network computing, then we can expect that most physical data flows will be implemented with some type of GUI technology. The physical data flow constructs would look something like the following. For the physical output, notice that two formats of the physical output are possible. We could have added the junction symbols (Chapter 9) to make the flows mutually contingent (both required) or mutually exclusive (either/or, but not both).

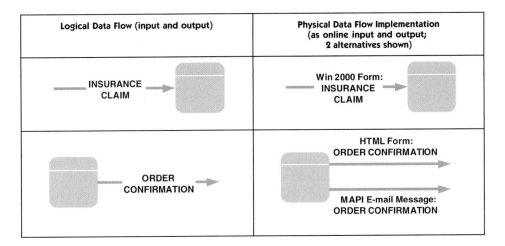

Remote Batch   Remote batch combines the best aspects of batch and online inputs and outputs. Distributed online computers handle data input and editing. Edited transactions are collected into a batch file for later transmission to host computers that process the file as a batch. Results are usually transmitted as a batch back to the original computers.

Remote batch is hardly a new alternative, but personal computers have given the option new life. For example, one of the authors' colleagues uses a Microsoft *Access* program to input and test the feasibility of a schedule of classes for his academic department each semester. When finished, he generates a comma delimited file to transmit to the academic scheduling unit for batch processing. The entire physical input model looks something like this:

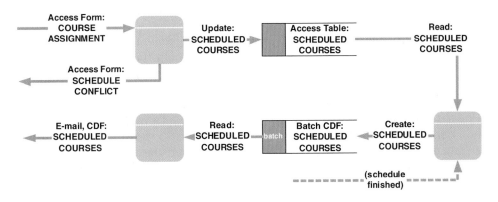

Remote batch using PCs should get another boost with the advances in hand-held and subnotebook computer technology. These four-ounce to four-pound computers can be used to collect batches of everything from inventory counts to mortgage applications. The inputs are remotely batched on the device for later transmission as a batch.

Keyless Data Entry (and Automatic Identification)   Keying in data has always been a major source of errors in computer inputs (and inquiries). Any technology that reduces or eliminates the possibility of keying errors should be considered for system design. In batch systems, keying errors can be eliminated through optical character reading (OCR) and optical mark reading (OMR) technology. Both are still viable options for input design. The physical data flow construct is shown below.

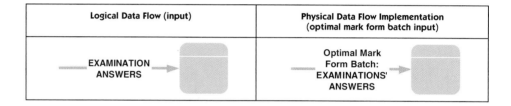

The real advances in keyless data entry are coming for online systems in the form of auto-identification systems. For example, bar-coding schemes (such as the Universal Product Codes that are common in the retail industry) are widely available for many modern applications. For example, Federal Express creates a bar code–based label for all packages when you take the package to a center for delivery. The bar codes can be read and traced as the package moves across the country to its final destination. Bar code technology is being constantly improved to compress greater amounts of data into smaller labels. The physical data flow construct is shown below. (The receiving physical process would be named for the function it performs.)

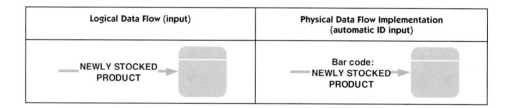

Pen Input   As pen-based operating systems (e.g., the *Palm OS* and Microsoft's *Windows Mobile*) become more widely available and used, and the tools for building pen-based applications become available and standardized, we expect to see more system designs that exploit this technology.

Some businesses already use this technology for remote data collection. For example, UPS uses pen-based notebook systems to help track packages through the delivery system. The driver calls up the package tracking number on the special tablet computer. The customer signs the pad in the designated area. When the driver returns to the truck and places the tablet computer back in its docking cradle, the updated delivery data is transmitted by cellular modem to the distribution center where the package tracking system updates the database (ultimately enabling the shipper to know that you have received the package via a simple Web inquiry).

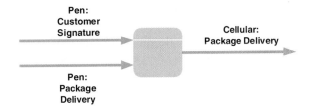

Electronic Messaging and Work Group Technology   Electronic mail has grown up! No longer merely a way to communicate more effectively, information systems are being designed to directly incorporate the technology. For example, Microsoft *Exchange Server* and IBM *Lotus Notes* allow for the construction of intelligent electronic forms that can be integrated into any application. Basic messaging services can also be integrated into applications.

For example, any employee via an e-mail-based form could initiate travel requests. The system takes the data submitted on the form and follows predefined rules to automatically route the request to the appropriate decision makers. For example, less expensive travel requests might be routed directly to a business officer. More expensive requests might be routed first to a department head for approval and then to a business officer. Eventually, approved forms can be automatically input to the appropriate reimbursement processing information system for normal processing. And at each step, the messaging system automatically informs the initiator of progress via e-mail. A physical DFD that included an e-mail message implementation was presented earlier.

Electronic Data Interchange   Businesses that operate in many locations and businesses that seek more efficient exchange of transactions and data with other businesses often utilize electronic data interchange. **Electronic data interchange (EDI)** is the standardized electronic flow of business transactions or data between businesses. Typically, many businesses must commit to a data format to make EDI feasible.

With EDI, a business can eliminate its dependence on paper documents and mail. For example, most colleges now accept SAT or ACT test scores via EDI from national testing centers. This has been made possible because college registrars have agreed to a standard format for these test scores.

electronic data interchange (EDI) the standardized electronic flow of business transactions or data between businesses.

Logical Data Flow (input)	Physical Data Flow Implementation (automatic ID input)
STUDENT APTITUDE SCORE	EDI: STUDENT APTITUDE SCORES

Imaging and Document Interchange   Another emerging I/O technology is based on image and document interchange. This is similar to EDI except that the actual images of forms and data are transmitted and received. It is particularly useful in applications in which the form images or graphics are required. For example, the insurance industry has made great strides in electronically transmitting, storing, and using claims images. Other imaging applications combine data with pictures or graphs. For example, a law enforcement application can store, transmit, and receive photographic images and fingerprints.

Middleware   Most of the above subsections focused on input and output—the user interface. But many system designs require process-to-process physical data flows.

middleware utility software that enables communication between different processors in a system.

Earlier in this chapter, we described various client/server and network computing scenarios that automatically include process-to-process data flows because clients and servers must talk to one another. They do this through middleware. **Middleware** is utility software that enables communication between different processors in a system. It may be built into the respective operating systems or added through purchased middleware products. Middleware products allow the programmers to ignore underlying communication protocols.

Middleware is said to be the "slash" in "client/server." There are three classes of middleware that happen to correspond to the middle three layers of our distributed systems framework—presentation logic, application logic, and data manipulation logic:

- *Presentation middleware* allows a programmer to build user interface components that can talk to Web browsers or a desktop GUI. For example, HTTP allows the programmer to communicate with a Web browser through a standard application programmer interface (API).
- *Application middleware* enables two programmer-written processes on different processors to communicate with one another in whatever way is best suited to the overall application. Application middleware is essential to multitier application development. Examples of application middleware are numerous: remote procedure calls (RPCs), message queues, and object request brokers.
- *Database middleware* allows a programmer to pass SQL commands to a database engine for processing through a standard API.

    Another common type of middleware is ODBC (object database connectivity) and JDBC (Javabean database connectivity), which automatically translate the SQL commands of one database server for use on a different database server (for example, *Oracle* to *SQL Server*, or vice versa).

On a physical data flow diagram, middleware can be depicted by specifying the middleware class name on the physical data flow (e.g., ODBC).

> Process Architectures—The Software Development Environment

software development environment (SDE) a language and tool kit for developing applications.

The process architecture of an application is defined in terms of the software languages and tools that will be used to develop the business logic and application programs for that process. Typically, this is expressed as a menu of choices because different software development environments are suited to different applications. A **software development environment (SDE)** is a language and tool kit for constructing information system applications. One way to classify SDEs is according to the type of client/server or network computing architecture they support.

SDEs for Centralized Computing and Distributed Presentation Not that long ago, the software development environment for centralized computing was very simple. It consisted of the following:

- An editor and compiler, usually *COBOL,* to write programs.
- A transaction monitor, usually *CICS,* to manage any online transactions and terminal screens.
- A file management system, such as *VSAM,* or a database management system, such as *DB2,* to manage stored data.

That was it! Because all these tools executed on the mainframe, only that computer's operating system (more often than not, *MVS*) was critical.

The personal computer brought many new *COBOL* development tools down to the mainframe. A PC-based *COBOL* SDE such as the Micro Focus *COBOL Workbench* usually provided the programmer with more powerful editors and testing and debugging tools at the workstation level. A programmer could do much of the development work at that level and then upload the code to the central computer for system testing, performance tuning, and production. Frequently, the SDE could be interfaced with a CASE tool and code generator to take advantage of process models developed during systems analysis.

Eventually, SDEs provided tools to develop distributed presentation client/server systems. For example, the Micro Focus *Dialog Manager* provided *COBOL Workbench* users with tools to build *Windows*-based user interfaces that could cooperate with the *CICS* transaction monitors and the mainframe *COBOL* programs.

SDEs for Two-Tier Client/Server   Today the typical SDE for two-tiered client/server applications (also called distributed data) consists of a client-based programming language with built-in SQL connectivity to one or more server database engines. Examples of two-tiered client/server SDEs include Sybase's *PowerBuilder,* Microsoft's *Visual Studio,* and Borland's *Delphi* (Client/Server Edition). Typically, these SDEs provide the following:

- Rapid application development (RAD) for quickly building the graphical user interface that will be replicated and executed on all the client PCs.
- Automatic generation of the template code for the above GUI and associated system events (such as mouse-clicks, keystrokes, etc.) that use the GUI. The programmer only has to add the code for the business logic.
- A programming language that is compiled for replication and execution on the client PCs.
- Connectivity (in the above language) for various relational database engines and interoperability with those engines. Interoperability is achieved by including SQL database commands (e.g., to create, read, update, delete, and sort records) that will be sent to the database engine for execution on the server.
- A sophisticated code testing and debugging environment for the client.
- A system testing environment that helps the programmer develop, maintain, and run a reusable test script of user data, actions, and events against the compiled programs to ensure that code changes do not introduce new or unforeseen problems.
- A report writing environment to simplify the creation of new end-user reports off a remote database.
- A help authoring system for the client PCs.

Today most of these tools come in the bundled SDE, but independent software tool vendors have emerged to produce replacement tools that often provide still greater functionality and/or productivity than those provided in the basic SDE. To learn more about such add-on tools, search the Internet for Programmers Paradise, a software development tool Web storefront.

Some of the process logic of any two-tiered client/server application can be offloaded to the database server in the form of stored procedures. In this case, stored procedures are written in a superset of the SQL language. These procedures are then "called" from the client for execution on the server. Different experts seem to love or hate stored procedures. On the plus side, stored procedures can be made to better enforce data integrity in database tables. They are reusable and verifiable. On the negative side, they blur the distinction between the application and data manipulation layers of our framework—they are application logic that executes on the database servers. Many designers prefer a more cohesive design strategy called *clean layering.* **Clean layering** requires that the presentation, application, and data layers of an application be physically separated. Clean layering is said to allow components of each layer to be revised and enhanced without affecting other layers in the system.

clean layering a design strategy that requires that presentation, application, and data layers be physically separated.

SDEs for Multitier Client/Server   The current state of the art in enterprise application development is occurring in SDEs for three-tiered (and beyond) client/server architectures. Unlike two-tiered applications, *n*-tiered applications must support more than 100 users with mainframelike transaction response time and throughput, with

100 gigabyte or larger databases. While the two-tiered SDEs described earlier are trying to expand in this market, a different class of SDEs currently dominates the market. Typically, the SDEs in this class must provide all the capabilities typically associated with two-tiered SDEs plus the following:

- Support for heterogeneous computing platforms, both client and server.
- Code generation and programming for both clients and servers.
- A strong emphasis on reusability using software application frameworks, templates, components, and objects.
- Bundled minicase tools for analysis and design that interoperate with code generators and editors.
- Tools that help analysts and programmers partition application components between the clients and servers.
- Tools that help developers deploy and manage the finished application to clients and servers. This generally includes security management tools.
- The ability to automatically scale the application to larger and different platforms, client and server.
- Sophisticated software version control and application management.

Examples of *n*-tiered client/server SDEs include Dynasty's *Dynasty*, and IBM's *VisualAge* (a family of products). Again, a large number of independent software tool vendors are building add-on and replacement tools for these SDEs.

SDEs for Internet and Intranet Client/Server   Rapid application development tools are emerging to enable client/server Internet and intranet applications. Most of these languages are built around four core standard technologies:

HTML (hypertext markup language)—the language used to construct most Internet and intranet page content and hyperlinks.

XML (extensible markup language)—an extensible language for transporting data and properties across the Web.

CGI (Computer Gateway Interface)—a standard for publishing graphical World Wide Web components, constructs, and links.

Java—a general-purpose programming language for creating platform-independent programs, servlets, and applets that can execute from within a browser's *Java Virtual Machine*.

Examples of *Java*-specific SDEs include IBM's *WebSphere* and Borland's *Jbuilder*. These SDEs can create Internet, intranet, and non-Internet/intranet applications. Virtually all existing two-tiered and *n*-tiered SDEs are also evolving to support *HTML, XML, CGI*, and *Java*.

Application Architecture Strategies for Systems Design

Regardless of what it is called, all information systems have an application architecture. Different organizations apply different strategies to determining application architecture. Let's briefly classify the two most common approaches.

> The Enterprise Application Architecture Strategy

In the enterprise application architecture strategy, the organization develops an enterprisewide information technology architecture to be followed in all subsequent information systems development projects. This IT architecture defines the following:

- The approved network, data, interface, and processing technologies and development tools (inclusive of hardware and software, and clients and servers).

- A strategy for integrating legacy systems and technologies into the application architecture.
- An ongoing process for continuously reviewing the application architecture for currency and appropriateness.
- An ongoing process for researching emerging technologies and making recommendations for their inclusion in the application architecture.
- A process for analyzing requests for variances from the approved application architecture.

An initial application architecture is usually developed as a separate project or as part of a strategic information systems planning project. The ongoing maintenance of the application architecture is usually assigned to a permanent information technology research group or to an enterprise application architecture committee.

Subsequent to the approval of the application architecture, every information system development project is expected to use or choose technologies based on that architecture. In most cases, this greatly simplifies the architecture phase of a system development methodology. You simply select from the approved technologies according to the architecture's rules or guidelines.

Of course, even if a technology is approved in the application architecture, it is subject to a feasibility analysis, as described in the next subsection.

> The Tactical Application Architecture Strategy

In the absence of an enterprisewide application architecture, each project must define its own architecture for the information system being developed. There still may exist some sort of information technology research and deployment group.

While the proposed application architecture for any new information system may be influenced by existing technologies, the developers usually have somewhat greater latitude in requesting new technologies. Of course, the final decision must be defended and approved as feasible. IT feasibility usually includes the following aspects:

- *Technical feasibility*—This can be either a measure of a technology's maturity, a measure of the technology's suitability to the application being designed, or a measure of the technology's ability to work with other technologies.
- *Operational feasibility*—This is a measure of how comfortable the business management and users are with the technology and how comfortable the technology managers and support personnel are with the technology.
- *Economic feasibility*—This a measure of both whether or not the technology can be afforded and whether it is cost-effective, meaning the benefits outweigh the costs.

Feasibility criteria and techniques for measuring them were covered in Chapter 11.

Modeling the Application Architecture of an Information System

The use of logical DFDs to model process requirements is a fairly accepted practice. However, the transition from analysis-oriented logical DFDs to design-oriented physical DFDs has historically been somewhat mysterious and elusive. We desire a high-level general design that can serve as an application architecture for the system and a general design for the processes that make up the system. At the same time, we don't want to get caught up in a counterproductive modeling exercise that slows our progress in systems design and rapid application development. Simply stated, we want a blueprint to guide us through detailed design and construction. The blueprint will identify design units for detailed specification or rapid development, whichever is most productive in our project.

> Drawing Physical Data Flow Diagrams

The mechanics for drawing physical DFDs are virtually identical to those for logical DFDs. The rules of correctness are also identical. An acceptable design results in:

- A system that works.
- A system that fulfills user requirements (specified in the logical DFDs).
- A system that provides adequate performance (throughput and response time).
- A system that includes sufficient internal controls (to eliminate human and computer errors, ensure data integrity and security, and satisfy auditing constraints).
- A system that is adaptable to ever-changing requirements and enhancements.

We could develop a single physical DFD for the entire system or a set of physical DFDs for the target system. Our methodology suggests the following:

- A physical data flow diagram should be developed for the network architecture. Each process on this diagram is a physical process<u>or</u> (client or server) in the system. Each server is its own processor; however, it is usually impractical to show each client. Instead, each class of clients (e.g., an order entry clerk) is represented by a single processor.
- For each processor on the above model, a physical data flow diagram should be developed to show the event processes (see Chapter 9) that will be assigned to that processor. It is possible that you would choose to duplicate some event processes on multiple processors. For instance, orders may be processed on each region's servers and clients.
- For all but the simplest event processes, they should be factored into design units and modeled as a single physical data flow diagram. A **design unit** is a self-contained collection of processes, data stores, and data flows that share similar design attributes. A design unit serves as a subset of the total system whose inputs, outputs, files and databases, and programs can be designed, constructed, and unit tested as a single subsystem.

design unit a self-contained collection of processes, data stores, and data flows that share similar design attributes.

An example would be a set of processes (one or more) to be designed as a single program. The design unit could then be assigned to a single programmer (or team) who (which) can work independently of other programmers and teams without adversely affecting the work of the other programmers. The implemented units would then be assembled into the final application system. Design units can also be prioritized for implementing versions of a system.

> Prerequisites

Let's set the table by describing the prerequisites to creating physical DFDs. They include:

- A logical data model (entity relationship diagram created in Chapter 8).
- Logical process models (data flow diagrams created in Chapter 9).
- Repository details for all of the above.

Given these models and details, we can distribute data and processes to create a general design. Your general design will normally be constrained by one or more of the following:

- Architectural standards that predetermined the choice of database management systems, network topology and technology, user interface(s), and/or processing methods.
- Project objectives that were defined at the beginning of systems analysis and refined throughout systems analysis.

- The feasibility of chosen or desired technology and methods. (Feasibility analysis techniques were covered in Chapter 11.)

Within any restrictions of those constraints, the ensuing techniques can be applied.

> The Network Architecture

The first physical DFD to be drawn is the network architecture DFD. A *network architecture DFD* is a physical data flow diagram that allocates processors (clients and servers) and devices (e.g., machines and robots) to a network and establishes (1) the connectivity between the clients and the servers and (2) where users will interact with the processors (usually only the clients).

To identify the processors and their locations, the developer utilizes two resources:

- If an enterprise information technology architecture exists, that architecture likely specifies the client/server vision that should be targeted.
- The advice of competent network managers and/or specialists should be solicited to determine what's in place, what's possible, and what impact the system may have on the computer network.

Network architecture DFDs (see Figure 13-11) need to be labeled to show somewhat different information than normal DFDs. They don't show specific data flows

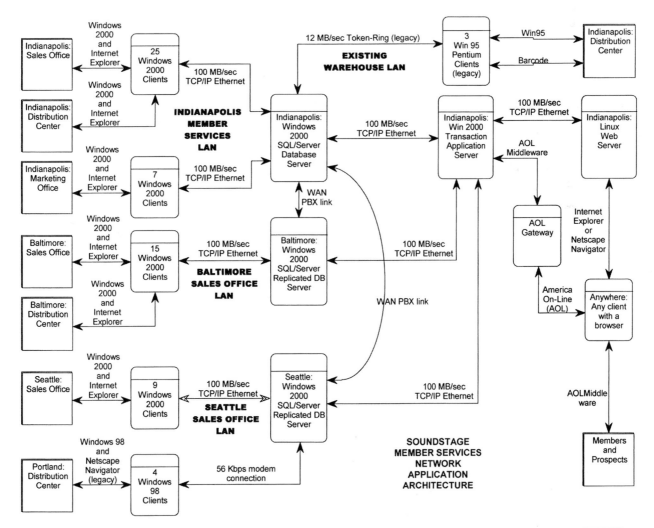

FIGURE 13-11 Network Architecture DFD

per se. Instead, they show highways over which data flows may travel in either direction. Also, network topology DFDs indicate the following:

- *Servers and their physical locations*—Servers are not always located at the sites indicated on a location connectivity diagram. Network staff access to servers is usually an issue. Some network management tasks can be accomplished remotely, and some tasks also require hands-on access.
- *Clients and their physical locations*—In this case, the location connectivity diagram is useful in identifying "groups" of like users (e.g., ORDER CLERKS, SALES REPRESENTATIVES, etc.) who will be serviced by similar clients. A single processor should represent the entire group at a single location. The same group may be replicated in multiple locations. For example, you would expect each SALES REGION to have similar types of employees.
- *Processor specifications*—The repository descriptions of processors can be used to define processor specifications such as RAM, hard-disk capacity, and display.
- *Transport protocols*—Connections are labeled with transport protocols (e.g., TCP/IP) and other relevant physical parameters.

The network topology DFD can be used to either design a computer network or document the design of an existing computer network. In either case, the network is being modeled so that we can subsequently assign information system processes, data stores, and data flows to servers on the network.

> Data Distribution and Technology Assignments

The next step is to distribute data stores to the network processors. The required logical data stores are already known from systems analysis—as data stores on the logical DFDs (Chapter 9) or as entities on the logical ERDs (Chapter 8). We need only determine where each will be physically stored and how they will be implemented.

To distribute the data and assign their implementation methods, the developers utilize three resources:

- If available, the data distribution matrices from systems analysis (Chapters 8 and 9) model the data needs at business locations from a technology independent perspective.
- If an enterprise information technology architecture exists, that architecture likely specifies the database vision and technologies that should be targeted.
- The advice of data and database administrators should be solicited to determine what's in place, what's possible, and what impact the database may have on the overall system.

The distribution options were described earlier in the chapter and are summarized as follows:

- *Store all data on a single server.* In this case, the database (consisting of multiple tables) should be named, and that named database and its implementation method (e.g., Oracle: dbmemberServices) should be added to the physical DFD and connected to the appropriate processor.
- *Store specific tables on different servers.* In this case, and for clarity's sake, we should record each table as a data store on the physical DFD and connect each to the appropriate server.
- *Store subsets of specific tables on different servers.* In this case we record the tables exactly as above except that we indicate which tables are subsets of the total set of records. For example, the label DB2: ORDERS TABLE (REG SUBSET) would indicate that a subset of all orders for a region is stored in a DB2 database table.
- *Replicate (duplicate) specific tables or subsets on different servers.* In this case, replicated data stores are shown on the physical DFD. One copy of any replicated table is designated as the MASTER, and all other copies are designated as COPY or REPLICANT.

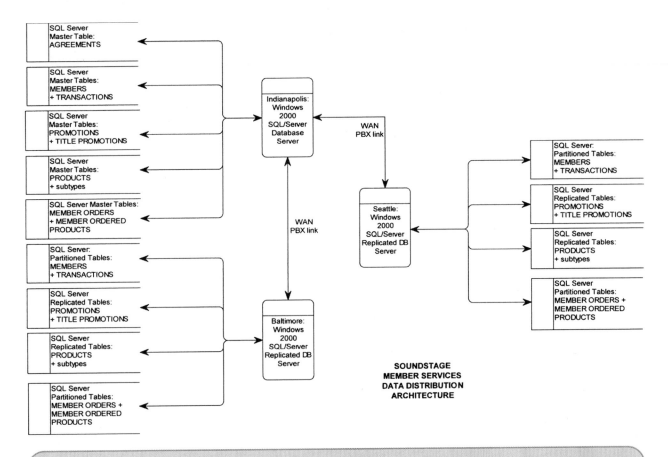

FIGURE 13-12 Data Distribution and Technology Assignments for SoundStage

Why distribute data storage? There are many possible reasons. First, some data instances are of local interest only. Second, performance can often be improved by subsetting data to multiple locations. Finally, some data needs to be localized to assign custodianship of that data. The data distribution and technology assignments for the SoundStage case study are shown in Figure 13-12.

Data distribution decisions can be very complex—normally the decisions are guided by data and database professionals and taught in data management courses and textbooks. In this book we want to consider only how to document the partition and duplication decisions.

> Process Distribution and Technology Assignments

Information system processes can now be assigned to processors as follows:

- For two-tiered client/server systems, all the logical event diagrams (Chapter 6) are assigned to the client.
- For three-tiered client/server and network computing systems, you must closely examine each event's primitive (detailed) data flow diagram. You need to determine which primitive processes should be assigned to the client and which should be assigned to an application server. In general, data capture and editing are assigned to clients and other business logic is assigned to servers. If you partition different aspects of a logical DFD to different clients and servers, you should draw separate physical DFDs for the portions on each client and server.

After partitioning, each physical DFD corresponds to a design unit for a given business event. (Business events, or use cases, were discussed in Chapter 7.) For each of these design units, you must assign an implementation method, the SDE that will be used to implement that process. You must also assign implementation methods to the data flows.

SoundStage's Member Services system will be implemented with a multitiered client/server and network computing architecture. A sample DFD for one event to be assigned to a client is shown in Figure 13-13. Notice that the data stores are shown

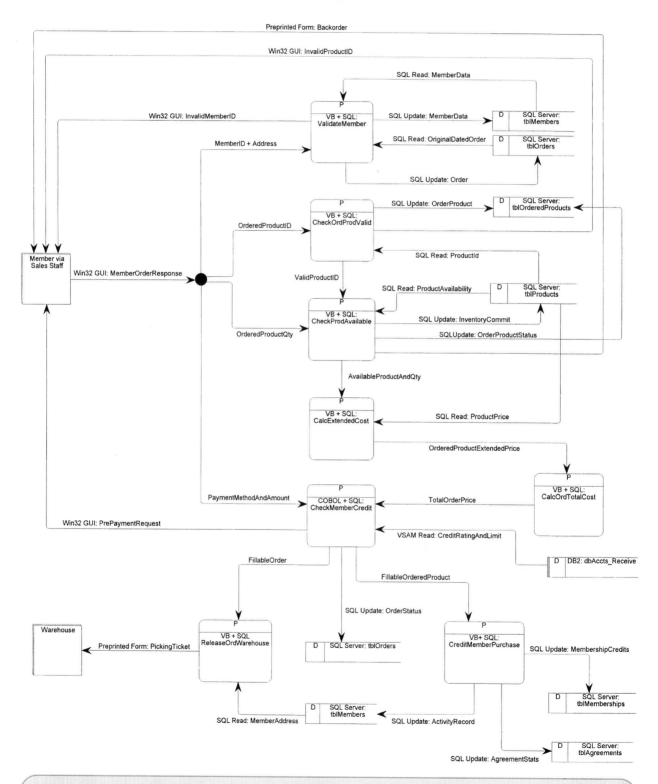

FIGURE 13-13 A Physical DFD for an Event

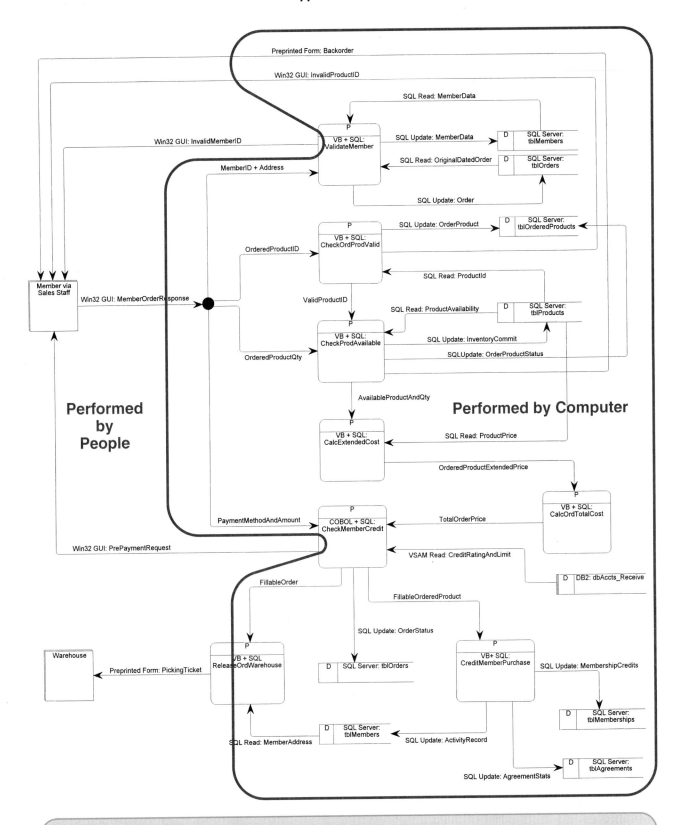

FIGURE 13-14 The Person/Machine Boundary

even though we know they have been partitioned to a database server. This is for the benefit of the programmers who must implement the DFD.

> The Person/Machine Boundaries

The last step of process design is to factor out any portion of the physical DFDs that represent manual, not computerized, processes. This is sometimes called establishing a person/machine boundary. Establishing a person/machine boundary is not difficult, but it is not as simple as you might first think. The difficulty arises when the person/machine boundary cuts through a logical process—in other words, part of the process is to be manual and part is to be computerized. This situation is common on logical DFDs because they are drawn without regard to implementation alternatives.

Figure 13-14 adds the person/machine boundary to a physical DFD. Notice that our boundary cuts through several processes, including the CHECK MEMBER CREDIT process. The solution to this process requires two steps:

1. The manual process portions are pulled out as a separate design unit (see Figure 13-15). All these processes are completely manual. The interfaces of the manual design units to the computerized processes (on Figure 13-14) are depicted as external agents. Ultimately, the manual processes in the design unit must be clearly described to those people who will have to perform them.
2. If necessary, the processes on the original diagram should be renamed to reflect only the computerized portion. (In practice, the processes were already named that way.)

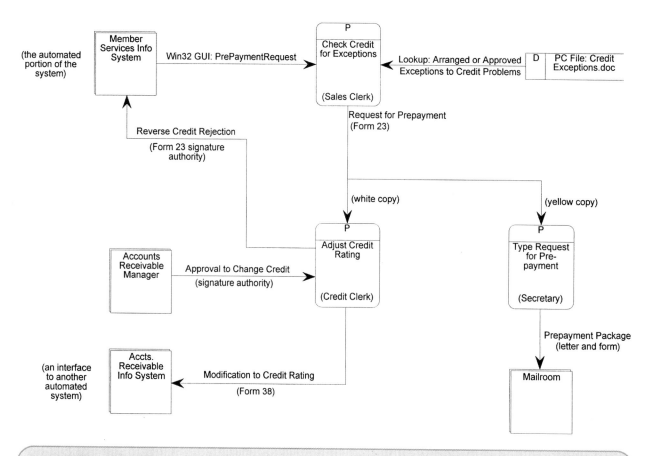

FIGURE 13-15 A Manual Design Unit

In this chapter, you have learned how to outline the design of a new information system to fulfill the requirements identified and modeled during systems analysis. This general design for the new system will guide the detailed design and construction of that system.

Most readers will now progress to the detailed design chapters that build on the general design for the new system. For most of you, we recommend you start with Chapter 14, "Database Design." Most design-by-prototyping and rapid application development techniques are absolutely dependent on the existence of the planned information system's database. Databases must be carefully designed to ensure adaptability and flexibility during the system's lifetime. Thus, Chapter 14 is the best place to begin your study of detailed design. Subsequently, you can move on to chapters that cover other aspects of detailed design, including inputs, outputs, and programs.

 ## Chapter Review

1. Physical data flow diagrams model the technical and human design decisions to be implemented as part of an information system. They communicate technical choices and other design decisions to those who will actually construct and implement the system.

2. An information technology architecture defines the technologies to be used by one, more, or all information systems. There are four categories of technology architectures: network, data, interface, and process.

3. A distributed system is one in which the components of an information system are distributed to multiple locations in a computer network. The five layers of distributed systems architecture are (*a*) presentation, (*b*) presentation logic, (*c*) application logic, (*d*) data manipulation, and (*e*) data.

4. A local area network is a set of client computers connected to one or more servers through either cable or wireless connections over relatively short distances.

5. A file server system is a LAN-based solution in which a server computer hosts only the data layer. All other layers are implemented on the client.

6. The prevailing computing model is currently client/server in which the presentation, presentation logic, application logic, data manipulation, and data layers are distributed between client PCs and one or more servers. Clients are classified by their power as thin or fat. Servers are dedicated to functions such as database, transactions, applications, messaging or work group, or Web.

7. Distributed presentation, distributed data, and distributed data and logic are types of client/server systems.

8. Network computing uses Internet technology to build Internet or intranet applications.

9. Data storage is typically implemented using distributed relational database technology that either partitions data to different servers or replicates data on multiple servers.

10. User interface options include batch, online, remote batch, keyless data entry (including optical character/mark and bar-coding methods), pen input, electronic messaging, electronic data interchange, and imaging.

11. System interfacing is typically implemented using middleware, software that enables processes to communicate with one another.

12. Processes are implemented using highly integrated tool kits called software development environments.

13. Application architectures may be developed and enforced strategically, or they may tactically evolve on a project-by-project basis.

Review Questions

1. In traditional structured analysis and design, what system models are developed, and in what order?
2. Why is the complete structured analysis and design methodology seldom employed anymore?
3. When a logical process is divided into multiple physical processes, or if more physical processes are added, what is it important for designers to check?
4. Why is the number of physical processes shown on a physical DFD generally greater than the number of the logical processes?
5. What does a physical data flow represent?
6. What type of data store is often overlooked by designers in conducting systems design?
7. Although centralized systems are less complex and easier to implement, distributed systems have

pretty much taken over from centralized systems. What were some of the reasons for this?

8. What is the difference between the presentation layer and the presentation logic layer?
9. What is a file server system, and what kind of network environment does it use?
10. What are some of the inherent limitations and disadvantages of a file server system?
11. What is the difference between a thin client and a fat client?
12. What is the network architecture used in e-commerce? Please explain how each layer is related.
13. What is the sequence of high-level tasks for modeling the application architecture of an information system?

Problems and Exercises

1. You are in the middle of the system design phase for a project to develop a corporate intranet, and the project team is holding a planning meeting. One of the system users on the project, who has said very little during the meeting, finally speaks up and says, "All you techies keep talking about the application architecture we're going to be designing. I don't have a clue what you're talking about." Define and explain what application architecture is to the nontechnical system users in the group.
2. What is the purpose of the physical data flow diagram? In general, how are they different from logical data flow diagrams? What basic shapes and connections do physical DFDs use? Are physical DFDs a legacy design tool, or are they still a viable tool in today's object-oriented world?
3. You are working on a project to design a new order system for a distributor of auto supplies. You are developing the physical DFD, and one of the logical processes is "Check Inventory in Stock." If this process is to be performed both by people and by computer, how would you show this as a physical process? (Note: Use the diagram formats shown after Figure 13-1.) What if the process is completely done by computer, but using different technologies?
4. Explain data replication and its purpose. In what type of database system would you find data replication?
5. Complete the following sentences:

   A _____ system is also called a distributed data and application client/server system or

_____. The _____ resides on an _____, the _____ resides on the client server, and the data and _____ on the _____.

A _____-tiered system is also called _____, and the _____ resides on the clients, and the data and data manipulation layers on the server.

A _____ client/server system please the presentation and the presentation logic layers on the _____, and the _____ on the server.

6. What do file server systems and client/server systems have in common? What is different? What are the most important advantages of a client/server solution?
7. You are working in the IT shop of a rapidly growing organization that is planning to implement a new client/server system. Initially, there will be slightly fewer than 100 clients, with a substantial amount of data input and data analysis activity across the network. The business drivers are to be able to get data in and to "crunch" it quickly. The budget for the project is robust and allows for the purchase of powerful workstations and personal computers. The designers on the team are pretty well evenly divided between a two-tiered and three-tiered client/server architecture. They are looking to you for advice. What would you recommend? Why?

8. Internet technology has grown at an explosive rate over the past decade. In the view of many, network computing architectures represents a major move in a radically different direction away from client/server architectures. Why is this?

9. Match the terms in the first column with the definitions or examples in the second column.

1.	Thin client	A.	Patient Treatment Records
2.	Logical Data Store	B.	Data input screen
3.	Groupware server	C.	SQL Insert: New Account
4.	Processor	D.	Data terminal
5.	Transaction server	E.	SAS File: Waiting List Report
6.	Mainframe	F.	CORBA
7.	Presentation Layer	G.	Report-formatting application
8.	Physical data flow	H.	Distributed system
9.	Physical Data Store	I.	Centralized system
10.	Presentation Logic Layer	J.	Microsoft Exchange
11.	Wide Area Network (WAN)	K.	Tuxedo
12.	Object-sharing standard	L.	Customer
13.	Application logic layer	M.	Statistical analysis application

10. Batch processing has been in use since the 1950s and many people consider batch processing to be an obsolete method of processing data. But if batch processing is obsolete, why then are new batch processing applications still being developed?

11. Match the terms in the first column with the definitions or examples in the second column.

1.	Two-tiered client/ server SDE	A.	HTTP
2.	Design unit	B.	CICS
3.	EDI	C.	Physically separated presentation, data, and application layers
4.	Application middleware	D.	Employee's "start page"
5.	Multitier client/ server SDE	E.	Determining distribution of application components
6.	Virtual business	F.	*Windows CE*
7.	Intranet Client/. Server SDE	G.	*Allegris*
8.	Partitioning	H.	*PowerBuilder*
9.	Presentation middleware	I.	Self-contained collection of data flows, stores, and processes
10.	Clean layering	J.	Object request brokers
11.	Pen-input	K.	Online commercial banking
12.	Intranet portal	L.	*XML*
13.	Transaction monitor	M.	Amazon.com

12. You are working on a complex project to implement an enterprise-level information system in your organization. You are almost finished with creating the physical data flow diagram from the logical DFD, and realize there are a number of manual processes intertwined with computerized processes. What caused this to occur? How should you show the manual processes on the physical DFD, or do you need to show them at all?

13. You have been given a set of physical DFDs for a new system to review for acceptability. What questions should you ask yourself when reviewing them?

Projects and Research

1. Although your friends tease you about it, you are an unabashed collector of vintage folk songs from the 1950s and 1960s. Your collection now totals several thousand recordings in various formats. To help keep better track of the recordings, you have decided to develop a simple inventory system in Microsoft *Access*. You want to be able to add new recordings to the system, update information on the ones that you have, search on multiple fields for a particular recording or artist, and generate various reports. Design the system, using the techniques learned to date, then draw a context data flow diagram and logical data flow diagram.

2. Many organizations have implemented intranets. Contact or visit several local organizations in the public and private sector that have intranets. Find the unit or person who is responsible for the organization's intranet, and discuss its application architecture, features, policies, issues, and so on.

   a. Describe each of the organizations you contacted.

   b. Describe each of their intranets and how they are used by employees.

   c. Are they primarily informational intranets, or are any of them being used as a portal where everything—their desktop applications and any information system applications they use for their job—runs from the intranet browser?

d. Who "owns" the intranet in the organization, and who is responsible for posting content or keeping it current?

e. Do your discussions make you feel that the organizations' employees and intranet owners understand the potential of their intranet and are using it to its full potential? Explain your -answer.

3. You are a consultant who has been hired by a company to help update its IT architecture plan. The company manufactures electric generators and has sales offices and service facilities throughout North America and Central America. One of your tasks is to recommend a data distribution strategy to the company. Although you are familiar with the principles of application architectures and the methods for documenting them, it has been a while since you've been in a position of recommending a data distribution approach. Research on the Web or in your school library the questions you should ask and the criteria to use in order to make an appropriate recommendation regarding the data distribution approach the company should take.

4. Mainframe computers once dominated information technology, but have slipped into the shadow of other technologies such as client/server systems. Every so often there seems to be a spate of articles reporting the final death of the mainframe. But just as often there seems to be another burst of articles reporting the resurgence of the mainframe. Research the topic of trends in mainframe computing in your school library, on the Internet, and/or with some experienced IT managers.

   a. Describe your research sources and their positions.

   b. Were you able to find "hard" information, such as the number of sales per year of mainframes?

   c. Did you find any differences in the usage of mainframe computers between the public and private sectors? Describe.

d. On the basis of the information you found, what conclusions would you draw about the state of mainframe computing today?

e. What about 10 years from now? What role, if any, do you think that mainframe computing will play in the public and private sectors? Support your answer.

5. New development tools to enable client/server Internet applications seem to be emerging almost every day (a slight exaggeration). Research one of these new development tools, and the core technology, such as *.NET* or *XML*, around which it is built. Then prepare an analysis for your chief information officer (real or hypothetical) evaluating this new tool and/or technology and the potential for usage by your organization. Note: since the target audience for this analysis is an executive, your document should touch on the salient points at an appropriately high level, though references or links to detailed technical documentation should also be included.

6. Visit or study a large corporation or government agency in your area, and ask about its application architectures. See if you can obtain a copy of its IT architecture plan or an equivalent document.

   a. Describe the organization you studied.

   b. Describe its application architecture(s). What is its predominant application architecture?

   c. What type of Internet-based computing architecture(s) does it use? How large a role does it play?

   d. Does the organization still use mainframe or minicomputer technology? If so, describe how it is currently used.

   e. What changes in application architecture technology does it envision will take place over the next five years. What is the organization's strategy for dealing with these technological changes? Do you think its strategy will be effective?

   f. Draw a high-level context diagram of its overall information system architecture.

Minicases

1. Consider an e-commerce site that you previously researched for Wow Munchies (the grocery store). Make any necessary assumptions, and conduct any new research that is needed. Create a network computing architecture model for Wow Munchies. State your assumptions, along with any background research you have done, in a short paper.

2. You will find, as a systems analyst, that the person who designs a system is often not the one who develops the program for it. On page 504, the book

states that an acceptable design of a physical data flow results in:

   a. A system that works

   b. A system that fulfills user requirements.

   c. A system that provides adequate performance.

   d. A system that includes sufficient internal controls.

   e. A system that is adaptable to ever-changing requirements.

Find an example of a system DFD design that *seemed* acceptable, but did not result in a system that met the stated requirements. Was the problem one of design or technical complications? Did the programmers understand what the analysts were requesting? Share with the class.

3. You are on a team that has been directed to design a system for VideoStore, a movie rental company. This company rents movies only in stores (not online) and has about 10 stores throughout the state of Ohio. Discuss, step by step, how you would go through the life cycle process up to and including the design phase to create this system. Be as detailed as you can, and use specific examples.

4. Return to the Chapter 6 material on PIECES and the material in Chapter 11 on the candidate systems matrix. What are the strengths of each? Utilize them together to consider three potential systems for the video rental store you researched in the previous problem. How does using multiple-perspective matrices give you a more thorough view of system options?

Suggested Readings

Berstein, Phillip, and Eric Newcomer. *Principles of Transaction Processing: For the Systems Professional.* San Francisco: Morgan Kaufman Publishers, 1997. This book covers virtually every transaction processing model, transaction monitor, and transaction server currently implemented.

Gane, Chris, and Trish Sarson. *Structured Systems Analysis: Tools and Techniques.* Englewood Cliffs, NJ: Prentice Hall, 1979. This classic on process modeling became the basis of physical data flow diagrams.

Goldman, James. *Applied Data Communications: A Business-Oriented Approach,* 2nd ed. New York: John Wiley & Sons, 1998. Our colleague at Purdue has written an excellent textbook for those seeking to learn about data communications and networking from a business perspective.

Goldman, James; Phillip Rawles; and Julie Mariga. *Client/Server Information Systems: A Business-Oriented Approach.* New York: John Wiley & Sons, 1999. Our colleagues have written an outstanding textbook that introduces students to information technology architecture for information systems.

Kara, Dan. "Why Partition? Multitiered Application Architecture." In *Application Development Trends.* Natick, MA: Software Productivity Group, May 1997, pp. 38–46. This article stimulated our interest and research on the need to develop and teach partitioning techniques as part of this book.

Kara, Daniel A., et al. "Enterprise Application Development: Seminar Notes." Chicago: Software Productivity Group, November 12, 1996. This seminar and the writings of the SPG have strengthened our understanding of two-tiered and *n*-tiered software application development techniques and technologies.

Orfali, Robert; Dan Harkey; and Jeri Edwards. *Client/Server Survival Guide,* 3rd ed. New York: John Wiley & Sons, 1999. This professional reference manual has served us well for three editions of client/server technology and terminology evolution.

Renaud, Paul. *Introduction to Client/Server Systems,* 2nd ed. New York: John Wiley & Sons, 1996. This is another reference book on the primary distributed computing architecture of our time.

Smith, Patrick, and Steve Guengerich. *Client/Server Computing,* 2nd ed. Indianapolis, IN: SAMS Publishing, 1994. This professional book has been used to teach the basics of client/server technology and architecture to our students at Purdue. Given the rapid evolution of this technology, there may now exist a third edition. Check out the technology case studies in the appendixes.

Theby, Stephen E. "Derived Design: Bridging Analysis and Design." McDonnell Douglas Professional Services: Improved System Technologies, 1987. The techniques described in this paper are the basis for a phase in STRADIS (Structured Analysis, Design, and Implementation of Information Systems), a systems development methodology. The technique was altered and simplified to make it suitable to the level of this textbook. As authors, we were quite impressed with the full derived design technique as advocated in the STRADIS methodology.

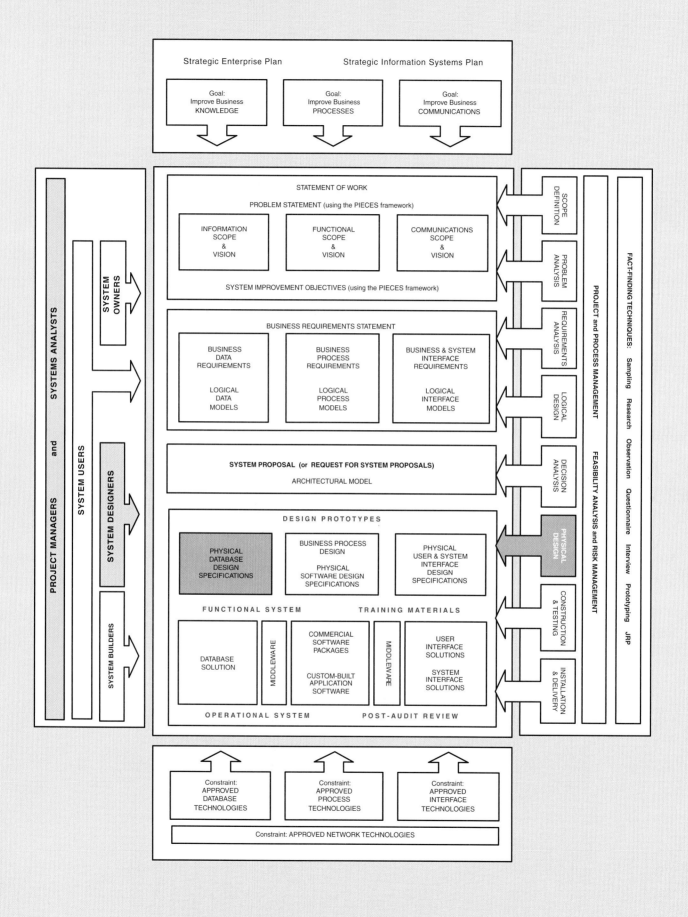

Strategic Enterprise Plan   Strategic Information Systems Plan

Goal:
Improve Business
KNOWLEDGE

Goal:
Improve Business
PROCESSES

Goal:
Improve Business
COMMUNICATIONS

STATEMENT OF WORK

PROBLEM STATEMENT (using the PIECES framework)

INFORMATION
SCOPE
&
VISION

FUNCTIONAL
SCOPE
&
VISION

COMMUNICATIONS
SCOPE
&
VISION

SYSTEM IMPROVEMENT OBJECTIVES (using the PIECES framework)

BUSINESS REQUIREMENTS STATEMENT

BUSINESS
DATA
REQUIREMENTS

BUSINESS
PROCESS
REQUIREMENTS

BUSINESS & SYSTEM
INTERFACE
REQUIREMENTS

LOGICAL
DATA
MODELS

LOGICAL
PROCESS
MODELS

LOGICAL
INTERFACE
MODELS

SYSTEM PROPOSAL (or REQUEST FOR SYSTEM PROPOSALS)

ARCHITECTURAL MODEL

DESIGN PROTOTYPES

PHYSICAL
DATABASE
DESIGN
SPECIFICATIONS

BUSINESS PROCESS
DESIGN

PHYSICAL
SOFTWARE DESIGN
SPECIFICATIONS

PHYSICAL
USER & SYSTEM
INTERFACE
DESIGN
SPECIFICATIONS

FUNCTIONAL SYSTEM   TRAINING MATERIALS

DATABASE
SOLUTION

MIDDLEWARE

COMMERCIAL
SOFTWARE
PACKAGES

CUSTOM-BUILT
APPLICATION
SOFTWARE

MIDDLEWARE

USER
INTERFACE
SOLUTIONS

SYSTEM
INTERFACE
SOLUTIONS

OPERATIONAL SYSTEM   POST-AUDIT REVIEW

SYSTEM OWNERS

SYSTEM USERS

SYSTEM DESIGNERS

SYSTEM BUILDERS

SYSTEMS ANALYSTS

PROJECT MANAGERS   and

SCOPE DEFINITION

PROBLEM ANALYSIS

REQUIREMENTS ANALYSIS

LOGICAL DESIGN

DECISION ANALYSIS

PHYSICAL DESIGN

CONSTRUCTION & TESTING

INSTALLATION & DELIVERY

PROJECT and PROCESS MANAGEMENT

FEASIBILITY ANALYSIS and RISK MANAGEMENT

FACT-FINDING TECHNIQUES: Sampling   Research   Observation   Questionnaire   Interview   Prototyping   JRP

Constraint:
APPROVED
DATABASE
TECHNOLOGIES

Constraint:
APPROVED
PROCESS
TECHNOLOGIES

Constraint:
APPROVED
INTERFACE
TECHNOLOGIES

Constraint: APPROVED NETWORK TECHNOLOGIES

14

Database Design

Chapter Preview and Objectives

Data storage is a critical component of most information systems. This chapter teaches the design and construction of physical databases. You will know that you have mastered the tools and techniques of database design when you can:

∎ Compare and contrast conventional files and modern, relational databases.

∎ Define and give examples of fields, records, files, and databases.

∎ Describe a modern data architecture that includes files, operational databases, data warehouses, personal databases, and work group databases.

∎ Compare the roles of systems analyst, data administrator, and database administrator as they relate to databases.

∎ Describe the architecture of a database management system.

∎ Describe how a relational database implements entities, attributes, and relationships from a logical data model.

∎ Transform a logical data model into a physical, relational database schema.

∎ Generate SQL code to create the database structures in a schema.

Introduction

In the decision analysis phase of the SoundStage Member Services system project it was decided to implement the data for the system in *SQL Server*. Now in the physical design phase, Bob Martinez has been working on the physical design of the database.

To maximize throughput, the entire database will be replicated at each distribution center. Each instance of the database will be stored on a Dell PowerEdge server with quad Xeon processors and RAID level 5 hard drives. Fortunately, *SQL Server* has built-in capability to synchronize replicated data.

Bob refined the normalized entity relationship diagram he created during the logical design phase. Using the CASE tool, *System Architect,* he revised table and field names according to accepted SoundStage naming conventions. He created indexes on all key fields as well as nonkey fields wth subsetting criteria requirements. He created primary key and foreign key constraints on the tables. He also created other constraints to implement business rules that require default values for some fields, require non-null values in some fields, or limit field entries to a certain domain of values.

System Architect automatically generated the SQL code that will be used to construct the actual database. In the meantime, Bob used that SQL code to create a desktop prototype of the database in the Microsoft Data Engine (MSDE), an *SQL Server*-compatible development database engine. The prototype will give developers something they can test their SQL and programs against.

Conventional Files versus the Database

file a collection of similar records.

database a collection of interrelated files.

All information systems create, read, update, and delete (sometimes abbreviated *CRUD*) data. This data is stored in files and databases. A **file** is a collection of similar records. Examples include a CUSTOMER FILE, ORDER FILE, and PRODUCT FILE. A **database** is a collection of interrelated files. The key word is *interrelated*. A database is *not* merely a collection of files. The records in each file must allow for relationships (think of them as "pointers") to the records in other files. For example, a SALES database might contain order records that are linked to their corresponding CUSTOMER and PRODUCT records.

Let's compare the file and database alternatives. Figure 14-1 illustrates the fundamental difference between the file and database environments. In the file environment, data storage is built around the applications that will use the files. In the database environment, applications will be built around the integrated database. Accordingly, the database is not necessarily dependent on the applications that will use it. In other words, given a database, new applications can be built to share that database. Each environment has its advantages and disadvantages.

As shown in the chapter home page, this chapter is concerned with the design and (initial) construction of the database for an information system. The prerequisite is a data (requirements) model from Chapter 8. The deliverables are a database (design) schema and database (definition) program.

> The Pros and Cons of Conventional Files

In most organizations, many existing information systems and applications are built around conventional files. You may already be familiar with various conventional file organizations (e.g., indexed, hashed, relative, and sequential) and their access methods (e.g., sequential and direct) from a programming course. These conventional files will likely be in service for quite some time.

Conventional files are relatively easy to design and implement because they are normally designed for use with a single application or information system, such as ACCOUNTS RECEIVABLE or PAYROLL. If you understand the end user's output requirements for that

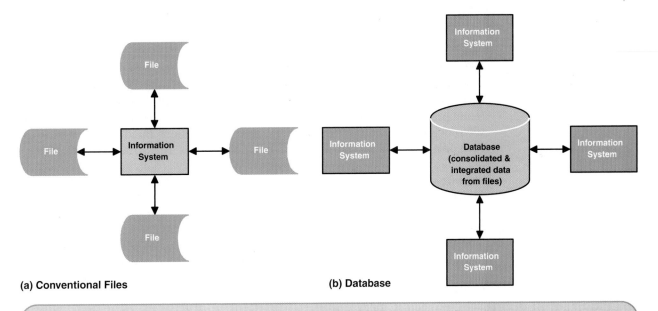

(a) Conventional Files

(b) Database

FIGURE 14-1 Conventional Files versus the Database

system, you can easily determine the data that will have to be captured and stored to produce those outputs and define the best file organization for those requirements.

Historically, another advantage of conventional files has been processing speed. They can be optimized for the access of the application. At the same time, they can rarely be optimized for shared use by different applications or systems. Still, files have generally outperformed their database counterparts; however, this limitation of database technology is rapidly disappearing thanks to cheaper and more powerful computers and more efficient database technology.

Conventional files also have numerous disadvantages. Duplication of data items in multiple files is normally cited as the principal disadvantage of file-based systems. Files tend to be built around single applications without regard to other (future) applications. Over time, because many applications have common data needs, the common data elements get stored redundantly in many different files. This duplicate data results in duplicate inputs, duplicate maintenance, duplicate storage, and possibly data integrity problems (different files showing different values for the same data item).

And what happens if the data format needs to change? Consider the problem faced by many firms if all systems must support a nine-digit zip code or four-digit years. Because these fields may be stored in many files (with different names), each file would have to be studied and identified. Subsequently, all of the programs that use these zip code and date fields would have to be changed.

A significant disadvantage of files is their inflexibility and nonscalability. Files are typically designed to support a single application's *current* requirements and programs. Future needs—such as new reports and queries—often require that these files be restructured because the original file structure cannot effectively or efficiently support the new requirements. But if we elect to restructure those files, all programs using those files would also have to be rewritten. In other words, the current programs have become dependent on the current files, and vice versa. This usually makes reorganization impractical; therefore, we elect to create new, redundant files (same data, structured differently) to meet the new requirements. But that exacerbates the aforementioned redundancy problem. Thus, the inflexibility and redundancy problems tend to complicate one another!

As legacy file-based systems and applications become candidates for reengineering, the trend is overwhelmingly in favor of replacing file-based systems and applications with database systems and applications. For that reason, we have elected to focus this chapter on database design.

> The Pros and Cons of Databases

We've already stated the principal advantage of databases—the ability to share the same data across multiple applications and systems. A common misconception about the database approach is that you can build a single superdatabase that contains all data items of interest to an organization. This notion, however desirable, is not currently practical.[1] The reality of such a solution is that it would take forever to build such a complex database. Realistically, most organizations build several databases, each one sharing data with several information systems. Thus, there will be *some* redundancy between databases. However, this redundancy is both greatly reduced and, ultimately, controlled.

Database technology offers the advantage of storing data in flexible formats. This is possible because databases are defined separately from the information systems and application programs that will use them. Theoretically, this allows us to use the data in ways not originally specified by the end users. Care must be taken to truly achieve this *data independence.* If the database is well designed, different combinations of the same data can be easily accessed to fulfill future report and query needs. The database scope can even be extended without changing existing programs that use it. In other words, new fields and record types can be added to the database without affecting current programs.

Database technology provides superior *scalability,* meaning that the database and the systems that use it can be grown or expanded to meet the changing needs of an organization. Database technology provides better technology for client/server and network computing architectures. Such architectures typically require that the database run on its own server. Client/server and network computing was covered in Chapter 13.

On the other hand, database technology is more complex than file technology. Special software, called a *database management system (DBMS),* is required. While a DBMS is still somewhat slower than file technology, the performance limitations are rapidly disappearing. Considering the long-term benefits described earlier, most new information systems development is using database technology.

But the advantages of data independence, greatly reduced data redundancy, and increased flexibility come at a cost. Database technology requires a significant investment. The cost of developing databases is higher because analysts and programmers must learn how to use the DBMS. Finally, to achieve the benefits of database technology, analysts and database administrators and experts must adhere to rigorous design principles.

Another potential problem with the database approach is the increased vulnerability inherent in the use of shared data. You are placing all your eggs in one basket. Therefore, backup and recovery and security and privacy become important issues in the world of databases.

Despite the problems discussed, database usage is growing by leaps and bounds. The technology will continue to improve, and performance limitations will all but disappear. Design methods and tools will also improve. For these reasons, this chapter will focus on database design as an important skill for systems analysts.

Database Concepts for the Systems Analyst

We should begin with a disclaimer. Many of the concepts and issues that are important to database design are also taught in database and data management courses. Most information systems curricula include at least one such course. It is not our intent in this chapter to replace that course. Students of information systems should actively seek out courses that *focus* on data management and database techniques;

---

[1]Enterprise resource planning (ERP) applications provide a common, customizable database that truly supports almost all the core, operational, and managerial data required in many organizations. For example, *SAP R/3* provides several thousand tables.

those courses will cover many more relevant technologies and techniques than we can cover in this single chapter.

That said, we will first introduce (or, for some of you, review) the database concepts and issues that are pertinent to the systems analyst's responsibilities in information system design. Although the chapter focus is on database design, experienced readers will immediately notice that many of the concepts transcend the choice between files and databases.

> Fields

Fields are common to both files and databases. A **field** is the physical implementation of a data attribute (introduced in Chapter 8). Fields are the smallest unit of *meaningful* data to be stored in a file or database. There are four types of fields that can be stored: *primary keys, secondary keys, foreign keys,* and *descriptive fields.*

A **primary key** is a field whose values identify one and only one record in a data entity. (This concept was introduced previously in Chapter 8.) For example, CUSTOMER NUMBER uniquely identifies a single CUSTOMER record in a database, and ORDER NUMBER uniquely identifies a single ORDER record in a database. Also recall from Chapter 8 that a primary key might be created by combining two or more fields (called a *concatonated key*).

A **secondary key** is an *alternate* identifier for a database. A secondary key's value may identify either a single record (as with a primary key) or a subset of all records (such as all ORDERS that have the ORDER STATUS of back-ordered). A single file in a database may have only one primary key, but it may have several secondary keys. To facilitate searching and sorting, an *index* is frequently created for keys.

Foreign keys (also introduced in Chapter 8) are pointers to the records of a different file in a database. Foreign keys enable the database to link the records of one type to those of another type. For example, an ORDER RECORD contains the foreign key CUSTOMER NUMBER to "identify" or "point to" the CUSTOMER record that is associated with the ORDER. Notice that a foreign key in one file requires the existence of the corresponding primary key in another table—otherwise, it does not point to anything! Thus, the CUSTOMER NUMBER in an ORDERS file requires the existence of a CUSTOMER NUMBER in the CUSTOMERS file in order to link those files.

A **descriptive field** is any other (*nonkey*) field that stores business data. For example, given an EMPLOYEES file, some descriptive fields include EMPLOYEE NAME, DATE HIRED, PAY RATE, and YEAR-TO-DATE WAGES.

The business requirements for both keys and descriptors were defined when you performed data modeling in systems analysis (Chapter 8).

> Records

Fields are organized into records. Records are common to both files and databases. A **record** is a collection of fields arranged in a predefined format. For example, a CUSTOMER RECORD may be described by the following fields (notice the common notation):

> CUSTOMER (NUMBER, LAST-NAME, FIRST-NAME, MIDDLE-INITIAL, POST-OFFICE-BOX-NUMBER, STREET-ADDRESS, CITY, STATE, COUNTRY, POSTAL-CODE, DATE-CREATED, DATE-OF-LAST-ORDER, CREDIT-RATING, CREDIT-LIMIT, BALANCE, BALANCE-PAST-DUE . . .)

During systems design, records will be classified as either fixed-length or variable-length records. Most database technologies impose a *fixed-length record structure,* meaning that each record instance has the same fields, same number of fields, and same logical size. Some database systems will, however, compress unused fields and values to conserve disk storage space. The database designer must generally understand and specify this compression in the database design.

In your prior programming courses (especially *COBOL*), you may have encountered *variable-length record structures* that allow different records in the same file to

---

field the smallest unit of meaningful data to be stored in a file or database.

primary key a field or group of fields that uniquely identifies a record.

secondary key a field that identifies a single record or a subset of related records.

foreign key a field that points to records in a different file in a database.

descriptive field a nonkey field.

record a collection of fields arranged in a predetermined format.

have different lengths. For example, a variable-length order record might contain certain common fields that occur once for every order (e.g., ORDER NUMBER, ORDER DATE, and CUSTOMER NUMBER) and other fields that repeat some number of times based on the number of products sold on the order (e.g., PRODUCT NUMBER and QUANTITY ORDERED). Database technologies typically disallow (or at least discourage) variable-length records. This is not a problem, as we'll show later in the chapter.

When a computer program reads a record from a database, it actually retrieves a group or *block* (or *page*) of records at a time. This approach minimizes the number of actual disk accesses. A **blocking factor** is the number of *logical records* included in a single read or write operation (from the computer's perspective). A block is sometimes called a *physical record*. Today, the blocking factor is usually determined and optimized by the chosen database technology, but a qualified database administrator may be allowed to fine-tune that blocking factor for performance. Database tuning considerations are best deferred to a database course or textbook.

blocking factor the number of logical records included in a single read or write operation.

> Files and Tables

Similar records are organized into groups called *files*. In database systems, a file is frequently called a *table*. A **file** is the set of all occurrences of a given record structure. A **table** is the *relational* database equivalent of a file. Relational database technology will be introduced shortly. Some types of conventional files and tables are:

file the set of all occurrences of a given record structure.

table the relational database equivalent of a file.

- **Master files** or tables contain records that are relatively permanent. Thus, once a record has been added to a master file, it remains in the system indefinitely. The values of fields for the record will change over its lifetime, but the individual records are retained indefinitely. Examples of master files and tables include CUSTOMERS, PRODUCTS, and SUPPLIERS.

master file a table containing records that are relatively permanent.

- **Transaction files** or tables contain records that describe business events. The data describing these events normally has a limited useful lifetime. For instance, an INVOICE record is ordinarily useful until the invoice has been paid or written off as uncollectible. In information systems, transaction records are frequently retained *online* for some period of time. Subsequent to their useful lifetime, they are *archived* off-line. Examples of transaction files include ORDERS, INVOICES, REQUISITIONS, and REGISTRATIONS.

transaction file a table containing records that describe business events.

- **Document files** and tables contain stored copies of historical data for easy retrieval and review without the overhead of regenerating the document.

document file a table containing historical data.

- **Archival files** and tables contain master and transaction file records that have been deleted from online storage. Thus, records are rarely deleted; they are merely moved from online storage to off-line storage. Archival requirements are dictated by government regulation and the need for subsequent audit or analysis.

archival file a table containing master and transaction file records that have been deleted from online storage.

- **Table look-up files** contain relatively static data that can be shared by applications to maintain consistency and improve performance. Examples include SALES TAX TABLES, ZIP CODE TABLES, and INCOME TAX TABLES.

table look-up file a table containing relatively static data that can be shared.

- **Audit files** are special records of updates to other files, especially master and transaction files. They are used in conjunction with archival files to recover "lost" data. Audit trails are typically built into better database technologies.

audit file a table containing records of updates to other files.

In the not-too-distant past, file design methods required that the analyst specify precisely how the records in a database should be sequenced (called *file organization*) and accessed (called *file access*). In today's database environment, the database technology itself usually predetermines and/or limits the file organization for all tables contained in the database. Once again, a trained database administrator may be given some control over organization, storage location, and access methods for the purpose of performance tuning.

> Databases

As described earlier, stand-alone, application-specific files were once the lifeblood of most information systems; however, they are being slowly but surely replaced with databases. Recall that a database may loosely be thought of as a set of interrelated files. By interrelated, we mean that records in one file may be associated or linked with the records in a different file.

For example, a STUDENT record may be linked to all of that student's COURSE records. In turn, a COURSE record may be linked to the STUDENT records that indicate completion of that course. This two-way linking and flexibility allow us to eliminate *most* of the need to redundantly store the same fields in the different record types. Thus, in a very real sense, multiple files are consolidated into a single file—the database.

The idea of relationships between different collections of data was introduced in Chapter 8. In that chapter, you learned to discover a system's data requirements and model those requirements as *entities* and *relationships*. The database now provides for the technical implementation of those entities and relationships.

So many applications are now being built around database technology that database design has become an important skill for the analyst. The history of information systems has led to one inescapable conclusion:

> *Data is a resource that must be controlled and managed!*

Data Architecture   Data becomes a business resource in a database environment. Information systems are built around this resource to give computer programmers and end users flexible access to data. A business's **data architecture** defines how that business will develop and use both files and databases to store all of the organization's data, which file and database technology is to be used, and what kind of administrative structure will be set up to manage the data resource.

Figure 14-2 illustrates the data architecture into which many companies have evolved. As shown in the figure, most companies still have numerous conventional file-based information system applications, most of which were developed before the emergence of high-performance database technology. In many cases, the processing efficiency of these files or the projected cost of redesigning these files has slowed conversion of the systems to database.

As shown in Figure 14-2, **operational** (or *transactional*) **databases** are developed to support day-to-day operations and business transaction processing for major information systems. These systems are developed (or purchased) over time to replace the conventional files that formerly supported applications. Access to these databases is limited to computer programs that use the DBMS to process transactions, maintain the data, and generate regularly scheduled management reports. Some query access may also be provided.

Many information systems shops hesitate to give end users access to operational databases for queries and reports. The volume of unscheduled reports and queries could overload the computers and hamper business operations that the databases were intended to support. Instead, data warehouses are developed, possibly on separate computers.

Data warehouses store data extracted from the operational databases. Query tools and decision support tools are then used to generate reports and analyses off these data warehouses. These tools often allow users to extract data from both conventional files and operational databases. This is sometimes called *data mining*.

Figure 14-2 also shows *personal* and *work group* (or *departmental*) *databases*. Personal computer and local network database technology has rapidly matured to allow end users to develop personal and departmental databases. These databases may contain unique data, or they may import data from conventional files, operational databases, and/or data warehouses. Personal databases are built using PC database technology such as *Access, dBASE,* and *Visual FoxPro.*

---

data architecture a definition of how files and databases are to be developed.

operational database a database that supports day-to-day operations and transactions for an information system. Also called *transactional database.*

data warehouse a database that stores data extracted from operational databases.

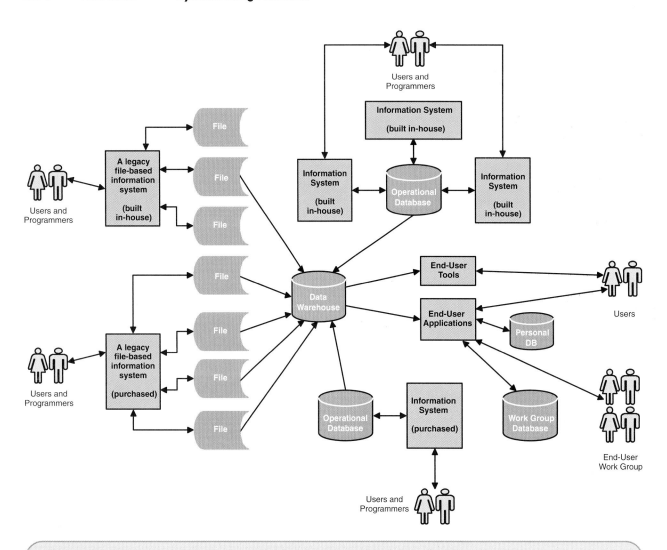

FIGURE 14-2 A Typical, Modern Data Architecture

data administrator a database specialist responsible for data planning, definition, architecture, and management.

database administrator a specialist responsible for database technology, design, construction, security, backup and recovery, and performance tuning.

database architecture the database technology used to support data architecture.

Contemporary data architecture also allows for Internet-enabled database technology. For example, *Oracle 10g* provides special tools and facilities for Web-enabling a database.

Admittedly, this overall scenario is advanced, but many firms are currently using variations of it. To manage the enterprisewide data resource, a staff of database specialists may be organized around the following administrators: A **data administrator** is responsible for the data planning, definition, architecture, and management. One or more **database administrators (DBAs)** are responsible for the database technology, database design and construction consultation, security, backup and recovery, and performance tuning. In smaller businesses, these roles may be combined or assigned to one or more systems analysts.

Database Architecture So far, we have made several references to the *database technology* that makes the above data architecture possible. **Database architecture** refers to the database technology, including the database engine, database utilities, database CASE tools for analysis and design, and database application development tools. The control center of a database architecture is its database management system.

A **database management system (DBMS)** is specialized computer software, available from computer vendors, that is used to create, access, control, and manage the database. The core of the DBMS is often called its *database engine*. The engine responds to specific commands to create database structures and then to create, read, update, and delete records in the database. The database management system is purchased from a database technology vendor such as Oracle, IBM, Microsoft, or Sybase.

Figure 14-3 depicts a typical database management system architecture. A systems analyst or database analyst designs the structure of the data in terms of record types, fields contained in those record types, and relationships that exist between record types. These structures are defined to the database management system using its data definition language. **Data definition language (DDL)** is used by the DBMS to physically establish those record types, fields, and structural relationships. Additionally, the DDL defines views of the database. Views restrict the portion of a database that may be used or accessed by different users and programs.

> **database management system (DBMS)** special software used to create, access, control, and manage a database.

> **data definition language (DDL)** a language used by a DBMS to define a database or a view of a database.

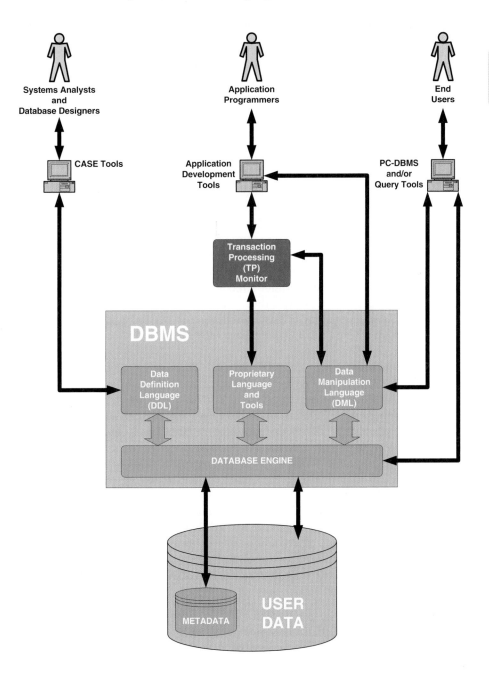

FIGURE 14-3

A Typical Database Management System Architecture

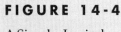

FIGURE 14-4

A Simple, Logical Data Model

Most database management systems store both *user data* and *metadata*—the data (or specifications) about the data—such as record and field definitions, synonyms, data relationships, validation rules, help messages, and so forth. Some metadata is stored in the actual database, while other metadata is stored in CASE tool repositories.

To help design databases, CASE tools may be provided either by the database technology vendor (e.g., Oracle's *Designer*) or from a third-party CASE tool vendor (Popkin's *System Architect,* Microsoft's *Visio Enterprise,* or Computer Associates' *ERwin,* etc.).

The database management system also provides a data manipulation language to access and use the stored data in applications. A **data manipulation language (DML)** is used to create, read, update, and delete records in the database and to navigate between different records and types of records—for example, from a CUSTOMER record to the ORDER records for that customer. The DBMS and DML hide the details concerning how records are organized and allocated to the disk. In general, the DML is very flexible in that it may be used by itself to create, read, update, and delete records or its commands may be "called" from a separate host programming language such as *COBOL, Visual Basic,* or *Java.*

Many DBMSs don't require the use of a DDL to construct the database or a DML to access the database. Instead (or in addition), they provide their own proprietary tools and commands to perform those tasks. This is especially true of PC-based DBMSs such as Microsoft *Access. Access* provides a simple graphical user interface to create the tables and both a form-based environment and scripting language (*Visual Basic for Applications*) to access, browse, and maintain the tables.

Many DBMSs also include proprietary report-writing and inquiry tools to allow users to access and format data without directly using the DML. Many high-end DBMSs are designed to interact with popular third-party transaction processing monitors.

All of the above technology is illustrated in Figure 14-3. Today, almost all new database development is using relational database technology.

Relational Database Management Systems   There are several types of database management systems. They can be classified according to the way they structure records. Early database management systems organized records in hierarchies or networks implemented with indexes and linked lists. Today, most successful database management systems are based on relational technology. **Relational databases** implement data in a series of two-dimensional tables that are "related" to one another via foreign keys. Each table (sometimes called a *relation*) consists of named columns (which are fields or attributes) and any number of unnamed rows (which correspond to records).

Figure 14-4 illustrates a logical data model. Figure 14-5 is the physical, relational database implementation of that data model (called a *schema*). In a relational database, files are seen as simple two-dimensional tables, also known as relations. The rows are records. The columns correspond to fields.

The following shorthand notation for tables is commonly encountered in systems design and database books.

CUSTOMERS (<u>CUSTOMER-NUMBER</u>, CUSTOMER-NAME, CUSTOMER-BALANCE, . . .)

ORDERS (<u>ORDER-NUMBER</u>, CUSTOMER-NUMBER (FK), . . .)

ORDERED-PRODUCTS (<u>ORDER-NUMBER</u> (FK), <u>PRODUCT-NUMBER</u> (FK), QUANTITY-ORDERED, . . .)

PRODUCTS (<u>PRODUCT-NUMBER</u>, PRODUCT-DESCRIPTION, QUANTITY-IN-STOCK, . . .)

---

data manipulation language (DML) a DBMS language used to create, read, update, and delete records.

---

relational database a database that implements data as a series of two-dimensional tables that are related via foreign keys.

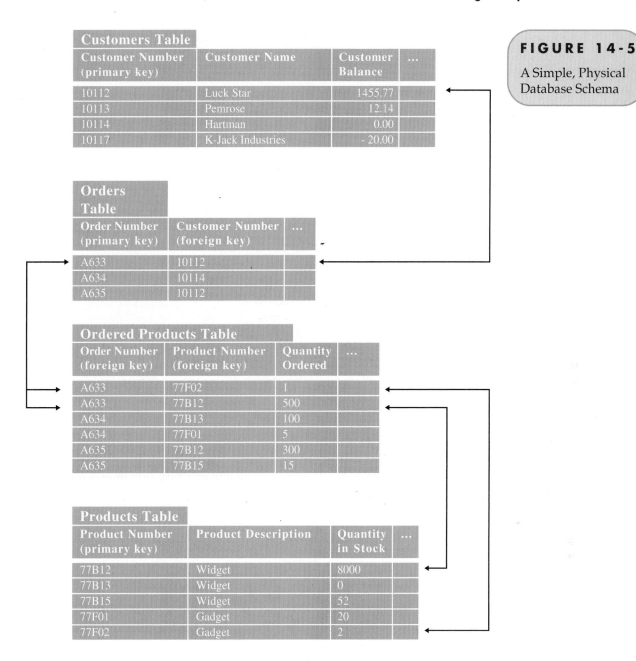

FIGURE 14-5
A Simple, Physical Database Schema

Both the DDL and DML of most relational databases is called *SQL* (pronounced "S-Q-L" by some and "sequel" by others). SQL supports complete database creation, maintenance, and usage. To access data in tables and records, SQL provides the following basic commands:

- SELECT specific records from a table based on specific criteria (e.g., SELECT CUSTOMER WHERE BALANCE > 500.00).
- PROJECT out specific fields from a table (e.g., PROJECT CUSTOMER TO INCLUDE ONLY CUSTOMER-NUMBER, CUSTOMER-NAME, BALANCE).
- JOIN two or more tables across a common field—a primary and foreign key (JOIN CUSTOMER AND ORDER USING CUSTOMER-NUMBER).

When used in combination, these basic commands can address most database requirements. A fundamental characteristic of SQL is that commands return a set of records, not necessarily just a single record (as in nonrelational database and file technology). SQL databases also provide commands for creating, updating, and deleting records, as well as sorting records.

FIGURE 14-6

User/Designer
Interface for a
Relational PC
DBMS (Microsoft
Access)

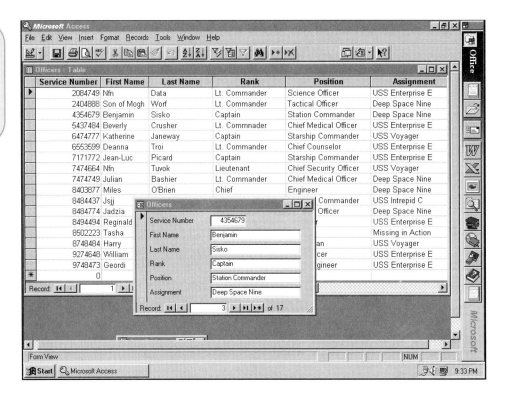

---

trigger a program embedded
within a table and is automatically invoked by updates to
another table.

---

stored procedures a program embedded in a table
that can be called from an
application program.

High-end relational databases also extend the SQL language to support triggers and stored procedures. **Triggers** are programs embedded within a table that are automatically invoked by updates to another table. For example, if a record is deleted from a PASSENGER AIRCRAFT table, a trigger can force the automatic deletion of all corresponding records in a SEATS table for that aircraft. **Stored procedures** are programs embedded within a table that can be called from an application program. For example, a data validation algorithm might be embedded in a table to ensure that new and updated records contain valid data *before* they are stored. Stored procedures are written in a proprietary extension of SQL such as Microsoft's *Transact SQL* or Oracle's *PL/SQL.*

Both triggers and stored procedures are reusable because they are stored with the tables themselves (as metadata). This eliminates the need for application programmers to create the equivalent logic within each application that uses the tables.

All high-end relational database management systems (e.g., *Oracle, UDB/DB2,* and *SQL Server*) and many personal computer relational database management systems (such as Microsoft *Access*) support the SQL language standards.

Examples of high-performance relational DBMSs include Oracle Corporation's *Oracle,* IBM's *DB2,* Microsoft's *SQL Server* (being used in the SoundStage project), and Sybase Corporation's *Sybase.* Many of these databases run on mainframes, minicomputers, and network database servers. Additionally, most personal computer DBMSs are relational (or at least partially so). Examples include Microsoft's *Access* and *Visual Foxpro.* These database engines can run on both stand-alone personal computers and local area network file servers. Figure 14-6 illustrates a relational database management system's user interface.

Prerequisite for Database Design—Normalization

In Chapter 8 you learned how to model data requirements for an information system. That model took the form of a fully attributed entity relationship diagram and a repository of metadata. Chapter 8 also taught a technique called *data analysis* or *normalization.* This technique was used to produce a data model that meets the following quality criteria:

- *A good data model is simple.* As a general rule, the data attributes that describe an entity should describe only that entity.
- *A good data model is essentially nonredundant.* This means that each data attribute, other than foreign keys, describes at most one entity.
- *A good data model should be flexible and adaptable to future needs.* In the absence of this criteria, we would tend to design databases to fulfill only *today's* business requirements.

So how do we achieve the above goals? How can you design a database that can adapt to future requirements that you cannot predict? The answer lies in data analysis.

Recall that normalization is a three-step technique that places the data model into first normal form, second normal form, and third normal form. Database design should proceed only if the underlying logical data model is in at least 3NF. For a more detailed explanation, we encourage you to review Chapter 8.

Conventional File Design

The focus of this chapter is on database design; however, we would be remiss to not say a few words about conventional file design. First, file design is simplified because of its orientation to a single application. Typically, the output and input designs (Chapters 15 and 16) would be completed first since the file design is dependent on supporting those application requirements.

Most fundamental entities from the data model would be designed as master or transaction records. The master files are typically fixed-length records. Associative entities from the data model are typically joined into the transaction records to form variable-length records (based on the one-to-many relationships). Other types of files (not represented in the data model) are added as necessary.

Two important considerations of conventional file design are *file access* and *organization.* The systems analyst usually studies how each program will access the records in the file (sequentially or randomly) and then selects an appropriate file organization (e.g., sequential, indexed, hashed, etc.). In practice, many systems analysts select an indexed sequential (or ISAM/VSAM) organization to support the likelihood that different programs will require different access methods into the records.

Modern Database Design

The design of any database will usually involve the DBA and database staff. They will handle the technical details and cross-application issues. Still, it is useful for the systems analyst to understand the basic design principles for relational databases.

The design rules presented here are, in fact, guidelines. We cannot cover every idiosyncrasy. Also, because SoundStage has elected to use Microsoft's *SQL Server* as its database management system, our design will be constrained by that technology. Each relational DBMS presents its own capabilities and constraints. Fortunately, the guidelines presented here are fairly generic and applicable to most DBMS environments. Database courses and textbooks tend to cover a wider variety of technology and issues.

Computer-assisted systems engineering (CASE) has been a continuing theme throughout this book. There are specific CASE products that address database analysis and design (e.g., Computer Associates' *ERwin*). Also, most general-purpose CASE tools now include database design tools. In this example, we continued to use Popkin's *System Architect* CASE product for the SoundStage case study. Finally, most CASE tools (including *System Architect*) can automatically generate SQL code to construct the database structures for the most popular database management systems. This code generation capability is an enormous time-saver.

> Goals and Prerequisites of Database Design

The goals of database design are as follows:

- A database should provide for the efficient storage, update, and retrieval of data.
- A database should be reliable—the stored data should have high integrity to promote user trust in the data.
- A database should be adaptable and scalable to new and unforeseen requirements and applications.
- A database should support the business requirements of the information system.

The system's logical data model—in our case, a fully attributed and normalized entity relationship diagram (ERD)—serves as the prerequisite. This model, from Chapter 8, is reproduced in Figure 14-7. Every attribute in that model must be defined as to its data type, domain, and default. These properties were also covered in Chapter 8.

> The Database Schema

database schema a model or blueprint representing the technical implementation of a database.

The design of a database is depicted as a special model called a database schema. A **database schema** is the *physical* model or blueprint for a database. It represents the technical implementation of the logical data model. (*System Architect* calls it a *physical data model*.)

> NOTE: We should acknowledge some potentially confusing terminology here. We are using the terms *logical* and *physical* in a manner consistent with earlier chapters in this book. Unfortunately, most database books use the terms *conceptual* (our *logical*) and *logical* (our *physical*). We apologize for this unavoidable industry confusion.

A relational database schema defines the database structure in terms of tables, keys, indexes, and integrity rules. A database schema specifies details based on the capabilities, terminology, and constraints of the chosen database management system. Each DBMS supports different data types, integrity rules, and so forth.

The transformation of the logical data model into a physical relational database schema is governed by some fairly generic rules and options. These rules and guidelines are summarized as follows:

1. Each fundamental, associative, and weak entity is implemented as a separate table. Table names may have to be formatted according to the naming rules and size limitations of the DBMS. For example, a logical entity named MEMBER ORDERED PRODUCT might be changed to a physical table named tblMemberOrdProd. The prefix and compression of spaces is consistent with contemporary naming standards and guidelines in modern programming languages.
   a. The primary key is identified as such and implemented as an index into the table.
   b. Each secondary key is implemented as its own index into the table.
   c. An index should be created for any nonkey attributes that were identified as subsetting criteria requirements (Chapter 8).
   d. Each foreign key will be implemented as such. The inclusion of these foreign keys implements the relationships on the data model and allows tables to be joined in SQL and application programs.
   e. Attributes will be implemented with fields. These fields correspond to columns in the table. The following technical details must usually be specified for each attribute. (These details may be automatically inferred by the CASE tool from the logical descriptions in the data model.)
      Field names may have to be shortened and reformatted according to DBMS constraints and internal rules. For example, in the logical data model,

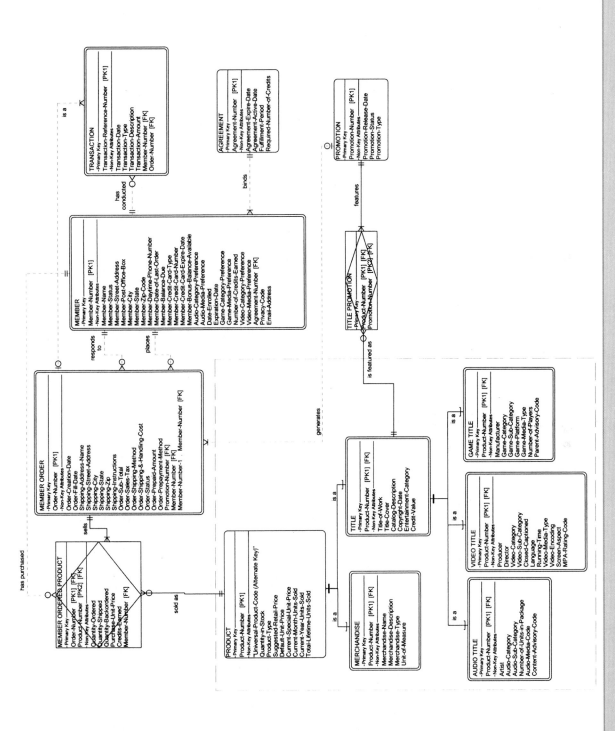

F I G U R E   1 4 - 7   SoundStage Logical Data Model in Third Normal Form

most attributes might be prefaced with the entity name (e.g., MEMBER NAME). In the physical database, we might simply use NAME.

 i. *Data type.* Each DBMS supports different data types and terms for those data types. Figure 14-8 shows different physical data types for a few different database management systems.

 ii. *Size of the field.* Different DBMSs express precision of real numbers differently. For example, in *SQL Server,* a size specification of NUMBER (3,2) supports a range from −9.99 to 9.99.

 iii. *NULL or NOT NULL.* Must the field have a value before the record can be committed to storage? Again, different DBMSs may require different reserved words to express this property. By definition, primary keys can never be allowed to have NULL values.

 iv. *Domains.* Many database management systems can automatically edit data to ensure that fields contain legal data. This can be a great benefit to ensuring data integrity independent from the application programs. If the programmer makes a mistake, the DBMS catches the mistake. But for DBMSs that support data integrity, the rules must be precisely specified in a language that is understood by the DBMS.

 v. *Default.* Many database management systems allow a default value to be automatically set in the event that a user or programmer creates a record containing fields with no values. In some cases, NULL serves as the default.

 vi. Again, many of the above specifications were documented as part of a complete logical data model. If that data model was developed with a CASE tool, the CASE tool may be capable of automatically translating the data model into the physical language of the chosen database technology.

2. Supertype/subtype entities present additional options as follows:
 a. Each supertype and subtype can be implemented with a separate table (all having the same primary key).
 b. Alternatively, if the subtypes are of *similar size* and data content, a database administrator may elect to collapse the subtypes into the supertype to create a single table. This presents certain problems for setting defaults and checking domains. In a high-end DBMS, these problems can be overcome by embedding the default and domain logic into stored *procedures* for the table.
 c. Alternatively, the supertype's attributes could be duplicated in a table for each subtype.
 d. Some combination of the above options could be used.

3. Evaluate and specify referential integrity constraints (described in the next section).

The SoundStage database schema was automatically generated from the logical data model by our CASE tool *System Architect.* It is illustrated in Figure 14-9. We call your attention to the following numbered bullets on the figure:

❶ Each rounded rectangle defines a table. The named rows in the rectangle actually correspond to the named columns that will be created for the table.

❷ SoundStage has defined a standard naming convention for tables and columns. The conventions are based on the programming guidelines called *Hungarian Notation.* Each object is named without spaces, dashes, or underscores. And each object is given a prefix that defines all similar objects. For database objects, the following standards were used:

tbl   Indicates a database table.
col   Indicates a column in the table.

FIGURE 14-8 Partial List of Physical Data Types for Different Database Technologies

Logical Data Type (to be stored in field)	Physical Data Type Microsoft *Access*	Physical Data Type Microsoft *SQL Server*	Physical Data Type *Oracle*
Fixed-length character data (*use for fields with relatively fixed length character data*)	TEXT	CHAR (size) *or* character (size)	CHAR (SIZE)
Variable-length character data (*use for fields that require character data but for which size varies greatly—such as ADDRESS*)	TEXT	VARCHAR (max size) *or* character varying (max size)	VARCHAR2 (max size)
Very long character data (*use for long descriptions and notes—usually no more than one such field per record*)	MEMO	TEXT	LONG VARCHAR *or* LONG VARCHAR2
Integer number	NUMBER	INT (size) *or* integer *or* smallinteger *or* tinyinteger	INTEGER *or* NUMBER (size) *or* smallint or byte
Decimal number	NUMBER	DECIMAL (size, decimal places) *or* NUMERIC (size, decimal places)	DECIMAL (size, decimal places) *or* NUMBER (size, decimal places) *or* NUMBER
Financial number	CURRENCY	MONEY *or* SMALLMONEY	*see decimal number*
Date (with time)	DATE/TIME	DATETIME *or* SMALLDATETIME *Depending on precision needed*	DATE
Current time (*use to store the date and time from the computer's system clock*)	*not supported*	TIMESTAMP	TIMESTAMP
Yes or No; or True or False	YES/NO	BIT	*use CHAR(1) and set a yes or no domain*
Image	OLE OBJECT	IMAGE	LONGRAW
Hyperlink	HYPERLINK	VARBINARY	RAW
Can designer define new data types?	*NO*	*YES*	*YES*

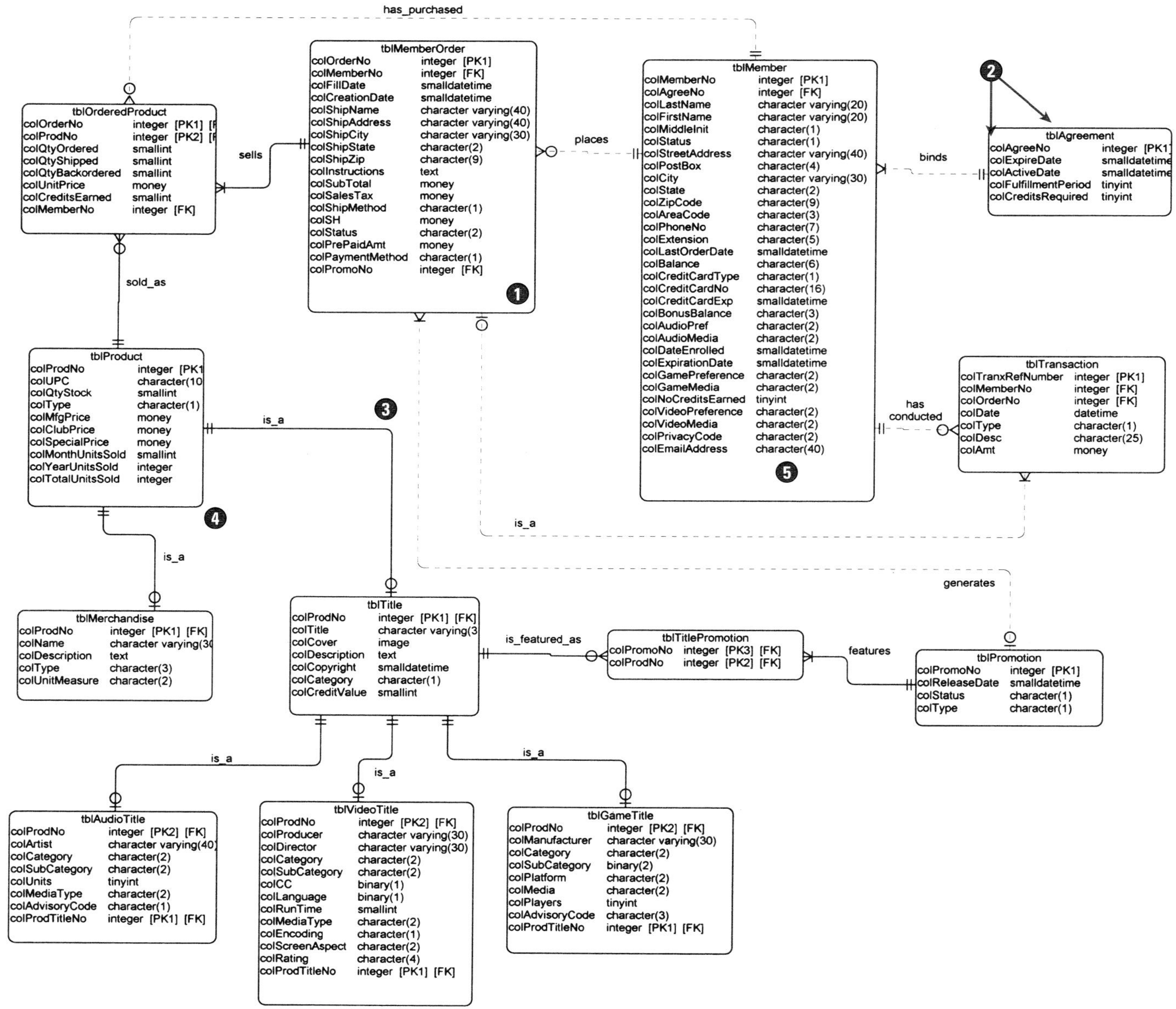

FIGURE 14-9 Initial SoundStage Physical Database Schema

Although not depicted on the schema, other common database prefixes may be included in the schemas underlying data dictionary (repository) such that those prefixes may be used to generate correct code. Possibilities include:

db   Indicates the database itself.
idx   Indicates an index built for a table.
dom   Indicates a domain that can be applied to one or more fields.

❸ Logical relationships, both identifying and nonidentifying, are transformed in *constraints* that are implemented using the foreign keys.

❹ We elected to make each supertype and subtype entity in the logical, generalization hierarchy into its own physical table. (This was the default option for *System Architect*'s physical data model generator.)

❺ Notice that *System Architect* automatically inferred physical data types for each field based on (1) the selection of Microsoft *SQL Server* as the target database management system and (2) the logical data types we had defined for each entity's attributes during systems analysis. The generated physical data types can be changed to reduce storage space required, improve data integrity, or better represent all the possible values included in the domain.

Although not depicted on the database schema, the schema generator also creates an index for each primary key indicated in the schema. You can add additional indexes for unique secondary keys (such as the Universal Product Code, or UPC, field in tblProduct) or for any nonkey attribute that can be used to subset all records in a table (such as tblTransaction.colType). These indexes can improve the performance of the final database.

Some CASE tools generate database schemas with considerably more detail than our example. For example, some database schemas indicate for each field whether or not the field must take on a value:

- NULL means the field does not have to have a value.
- NOT NULL means the field must have a value. Because primary keys are used to uniquely access records, no PK field may take on NULL values.

Would you ever want to compromise the third normal form entities when designing the database? For example, would you ever want to combine two third-normal-form entities into a single table (that would, by default, no longer be in third normal form)? Usually not! Although a database administrator may create such a compromise to improve database performance, he or she should carefully weigh the advantages and disadvantages. Although such compromises may mean greater convenience through fewer tables or better overall performance, such combinations may also lead to the possible loss of data independence—should future new fields necessitate re-splitting the table into two tables, programs will have to be rewritten. As a general rule, combining entities into tables is not recommended.

> Data and Referential Integrity

Database integrity is about trust. Can the business and its users trust the data stored in the database? Data integrity provides necessary internal controls for the database. There are at least three types of data integrity that must be designed into any database.

Key Integrity   Every table should have a primary key (which may be concatenated). The primary key must be controlled such that no two records in the table have the same primary-key value. (Note that for a concatenated key, the concatenated value must be unique—not the individual values that make up the concatenation.)

Also, the primary key for a record must never be allowed to have a NULL value. That would defeat the purpose of the primary key, to uniquely identify the record.

If the database management system does not enforce these rules, other steps must be taken to ensure them. Most DBMSs do enforce key integrity.

Domain Integrity   Appropriate controls must be designed to ensure that no field takes on a value that is outside the range of legal values. For example, if GRADE POINT AVERAGE is defined to be a number between 0.00 and 4.00, then controls must be implemented to prevent negative numbers and numbers greater than 4.00.

Not long ago, application programs were expected to perform all data editing. Today, most database management systems are capable of enforcing domain rules. For the foreseeable future, the responsibility for data editing will continue to be shared between the application programs and the DBMS.

Referential Integrity   The architecture of relational databases implements relationships between the records in tables via *foreign keys*. The use of foreign keys increases the flexibility and scalability of any database, but it also increases the risk of referential integrity errors. A **referential integrity** error exists when a foreign-key value in one table has no matching primary-key value in the related table. For example, an INVOICES table usually includes a foreign key, CUSTOMER NUMBER, to "reference back to" the matching CUSTOMER NUMBER primary key in the CUSTOMERS table. What happens if we delete a CUSTOMER record? There is the *potential* that we may have INVOICE records whose CUSTOMER NUMBER has no matching record in the CUSTOMERS table. Essentially, we have compromised the referential integrity between the two tables.

> **referential integrity** the assurance that a foreign-key value in one table has a matching primary-key value in the related table.

How do we prevent referential integrity errors? One of two things should happen. When considering the deletion of CUSTOMER records, either we should automatically delete all INVOICE records that have a matching CUSTOMER NUMBER (which doesn't make much business sense) or we should disallow the deletion of the CUSTOMER record until we have deleted all INVOICE records.

Referential integrity is specified in the form of deletion rules as follows:[2]

- *No restriction*—Any record in the table may be deleted without regard to any records in any other tables.

  In looking at the final SoundStage data model, we could not apply this rule to any table.
- *Delete:Cascade*—A deletion of a record in the table must be automatically followed by the deletion of matching records in a related table. Many relational DBMSs can automatically enforce delete:cascade rules using triggers.

  In the SoundStage data model, an example of a valid delete:cascade rule would be from MEMBER ORDER to MEMBER ORDERED PRODUCT. In other words, if we delete a specific MEMBER ORDER, we should automatically delete all matching MEMBER ORDERED PRODUCTS for that order.
- *Delete:Restrict*—A deletion of a record in the table must be disallowed until any matching records are deleted from a related table. Again, many relational DBMSs can automatically enforce delete:restrict rules.

  For example, in the SoundStage data model, we might specify that we should disallow the deletion of any PRODUCT as long as there exists MEMBER ORDERED PRODUCTS for that product.
- *Delete:Set null*—A deletion of a record in the table must be automatically followed by setting any matching keys in a related table to the value NULL. Again, many relational DBMSs can enforce such a rule through triggers.

  The Delete:Set null option was not used in the SoundStage data model. It is used only when you are willing to delete a master table record but you don't want to delete corresponding transaction table records for historical reasons. By setting the foreign key to NULL, you are acknowledging that the record does not point back to a corresponding master record, but at least you don't have it pointing to a nonexisting master record.

The final database schema, complete with referential integrity rules, is illustrated in Figure 14-10. This is the blueprint for writing the SQL code (or equivalent) to create the tables and data structures.

---

[2]Knowledgeable database students know that there are also insertion and update rules for referential integrity. A full discussion of these rules is deferred to database courses and textbooks.

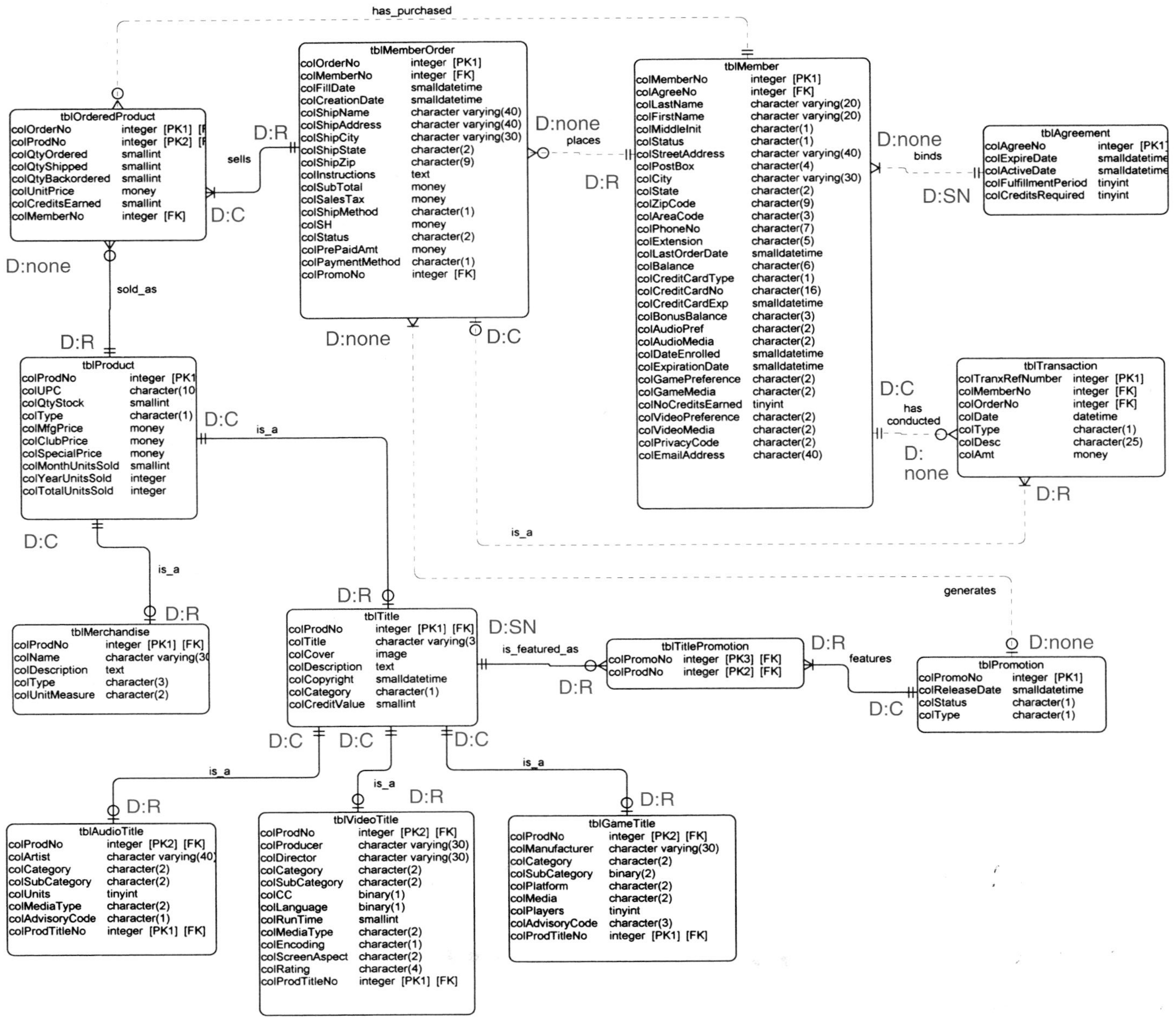

FIGURE 14-10 Final SoundStage Physical Database Schema

> Roles

Some database standards insist that no two fields have *exactly* the same name. This constraint simplifies documentation, help systems, and metadata definitions. This presents an obvious problem with foreign keys. By definition, a foreign key must have a corresponding primary key. During *logical* data modeling, using the same name suited our purpose of helping the users understand that the foreign keys allow us to match related records in different entities. But in a *physical* database, it is not always necessary or even desirable to have these redundant field names in the database.

role name a foreign key name that reflects the purpose it serves in a table.

To fix this problem, foreign keys can be given role names. A **role name** is an alternate name for a foreign key that clearly distinguishes the purpose that the foreign key serves in the table. For example, in the SoundStage database schema, PRODUCT_NUMBER is a primary key for the PRODUCTS table and a foreign key in the MEMBER ORDERED PRODUCTS table. The name should not be changed in the PRODUCTS table. But it may make sense to rename the foreign key to ORDERED_PRODUCT_NUMBER to more accurately reflect its role in the MEMBER ORDERED PRODUCTS table.

The decision to require role names or not is usually established by the data or database administrator.

> Database Distribution and Replication

In Chapter 8, "Data Modeling and Analysis," we briefly introduced the concept of logical data distribution analysis. *Data distribution analysis* establishes which business locations need access to which logical data entities and attributes.

We used a simple matrix in Chapter 8 to map entities and attributes to locations. Many CASE tools, including *System Architect,* include facilities for building such a matrix. We should give some consideration now to the impact of data distribution analysis on database design.

In today's multitier, client/server, network-centric world, information systems and databases are rarely centralized. Instead, they are distributed across a network that may span many buildings, cities, states, or countries. Accordingly, we may need to partition, distribute, or replicate all or part of a database design to different physical database servers in different physical locations. Basically, we need to perform a physical database distribution analysis that takes into consideration what we learned during our logical data distribution analysis.

Essentially, we have a number of distribution options available to us:

- *Centralization* of the database. In other words, we would implement the database on a single server regardless of the number of physical locations that may require access to it. This solution is simple and the easiest to maintain; however, it violates a data management rule that has become important to many data administrators and users—data should be located as closely as possible to its users.
- *Horizontal distribution* of the data. In this option, each table (or entire rows in a table) would be assigned to different database servers and locations. This option results in efficient access and security because each location has only those tables and rows required for that location. Unfortunately, data cannot always be easily recombined for management analysis across sites.
- *Vertical distribution* of the data. In this option, specific columns of tables are assigned to specific databases and servers. The advantages and disadvantages are very similar to that of horizontal distribution.
- *Replication* of the data. Replication refers to the physical duplication of entire tables to multiple locations. Most high-end, enterprise database management systems include replication technology that coordinates updates to the duplicated tables and records to maintain data integrity. This solution offers performance and accessibility advantages and reduces network traffic, but it also increases the complexity of data integrity and requires more physical storage capacity.

These alternatives are not mutually exclusive. The designer must carefully plan degrees of data distribution and replication.

Given our physical database schema, we can define *views* that correspond to specific geographic locations (and subviews for different users and applications). A database view may be very selective. It may include a specific subset of tables, a specific subset of columns in tables, or even a specific subset of records in tables. Each view must be carefully synchronized with the master database schema such that changes to the master schema can, if appropriate, be propagated to the views. CASE tools can be very helpful in defining views and keeping all views in sync.

For the SoundStage project, we plan to replicate the entire database in each of three cities. The data integrity for common tables will be implemented using *SQL Server's* replication technology. The systems analyst will not typically program the replication rules. A qualified database analyst or administrator will do that. Since we will implement the entire physical database schema on each city's server, there is no need to define views for our project.

> Database Prototypes

Prototyping is not an alternative to carefully thought-out database schemas. On the other hand, once the schema is completed, a prototype database can usually be generated very quickly. Most modern DBMSs include powerful, menu-driven database generators that automatically create a DDL and generate a prototype database from that DDL. A database can then be loaded with test data that will prove useful for prototyping and testing outputs, inputs, screens, and other systems components.

> Database Capacity Planning

A database is stored on disk. Ultimately, the database administrator will want an estimate of disk capacity for the new database to ensure that sufficient disk space is available. Database capacity planning can be calculated with simple arithmetic as follows. This simple formula ignores factors such as packing, coding, and compression, but by leaving out those possibilities, you are adding slack capacity.

1. For each table, sum the *field* sizes. This is the *record size* for the table. Avoid the implications of compression, coding, and packing—in other words, assume that each stored character and digit will consume one byte of storage. Note that formatting characters (e.g., commas, hyphens, slashes) are almost never stored in a database. Those formatting characters are added by the application programs that will access the database and present the output to the users.
2. For each table, multiply the *record size* times the number of entity instances to be included in the table. It is recommended that growth be considered over a reasonable time period (e.g., three years). This is the *table size.*
3. Sum the *table sizes.* This is the *database size.*
4. Optionally, add a slack capacity buffer (e.g., 10 percent) to account for unanticipated factors or inaccurate estimates above. This is the *anticipated database capacity.*

> Database Structure Generation

CASE tools are frequently capable of generating SQL code for the database directly from a CASE-based database schema. This code can be exported to the DBMS for compilation. Even a small database such as the SoundStage model can require 50 pages or more of SQL data definition language code to create the tables, indexes, keys, fields, and triggers. Clearly, a CASE tool's ability to automatically generate syntactically correct code is an enormous productivity advantage. Furthermore, it almost always proves easier to modify the database schema and regenerate the code than to maintain the code directly. Figure 14-11 is a two-page sample of code generated by *System*

```
/* SQL Product = SQL Server V7 */

CREATE DATABASE dbMemberServices

CREATE TABLE tblAgreement(
      colAgreementNo                integer NOT NULL,
      colExpireDate                 smalldatetime NULL,
      colActiveDate                 smalldatetime NOT NULL,
      colFulfillmentPeriod          tinyint NOT NULL,
      colCreditsRequired            tinyint NOT NULL)
go

ALTER TABLE tblAgreement ADD CONSTRAINT AGREEMENT_PK
   PRIMARY KEY  (colAgreementNo)
go

CREATE TABLE tblAudioTitle(
      colArtist                     character varying(40) NULL,
      colCategory                   character(2) NOT NULL,
      colSubCategory                character(2) NOT NULL,
      colUnits                      tinyint NOT NULL          DEFAULT 1,
      colMediaType                  character(2) NOT NULL,
      colAdvisoryCode               character(1) NULL,
      colProductNo                  integer NOT NULL)
go

sp_bindrule domAudioCategory, 'tblAudioTitle.colCategory'
go

sp_bindrule domAudioCategory, 'tblAudioTitle.colSubCategory'
go

sp_bindefault defOne, 'tblAudioTitle.colUnits'
go

sp_bindrule domPositive, 'tblAudioTitle.colUnits'
go

sp_bindrule domAudioMedia, 'tblAudioTitle.colMediaType'
go

sp_bindrule domContentAdvisory, 'tblAudioTitle.colAdvisoryCode'
go

ALTER TABLE tblAudioTitle ADD CONSTRAINT AUDIO_TITLE_PK
   PRIMARY KEY  (colProductNo)
go

CREATE TABLE tblGameTitle(
      colManufacturer               character varying(30) NOT NULL,
      colCategory                   character(2) NOT NULL,
      colSubCategory                binary(2) NULL,
      colPlatform                   character(2) NOT NULL,
      colMedia                      character(2) NOT NULL,
      colPlayers                    tinyint NOT NULL,
      colAdvisoryCode               character(3) NULL,
```

FIGURE 14-11   Partial SQL Code to Construct the SoundStage Database

```
        colProductNo                      integer NOT NULL)
go

sp_bindrule domGameCategories, 'tblGameTitle.colCategory'
go

sp_bindrule domGameCategories, 'tblGameTitle.colSubCategory'
go

sp_bindrule domGamePlatforms, 'tblGameTitle.colPlatform'
go

sp_bindrule domGameMediaTypes, 'tblGameTitle.colMedia'
go

sp_bindefault defOne, 'tblGameTitle.colPlayers'
go

ALTER TABLE tblGameTitle ADD CONSTRAINT GAME_TITLE_PK
   PRIMARY KEY  (colProductNo)
go

CREATE TABLE tblMember(
        colMemberNo                       integer NOT NULL,
        Last_Name                         character varying((20)) NOT NULL,
        First_Name                        character varying((19)) NOT NULL,
        Middle_Initial                    character((1)) NOT NULL,
        colStatus                         character(1) NOT NULL,
        colStreetAddress                  character varying(40) NOT NULL,
        colPostBox                        character(4) NULL,
        colCity                           character varying(30) NOT NULL,
        colState                          character(2) NOT NULL,
        colZipCode                        character(9) NOT NULL,
        Area_Code                         character((3)) NOT NULL,
        Phone_Number                      character((7)) NOT NULL,
        Extension                         character((5)) NOT NULL,
        colLastOrderDate                  smalldatetime NULL,
        colBalance                        character(6) NOT NULL        DEFAULT 0.00,
        colCreditCardType                 character(1) NULL,
        colCreditCardNo                   character(16) NULL           UNIQUE,
        colCreditCardExp                  smalldatetime NULL,
        colBonusBalance                   character(3) NULL            DEFAULT 0,
        colAudioPref                      character(2) NOT NULL,
        colAudioMedia                     character(2) NOT NULL,
        colDateEnrolled                   smalldatetime NOT NULL,
        colExpirationDate                 smalldatetime NULL,
        colGamePreference                 character(2) NOT NULL,
        colGameMedia                      character(2) NOT NULL,
        colNoCreditsEarned                tinyint NOT NULL,
        colVideoPreference                character(2) NOT NULL,
        colVideoMedia                     character(2) NOT NULL,
        colAgreementNo                    integer NOT NULL,
        colPrivacyCode                    character(2) NULL,
        colEmailAddress                   character(40) NULL)
go
```

FIGURE 14-11 Concluded

The Next Generation: Database Design

Relational database technology is widely deployed and used in contemporary information systems shops. The skills taught in this chapter will remain viable well into the foreseeable future. But one new technology that is slowly emerging could ultimately change the landscape dramatically—*object* database management systems.

Object database management systems store true objects—that is, encapsulated data and all of the processes that can act on that data. Because relational database management systems are so widely used, we don't expect this change to happen quickly. Furthermore, the relational DBMS vendors are not likely to give up their market share without a fight. It is expected that these vendors either will build object technology into their existing relational DBMSs or will create new, object DBMSs and provide for the transition between relational and object models. Regardless, this is one technology to keep an eye on.

Architect from the SoundStage database schema. The SoundStage example actually generated more than 30 pages of SQL code to create the database in Figure 14-9. Can you imagine the effort required to hand-code that much SQL (with reasonable accuracy)? Clearly, CASE tool generation of SQL code can be very productive—however, the code generated is only as good and complete as the data model.

Let's begin with the obvious! If you have information systems career aspirations, you had better plan to take one or more true database courses. The topics presented in this chapter represent only the tip of the iceberg as it relates to database technology, development, and management. Most IS curricula include at least one good database or data management course to add value to your education. Take it!

You have only begun your journey through system design. The database is the *brain* of a new system or application. The subsequent chapters focus on the design of other crucial body parts. Chapters 15 through 17 teach input, output, and user interface design, respectively. Think of inputs, outputs, and interfaces as the *soul* of the system.

Chapter Review

1. The data captured by an information system is stored in files and databases. A file is a collection of similar records. A database is a collection of interrelated files.

2. Many legacy systems were built with file technology. Because files were built for specific applications, their design was optimized for those applications. This close relationship between the files and their applications made it difficult to restructure the files to meet future requirements. And because many applications use the same data, it is not uncommon to find redundant files with data values that do not always match.

3. As the above legacy systems are slowly reengineered, they are usually converted to database technology. Well-designed databases share nonredundant data and overcome all the limitations of conventional files.

4. Database design is the process of translating logical data models (Chapter 8) into physical database schemas.

5. The smallest unit of meaningful data that can be stored is called a field. There are four types of fields:

   a. A primary key is a field that uniquely identifies one and only one record in a file or table.
   b. A secondary key is a field that may either uniquely identify one and only one record in a file or table or identify a set of records with some common, meaningful characteristic.
   c. A foreign key is a field that points to a related record in a different table.
   d. All other fields are called descriptive fields.

6. Fields are organized into records, and similar records are organized into files or tables.

7. A database is a collection of tables (files) with logical pointers that relate records in one table to records in a different table.

8. The data architecture that has evolved in most organizations includes conventional files, operational databases, data warehouses, and personal and work group databases. To coordinate this complex infrastructure, many organizations assign a data administrator to plan and manage the overall data resource and database administrators to implement and manage specific databases and database technologies.

9. A database architecture is built around a database management system (DBMS) that provides the technology to define the database structure and then to create, read, update, and delete records in the tables that make up that structure. A DBMS provides a data language to accomplish this. That language provides at least two components:

   a. A data definition language to create and maintain the database structure and rules.
   b. A data manipulation language to create, read, use, update, and delete records in the database.

10. Today, relational database management systems are used to support the development and reengineering of the overwhelming number of information systems. Relational databases store data in a collection of tables that are related via foreign keys.

    a. The data definition and manipulation languages of most relational DBMSs are consolidated into a standard language known as SQL.
    b. High-end relational database management systems support triggers and stored procedures, programs that are stored with the tables and callable from other SQL-based programs.

11. Data analysis and normalization are techniques for removing impurities from a data model as a preface to designing the database. These impurities can make a database unreliable, inflexible, and nonscalable.

12. Distribution and replication decisions should be made before database design. Each unique database should be represented by its own logical data submodel.

13. A database schema is the physical model for a database based on the chosen database technology. The rules for transforming a logical data model into a physical database schema are generalized as follows:

    a. Each entity becomes a table.
    b. Each attribute becomes a field (column in the table).
    c. Each primary and secondary key becomes an index into the table.
    d. Each foreign key implements a possible relationship between instances of the table.

14. Database integrity should be checked and, if necessary, improved to ensure that the business and its users can trust the stored data.

    a. Key integrity ensures that every record will have a unique, non-NULL primary-key value.
    b. Domain integrity ensures that appropriate fields will store only legitimate values from the set of all possible values.
    c. Referential integrity ensures that no foreign-key value points to a nonexistent primary-key

value. A deletion rule should be specified for every relationship with another table. The deletion rules either cascade the deletion to related records in other tables, disallow the deletion until related records in other tables

are first deleted, or allow the deletion but set any foreign keys in related tables to NULL.

15. SQL-DDL is written or generated to create the database.

Review Questions

1. What does the acronym CRUD represent?
2. In looking at Figure 14-1 in the textbook, what can you conclude regarding the characteristics of conventional files and databases?
3. Why do conventional files tend be have duplication of data?
4. Why are conventional files easy to design and implement?
5. What is a common misconception about databases?
6. Why is storing data in a database riskier than storing it in a file?
7. What is a secondary key?
8. What is a fixed-length record structure?
9. What are some common types of conventional files and tables?
10. In comparing operational databases and data warehouses, which generally has fewer CRUD activities? Why?
11. What is a database engine?
12. What is metadata? If database administrators need to define metadata, what kind of language should they use (DDL or DML)? Why?
13. What is a relational database?
14. What is the difference between a relational database schema and a database schema?
15. What are the common deletion rules to enforce referential integrity?

Problems and Exercises

1. When an organization decides to replace a legacy system, it usually chooses a contemporary database system over a traditional file-based system. But each type of system has its own advantages and disadvantages. Identify whether each characteristic listed below generally belongs to a file-based system or to a database system.

   • High cost of development.
   • Generally designed to be used with a single system or application.
   • Greater data privacy concerns.
   • Controlled redundancy.
   • Suboptimal performance for shared use by multiple systems.
   • Tends to be slower.
   • Data formats are flexible.
   • Identifying data elements is relatively quick and straightforward.
   • Designed to support current requirements.
   • Higher training costs.
   • Records are linked to related records.
   • Rigorous design standards.
   • Optimized single application processing speed.
   • Tendency towards data redundancy.
   • Increased vulnerability.

   • Data storage is built around the hub of the information system.
   • "Silo" effect.
   • Scalable.

2. Although database systems have become the systems of choice for new and reengineered systems, are there any situations where a traditional file-based system might be chosen instead? Explain your answer.

3. The textbook states that "data is a resource that must be controlled and managed." Explain this statement, and indicate whether you agree or not, and why.

4. Consider a local car dealership that has been in business for fifteen years or so. In addition to selling and leasing new and used cars, the dealership has a parts department, service department, and auto body department. It was one of the first car dealerships to automate in the mid-70s, and while most of the information systems have been periodically updated, a few of the original systems are still in use, including some stand-alone systems. About five years ago, the dealership began to gradually replace some of the traditional file-based systems with a relational database system, but the

conversion is far from over. Last year, the owner read in an airline magazine about data warehouses, and hired a local systems integrator to install one, which is in progress. The dealership has a couple of jack-of-all-trades IT people on staff, but most of the development work is contracted out. Given this scenario, draw a high-level diagram of what the dealership's data architecture might look like; use Figure 14-2 as an example.

5. One of the dealership's legacy stand-alone systems in the preceding exercise is the salesperson work schedule system. This system was developed in the 1980s on a single PC, using *dBASE III,* to keep track of each salesperson's daily and weekly work schedule. The dealership plans to retire this antiquated system and incorporate it into a relational database system. What are two or three tables you might expect to find in a scheduling system? Using Figure 14-5 as an example, show these tables in a simple, physical database schema. Include the primary key, any foreign keys, one or two nonkey attributes, and a few values for each object.

6. Match the following terms in the first column with the definitions or examples in the second column.

1. DBMS	a. Physical implementation of a database
2. Transaction file	b. Federal register of country codes
3. Data warehouse	c. 5NF
4. Primary key	d. Microsoft *Access*
5. SQL	e. Delete:Set null
6. CASE tool	f. Sybase *IQ*
7. Hungarian notation	g. SSN
8. Normalization	h. Daily hospital admissions file
9. Table look-up file	i. *Oracle 10g*
10. Referential integrity	j. Standard naming convention for tables
11. Schema	k. *System Architect*
12. Personal database	l. ALTER TABLE

7. You are a systems designer and have a friend who owns a small bookstore and mail-order business specializing in rare books and first editions. Total sales average about a dozen books per day, store and mail order combined. Your friend wants to start selling books over the Web also, and has come to you for advice. He has heard that he should go with a relational database and that *Oracle* is the best one to get. Do you concur? If not, what would you recommend?

8. In transforming a logical data model into a physical relational database schema, what options does the database administrator have in how supertype/subtype entities are implemented?

9. You need to calculate the anticipated database capacity for constructing a database with four tables, as shown below:

Table 1	Table 2	Table 3	Table 4
Field 1: 32 char	Field 1: 5 char	Field 1: 2 char	Field 1: 16 char
Field 2: 15	Field 2: 7	Field 2: 7	Field 2: 30
Field 3: 7	Field 3: 13	Field 3: 8	Field 3: 12
Field 4: 12	Field 4: 6	Field 4: 4	Field 4: 7
Field 5: 9	Field 5: 4		Field 5: 54
Field 6: 12			Field 6: 3
Field 7: 6			Field 7: 1
Field 8: 2			

Table 1 will be initially loaded with 200,000 records, Table 2 with 100,000 records, and the other two tables with 40,000 records each. The expected rate of growth in the number of records is 20 percent each year for three years.

According to the database capacity planning steps described in the chapter, what is the anticipated database capacity?

10. You are studying the design documentation for an extremely large information system used by your organization. As expected, all entities on the logical data model are in third normal form. But in comparing the logical data model to the physical database schema, you notice that two tables shown separately on the logical data model have been combined into one table on the physical database schema. There are no notes explaining this, and you can't ask the database administrator who signed off on the design documents but is no longer with the company. What might the reason have been for compromising third normal form? What are the potential consequences?

11. Explain the concept of referential integrity, and give an example. What is a referential integrity error? Provide an example, and explain the possible consequences of a referential integrity error.

12. The deletion rules for enforcing referential integrity include both Delete:Cascade and Delete:Restrict. In general, what criteria should a DBA use in deciding whether to use a Delete:Cascade or Delete:Restrict rule for deletion?

13. Database centralization is one of a number of distribution options, but it violates the rule that data should be managed and stored in close proximity to its users. Discuss the reasons why many data administrators and users consider this rule to be important. Given the growth of Web-based applications and global information systems, is this rule still important and/or viable?

Projects and Research

1. Interview the chief information officers or IT managers of several organizations that use relational database technologies. Ask them about their experience in moving from traditional flat file information systems to relational database systems.

   a. Describe the CIOs you interviewed and their organizations.
   b. Describe their database environment.
   c. When did their organizations move from traditional file systems to relational database systems?
   d. What type of problems did they encounter, and how were they resolved (or were they resolved)?

   If they had to do it again, what would they do differently?

2. More often than not, database environments in an organization reflect data structures that have been developed over a period of years, sometimes haphazardly, and that often reflect a variety of architectural styles and structures. Look at the existing database environment in your organization or school or in a local company.

   a. What is the age of the oldest information system used in the organization?
   b. What is the age of the newest?
   c. Does the organization have both traditional file-based systems and modern relational database systems?
   d. What are the different user systems, end-user tools, end-user applications? Any data warehouses? Any Web-based applications?
   e. Draw a data architecture diagram based upon the diagram shown in Figure 14-2.

3. Modern databases require a high level of skill and knowledge to be adequately supported. See if you can interview the database administrators in three or four local organizations which use contemporary database systems:

   a. Describe the systems which are managed by the database administrators you interviewed.
   b. On average, how much experience and training did they have in information technology before becoming database administrators?
   c. On average, how much time do they spend in training, formal or informal, per year keeping their skills current? Do they feel they receive enough training?
   d. Compute the average cost of annual training per database administrator. Include both direct costs for training and indirect "lost opportunity" costs. If actual costs are not known, use a direct training cost of $500/day and average DBA salary and benefits of $75 per hour.

   e. Do you think training costs for DBA administrators are higher than for administrators of flat file systems? Why or why not?

4. After you talk to the database administrators regarding training, also ask them about:

   a. Normalization—do they generally normalize to third normal form? Higher?
   b. In general, how much time is spent on normalization for a large or enterprise-level database?
   c. What are the three biggest chronic problems faced by each of the database administrators?
   d. How often do they have to modify and/or update the database schema? Do they have a formal process for identifying and making updates, or is it done on an ad hoc basis?
   e. If they were the CIO in their organization, what would they change?

5. CASE tools, such as *System Architect,* are used for database development and support. Search on the Web and in trade journals for some of the popular CASE tools currently in use.

   a. What CASE tools did you find, and who are their manufacturers?
   b. What is the number of installed bases, or IT shops, using each of these CASE tools?
   c. What is the range of cost for these CASE tools? Do you think they are cost-effective?
   d. Compare and contrast these CASE tools in terms of their features and capabilities.
   e. Which one would you use if you were a DBA and cost was not a concern? Why?

6. Currently, relational database technology is probably the most prevalent database technology used in modern information technology shops. But database technology is an evolving field and new technologies are being developed constantly. Search on the Web or in your school library for articles on emerging database technologies, such as object database management systems. Make sure to include white papers from companies such as Oracle, Sybase, IBM, and Microsoft.

   a. What articles and papers did you find?
   b. What are some of the new database technologies that are entering the market or that are currently under development?
   c. Compare and contrast each of them with contemporary relational database technology.
   d. What influence does the growth of Internet applications have on these database technologies?
   e. Which one, if any, do you feel is a serious contender to replace relational database technology? Why?

Minicases

With your professor's help, liaison with a team from either a systems analysis or computer programming class at another school. Your assignment, to complete together, is to build a suitable Web page for a small business or nonprofit organization of your choosing. You will be graded on completeness, functionality, professionalism, and teamwork. A communication suggestion is to utilize e-mail as much as possible.

1. Meet the team at the other school via the phone, virtual meeting environment, discussion board, or e-mail per your professor's instructions. If you use a virtual meeting environment, you may need to install and learn how to use the appropriate software. Determine and establish team guidelines and rules for:

   - Deadlines—how will the team handle a slipped deadline by one of the team members? Who will be in charge of setting the time line?
   - Communication—How will you communicate? How often? Do some of your members communicate better using one method over the others (i.e., preferences)?
   - Miscommunication of Personal Differences— How will you address miscommunication among members or arising personal differences?
   - Expectations—What are the team's expectations for quality? Behavior? What will you do if someone does not perform up to expectations?

Submit to your professor an agreement, signed by each of the team members, concerning how these matters will be addressed.

2. Meet local small business owners or representatives of nonprofit organizations. Find a company or organization that will host your team to produce a Web site for them (nonmonetarily, of course). Find out from your school's risk management or legal department what paperwork is necessary for you and your "client" to complete. (Why is this necessary?)

3. Determine the business's or organization's requirements through interviews, forms, surveys, JAD, and the like, and create the appropriate models and studies for the Web site. Don't forget to consider costs, legal issues, and specific company needs in your models and paper. You will be graded on completeness, correctness, clarity, and professionalism.

4. Create the Web site using appropriate technologies, getting a domain name, and so forth. Set up e-mail with the domain name, as well. If it is appropriate for the company to have shopping cart and online payment capabilities, make sure that those are fully functional. Stress test the site by exchanging URL's with another team *before* you submit it to your professor *or* the client.

Suggested Readings

Bruce, Thomas. *Designing Quality Databases with IDEF1X Information Models.* New York: Dorset House Publishing, 1992. This has rapidly become our favorite practical database design book. Incidentally, the foreword was written by John Zachman, whose *Framework for Information Systems Architecture* inspired our own information system building blocks framework.

McFadden, Fred; Jeffrey Hoffer; and Mary Prescott. *Modern Database Management*, 5th ed. Reading, MA: Addison-Wesley, 1994. For those seeking to expand their overall data management and database education, this is one of the most popular introductory textbooks on the market and our own favorite. These authors do a particularly thorough job of explaining distributed database design (in much greater detail than is possible in our book).

Teorey, Toby. *Database Modeling & Design: The Fundamental Principles,* 2nd ed. San Francisco: Morgan Kaufman Publishers, 1990. This is our favorite database design conceptual book. Appendix A provides a concise review of the SQL language.

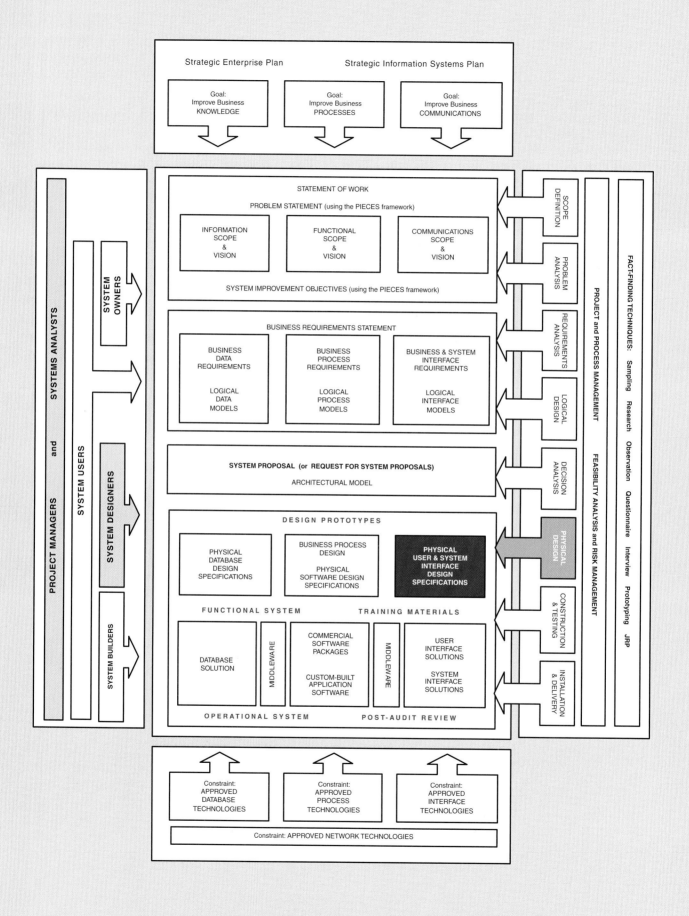

15

Output Design and Prototyping

Chapter Preview and Objectives

In this chapter you will learn how to design and prototype computer outputs. You will know how to design and prototype outputs when you can:

▌ Distinguish between internal, external, and turnaround outputs.

▌ Differentiate between detailed, summary, and exception reports.

▌ Identify several output implementation methods.

▌ Differentiate among tabular, zoned, and graphic formats for presenting information.

▌ Distinguish among area, bar, column, pie, line, radar, donut, and scatter charts and their uses.

▌ Describe several general principles that are important to output design.

▌ Design and prototype computer outputs.

Output and input design represent something of a "chicken or egg" sequencing problem. Which do you do first? In this edition, we present output design first. Classic system design prefers this approach as something of a system validation test—design the outputs and then make sure the inputs are sufficient to produce the outputs. In practice, this sequencing of tasks becomes less important because modern systems analysis techniques sufficiently predefine logical input and output requirements. You and your instructor may safely swap Chapters 15 and 16 if you prefer.

Introduction

Bob Martinez was glad that as part of the data design step he created a prototype database in the Microsoft Data Engine (MSDE). It is really coming in handy now as he designs reports for the system.

Bob created a simple Microsoft *Access* database, connected it to the MSDE database, and entered some sample data. Then working from discovery prototypes created in the analysis phase and use-case narratives, he used the *Access Report Wizards* to create a rough prototype for each printed report. With feedback from the users, he refined the report designs through several iterations. For reports that called for user-entered customization parameters, Bob created *Access* forms to simulate the customization interface.

Of course, the actual system will use neither Microsoft *Access* nor MSDE. But when programmers eventually get into the actual system construction, these reports and forms will guide their work and assure that the actual system meets all user requirements.

Output Design Concepts and Guidelines

Outputs present information to system users. Outputs are the most visible component of a working information system. As such, they are often the basis for the users' and management's final assessment of the system's value. During requirements analysis, you defined *logical* output requirements. During decision analysis, you *may* have considered different physical implementation alternatives. In this chapter, you will learn how to physically design the outputs.

Today, most outputs are designed by rapidly constructing prototypes. These prototypes may be simple computer-generated mock-ups with dummy data, or they may be generated from prototype databases such as Microsoft *Access,* which can be rapidly constructed and populated with test data. These prototypes are rarely fully functional. They won't contain security features or optimized data access that will be necessary in the final version of a system. Furthermore, in the interest of productivity, we may not include every button or control feature that would have to be included in a production system.

During requirements analysis, outputs were modeled as data flows that consist of data attributes. Even in the most thorough of requirements analysis, we will miss requirements. Output design may introduce new attributes or fields to the system.

We begin with a discussion of types of outputs. Outputs can be classified according to two characteristics: (1) their distribution and audience and (2) their implementation method. Figure 15-1 illustrates this taxonomy. The characteristics are discussed briefly in the following sections.

> Distribution and Audience of Outputs

One way to classify outputs is according to their distribution inside or outside the organization and the people who read and use them. **Internal outputs** are intended for the system owners and system users within an organization. They only rarely find their way outside the organization. Internal outputs support either day-to-day business operations or management monitoring and decision making. Figure 15-2 illustrates three basic subclasses of internal outputs:

internal output an output for system owners and users within an organization.

- **Detailed reports** present information with little or no filtering or restrictions. The example in Figure 15-2(a) is a listing of all purchase orders that were generated on a particular date. Other examples of detail reports would be a detailed listing of all customer accounts, orders, or products in inventory. Some detailed reports are historical. Other detailed reports are regulatory, that is, required by government.

detailed report an internal output that presents information with little or no filtering.

FIGURE 15-1   A Taxonomy for Computer-Generated Outputs

Distribution / Delivery	Internal Output (reporting)	Turnaround Output (external; then internal)	External Output (transactions)
Printer	Detailed, summary, or exception information printed on hard-copy reports for internal business use. Common examples: management reports	Business transactions printed on business forms that will eventually be returned as input business transactions. Common examples: phone bills and credit card bills	Business transactions printed on business forms that conclude the business transactions. Common examples: paychecks and bank statements
Screen	Detailed, summary, or exception information displayed on monitors for internal business use. Reports may be tabular or graphical. Examples: online management reports and responses to inquiries	Business transactions displayed on monitors in forms or windows that will also be used to input other data to initiate a related transaction. Examples: Web-based display of stock prices with the point-and-click purchase option	Business transactions displayed on business forms that conclude the business transactions Examples: Web-based report detailing banking transactions
Point-of-sale terminals	Information printed or displayed on special-purpose terminals dedicated to specific internal business functions. Includes wireless communication information transmission. Examples: end-of-shift cash register balancing report	Information printed or displayed on a special-purpose terminal for the purpose of initiating a follow-up business transaction. Examples: grocery store monitor that allows customer to monitor scanned prices, to be followed by input of debit or credit card payment authorization	Information printed or displayed on special-purpose terminals dedicated to customers. Examples: account balances display at an ATM machine or printout of lottery tickets; also, account information displayed via television over cable or satellite
Multimedia (audio or video)	Information transformed into speech for internal users. Not commonly implemented for internal users	Information transformed into speech for external users who respond with speech or tone input data. Examples: telephone touch-tone class schedule as part of course registration system	Information transformed into speech for external users. Examples: movie trailer for prospective online buyers of DVDs or telephone response to mortgage payoff query
E-mail	Displayed messages related to internal business information. Examples: e-mail messages announcing availability of new online business report	Displayed messages intended to initiate business transaction. Examples: e-mail messages whose responses are required to continue processing a business transaction	Messages related to business transactions. Examples: e-mail message confirmations of business transactions conducted via e-commerce on the Web
Hyperlinks	Web-based links to internal information that is enabled via HTML or XML formats. Examples: integration of all information system reports into a Web-based archival system for online archival and access	Web-based links incorporated into Web-based input pages to provide users with access to additional information. Examples: on a Web auction page, hyperlinks into a seller's performance history with an invitation to add a new comment	Web-based links incorporated into Web-based transactions Examples: hyperlinks to privacy policy or an explanation of how to interpret or respond to information in a report or transaction
Microfiche	Internal management reports archived to microfilm that requires minimal physical storage space. Examples: computer output on microfilm (COM)	Not applicable unless there is an internal need to archive turnaround documents. Examples: computer output on microfilm (COM)	Not applicable unless there is an internal need for copies of external reports. Examples: computer output on microfilm (COM)

FIGURE 15-2a and 15-2b

Levels of Report Detail

(a) Detailed reports

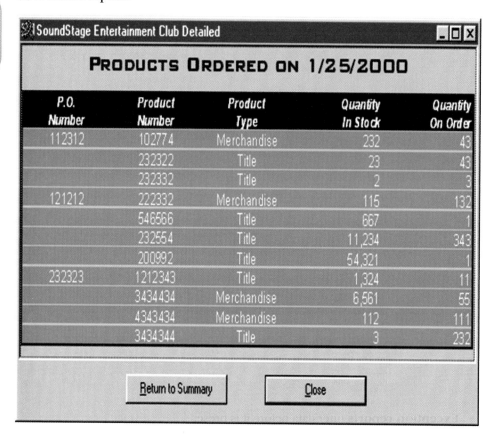

(b) Summary reports

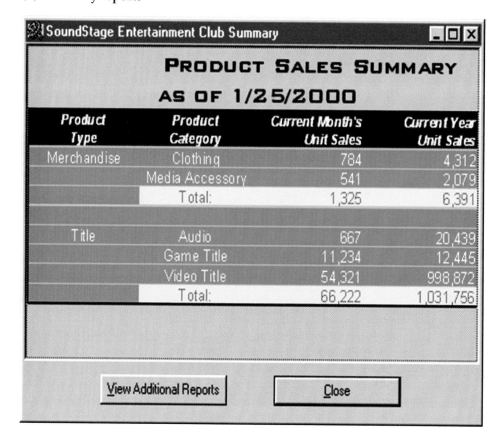

(c) Exception reports

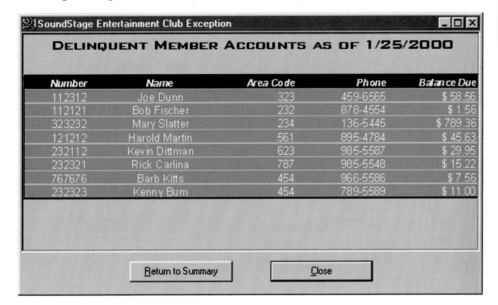

FIGURE 15-2c

Levels of Report
Detail

- **Summary reports** categorize information for managers who do not want to wade through details. The sample report in Figure 15-2(b) summarizes the month's and year's total sales by product type and category. The data for summary reports is typically categorized and summarized to indicate trends and potential problems. The use of graphics (charts and graphs) on summary reports is also rapidly gaining acceptance because they more clearly summarize trends at a glance.
- **Exception reports** filter data before it is presented to the manager as information. Exception reports include only exceptions to some condition or standard. The example in Figure 15-2(c) depicts the identification of delinquent member accounts. Another classic example of an exception report is a report that identifies items that are low in stock.

The opposite of internal outputs is external outputs. **External outputs** leave the organization. They are intended for customers, suppliers, partners, and regulatory agencies. They usually conclude or report on business transactions. Examples of external outputs are invoices, account statements, paychecks, course schedules, airline tickets, boarding passes, travel itineraries, telephone bills, purchase orders, and mailing labels.

Figure 15-3 illustrates a sample external output for SoundStage Entertainment Club. This sample, like many external outputs, is initially created as a blank, preprinted form that is designed and duplicated by forms manufacturers for use with computer printers.

Some outputs are both external and internal. They begin as external outputs that exit the organization but ultimately return (in part or in whole) as an internal input. **Turnaround outputs** are those external outputs that eventually reenter the system as inputs. Figure 15-4 demonstrates a turnaround document. Notice that the invoice has upper and lower portions. The top portion is to be detached and returned with the customer payment as an input.

summary report an internal output that categorizes information for managers.

exception report an internal output that filters data to present information that reports exceptions to some condition or standard.

external output an output that leaves the organization.

turnaround output an external output that may reenter the system as an input.

> Implementation Methods for Outputs

We assume you are familiar with different output devices, such as printers, plotters, computer output on microfilm (COM), and PC display monitors. These are standard topics in most introductory information systems courses. In this chapter, we are more concerned with the actual output than with the device. A good systems analyst will consider all available options for implementing an output. Let's briefly examine implementation methods and formats. You should continue to reference Figure 15-1 as we complete this introduction to the output taxonomy.

SoundStage Entertainment Club
Fax 317-494-5222

The following number must appear on all related correspondence, shipping papers, and invoices:
P.O. NUMBER: 712812

To:
CBS Fox Video Distribution
26253 Rodeo DR
Hollywood, CA

Ship To:
SoundStage Entertainment Club
Shipping/Receiving Station
Building A
2630 Darwin Drive
Indianapolis, IN 45213

P.O. DATE	REQUISITIONER	SHIP VIA	F.O.B. POINT	TERMS
5-3-03	LDB	UPS		Net 30

QTY	DESCRIPTION	UNIT PRICE	TOTAL
20000	Star Wars: The Phantom Menace (VHS)	15.99	319,800.00
3000	Star Wars: The Phantom Menace (DVD Dolby Digital)	19.99	59,970.00
500	Star Wars: The Phantom Menace (DVD DTS)	24.99	12,495.00
8000	Star Wars: The Phantom Menace (PlayStation II)	16.99	135,920.00
400	Star Wars: The Phantom Menace Soundtrack (CD)	16.99	6,796.00
600	Star Wars: The Phanton Menace Theater Poster	4.99	2,994.00

Subtotal	537,975.00
Tax	37,658.25
Total	575,633.25

1. Please send two copies of your invoice.

2. Enter this order in accordance with the prices, terms, delivery method, and specifications listed above.

3. Please notify us immediately if you are unable to ship as specified.

Madge Worthy   5-4-03
Authorized by   Date

Printed Output   The most common medium for computer outputs is paper—printed output. Currently, paper is the cheapest medium we will survey. Although the paperless office has been predicted for many years, it has not yet become a reality. Perhaps there is a psychological dependence on paper as a medium. In any case, paper output will be with us for a long time.

Printed output may be produced on impact printers, but increasingly it is printed on laser printers, which have become increasingly cost-effective. Internal outputs are typically printed on blank paper (called *stock paper*). External outputs and turnaround documents are printed on *preprinted forms.* The layout of preprinted forms (such as blank checks and W-2 tax forms) is predetermined, and the blank documents are mass-produced. The preprinted forms are run through the printer to add the variable business data (such as *your* paycheck and W-2 tax form).

Perhaps the most common format for printed output is tabular. **Tabular output** presents information as columns of text and numbers. Most of the computer programs you've written probably generated tabular reports. The sample detailed, summary, and exception reports illustrated earlier in the chapter (Figure 15-2) were all tabular.

An alternative to tabular output is zoned output. **Zoned output** places text and numbers into designated areas or boxes of a form or screen. Zoned output is often used

tabular output an output that presents information as columns of text and numbers.

zoned output an output that presents text and numbers in designated areas of a form or screen.

SoundStage Entertainment Club

2630 Darwin Drive - Bldg B
Indianapolis, IN 45213
317 496 0998 fax 317 494 0999

Invoice No.   301231

INVOICE

Customer

Name	KATRINA SMITH
Address	3019 DURAC DR
City	LITTLE ROCK State AR ZIP 42653
Phone	502-430-4545

Due Date	2/24/03
Order No.	346910

Payment Amt

....

Detach and return top portion with payment

Qty	Description	Unit Price	TOTAL
1	EAGLES HELL FREEZES OVER (DVD DD)	$19.99	$19.99
1	THE GRAMMY BOX (CD) ***COUNTS AS 3 CREDITS	$21.99	$21.99
1	GONE WITH THE WIND DIRECTORS CUT (DVD DS)	$17.99	$17.99
1	SIXTH SENSE (VHS)	FREE SS CR	$0.00
1	A BUG'S LIFE (VHS)	FREE SS CR	$0.00
1	NASCAR 2000 (VHS) *** CLOSEOUT (NO SS CR)	$9.99	$9.99

10 SOUNDSTAGE CREDITS WERE USED TO PAY
FOR PART OF THIS PURCHASE

WE APPRECIATE THE FINE MANNER IN WHICH YOU
HAVE PAID ON YOUR ACCOUNT. IN APPRECIATION
WE HAVE ADDED 7 SOUNDSTAGE CREDITS TO
YOUR ACCOUNT

YOU CAN EARN 7 CREDITS BY PAYING THIS
INVOICE BY THE DUE DATE

SubTotal	$69.96
Shipping & Handling	$7.00
Taxes	$2.95
TOTAL	$79.91

Payment Details

○ Cash
○ Check
○ Credit Card

Name _____
CC # _____
       Expires _____

Office Use Only

Please return top portion invoice with payment. Make checks payable to:
SoundStage Entertainment Club.

RETURN TOP PORTION WITH PAYMENT

FIGURE 15-4 Typical Turnaround Document

in conjunction with tabular output. For example, an order output contains zones for customer and order data in addition to tables (or rows of columns) for ordered products.

Screen Output   The fastest-growing medium for computer outputs is the online display of information on a visual display device, such as a CRT terminal or PC monitor. The pace of today's economy requires information on demand. Screen output is most suited to this requirement.

While screen output provides the system user with convenient access to information, the information is only temporary. When the information leaves the screen, that information is lost unless it is redisplayed. For this reason, printed output options are usually added to screen output designs.

Thanks to screen output technology, tabular reports—especially summary reports—can be presented in graphical formats. **Graphic output** is the use of a pictorial chart to convey information in ways that demonstrate trends and relationships not easily seen in tabular output.

> **graphic output** an output that uses a pictorial chart to convey information.

To the system user, a picture can be more valuable than words. There are numerous types and styles of charts for presenting information. Figure 15-5 summarizes various types of charts that can be output with today's technology. Report writing technology and spreadsheet software can quickly transform tabular data into charts that enable the reader to more quickly draw conclusions.

The popularity of graphic output has also been stimulated by the availability of low-cost, easy-to-use graphics printers and software, especially in the PC industry. Later in this chapter we will show you an alternative graphic design for a SoundStage output.

Point-of-Sale Terminals   Many of today's retail and consumer transactions are enabled or enhanced by point-of-sale (POS) terminals. The classic example is the automated teller machine (ATM). POS terminals are both input and output devices. In this chapter, we are interested only in the output dimension. ATMs display account balances and print transaction receipts. POS cash registers display prices and running totals as bar codes are scanned, and they also produce receipts. Lottery POS terminals generate random numbers and print tickets. All are examples of outputs that must be designed.

Multimedia   *Multimedia* is a term coined to collectively describe any information presented in a format other than traditional numbers, codes, and words. This includes graphics, sound, pictures, and animation. It is usually presented as a contemporary extension to screen output. Increasingly, multimedia output is being driven by the transition of information systems applications to the Internet and intranets.

We've already discussed graphical output. But other multimedia formats can be integrated into traditional screen designs. Many information systems offer film and animation as part of the output mix. Product descriptions as well as installation and maintenance instructions can be integrated into online catalogs using multimedia tools. Sound bites can also be integrated.

But multimedia output is not dependent on screen display technology. Sound, in the form of telephone touch-tone–based systems, can be used to implement an interesting output alternative. Many banks offer their customers touch-tone access to a wide variety of account, loan, and transaction data.

E-mail   E-mail has transformed communications in the modern business world, if not society as a whole. New information systems are expected to be message-enabled. How does this impact output design? Transactional systems are increasingly Web-enabled. When you purchase products over the Web, you almost always receive automated e-mail output to confirm your order. Follow-up e-mail may inform you of order fulfillment progress and initiate customer follow-up (a form of turnaround output).

Internal outputs may also be e-mail–enhanced. For example, a system can push notification of the availability of new reports to interested users. Only those users who truly need the report will access the report and print it. This can generate a significant cost savings over mass distribution.

	Sample	Selection Criteria
Line Chart		**Line charts** show one or more series of data over a period of time. They are useful for summarizing and showing data at regular intervals. Each line represents one series or category of data.
Area Chart		**Area charts** are similar to line charts except that the focus is on the area under the line. That area is useful for summarizing and showing the change in data over time. Each line represents one series or category of data.
Bar Chart		**Bar charts** are useful for comparing series or categories of data. Each bar represents one series or category of data.
Column Chart		**Column charts** are similar to bar charts except that the bars are vertical. Also, a series of column charts may be used to compare the same categories at different times or time intervals. Each bar represents one series or category of data.
Pie Chart		**Pie charts** show the relationship of parts to a whole. They are useful for summarizing percentages of a whole within a single series of data. Each slice represents one item in that series of data.
Donut Chart		**Donut charts** are similar to pie charts except that they can show multiple series or categories of data, each as its own concentric ring. Within each ring, a slice of that ring represents one item in that series of data.
Radar Chart		**Radar charts** are useful for comparing different aspects of more than one series or category of data. Each data series is represented as a geometric shape around a central point. Multiple series are overlaid so that they can be compared.
Scatter Chart		**Scatter charts** are useful for showing the relationship between two or more series or categories of data measured at uneven intervals of time. Each series is represented by data points using either different colors or bullets.

FIGURE 15-5 Chart Types and Selection Criteria

Hyperlinks   Many outputs are now Web-enabled. Many databases and consumer ordering systems are now Web-enabled. Web hyperlinks allow users to browse lists of records or search for specific records and retrieve various levels of detailed information on demand. Obviously, this medium can and is extended to computer inputs.

Technology exists to easily transform internal reports into HTML or XML formats for distribution via intranets. This reduces dependence on printed reports and screen reports that require a specific operating system or version (such as *Windows*). Essentially, all the recipient requires is a current browser that can run on any computer platform (*Windows, Mac, Linux,* or *UNIX*).

But Web-enabled output goes beyond presenting traditional outputs via the Internet and intranets. Many businesses have invested in Web-based internal report systems that consolidate weeks, months, and years of traditional internal reports into an organized database from which the reports can be recalled and displayed or printed. These systems don't create new outputs. They merely reformat previous reports for access via a browser. Think of it as an on-demand, Web-enabled report archival system. Examples of such reporting systems include DataWatch *Monarch/ES* and NSA *Report.Web.*

Microfilm   Paper is bulky and requires considerable storage space. To overcome the storage problem, many businesses use microfilm as an output medium. The first film medium is microfilm. More commonly, they turn to *microfiche,* small sheets of *microfilm* capable of storing dozens or hundreds of pages of computer output. The use of film presents its own problems; microfiche and microfilm can be produced and read only by special equipment.

This completes our introduction to output concepts. If you study Figure 15-1 carefully, you can see that implementation and distribution options can be combined to develop very creative, user-friendly, and exciting outputs.

How to Design and Prototype Outputs

In this section, we'll discuss and demonstrate the process of output design and prototyping. We'll introduce some tools for documenting and prototyping output design, and we'll also apply the concepts you learned in the last section. We will demonstrate how automated tools can be used to design and prototype outputs and layouts to system users and programmers. As usual, each step of the output design technique will be demonstrated using examples drawn from our SoundStage Entertainment Club case study.

> Automated Tools for Output Design and Prototyping

In the not-too-distant past, the primary tools for output design were *printer spacing charts* (see Figure 15-6) and *display layout charts*. Today, this approach is not practiced much. It is a tedious process that is not conducive to today's preferred prototyping and rapid application development strategies, which use automated tools to accelerate the design process.

Before the availability of automated tools, analysts could sketch only rough drafts of outputs to get a feel for how system users wanted outputs to look. With automated tools, we can develop more realistic prototypes of these outputs. Perhaps the least expensive and most overlooked prototyping tool is the common spreadsheet. Examples include Lotus *1-2-3* and Microsoft *Excel.* A spreadsheet's tabular format is ideally suited to the creation of tabular output prototypes. And most spreadsheets include facilities to quickly convert tabular data into a variety of popular chart formats. Consequently, spreadsheets provide an unprecedented way to quickly prototype graphical output for system users.

Arguably, the most commonly used automated tool for output design is the PC-database application development environment. Many of you have no doubt learned Microsoft *Access* in either a PC literacy or database development course. While *Access* is not powerful enough to develop most enterprise-level applications, you may be

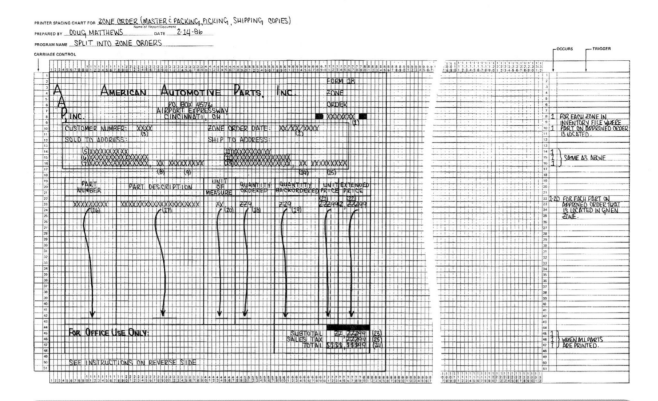

FIGURE 15-6   Printer Spacing Chart

surprised at how many designers use *Access* to prototype such applications. First, it provides rapid development tools to quickly construct a single-user (or few-user) database and test data. That data can subsequently feed the output design prototypes to increase realism. Designers can use *Access*'s report facility to lay out proposed output designs and test them with users.

Many CASE tools include facilities for report and screen layout and prototyping using the project repository created during requirements analysis. *System Architect's* screen design facility is demonstrated in Figure 15-7.

The above automated tools have significantly accelerated and enhanced the output design process. But the ultimate output design process would not only prototype the output's design but also serve as the final implementation of that output. This more sophisticated solution is found in report writing output tools such as Business Objects' *Crystal Reports* and Actuate's *e.Reporting Suite*. These products create the actual "code" to be integrated in the operational information system. Figure 15-8 illustrates two screens from the *Crystal Reports* tool being used to create a SoundStage report from a prototype database.

> Output Design Guidelines

Many issues apply to output design. Most are driven by human engineering concerns—the desire to design outputs that will support the ways in which system users work. The following general principles are important for output design:

1. *Computer outputs should be simple to read and interpret.* These guidelines may enhance readability:
   a. Every output should have a title.
   b. Every output should be dated and time-stamped. This helps the reader appreciate the currency of information (or lack thereof).
   c. Reports and screens should include sections and headings to segment information.

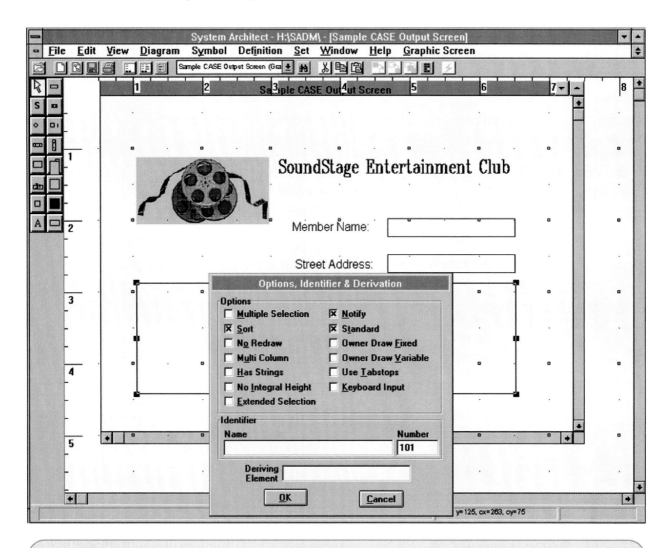

FIGURE 15-7   CASE Tool for Output Design

    d. In form-based outputs, all fields should be clearly labeled.

    e. In tabular-based outputs, columns should be clearly labeled.

    f. Because section headings, field names, and column headings are sometimes abbreviated to conserve space, reports should include or provide access to legends to interpret those headings.

    g. Only required information should be printed or displayed. In online outputs, use information hiding and provide methods to expand and contract levels of detail.

    h. Information should never have to be manually edited to become usable.

    i. Information should be balanced on the report or display—not too crowded, not too spread out. Also, provide sufficient margins and spacing throughout the output to enhance readability.

    j. Users must be able to easily find the output, move forward and backward, and exit the report.

    k. Computer jargon and error messages should be omitted from all outputs.

  2. *The timing of computer outputs is important.* Output information must reach recipients while the information is pertinent to transactions or decisions. This can affect how the output is designed and implemented.

  3. *The distribution of (or access to) computer outputs must be sufficient to assist all relevant system users.* The choice of implementation method affects distribution.

(a)

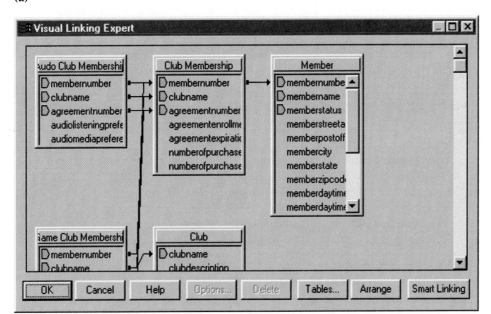

FIGURE 15-8

Report Writer Tool
for Report Design

(b)

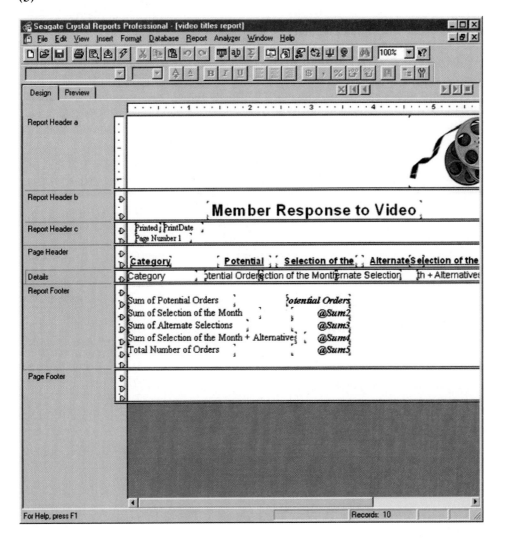

4. *The computer outputs must be acceptable to the system users who will receive them.* An output design may contain the required information and still not be acceptable to the system user. To avoid this problem, the systems analyst must understand how the recipient plans to use the output.

> The Output Design Process

Output design is not a complicated process. Some steps are essential, and others are dictated by circumstances. The steps are:

1. Identify system outputs and review logical requirements.
2. Specify physical output requirements.
3. As necessary, design any preprinted external forms.
4. Design, validate, and test outputs using some combination of:
   a. Layout tools (e.g., hand sketches, printer/display layout charts, or CASE).
   b. Prototyping tools (e.g., spreadsheet, PC DBMS, 4GL).
   c. Code-generating tools (e.g., report writer).

In the following subsections, we examine these steps and illustrate a few examples from the SoundStage project.

Step 1: Identify System Outputs and Review Logical Requirements   Output requirements should have been defined during requirements analysis. Physical data flow diagrams (or design units, both described in Chapter 13) are a good starting point for output design. Those DFDs identify both the net outputs of the system (process-to-external agent) and the implementation method.

Depending on your system development methodology and standards, each of these net output data flows may also be described as a logical data flow in a data dictionary or repository (see data structures, Chapter 9). The data structure for a data flow specifies the attributes or fields to be included in the output. If those requirements are specified in the relational algebraic notation, you can quickly determine which fields repeat, which fields have optional values, and so on. Consider the following data structure:

Data Structure Defining Logical Requirements	Comment
INVOICE = INVOICE NUMBER	← Unique identifier of the output.
+ INVOICE DATE	← One of many fields that must take on a value. Lack of parentheses indicates a value is required.
+ CUSTOMER NUMBER	
+ CUSTOMER NAME	
+ CUSTOMER BILLING ADDRESS = ADDRESS >	← Pointer to a related definition.
+ 1 {SERVICE DATE +	← Begins group of fields that repeats 1 − n times.
SERVICE PROVIDED +	
SERVICE CHARGE} N	
+ PREVIOUS BALANCE DUE	← More required fields with single values.
+ PAYMENTS RECEIVED	
+ TOTAL NEW SERVICE CHARGES	
+ INTEREST CHARGES	
+ NEW BALANCE DUE	
+ MINIMUM PAYMENT DUE	
+ PAYMENT DUE DATE	
+ (DEFAULT CREDIT CARD NUMBER)	
+ ([CREDIT MESSAGE | PAYMENT MESSAGE])	← Field does not have to have value.
	← Field does not have to have value, but if it does, it will provide only one of two possible field options.

Without such precise requirements, discovery prototypes may exist that were created during requirements analysis. In either case, a good requirements statement should be available in some format.

Step 2: Specify Physical Output Requirements   Recall that the decision analysis phase should have established some expectation of how most output data flows will eventually be implemented. Relative to outputs, the decisions were made by determining the best medium and format for the design and implementation based on:

- Type and purpose of the output.
- Operational, technical, and economic feasibility.

Because feasibility is important to more than just outputs, the techniques for evaluating feasibility were covered separately (in Chapter 10). The first set of criteria, however, is described in the following list:

- Is the output for internal or external use?
- If it's an internal output, is it a detailed, summary, or exception report?
- If it's an external report, is it a turnaround document?

After assuring yourself that you understand what type of report the output is and how it will be used, you need to address several design issues:

1. What implementation method would best serve the output? Various methods were discussed earlier in the chapter. You will have to understand the purpose or use of the output to determine the proper method. You can select more than one method for a single output—for instance, screen output with optional printout. Clearly, these decisions are best addressed with the system users.
   a. What would be the best format for the report? Tabular? Zoned? Graphic? Some combination?
   b. If a printout is desired, you must determine what type of form or paper will be used. Stock paper comes in three standard sizes (all specified in inches): 8½ × 11, 11 × 14, and 8½ × 14 inches. You need to determine the capabilities and limitations of the intended printer.
   c. For screen output, you need to understand the limitations of the users' display devices. Despite the increase in larger 19- and 21-inch high-resolution monitors, most users still have 15- and 17-inch displays and have their screen resolution set as low as 640 × 480 pixels (especially as you reach out directly to consumers in e-commerce applications). It is still recommended that screen outputs (including forms or pages within your application) be designed for the lowest common denominator.
   d. Form images can be stored and printed with modern laser printers, thereby eliminating the need for dealing with forms manufacturers in some businesses.
2. How frequently is the output generated? On demand? Hourly? Daily? Monthly? For scheduled outputs, when do system users need the report?
   a. Users generate many reports on demand. It can be helpful to use automated e-mail to notify users that new versions are available.
   b. If reports are to be printed by the information services department, they must be worked into the information systems operations schedule. For instance, a report the system user needs by 9:00 A.M. on Thursday may have to be scheduled for 5:30 A.M. Thursday. No other time may be available.
3. How many pages or sheets of output will be generated for a single copy of a printed output? This information may be necessary to accurately plan paper and forms consumption.
4. Does the output require multiple copies? If so, how many?
   a. Impact printers are usually required to print all copies of a multicopy form at the same time.
   b. Laser printers can print multiple copies of a form only one after the other. This means that if the copies are different in color or fields, the preprinted forms must be collated before final printing.

5. For printed outputs, have distribution controls been finalized? For online outputs, access controls should be determined.

These design decisions should be recorded in the data dictionary/project repository. Let's consider an example from our SoundStage Entertainment Club case.

One output for SoundStage is the MEMBER RESPONSE SUMMARY REPORT. This report was requested to provide internal management with information regarding customer responses to the monthly promotional offers. The following design requirements were established:

1. The manager will request the report from his or her own workstation. It was determined that the information should be presented as a screen output in both tabular and graphical formats (to be determined via prototypes).
   a. All managers have 17-inch or larger display monitors.
   b. Managers should have the option of obtaining a laser printer output via their LAN configuration. Printouts should be on 8½- × 11-inch stock paper.
2. Managers must be able to display the report on demand. Managers have requested automatic e-mail notification of the availability of any newly generated version of the report. A hyperlink to the latest version of the report should also be made available in the standard home page of every Member Services manager, level 3 and above.
3. Graphical output should be displayable in a single screen and printable on a single page. Tabular data may be printed on one to two pages. The volume of pages is not considered significant for this report.
4. The report must be restricted in access to managers whose network accounts carry level-3 or higher account privileges. The report should include a "Confidential" watermark and a message that prohibits external distribution or information sharing without the written permission of Internal Audit.

Step 3: Design Any Preprinted Forms   External and turnaround documents are separated here for special consideration because they contain considerable constant and preprinted information that must be designed before designing the final output. In most cases, the design of a preprinted form is subcontracted to a forms manufacturer. The business, however, must specify the design requirements and carefully review design prototypes. The design requirements address issues such as the following:

- What preprinted information must appear on the form? This includes contact information, headings, labels, and other common information to appear on all copies of the form.
- Should the form be designed for mailing? If so, address locations become important based on whether or not windowed envelopes will be used.
- How many forms will be required for printing each day? Week? Month? Year?
- What will be the form's size? Form size, along with volume (above), can impact mailing costs.
- Will the form be perforated to serve as a turnaround document? Also, for turnaround documents the location of the address becomes more critical because the return address for the external output becomes the mailing address for the returned document.
- What legends, policies, and instructions need to be printed on the form (both front and back)?
- What colors will be used, and for which copies?

For external documents, there are also several alternatives. Carbon and chemical carbon are the most common duplicating techniques. Selective carbons are a variation whereby certain fields on the master copy will not be printed on one or more of the remaining copies. The fields to be omitted must be communicated to the forms manufacturer. Two-up printing is a technique whereby two sets of forms, possibly including carbons, are printed side by side on the printer.

A SoundStage preprinted output form was previously displayed as Figure 15-3.

Step 4: Design, Validate, and Test Outputs   After design decisions and details have been recorded in the project repository, we must design the actual format of the report. The format or layout of an output directly affects the system user's ability to read and interpret it. The best way to lay out the format is to sketch or, better still, generate a sample of the report or document. We need to show that sketch or prototype to the system user, get feedback, and modify the sample. It's important to use realistic or reasonable data and demonstrate all control breaks.

The most important issue during the design step is format. Figure 15-9 summarizes a number of design issues and considerations for printed and tabular reports. Many of these considerations apply equally to screen outputs. Also, screen output offers a number of special considerations that are summarized in Figure 15-10.

The SoundStage management expressed concern that the MEMBER RESPONSE SUMMARY output could potentially become too lengthy. Often the manager is interested in seeing only information pertaining to member responses for one or a few different product promotions. Thus, it was decided that the manager needed the ability to "customize" the output. The screens used to allow the manager to specify the customization desired should be prototyped as well as the report and graph containing the actual information. Figure 15-11(a) shows the prototype of the screen the user

FIGURE 15-9   Tabular Report Design Principles

Design Issue	Design Guideline	Examples
Page size	At one time, most reports were printed on oversized paper. This required special binding and storage. Today, the page sizes of choice are standard (8½″ × 11″) and legal (8½″ × 14″). These sizes are compatible with the predominance of laser printers in the modern business.	Not applicable.
Page orientation	Page orientation is the width and length of a page as it is rotated. The *portrait* orientation (e.g., 8½ W × 11 L) is often preferred because it is oriented the way we orient most books and reports; however, *landscape* (e.g., 11 W × 8½ L) is often necessitated for tabular reports because more columns can be printed.	portrait landscape
Page headings	Page headers should appear on every page. At a minimum, they should include a recognizable report title, date and time, and page numbers. Headers may be consolidated into one line or use multiple lines	JAN 4, 2001 PAGE 4 of 6 OVERSUBSCRIPTIONS BY COURSE
Report legends	A legend is an explanation of abbreviations, colors, or codes used in a report. In a printed report, a legend can be printed on only the first page or on every page. On a display screen, a legend can be made available as a pop-up dialogue box.	REPORT LEGEND SEATS NUMBER OF SEATS IN THE CLASSROOM LIM COURSE ENROLLMENT LIMIT REQ NUMBER OF SEATS REQUESTED BY DEPARTMENT RES NUMBER OF SEATS RESERVED FOR DEPARTMENT USED NUMBER OF SEATS USED BY DEPARTMENT AVL NUMBER OF SEATS AVAILABLE FOR DEPARTMENT OVR NUMBER OF OVERSUBSCRIPTIONS FOR DEPARTMENT

FIGURE 15-9 Concluded

Design Issue	Design Guideline	Examples
Column headings	Column headings should be short and descriptive. If possible, avoid abbreviations. Unfortunately, this is not always possible. If abbreviations are used, include a legend (see "Report legends").	Self-explanatory.
Heading alignments	The relationship of column headings to the actual column data under those headings can greatly affect readability. Alignment should be tested with users for preferences, with a special emphasis on the risk of misinterpretation of the information. See examples for possibilities (which can be combined).	Left justification (good for longer and variable-length fields): NAME == XXXXXXX X XXXXXXXX XXXXXX Right justification (good for some numeric fields, especially monetary fields); be sure to align decimal points: AMOUNT ==================== $$$,$$$.¢¢ Center (good for fixed-length fields and some moderate-length fields): STATUS ========== XXXX XXXX
Column spacing	The spacing between columns impacts readability. If the columns are too close, users may not properly differentiate between the columns. If they are spaced too far apart, the user may have difficulty following a single row all the way across a page. As a general rule of thumb, place 3–5 spaces between each column.	Self-explanatory.
Row headings	The first one or two columns should serve as the identification data that differentiates each row. Rows should be sequenced in a fashion that supports their use. Frequently rows are sorted on a numerical key or alphabetically.	By number: STUDENT ID STUDENT NAME ==================== ================================ 999–38–8476 MARY ELLEN KUKOW 999–39–5857 By alpha: SERVICE CANCEL SUBSCR TOTAL =========== =========== =========== ========= HBO 45 345 7665
Formatting	Data is often stored without formatting characters to save storage space. Outputs should reformat that data to match the users' norms.	As stored: As output: 307877262 307–87–7262 8004445454 (800) 444–5454 02272000 Feb 27, 2000

Control breaks	Frequently, rows represent groups of meaningful data. Those groups should be logically grouped in the report. The transition from one group to the next is called a *control break* and is frequently followed by subtotals for the group.	RANK NAME SALARY
		======= =========================== ===========
		CPT JANEWAY, K 175,000
		CPT KIRK, J 225,000
		CPT PICARD, J 200,000
		CPT SISKO, B 165,000
		    --------- 	
		    CAPTAINS TOTAL 765,000 ☜ a control break
		
		LTC CHAKOTAY 110,000
		LTC DATA 125,000
		LTC RIKER, W 140,000
		LTC SPOCK, S 155,000
		    --------- 	
		    EXEC OFFCR TOTAL 530,000

End of report	The end of a report should be clearly indicated to ensure that users have the entire report.	*** END OF REPORT ***

FIGURE 15-10 Screen Output Design Principles

Screen Design Consideration	Design Guidelines
Size	Different displays support different resolutions. The designer should consider the "lowest common denominator." The default window size should be less than or equal to the worst resolution display in the user community. For instance, if some users have only a 640 × 480 pixel resolution display, don't design windows to open at an 800 × 600 pixel resolution.
Scrolling	Online outputs have the advantage of not being limited by the physical page. This can also be a disadvantage if important information such as column headings scrolls off the screen. If possible, freeze important headings at the top of a screen.
Navigation	Users should always have a sense of where they are in a network of online screens. Given that, users also require the ability to navigate between screens. WINDOWS: Outputs appear in windows called *forms*. A form may display one record or many. The scroll bar should indicate where you are in the report. Buttons are frequently provided to move forward and backward through records in the report and to exit the report. INTERNET: Outputs appear in windows called *pages*. A page may display one record or many. Buttons or hyperlinks may be used to navigate through records. Custom search engines can also be used to navigate to specific locations within a report.
Partitioning	WINDOWS: *Zones* are forms within forms. Each form is independent of the other but can be related. The zones can be independently scrollable. The Microsoft *Outlook* bar is one example. Zones can be used for legends or control breaks that take the user to different sections within a report. INTERNET: *Frames* are pages within pages. Users can scroll independently within pages. Frames can enhance reports in many ways. They can be used for a legend, table of contents, or summary information.
Information hiding	Online applications such as those that run under *Windows* or within an Internet browser offer capabilities to hide information until it either is needed or becomes important. Examples of such information hiding include: • Drill-down controls that show minimal information and provide readers with simple ways to expand or contract the level of detail displayed. — In *Windows* outputs the use of a small plus or minus sign in a small box to the left of a data record offers the option of expanding or contracting the record into more or less detail. All of this expansion and contraction occurs within the output's window. — In *intranet* applications, any given piece of summary information can be highlighted as a hyperlink to expand that information into greater detail. Typically, the expanded information is opened in a separate window so that the reader can use the browser's forward and backward buttons to switch between levels of detail. • Pop-up dialogue boxes may be triggered by information.
Highlighting	Highlighting can be used in reports to call users' attention to erroneous data, exception data, or specific problems. Highlighting can also be a distraction if misused. Ongoing human factors research will continue to guide our future use of highlighting. Examples of highlighting include: • Color (avoid colors that color-blind persons cannot distinguish). • Font and case (changing case can draw attention). • Justification (left, right, or centered). • Hyphenation (not recommended in reports). • Blinking (can draw attention or become annoying). • Reverse video.
Printing	For many users, there is still comfort in printed reports. Always provide users the option to print a permanent copy of the report. For Internet use, reports may need to be made available in industry-standard formats such as Adobe *Acrobat*, which allows users to open and read the reports using free and widely available software.

FIGURE 15-11

Report Customiza-
tion and Tabular
Report Prototypes

(a) Report customization prototype

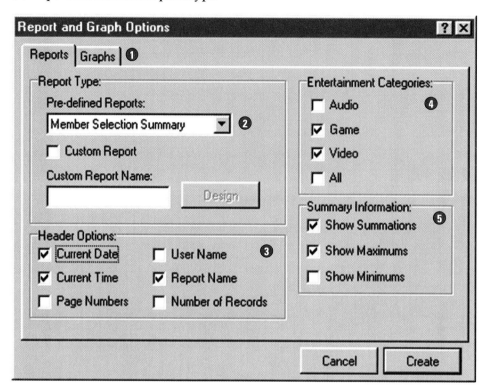

(b) Tabular report prototype

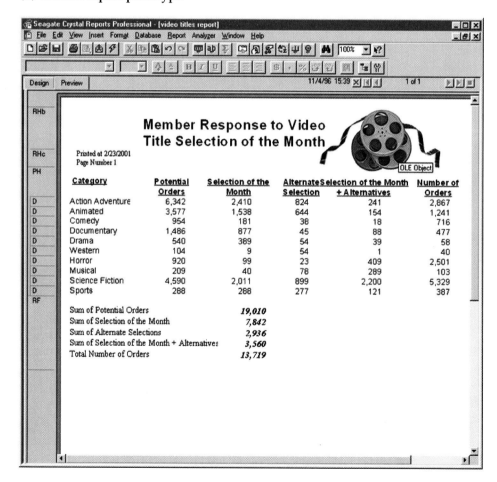

can use to choose a particular report (or graph) and customize its content. The following points should be noted:

❶ A tab dialogue box is used to allow the user to select between obtaining a report and obtaining a graph. A tab control is used to present a series of related information. If the user clicks on the tab labeled "Graphs," information would be displayed for customizing the output as a graph.

❷ A drop-down list is used to select the desired report. The user can click on the downward arrow to obtain a list of possible reports to choose from.

❸ The user is provided with a series of check boxes that correspond to general options for customizing the selected report. The user simply "checks" the options he wishes to have on the report.

❹ A group of check boxes is also used to allow the user to select one or more product categories she wishes to include on the report.

❺ Once again, a group of check boxes is used to allow the user to further customize the report. Here the user is allowed to indicate the type of summary information or totals desired for each product category.

Let's now look at a prototype of the report that will result from the previous report customization dialogue. Figure 15-11(b) is a prototype of a screen output version of the actual report. Examine the content and appearance of the tabular design. Notice that the user is allowed to scroll vertically and horizontally to view the entire report. In addition, buttons are provided to allow the user to toggle forward and backward to view different report pages.

Finally, let's look at a prototype of a graphic version of the MEMBER RESPONSE SUMMARY output (see Figure 15-12). Note the following:

- The graph is clearly labeled along the vertical and horizontal axes.
- A legend has been provided to aid in interpreting the graph bars.

When you are prototyping outputs, it is important to involve the user to obtain feedback. The user should be allowed to actually "exercise" or test the screens. Part of that experience should involve demonstrating how the user may obtain appropriate help or instructions, drill-down to obtain additional information, navigate through

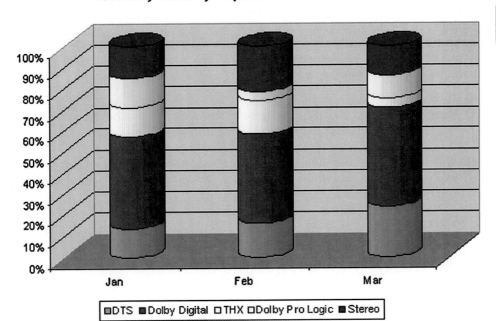

Quarterly Sales by Popular DVD Audio Format

□ DTS   ■ Dolby Digital   □ THX   □ Dolby Pro Logic   ■ Stereo

FIGURE 15-12

Graphical Report Prototype

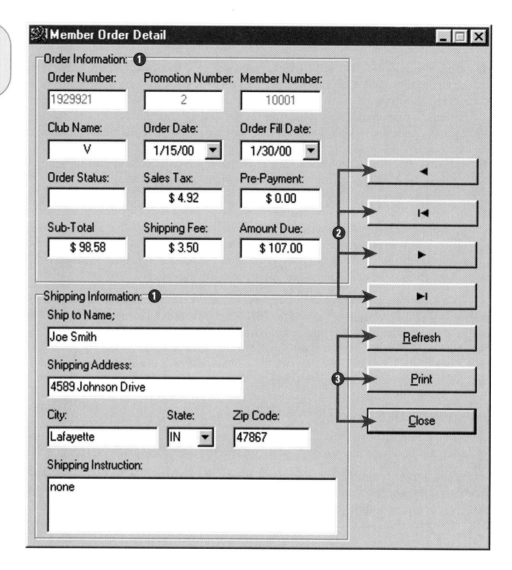

pages, request different formats that are available, size the outputs, and perform test customization capabilities. All features should be demonstrated or tested.

Thus far, we have presented samples of only a tabular and a graphical report. Another type of output is a record-at-a-time report. Users can browse forward and backward through individual records in a file. A sample screen for a record-at-a-time output is shown in Figure 15-13. We call your attention to the following:

❶ Each field is clearly labeled.

❷ Buttons have been added for navigation between records. The almost universally accepted buttons are for FIRST RECORD, NEXT RECORD, PREVIOUS RECORD, and LAST RECORD.

❸ We added buttons for the user to get a printed copy of the output, as well as to exit the report when finished. (Consistent with prototyping, the programmer will write the code for exiting later.)

> Web-Based Outputs and E-Business

The last output design considerations we want to address concern Web-based outputs. The SoundStage project will add various e-commerce and e-business capabilities to the Member Services information system. Some of these capabilities will affect output design.

One logical output requirement for the project is catalog browsing. Members should be able to browse and search catalogs, presumably as a preface to placing

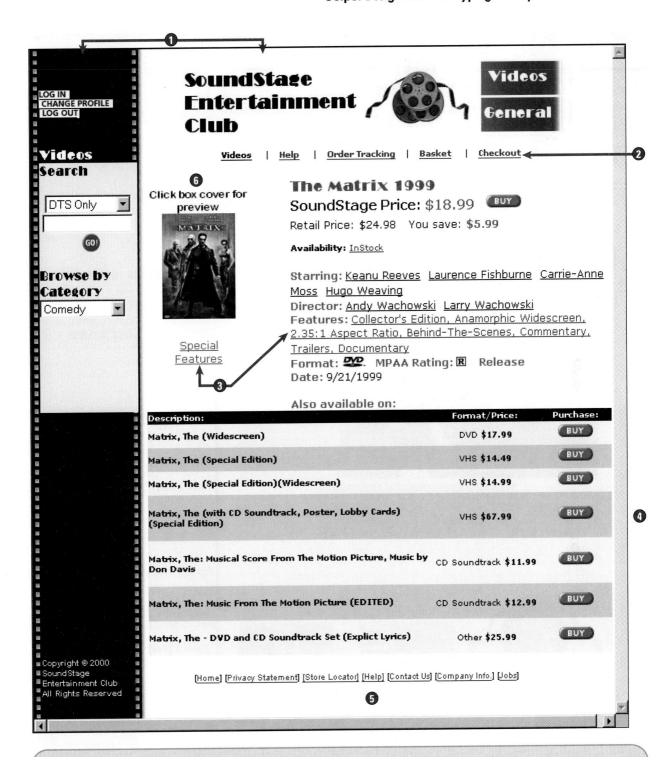

FIGURE 15-14 Web Database Output

orders. The catalog itself is the output. Figure 15-14 is a prototype screen for the physical catalog output. Note the following:

❶ This output uses frames to allow the user to focus separately on navigation and output.

❷ The screen uses hyperlinks to provide navigation through complex menu structures that are related to the output.

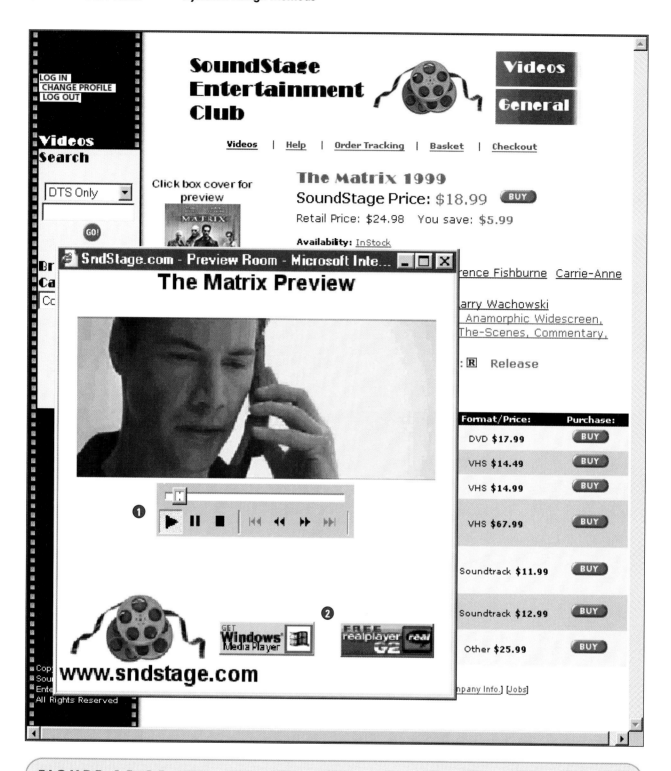

FIGURE 15-15 Windows *Media Player* Output

❸ Hyperlinks also allow the user to get additional information. This functionality is referred to as "drill-down."

❹ Shading is used to separate each detail line. This practice reflects the more artistic approach used to design Web-based outputs. Also, the "BUY" buttons have effectively transformed this output into a trigger for subsequent inputs. This is the e-commerce virtual equivalent of a turnaround document!

⑤ Most Web-based output screen designs require standard footers on the screen to provide additional navigation.

⑥ A picture can be a selectable object. In this case it represents another type of drill-down where the user is able to obtain additional information.

Another output requirement is to allow members to play video trailers and audio sound bites for products to preview candidate purchases. The preview will be triggered by a hyperlink in the previous screen, and it will activate a multimedia player as shown in Figure 15-15 on the previous page. Such output extensions are expected to become the norm as Internet- and intranet-based applications grow in popularity.

❶ Web-based outputs frequently use plug-ins. This output screen has the standard buttons associated with a typical audio or video player.

❷ Web-based outputs also commonly provide appropriate plug-ins or plug-in versions needed for the session.

Learning Roadmap

This chapter provided a detailed overview of the design and prototyping of computer outputs for a systems development project. It is recommended that you now complete Chapter 16 and not skip to Chapter 17. Chapter 16 deals with designing and prototyping an application's inputs. Chapter 17 deals with designing and prototyping an application's overall interface. As such, Chapter 17 involves tying together and presenting the applications functions addressed in Chapters 15 and 16.

1. Outputs can be classified according to two characteristics:

   a. Their distribution inside or outside the organization and the people who read and use them.

   b. Their implementation method.

2. Internal outputs are intended for the system owners and users within an organization. They only rarely find their way outside the organization. There are three subclasses of internal outputs:

   a. Detailed reports—present information with little or no filtering or restrictions.

   b. Summary reports—categorize information for managers who do not want to wade through details.

   c. Exception reports—filter data before it is presented to the manager as information.

3. External outputs leave the organization. They are intended for customers, suppliers, partners, and regulatory agencies. They usually conclude or report on business transactions.

4. Some outputs are both external and internal. They begin as external outputs that exit the organization but return in part or in whole.

5. Turnaround outputs are those external outputs that eventually reenter the system as inputs.

6. A good systems analyst will consider all available options for implementing an output. Several methods and formats exist:

   a. The most common medium for computer outputs is paper—printed output. Internal outputs are typically printed on blank paper (called stock paper). External outputs and turnaround documents are printed on preprinted forms.

      i) Perhaps the most common format for printed output is tabular. Tabular output presents information as columns of text and numbers.

      ii) An alternative to tabular output is zoned output. Zoned output places text and numbers into designated areas or boxes of a form or screen.

   b. Screen output is most suited to the pace of today's economy, which requires information on demand. Screen output technology allows reports to be presented in graphical formats. Graphic output is the use of a pictorial chart to convey information in ways that demonstrate trends and relationships not easily seen in tabular output.

   c. Many of today's retail and consumer transactions are enabled or enhanced by point-of-sale (POS) terminals.

   d. *Multimedia* is a term coined to collectively describe any information presented in a format other than traditional numbers, codes, and words. This includes graphics, sound, pictures, and animation.

   e. E-mail is becoming a very popular output medium as a means of reaching large audiences and generating significant cost savings.

   f. Web hyperlinks allow users to browse lists of records or search for specific records and retrieve various levels of detailed information on demand.

   g. Paper requires considerable storage space. To overcome the storage problem, many businesses use microfilm as an output medium.

7. The most commonly used automated tool for output design is the PC-database application development environment. Many CASE tools also include facilities for report and screen layout and prototyping using the project repository created during requirements analysis.

8. The following general principles are important for output design:

   a. Computer outputs should be simple to read and interpret.

   b. The timing of computer outputs is important—their recipients must receive output information while the information is pertinent to transactions or decisions.

   c. The distribution of (or access to) computer outputs must be sufficient to assist all relevant system users.

   d. The computer outputs must be acceptable to the system users who will receive them.

9. Output design is not a complicated process. Some steps are essential, and others are dictated by circumstances. The steps are:

   a. Identify system outputs and review logical requirements.

   b. Specify physical output requirements.

   c. As necessary, design any preprinted external forms.

   d. Design, validate, and test outputs using some combination of:

      i) Layout tools (e.g., hand sketches, printer/display layout charts, or CASE).

      ii) Prototyping tools (e.g., spreadsheet, PC DBMS, 4GL).

      iii) Code-generating tools (e.g., report writer).

Review Questions

1. What are some of the characteristics of prototypes?
2. How are outputs classified?
3. What is the difference between the summary report and the exception report?
4. What are some examples of external reports?
5. What is the difference between tabular output and zoned output?
6. Why are printed reports needed in addition to the screen outputs?
7. What are some of the examples of pictorial charts?
8. Why should graphic outputs be used?
9. What are some of the output design guidelines?
10. What are the steps basic for designing output?
11. What are the two most important kinds of criteria that analysts should consider when they specify physical output requirements? Why are they important?
12. What are some of the design issues that analysts need to consider?
13. What are preprinted forms for?
14. What is the advantage of using frames when displaying information on the Internet?

Problems and Exercises

1. One hundred years ago, if you were designing a report, what different delivery methods and media were available? What about 50 years ago? Today? What do you think has been the biggest change in reports over the past 100 years?
2. You are working as a systems designer for the county Department of Social Services. The director of the county child protection agency is concerned about the agency's caseload and the length of time that cases remain open. The agency's objective is to have no open cases older than 60 days, and preferably none older than 30 days. The director wants a monthly report showing the number of cases, by age, for each of the 12 child protection workers in the agency. What subclass of report should you design? Should the output format be tabular or zoned? Describe the data structure defining logical requirements for the report. Use the format described in the chapter.
3. Use the information in the preceding question to create a prototype of the report; use an automated tool such as Microsoft *Access* (or if you prefer, you can create a prototype the old-fashioned way). Populate the report with several sample records, in alphabetical order by worker last name.
4. The director of the child protection agency is pleased with the report, but would also like to see it in graphic format. What chart type(s) would be inappropriate? Why? What chart type(s) would be appropriate for this type of report? Why? Which one do you think would be the best one? Why?
5. What subclass of report would you design for the sales manager of a car dealership whose job it is to review vehicle sales each week and year to date? What data elements should be included in the report? What should you ask the sales manager before you design the report?
6. The sales manager also has to know, on a weekly basis, who didn't make their sales quota for the previous week and/or for the year to date. What subclass of report is needed in this situation? What data elements would you include, and how would you group them?
7. Match the definitions or examples in the first column with the terms in the second column.

A. On-demand Web-enabled report archival system	1. Detailed report
B. ATM	2. Display layout chart
C. Traditional output medium	3. External output
D. Report of Delinquent Accounts	4. Scatter chart
E. Transition between different data groups	5. Summary report
F. Report of Vehicles in Inventory	6. Control break
G. "Buy" button on Web site	7. Zoned output
H. Screen design tool seldom used anymore	8. Turnaround output

I. Sales order

J. Shows relationship between two or more series of data

K. Quarterly Report of Sales By Region

L. Sales receipt

9. Exception report

10. DataWatch *Monarch/ES*

11. POS Terminal

12. Microfilm

8. The sales manager has asked you to develop an automated chart to show the company's annual sales by quarter for the past five years. The manager considers bar charts boring and wants to use a pie chart, instead, to show the five-year sales report in an easy-to-read report. What are your thoughts about the manager's idea? Explain.

9. The director of the Child Protection Agency is very pleased with the summary report that you designed in Question 2. To help the child protection workers manage and prioritize their caseload, the director would now like you to design a report that would go to each worker, showing their open cases, including the age of each case. Further, the report should be a turnaround document, where the child protection workers can provide status on each of their open cases, including the estimated date of completion. What subclass of report is needed? What data elements are needed? In what order should the cases be listed? Create a prototype design for this report.

10. Complete the sentences below.

   a. The purpose of outputs is to present _____ to system users. Because they are the most _____ part of an information system, system users and owners often base the _____ of an information system on the outputs.

   b. In designing outputs, a good place to begin is with the _____, because they identify both the _____ and the _____ method.

   c. Outputs can be categorized by two characteristics: (1) by their _____ and _____, and (2) by their _____.

   d. In a report, _____ often occur at _____, which are used to transition from one _____ of data to the next one.

   e. In a tabular report, _____ is influenced by column _____, which generally should be 3-5 _____.

11. You are a systems designer working in the IT division of a large manufacturing corporation with plants throughout the country. The CIO mentions to you in passing that the vice president of marketing wants a new executive-level report showing daily production by region and by office in order to review production levels and fix problems quickly. Your CIO tells you to have a preliminary design and prototype ready the day after tomorrow. On the basis of the information you have been given, what type of report is needed? Is it for internal or external use? Assuming that the corporation's information system already captures the data needed for this report, what are some of the remaining design issues?

12. In the preceding scenario, what common tool could you use if your organization doesn't use CASE tools or dedicated report-writing tools for screen layout and prototyping? For an executive-level report, what are the most critical principles to apply in designing the output? (Remember, your future with the company may depend upon knowing and being able to apply these principles.)

13. You have volunteered to work on the Web site of your local library. The library plans to develop an online catalog of books that can be reserved by library patrons from their home computers via the Internet. Many of these patrons are senior citizens. What are some of the screen design issues that should be taken into consideration?

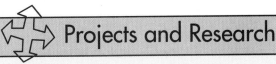

Projects and Research

1. In the 1990s (and even before), there was a great deal of discussion regarding the paperless office. Some industry pundits and futurists predicted that within a short period of time, paper would become a legacy product in many organizations. Yet today the reality seems quite different and, in fact, businesses are consuming and churning out more paper than ever. Do some research on the Web for both contemporary and past articles on this subject.

   a. Describe the articles that you found.
   b. Compare and contrast their viewpoints.
   c. Contact a large organization and a government agency in your area. Do either consider the paperless office to be an objective? If so, what are their plans for achieving it, and is progress being made?
   d. What about your own organization or school?
   e. What is your own position on this subject? Do you think a paperless office is a viable concept? Why or why not?

2. Designing a form or interface screen has been compared to watching an Olympic gymnast: It looks deceptively simple until you actually try to do it yourself. Consider the following questions:

   a. On the basis of your own experience, as well as your readings from this and other textbooks, what makes one form or interface screen "good" and another one "bad"?
   b. Pick a form or interface screen that you feel is particularly horrible. Describe why.
   c. Redesign the form or screen into one that you feel is "good."
   d. Have a couple of fellow students or associates compare and critique the "before" and "after" versions of the form or screen. How did they rate your "after" version compared to the "before" version?
   e. Can you have a well-designed form or screen if the data itself that is to be captured is not well designed? Why or why not?
   f. In today's global village, would a form or interface screen design that is considered good in one culture be considered universally good? How much of an influence do cultural differences have upon design?

3. Although it is probably a cliché to say we are living in a time of unprecedented technological change, it is difficult to truly comprehend the enormous changes that have taken place in a very short time and their impact upon us. To help get a sense of these changes, consider the following questions:

   a. Identify the different forms of output methods developed in the past 1,000 years, and draw them on a time line. How many output methods did you identify, and how many of them were commercially available in the last 50 years? In the last 25? In the last 10?
   b. What is the earliest version of a turnaround document that you can find?
   c. According to your research, when did microfilm become widely available? What was its impact upon private- and public-sector organizations?
   d. What about screen output? When did PC monitors come into widespread use? What has been their impact upon private- and public-sector organizations?
   e. Of all the output methods in use today, which one do you think had the most significant impact upon governments and cultures? Why?

4. Predictions of future technological advances and breakthroughs are notoriously unreliable. For example, in the 1960s some futurists predicted that we would all soon be commuting to work in our own personal aircraft or driving atomic-powered cars.

   a. In the 1980s, there was a good deal of speculation that holograms might soon become a common and revolutionary form of output. Research this topic—what did you find?
   b. Research recent articles describing new output methods that are still in the concept stage or on the (virtual) drawing board, but that industry pundits predict we may see in the not-too-distant future. Describe what you find.
   c. What do *you* predict will be the next breakthrough in output methods?
   d. If your prediction is correct, what is the potential impact upon what you do, as a systems analyst or designer, or how you do it?
   e. Do you think that most systems analysts and designers should pay attention to new technologies that are not yet commercially available even though many of them never pan out? Should they just sit back and wait to see what appears on the market?

5. Many organizations have implemented a company intranet. But it appears that relatively few (at least as of this date) have integrated their intranet with desktop productivity tools, such as Microsoft *Office,* e-mail and calendaring, and the specific data input/output applications used by employees.

   a. Contact several local private- and public-sector organizations. Do they currently have an intranet implemented?
   b. Describe how the intranets are being used and what features they have. Are any of these intranets integrated with desktop productivity tools and/or applications used by the organization?
   c. See if you can take a look at several of these intranets. Are their interface screens well designed? What, if any, are the differences between intranets and Internet applications that need to be taken into consideration when designing the screen interfaces for an intranet?
   d. What if you had the opportunity to design a fully integrated intranet for your organization? What features and functionality would you include?
   e. Create a prototype design for your intranet.

6. In today's global economy environment, information on demand is the expected norm. This is a very recent development that has had a profound impact on organization and individuals. Consider the following questions:

   a. In 1800, if a merchant company in Europe sent a turnaround report to its agent in New York, how long would the company expect that it would take—at a minimum—to receive the return report?
   b. What about in 1900?
   c. In 1950?
   d. In 2005—at a maximum?
   e. Describe what you believe to be the most significant impact this change in reporting speed has had both on organizations and on individuals.
   f. Is this extraordinary change in the speed of reporting—and in expectations—good or bad? For the organization? For an individual employee?

Minicases

1. Collect an example of a detailed report, a summary report, and an exception report. Submit them, along with a brief description of the information in them, to your professor. What were the similarities between the reports? The differences?
2. An online form can be set up to "dump" the form contents into an e-mail and send it to a specific e-mail account. Find the code snippets to do this, and create a simple online form that will send contents to your e-mail address. Fill out and submit your form at least one time. Forward the e-mail you have received with form contents to your professor, along with the URL of your form.
3. Information should only be inputted into an information system *one time*. After that, information should be shared digitally across departments, with no need for reentry of data. Why is this? What type of common data formatting problems do you think system designers run into when they set up online forms (such as the one you did in Question 2) that send data directly to a database?
4. Find an example of a really well designed output (could be a form, report, e-mail, etc.). Then find an example of a poorly designed one. Present both to your class. Lead a discussion on improving the poorly designed example, using specific attributes of excellence from the well-designed output. You will be graded on your ability to engage the class and work as a team member to improve the output medium.

Suggested Readings

Andres, C. *Great Web Architecture.* Foster City, CA: IDG Books Worldwide, 1999. Books on effective Web interface design are beginning to surface. The science of human engineering for Web interfaces has not yet progressed as far as client/server interfaces (e.g., *Windows*). Here is an early title that explores many dimensions of Web architecture and interfaces using real-world examples.

Application Development Strategies (monthly periodical). Arlington, VA: Cutter Information Corporation. This is our favorite theme-oriented periodical that follows system development strategies, methodologies, CASE, and other relevant trends. Each issue focuses on a single theme. This periodical will provide a good foundation for how to develop prototypes.

Galitz, W. O. *User-Interface Screen Design.* New York: John Wiley & Sons, 1993. This is our favorite book on overall user interface design.

Shelly, G., T. Cashman, and H. Rosenblatt. *Systems Analysis and Design,* 3rd ed. Cambridge, MA: Course Technology, 1998. We mention our competitors for their excellent coverage of tabular, printed output design. They afford many more pages of coverage and examples than we could in our latest edition.

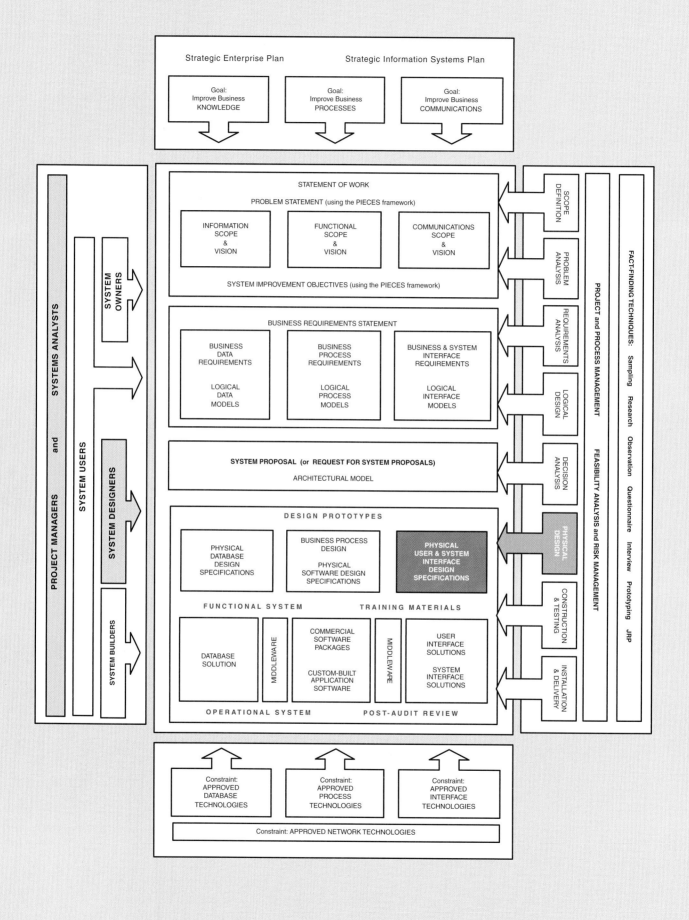

16

Input Design and Prototyping

Chapter Preview and Objectives

In this chapter you will learn how to design computer inputs. It is the second of three chapters that address the design of online systems using a graphical user interface for either client/server or Web-based systems. You will know how to design inputs when you can:

▎ Define the appropriate format and media for a computer input.

▎ Explain the difference between data capture, data entry, and data input.

▎ Identify and describe several automatic data collection technologies.

▎ Apply human factors to the design of computer inputs.

▎ Design internal controls for computer inputs.

▎ Select proper screen-based controls for input attributes that are to appear on a GUI input screen.

▎ Design a Web-based input interface.

Output and input design represent something of a "chicken or egg" sequencing problem. Which do you do first? In this edition, we present output design first. Classic system design prefers this approach as something of a system validation test—design outputs and then make sure the inputs are sufficient to produce the outputs. In practice, this sequencing of tasks becomes less important because modern systems analysis techniques sufficiently predefine logical input and output requirements. You and your instructor may safely swap Chapters 15 and 16 if you prefer.

Introduction

Bob Martinez has been given the assignment of prototyping the Web-based member order entry screen for the SoundStage Member Services system project (you can see his work in Figures 16-11 and 16-12). This was clearly the most fun part of the system project for Bob. A graphic design consulting firm had created the overall look and feel. But it was up to Bob to create the actual prototype.

He decided to use *Visual Studio .NET* just because he was comfortable with its GUI designer. Bob was able to pretty quickly put together the Web pages so that they looked as if they were a real shopping cart. Of course, it didn't have any real programming code. Whatever you searched for, you got *The Matrix* (Bob's all-time favorite movie), and the data grid shown at the bottom of Figure 16-11 had been filled in by hand, not by the database. But the layout included all requirements that had previously been identified. And all the links worked, though they all went to an "Under Construction" page.

Both SoundStage employees and selected members exercised the Web pages. Having gotten generally favorable feedback, Sandra is now having Bob refine and expand the prototype. Then it will go back to the users for more testing, and then more refinement, until they have a Web design that really can be programmed.

Input Design Concepts and Guidelines

"Garbage in! Garbage out!" This overworked expression is no less true today than it was when we first studied computer programming. Management and users make important decisions based on system outputs (Chapter 15). These outputs are produced from data that is either input or retrieved from databases. And any data in the databases must have been first input. In this chapter, you are going to learn how to design computer inputs. Input design serves an important goal—capture and get the data into a format suitable for the computer.

Today most inputs are designed by rapidly constructing prototypes. These prototypes may be simple computer-generated mock-ups, or they may be generated from prototype database structures such as those developed for Microsoft *Access*. These prototypes are rarely fully functional. They won't contain security features, data editing, or data updates that will be necessary in the final version of a system. Furthermore, in the interest of productivity, they may not include every button or control feature that would have to be included in a production system.

During requirements analysis, inputs were modeled as data flows that consist of data attributes. Even in the most thorough of requirements analysis, we will miss requirements. Input design may introduce new attributes or fields to the system. This is especially true if output design introduced new attributes to the outputs—the inputs must always be sufficient to produce the outputs!

We begin with a discussion of types of inputs. Inputs can be classified according to two characteristics: (1) how the data is initially captured, entered, and processed and (2) the method and technology used to capture and enter the data. Figure 16-1 illustrates this taxonomy. The characteristics are discussed briefly in the following sections.

> Data Capture, Data Entry, and Data Processing

When you think of "input," you usually think of input devices, such as keyboards and mice. But input begins long before the data arrives at the device. To actually input business data into a computer, the systems analyst may have to design source documents, input screens, and methods and procedures for getting the data into the computer (from customer to form to data entry clerk to computer).

FIGURE 16-1   An Input Taxonomy

Process Method	Data Capture	Data Entry	Data Processing
Keyboard	Data is usually captured on a business form that becomes the source document for input. Data can be collected real-time (over the phone).	Data is entered via keyboard. This is the most common input method but also the most prone to errors.	OLD: Data can be collected into batch files (disk) for processing as a batch. NEW: Data is processed as soon as it has been keyed
Mouse	Same as above.	Used in conjunction with keyboard to simplify data entry. Mouse serves as a pointing device for a screen. Can be with geographical user interfaces to reduce errors through point-and-click choices.	Same as above, but the use of a mouse is most commonly associated with online and real-time processing.
Touch screen	Same as above.	Data is entered on a touch screen display or handheld device. Data entry users either touch commands and data choices or enter data using handwriting recognition.	On PCs, touch screen choices are processed same as above. On handheld computers, data is stored on the handheld for later processing as a remote batch.
Point of sale	Data is captured as close to the point of sale (or transaction) as humanly possible. No source documents.	Data is often entered directly by the customer (e.g., ATM) or by an employee directly interacting with the customer (e.g., retail cash register). Input requires specialized, dedicated terminals that utilize some combination of the other techniques in this table.	Data is almost always processed immediately as a transaction or inquiry.
Sound	Data is captured as close to the source as possible, even when the customer is remotely located (e.g., at home or place of employment).	Data is entered using touch-tones (typically from a telephone). Usually requires fairly rigid command menu structure and limited input options.	Data is almost always processed immediately as a transaction or inquiry.
Speech	Same as sound.	Data (and commands) is spoken. This technology is not as mature and is much less reliable and common than other techniques.	Data is almost always processed immediately as a transaction or inquiry.
Optical mark	Data is recorded on optical scan sheets as marks or precisely formed letters, numbers, and punctuation. This is the oldest form of automatic data capture.	Eliminates the need for data entry. (Very commonly used in education for test scoring, course evaluations, and surveys.)	Data is almost always processed as a batch.
Magnetic ink	Data is usually prerecorded on forms that are subsequently completed by the customer. The customer records additional data on the form.	A magnetic ink reader reads the magnetized data. The customer-added data must be entered using another input method. This technique is used in applications requiring high accuracy and security, the most common of which is bank checks (for check number, account number, bank ID).	Data is almost always processed as a batch.
Electromagnetic	Data is recorded directly on the object to be described by data.	Data is transmitted by radio frequency.	Data is almost always processed immediately.
Smart card	Data is recorded directly on a device to be carried by the customer, employee, or other individual that is described by that data.	Data is read by smart card readers.	Data is almost always processed immediately.
Biometric	Unique human characteristics become data.	Data is read by biometric sensors. Primary applications are security and medical monitoring.	Data is processed immediately.

This brings us to our first fundamental question. What is the difference between data capture and data entry? Data happens! It accompanies business events called *transactions.* Examples include ORDERS, TIME CARDS, RESERVATIONS, and the like. We must determine when and how to capture the data when "it happens."

data capture the identification and acquisition of new data.

Data capture is the identification and acquisition of new data. *When* is easy! It's always best to capture the data as soon as possible after it originates. *How* is another story! Historically, special paper forms called *source documents* were used. **Source documents** are forms used to record business transactions in terms of data that describes those transactions.

source document a form used to record data about a transaction.

Display screens that can duplicate the appearance of almost any paper-based form are gradually replacing the paper forms. This trend is being accelerated by Web-based e-commerce and e-business. Still, business forms are commonly used as source documents for data entry. Design of source documents requires care. The layout and readability will affect the speed of data entry.

data entry the process of translating data into a computer-readable format.

Data entry is the process of translating the source data or document into a computer-readable format. Because data entry used to be 100 percent keyboard-based, businesses employed armies of data entry clerks. As online computing became more common, the responsibility for data entry shifted directly to system users. Today another transformation is occurring. Thanks to personal computers and the Internet, some data entry has shifted directly to the consumer. In all cases, data entry produces input for data processing.

Entered data must subsequently be processed—data processing. In this chapter, we are not concerned with how the data is transformed into outputs. But we are interested in the timing of input processing. When does the input data get processed?

Batch Processing   Batch processing used to be the dominant form of data processing. In **batch processing,** the entered data is collected into files called *batches.* Each file is processed as a batch of many transactions. Contrary to popular belief, some data is still processed in batches. Time cards are the classic example. Most batches are recorded as disk files (hence the term *key-to-disk*). Some older systems may still record batches on magnetic tape (*key-to-tape*).

batch processing a data processing method whereby data about many transactions is collected as a single file which is then processed.

Online Processing   Today most (but not all) information systems have been converted to online processing. In **online processing,** the captured data is processed immediately. Initially, data was entered at terminals. Today, that same data is captured on PCs and workstations to take advantage of their ability to perform some of the data validation and editing before it gets sent to the server computers. Because of PCs, we rarely hear the term *online processing* anymore. We usually hear the term *client/server,* where the PC is the client.

online processing a data processing method whereby data about a single transaction is processed immediately.

Most of today's applications present the user with a PC-based *graphical user interface (GUI).* Microsoft *Windows* is the dominant GUI in today's businesses. But the emergence of the Web as a platform for Internet and intranet applications may make a Web browser the most important user interface in the future. Microsoft *Internet Explorer* and Mozilla *Firefox* are the dominant browser interfaces in today's market. This chapter will address input design techniques for both the *Windows* client/server interface and the browser interface.

Remote Batch   Batch and online represent extremes on the processing spectrum. A combination solution also exists—the remote batch. In **remote batch processing,** data is entered using online editing techniques; however, the data is collected into a batch instead of being immediately processed. Later, the batch is processed.

remote batch processing a data processing method whereby data is entered online, collected as a batch, and processed at a later time.

Modern remote batch can take several forms. A simple example uses a PC-based front-end application to capture and store the data. The data can later be transmitted across a network for batch processing. A more contemporary example of remote batch processing uses disconnected laptop or handheld computers (or devices) to collect data for later processing. If you've recently received a package from UPS or Federal Express, you've seen such devices used by the drivers to record pickups and deliveries.

Now that we've covered the basic data capture, data entry, and data processing techniques, we can more closely examine the input methods shown as rows in Figure 16-1.

> Input Methods and Implementation

Different input devices, such as keyboards and mice, are covered in most introductory information systems courses. In this section, we are more interested in the method and its implementation than in the technology. In particular, we are interested in how the choice of a method affects data capture, entry, and processing as described in the previous section. You should continue to study Figure 16-1 as we introduce these methods.

Keyboard   Keyboard data entry remains the most common form of input. Unfortunately, it requires the most data editing because people make mistakes keying data from source documents. Fortunately, graphical user interfaces such as Microsoft *Windows* and Web browsers now make it possible to design online screens that reduce errors by forcing correct choices on the user. We will explore several useful GUI controls for such interfaces in the next section.

Mouse   A mouse is a pointing device used in conjunction with graphical user interfaces. The mouse has made it easy to navigate online forms and click on commands and input options. For example, the legitimate values for an attribute can be recorded on a screen as "clickable" boxes or buttons that eliminate the need to key in that data. This results in fewer data entry errors. We will explore mouse-based controls in our input designs for this chapter.

Touch Screen   An emerging technology that will greatly impact input design in the near future is the touch screen display. Such displays are common in handheld and palmtop computers that are finding their way into countless information system applications. A Symbol Technologies handheld computer based on the *Palm Operating System* is shown in the margin. Such devices simplify many data collection activities in a warehouse and on a manufacturing shop floor. Touch screen buttons can be programmed to collect the data. Most such devices support handwriting recognition as well. The Symbol Technologies unit depicted also can scan and read bar codes (discussed shortly).

Point of Sale   Point-of-sale (POS) terminals have been with us for some time. They have all but replaced old-fashioned cash registers. These terminals capture data at the point of sale and provide time-saving ways to enter data, perform transactional calculations, and produce some output. Like the handhelds just described, most can scan and read bar codes to eliminate keying errors. Automatic teller machines (ATMs), another form of POS terminal, are operated directly by the consumer.

A Handheld
Computer

Sound and Speech   Sound represents another form of input. You might have used a touch-tone telephone-based system to register for this course. Such tone-based systems require special input/output technology that drives the design. Those systems are beyond the scope of this book.

A more sophisticated form of this input method uses voice recognition technology to make it possible to input data. Currently this technology is relatively immature and unreliable. It is best utilized to input commands, not data. But the time may come when voice recognition technology replaces the keyboard as the principal means by which we enter data.

The remaining input methods are broadly classified as *automatic data capture (ADC)*. With advancements in today's input technology, we can eliminate much (and sometimes all) human intervention associated with the input methods discussed in the previous section. By eliminating human intervention we can decrease the time delay and errors associated with human interaction.

Optical Mark   *Optical mark recognition (OMR)* technology for input has existed for several decades. It is primarily batch processing–oriented. The classic example is the optical mark forms used for objective-based questions (e.g., multiple choice) on examinations. The technology is also useful in surveys and questionnaires or any other application where the number of possible data values is relatively limited and highly structured. Most applications that could benefit from this input method have probably already exploited it.

Optical character recognition (OCR) is less prevalent despite its maturity. It requires that the user or customer carefully handwrite input data on a business form. If the letters and numbers are properly scribed, an OCR reader can process the forms without human intervention. Obviously, this depends on the handwriting of the user or customer. But it does work. Columbia House Record Club used to use an OCR form for customer responses to orders. Like most OCR applications, the number of fields to be input was very small (reducing the possibility of errors). Processing methods must be implemented for any inputs rejected due to illegibility.

Today the most prevalent form of optical technology involves bar coding. *Bar codes* are on almost every product we buy, but bar-coding technology is not limited to retail sales. You can create bar codes for almost any business application. You can even integrate bar codes into *Windows*-based applications, as shown in Figure 16-2.

Magnetic Ink   Magnetic ink ADC technology is one you will likely recognize. It usually involves using magnetic stripe cards, but it also may include the use of magnetic ink character recognition (MICR). Over 1 billion magnetic stripe cards are in use today! They have found their way into a number of business applications, such as credit card transactions, building security access control, and employee attendance tracking. MICR is most widely used in the banking industry.

FIGURE 16-2

Bar Codes in a *Windows* Application

Electromagnetic Transmission   Electromagnetic ADC technology is based on the use of radio frequency to identify physical objects. This technology involves attaching a tag and antenna to the physical object that is to be tracked. The tag contains memory that is used to identify the object being tracked. The tag can be read by a reader whenever the object resides within the electromagnetic field generated by the reader. This identification technology is becoming very popular in applications that involve tracking physical objects that are out of sight and on the move. For example, electromagnetic ADC is being used for public transportation tracking and control, tracking manufactured products, and tracking animals, to name a few.

Smart Cards   Smart card technology has the ability to store a massive amount of information. Smart cards are similar to, but slightly thicker than, credit cards. They also differ in that they contain a microprocessor, memory circuits, and a battery. Think of it as a credit card with a computer on board. They represent a portable storage medium from which input data can be obtained. While this technology is only beginning to make inroads in the United States, smart cards are used on a daily basis by over 60 percent of the French population. Smart card applications are particularly promising in the area of health records, where a person's blood type, vaccinations, and other past medical history can be made readily available. Other uses may include such applications as passports, financial information for point-of-sale transactions, and pay television, to name a few. Another future application could be a combination debit card that automatically maintains and displays your account balance. A smart card used in a security application is shown in the margin.

A Smart Card

Biometric   Biometric ADC technology is based on unique human characteristics or traits. For example, an individual can be identified by his or her unique fingerprint, voice pattern, or pattern of certain veins (retina or wrist). Biometric ADC systems consist of sensors that capture an individual's characteristic or trait, digitize the image pattern, and then compare the image to stored patterns for identification. Biometric ADC is popular because it offers the most accurate and reliable means for identification. This technology is particularly popular for systems that require security access.

> System User Issues for Input Design

Because inputs originate with system users, human factors play a significant role in input design. Inputs should be as simple as possible and be designed to reduce the possibility of incorrect data being entered. The needs of system users *must* be considered. With this in mind, several human factors should be evaluated.

The volume of data to be input should be minimized. The more data that is input, the greater the potential number of input errors and the longer it takes to input that data. Thus, numerous considerations should be given to the data that is captured for input. These general principles should be followed for input design:

- *Capture only variable data.* Do not enter constant data. For instance, when deciding what elements to include in a SALES ORDER input, we need PART NUMBERS for all parts ordered. However, do we need to input PART DESCRIPTIONS for those parts? PART DESCRIPTION is probably stored in a database table. If we input PART NUMBER, we can look up PART DESCRIPTION. Permanent (or semipermanent) data should be stored in the database. Of course, inputs must be designed for maintaining those database tables.
- *Do not capture data that can be calculated or stored in computer programs.* For example, if you input QUANTITY ORDERED and PRICE, you don't need to input EXTENDED PRICE, which is equal to QUANTITY ORDERED × PRICE. Another example is incorporating FEDERAL TAX WITHHOLDING data in tables (arrays) instead of keying in that data every time.
- *Use codes for appropriate attributes.* Codes were introduced earlier. Codes can be translated in computer programs by using tables.

If source documents are used to capture data, they should be easy for system users to complete and subsequently enter into the system. The following suggestions may help:

- *Include instructions for completing the form.* Remember that people don't like to have to read instructions printed on the back side of a form.
- *Minimize the amount of handwriting.* Many people suffer from poor penmanship. The data entry clerk or CRT operator may misread the data and input incorrect data. Use check boxes wherever possible so that the system user only needs to check the appropriate values.
- *Data to be entered (keyed) should be sequenced so that it can be read like this book, top to bottom and left to right.* Figure 16-3(a) demonstrates a good flow. The system user should not have to move from right to left on a line or jump around on the form, as shown in Figure 16-3(b), to enter data.

FIGURE 16-3

Good and Bad Flow in a Form

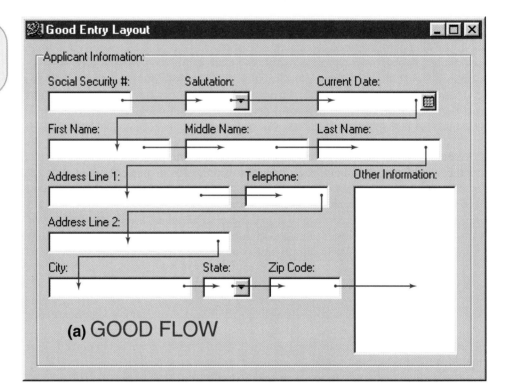

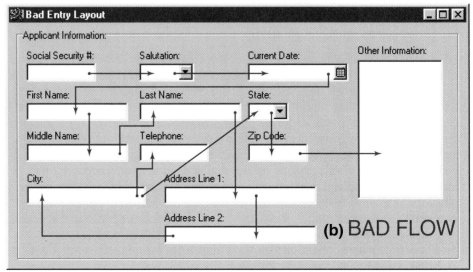

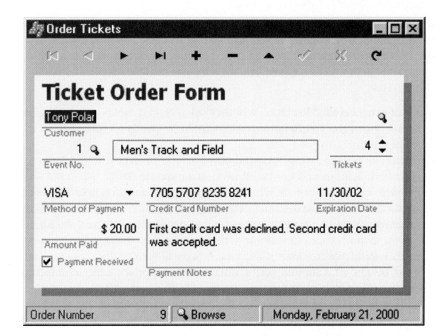

FIGURE 16-4

Metaphoric Screen
Design

- *When possible, use designs based on known metaphors.* The classic example of this is the personal finance application *Quicken.* The program's ease of use is greatly enhanced by its on-screen re-creation of the checkbook metaphor. The user writes checks by filling in a graphical representation of the check. And the check register looks exactly like its paper equivalent. Not all inputs lend themselves to metaphors, but some are greatly enhanced by the imitation (see Figure 16-4).
- There are several other guidelines and issues specific to data input for GUI screen designs. We'll introduce these guidelines, as appropriate, when we discuss GUI controls for input design later in this chapter, as well as in the chapters on output design and user interface design.

> Internal Controls—Data Editing for Inputs

Internal controls are a requirement in all computer-based systems. Internal input controls ensure that the data input to the computer is accurate and that the system is protected against accidental and intentional errors and abuse, including fraud. The following internal control guidelines are offered:

1. The number of inputs should be monitored. This is especially true with the batch method, because source documents may be misplaced, lost, or skipped.

   — In batch systems, data about each batch should be recorded on a batch control slip. Data includes BATCH NUMBER, NUMBER OF DOCUMENTS, and CONTROL TOTALS (e.g., total number of line items on the documents). These totals can be compared with the output totals on a report after processing has been completed. If the totals are not equal, the cause of the discrepancy must be determined.

   — In batch systems, an alternative control would be one-for-one checks. Each source document would be matched against the corresponding historical report detail line that confirms the document has been processed. This control check may be necessary only when the batch control totals don't match.

   — In online systems, each input transaction should be logged to a separate audit file so that it can be recovered and reprocessed if there is a processing error or if data is lost.

2. Care must also be taken to ensure that the data is valid. Two types of errors can infiltrate the data: data entry errors and invalid data recorded by system users. Data entry errors include copying errors, transpositions (typing 132 as 123), and slides (keying 345.36 as 3453.6). The following techniques are widely used to validate data:

— *Existence checks* determine whether all <u>required</u> fields on the input have actually been entered. Required fields should be clearly identified as such on the input screen.
— *Data-type checks* ensure that the correct type of data is input. For example, alphabetic data should not be allowed in a numeric field.
— *Domain checks* determine whether the input data for each field falls within the legitimate set or range of values defined for that field. For instance, an upper-limit range may be put on PAY RATE to ensure that no employee is paid at a higher rate.
— *Combination checks* determine whether a known relationship between two fields is valid. For instance, if the VEHICLE MAKE is Pontiac, then the VEHICLE MODEL must be one of a limited set of values that comprises cars manufactured by Pontiac (Firebird, Grand Prix, and Bonneville, to name a few).
— *Self-checking digits* determine data entry errors on primary keys. A *check digit* is a number or character that is appended to a primary key field. The check digit is calculated by applying a formula, such as Modulus 11, to the actual key. The check digit verifies correct data entry in one of two ways. Some data entry devices can automatically validate data by applying the same formula to the data as it is entered by the system user. If the check digit entered doesn't match the check digit calculated, an error is displayed. Alternatively, computer programs can also validate check digits by using readily available subroutines.
— *Format checks* compare data entered against the known formatting requirements for that data. For instance, some fields may require leading zeros, while others don't. Some fields use standard punctuation (e.g., Social Security numbers or phone numbers). A value "A4898 DH" might pass a format check, while a similar value "A489 ID8" would not.

In Chapter 14, you learned that most database management systems perform data validation checks similar to those described in the above list. So why do we need input controls? Simple! Most applications today are networked. Erroneous data is both a network traffic bottleneck and a detractor for transaction throughput and response time. It is always best to capture and correct input errors as close as possible to the source—hence the emphasis on input controls and validation.

GUI Controls for Input Design

As mentioned earlier, most new applications being developed today include a graphical user interface (GUI). Most are based on Microsoft *Windows,* but the pervasive adoption of the Internet, combined with Web-based e-commerce, is quickly driving some interfaces to the Web browser. While GUI designs provide a more user-friendly interface, they also present more complex design issues than their predecessors. This chapter will not attempt to address all the GUI design issues; entire books have been written on the subject. Several of our favorites are listed in the Suggested Readings.

Rather, this chapter will focus on selecting the proper screen-based controls for entering data on a GUI screen. Think of controls as "widgets" for building a user interface. They are included in most contemporary application development environments such as Microsoft's *Access* and *Visual Studio .NET,* Sybase's *PowerBuilder,* InPrise's *JBuilder,* Symantec's *Visual Café,* IBM's *Visual Age,* and many others. Many of these tools share controls (and code) via the repository. This approach is called *repository-based programming.*

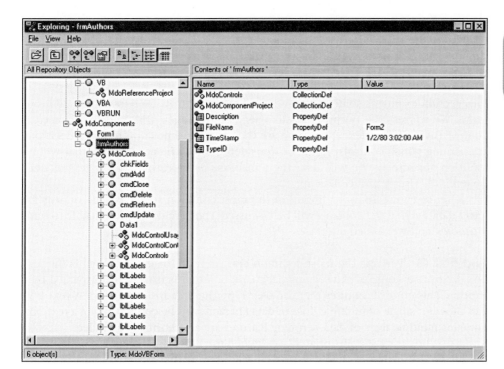

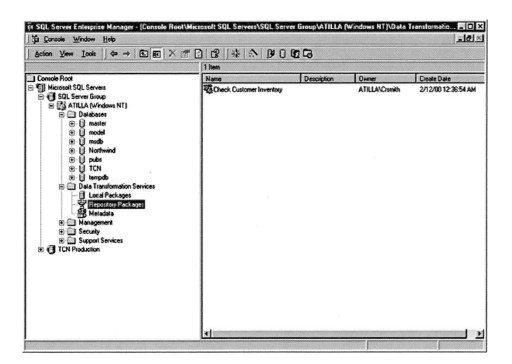

Figure 16-5 illustrates access to a repository that contains input controls and code. The approach is based on the object-oriented and component-based programming techniques that have become pervasive in application development. This figure depicts controls that could be used by various systems analysts or programmers to prototype an interface. The developers can, in a single location, define most of the properties and constraints for a reusable field and the data validation code for that field. Once defined, the object or control can be used by any number of other systems analysts and programmers in the organization. This repository-based approach guarantees that every instance of the field will be used in a consistent manner. Furthermore, the repository entries can be changed if business rules dictate, and no additional changes to the applications will be required.

> Common GUI Controls for Inputs

This section examines some of the most common controls used in GUI-based input forms. We address the purpose, advantages, disadvantages, and guidelines for each control. Given this understanding, we are then in a good position to make decisions concerning which controls should be considered for each data attribute that will be input on our screens. We will defer the transitions between our screen designs until Chapter 17, "User Interface Design."

Refer to Figure 16-6 as a library of the most common screen-based controls for input data. Each of the controls will be discussed. They are equally applicable to both *Windows*- and Web-based interfaces.

Text Box ❶   Perhaps the most common control used for input of data is the text box. A *text box* consists of a rectangular-shape box that is usually accompanied by a caption. This control requires that the user type the data inside the box. A text box can allow for single or multiple lines of data characters to be entered. When a text box contains multiple lines of data, scrolling features are also normally included.

A text box is most appropriately used when the input data values are unlimited in scope and the analyst is unable to provide the users with a meaningful list of values from which they can select. For example, a single-line text box would be an appropriate control for capturing a new customer's last name because the possibilities for the customer's last name are virtually impossible to predetermine. A text box would also be appropriate for capturing data about shipping instructions that describe a particular order placed by a customer. Once again, the possible values for shipping instructions are virtually unlimited. In addition, the multiple-line text box would be appropriate due to the unpredictable length of the shipping instructions. In cases where the text box is not large enough to view the entire input data values, the text box may use scrolling and word-wrap features.

FIGURE 16-6

Common GUI
Input Controls

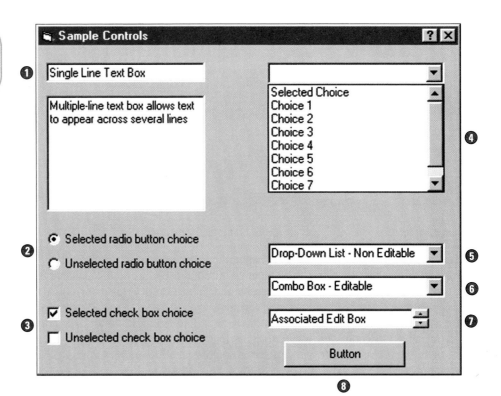

Numerous guidelines should be followed when using a text box on an input screen. A text box should be accompanied by a descriptive, meaningful caption. Avoid using abbreviations for captions. Only the first character of the caption's text should be capitalized.

The location of the caption is also significant. The user should be able to clearly associate the caption with the text box. Therefore, the caption should be located to the left of the actual text box or left-aligned immediately above the text box. Finally, it is also generally accepted that the caption be followed by a colon to help the user visually distinguish the caption from the box.

Generally, the size of the text box should be large enough for all characters of fixed-length input data to be entered and viewed by the user. When the length of the data to be input is variable and could become quite long, the text box's scrolling and word-wrapping features should be applied.

Radio Button ❷   Radio buttons provide the user with an easy way to quickly identify and select a particular value from a value set. A *radio button* consists of a small circle and an associated textual description that correspond to the value choice. The circle is located to the left of the textual description of the value choice. Radio buttons normally appear in groups—one radio button per value choice. When a user selects the appropriate choice from the value set, the circle corresponding to that choice is partially filled to indicate it has been selected. When a choice is selected, any default or previously selected choice's circle is deselected. Radio buttons also give the user the flexibility of selecting via the keyboard or mouse.

Radio buttons are most appropriate when a user may be expected to input data that has a limited predefined set of mutually exclusive values. For example, a user may be asked to input an ORDER TYPE and GENDER. Each of these has a limited, predefined, mutually exclusive set of valid values. For example, when the users are to input an ORDER TYPE, they might be expected to indicate one and only one value from the value set "regular order," "rush order," or "standing order." For GENDER, the user would be expected to indicate one and only one value from the set "female," "male," or "unknown."

There are several guidelines to consider when using radio buttons as a means for data input. First, radio buttons should present the alternatives vertically aligned and left-justified to aid the user in browsing. If necessary, the choices can be presented where they are aligned horizontally, but adequate spacing should be used to help visually distinguish the choices. Also, the group of choices should be visually grouped to set them off from other input controls appearing on the screen. The grouping should also contain an appropriate meaningful caption. For example, radio buttons for male, female, and unknown might be vertically aligned and left-justified with the heading/caption "Gender" left-justified above the set.

The sequencing of the choices should also be given consideration. The larger the number of choices, the more thought should be given to the ease of scanning and identifying the choices. For example, in some cases it may be more natural for the user to locate choices that are presented in alphabetical order. In other cases, the frequency in which a value is selected may be important in regard to where it is located in the set of choices.

Finally, it is not recommended that radio buttons be used to select the value for an input data whose value is simply a yes/no (or on/off state). Instead, a check box control should be considered.

Check Box ❸   As with text boxes and radio buttons, a *check box* also consists of two parts. It consists of a square box followed by a textual description of the input field for which the user is to provide the yes/no value. Check boxes provide the user with the flexibility of selecting the value via the keyboard or mouse. An input data field whose value is yes is represented by a square that is filled with a "✓." The absence of a "✓" means the input field's value is no. The user simply toggles the input field's value from one value/state to the other as desired.

Often a user needs to input a data field whose value set consists of a simple yes or no value. For example, a user may be asked for a yes/no value for such items as the following input data: CREDIT APPROVED? SENIOR CITIZEN? HAVE YOU EVER BEEN CONVICTED OF FRAUD? and MAY WE CONTACT YOUR PREVIOUS EMPLOYER? In each situation a check box control could be used. A check box control offers a visual and intuitive means for the user to input such data.

The previous example represented a simplified scenario for the use of a stand-alone check box. On a single input screen it may be desirable to ask a user to enter values for a number of related input fields having a yes/no value. For example, a receptionist at a health clinic may be entering data from a completed patient form. On a section of that form, the patient may have been asked about a number of illnesses. The patient may have been asked about his or her past medical history and instructed to "check all that apply" from a list of types of various illnesses. If properly designed, the receptionist's input screen would represent each illness as a separate input field using a check box control. The controls would be physically associated into a group on the screen. The group would also be given an appropriate heading/caption. Recognize that even though the check boxes may be visually grouped on the screen, each check box operates as a separate independent input field.

Following these recommended guidelines will improve the use of check box controls. Make sure the textual description is meaningful to the user. Look for opportunities to group check boxes for related yes/no input fields and provide a descriptive group heading.

To aid in the user's browsing and selecting from a group of check boxes, arrange the group of check box controls so that they are aligned vertically and left-justified. If necessary, align horizontally and be sure to leave adequate space to visually separate the controls from one another. Finally, provide further assistance to the user by appropriately sequencing the input fields according to their textual description. In most cases, where the number of check box controls is large, the sequencing should be alphabetical. In cases where the text description represents dollar ranges or some other measurement, the sequencing may be according to numerical order. In other cases, such as those where a very limited number of controls are grouped, the basis for sequencing may be according to the frequency that a given input data field's yes/no value is selected. (All input data fields represented using a check box have a default value—either checked or unchecked.)

List Box ❹   A list box is a control that requires that the user select a data item's value from a list of possible choices. The *list box* is rectangular and contains one or more rows of possible data values. The values may appear as either a textual description or a graphical representation. List boxes having a large number of possible values may include scroll bars for navigating through the list of choices.

It is also common for a list box's row to contain more than one column. For example, a list box could simply contain rows having a single column of permissible values for an input data item called JOB CODE. However, it may be asking too much to expect the user to recognize what each job code actually represents. In this case, to place the values of JOB CODE into a meaningful perspective, the list box could include a second column containing the corresponding JOB TITLE for each job code.

How does one choose between a radio button and a list box control? Both controls are useful in ensuring that the user enters the correct value for a data item. Both are also appropriate when it is desirable to have the value choices constantly visible to the user.

The decision is normally driven by the number of possible values for the data item and the amount of screen space available for the control. Scrolling capabilities make list boxes appropriate for use in cases where there is limited screen space available and the input data item has a large number of predefined, mutually exclusive values from which to choose.

There are several guidelines to consider when using a list box as a means for data input. A list box should be accompanied by a descriptive caption. Avoid using abbreviations for captions, and capitalize only the first character of the caption's text. It is also generally accepted that the caption be followed by a colon to help the user visually distinguish the caption from the box.

The location of the caption is also significant. The user should be able to clearly associate the caption with the list box. Therefore, the caption should appear left-justified immediately above the actual list box.

There are also several guidelines relating to the list box. First, it is recommended that a list box contain a highlighted default value. Second, consider the size of the list box. Generally, the width of the list box should be large enough for most characters of fixed-length input data to be entered and viewed by the user. The length of the box should allow for at least three choices and be limited in size to containing about seven choices. In both cases, scrolling features should be used to suggest that additional choices are available to the user.

If graphical representations are used for value choices, make sure the graphics are meaningful and truly representative of the choice. If textual descriptions are used, use mixed-case letters and ensure that the descriptions are meaningful. It is important that these decisions or judgments be based on the perspective and opinions of the user!

You should also give careful thought to the ease with which a user can scan and identify the choices appearing in the list box. The list of choices should be left-justified to aid in browsing. Be sure to involve the user when addressing the order in which choices will appear in the list. In some cases, it may be natural to the user if the list of choices appears in alphabetical order. In other cases, the frequency in which a value is selected may be important in regard to where it is located in the list.

Drop-Down List ❺   A drop-down list is another control that requires the user to select a data item's value from a list of possible choices. A *drop-down list* consists of a rectangular selection field with a small button connected to its side. The small button contains the image of a downward-pointing arrow and bar. This button is intended to suggest to the user the existence of a hidden list of possible values for a data item.

When requested, the hidden list appears to "drop or pull down" beneath the selection field to reveal itself to the user. The revealed list has characteristics similar to the list box control mentioned in the previous section. When the user selects a value from the list of choices, the selected value is displayed in the selection field and the list of choices once again becomes hidden from the user.

A drop-down list should be used in cases where the data item has a large number of predefined values and screen space availability prohibits the use of a list box. One disadvantage of a drop-down list is that it requires extra steps by the user, in comparison to the previously mentioned controls.

Many of the guidelines for using list boxes directly apply to drop-down lists. One exception is the placement of the caption. The caption for a drop-down list is generally either left-aligned immediately above the selection field portion of the control or located to the left of the control.

Combination Box ❻   A *combination box,* often simply called a combo box, is a control whose name reflects the fact that it combines the capabilities of a text box and list box. A combo box gives the user the flexibility of entering a data item's value (as with a text box) or selecting its value from a list (as with a list box).

At first glance, a combo box closely resembles a drop-down list control. Unlike the drop-down list control, however, the rectangular box can serve as an entry field for the user to directly enter a data item's value. Once the small button is selected, a hidden list is revealed. The revealed list appears slightly indented beneath the rectangular entry field.

When the user selects a value from the list of choices, the selected value is displayed in the entry field and the list of choices once again becomes hidden from the user.

A combo box is most appropriately used where screen space is limited and it is desirable to provide the user with the option of selecting a value from a list or typing a value that may or may not appear as an option in the list.

The same guidelines for using drop-down lists directly apply to combo boxes.

Spin Box ❼ A *spin box* is a screen-based control that consists of a single-line text box followed immediately by two small buttons. The two buttons are vertically aligned. The top button has an arrow pointing upward, and the bottom button has an arrow pointing down. This control allows the user to enter data directly into the associated text box or to select a value by clicking on the buttons to scroll (or "spin") through a list of values. The buttons have a unit of measure associated with them. When the user clicks on one of the arrow buttons, a value will appear in the text box. The value in the text box is manipulated by clicking on the arrow buttons. The upward pointing button will increase the value in the text box by a unit of measure, whereas the downward pointing button will decrease the value in the text box by the same unit of measure.

A spin box is most appropriately used to allow the user to make an input selection by using the buttons to navigate through a small set of meaningful choices or by directly keying the data value into the text box. The data values for a spin box should be capable of being sequenced in a predictable manner.

Spin boxes should contain a label or caption that clearly identifies the input data item. This label should be located to the left of the text box or left-aligned immediately above the text box portion of the control. Finally, spin boxes should always contain a default value in the text box portion of the control.

Buttons ❽ Strictly speaking, buttons are not input controls. They do not contribute to the selection of or input of actual data. Nonetheless, input form design is incomplete without them. *Buttons* serve several purposes. They allow a user to commit all of the data to be processed, or cancel a transaction, or get help. They can be used to navigate between instances of the same form.

Many more screen-based controls are available for designing graphical user interfaces. The above are the most common controls for capturing input data. There are others, and you should become familiar with them and their proper usage for inputting data. In later chapters you will be exposed to several other controls used for other purposes. Keep on top of developments in the area of GUI as new controls are sure to be made available.

> Advanced Input Controls

Figures 16-7(a) and (b) illustrate additional controls for data input. These advanced controls can be used in *Windows* interfaces to create a more sophisticated look and feel. Equivalent controls are likely available for Web-based applications, but most Web-based e-commerce applications aspire to simpler formats. We will not discuss these controls in detail, but they are summarized as follows:

- *Drop-down calendar.* A field is illustrated in Figure 16-7(a). Clicking the down arrow next to the date creates the pop-up calendar shown in Figure 16-7(b). The familiar calendar is another example of metaphoric design.
- *Slider edit calendar.* This is a nonnumeric means of selecting a value.
- *Masked edit control.* This control builds the format checks described earlier right into the field.
- *Ellipsis control.* Clicking on the three dots causes a pop-up dialogue to appear for data entry. It might be used for a field that consists of several parts (such as an address—street, city, state, and zip code.

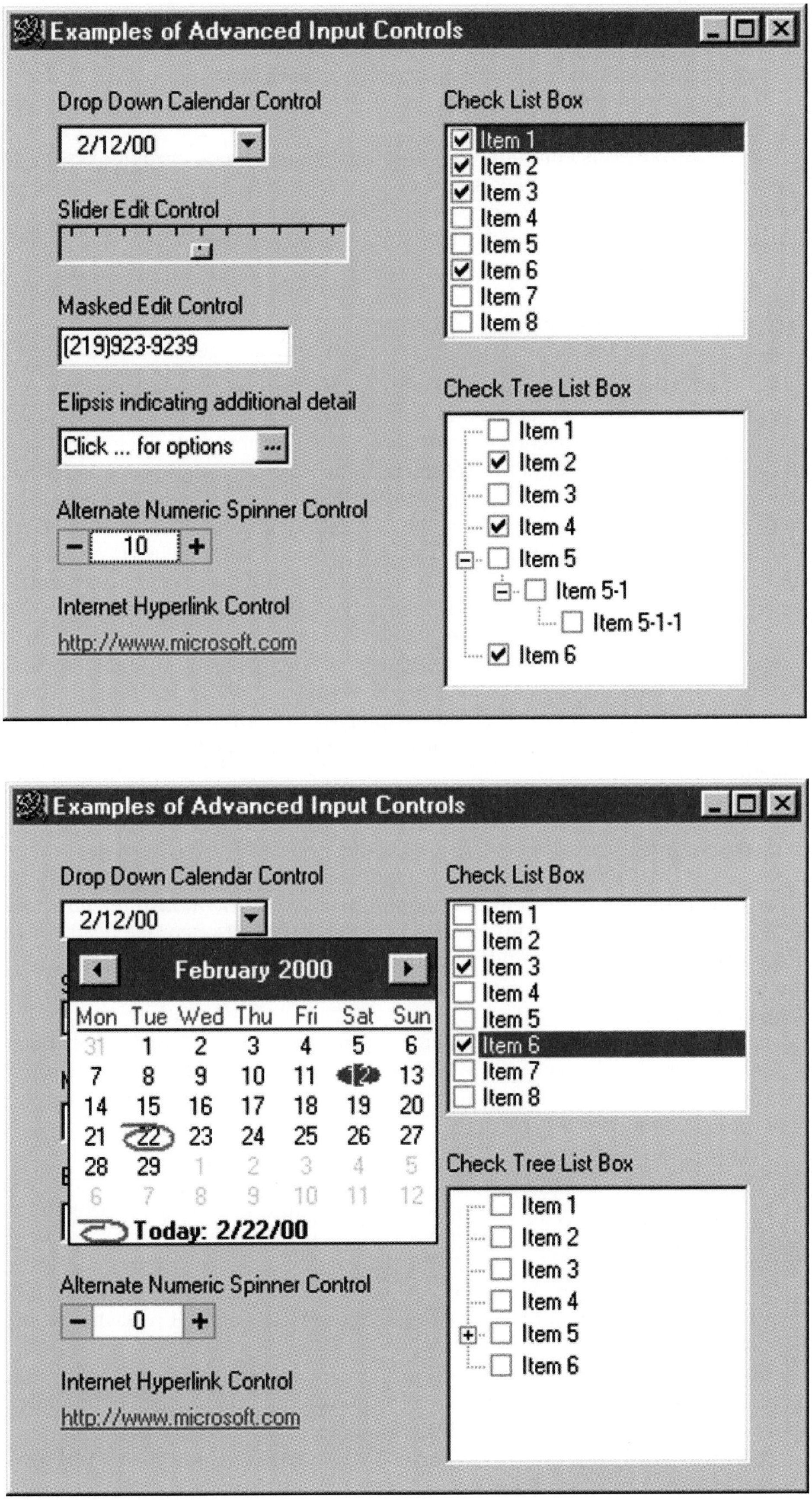

FIGURE 16-7

Advanced GUI Input Controls

- *Alternate numeric spinner.* This is a different type of input spinner.
- *Internet hyperlink.* Similar in function to a button, a hyperlink can be linked to Web pages, but it can also be linked to other *Windows* forms. This is an effective way to hide related input forms that do not apply to all or most users.
- *Check list box.* This control is useful for combining several check boxes in situations where several boxes may be applicable.
- *Check tree list box.* This control is useful for presenting data options that need to be hierarchically organized into a treelike structure.

How to Design and Prototype Inputs

How do you design online inputs? Traditionally, designers were concerned with the overall content, appearance, and functionality of the input screen—in relative isolation of other screens that needed to be designed. The designers knew they would simply design a subsequent set of menu screens from which the users would select an option that would lead them to the appropriate input screen. Simple enough. However, given today's graphical environments, there is an emphasis on developing an overall system that blends well into the user's workplace environment. This emphasis rarely results in a hierarchical, menu-driven application interface that characterized the more traditional text- or command-based applications of old.

The following sections will demonstrate how the first stage of input design is completed. We will draw on examples from our SoundStage case study. We will examine both client/server, *Windows*-based inputs and Web-based, e-commerce inputs that run in a browser. Later, in Chapter 17, we will integrate the outputs (from Chapter 15) and inputs from this chapter into an overall user interface and dialogue.

> Automated Tools for Input Design and Prototyping

In the recent past, the primary tools for input design were *record layout charts* and *display layout charts.* Today, this "sketching" approach is not often practiced. It is a tedious process that is not conducive to today's preferred prototyping and rapid application development strategies, which use automated tools to accelerate the design process.

Before the availability of automated tools, analysts could sketch only rough drafts of inputs to get a feel for how system users wanted outputs to look or how the batch records would be structured. With automated tools, we can develop more realistic prototypes of these inputs.

Arguably, the most commonly used automated tool for input design is the PC-database application development environment. While Microsoft *Access* is not powerful enough to develop most enterprise-level applications, you may be surprised at how many designers use *Access* to prototype such applications. Given a database structure (easily specified in *Access*), you can quickly generate or create forms for inputting data. You can include most of the GUI controls we described in this chapter. The users can subsequently exercise those forms and tell you what works and what doesn't.

Many CASE tools include facilities for report and screen layout and prototyping using the project repository created during requirements analysis. *System Architect's* screen design facility was previously demonstrated in Chapter 15, Figure 15-7.

Most GUI-based programming languages, such as *Visual Basic,* can be easily used to construct nonfunctional prototypes of inputs. The key term here is *nonfunctional.* The forms will look real, but there will be no code for implementing any of the buttons or fields. That is the essence of <u>rapid</u> prototyping.

> The Input Design Process

Input design is not a complicated process. Some steps are essential, and others are dictated by circumstances. The steps are:

1. Identify system inputs and review logical requirements.
2. Select appropriate GUI controls.
3. Design, validate, and test inputs using <u>some combination of</u>:
   a. Layout tools (e.g., hand sketches, printer/display layout charts, or CASE).
   b. Prototyping tools (e.g., spreadsheet, PC DBMS, 4GL).
4. If necessary, design the source document.

In the following subsections, we examine these steps and illustrate a few examples from the SoundStage project.

Step 1: Identify System Inputs and Review Logical Requirements Input requirements should have been defined during requirements analysis. Physical data flow diagrams (or design units; both described in Chapter 13) are a good starting point for input design. Those DFDs identify both the net outputs of the system (external agent to process) and the implementation method.

 Your system development methodology and standards will determine whether each of these net input data flows may also be described as a logical data flow in a data dictionary or repository (see Chapter 9). The data structure for a data flow specifies the attributes or fields to be included in the output. If those requirements are specified in the relational algebraic notation, you can quickly determine which fields repeat, which fields have optional values, and so on. Consider the following data structure:

Data Structure Defining Logical Requirements	Comment
ORDER = <u>ORDER NUMBER</u>	← Unique identifier of the output.
+ ORDER DATE	← One of many fields that must take on a value. Lack of parentheses indicates a value is required.
+ CUSTOMER NUMBER	
+ CUSTOMER NAME	
+ CUSTOMER SHIP ADDRESS = ADDRESS >	← Pointer to a related definition.
+ (CUSTOMER BILLING ADDRESS = ADDRESS >)	
+ 1 {PRODUCT NUMBER + QUANTITY ORDERED} N	← A group of fields that repeats 1 − n times. Parentheses indicate optional value.
+ (DEFAULT CREDIT CARD NUMBER)	← An optional field, meaning one that does not have to have a value.

In the absence of such precise requirements, there may exist discovery prototypes that were created during requirements analysis. In either case, a good requirements statement should be available in some format.

 Input requirements specified during requirements analysis for the SoundStage case study were reviewed, and it was determined that three inputs pertained to the subject VIDEOTAPE. It was also determined that a single input screen could be used to support the three inputs—NEW VIDEO TITLE, DISCONTINUED VIDEO TITLE, and VIDEO TITLE UPDATE. The data content for the three inputs should capture or display the following data:

PRODUCT NUMBER +
UNIVERSAL PRODUCT CODE +
QUANTITY IN STOCK +
PRODUCT TYPE +
MANUFACTURER'S SUGGESTED RETAIL UNIT PRICE +
CLUB DEFAULT UNIT PRICE +
CURRENT SPECIAL UNIT PRICE +

> CURRENT MONTH UNITS SOLD +
> CURRENT YEAR UNITS SOLD +
> TOTAL LIFETIME UNITS SOLD +
> TITLE OF WORK +
> CATALOG DESCRIPTION +
> COPYRIGHT DATE +
> CREDIT VALUE +
> PRODUCER +
> DIRECTOR +
> VIDEO CATEGORY

The attributes PRODUCT NUMBER, MONTHLY UNIT SALES, YEAR UNIT SALES, and TOTAL UNIT SALES are not to be entered by the user. Rather, these attributes are to be automatically generated by the system. Also, for the TITLE COVER, the user will be expected to simply specify a bitmap file that will contain an actual image of the new video title.

Step 2: Select Appropriate GUI Controls   Now that we have an idea of the content for our input, we can address the proper screen-based control to use for each attribute to appear on our screen. Using the repository-based programming approach, we would first check to see if such decisions and other attribute characteristics have already been made and recorded as repository entries. If so, we would simply reuse the repository entries that correspond to the attributes we will use on our input screens. In cases where there is no repository entry, we will have to simply create them.

To choose the correct control for our attributes, we must begin by examining the possible values for each attribute. Here are some preliminary decisions regarding our input attributes identified in the previous step:

- PRODUCT NUMBER, CURRENT MONTH UNITS SOLD, CURRENT YEAR UNITS SOLD, TOTAL LIFETIME UNITS SOLD, UNIVERSAL PRODUCT CODE, MANUFACTURER'S SUGGESTED RETAIL UNIT PRICE, CLUB DEFAULT UNIT PRICE, CURRENT SPECIAL UNIT PRICE, PRODUCER, and DIRECTOR attributes all have input data values that are unlimited in scope or noneditable. Since the designer is unable to provide the user with a meaningful list of values from which to choose, a single-line text box was chosen. Since the attribute CATALOG DESCRIPTION also fits this criteria, a multiple-line text box (referred to as a "memo box" by some products) was selected.
- PRODUCT TYPE, LANGUAGE, VIDEO ENCODING, SCREEN ASPECT, and VIDEO MEDIA TYPE all contain a limited predefined set of values. Therefore, it was determined that radio buttons would be the preferred screen-based control for these input items.
- It was determined that CLOSED CAPTION? is an input attribute that contains a yes/no value. Therefore, a check box was selected as the control for this attribute.
- QUANTITY IN STOCK, RUNNING TIME, COPYRIGHT DATE, and CREDIT VALUE contain data values that can be sequenced in a predictable manner. Thus, a spin box with an associated text box would be a good choice for these attributes.
- The attributes VIDEO CATEGORY and VIDEO SUBCATEGORY contain a large number of predefined values. With so many attributes to display on our screen, it was determined that a drop-down list would be the best control choice.
- TITLE COVER presented an interesting challenge. Its value is actually a drive, directory, and name of a file that contains a bitmap image of the cover of the video title. This attribute will make use of an advanced control called an *image box* to store a picture of the video title cover. When this object is selected by the user, a set of controls and special dialogue (user interaction) will be used to capture the input for this item. We'll illustrate this input later in step 3.

Once again, there are many other screen-based controls that could be used to input data. Our examples focus on the most commonly used controls. How well you complete this activity will be a function of how knowledgeable you are about these common controls and other more advanced controls.

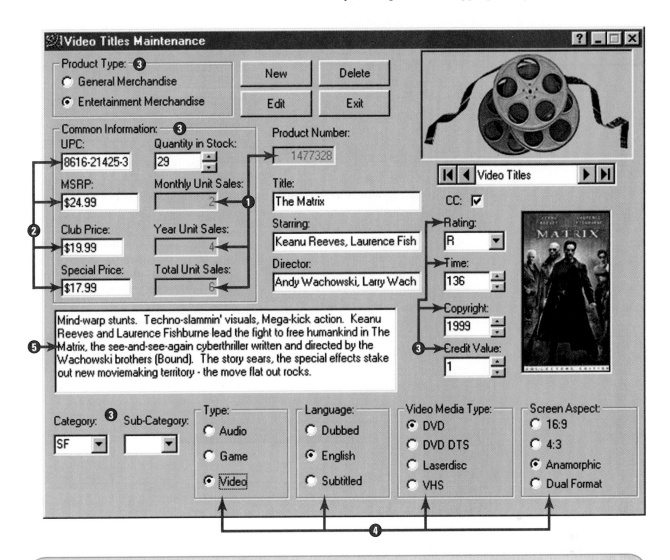

FIGURE 16-8   Input Prototype for Video Title Maintenance

Step 3: Design, Validate, and Test Inputs   This step involves developing prototype screens for users to review and test. Their feedback may result in the need to return to steps 1 and 2 to add new attributes and address their characteristics.

Let's take a look at a couple of SoundStage screen prototypes. Figure 16-8 represents a possible prototype screen for handling NEW VIDEO TITLE, DISCONTINUED VIDEO TITLE, and VIDEO TITLE UPDATE. The logo appearing in the upper-right portion of the screen was included to adhere to a company standard—all screens must display the company logo. The buttons also appearing in the upper center and right portion of the screen were added because of the decision to combine the three inputs into a single screen. They were needed to give the user the option of selecting the desired type of input and record action. We will discuss these buttons and other command and navigation controls and their use in Chapter 17.

Note the following issues in Figure 16-8:

❶ The PRODUCT NUMBER, MONTHLY UNIT SALES, YEAR UNIT SALES, and TOTAL UNIT SALES are screened in a special color as a visual clue to the user that these fields are locked and the user cannot enter data into them. These fields are automatically generated by the system. Other fields appearing on the screen have a white background as a visual clue that they can be edited.

❷ Edit masks were specified for these input fields. The UNIVERSAL PRODUCT CODE field contains dashes in specified locations. The user does not actually enter these dashes. Rather, the user simply types in the numbers, and afterward the entire content is redisplayed according to the specified edit mask. The same is true for the MANUFACTURER'S SUGGESTED RETAIL PRICE, CLUB DEFAULT UNIT PRICE, and CURRENT SPECIAL UNIT PRICE fields. For example, in either of these three fields the user could type the number 9 and press enter, and the content would be redisplayed (according to the edit mask) with a dollar sign and decimal point.

❸ Each field on a screen has been given a label that is meaningful to the users. Feedback from users indicated "CC" was a commonly recognized abbreviation for "closed caption." Also, the users indicated that a label was not necessary for CATALOG DESCRIPTION.

❹ Related radio buttons have been arranged in a group box that contains a descriptive label. Group boxes are frequently used to visually associate a variety of controls that are related. For example, the fields inside the group box labeled "Common Information" were grouped because the user associates these attributes with any type of SoundStage product. Also, each label that corresponds to a radio button option is not what is actually input and stored in the database. Rather, what you see is the meaning of the value. The actual value that is stored is a code. For example, the code value E would actually be stored instead of "English" if the user selects the radio button labeled "English" for the attribute LANGUAGE.

❺ The multiple-line text box has a vertical scroll bar feature if the text fills the text box. This is a visual clue that there is additional text not appearing inside the CATALOG DESCRIPTION field.

In prototyping input screens, you need to let the user exercise or test the screens. Part of that experience should involve demonstrating how the user may obtain appropriate help or instructions. New versions of Microsoft products use what are called "tooltips" to provide a brief description of buttons and boxes that appear on a screen. The tooltip description displays when the user positions the mouse over the top of the object. Also, the F1 key is universally accepted as initiating context-sensitive help. A help button is another option. Whichever approach(es) you use, it is not necessary to actually implement the help in a prototype.

Finally, prototypes need not display all details to a user unless they are requested (or triggered by a user action). For example, the drop-down list for Motion Picture Association of America RATING code displays only a default value. However, the downward-pointing arrow is a visual clue that a list box containing possible values exists. The list box may be viewed by simply clicking on the downward pointing arrow. The result of that action is illustrated in the margin.

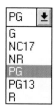

A Drop-Down Menu

The previous example was fairly simple because it contained only data that might be updated in one database table. But what if an input includes data to be updated in more than one table? And suppose there is a one-to-many relationship between the tables. Consider MEMBER ORDER, which has a one-to-many relationship to MEMBER ORDERED PRODUCTS. How do we design a single input to capture the data for both tables?

Figure 16-9 represents a prototype screen for entering MEMBER and MEMBER ORDERED PRODUCTS on a single form. The form is segmented into two windowpanes. MEMBER data is in the top pane, and MEMBER ORDERED PRODUCT data is in the bottom pane. You may be wondering what happens if the number of MEMBER ORDERED PRODUCTS exceeds the space allotted for that pane. In other words, where is the scroll bar for the bottom pane? Many *Windows* GI controls are "intelligent." If the number of rows in the bottom pane exceeds the space, a vertical scroll bar will automatically appear.

As one last *Windows* example, Figure 16-10 shows a single-screen design that consolidates three different or similar inputs from our data flow diagrams: NEW MEMBER, MEMBER CANCELLATION, and MEMBER UPDATE. This form also uses the standard input controls

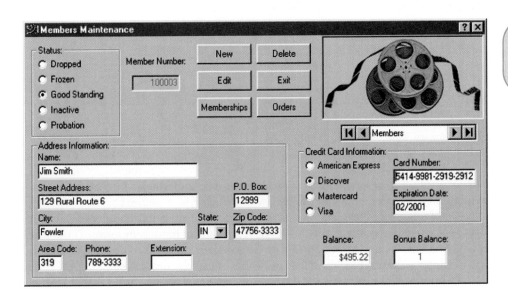

that we've discussed in this chapter. The consolidation of logical and physical data
flows into single-screen designs is very common.

Step 4: If Necessary, Design the Source Document   If a source document will be
used to capture data, we must also design that document. The source document is for
the system user. In its simplest form, the prototype may be a simple sketch or an industrial artist's rendition.

A well-designed source document will be divided into zones. Some zones are
used for identification; these include company name, form name, official form number, date of last revision (an important attribute that is often omitted), and logos.
Other zones contain data that identifies a specific occurrence of the form, such as
form sequence number (possibly preprinted) and date. The largest portion of the
document is used to record transaction data. Data that occurs once and data that re-
peats should be logically separated. Totals should be relegated to the lower portion of
the form because they are usually calculated and, therefore, not input. Many forms

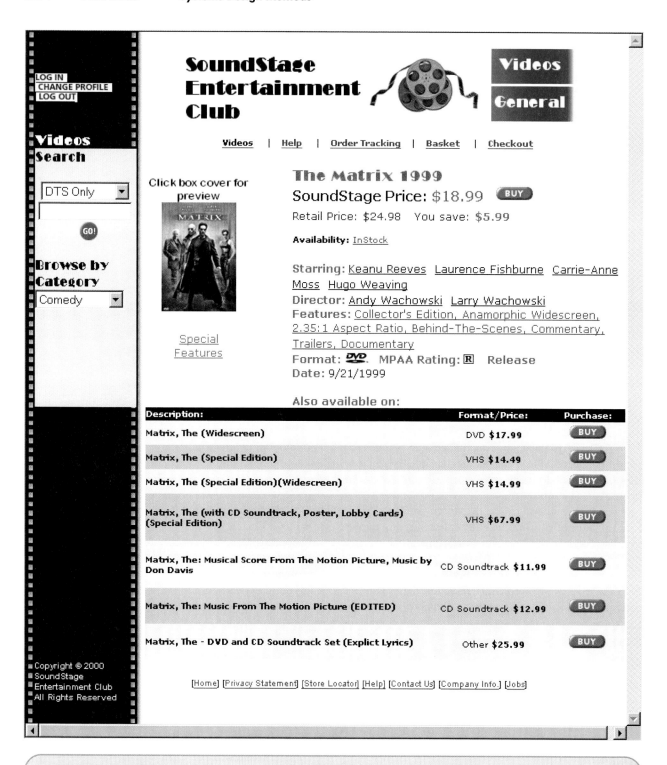

FIGURE 16-11 Input Prototype for Web Shopping Cart

include an authorization zone for signatures. Instructions should be placed in a convenient location, preferably not on the back of the form.

Prototyping tools have become more advanced in recent years. Spreadsheet programs such as Microsoft's *Excel* can make very realistic models of forms. These tools give you outstanding control over font styles and sizes, graphics for logos, and the like. Laser printers can produce excellent printouts of the prototypes.

Another way to prototype source documents is to develop a rough model using a word processor. Pass the model to one of the growing number of desktop publishing systems that can transform the rough model into impressive-looking forms (so impressive, in fact, that some companies now develop forms this way instead of subcontracting their design to a forms manufacturer).

> Web-Based Inputs and E-Business

The last input design considerations we want to address concern Web-based outputs. The SoundStage project will add various e-commerce and e-business capabilities to the Member Services information system. Some of these capabilities will require Web-based inputs that must be designed.

One logical output requirement for the project is Web-based MEMBER ORDER. We just showed you the client/server version. Now let's look at the Web-based version. It is common to present a Web storefront (Figure 16-11 on page 604). In addition to providing the member with information about SoundStage products (an output), the member can click the "buy" button to initiate a purchase. That takes the member to what has become a common metaphor screen in e-commerce applications, the *shopping cart* screen (see Figure 16-12). Web interfaces tend to be somewhat more artistic

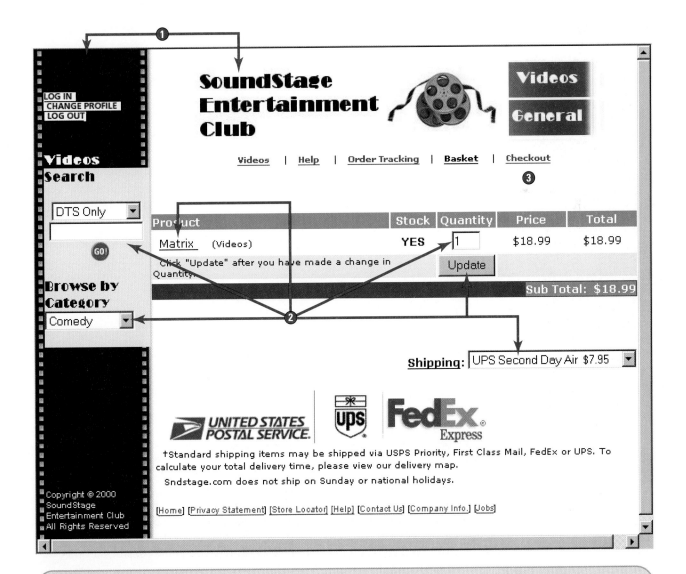

FIGURE 16-12   Input Prototype for Web Shopping Cart

than *Windows* interfaces. Perhaps that is part of the appeal. The interface needs to be visually appealing to entice the customer to purchase products in the absence of a verbal sales pitch. In Figure 16-12:

❶ The shopping cart "frame" is independent of the general navigation frame (on the left). The latter allows the user to search and browse the entire Web site, hopefully to find additional products to add to the shopping cart.

❷ Buttons, text boxes, hyperlinks, drop-down boxes, and other common controls are here applied to a Web interface instead of a *Windows* interface.

❸ A checkout hyperlink sends the member to the next "page" to complete the transaction.

The Web interface offers several advantages such as the automatic ability for members to use their forward and backward buttons to navigate different inventory and order pages at the Web site.

Learning Roadmap

This chapter provided a detailed overview of the input design tasks of a project. If you haven't covered output design, you should go back and read Chapter 15 next. Otherwise, the next logical chapter is Chapter 17, "User Interface Design." User interface design ties the input and output screens together into an overall user experience. As we did in this chapter, we will address both client/server, *Windows*-based user interfaces and Web-based, e-business solutions.

Chapter Review

1. Several concepts are important to input design. One of the first things you must learn is the difference between data capture, data entry, and data processing. Alternative input media and methods must also be understood before designing the inputs. And because accurate data input is so critical to successful processing, file maintenance, and output, you should also learn about human factors and internal controls for input design.

2. Data happens! It accompanies business events called transactions. Examples include orders, time cards, reservations, and the like. This is an

important concept because system designers must determine when and how to capture the data. The designer must understand the difference between the following:

a. Data capture is the identification and acquisition of new data.

b. A source document is a paper form used to record business transactions in terms of data that describe those transactions.

c. Data entry is the process of translating the source data into a computer-readable format. That format may be a magnetic disk, an optical mark form, a magnetic tape, or a floppy diskette, to name a few.

3. Data must be processed using one of the following techniques:

a. In batch processing, the entered data is collected into files called batches that are processed later.

b. In online processing, the captured data is processed immediately.

c. In remote batch processing, data is entered using online editing techniques; however, the data is collected into batches for later processing.

4. The systems analyst usually selects the method and medium for all inputs. Input methods include:

— Keyboard	— Optical mark
— Mouse	— Magnetic ink
— Touch screen	— Electromagnetic signature
— Point of sale	— Smart cards
— Sound and speech	— Biometrics

5. Most new applications being developed today consist of screens having a "graphical"-looking appearance. This type of appearance is referred to as a graphical user interface (GUI).

6. Inputs should be as simple as possible and designed to reduce the possibility of incorrect data being entered. Furthermore, the needs of data entry clerks must also be considered. With this in mind, system designers should understand human factors that should be evaluated during input design.

7. Input controls ensure that the data input to the computer is accurate and that the system is protected against accidental and intentional errors and abuse, including fraud.

8. When designing input screens for an application that will contain a GUI appearance, the designer must be careful to select the proper control object for each input attribute. Each control serves a specific purpose, has certain advantages and disadvantages, and should be used according to guidelines. Some of the most commonly used screen-based controls for inputting data include text box, radio button, check box, list box, drop-down list, combination box, and spin box.

Review Questions

1. What is the goal of input design?
2. What is the relationship between source documents and data entry?
3. What is the next step after data is entered? What are the different methods used for this step, and how are they different in terms of the timing?
4. What are the different input methods described in the textbook?
5. What is the difference between OMR and OCR?
6. For biometric input, what and how is data entered and processed into the information systems?
7. Why is smart card technology able to store a tremendous amount of information? What are some examples of the applications of smart card technology?
8. Why are human factors important in input design? What principles need to be considered in input design?
9. What are some of the techniques used to validate data?
10. Under what circumstances should we choose to use radio buttons or check boxes?
11. What are the similarities of a drop-down list and a combo box?
12. What are some advanced input controls suggested in the textbook?
13. What are the steps for input design process?
14. What is a well-designed source document?
15. What are the challenges facing Web interfaces compared to *Windows* interfaces when designing input suggested in the textbook?

Problems and Exercises

1. What overriding goal should every systems designer, in performing input design, never lose sight of?

2. The owner of a chain of fast-food sit-down restaurants has hired your company to design a method to get orders to customers faster, with less labor, but without any loss of quality. Currently, the fast-food restaurants use the conventional method of having customers wait in line to order and pay; then the order is printed out and given to the food preparation specialist (chef). Can you think of how technology might be used to meet these objectives?

3. Look at the following portion of a data input screen used by the technicians in a company to order parts from the company warehouse. Is there anything wrong with the design of this input screen?

   ENTER TECHNICIAN #: ■■ ENTER TECHNICIAN NAME: ■■■■■
   ENTER PART #: ■■■■■■ ENTER PART DESCRIPTION: ■■■■■

4. Point of sale (POS) terminals, such as those used in ATMs, gas stations, and in stores, have become extremely common due to their convenience and versatility. But in terms of human interface design, their input methods sometimes leave something to be desired. What are the areas where you think improvement is needed?

5. Answer the following true/false questions: Qualify or explain your answers as needed.

   a. System users tend to be confused by data entry codes, and frequently enter the wrong code; therefore, their use should be avoided.

   b. Batch processing is still a viable data processing process.

   c. There is little correlation in terms of data accuracy between the point at which data originates and the length of time before the data is captured.

   d. The computer mouse was invented in order to optimize the use of graphical-user interfaces on personal computers using *Windows* or *Apple* operating systems.

   e. Using metaphor-based screen design is considered too "cute" and unprofessional.

   f. Radio buttons are best used only when there is a very small number of previously defined values that have no commonality.

6. Think about the best and the worst data input screens that you have used, heard about, and/or worked on. Using your own experience, as well as this chapter, list at least five input screen requirements and/or principles (other than the one described in Question 3) that you consider to be important. Explain why you selected each of them.

7. You are designing an input screen for a client treatment data system which will be used by a county in your state. Alcohol and drug treatment providers will enter demographic and treatment data on their clients, then send it to the county Department of Behavioral Health. The system will run on a client/server network using a relational DBMS. Business rules require an entry for all data elements; there are no discretionary or optional fields. Data will include a mixture of alphanumeric data, numeric data used for calculations, and dates. Some of the data fields will have interdependencies between them, such as a field for gender and another field for whether the client is pregnant. Unique client identifiers will be used as keys and will be generated by the system.

   Where should you design the data input controls and edits—at the client-side or server-side of the network? Why? What types of edits and validation checks should be included?

8. Match the definitions or examples in the first column with the terms in the second column:

   A. Magnetic stripe card

   B. Single-line text box with two vertically aligned buttons

   C. FedEx pickup and delivery data processing method

   D. Text description of value choice associated with circle

   E. Voice recognition system

   F. Identification and acquisition of new data

   G. Input device most prone to data entry errors

   H. Paper form used to record business transaction(s)

   I. Example of metaphor-based screen design

   1. Slider edit calendar

   2. Data capture

   3. OMR

   4. Quicken

   5. Microsoft *Visual Basic*

   6. Type of ADC technology

   7. Check digit formula

   8. Radio button

   9. Remote batch processing

J. Optical mark forms used for objective question exams

K. Nonnumeric method for value selection

L. Technique to determine primary key data entry errors

M. Example of repository-based programming approach

10. Biometric ADC system

11. Source document

12. Keyboard

13. Spin box

9. Text boxes may be the most frequently used control for data input in GUI interfaces. What conventions and guidelines should the system designer follow when designing input screens that include text boxes?

10. Fill in the blanks for the following statements:

a. Source documents should be portioned into different zones for _____ data, _____ data, _____, and depending upon the form, for data that identifies a unique _____ of the form.

b. After _____ system inputs and reviewing _____, the next step in the _____ process is to _____.

c. _____ GUI controls allow a _____ to automatically appear if the number of _____ exceeds the allotted space for that _____.

d. Although technically _____ are _____ input controls, they enable users to _____ to or _____ a transaction.

e. The _____ data input control is the _____, which is best used when the scope of input data values is _____.

11. What basic questions should you ask yourself in deciding on the best GUI control to use for each data attribute to be captured and input? Provide examples of data attributes, and instructions on matching the data attribute with the GUI control that is the best in that situation.

12. Designs based on generic, easily identifiable metaphors are generally well received by system users, particularly novice ones, because their familiarity enhances the perception of ease-of-use and user-friendliness. Your company wants to replace its paper telephone message forms with an electronic version that can be sent as an e-mail attachment. Create a metaphoric screen design for one of the common paper telephone message forms. (Hint: This will not require a screen design tool, but can be designed in Microsoft *Word* or *Excel*. Create your own, rather than using one of the templates that are commonly available.)

13. As the Internet continues to grow in its business influence, it is increasingly being used for applications that were formerly client/server-based applications. What are some of the similarities in data input design between Web-based applications and client/server-based applications? What are some of the differences? Give examples.

Projects and Research

1. As the textbook mentions, many organizations once employed huge numbers of clerks to perform data entry. As personal computers and online computing became more common, system users began assuming responsibility for data entry, and the ranks of data entry clerks shrank dramatically in most organizations. Today the explosive growth of the Internet is having a similarly profound impact upon the organizational structure of most large companies and government agencies.

   a. Research the transformation of responsibility for data entry, from data entry clerks to system users, that occurred in the 1980s and 1990s. Discuss the issues faced by these organizations and their employees.

   b. Research the transformation that is taking place today as the Internet facilitates the growth of customer-based data entry. What are the implications? What is the impact upon companies' organizational structure, its employees, and its customers?

   c. Research a large company or government agency in your area. Compare its organizationa structure 25 years ago to 10 years ago and to today. What relevant changes did you find?

   d. Research articles in business and IT journals regarding the impact technology is having upon organizations in the private and public sectors. What are some of the predictions regarding what the next transformation in data entry will be? Do you think these predictions will come true?

   e. Overall, are we better off or worse off for these changes?

2. Despite the inroads being made by other input methods, the venerable keyboard is still far and away the most common method used for data input, and it is likely to stay that way for a while. But could the current basic keyboard design itself be improved? Research the history of the keyboard, as well as recent developments in keyboard design, then consider the following questions:

   a. What is the basis for the "QWERTY" layout?
   b. Are more efficient keyboard layouts available? If so, why aren't they being widely used?
   c. Keyboards are a form of tactile input. Are there methods of tactile data input other than pressing a key which represents a letter or number?
   d. What about the impact of repetitive stress disorders such as carpal tunnel syndrome, which can be caused or exacerbated by keyboards? What is the estimated annual cost in lost productivity and medical care?
   e. If you were asked to redesign the keyboard, what changes would you make? Why?

3. Voice recognition technology is used fairly frequently for entering commands or responding to automated questions data over the telephone. Also, some technology experts predict that voice recognition will one day replace keyboards for data entry; currently there are very few applications that use voice recognition technology for data entry. Research recent developments in voice recognition technology and respond to the questions below:

   a. What articles did you find? What are their viewpoints regarding voice recognition technology?
   b. What is the current state of the art of voice recognition technology?
   c. Do you think that voice recognition technology has finally matured to the point where it can soon be a viable option for keyboard data entry? Or does it still need some further maturing? Or does it represent a technological dead end? Explain.
   d. If you have access to applications that include voice recognition software for data or text entry (or can download a free trial copy without violating any licensing or usage restrictions), try using them. How would you evaluate them?

4. A number of emerging technologies are classified as automatic data capture (ADC) technologies. In essence, they get people out of the data input loop. One of these technologies is radio frequency identification (RFID) technology, which is quickly becoming very common. Research RFID technology and business applications on the Web and/or in your school's library.

   a. Explain how RFID devices work.
   b. How is RFID currently being used in the private and public sectors?
   c. What are their main advantages? Their disadvantages?
   d. What are some of the social, economic, and political implications regarding RFID technology?
   e. What will be some of the applications to which RFID technology will put to use 10 years from now?

5. Generally, one doesn't have to look too far in most organizations to find a poorly designed source document or data input screen, or both. Contact a nonprofit or similar type of organization and volunteer to review their forms and data input screens, and to redesign one or two of them.

   a. Describe the organization for which you did the volunteer work.
   b. Did you find any design problems in either the source documents or the input screens? Describe the design problems.
   c. Describe the source document and/or input screen that you redesigned. Include a sample if possible.
   d. What changes did you make, and what process did you follow in making the changes? Include a sample.
   e. What challenges did you face in redesigning the form and/or screen?
   f. Was the organization pleased with your redesign? Is it using the new form and screen?

6. Your Web design company has been hired by a supermarket chain to develop a short Web-based survey on what customers like and don't like about the supermarket chain. The objective is to have an employee in each store walking through the store with a laptop and randomly choosing customers to survey regarding their likes and dislikes. Customers will be asked three to five questions. Survey data will be entered directly into the laptop via a Web-based application.

   a. What are some of the high-level considerations that need to be addressed before designing the survey form?
   b. What questions would you include in the survey?
   c. What other data would you include?
   d. Design a prototype of the form.
   e. Was this easier or harder to do than you thought? Describe any challenges you faced that you didn't anticipate.

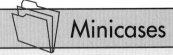

1. Input design affects not only the ease of use of a system, but also the security. Find examples of how systems can be crashed by the characters that are input into text boxes. Find out how to address these security loopholes or concerns. Prepare a short paper on this material and present to the class.

2. In minicase 2 from the previous chapter, you created an online form. Did you appropriately use the different methods for inputs, such as input boxes, radio buttons, drop downs, and the like? What would you change? Why?

3. Make the changes you suggested in minicase 2. Submit screen shots of the form both before and after you made the changes. Be sure to include the URL of your form, so that the professor can check your work.

4. Research input methods for blind users. Write a short paper briefly describing these input methods and how you can integrate them into an information system.

Suggested Readings

Andres, C. *Great Web Architecture*. Foster City, CA: IDG Books Worldwide, 1999. Books on effective Web interface design are beginning to surface. The science of human engineering for Web interfaces has not yet progressed as far as client/server interfaces (e.g., *Windows*). Here is an early title that explores many dimensions of Web architecture and interfaces using real-world examples.

Application Development Strategies (monthly periodical). Arlington, MA: Cutter Information Corporation. This is our favorite theme-oriented periodical that follows system development strategies, methodologies, CASE, and other relevant trends. Each issue focuses on a single theme. This periodical will provide a good foundation for how to develop input prototypes.

Dunlap, Duane. *Understanding and Using ADC Technologies. A White Paper for the ADC Industry*. A SCAN TECH 1995 Presentation. October 23, 1995, Chicago. We are indebted to our friend and colleague. Professor Dunlap is a leader in the field of ADC. This paper was the basis for much of our discussion on the trends in ADC technology.

Fitzgerald, Jerry. *Internal Controls for Computerized Information Systems*. Redwood City, CA: Jerry Fitzgerald & Associates, 1978. This is our reference standard on the subject of designing internal controls into systems. Fitzgerald advocates a unique and powerful matrix tool for designing controls. This book goes far beyond any introductory systems textbook; it is must reading.

Galitz, W. O. *User-Interface Screen Design*. New York: John Wiley & Sons, 1993. This is our favorite book on overall user interface design. The author offers several flowcharts of the decision process in applying GUI controls to inputs.

Kozar, Kenneth. *Humanized Information Systems Analysis and Design*. New York: McGraw-Hill, 1989. A good user-oriented treatment of input design.

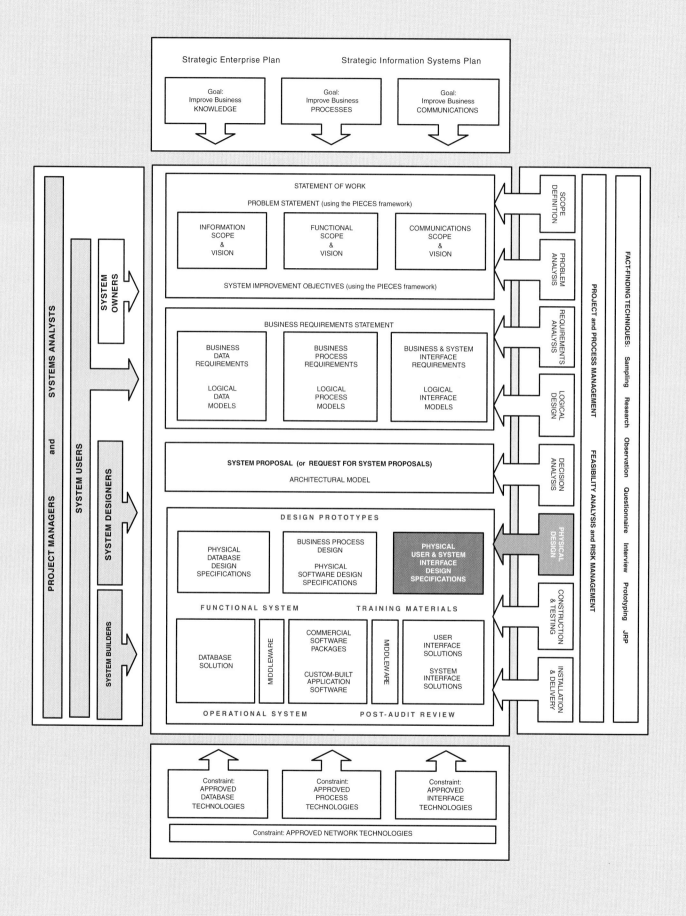

17

User Interface Design

Chapter Preview and Objectives

In this chapter you will learn how to design and prototype the user interface for a system. The user interface should provide a friendly means by which the user can interact with the application to process inputs and obtain outputs. In Chapters 15 and 16, you learned how to design and prototype outputs and inputs. User interface design and prototyping address the overall presentation of the application and may require revisions to the preliminary output and input prototypes. Today there are two commonly encountered interfaces: terminals (or microcomputers behaving as terminals) used in conjunction with mainframes and the more common display monitors connected to microcomputers. There are also several strategy styles for designing the user interface for systems. You will know that you've mastered user interface design when you can:

■ Distinguish between different types of computer users and design considerations for each.

■ Identify several important human engineering factors and guidelines and incorporate them into a design of a user interface.

■ Integrate output and input design into an overall user interface that establishes the dialogue between users and computer.

■ Understand the role of operating systems, Web browsers, and other technologies for user interface design.

■ Apply appropriate user interface strategies to an information system. Use a state transition diagram to plan and coordinate a user interface for an information system.

■ Describe how prototyping can be used to design a user interface.

Introduction

Designing the user interface for the in-house client/server part of the SoundStage Member Services system has been a longer process than Bob Martinez anticipated. Some of the final design is shown in Figures 17-17, 17-18, and 17-19, but it didn't start out looking like that. Bob's first design used icons instead of text for all menu buttons. For instance, a picture of a stack of papers indicated "Orders," and a picture of balloons indicated "Promotions." But feedback from users testing the interface revealed that many of them had no clue what those icons meant and wasted time trying to figure them out. So a decision was made to switch to text labels. Various kinds of controls were tested for the screens shown in Figures 17-18 and 17-19. Each iteration went through user testing, leading to revisions and more user testing. It was all a lot of work. But Bob knew this work on the front end would be much less than the work of answering thousands of phone calls from confused and upset users if they implemented a user interface that was lacking.

User Interface Design Concepts and Guidelines

In the two previous chapters, we addressed output and input design. In this chapter, we integrate output and input design into an overall user interface that establishes the dialogue between users and computer. The dialogue determines everything, from starting the system or logging into the system, to setting options and preferences, to getting help. And the presentation of the outputs and inputs is also part of the interface. We need to examine the screen-to-screen transitions that can occur. In client/server applications (e.g. network-based *Windows*) and Web applications (e.g., Internet- or intranet-based browsers), the user has many alternative paths through menus, hyperlinks, dialogues, and the like. This makes for very accommodating and friendly user interfaces, but it greatly complicates design and programming.

Today most user interfaces are designed by rapidly constructing prototypes. These prototypes are generated using rapid application development environments such as Microsoft's *Visual Studio,* Borland's *JBuilder* (for *Java*), or IBM's *VisualAge* (for various languages). These prototypes are rarely <u>fully</u> functional, but they do contain enough functionality to demonstrate the interface. For example, a help system prototype may be functional to the extent that it calls up a few sample screens to demonstrate levels of assistance. Or a security system might have just enough functionality to demonstrate representative log-in errors even though it is not actually authenticating users. When we get to the construction phase of the life cycle, programmers and analysts will complete the functionality.

We begin our study by examining types of users, human factors, and human engineering guidelines that affect user interface design.

> Types of Computer Users

Nowhere are human factors as important as they are in user interface design. Just ask the typical systems analyst who spends half the day answering phone calls from system users who are having difficulty using information systems and computer applications. The overriding consideration of user interface design is the same as that for business and technical writing—understand your audience. In interface design, the audience is the SYSTEM USER.

On the chapter home page at the beginning of this chapter, we highlighted SYSTEMS ANALYSTS, SYSTEM DESIGNERS, and SYSTEM BUILDERS as the stakeholders that actually <u>perform</u> user interface design, but we also highlighted SYSTEM USERS as the stakeholders for which the user interface design is intended. Users must <u>test</u> and <u>evaluate</u> that design. The chapter home page also illustrates that user interface design is performed during the DESIGN and CONSTRUCTION phases and that the activities result in the INTERFACE SPECIFICATIONS (prototypes) and PROGRAM building blocks.

SYSTEM USERS can be broadly classified as either expert or novice—and either nondiscretionary or discretionary.

An **expert user** is an experienced computer user who has spent considerable time using specific application programs. The use of a computer is usually considered nondiscretionary. In the mainframe computing era, this was called a *dedicated user.* Expert users generally are comfortable with (but not necessarily experts in) the application's operating environment (e.g., *Windows* or a Web browser). They have invested time in learning to use the computer. They will invest time in overcoming less-than-friendly user interfaces. In general, they have memorized routine operations to an extent that they neither seek nor want excessive computer feedback and instructions. They want to be able to accomplish their task in as few actions and keystrokes as possible.

The **novice user** (sometimes called a *casual user*) is a less experienced computer user who will generally use a computer on a less frequent, or even occasional, basis. The use of a computer <u>may</u> be viewed as discretionary (although this is becoming less and less true). Stated simply, the novice users need more help than the expert users. Help takes many forms, including menus, dialogues, instructions, and help screens. Most managers, despite their increasing computer literacy, fall into the novice category. They are paid to recognize and solve problems, exploit opportunities, and create plans and manage the vision—not to learn and use computers. Computers are considered tools by modern managers. When the need arises, they want to realize their benefit as quickly as possible and move on.

Expert and novice users are actually extremes on the continuum of all users. The totally novice user who hasn't used a computer is becoming less common. Few college curricula don't require computer literacy for all majors, and students in all majors have discovered the value of increased interdisciplinary computer expertise (sometimes called *informatics*). Novice users also usually graduate to expert users through practice and experience. The net societal impact of the Internet is that more people are becoming increasingly comfortable with computers—creating a class of users that is less novice and more expert with each passing year. Is it any wonder that user interface design is racing toward Web browser-like interfaces, even within *Windows* applications?

It is difficult to imagine today's young students and professionals being uncomfortable with computers. Regardless, most of today's systems are designed for the novice system user but adapting to the expert user. The focus is on user friendliness or human engineering.

> Human Factors

Before designing user interfaces, you may find it useful to understand the elements that frequently cause people to have difficulty with computer systems. Our favorite user interface design expert, Wilbert Galitz (see the Suggested Readings) offers the following interface problems:

- Excessive use of computer jargon and acronyms.
- Nonobvious or less-than-intuitive design.
- Inability to distinguish between alternative actions ("What do I do next?").
- Inconsistent problem-solving approaches.
- Design inconsistency.

According to Galitz, these problems result in confusion, panic, frustration, boredom, misuse, abandonment, and other undesirable consequences.

To solve these problems, Galitz offers the following overriding "commandments" of user interface design:

- *Understand your users and their tasks.* This becomes increasingly difficult as we extend our information systems to implement business-to-consumer (B2C) and business-to-business (B2B) functionality using the Internet.

---

expert user an experienced computer user.

novice user an inexperienced or casual computer user.

- *Involve the user in interface design.* Find out what the users like and dislike in their current applications. Involve them in screen design and dialogue from the beginning. This commandment is easily enabled with today's PC-database and rapid application development technology.
- *Test the system on actual users.* Observation and listening are the key skills here. After initial training, try to avoid excessive coaching and forcing users to learn the system. Instead, observe their actions and mistakes, and listen to their comments and questions to better understand their interaction with the user interface.
- *Practice iterative design.* The first user interface will probably be unsatisfactory. Expect any user interface design to go through multiple design iterations and testing. When is the interface finished? Probably never! But Galitz suggests that a good goal is that 95 percent of the typical users (be they novice or expert) can perform intended tasks (be they routine or less common) without difficulty or help.

> Human Engineering Guidelines

Given the type of user, a number of important human engineering factors should be incorporated into the design:

- *The system user should always be aware of what to do next.* The system should always provide instructions on how to proceed, back up, exit, and the like. Several situations require some type of feedback:
  — *Tell the user what the system expects right now.* This can take the form of a simple message such as READY, TYPE COMMAND, SELECT ONE OR MORE OPTIONS, or TYPE DATA.
  — *Tell the user that data has been entered correctly.* This can be as simple as moving the cursor to the next field in a form or displaying a message such as DATA OK.
  — *Tell the user that data has not been entered correctly.* Short, simple messages about the correct format are preferred. Help functions can supplement these messages with more extensive instructions and examples.
  — *Explain to the user the reason for a delay in processing.* Some actions require several seconds or minutes to complete. Examples include sorting, indexing, printing, and updating. Simple messages such as SORTING—PLEASE STAND BY, or INDEXING—THIS MAY TAKE A FEW MINUTES or PLEASE WAIT tell the user that the system has not failed. The *Windows* hourglass or the *Internet Explorer* revolving globe are iconic clues that processing is occurring.
  — *Tell the user that a task was completed or was not completed.* This is especially important in the case of delayed processing, but it is also important in other situations. A message such as PRINTING COMPLETE or PRINTER NOT READY—TRY AGAIN OR CONTACT YOUR NETWORK ADMINISTRATOR will suffice.
- *The screen should be formatted so that the various types of information, instructions, and messages always appear in the same general display area.* This way, the system user knows approximately where to look for specific information. In most windowing environments, standards often dictate the location of status messages or pop-up dialogue windows.
- *Messages, instructions, or information should be displayed long enough to allow the system user to read them.* Most experts recommend that important messages be displayed until the user acknowledges them.
- *Use display attributes sparingly.* Display attributes, such as blinking, highlighting, and reverse video, can be distracting if overused. Judicious use

allows you to call attention to something important—for example, the next field to be entered, a message, or an instruction.

- *Default values for fields and answers to be entered by the user should be specified.* In windowing environments, valid values are frequently presented in a separate window or dialogue box as a scrollable region. The default value, if applicable, should usually be first and clearly highlighted.
- *Anticipate the errors users might make.* System users will make errors, even when given the most obvious instructions. If it is possible for the user to execute a dangerous action, let it be known (for example, a message or dialogue box could read ARE YOU SURE YOU WANT TO DELETE THIS FILE?). An ounce of prevention goes a long way!
- *With respect to errors, a user should not be allowed to proceed without correcting an error.* Instructions (and examples) on how to correct the error should be displayed. The error can be highlighted with sound or color and then explained in a pop-up window or dialogue box. A HELP option can be defined to trigger display of additional instructions.
- *If the user does something that could be catastrophic, the keyboard should be locked to prevent any further input, and an instruction to call the analyst or technical support should be displayed.*

> Dialogue Tone and Terminology

The overall flow of screens and messages is called a **dialogue.** The tone and terminology of a dialogue are very important human factors in user interface design. With respect to the tone of the dialogue, the following guidelines are offered:

- *Use simple, grammatically correct sentences.* It is best to use conversational English rather than formal, written English.
- *Don't be funny or cute!* When someone has to use the system 50 times a day, the intended humor quickly wears off.
- *Don't be condescending.* Don't insult the intelligence of the system user. For instance, don't offer repeated praise or rewards.

With respect to the terminology used in a computer dialogue, the following suggestions may prove helpful:

- *Don't use computer jargon.*
- *Avoid most abbreviations.* Abbreviations assume that the user understands how to translate them. Check first!
- *Use simple terms.* Use NOT CORRECT instead of INCORRECT. There is less chance of misreading or misinterpretation.
- *Be consistent in your use of terminology.* For instance, don't use both EDIT and MODIFY to mean the same action.
- *Carefully phrase instructions—use appropriate action verbs.* The following recommendations should prove helpful:
  — Use SELECT or CHOOSE instead of PICK when referring to a list of options. Be sure to indicate whether the user can select only one or more than one option from the list of available options.
  — Use TYPE, not ENTER, to request the user to input specific data or instructions. The term ENTER may be confused with the enter key.
  — Use PRESS, not HIT or DEPRESS, to refer to keyboard actions. Whenever possible, refer to keys by the symbols or identifiers that are actually printed on the keys. For instance, the ↵ symbol is used on some keyboards to designate the RETURN or ENTER key.
  — When referring to the on-screen mouse cursor, use the term POSITION THE CURSOR, not POINT THE CURSOR.

dialogue the overall flow of screens and messages for a application.

User Interface Technology

Most of today's user interfaces are graphical. The basic structure of the *graphical user interface* (or *GUI*) is provided within either the computer operating system or the Internet browser of choice. In *client/server information systems,* the user interface client is implemented to execute within the PC operating system. In *Internet and intranet information systems,* the user interface is implemented to execute within the PC's Web browser (which, in turn, executes within the PC operating system).

> Operating Systems and Web Browsers

The dominant GUI-based operating system for today's client computers (as in a client/server network) is Microsoft *Windows* (various versions). Apple's *Macintosh* and the various flavors of *UNIX* (including *Linux*) also hold market share. For the growing numbers of handheld and palm-top client computers, the current dominant operating system is Palm's *Palm OS.* Microsoft *Windows Mobile* also holds a percentage of that market.

Increasingly, the operating system is not the key technology factor in user interface design. Internet and intranet applications run within a Web browser. Most browsers run in many operating systems, making it possible to design a user interface that is less dependent on the computer itself. The advantages of this computer *platform independence* should be obvious. Instead of writing a user interface for each anticipated computing platform and operating system, you write it for one or two browsers. As we go to press, the dominant Web browsers are Microsoft *Internet Explorer* and Mozilla *Firefox,* but version problems within browsers can exist in the user community.

In addition to the operating systems and browsers, the overall design of a user interface is enhanced or restricted by the available features of the users' display monitor, keyboard, and pointing devices. Let's briefly examine some of the other considerations.

> Display Monitor

The size of the display area is critical to user interface design. Not all displays are PC monitors! A number of non-PC *terminals* still exist. Terminals are non-PC displays that merely display data and information transmitted by a remote computer, usually a mainframe. And while many terminals have been replaced by PCs, users are still frequently forced to interface with the legacy mainframe applications using *terminal emulators* that open a window on the screen that still displays information and instructions in the original, pre-*Windows* terminal format. For these terminals and terminal emulators, the two most common display areas were 25 lines by 80 columns and 25 lines by 132 columns.

Fortunately, the personal computer monitor has replaced most terminals, and most newer and reengineered applications are being written to a graphical interface. For PC monitors, we don't measure the display in terms of lines and columns. And while diagonal measures such as inches are often quoted, the more relevant measure is graphical resolution. Graphical resolution is measured in pixels, the number of distinct points of light displayed on the screen. Today's most common resolution is 1,024,000 horizontal pixels by 800,000 vertical pixels in a 17-inch diagonal display. Larger display sizes support even more pixels; however, the designer should generally design the user interface with the assumption of the lowest common or reasonable denominator.

Obviously, handheld and palm-top computers and specialized terminal displays (such as those in cash registers and ATMs) support much smaller displays that must be considered in user interface design.

The manner in which the display area is shown to the user is controlled by both the technical capabilities of the display and the operating system capabilities. Paging and scrolling are the two most common approaches to showing the display area to the user. **Paging** displays a complete screen of characters at a time. The complete display area is known as a *page* (or *screen*). The page is replaced on demand by the next or previous page, much like turning the pages of a book. **Scrolling** moves the displayed information up or down on the screen, one line at a time. This is similar to the way movie and television credits scroll up the screen at the end of a movie. Once again, PC displays offer a wider range of paging and scrolling options.

paging displaying a complete screen of characters at a time.

scrolling displaying information up or down a screen, one line at a time.

> Keyboards and Pointers

Most (but not all) terminals and monitors are integrated with keyboards. The obvious exception is palm computers such as the PalmPilot. The critical features of the keyboard include character set and function keys.

The character set of most PC keyboards is fairly standard. These character sets can be extended with software to support additional characters and symbols. For specialized terminals or workstations, the manufacturer can design custom keyboards. Most keyboards contain special keys called *function keys*. PC keyboards usually have 12 such function keys. Terminals have been known to include as many as 32 function keys. **Function keys** (usually labeled F1, F2, and so on) can be used to program certain common, repetitive operations in a user interface (for example, HELP, EXIT, and UPDATE). In an operating system, function keys are often predefined, but application developers can customize them for specific systems. Function keys should be used consistently. That is, any information system's programs should consistently use the same function keys for the same purposes. For example, F1 is commonly used as the help key in both operating systems and applications.

function keys a series of special keyboard keys used to program special operations.

Most GUIs (including operating systems and browsers) use pointing devices including mice, pens, and touch-sensitive screens. Obviously, the most common pointer is the mouse. A **mouse** is a small hand-size device that sits on a flat surface near the terminal. It has a small roller on the underside. As you move the mouse on the flat surface, it causes the pointer to move across the screen. Buttons on the mouse allow you to select objects or commands to which the cursor has been moved. Driven by the need to scroll through Web pages and other documents, many mice now include a wheel that allows a user to more easily scroll through pages and documents without using the scroll bars.

mouse a device used to cause a pointer to move across a display screen.

Pens are becoming important in applications that use handheld devices (such as PalmPilots). Because such devices frequently don't include keyboards, the user interface may need to be designed to allow "typing" on a keyboard displayed on the screen or using a handwriting standard such as *Graffiti* or *Jot*. Prebuilt components exist to implement these common features.

As previously noted, the most common user interface is graphical—either *Windows*-based or Web browser–based. The remainder of this chapter will focus on graphical user interface design.

Graphical User Interface Styles and Considerations

User interface design is the specification of a dialogue or conversation between the system user and the computer. This dialogue generally results in data input and information output. There are several styles of graphical user interfaces. Traditionally these styles were viewed as alternatives, but they are increasingly blended. This section presents an overview of several different styles or strategies used for designing graphical user interfaces and how they are being incorporated into today's applications. We will demonstrate these styles with popular software applications.

> Windows and Frames

The basic construct of a GUI (both operating system– and browser-based) is the *window*. A window is a rectangular, bordered area. A title (and optionally a file name) is displayed at the top of each window.

A window can be smaller or larger than the actual display monitor's viewable area. It usually includes standardized controls in the upper right-hand corner to *maximize* itself to the display screen's size, *minimize* itself to an icon (at the bottom of the screen), toggle to a previous size, and *exit* (or *close*).

The file, form, or document displayed within a window may or may not fit in that window. When the file, form, or document exceeds the window size, *scroll bars* on the right-hand side and bottom of the window are used to navigate that file, form, or document and indicate the current position of the cursor relative to the entire file, form, or document.

A window may be divided into zones called *frames*. Each frame can act independently of the other frames in the same window, using features such as paging, scrolling, display attributes, and color. Each frame can be defined to serve a different purpose. Frames are common in both *Windows* and Web browsers.

Within a window or frame, you can use all of the *user interface controls* that were used in the previous two chapters (such as *text boxes, radio buttons, check boxes, drop-down lists, buttons,* etc.). Additionally, many other user interface types of controls will be introduced later in the chapter.

Finally, a window frequently has a *task bar* or tray across the bottom of the window. This task bar can be used to display messages, progress, or special tools (to be discussed later).

> Menu-Driven Interfaces

menu driven a dialogue strategy that requires that the user select an action from a menu of choices.

The oldest and most commonly employed dialogue strategy is menu selection. Different types of menus cater to novice and expert users. **Menu-driven** strategies require that the user select an action from a menu of alternatives.

Menu-driven dialogues actually predate GUIs. A typical pre-GUI hierarchical menu is illustrated in Figure 17-1. Menu options can be logically grouped into high-level options to simplify presentation. As shown in the figure, if the main menu option DISPLAY WARRANTY REPORTS is selected, the submenu WARRANTY SYSTEM REPORT MENU will appear. Then, if the PART WARRANTY SUMMARY option is selected, the report customization and report screens are displayed in sequence. There is no technical limit to how deeply hierarchical menus can be nested. However, the deeper the nesting, the greater the need for direct paths to deeply rooted menu options for the expert user, who may find navigating through multiple levels annoying (called *screen thrashing*). And most users also require ways to escape back to the main or higher-level submenus without backtracking through each of the original screens.

Pre-GUI hierarchical menus were relatively easy to design. A dialogue chart such as the one shown in Figure 17-2 (taken from an earlier edition of our book) was used to map the screen-to-screen transitions and ensure consistency and completeness. But the arrival of graphical user interfaces greatly complicated menu design.

In operating system GUIs such as Microsoft *Windows,* user dialogues are not hierarchical. Think about it! Between the time you start and exit a *Windows* application, such as your word processor, the number of different actions and paths you take through your application is seemingly endless. The dialogue is hardly hierarchical! Such dialogue design cannot be modeled as easily by hierarchically based dialogue charts. Let's examine how GUI menus work.

Pull-Down and Cascading Menus In a GUI, menus are usually implemented with pull-down and cascading menus from a *menu bar* as shown in Figure 17-3(a). Each menu option is actually a group of related commands and actions. A menu template is shown in

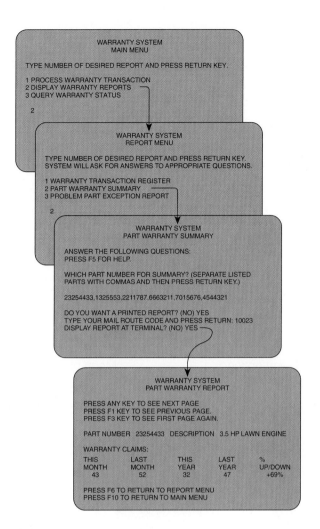

FIGURE 17-1

A Classical Hierarchical Menu Dialogue

the margin. Many of these menu groups are common to many or all applications. For example, *Windows*-based applications typically include the following menu groups:

File	Edit	View	Format	Tools	Actions	Window	Help

Users can select a menu group using either the mouse or a keyboard shortcut (e.g., simultaneously pressing the Alt-key plus the <u>underlined letter</u>, called a *mnemonic, shortcut,* or *hot key*).

Each menu group has its own *pull-down menu.* When the user selects a group from the menu bar, a submenu is pulled down:

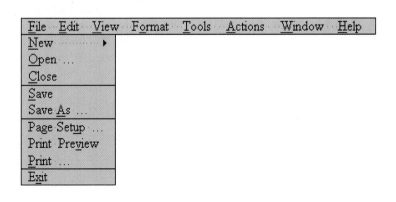

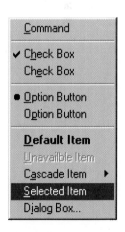

Menu Template

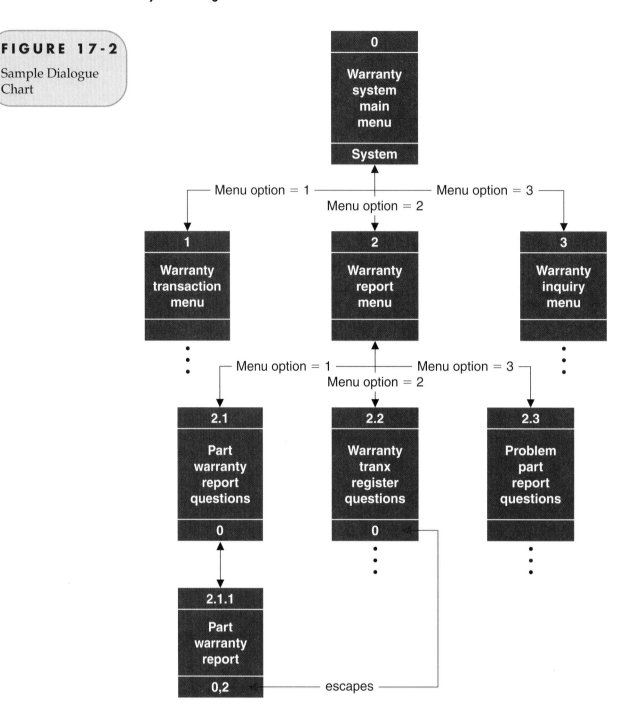

Notice that the submenu choices may be subgrouped by horizontal lines (e.g., grouping all SAVE or PRINT submenu commands). In some cases, a named submenu action is followed by ellipses (three dots) indicating that a *dialogue box* (window) (see Figure 17-3b) will subsequently appear (pop up) to present additional options or collect additional instructions. In other cases, a named submenu action will have a small arrow indicating yet another submenu. This is called a *cascading menu.*

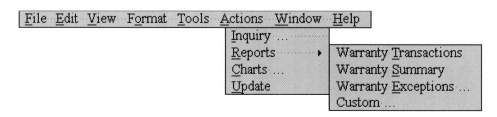

(a)

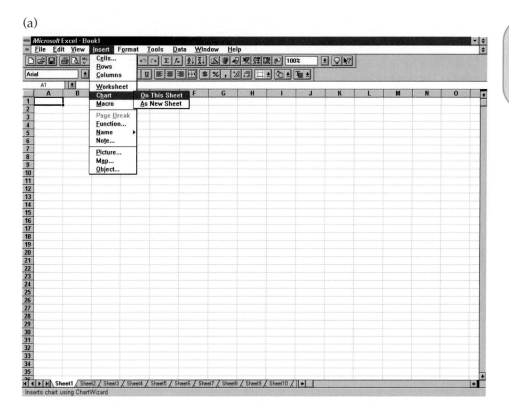

FIGURE 17-3

(a) Pull-Down and
Cascading Menus
(b) Dialogue Box

(b)

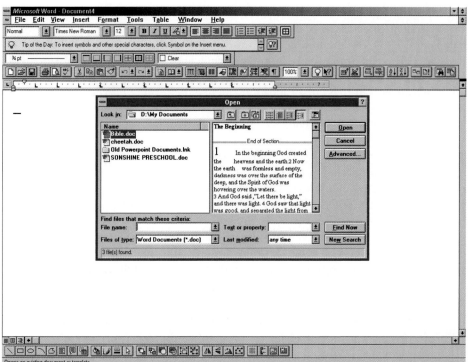

Tear-Off and Pop-Up Menus   Not all menus are relegated to the menu bar. Some
GUIs allow *tear-off menus.* With a tear-off menu, the user can select a drop-down
menu or cascaded menu, "drag it off" the menu bar, and relocate it elsewhere on the
screen. This is especially useful if the menu must be continually used. Only a copy of
the original menu is actually torn off.

A *pop-up menu* is context-sensitive and dependent on a pointing device. Activated by the user's clicking of the right mouse button, a menu pops up from nowhere (see Figure 17-4). The menu that pops up depends on the location of the cursor on the screen. The cursor may be pointing to a blank area, a field, a cell, a word, or an object. The right-button click will bring up a menu displaying only those actions that apply to whatever is at that cursor location—hence the term *context sensitive.* Pop-up menus may also cascade. Pop-up menus are primarily for expert users because there is no visual clue to their presence.

Toolbar and Iconic Menus   *Toolbars* consist of *icons* (pictures) that represent menu shortcuts for actions and commands that are normally embedded in the drop-down and cascading menus (see Figure 17-5). In *Windows* applications, a toolbar of commonly used actions is found immediately beneath the menu bar. The user can click on any of these tools or icons to immediately invoke that action without going through the menus. Toolbars can be created for any application. Application developers can provide users with some flexibility for customizing those toolbars.

While the default location for most toolbars is immediately under the menu bar, many applications allow toolbars to be relocated to the left, right, or bottom of the window at the convenience of the user. This is called *docking* the toolbar. Also, some toolbars can be made to *float* (or move) within any convenient location <u>inside</u> the window.

> NOTE: In Web-based applications, the toolbar is provided by the browser and cannot be customized to specific applications. The most important icons on the browser toolbar are the PAGE FORWARD, PAGE BACKWARD, and HOME PAGE icons that are standard to all Web-based Internet and intranet navigation.

Iconic menus use pictures to represent menu options in the main <u>body</u> of the window. In *Windows* applications, these iconic menus are frequently used to provide a control center (of main functions and activities) for a computer application or to

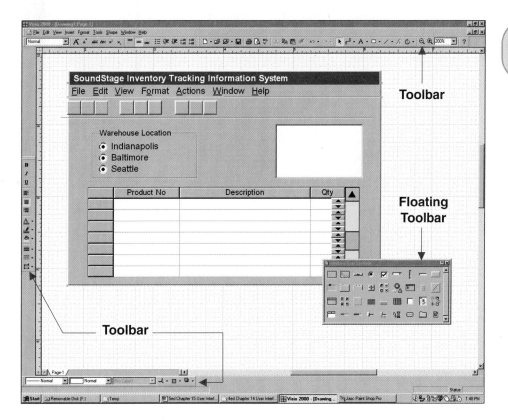

FIGURE 17-5
Toolbars

document the business steps in using a computer application. Figure 17-6 demonstrates an iconic menu. Each button represents an intuitive menu choice.

Iconic menus are very popular in Web-based applications because those applications run in the browser—browsers do not allow the developer to alter the menu commands in the browser's menu bar. Instead, Web applications frequently use clickable pictures, icons, and buttons to represent the menu options.

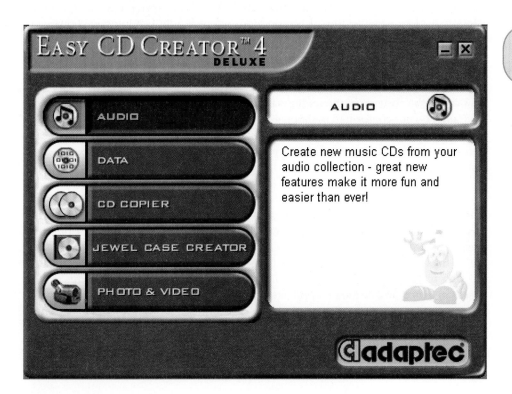

FIGURE 17-6
Iconic Menu

FIGURE 17-7

Consumer-Style Interface

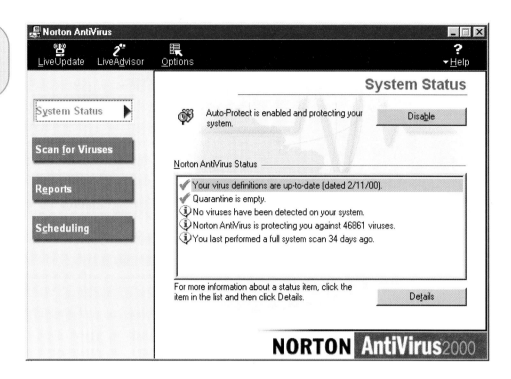

The popularity of Weblike interfaces is significantly influencing *Windows* user interface design. Most client/server information systems have implemented the client user interface to emulate the user's most commonly used PC tools such as the word processor and spreadsheet. The user is familiar with those tools; therefore, it makes sense to design other applications to mimic those menus, toolbars, dialogue boxes, and the like. But the popularity of Web-based applications has given rise to a new consumer-style interface.

Like Web pages, *consumer-style interfaces* for *Windows* applications are somewhat more artistic. While menu bars may still be used, the primary look and feel of the window is more Weblike; thus, it is more consumer-friendly. The interface consists of clickable icons and buttons that replace more traditional *Windows* menu approaches. When not overly complicated, this can be a friendlier "face" for *Windows* applications than is traditionally seen in applications such as Microsoft *Office* and Lotus *Smart-Suite*. A consumer-style *Windows* interface is illustrated in Figure 17-7.

Notice the absence of the traditional menu bar. Also notice that the buttons (in the left frame) do not conform to traditional *Windows* size and style. The background image is more artistic, as is the use of fonts and color. (Many organizations include a graphic designer as part of the team to develop consumer-style interfaces.) We expect such consumer-friendly styles to be embraced by future *Windows*-based information systems.

Hypertext and Hyperlink Menus *Hypertext* and *hyperlinks* are products of contemporary Web-based user interfaces. Hypertext and hyperlinks were originally created to navigate within and between Web pages and sites. A word, term, or phrase is marked as a hyperlink (usually formatted as underlined text, usually with color). Clicking on the hyperlink navigates the user to the associated page (or bookmark in a page).

This technology can be easily extended and adapted to implement menus in Web-based Internet and intranet applications. Because these applications run in the browser, and because the browser's menu bar and commands are fixed, we cannot easily implement custom menus as we do in *Windows* applications. Instead, we use hypertext and hyperlinks to implement those menus in the body of the Web page. Each menu option is a hypertext phrase (or a hyperlinked icon or button) that invokes actions or forms on other Web pages. Essentially, this approach creates

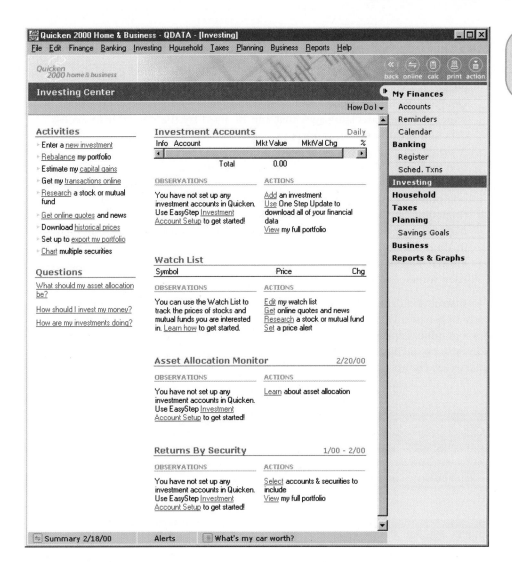

FIGURE 17-8

Hybrid *Windows/* Web Interface

hierarchical menu structures similar to those that were introduced earlier in Figure 17-1. It is something of an irony that menu design for Web-based applications is being driven by an approach that returns to a style that was extensively used in legacy mainframe applications!

Hypertext and hyperlinks are no longer exclusive to Internet and intranet applications. Many contemporary *Windows* applications have embraced the popularity of the Web by presenting a *hybrid Windows/Web* user interface. For example, Figure 17-8 demonstrates the user interface for Intuit's *Quicken,* the popular personal finance program. With its many hyperlinks, it looks like a Web page. While it does include many optional Web-enabled features, it is actually a *Windows* application! The first clue is that it runs in its own window, not the browser's window. It also has its own *Windows* menu bar, complete with all the custom pull-down and cascading menus that are common to *Windows* applications. We expect this hybrid interface to become increasingly pervasive as businesses embrace the Internet and intranets as the fundamental foundation for all information systems.

> Instruction-Driven Interfaces

Instead of menus, or in addition to menus, some applications are designed using a dialogue based on an *instruction set* (also called a *command language interface*). Because the user must learn the syntax of the instruction set, this approach is most suitable for expert users. Three types of syntax can be defined. Determining which type should be used depends on the available technology:

- A *language-based syntax* is built around a widely accepted command language that can be used by the user to invoke actions. Examples include *Query by Example (QBE)* and *Structured Query Language (SQL),* both of which are database languages that can be used by the end user to access data and create custom reports.
- A *mnemonic syntax* is built around commands defined for custom information system applications. Users are provided with a screen console in which they can enter commands that will invoke actions and responses from the computer user. Ideally, the commands should be meaningful to the user (including any abbreviations allowed).
- *Natural language syntax* allows users to enter questions and commands in their native language. The system interprets these commands against a known syntax and requests clarification if it doesn't understand what the user wants.

Instruction-driven styles were common to legacy mainframe applications and early DOS-based PC applications. But this style of interaction can still be found in today's graphical applications. For example, Microsoft's *Access* database product contains a query facility that allows the developer to visually (point and click) develop a query (see Figure 17-9). The developer simply selects from database tables, columns,

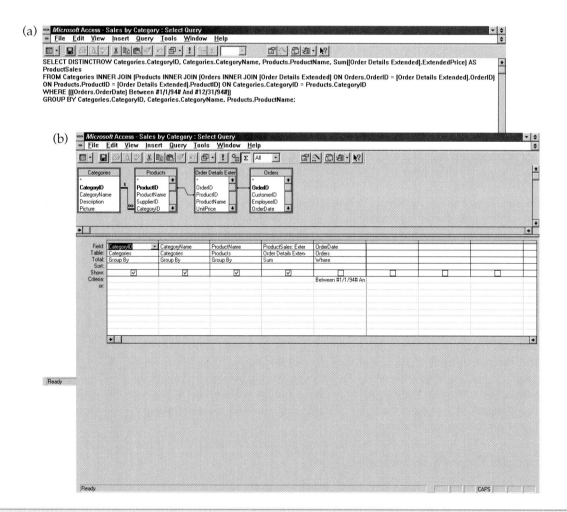

FIGURE 17-9 Instruction-Driven Interface

and rows to include in a query, as shown in Figure 17-9(a). Then, if desired, the developer can view and edit the command-level SQL code that implements the query, as shown in Figure 17-9(b). Once again, the instruction set approach requires a degree of user expertise, experience, and know-how.

> Question-Answer Dialogues

A *question-answer dialogue* style is primarily used to supplement either menu-driven or instruction-driven dialogues. Users are prompted with questions to which they supply answers. The simplest questions involve YES or NO answers—for instance:

DO YOU WANT TO SEE ALL PARTS? [NO].

Notice how the user was offered a default answer! Questions can be more elaborate. For example, the system could ask:

WHICH PART NUMBER ARE YOU INTERESTED IN?

This strategy requires that you consider all possible correct answers and deal with the actions to be taken if incorrect answers are entered. Question-answer dialogue is difficult because you must try to consider everything that the system user might do wrong!

Question-answer dialogues are very popular in Web-based applications. For example, a car reservation system may ask a series of questions to define what type of car and rental agreement you require:

WHERE DO YOU WANT TO PICK UP YOUR RENTAL VEHICLE?
WHERE DO YOU PLAN TO RETURN YOUR RENTAL VEHICLE?
WHAT IS THE PICKUP DATE AND TIME?
WHAT IS THE RETURN DATE AND TIME?
WHAT TYPE AND SIZE OF VEHICLE DO YOU NEED?
DO YOU HAVE ANY PROMOTIONAL COUPONS? . . .

A drop-down list of alternative answers may accompany each question. Together, these questions and answers define a business transaction.

> Special Considerations for User Interface Design

In addition to establishing a user interface style, analysts must address certain special considerations for user interface design. How will users be recognized and authenticated to use the system? Are there any security or privacy considerations to be accommodated in the user interface? Finally, how will users get help via the user interface?

Internal Controls—Authentication and Authorization   In most environments, system users must be authenticated and authorized by the system before they are permitted to perform certain actions. In other words, system users must "log into" the system. Most log-ins require both a USER ID and a PASSWORD. System users should not be required to learn and memorize multiple USER IDS and PASSWORDS. Ideally, they should be required to use the same log-in as is used for their local area network account. (*Windows XP, NT,* and *2000* allow for this authentication to occur without the need to retype either field.)

Figure 17-10(a) demonstrates the user interface for the SoundStage log-in. The USER ID and PASSWORD will be authenticated against the network accounts file. Notice that the password is printed as asterisks as the user types it in, a common security and privacy measure. Should the user ID or password fail to be authenticated, the security authorization dialogue in Figure 17-10(b) will be displayed.

Authentication is only half of the solution. Once authenticated, the user's access and service privileges for this information system must be established. There are many models for establishing and managing privileges. An important guideline is to assign privileges to *roles,* not to individuals. In most businesses, people change jobs routinely—they are reassigned and promoted to new job responsibilities and roles. Also, job descriptions and

(a)

(b)

roles change from time to time. Finally, people leave the business and some are terminated. For all of these reasons, privileges should be assigned to roles. Then it is a simple matter of identifying the roles that any USER ID can assume.

For each role, the specific privileges that should be assigned to the role need to be defined. Privileges may include permission to read specific tables or views; permission to create, update, or delete records (rows) in specific tables or views; permission to generate and view specific reports; permission to execute specific transactions; and the like. Although not technically part of the interface, defining these roles and permissions is needed both to design an appropriate log-in interface and to functionally specify the complete authentication and authorization security model for the system.

Different user views could actually be applied to customize the user interface for different categories of users. For example, it is fairly easy to "ghost" (change the font from black to gray) and disable those menu options and dialogue boxes that are to be restricted from certain classes of system users.

With the emergence of e-commerce, consumers and other businesses must have confidence that we are who we claim to be. Consumers may be providing credit card numbers and other private information for transmission over the Internet. For this reason, SoundStage purchased a Web certification to authenticate itself to its club members and prospective members. At any time, using the browser interface, SoundStage members can view the authentication certificate in Figure 17-11. With this certification, the SoundStage Web site will display a "Secure Server Certification" icon (see margin—the *padlock*) that will tell consumers their data will be encrypted (securely scrambled) to ensure that their credit card and personal data is not being intercepted or accessed by others when passed along the network.

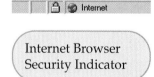

Internet Browser
Security Indicator

Online Help   Nobody wants to read the manuals anymore! At least that's the way it seems. And to some degree, it's justified. Manuals are essentially sequential files of

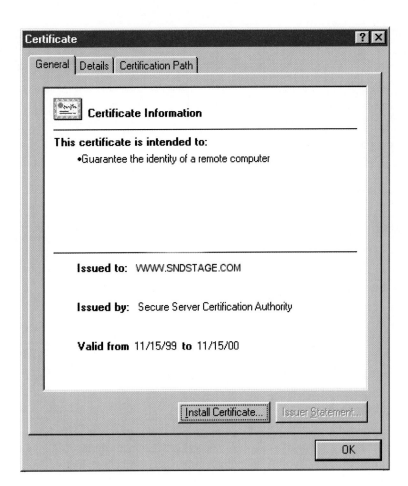

FIGURE 17-11

Server Security
Certificate

information. People want immediate, direct access to context-sensitive help, that is, help that is smart enough to figure out what they might be trying to do. There is definitely a trend toward building help systems and tutorials directly into the application. Online help becomes part of the user interface.

The general-purpose help for an application is built into the Help menu for *Windows* applications. For Web applications, help is usually built as separate pages, usually "pop-up" pages in separate windows, so that the user can also remain focused on the page that initiated the need for help.

Today, HTML (Hypertext Markup Language) is gradually becoming the universal language for constructing help systems for graphical user interfaces—both Web and *Windows* applications. For example, the entire help system for Microsoft *Office* is now written in HTML.

The design, construction, and testing of a help system is simplified by today's automated tools. A complete *help system* includes a table of contents, numerous instructions, examples, and a thorough index. Many *help authoring packages,* such as Macromedia's *RoboHelp,* leverage the help author's word processor to help with the planning, outlining, writing, indexing, and hypertext-linking aspects of authoring a complete help system.

A well-designed help system will implement a wide range of help elements. Perhaps the most commonly encountered types of help are those that users must initiate. As mentioned earlier, the F1 function key is almost universally accepted as a help request command. Likewise, a standard Help menu bar option is commonly used to organize and present different types and levels of help in most *Windows*-based applications (commercial or custom-built). Finally, as is illustrated in Figure 17-12, *Windows* and Web-based interfaces frequently use *tool tip* controls to provide pop-up help associated with specific tool and object icons. Tool tips appear when the user

FIGURE 17-12
Help Tool Tip

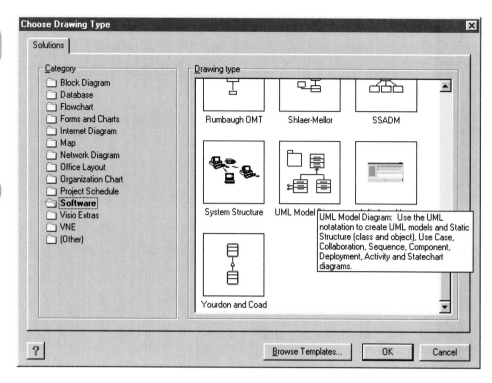

Help Agent

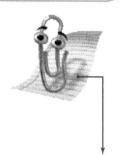

What would you like to do?

- Change the default chart type
- Troubleshoot charts
- Change the display of chart labels, data tables, legends, gridlines, or axes
- Change the view of a 3-D chart
- Change chart labels, titles, and other text
- Change the way data is plotted

How do I customize the X-Axis on my chart

| Options | Search |

Natural Language Processing

agent reusable software object that can operate across different applications and networks.

momentarily positions the cursor over the icon (or object) on the screen. Tool tips are appropriate for all icons because the user interface designer can never be assured that the image or label appearing on an icon is going to be meaningful to the system user.

Two additional and common help features particularly effective for the more novice user are help wizards and help agents (or assistants). As is illustrated in Figure 17-13(a), a *help wizard* guides the users through complex processes by presenting a sequence of dialogue boxes that require user input and system feedback. We call your attention to the following:

1 As is typical of help wizards, the dialogue usually includes a series of instructions or questions for the user to respond to.

2 The wizard contains explanations to aid in the user's understanding and decision making.

3 The wizard also provides a button for requesting more detailed help in completing the task.

4 The "Next" button suggests additional or subsequent steps to be supported by the help wizard. (The "Next" button is usually changed to "Finish" once a sequence of dialogue boxes is complete.) Figure 17-13(b) shows the resulting screen and subsequent step supported by the help wizard.

Microsoft and third-party software control vendors actually sell wizards to help developers construct wizards!

Agents are another technology with applications to help systems. **Agents** are reusable software objects that can operate across different software applications and even across networks. Microsoft's *help agent* (referred to as an *assistant*) provides a common help assistant in *Office* applications. In its default form, it presents itself as an animated paper clip (see margin). (Microsoft's help agent can be programmed into custom applications, both for *Windows* and for the *Internet Explorer* Web browser.) A single user click on this help agent initiates help.

The Microsoft help agent is complemented by natural language processing technology (see margin) that allows the user to write an inquiry in natural language phrases that are interpreted by the agent to present the most likely help responses. The user can then select one of those responses or enter into the more detailed help index.

(a)

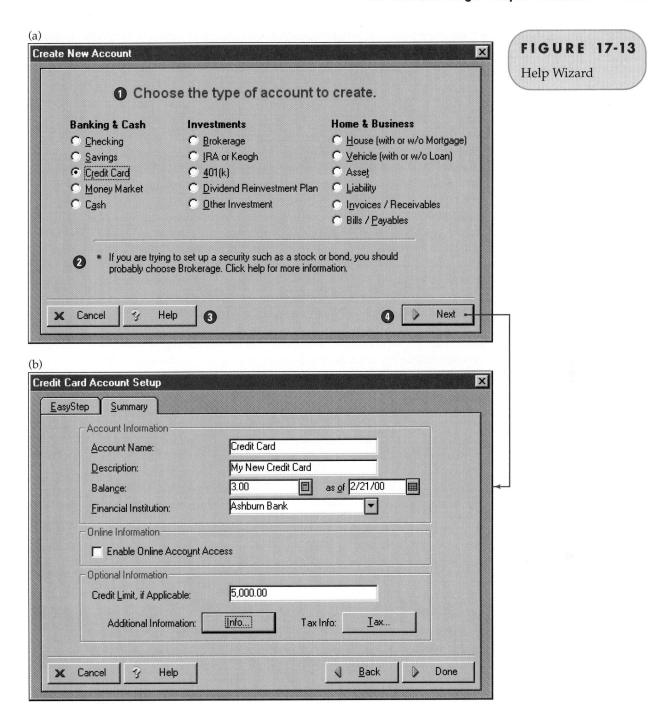

FIGURE 17-13

Help Wizard

(b)

The overriding theme for designing a good help system is that the designer should anticipate system user errors. When designing the user interface to report such errors, the designer should always provide the system user with help to resolve the error. After leaving any help session, users should always be returned to where they were in the application before requesting or receiving the help.

How to Design and Prototype a User Interface

Today's graphical environments create an emphasis on developing an overall system that blends well into the user's workplace environment. The following sections will demonstrate how to design a user interface for a graphical environment. We will draw on examples from the SoundStage case study. We will examine both client/server, *Windows*-based inputs and Web-based, e-commerce inputs that run in a browser.

> Automated Tools for User Interface Design and Prototyping

The automated tools for supporting user interface design and prototyping are the same as the tools we identified in Chapters 15 and 16 for output and input design. The most commonly used automated tool for user interface design is the PC-database application development environment. Most PC-database products such as Microsoft's *Access* are not powerful enough to develop most enterprise-level applications, but they are more than adequate to use in prototyping an application's user interface screens. Given a database structure (easily specified in *Access*), you can quickly generate or create forms for inputting data. You can include most of the GUI controls we described in this chapter. The users can subsequently exercise those forms and tell you what works and what doesn't.

Many CASE tools also include facilities for screen layout and prototyping using the project repository created during requirements analysis. *System Architect's* screen design facility was previously demonstrated in Chapter 15, Figure 15-7.

Most GUI-based application development environments, such as Microsoft's *Visual Studio,* can be easily used to construct nonfunctional prototypes of user interface screens. The key term here is *nonfunctional.* The forms will look real, but there will be no code to implement any of the buttons or fields. That is the essence of rapid prototyping. For example, Figure 17-14 demonstrates a *Visual Studio* dialogue for building a simple menu.

In Chapter 16 we introduced a number of input controls that could be included in any window. The number of controls available to the interface designer is limited only by the applications development environment that will be used to construct the interface. Figure 17-15 illustrates a few additional controls that are available in the *Visual Studio* environment, including outlook bars, sortable columns with headings, gauge controls, directory list boxes, and noninput drop-down lists.

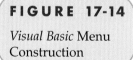

FIGURE 17-14

Visual Basic Menu Construction

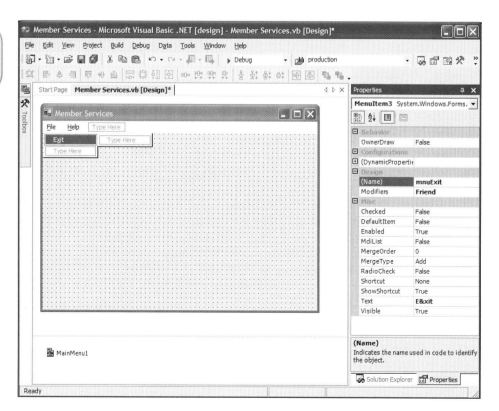

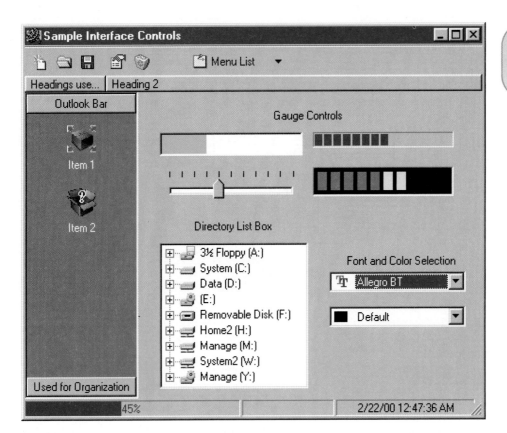

FIGURE 17-15

Additional User
Interface Controls

> The User Interface Design Process

User interface design is not a complicated process. The basic steps involved are:

1. Chart the user interface dialogue.
2. Prototype the dialogue and user interface.
3. Obtain user feedback.
4. If necessary, return to step 1 or 2.

In reality, the steps are not strictly sequential in practice. Instead, the steps are iterative—for example, as prototypes are developed, they are reviewed by the system users, who provide feedback that may require revisions or a new prototype. In the following subsections, we examine these steps in a single iteration and illustrate a few examples from the SoundStage project.

Step 1: Chart the Dialogue   A typical user interface may involve many possible screens (which may consist of several windows), perhaps hundreds! Each screen can be designed and prototyped. But what about the coordination of these screens?

Screens typically occur in a specific order. You may also be able to toggle among the screens. Additionally, some screens may appear only under certain conditions. To make matters even more difficult, some screens may occur repetitively until some condition is fulfilled. This sounds almost like a programming problem, doesn't it? We need a tool to coordinate the screens that can occur in a user interface. A **state transition diagram (STD)** is used to depict the sequence and variations of screens that can occur when the system user sits at the terminal. (The authors are using the term *screen* in a general sense. When graphical interfaces are being designed, the term may refer to an entire display screen, a window, or a dialogue box.) You can think of it as a road map. Each screen is analogous to a city. Not all roads go through all cities. Rectangles are used to represent display screens. Arrows represent the flow of control and the triggering event causing the screen to become active or receive focus.

state transition diagram (STD) a tool used to depict the sequence and variation of screens that can occur during a user session.

FIGURE 17-16

SoundStage Partial
State Transition
Diagram

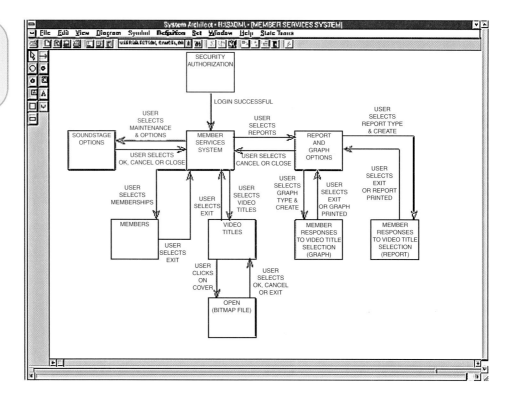

The rectangles describe only what can appear during the dialogue. The direction of the arrows indicates the order in which these screens occur. A separate arrow, each with its own label, is drawn for each direction because different actions trigger flow of control from and flow of control to a given screen.

Let's examine a dialogue that is under construction for the SoundStage project (see Figure 17-16). The <u>partially completed</u> SoundStage state transition diagram is being developed using a CASE product, Popkin's *System Architect*. Note the following:

❶ The partial state transition diagram includes references to some of the Sound-Stage input screens developed in Chapter 16.

❷ The diagram also includes references to some of the output screens designed in Chapter 15.

❸ The MEMBER SERVICES SYSTEM screen will be a new screen that will need to be designed and prototyped. This screen will serve as the application's main window. It will play a major role in providing the user with the ability to get access to the system's input and output screens, which were designed earlier. It will also provide the user with the ability to complete a number of additional functions (beyond input and output processing) that are commonly established during user interface design. It will be accessible only when the users have first been provided with the SECURITY AUTHORIZATION screen and have successfully logged into the system.

❹ The SOUNDSTAGE OPTIONS screen is another new screen to be created. This screen will allow users to set various user options and defaults to be used during their session—for example, selecting a printer, zooming, and many other options.

State transition diagrams such as the one presented in Figure 17-16 can become quite large, especially when all input, output, help, and other screens are added to the diagram. Therefore, it is common to partition the diagram into a set of separate simpler and easier-to-read diagrams.

Step 2: Prototype the Dialogue and User Interface   Recall that we have some new screens to design and prototype. Some of these new screens were identified to

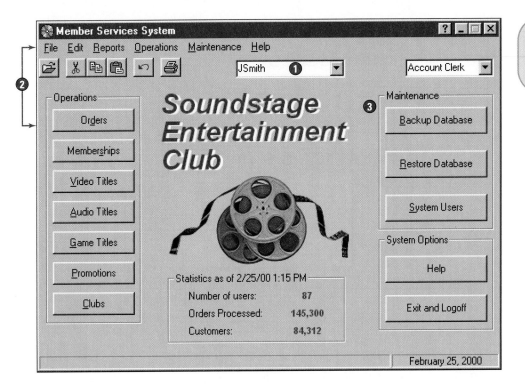

bring together the application and its input and output screens that were designed earlier. Some screens were identified to provide the users with some flexibility in customizing the application's interaction to suit their own preferences. Still others may have been identified to deal with system controls, such as backup and recovery.

Let's look at some new screens that were to be created for the SoundStage Member Services System. System users would first be presented with the authentication log-in screen that was discussed earlier in the chapter (see Figure 17-10). According to the state transition diagram, the successful log-in of a user results in the SoundStage Member Services System main menu screen depicted in Figure 17-17. Notice the following:

❶ The users and their access privileges are confirmed. Based on the users' access privilege, certain functions will be enabled and disabled.

❷ Through a menu bar selection *or* through a vertical menu of buttons, the user can complete common Member Services business operations. These buttons will lead to screens that allow the user to process appropriate transactions via input screens designed and prototyped earlier. Text labels were used for buttons because the analyst was unable to establish icons (pictures) that all users could readily identify with as a representation of the operations. The menu bar and buttons contain hot keys to provide the user with the flexibility of selecting via the keyboard or mouse. A group box was used to visually associate the buttons that represent related operations.

❸ The user has the ability to complete various routine maintenance operations.

Via the menu bar of the MEMBER SERVICES SYSTEM screen, users can choose to set options for their work session. This new screen is depicted in Figure 17-18.

❶ This screen utilizes *tabs* as a means of allowing the user to alter four related sets of options.

❷ A *slider* control is used to allow the user to adjust the priority for background queries. This control is often used for items whose values are best presented as a spatial representation and when an approximate rather than precise value is sufficient.

FIGURE 17-18

SoundStage Options
and Preferences
Screen

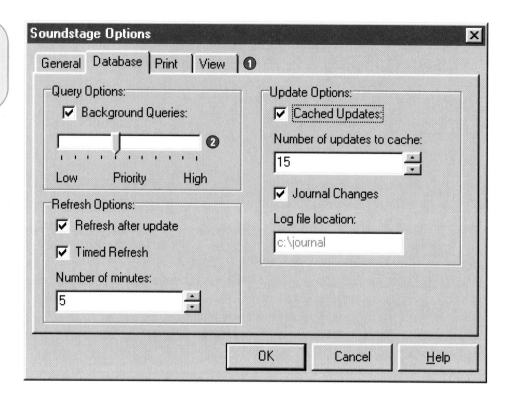

In reality, the analyst would need to prototype the content and appearance of the "General," "Print," and "View" tabs as well as the "Database" tab. According to the state transition diagram, this screen will return control to the parent window, MEMBER SERVICES SYSTEM.

According to the state transition diagram, system users are also to be provided with the opportunity to specify report customization preferences. Figure 17-19 depicts a prototype screen that allows SoundStage users to choose a particular report (or graph) and customize its content.

FIGURE 17-19

SoundStage Report
Customization
Dialogue Screen

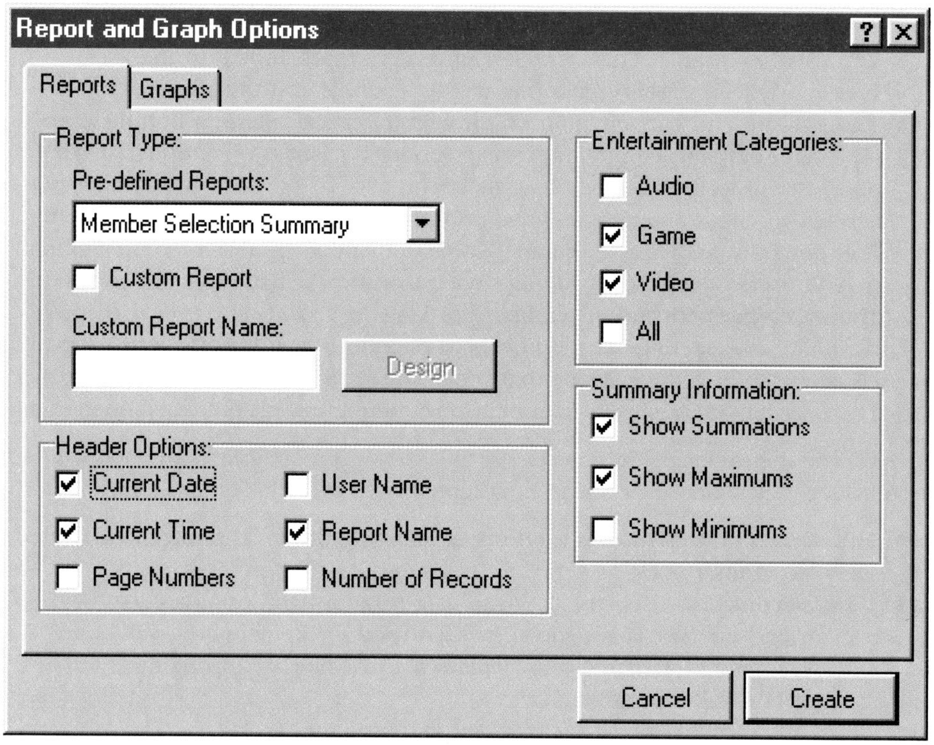

Study the state transition diagram and the screens that we just examined to see how this portion of the overall system dialogue would work. By studying the entire collection of screens, you may discover the need to revise some screens. Such issues as color, naming consistencies of common buttons and menu options, and other look-and-feel conflicts may need to be resolved. Once again, adherence to any standards governing GUIs should be confirmed.

Step 3: Obtain User Feedback   Exercising (or testing) the user interface is a key advantage of all the prototyping environments we have alluded to throughout this chapter. *Exercising* (or *testing*) *the user interface* means that system users experiment with and test the interface design before extensive programming and actual implementation of the working system. Analysts can observe this testing to improve on the design.

In the absence of prototyping tools, the analyst should at least simulate the dialogue by walking through the screen sketches with system users. User feedback is essential in user interface design. The analyst should encourage the user to participate in testing the application's interface. Finally, the analyst should expect to revisit steps 1 and 2 as needed changes become known.

This chapter provided a detailed overview of user interface design and prototyping. If you are taking an object-oriented approach and have not already done so, you will want to proceed to Chapter 18. If you are taking a traditional approach, you will move to Chapter 19 and the construction phase. The construction phase will cover how to build and test the networks, databases, and programs for the new system. The implementation phase will cover how to conduct a system test, develop a conversion plan, install databases, train users, and complete the conversion from the old to the new system. You will learn about the construction and implementation phases in the next chapter. Before proceeding, we recommend that you first review Chapter 3 to see where software design falls in the overall systems development process.

Learning Roadmap

Chapter Review

1. User interface design is concerned with the dialogue between a user and the computer. It is concerned with everything from starting the system or logging into the system to the eventual presentation of desired outputs and inputs.

2. Most user interfaces are designed by rapidly constructing prototypes using rapid application development environments. Such prototypes are rarely fully functional.

3. Relative to user interface design, the system users can be broadly classified as either expert or novice:

   a. An expert user is an experienced computer user who will spend considerable time using specific application programs. The use of a computer is usually considered nondiscretionary.

b. The novice user is a less experienced computer user who will generally use a computer on a less frequent, or even occasional basis. The use of a computer <u>may</u> be viewed as discretionary.

4. Several human factors frequently cause people to have difficulty with computer systems, including these interface problems:

   a. Excessive use of computer jargon and acronyms.
   b. Nonobvious or less-than-intuitive design.
   c. Inability to distinguish between alternative actions ("What do I do next?").
   d. Inconsistent problem-solving approaches.
   e. Design inconsistency.

5. Galitz offers the following overriding "commandments" of user interface design:

   a. Understand your users and their tasks.
   b. Involve the users in interface design.
   c. Test the system on actual users.
   d. Practice iterative design.

6. Given the type of user for a system, there are a number of important human engineering factors that should be incorporated into the design:

   a. The system user should always be aware of what to do next.
   b. The screen should be formatted so that the various types of information, instructions, and messages always appear in the same general display area.
   c. Messages, instructions, or information should be displayed long enough to allow the system user to read them.
   d. Use display attributes sparingly.
   e. Default values for fields and answers to be entered by the user should be specified.
   f. Anticipate the errors users might make.
   g. A user should not be allowed to proceed without correcting an error.
   h. The system user should never get an operating system message or fatal error.
   i. If the user does something that could be catastrophic, the keyboard should be locked to prevent any further input, and an instruction to call the analyst or technical support should be displayed.

7. The overall flow of screens and messages is called a dialogue. With respect to the tone of the dialogue, the following guidelines are offered:

   a. Use simple, grammatically correct sentences.
   b. Don't be funny or cute.
   c. Don't be condescending.

With respect to the terminology used in a computer dialogue, the following suggestions may prove helpful:

   a. Don't use computer jargon.
   b. Avoid most abbreviations.
   c. Use simple terms.
   d. Be consistent in your use of terminology.
   e. Carefully phrase instructions—use appropriate action verbs.

8. Most of today's user interfaces are graphical. The basic structure of the graphical user interface (or GUI) is provided either within the computer operating system or in the Internet browser.

9. The overall design of a user interface is enhanced or restricted by the available features of the user's display monitor, keyboard, and pointing devices.

10. There are several styles of graphical user interfaces, including menu-driven, instruction-driven, and question-answer dialogues.

11. Menu-driven strategies require that the user select an action from a menu of alternatives. GUI menu implementation may include:

   a. Pull-down and cascading menus.
   b. Tear-off and pop-up menus.
   c. Toolbar and iconic menus.
   d. Hypertext and hyperlink menus.

12. Instruction-driven interfaces are designed using a dialogue based on an instruction set. Three types of syntax may be used for the instruction set:

   a. Language-based syntax, which is built around a widely accepted command language that can be used by the user to invoke actions.
   b. Mnemonic syntax, which is built around commands defined for the custom information system applications.
   c. Natural language syntax, which allows users to enter questions and commands in their own native language.

13. A question-answer dialogue style is primarily used to supplement either menu-driven or instruction-driven dialogues. Users are prompted with questions to which they supply answers.

14. Internal controls and online help are some special considerations that should go into user interface design.

15. User interface design consists of three iterative steps:

   a. Chart the user interface dialogue.
   b. Prototype the dialogue and user interface.
   c. Obtain user feedback.

Review Questions

1. Why should the system users be involved in the process of designing user interfaces?
2. Who are expert users? Why are they called expert?
3. Why can some user interfaces cause users to feel confused, panicky, or frustrated?
4. What does it mean to test the system on actual users?
5. What should we do to ensure the system users are aware of what to do in the system?
6. How should the interfaces handle errors?
7. What are some factors that should be considered in terms of the terminology used in computer dialogues?
8. Why are Web browsers becoming more important when designing applications?
9. Explain paging and scrolling.
10. What should we consider when we design function keys for our applications?
11. Why are pens used in applications?
12. What is the relationship between windows and frames?
13. What are characteristics of a pop-up menu?
14. What are steps of the user interface design process?
15. What is the tool used to facilitate the charting of the dialogue?

Problems and Exercises

1. A fellow designer has asked you to review the dialogue to be used in several screens for a new application. Do you think the following messages comply with the guidelines for tone and terminology? Explain your answer as needed.

   a. An error message that says DISCHARGE DATE MUST BE ON OR AFTER ADMISSION DATE.
   b. An instruction that says ENTER THE CLIENT'S NAME NOW.
   c. An error message that says DATA IS SO FAR OUT OF RANGE IT HAS LEFT THE SOLAR SYSTEM.
   d. A question that asks customers DO YOU WANT TO RDF THE ACR BEFORE "UCI'ING" CMIS?
   e. An error message that says DON'T WORRY—NOT EVERYONE GETS IT RIGHT THE FIRST TIME.

2. Menu-driven interfaces, although older than GUI interfaces, are still very common. What type of user dialogue strategy does a menu-driven interface employ? What is the main difference between menu-driven interfaces and GUI interfaces? What is the major advantage of a menu-driven interface, and what is its major disadvantage?

3. Select an application that is used in your organization or school and that has more than one user interface screen. Also, try to select an application that you haven't used before, or at least one that you haven't used for a while. Test the user interface design against each of the human engineering guidelines described in the chapter. Make sure to enter incorrect as well as correct data.

   Describe the application and user interface screens that you tested. How many guidelines did the application meet, and which ones were they? How many guidelines did it fail, and which ones were they?

4. Considering the application described in the preceding exercise, what changes would you make to the user interface screens to bring them into compliance with the human engineering guidelines described in the textbook? Be specific in your answers. Have a couple of fellow students and/or co-workers review your changes, one as a user and the other as a designer. Did they find new problems?

5. Answer the following true/false questions. Explain your answers as necessary.

   a. Different action verbs should be used in screen dialogue to describe required keyboard actions in order to add variety and interest.
   b. Most managers are expert users, because they need a high level of PC expertise in order to manage effectively.
   c. Organizations should expect that expert designers, who come highly recommended and who are at the top of their pay scale, will need to refine and modify their user interface designs several times before the result will be satisfactory to the organization.
   d. *Windows* user interface design often borrows from Web interface styles and techniques.
   e. Applications need only one type of help menu or dialogue.

f. Users appreciate clever or humorous screen messages.

g. The process for designing user interfaces is straightforward and easy to understand.

6. In designing user interfaces, consideration must be given to information security and privacy. Describe some of the guidelines and considerations that must be taken into account in building internal controls into the user interface design.

7. Fill in the blanks for the following statements:

a. Some _____ interfaces use a _____ syntax that allows users to ask questions in their _____.

b. The _____ step in user interface design is to _____ to the _____ steps as part of the _____ process until users are _____ with the design.

c. SQL uses _____ syntax which allows _____ users to _____ the database.

d. Good _____ guidelines include _____ possible user _____, and _____ the user of an action's _____ before the action is _____.

e. One guideline for establishing users' _____ privileges is to base their privileges upon _____, not _____.

8. Match the definitions or examples in the first column with the terms in the second column:

A. Frequently required multiple level menu navigation

1. Platform independent

B. An application's overall sequence of screens and messages

2. Consumer-style interface

C. Full screen approach to display area seen by user at a time

3. Terminal emulators

D. Information is moved up or down one line at a time

4. State transition diagram

E. Computer expertise in multiple related fields of study

5. Iconic menu

F. Windows screen employing artistic Web-like "face"

6. Screen thrashing

G. Independent zones within a window

7. Mnemonic syntax

H. User interfaces that are not dependent upon a specific OS

8. B2C

I. Graphic tool used to show screen variation and sequence

9. Dialogue

J. Software to display mainframe screen format in a window

10. Scrolling

K. Command language interface meaningful to user

11. Informatics

L. Functionality based upon business-to-consumer transactions

12. Paging

M. Pictorial representation of menu option in main window body

13. Frames

9. You have been asked to design a series of user interface screens that will be used by both employees and customers. You proudly roll out your prototypes, expecting accolades and high praise. Instead, the users who try out your prototype—customers and employees alike—are confused, frustrated, and even angry. What are some of the interface problems they might have encountered which caused them to react like this? When you go to your boss for advice, your boss asks you a series of questions regarding the steps you followed. What are the questions you'll probably be asked?

10. At one time, most software applications came with a thick users' manual. Many of these users' manuals have disappeared in favor of sophisticated online help systems and tutorials. If you are a systems designer designing the help system for a new application, what are some of the important considerations to keep in mind?

11. It is not uncommon for an application to use hundreds of screens, windows, and dialogue boxes in its user interface. Coordinating the order and conditions under which these appear can be a difficult process that is prone to error. To help coordinate and document this process, state transition diagrams (STD) are used to illustrate the conditions under which screens, windows, and dialogue boxes appear, as well as their sequence.

Take an application in your organization or at school with which you are familiar. Create a state transition diagram for a part of the system, using Figure 17-16 as an example.

12. Assume that you are part of a project team that has been hired by a company that is moving from mainframe technology to client/server technology. You are working on the user interface design. The company wants this application to set the tone for subsequent applications to be developed. You have been given free rein to develop the conventions and standards for the user interface screens that will be used for the look-and-feel of this and subsequent applications. Create a

one-page list of what you believe to be the most important conventions and standards.

13. GUI and Web applications provide users with a variety of paths through the different parts of the application. The price paid for this user-friendliness and accommodation is complicated design and programming. Is it possible to have user-friendly and accommodating interfaces that don't require complicated programming and design?

 Projects and Research

1. This chapter, as well as the preceding two chapters, focused on the capture and presentation of data and information. In 1983, Edward Tufte wrote a book called *The Visual Display of Quantitative Information.* Many leading journals consider this book to be *the* definitive work on this subject. Since that time, Tufte has written numerous other books and articles on the display of information and data, how it shapes the ways we think, and how it can have profound consequences. Go to Tufte's Web site at *http://www.edwardtufte.com/tufte* and read some of the articles and forum discussions. Tufte's Web site also contains numerous links to other sites; you may find them to be equally interesting and/or valuable.

   a. Describe some of the articles and their viewpoints that you found on the Web site.
   b. What is "chartjunk"?
   c. What does Tufte have to say about project management charts? Are his suggestions viable in your opinion?
   d. What is Tufte's viewpoint on Microsoft *PowerPoint?* Do you agree? Why or why not?
   e. Assume you are starting a career as a systems designer. After reading the material on Tufte's Web site, describe those concepts, guidelines, or viewpoints, if any, that you believe are absolutely vital for every systems designer to understand and apply.

2. The textbook references another author, Wilbert Galitz. Galitz is one of a number of contemporary writers (several of whom are referenced in the suggested reading section), who are recognized as leaders in the area of human interface design. Use the Internet to research recent articles and forum discussions on the topic of human interface design and human engineering guidelines.
   a. Describe the articles you found, including their authors and viewpoints
   b. Discuss and compare any contrasting viewpoints you found on this topic.
   c. What are the authors' thoughts on the trend towards *Windows* and browser interfaces converging? Do they feel that eventually there will be little, if any, distinction between the two?

   d. What are their predictions, if any, regarding technological innovations that may fundamentally change human interface design in the future?
   e. On the basis of your research, do you feel that research in the area of human engineering and human interface designs is about as sophisticated and advanced as it is going to get? Why or why not?

3. So far, we've researched the viewpoints of leaders in the field of information presentation and human interface design for computer systems. What about experts in other areas?

   a. Find and interview several local graphic artists and designers in your area. What are their thoughts and viewpoints regarding how graphics can be used as a method of communication?
   b. Were their viewpoints conceptually similar or different compared to those you researched in the previous questions?
   c. Did you come away with anything that might be of value to you as a systems designer?
   d. What about the field of industrial psychology? Do some research in this area, and/or interview any educators or professors at your school who are knowledgeable in this discipline. Were their viewpoints different or about the same as the others you've researched or interviewed?
   e. Can you think of any fields or disciplines that might be helpful for designing human interfaces? If so, research them and report what you consider to be relevant and valuable.

4. Automated screen design tools are becoming increasingly powerful and sophisticated. Use the Web to find several of the leading design tools. Go to their manufacturers' Web sites and review their features. If trial versions are available, download them.

   a. What automated screen design tools did you find? Who manufactures them?
   b. Compare and contrast their features and functionality. Describe their different features in a matrix.
   c. If you were an independent designer, which one, if any, would you choose? Why?

d. Would you expect to see a significant difference in your productivity by using one of these tools? How much of a difference.

e. Do you feel that using these tools would enhance or constrain your creativity? Explain your answer.

5. Frequently, there is talk of "redesigning" government and making it operate more like private business. This raises the question of whether there is a fundamental difference between the public and private sectors, and whether this may have an impact upon how systems should be designed differently, depending upon whether they are intended for a government or a private sector organization.

a. Survey system designers in both the private and public sectors. Ask them what their top issues and problems are in terms of designing human interfaces.

b. What differences did you find?

c. What similarities?

d. Do you think there is enough commonality such that the same set of guidelines can apply to both public- and private-sector agencies? Why or why not?

e. Given a choice and assuming that salary and benefits were the same, would you rather be a systems designer in a public agency? A private agency? Why or why not?

6. Designing interface screens for B2C and B2B Web sites is considered by some to have a fundamentally different objective compared with other types of interface screens. Specifically, the purpose of these Web sites is to entice consumers and businesses to purchase their products or services.

a. Research articles on this topic. What did you find, and what were the viewpoints?

b. Summarize the difference in outlook, if any, between designing conventional input/output human interface screens and designing screens for B2C and B2B Web sites?

c. Do you agree with these viewpoints? Why or why not?

d. What type of background do you feel would be more valuable for B2C and B2B Web sites—a background in systems design or one in marketing and advertising? Explain your answer.

Minicases

There is a discussion, starting on page 615, on the human factors and human engineering issues in user interface design. The spirit of the discussion is that it is imperative that we understand the *people* who will be using the system and that we create a system interface that *they* understand and can use. But this is not an academic issue; it is a people-skills and people-understanding issue.

1. Interview someone you do not know well who is a complete nontechie. Your goal is to understand that person and his or her computer needs and wants. Things you need to consider in your interview:

a. Understand that person as a *person:* who are they? What are their likes and dislikes? Do they have a spouse? Children? What about sports? Hobbies? Do they work inside or outside the home? If you are interviewing them in their own "space" (home, office, etc.), take note of the personal effects that are in view. What do these things tell you about that person?

b. Understand them as a computer user: What are their experiences with computers? What types of things have they used a computer for? What *wouldn't* they use a computer for? Is there

something they find computers particularly useful for? Something that is particularly frustrating?

c. What is their body language telling you as you ask these questions? Are they at ease with you? Make a note of their reactions to you, how you are dressed, what you have said, and your own body language.

2. Using the knowledge you gained from your interview in minicase 1, design an interface for the individual you interviewed. What interface design modifications are you making so that the program will fit the individual? Explain in detail, and submit the results to your professor.

3. Meet with the person you interviewed in minicase 1 and present them with the design prototype you created. Get their feedback on the design. Do they like it? Could they navigate the pages? What about the design of the inputs? Is there anything they would change? What do they specifically like and dislike about the interface you created? Again, watch their body language. Are they telling you everything? What is your body language telling them? Be aware of your influence on the situation. Document the interview and submit the results to your professor.

4. Based on your second interview (minicase 3), revise your interface design. Then submit your work from the previous three minicases and this revision in a professional and complete deliverable to your professor. Be sure to include a brief discussion of what you learned from the person you interviewed, and from this experience.

Suggested Readings

Andres, Clay. *Great Web Architecture.* Foster City, CA: IDG Books Worldwide. This is an interesting title. It uses a "design by example" approach based on input from "top Web architects" to illustrate and discuss Web-based systems, including many with e-commerce and e-business aspects. This is not an academic title, but it is nonetheless interesting.

Galitz, Wilbert. *User-Interface Screen Design.* New York: Wiley QED, 1993. Ignore the date. This book remains our favorite user interface design book because it is so conceptually and fundamentally sound. Galitz teaches workstation, PC, and mainframe interface design here, based on well-thought-out principles and guidelines. We can't wait for the update. Would that we could afford to develop an entire elective course built around this outstanding book!

Galitz, Wilbert. *It's Time to Clean Your Windows: Designing GUIs That Work.* New York: John Wiley & Sons, 1994. This is another excellent book that provides an unbiased reference on designing graphical interfaces.

Hix, Deborah, and H. Rex Hartson. *Developing User Interfaces: Ensuring Usability through Product & Process.* New York: John Wiley & Sons, 1993. John Wiley & Sons must have the corner on user interface design books. These authors have academic roots. The book is somewhat hard to read, but nonetheless very well organized and written. We especially like the integration with systems analysis and design.

Horton, William K. *Designing & Writing Online Documentation: Help Files to Hypertext.* New York: John Wiley & Sons, 1990. We were able to provide only cursory coverage of this important topic.

Mandel, Theo. *Elements of User Interface Design.* New York: John Wiley & Sons, 1997. Here is a somewhat newer and very comprehensive book that includes some of the early design foundations for the Web.

Martin, Alexander, and David Eastman. *The User Interface Design Book for the Applications Programmer.* New York: John Wiley & Sons, 1996.

Microsoft Corporation. *Microsoft Windows User Experience: Official Guidelines for User Interface Developers and Designers.* Redmond, WA: Microsoft Press, 1999. This is the official standard for designing *Windows* user interfaces. There are many insights to Microsoft's intentions for maximizing the user experience.

Schmeiser, Lisa. *Web Design Templates Sourcebook.* Indianapolis, IN: New Riders Publishing, 1997. This is not an academic title or even a traditional professional market title. It caught our eye because it uses an HTML-based template approach (over 300) to present designs that can ultimately evolve into finished products. From our perspective, this represents an intriguing twist on the prototyping model.

Weinschenk, Susan, and Sarah C. Yeo. *Guidelines for Enterprise-Wide GUI Design.* New York: John Wiley & Sons, 1995. Clearly, John Wiley & Sons is the market's leading publisher for this subject.

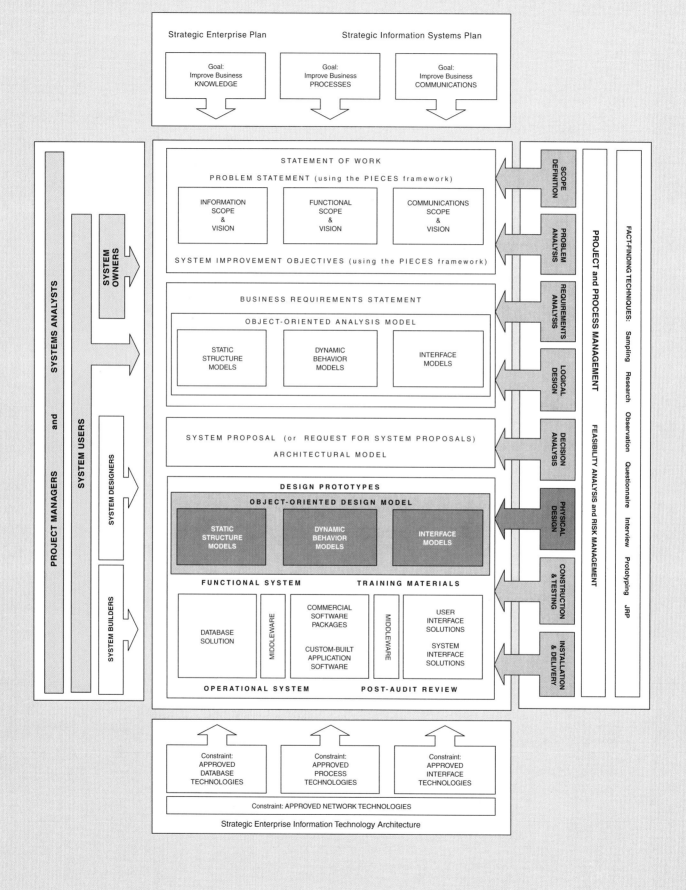

Strategic Enterprise Plan

Strategic Information Systems Plan

Goal:
Improve Business
KNOWLEDGE

Goal:
Improve Business
PROCESSES

Goal:
Improve Business
COMMUNICATIONS

SYSTEMS ANALYSTS

SYSTEM OWNERS

PROJECT MANAGERS and

SYSTEM USERS

SYSTEM DESIGNERS

SYSTEM BUILDERS

STATEMENT OF WORK

PROBLEM STATEMENT (using the PIECES framework)

INFORMATION
SCOPE
&
VISION

FUNCTIONAL
SCOPE
&
VISION

COMMUNICATIONS
SCOPE
&
VISION

SYSTEM IMPROVEMENT OBJECTIVES (using the PIECES framework)

BUSINESS REQUIREMENTS STATEMENT

OBJECT-ORIENTED ANALYSIS MODEL

STATIC
STRUCTURE
MODELS

DYNAMIC
BEHAVIOR
MODELS

INTERFACE
MODELS

SYSTEM PROPOSAL (or REQUEST FOR SYSTEM PROPOSALS)

ARCHITECTURAL MODEL

DESIGN PROTOTYPES

OBJECT-ORIENTED DESIGN MODEL

STATIC
STRUCTURE
MODELS

DYNAMIC
BEHAVIOR
MODELS

INTERFACE
MODELS

FUNCTIONAL SYSTEM   TRAINING MATERIALS

DATABASE
SOLUTION

MIDDLEWARE

COMMERCIAL
SOFTWARE
PACKAGES

CUSTOM-BUILT
APPLICATION
SOFTWARE

MIDDLEWARE

USER
INTERFACE
SOLUTIONS

SYSTEM
INTERFACE
SOLUTIONS

OPERATIONAL SYSTEM   POST-AUDIT REVIEW

SCOPE DEFINITION

PROBLEM ANALYSIS

REQUIREMENTS ANALYSIS

LOGICAL DESIGN

DECISION ANALYSIS

PHYSICAL DESIGN

CONSTRUCTION & TESTING

INSTALLATION & DELIVERY

PROJECT and PROCESS MANAGEMENT

FEASIBILITY ANALYSIS and RISK MANAGEMENT

FACT-FINDING TECHNIQUES: Sampling   Research   Observation   Questionnaire   Interview   Prototyping   JRP

Constraint:
APPROVED
DATABASE
TECHNOLOGIES

Constraint:
APPROVED
PROCESS
TECHNOLOGIES

Constraint:
APPROVED
INTERFACE
TECHNOLOGIES

Constraint: APPROVED NETWORK TECHNOLOGIES

Strategic Enterprise Information Technology Architecture

18

Object-Oriented Design and Modeling Using the UML

Chapter Preview and Objectives

This is the second of two chapters on object-oriented tools and techniques for systems development. This chapter focuses specifically on tools and techniques that are used during systems design. You will know object-oriented systems design when you can:

▌ Differentiate between entity, interface, control, persistence, and system classes.

▌ Understand the concepts of dependency and navigability.

▌ Define visibility and explain its three levels.

▌ Understand the concept of object responsibility and how it is related to message sending between object types.

▌ Describe the activities involved in object-oriented design.

▌ Differentiate between a design use-case narrative and an analysis use-case narrative.

▌ Describe CRC card modeling.

▌ Model class interactions with sequence diagrams.

▌ Construct a class diagram that reflects design specifics.

▌ Model object states with state machine diagrams.

▌ Understand the role of coupling and cohesion in object reuse.

▌ Describe the use of design patterns and two common design patterns.

▌ Differentiate between design patterns, object frameworks, and components.

▌ Understand the use of communication diagrams, component diagrams, and deployment diagrams.

Introduction

Chapters 13–17 showed Bob Martinez performing traditional structured design tasks to design the SoundStage Member Services system. How would his tasks have been different had the project followed an object-oriented approach?

In an object-oriented approach the systems analysis would still have had to design the application architecture as presented in Chapter 13, but it would have used different tools: namely, the component diagrams and deployment diagrams shown at the end of this chapter. They would have designed the program logic and structure using sequence diagrams, class diagrams, and state machine diagrams presented in this chapter. Assuming the data would be stored in a relational database (as is the case with most information systems), the tools presented in Chapter 14 would still be used, though they would have had to take a few additional steps to map the entity objects and their attributes to tables and fields in the database. They would also have still designed and prototyped the user interface using techniques presented in Chapters 15–17. In fact, activity diagrams with partitions are a useful tool for user interface design.

Coming out of object oriented analysis, Bob would have had activity diagrams, system sequence diagrams, and a class diagram of entity objects and their attributes. In the design phase, that class diagram has to be refined to include additional design objects and the assignment of behaviors and their parameters to objects. That process involves analyzing and designing object responsibilities and states using the tools and concepts presented in this chapter. The completed design documents (class diagram, sequence diagrams, machine state diagrams, etc.) could then be handed off to teams of programmers, who would program the objects with the specified behaviors and attributes.

The Design of an Object-Oriented System

object-oriented design (OOD) an approach used to specify the software solution in terms of collaborating objects, their attributes, and their methods.

In Chapter 10 we learned about object classes. So how are these classes put together into an application? What does an object-oriented system look like? In a pure object-oriented environment every piece of code exists inside an object class—all the user interface, all the program logic, everything. The application works by having classes send and receive messages from other classes. The goal of **object-oriented design (OOD)** is to specify the objects and messages of the system.

Figure 18-1 shows programming code for a Web page created in *C# .NET*. This Web page provides part of the user interface for the SoundStage system (notice the Text boxes, Labels, and Buttons). Near the top of the code we see "public class Login." This indicates that all the user interface code exists inside a class. Near the bottom of the screen, this user interface class creates an instance of the member class and calls it the validateLogin behavior (method) of that class. An object-oriented system is structured into at least three different types of object classes.

> Entity Classes

entity class an object class that contains business-related information and implements the analysis classes.

Entity classes usually correspond to items in real life (such as a MEMBER or ORDER) and contain information, known as *attributes,* that describes the different instance of the entity. They also encapsulate those behaviors (called *methods*) that maintain their information or attributes. These are the kinds of object classes we defined in Chapter 10. They are the heart of the system.

> Interface Classes

interface class an object class that provides the means by which an actor can interface with the system. Examples include a window, dialogue box, or screen. For nonhuman actors, an application program interface (API) is the interface class. Sometimes called a boundary class.

Users communicate with the system through the user interface, implemented as **interface classes.** The use-case functionality that describes the user directly interacting with the system should be placed in interface classes. The responsibility of an interface class is twofold:

FIGURE 18-1

An Object-Oriented Application

1. It translates the user's input into information that the system can understand and use to process the business event.
2. It takes data pertaining to a business event and translates the data for appropriate presentation to the user.

Each actor or user needs its own interface class to communicate with the system. In some cases, the user may need multiple interface classes. Take, for example, the ATM machine. Not only is there a display for presenting information, but there are also a card reader, money dispenser, and receipt printer. All of these would be considered interface object classes.

> Control Classes

Control classes implement the business logic or business rules of the system. Generally, each use case is implemented with one or more control classes. **Control classes** process messages from an interface class and respond to them by sending and receiving messages from the entity classes.

An object-oriented system could be implemented with just these three kinds of classes. But many methodologists include two other kinds of classes.

> Persistence Classes

The attributes of the entity classes are generally persistent, meaning they continue to exist beyond when the system is running. The functionality to read and write attributes in a database could be built into the entity classes. But if that functionality is put into separate persistence (or data access) classes, the entity classes are kept implementation neutral. That can allow the entity classes to be more reusable, a major goal of object-oriented design.

> System Classes

A final type of object class, the **system class,** isolates the other objects from operating system–specific functionality. If the system is ported to another operating system, only these classes and perhaps the interface classes have to be changed.

Why all these kinds of classes? Structuring the system this way makes the maintenance and enhancement of those classes simpler and easier.

control class an object class that contains application logic. Examples of such logic are business rules and calculations that involve multiple entity object classes. Control classes coordinate messages between interface classes and entity classes and the sequences in which the messages occur.

persistence class an object class that provides functionality to read and write persistent attributes in a database.

system class an object class that handles operating system–specific functionality.

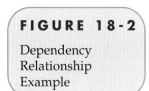

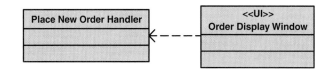

> Design Relationships

In object-oriented analysis, we concentrated on identifying the most common object relationships: associations, aggregation/composition relationships, and generalization/ specialization relationships. In object-oriented design, it is necessary to model more advanced relationships in order to accurately specify the software components. You will learn these relationships in the following sections.

Dependency Relationships A dependency relationship is used to model the association between two classes in two instances: (1) to indicate that when a change occurs in one class, it may affect the other class, and (2) to indicate the association between a persistent class and a transient class. Interface classes are typically transient and are modeled in this fashion. Draw your attention to Figure 18-2. In this example the ORDER DISPLAY WINDOW class is an interface class and is created to display the contents of an order. It is dependent on the PLACE NEW ORDER HANDLER class to map order information to it and to respond to events initiated from the interface. A dependency relationship is illustrated with a dashed arrow line.

Navigability As you learned in Chapter 10, by default, associations between classes are bidirectional, meaning that classes of one kind can navigate (send messages) to classes of the other kind. There may be times, though, when you want to limit the message sending to only one direction. For example, let's assume each system user must have a password, which the user must change every 30 days. Let's also assume that when a user changes passwords, the new one can't be a password he or she has used in the past six months. The model for this scenario is depicted in Figure 18-3. Given a USER, you'll want to find that user's current PASSWORD for authentication purposes or to change the current password. Thus, the USER class would send a message to the PASSWORD class. In most cases it wouldn't make sense that given a PASSWORD you would want to identify the corresponding USER. Navigability is illustrated with an arrowhead pointing only to the direction a message can be sent.

> Attribute and Method Visibility

visibility the level of access an external object has to an attribute or method.

How attributes and methods are accessed by other classes is defined by **visibility.** The UML provides three levels of visibility:

1. *Public*—denoted by the symbol "+."
2. *Protected*—denoted by the symbol "#."
3. *Private*—denoted by the symbol "−."

Public attributes can be accessed and public methods can be invoked by any other method in any other class. Protected attributes can be accessed and protected methods can be invoked by any method in the class in which the attribute or method is defined or in subclasses of that class. Private attributes can be accessed and private methods can be invoked only by any method in the class in which the attribute or method is defined. If a method needs to be invoked in response to a message sent by

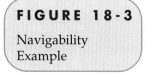

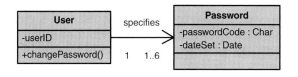

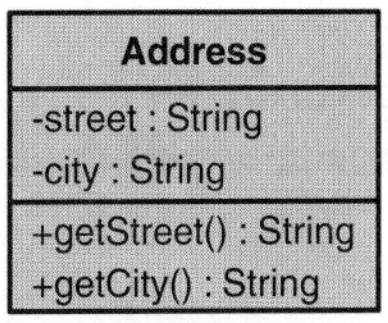

Address
-street : String
-city : String
+getStreet() : String
+getCity() : String

FIGURE 18-4

Visibility Example

another class, the method should be declared public. In most cases all attributes should be declared private to enforce encapsulation. Figure 18-4 depicts an example of denoting attribute and method visibility.

> Object Responsibilities

Recall that in object-oriented systems, objects encapsulate both data and behaviors. In design, we focus on identifying the behaviors a system must support and, in turn, designing the object **methods** for performing those behaviors. Along with behaviors, we determine the **object's responsibilities.**

In Chapter 10 you learned that objects have behaviors, or things that they can do. In object-oriented design it is important to recognize that an object has responsibility. Object responsibility is closely related to the concept of being able to send and/or respond to messages. Draw your attention to Figure 18-5. An ORDER object class has been assigned the responsibility of displaying a customer's order, but it needs help. First, it collaborates with the CUSTOMER class to get the customer data. Next, it collaborates with the MEMBER ORDERED PRODUCT class to get information about each product being ordered. The MEMBER ORDERED PRODUCT class cannot fulfill the entire request itself, so it needs to collaborate with the PRODUCT class to get detailed information about each product. Thus, when each class receives a message requesting a service, it has an obligation to respond to the message and fulfill the request.

A class responsibility is not the same thing as a class method. A class responsibility is implemented by the creation of one or more methods that may have to collaborate with other classes and methods, as presented above.

method the software logic that is executed in response to a message.

object responsibility the obligation that an object has to provide a service when requested and thus collaborate with other objects to satisfy the request if required.

The Process of Object-Oriented Design

In performing OOA, we defined use cases and identified objects based on ideal conditions and independent of any hardware or software solution. During object-oriented design, we want to refine those use cases and objects to reflect the actual environment of our proposed solution.

Object-oriented design includes the following activities:

1. Refining the use-case model to reflect the implementation environment.
2. Modeling class interactions, behaviors, and states that support the use-case scenario.
3. Updating the class diagram to reflect the implementation environment.

In the following sections we will review each of these activities to learn what steps, tools, and techniques are used to complete object-oriented design.

> Refining the Use-Case Model

In this iteration of use-case modeling, the use cases will be refined to include details of how the actor (or user) will actually interface with the system and how the system will respond to that stimulus to process the business event. The manner in which the user accesses the system—via a menu, window, button, bar code reader, printer, and so on—should be described in detail. The contents of windows, reports, and queries should also be specified within the use case. While refining use cases is often

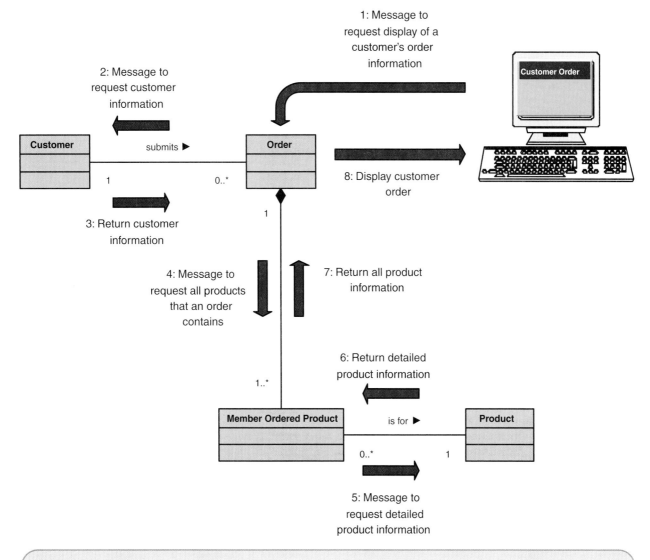

FIGURE 18-5 Object Responsibility

time-consuming and tedious, it is essential that they are completed. These use cases will be the basis on which subsequent user manuals and test scripts are developed during systems implementation. In addition, these use cases will be used by programmers to construct application programs during systems implementation.

In the following steps we will adapt each use case to the implementation environment or "reality" and document the results. It is important that each use case be highly detailed in describing the user interaction with the system. The refined use cases can then be used by the user to validate systems design and by the programmer for process and interface specifications.

Step 1: Transforming the "Analysis" Use Cases to "Design" Use Cases In Chapters 7 and 13 you learned how to do use-case modeling during systems analysis to document user requirements for a given business scenario. In this step, we refine each of those use cases to reflect the physical aspects of the implementation environment for our new system.

Figure 18-6 illustrates the refinement of the *Place New Order* use case that was originally defined during systems analysis. This version is identified as a design use case to distinguish it from the analysis version previously completed. We want to keep

Member Services System

Author(s): **K. Dittman**

Date: **11/21/02**

Version: **1.00**

Use Case Name:	Place New Order	❶ Use Case Type
Use Case ID:	MSS-SUC002.00	Business Requirements: ☐
Priority:	High	System Analysis: ☐
Source:	Requirement — MSS-R1.00 Requirements Use Case — MSS-BUC002.00	System Design: ☑
Primary Business Actor:	Club Member (Alias — Active Member, Member)	
Primary System Actor:	Club Member (Alias — Active Member, Member)	
Other Participating Actors:	• Warehouse (Alias — Distribution Center) (external receiver) • Accounts Receivable (external server)	
Other Interested Stakeholders:	• Marketing — interested in sales activity in order to plan new promotions. • Procurement — interested in sales activity in order to replenish inventory. • Management — interested in order activity in order to evaluate company performance and customer (member) satisfaction.	
Description:	This use case describes the event of a member submitting a new order for SoundStage products via the World Wide Web. The member selects the items he or she wishes to purchase. Once the member has completed shopping, the member's demographic information as well as account standing will be validated. Once the products are verified as being in stock, a packing order is sent to the distribution center for it to prepare the shipment. For any product not in stock, a back order is created. On completion, the member will be sent an order confirmation.	
Precondition:	The individual must be a registered user of the system. The member must have logged in to the system, and the member home page is being displayed.	
Trigger:	This use case is initiated when the member selects the option to enter a new order.	

Typical Course of Events:	Actor Action	System Response
	Step 1: The member clicks on the place new order icon (or link). ❷	**Step 2:** The system responds by displaying window *W11—Catalog Display*, a list of SoundStage products.* ❸ If the product list is greater than 50, which is the maximum number to be displayed on one page, the system calculates the number of pages required to display the products. The system then provides the member with the necessary navigational buttons, such as: [First], [Prev], [Next], [Last], and [1] [2] [3] [4], and so on.
	Step 3: The member scrolls through the available items by using the scroll bar buttons, the [Page Up] and [Page Down] keys, or the navigational controls specified in **step 2.** The member selects the ones he or she wishes to purchase by clicking the check box and entering the quantity to be ordered. ❹	**Step 4:** Once the member has completed making selections, the system retrieves the member's demographic information (shipping and billing addresses) and displays it in window *W02—Member Profile Display*. The system also prompts the member to make any required changes.
	Step 5: The member verifies demographic information (shipping and billing addresses). If no changes are necessary, the member clicks the [Continue] button.	**Step 6:** For each product ordered, the system verifies the product availability and determines an expected ship date, determines the price to be charged to the member, and determines the cost of the total order. If an item is not immediately available, it indicates that the product is back-ordered or that it has not been released for shipping (for preorders). If an item is no longer available, that is indicated also. The system then displays a summary of the order in window *W03—Order Summary Display*. The system also prompts the member to make any required changes.

FIGURE 18-6 Example of a Design Use Case

	Step 7: The member verifies the order. If no changes are necessary, the member clicks the [Continue] button. **Step 9:** The member responds by clicking the appropriate check box for the desired payment option. **Step 11:** The member verifies the order. If no changes are necessary, the member clicks the [Continue] button.	**Step 8:** The system checks the status of the member's account. If status is satisfactory, the system prompts the member to select the desired payment option (to be billed later or pay immediately with a credit card). **Step 10:** The system then displays a final summary of the order in window *W03—Order Summary Display*. The system also prompts the member to make any required changes. **Step 12:** The system records the order information (including back orders, if necessary). **Step 13:** Invoke abstract use case *MSS-AUC001.00 Determine Appropriate Distribution Center and Release Order to Be Filled.* **Step 14:** Once the order is processed, the system generates an order confirmation and displays it in window *W04—Order Confirmation Display*. The system also sends the confirmation by e-mail. Invoke abstract use case *MSS-AUC004.00 Send Electronic Member Correspondence.*
Alternate Courses:	**Alt-Step 3a:** If the member clicks on the item name, the system displays a pop-up window, *W15—Product Detail Display*, which contains all the product details, including a graphic of its cover. The member clicks the [Close] button to close the pop-up window. **Alt-Step 3b:** If member wants to perform keyword search, invoke abstract use case *MSS-AUC006.00 Search Product Catalog by Keyword*. **Alt-Step 5:** If member wants to change demographic information, invoke abstract use case *MSS-AUC007.00 Change Member Profile*. **Alt-Step 7:** If the order requires changes the member can delete any item no longer wanted by deselecting the check box by item and/or changing the order quantity. Once the member has completed the order changes, he or she clicks the [Update Order] button. The system reprocesses the order (**go to step 6**). If the member clicks the [Do More Shopping] button, **go to step 3.** If the member clicks the [Update Member Profile] button, invoke abstract use case *MSS-AUC007.00 Change Member Profile* and then **go to step 6.** **Alt-Step 8:** If the member's account is not in good standing, display to the member using window *W09—Member Account Status Display*, the account status, the reason the order is being held, and what actions are necessary to resolve the problem. In addition an e-mail is sent to the member with the same information. Invoke abstract use case *MSS—AUC004.00 Send Electronic Member Correspondence*. The system prompts the member to hold the order for later processing or cancel the order. If the member wishes to hold the order by clicking the [Save Order] button, the system records the order information, places it in hold status, and then displays the SoundStage main page, window *W00—Member Home Page*. If the member chooses to cancel the order by clicking the [Cancel Order] button, the system erases the inputted information, and then displays the SoundStage main page, window *W00—Member Home Page*. Terminate the use case. **Alt-Step 10:** If the member selects the option to pay by credit card, invoke abstract use case *MSS-AUC012.00 Pay by Credit Card*. If the member cannot pay by credit card, the system prompts the member to hold the order for later processing or cancel the order. If the member wishes to hold the order by clicking the [Save Order] button, the system records the order information, places it in hold status, and then displays the SoundStage main page, *window W00—Member Home Page*. If the member chooses to cancel the order by clicking the [Cancel Order] button, the system erases the inputted information and then displays the SoundStage main page, window *W00—Member Home Page*. Terminate the use case.	

FIGURE 18-6 Continued

	Alt-Step 11: If the order requires changes, the member can delete any item no longer wanted by deselecting the check box by item and/or changing the order quantity. Once the member has completed the order changes, he or she clicks the [Update Order] button. The system reprocesses the order (**go to step 6**). If the member clicks the [Do More Shopping] button, **go to step 3.** If the member clicks the [Update Member Profile] button invoke abstract use case *MSS-AUC007.00 Change Member Profile* and then **go to step 6.**
	Alt-Step 12: If all items ordered are on back order, the order is not released to the distribution center.
Conclusion:	This use case concludes when the member receives a confirmation of the order.
Postcondition:	The order has been recorded and, if the ordered products were available, released to the distribution center. For any product not available a back order has been created.
Business Rules:	• Member must have a valid e-mail address to submit online orders. • Member is billed for products only when they are shipped.
Implementation Constraints and Specifications:	• Use case must be available to the member 24/7. • Frequency—It is estimated that this use case will be executed 3,500 times per day. It should support up to 50 concurrent members.
Assumptions:	• Product can be transferred among distribution centers to fill orders. • Procurement will be notified of back orders by a daily report (separate use case). • The member responding to a promotion or a member using credits may affect the price of each ordered item. • The member can cancel the order at any time by clicking the [Cancel Order] button. If member hits [Back Page] button at any time, refresh current window.
Open Issues:	None

* Refer to user interface specification for window contents and specification.

FIGURE 18-6   Continued

the original analysis use cases separate from the refined design use cases to allow maximum flexibility in reusing use cases for variations of different physical implementations. We draw your attention to the following refinements to our use-case description in Figure 18-6:

❶ *Use-case type*—To reflect implementation details such as user interface constraints, tactical use cases called *system design use cases* are derived from the system analysis use cases.

❷ *Window controls*—In system design use cases, window controls such as icons, links, check boxes, and buttons are explicitly stated.

❸ *Window names*—The name of each user interface element (window name) is stated. If additional information about a user interface element exists, it is good practice to reference it. Otherwise, more detailed window specifications could be added to the use case.

❹ *Navigation instructions*—Directions on how the user navigates the user interface should be specified.

Step 2: Updating the Use-Case Model Diagram and Other Documentation to Reflect Any New Use Cases   After all the system analysis use cases have been transformed to system design use cases, it is quite possible that new use cases, use-case dependencies, or even actors have been discovered. It is very important that we keep our documentation accurate and current. Thus, in this step the use-case model diagram, the use-case dependency diagram, and the actor and use-case glossaries should be updated to reflect any new information introduced in step 1.

> Modeling Class Interactions, Behaviors, and States That Support the Use-Case Scenario

Step 1: Identify and Classify Use-Case Design Classes   In the previous section, we refined the use cases to reflect the implementation environment. In this activity we want to identify and categorize the design classes required by the functionality that was specified in each use case and identify the class interactions, their responsibilities, and their behaviors. See Figure 18-7.

❶ The Interface Classes column contains a list of classes mentioned in the use case that the users directly interface with, such as screens, windows, card readers, and printers. The only way an actor or user can interface with a system is via an interface class. Therefore, there should be at least one interface class per actor or user.

❷ The Controller Classes column contains a list of classes that encapsulate application logic or business rules. A use case should reveal one control class per unique user or actor.

❸ The Entity Classes column contains a list of classes that correspond to the business domain classes whose attributes were referenced in the use case.

Step 2: Identify Class Attributes   During both analysis and design, class attributes can be discovered. In efforts to transform analysis use cases into design use cases, we begin referencing the attributes in the use-case text. In this step, we examine each use case for additional attributes that haven't been previously identified, and we update our class diagram to include those attributes.

Step 3: Identify Class Behaviors and Responsibilities   Once we have identified all the objects needed to support the functionality of the use case, we shift our attention to defining the specific behaviors and responsibilities. This step involves the following tasks:

- Analyze the use cases to identify required system behaviors.
- Associate behaviors and responsibilities with classes.
- Model classes that have complex behavior.
- Examine the class diagram for additional behaviors.
- Verify classifications.

FIGURE 18-7   Interface, Control, and Entity Classes of *Place New Order* Use Case

❶ Interface Classes	❷ Controller Classes	❸ Entity Classes
W00-Member Home Page	Place New Order Handler	Billing Address
W02-Member Profile Display		Shipping Address
W03-Display Order Summary		Email Address
W04-Display Order Confirmation		Active Member
W09-Member Account Status Display		Member Order
W11-Catalog Display		Member Ordered Product
W15-Product Detail Display		Product
		Title
		Audio Title
		Game Title
		Video Title
		Transaction

FIGURE 18-8 Partial Summary of *Place New Order* Use-Case Behaviors

Behaviors	Automated/Manual	Class Type
Process new member order	Manual/Automated	Control
Click icon to place new order	Manual	
Retrieve product catalog information	Automated	Entity
Display W11-Catalog Display window	Automated	Interface
Scroll or page through catalog	Manual	
Select product to be ordered and enter quantity	Manual	
Retrieve member demographic information	Automated	Entity
Display W02 – Member Profile Display window	Automated	Interface
Verify member demographic information	Manual	
Validate quantity amount	Automated	Entity
Verify the product availability	Automated	Entity
Determines an expected ship date	Automated	Entity
Determine price of product	Automated	Entity
Determine cost of the total order	Automated	Entity
Display W03 – Order Summary Display window	Automated	Interface
Prompt user	Automated	Interface
Verify order information	Manual	
Check Status of member account	Automated	Entity
Prompt user for payment option	Automated	Interface
Store order information	Automated	Entity
Record back order information	Automated	Entity
Generate order confirmation	Automated	Entity
Display W04 – Order Confirmation Display	Automated	Interface
Click button or icon	Manual	

In Chapter 10 you learned that classes encapsulate data and behavior. Our first task in identifying the class behaviors and responsibilities is accomplished by once again examining our use case. The use-case description is examined to identify all *verb phrases*. Verb phrases suggest behaviors that are required to complete a use-case scenario. These verb phrases correlate to the system behaviors required to respond to the business event of a club member placing a new order. Each use case should be examined separately to identify behaviors associated with the use case.

Once the behaviors have been identified, our second task is to determine if the behaviors are manual or if they will be automated. If they are to be automated, they must be assigned to the appropriate object that will have the responsibility of carrying out that behavior. In Figure 18-8, which summarizes the *Place New Order* use-case behaviors, each verb phrase or behavior is listed along with its automated or manual designation. The third column lists the object type with which each behavior is associated.

In Figure 18-9, we have condensed the behavior list to show only the behaviors that need to be automated. Recall that the object types were defined earlier, in step 1. We will use the list in Figure 18-9 as the source of behaviors to be allocated in the next task.

The next task is to identify which behaviors should be associated with which class and to identify collaborations among those classes. One popular tool for that is the *class responsibility collaboration (CRC) card*.[1] A CRC card for the MEMBER ORDER

---

[1]CRC cards were pioneered by Kent Beck and Ward Cunningham.

FIGURE 18-9 Condensed Behavior List for *Place New Order*
Use Case

Behaviors	Class Type
Process new member order	Control
Retrieve product catalog information	Entity
Display W11-Catalog Display window	Interface
Retrieve member demographic information	Entity
Display W02 – Member Profile Display window	Interface
Validate quantity amount	Entity
Verify the product availability	Entity
Determines an expected ship date	Entity
Determine price of product	Entity
Determine cost of the total order	Entity
Display W03 – Order Summary Display window	Interface
Prompt user	Interface
Check Status of member account	Entity
Prompt user for payment option	Interface
Store order information	Entity
Record back order information	Entity
Generate order conformation	Entity
Display W04 – Order Confirmation Display	Interface

class is shown in Figure 18-10. A CRC card contains all use-case behaviors and responsibilities that have been associated with that class.

CRC cards can be developed and refined using an interactive process in which the cards are divided among a group of systems analysts or users. They then move through the steps of a use-case scenario, acting out the required collaborations using a spongy ball. The facilitator starts out by tossing the ball to the person holding the card of the class that is initially responsible for the scenario. That person describes the logic required to fulfill that responsibility. If the class needs information it doesn't have or must modify information it doesn't have, then the person tosses the ball to the peson holding the card with that information. The toss indicates a needed collaboration

FIGURE 18-10 CRC Card for MEMBER ORDER Class

Object Name: Member Order

Sub Object:

Super Object: Transaction

Behaviors and Responsibilities	Collaborators
Report order information	Member Ordered Product
Calculate subtotal cost	
Calculate total order cost	
Update order status	
Create Ordered Product	
Delete Ordered Product	

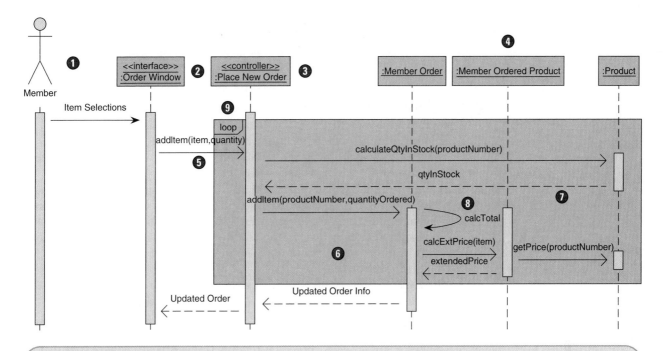

FIGURE 18-11 Sequence Diagram for Step 6 of the Place New Order Use Case

between the two classes, which is noted on the cards. When the scenario is completely acted out the ball is thrown back to the facilitator.

Analysis of the use-case scenarios may not reveal all behaviors for any given object type. On the other hand, by examining the class diagram, you may find additional behaviors (not mentioned in the use-case scenarios) that need to be assigned to an object class. For example, analyze the associations between the classes in Figure 18-12. How are those associations created or deleted? Which should be assigned that responsibility? As a rule, the class that controls the relationship should be responsible for creating or deleting the relationship. Draw your attention in Figure 18-12 to the association MEMBER ORDER and MEMBER ORDERED PRODUCT. By designing the system to have the MEMBER ORDER class have a behavior to "add ordered product," we have effectively given the MEMBER ORDER class control of creating this association. Also, recall from Chapter 10 that there are four "implicit" behaviors that can be associated with any object class: the abilities to create new instances, change its data or attributes, delete instances, and display information about the object class. While examining the use cases to identify and associate behaviors with object classes, we also focus on identifying the collaboration or cooperation that is necessary between classes. In Figure 18-10 the MEMBER ORDER class needs collaboration from the MEMBER ORDERED PRODUCT class to retrieve information about each of the products being ordered. Remember, if a class needs another class's attribute to accomplish a behavior, the collaborating class needs to have a behavior or method for providing that attribute.

Identifying the collaboration of object types is necessary to ensure that all use-case classes work in harmony to complete the processing required for the business event that triggers the use-case scenario.

Another tool for discovering and/or documenting class behaviors and responsibilities is a **sequence diagram.** In Chapter 10 we looked at system sequence diagrams, a high-level diagram that depicts the interaction between an actor and the system for a use-case scenario. A full-sequence diagram depicts the interaction between all the object classes involved in the scenario. A sequence diagram models the logic of a use case (or portion of a use case) by depicting the interaction of messages between objects in time sequence. The messages are arranged in time sequence from top to bottom.

sequence diagram a UML diagram that models the logic of a use case by depicting the interaction of messages between objects in time sequence.

A sequence diagram can be seen as a way to integrate the steps of a use case with the objects of a class diagram. It can be used as a communication tool with programmers to specify what methods (behaviors) to call in implementing a use case. Figure 18-11 shows one scenario for what is essentially step 6 of the *Place New Order* use case described in Figure 18-6. Figure 18-11 illustrates the following sequence diagram notations:

❶ *Actor*—the actor interacting with the user interface is shown with the use case actor symbol. Sometimes the actor is left off for the sake of simplicity. Sometimes the actor is represented with a box like the classes with a notation <<actor>>. The dashed vertical line extending downward from the actor indicates the life of the sequence.

❷ *Interface class*—the box indicates the user interface class code. To make sure there is no confusion as to what kind of class this is, <<interface>> is noted. As with many things in UML, whatever communicates best is right. The colon (:) is standard sequence diagram notation to indicate a running "instance" of the class. The dashed vertical line extending downward from the class indicates the life of the sequence.

❸ *Controller class*—every use case will have one or more controller classes, drawn with the same notation as the interface class and noted as <<controller>>.

❹ *Entity classes*—add boxes for each entity that needs to collaborate in the sequence of steps. Again, the colon (:) denotes an object instance, in other words, a specific order, specific product, and so forth.

❺ *Messages*—solid horizontal arrows indicate message inputs sent to the classes. Each message calls the behavior (or method) of the class to which the arrow points. The UML convention for messages is to begin the first word with a lowercase letter and append additional words with an initial uppercase letter and no space. In parentheses, include any parameters that need to be passed, following the same naming convention and separating individual parameters with commas.

❻ *Activation bars*—the bars that are set over the lifelines indicate the period of time during which each object instance exists. If you are familiar with any object-oriented programming language, you should recall instantiating objects to work with them in your program. The activation bars indicate the lifetime of an instance in RAM. Generally, objects are instantiated in response to messages. Persistent objects will, of course, continue to exist as stored data.

❼ *Return messages*—dashed horizontal arrows are return messages. Every behavior should return something, at least a true/false message indicating whether the behavior was successful. But for the sake of simplicity, return messages are often assumed and left off the sequence diagram.

❽ *Self-call*—an object can call its own method.

❾ *Frame*—we saw in Chapter 10 how to use a frame box in a system sequence diagram to indicate that one or more messages were optional (opt) steps. Here we use a frame to indicate that the controller needs to loop through all the items.

Let's walk through the sequence diagram shown in Figure 18-11. The Member makes his or her selections using the on-screen tools provided in the ORDER WINDOW (which is noted to be an interface class). The ORDER WINDOW then passes those selections with an item and quantity specification for each to the Controller class. The CONTROLLER loops through each of the items. The use case says that for each ordered item, the system must verify product availability. To do that the CONTROLLER sends a message to PRODUCT, calling its calculateQtyInStock method. We may have already identified calculateQtyInStock as a behavior of PRODUCT and so we can read it right off the class diagram and plug it in here. If it isn't a behavior already, then we can determine a need for its existence from this sequence diagram and then add it to the class diagram. Why would this behavior be assigned to PRODUCT? We see from Figure 18-11 that PRODUCT has a quantityInStock attribute, so it is the natural source of this information. PRODUCT returns quantityInStock to the CONTROLLER. The use case includes verbiage to handle items not in stock, but we are not following that scenario. This sequence diagram assumes all items are in stock.

Each in-stock item must be added to the order. Should that be a responsibility of MEMBER ORDER or MEMBER ORDERED PRODUCT? We see from Figure 18-12 that MEMBER ORDER

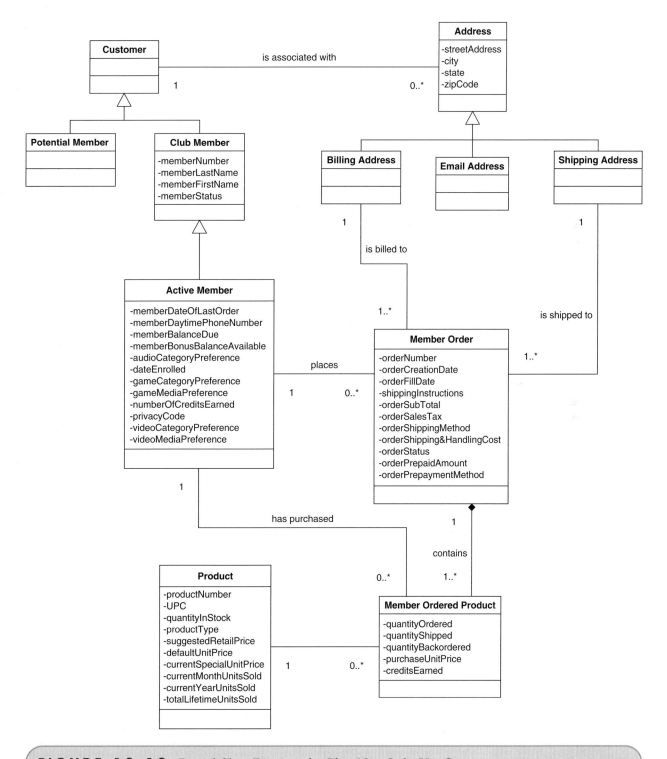

FIGURE 18-12   Partial Class Diagram for *Place New Order* Use Case

has a composition relationship to MEMBER ORDERED PRODUCT, making MEMBER ORDER responsible for the creation and deletion of instances. So we will have the CONTROLLER pass this message to MEMBER ORDER. As it adds an item, MEMBER ORDER needs to recalculate its total. So it calls one of its own methods (calcTotal). To do this calculation, it needs the extended price (quantity times price) of the new item, so it calls calcExtPrice of MEMBER ORDERED PRODUCT. That calculation needs price information, which is held by PRODUCT. So MEMBER ORDERED PRODUCT creates an instance of PRODUCT to look up the price. The extended price can then be passed back to MEMBER ORDER, which passes the entire order to the CONTROLLER. Finally, the CONTROLLER passes the order to the ORDER WINDOW for display.

From this we can determine what behaviors should be assigned to what classes and the parameters they will accept and return. Once the behaviors have been identified, documented, and associated to specific classes, then the class diagram can be updated to include those behaviors in the appropriate classes.

Before we move on, let's look at one other sequence diagram. Figure 18-13 shows a simple sequence diagram for the abstract use case Search Product Catalog by Keyword that is referred to in Alt-Step 3b of the *Place New Order* use case. When the member selects the Search by Keyword option and enters a keyword, the interface passes the request on to the controller. The controller calls the reportProduct method of PRODUCT, passing along the keyword. PRODUCT returns a collection of products that matches the keyword. If we were including persistence objects, we would show the data read statement going to the database.

The following are useful guidelines for constructing sequence diagrams:

- Identify the scope of the sequence diagram. You may wish to depict an entire use-case scenario or just one step.
- Draw the actor and interface class if your scope includes that.
- List the use-case steps down the left-hand side.
- Draw boxes for the controller class, and for each entity class that must collaborate in the sequence, based on the attributes it has or behaviors already assigned to it.

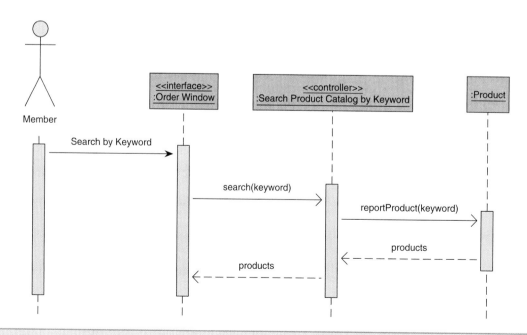

FIGURE 18-13 Sequence Diagram for *Search Product Catalog by Keyword* Use Case

Possible "States" of the Space Shuttle

FIGURE 18-14

Object State Example

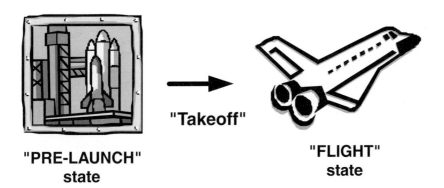

"Takeoff"

"PRE-LAUNCH" state

"FLIGHT" state

- Draw boxes for persistence and system classes if your scope includes that.
- Draw necessary messages and point each of them to the class that will fulfill the responsibility of responding to the message.
- Add activation bars to indicate the lifetime of each object instance.
- Add return messages that are needed for clarity.
- Add frames for loops, optional steps, alternate steps, and so on, as needed.

We will revisit sequence diagrams a bit later in this chapter when discussing design patterns.

Step 4: Model Object States   Our next task is to identify and model any object that has complex behavior based on the changes of its **state**. All objects are said to have state—the value of the object's attributes at one point in time. An object changes state when the value of one of its attributes changes. This change in state is triggered by a **state transition event.** Figure 18-14 shows the space shuttle resting on the launching pad in a state of *Pre-Launch*. After the shuttle takes off (an event), it changes state (state transition), and while it is in the air, it is in a state of *Flight*. We could have shown additional states such as *Landed, Checkout,* or *Refurbish* if the requirements specified this. Many of the objects in business systems have complex behaviors or go through many states and types of state.

A **state machine diagram** models the life cycle of a single object. It depicts the different states an object can have, the events that cause the object to change state over time, and the rules that govern the object's transition between states. In other words, it specifies from which state an object is allowed to transition to another state and under what conditions. A state machine diagram is constructed by performing the following activities:

- Identify the initial and final states (how is the object created and destroyed?).
- Identify other states an object may have during its lifetime.
- Identify triggers (events) that cause the object to leave a particular state.
- Identify state transition paths (when the object's state changes, what is the next state the object will be in?).

Figure 18-15 is a statechart diagram for the MEMBER ORDER object of the Member Services system. It begins with an initial state (solid circle) and transitions through a life cycle of different states (rounded-corner rectangles) until it reaches its final state

object state a condition of the object at one point in its lifetime.

state transition event an occurrence that triggers a change in an object's state through the updating of one or more of its attributes' values.

state machine diagram a UML diagram that depicts the combination of states that an object can assume during its lifetime, the events that trigger transitions between states, and the rules governing the objects transition. Also called a statechart diagram or state transition diagram.

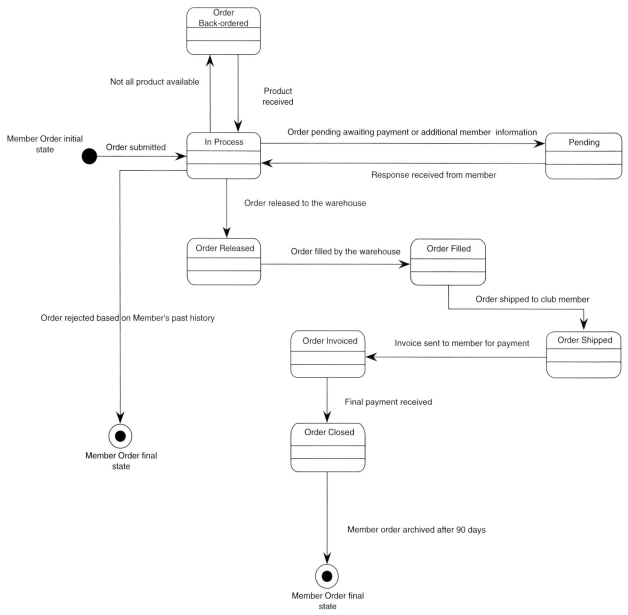

Constructed using Popkin Software's *System Architect*.

FIGURE 18-15 Member Order Statechart Diagram

(a solid circle inside of a hollow one). Each arrow represents an event that triggers the MEMBER ORDER to change from one state to another.

State machine diagrams are not required for all objects. Typically, a state machine diagram is constructed only for those objects that have clearly identifiable states and complex behavior. In our experience, any object that has an attribute called status is a good candidate for constructing a state machine diagram.

Finally, our last task is to verify the results from the previous tasks. This consists of conducting walkthroughs with the appropriate users. One verification approach that is commonly used is role playing. In **role playing,** the use-case scenarios are acted out by the participants. The participants may assume the role of an actor or an object type that collaborates to process a hypothetical business event. Message sending is simulated by using an item such as a ball that is passed (or

role playing the act of simulating object behavior and collaboration by acting out an object's behaviors and responsibilities.

sometimes thrown) between the participants. Role playing is quite effective in discovering missing objects and behaviors, as well as verifying the collaboration among objects.

> Updating the Object Model to Reflect the Implementation Environment

Once we have designed the objects and their required interactions, we can refine our class diagram to represent software classes in the application. A **design class diagram** typically includes the following:

design class diagram a diagram that depicts classes that correspond to software components that are used to build the software application.

- Classes.
- Associations and gen/spec and aggregation relationships.
- Attributes and attribute-type information.
- Methods with parameters.
- Navigability.
- Dependencies.

The following steps are used to transform the class diagram prepared in OOA to a design class diagram:

1. *Add design objects to diagram.* The entity, interface, and control objects that were previously identified should be added to the diagram. Because of diagram space and readability considerations, only the major interface objects should be included.
2. *Add attributes and attribute-type information to design objects.* OO programming languages allow the common attribute types such as Integer, Date, Boolean, and String(text), among others. OO languages also allow the definition of complex attribute types such as Address, Social Security Number, and Telephone Number; this is a powerful feature for a developer.
3. *Add attribute visibility.* Attributes can be defined as public, protected, or private.
4. *Add methods to design objects.* Define methods to get and update the values of all the attributes of each object. These types of methods are commonly referred to as "setters" and "getters" methods. It is common to exclude these methods from the design class diagram in order to save space and make the diagram more readable, because they always exist by default. Also, include methods to implement any previously identified responsibilities and behavior, such as creating or deleting class instances or forming or breaking class associations. Please note that method names are formatted based on the chosen programming language. How you format a method name in *Smalltalk* is different than how you do so in Java. In this textbook we will use the standard UML format of *methodName (parameterList)*.
5. *Add method visibility.* Methods can be defined as public, protected, or private.
6. *Add association navigability between classes.* Add navigability arrows to unidirectional associations to indicate the direction messages are sent between source and target classes.
7. *Add dependency relationships.* For any user interface class appearing on the diagram, draw a dependency line between it and the control object.

Figure 18-16 is a partial view of the SoundStage Member Services design class diagram. Please note the following:

❶ Visibility has been specified for each attribute. In this particular example all attributes are private, which is denoted by the symbol "−".
❷ Methods and their visibility have been specified.
❸ Navigability has been noted on some associations to indicate the passing of messages that go only one way.

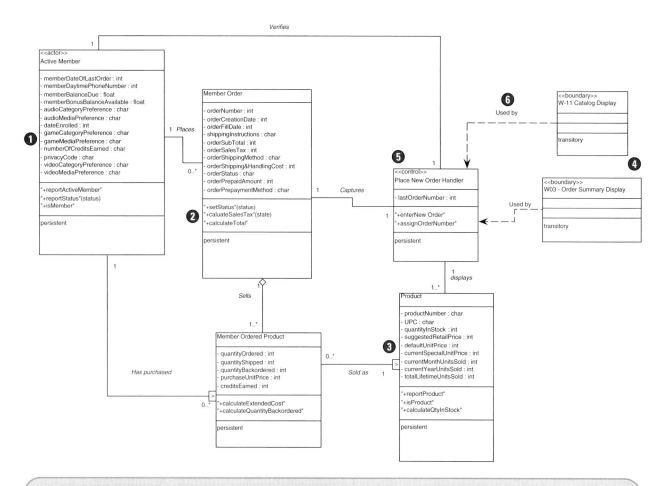

FIGURE 18-16 Partial Design Class Diagram for the *Place New Order* Use Case

❹ Interface classes have been added to show major user interface objects. In this particular software package these are considered boundary objects and are transitory in nature.

❺ A control class has been added to coordinate interactions between interface objects and entity objects. We have also given this control class the responsibility of assigning new order numbers, which are sequentially assigned.

❻ Interface objects are dependent on the control object.

Object Reusability and Design Patterns

coupling the degree to which one class is connected to or relies upon other classes.

cohesion the degree to which the attributes and behaviors of a single class are related to each other.

Look at the sequence diagram in Figure 18-17 and compare it to the sequence diagram shown in Figure 18-12. These are alternative ways of adding an item to an order. The major difference is that the design shown in Figure 18-17 places much more responsibility on the MEMBER ORDER entity class and much less on the PLACE NEW ORDER controller, MEMBER ORDERED PRODUCT, and PRODUCT. Essentially the controller passes the message to MEMBER ORDER and lets it act like a controller itself. We could design this interaction in other ways, as well. For instance the PLACE NEW ORDER controller could do all the work by passing messages directly to each class instead of going through MEMBER ORDER. So how do you know which is the best design?

The two overarching goals of object-oriented design are low coupling and high cohesion. **Coupling** is the degree to which one class is connected to or relies upon other classes. **Cohesion** is the degree to which all of the attributes and behaviors of

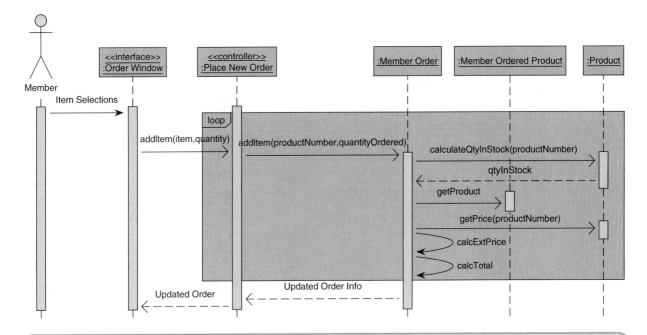

FIGURE 18-17   Alternate Sequence Diagram for Step 6 of the *Place New Order* Use Case

a single class are related to each other. By striving for high cohesion and low coupling, we want each class to focus on essentially one thing and for each class to be as independent as possible.

The reason behind the goals of high cohesion and low coupling is object reusability. Ideally, object classes created for one information system should be able to be reused in other information systems. That is why operating system–specific code and database-specific code are often designed into system and persistence classes. It allows, in theory, the entity classes to be reused with other databases and operating systems.

Several studies have documented the success of object reuse. An article that appeared in *ComputerWorld* tells how Electronic Data Systems (EDS) initiated two projects to develop the same system using two different programming languages.[2] One project used a traditional 3GL language called *PL/1,* and the other used *Smalltalk,* an object-oriented language. The results were impressive, as indicated in Table 18-1.

If our classes have high cohesion, meaning that they are essentially about one thing, then we are more likely to find reuse situations for them. If our classes have low coupling, meaning that they are relatively independent, then we can reuse one class or a few classes without having to import the entire class structure.

The design reflected in Figure 18-17 has higher coupling than the design in Figure 18-12. Why? Because MEMBER ORDER is related to (makes calls to) more classes. So we couldn't reuse MEMBER ORDER without also reusing both MEMBER ORDERED PRODUCT and PRODUCT. The Figure 18-17 design also has lower cohesion; MEMBER ORDER is responsible for getting the quantity in stock out of product. Should that really be its job?

Over the last few years the approaches that systems analysts use to design and develop software have radically changed with the advent of the Internet and Web-based applications. Many companies were under extreme pressure to provide a "Web-presence" in order to effectively compete with their competitors. With the business philosophy of "being first to market," developers were placed under extreme pressure to deliver functionality faster and faster, which often meant the quality of the product

---

[2]"White Paper on Object Technology: A Key Software Technology for the 90s," *ComputerWorld,* May 11, 1992.

TABLE 18-1	Comparison of an OO Language and a 3GL Language		
Programming Language	**Project Duration (calendar months)**	**Level of Effort (person-months)**	**Software Size (lines of code)**
PL/1	19	152	265,000
Smalltalk	3.5	10.4	22,000

was less than desirable. In their efforts to combat the quality problem and to achieve higher levels of reuse, developers began exploiting design patterns, object frameworks, and components.

Design Patterns

design pattern a common solution to a given problem in a given context, which supports reuse of proven approaches and techniques.

You probably have heard the phrase "Don't reinvent the wheel." Applied to software development it means don't write new software to solve a problem that someone else has already written software to solve correctly and efficiently. Many companies now take this approach with developing new applications. They would rather buy a software package off the shelf that meets the majority of their needs than build something from scratch. This approach ultimately saves time and money and makes good business sense if building the application would provide no competitive advantage, such as increased orders or greater market share. On a lesser scale, object-oriented developers look for the same reuse opportunities through the use of **design patterns.**

Over the course of many software projects, experienced developers collect a library of practices and routines, which worked well and correctly, that they can use over and over again in subsequent projects and even share with their fellow developers. These "development shortcuts," which are solutions to common design and programming problems, have come to be known as patterns. The goal of a pattern is not to discover or invent a new solution to a problem but to formally structure an existing solution to a common problem so that others may use it and take advantage of it. Figure 18-18 is an adaptation of the organizational pattern created by Martin

FIGURE 18-18

Organizational Pattern

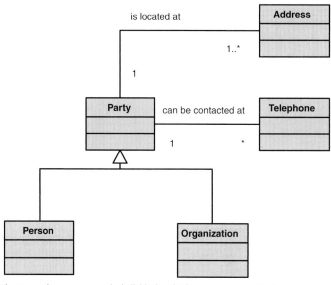

Example: An electric company's customers may be individuals or businesses. Many aspects of dealing with customers are the same, in which case they are treated as parties. Where they differ, they are treated through their subtype.

FIGURE 18-19 Gang-of-Four Patterns

Gang-of-Four Patterns		
Creational	**Structural**	**Behavioral**
Abstract factor	Adapter	Chain of responsibility
Builder	Bridge	Command
Factory method	Composite	Flyweight
Prototype	Decorator	Interpreter
Singleton	Façade	Iterator
	Proxy	Mediator
		Memento
		Observer
		State
		Strategy
		Template method
		Visitor

Fowler.[3] This pattern is very useful when your application has to work with organizational structures within a company or when individuals and companies can play the same role, such as customer.

There are two advantages to learning and using design patterns.

- They allow us to design information systems with the experiences of those who came before us rather than having to "reinvent the wheel."
- They provide designers a short-hand notation for discussing design issues. For instance, Bob Martinez might say to a colleague, "I know. Let's use a strategy pattern for promotions and build an adapter for integrating the Brand X sales tax class." You probably don't yet know what he's talking about. But if you read on, you will.

In 1995 Erich Gamma, Richard Helm, Ralph Johnson, and John Vlissides published *Design Patterns,* describing 23 patterns for OO design. The book quickly became know as the bible of design patterns, and the four authors became known collectively as the Gang of Four, often abbreviated as GOF. The 23 GOF patterns are divided into three categories (as shown in Figure 18-19): creational, structural, and behavioral. Creational patterns provide guidance for designing classes to instantiate new objects. Structural patterns provide guidance on how classes can be designed to form larger structures. Behavioral patterns provide guidance on the way in which classes interact to distribute responsibility. We will briefly discuss two sample patterns.

> The Strategy Pattern

SoundStage is always running promotions. When a member places an order, he or she may be using any one of a number of promotions. Some promotions are based on the total dollar amount of the order, some on the number of units, some on the kind of product ordered, some provide a percentage discount, some a dollar amount discount, and so on. The programming code to apply to each kind of promotion is significant. More importantly, it is constantly changing as the marketing people dream up new

---

[3]Martin Fowler is the chief scientist at ThoughtWorks, a cutting-edge consulting company. He has written several books and articles on OO development, and you can research his work at his Web site, martinfowler.com.

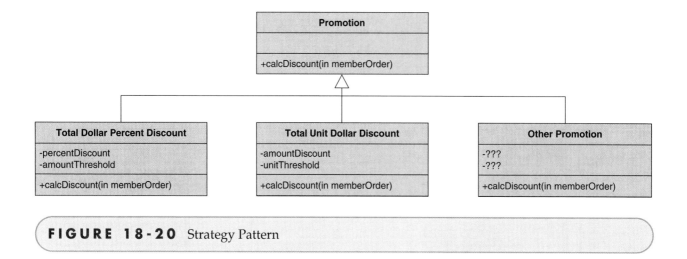

FIGURE 18-20 Strategy Pattern

promotions. How can our system incorporate existing and new promotions without constantly rewriting the controller classes?

Pattern:	Strategy
Category:	Behavioral
Problem:	How to design for varying and changing policy algorithms.
Solution:	Define each algorithm in a separate class with a common interface.

As illustrated in Figure 18-20, we can apply this pattern by creating various promotion classes that inherit from a supertype PROMOTION class. Each class has a standard interface method called calcDiscount, which returns the dollar amount that will be discounted when that promotion is applied to an order. The internal code to calculate each promotion will be entirely different for each promotion class. In fact, the MEMBER ORDER attributes needed for each calculation can even differ (number of units, total dollar amount, type of product, etc.). So we design the calcDiscount method to be passed to the entire MEMBER ORDER instance as a parameter. The promotion class can then do its job using any MEMBER ORDER attributes it needs.

> The Adapter Pattern

The SoundStage Member Services system has to calculate sales tax on orders. Keeping up on all the varying laws in each U.S. state and Canadian province is a daunting task. So SoundStage is going to buy prewritten tax calculation classes and plug them into the member services system. They found more than one vendor who could supply them with the classes, and each vendor's classes provide a different set of methods to call. They want to design the system so that if they ever change vendors, they have to change as little as possible in their system to work in the new classes.

Pattern:	Adapter
Category:	Structural
Problem:	How to provide a stable interface to similar classes with different interfaces.
Solution:	Add a class that acts as an adapter to convert the interface of a class into another interface that the client classes expect.

Figure 18-21 shows an implementation of the adapter pattern for SoundStage. We begin with the SALES TAX ADAPTER class. This provides an unchanging method (calcSalesTax) for the rest of the system to call. To integrate in the purchased class (BRAND X SALES TAX CALCULATOR) we write a new class (BRAND X ADAPTER) that inherits from SALES TAX ADAPTER class and includes all the code needed to call the purchased class. It translates (or adapts) the call from the system into a call that the purchased class can

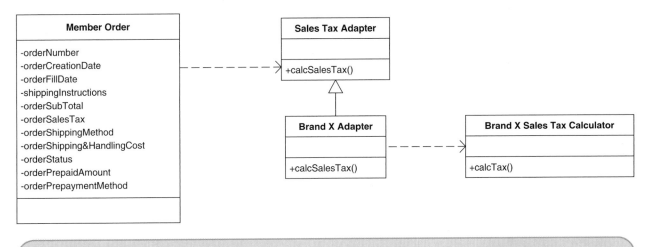

FIGURE 18-21 Adapter Pattern

accept. If we ever change vendors, then we have only to write a new adapter subtype; everything else stays the same.

Many Web sites are dedicated to the subject of patterns, and many excellent textbooks have been published containing tried-and-true patterns that have been developed by experts in the industry. Some are listed in the Suggested Readings at the end of the chapter.

> Object Frameworks and Components

Developers use **object frameworks** to take advantage of reusability and lessen development time. A framework is a subsystem of collaborating classes that provides a set of related services. Whereas patterns are design guidelines written on paper (or a Web page), frameworks are classes implemented in programming code and ready to call. One example is a calendar routine, used for calculating or displaying dates. Routines used for charting, printing, or any type of application utility would be good candidates for object frameworks. One of the more common object frameworks available is the software used to translate objects to relational tables and vice versa. This framework is required whenever your application is OO but the database you are using is non-OO, such as relational technology. By using object frameworks, developers can concentrate on developing the logic that is new or unique to the application, thus reducing the overall time required to build the entire system.

> **object framework** a set of related, interacting objects that provide a well-defined set of services for accomplishing a task.

Using **components,** the developer can easily package and distribute the programming code to others. A component represents a modular, physical element (EXE, DLL, or database) of the system that is replaceable. Components are sometimes thought of as "superobjects," but in reality a component consists of a set of related collaborating objects that have an interface and can be deployed as a single unit.

> **component** a group of objects packaged together into one unit. An example of a component is a dynamic link library (DLL) or executable file.

Additional UML Design and Implementation Diagrams

In Chapter 7 we were introduced to *use case diagrams*. In Chapter 10 we saw how to use *activity diagrams* and *class diagrams* in the analysis phase. In this chapter we saw how to build a design-level class diagram. We also saw in this chapter how to use *sequence diagrams* and *state machine diagrams* in the design phase. Sequence diagrams are useful when you want to study the behavior of several classes within a single use case. State machine diagrams are useful when you want to explore a single

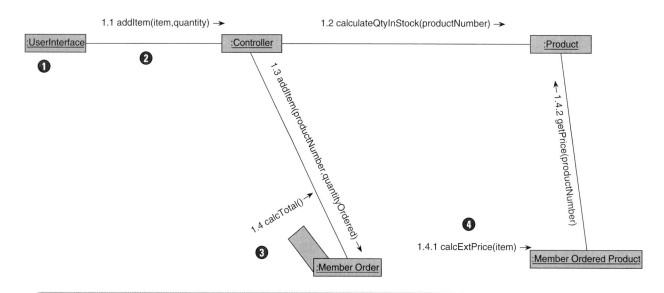

1.1 addItem(item,quantity) →   1.2 calculateQtyInStock(productNumber) →

:UserInterface ❶   :Controller ❷   :Product

1.3 addItem(productNumber,quantityOrdered) →

1.4 calcTotal() →

❸ :Member Order

1.4.2 getPrice(productNumber)

❹

1.4.1 calcExtPrice(item) →   :Member Ordered Product

FIGURE 18-22 Communication Diagram for Step 6 of the *Place New Order* Use Case

communication diagram
models the interaction of objects via messages, focusing on the structural organization of objects in a network format. Called a collaboration diagram prior to UML 2.0.

class across multiple use cases. You can also use activity diagrams in the design phase to graphically depict the sequential flow of activities of either a business process or a use case. In design they are very useful for modeling actions that will be performed when an operation is executing and the results of those actions—such as modeling the events that cause windows to be displayed or cleared.

That accounts for 5 of the 13 different diagrams of UML 2.0. What about the others? Briefly we will introduce you to three others. A **communication diagram** (called a collaboration diagram in earlier versions of UML) models the interaction of objects via messages. Thus, it is similar to a sequence diagram. But while a sequence diagram focuses on the timing or sequence of messages, a communication diagram focuses on the structural organization of objects in a network format. Figure 18-22 is a communication diagram that depicts the same interaction as the sequence diagram in Figure 18-12. Note the following on the diagram:

❶ *Class*—shown with a box symbol as in a sequence diagram but without the lifeline.
❷ *Messages*—show communication between the classes with arrows marking the direction and a notation of the method being called.
❸ *Self-calls*—can be shown as in sequence diagrams.
❹ *Numbering scheme*—Though this rule is often violated, the messages should be numbered with a nested scheme. This can be useful in depicting calls that are performed as part of a larger method. The numbering scheme here, for instance, makes it clear that steps 1.4.1 and 1.4.2 are done within step 1.4.

Since they accomplish much the same thing as sequence diagrams, when would you use one versus the other? Sequence diagrams are generally better when you want to emphasize the sequence of calls while communication diagrams are better when you want to emphasize the links. Overall, sequence diagrams are more popular. But since they are simpler and easier to draw on whiteboards, communication diagrams may be better tools for brainstorming alternative solutions.

component diagram
depicts the organization of programming code divided into components and how the components interact.

Component diagrams are implementation-type diagrams that are used to graphically depict the physical architecture of the software of the system. A single software component often implements a group of classes that form a cohesive subset of the system. Component diagrams can be used to show how programming code is divided into components and to depict the dependencies between those components.

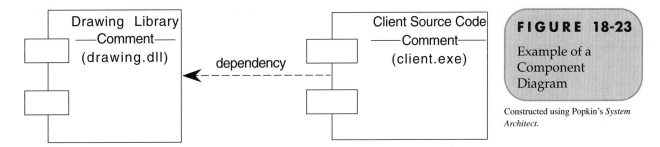

FIGURE 18-23

Example of a
Component
Diagram

Constructed using Popkin's *System Architect.*

Figure 18-23 is an example of a component diagram constructed using Popkin's *System Architect.* The component symbol shown here is a UML 1.X notation, which UML 2.0 has removed. It is, however, still in popular use.

Deployment diagrams are implementation-type diagrams that describe the physical architecture of the hardware and software in the system. They depict the software components, processors, and devices that make up the system's architecture. Figure 18-24 shows an example of a deployment diagram. Each box in the

deployment diagram
depicts the configuration of software components within the physical architecture of the system's hardware "nodes."

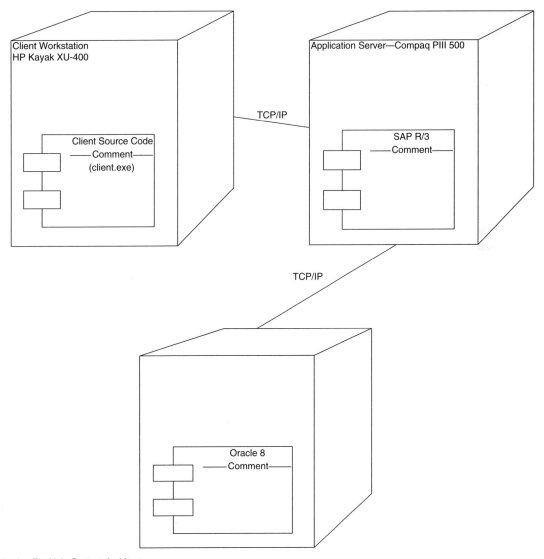

Constructed using Popkin's *System Architect.*

FIGURE 18-24 Sample Deployment Diagram

diagram is the symbol for a node, which in most cases is a piece of hardware. The hardware may be a PC, mainframe, printer, or even a sensor. Software that resides on the node is represented by the component symbol. The lines connecting the nodes indicate a communication path between the devices. In Figure 18-24 the connections are labeled with the type of communication protocols being used.

If you desire a more in-depth description of the purpose and use of the entire set of UML diagrams, there are excellent books available documenting the use of the UML. Many are listed in this chapter's Suggested Readings.

Learning Roadmap

This chapter provided an introduction to the object-oriented approach to systems design. Since prototyping is an integral part of object-oriented design, it is recommended that you learn about prototyping and user interfaces during system design. Thus, if you haven't already covered these topics, you should proceed to Chapters 15, 16, and 17 next.

To gain a better understanding of object-oriented design and its impact on the subsequent construction and implementation of a new system, it is recommended that you learn about object-oriented programming (OOP). Consider taking a course dealing with object-oriented programming. With the popularity of *Java* and the .NET framework, numerous books have been written on OOP. These books explain how the OO concepts presented in Chapters 10 and 18 are implemented in an object-oriented programming language environment.

Chapter Review

1. The approach of using object-oriented techniques for designing a system is referred to as object-oriented design.
2. Object-oriented design is concerned with identifying and classifying three class types, including interface, entity, and control object types. Interface and control object types are objects that are introduced as a result of implementation decisions that were made during systems design.

   a. Entity classes are identified during systems analysis and usually correspond to items in real life and contain information, known as attributes, that describes the different instances of the entity.

   b. Interface classes are introduced to represent a means through which the user will interface with the system. The responsibility of the interface class is twofold:

   i) It translates the user's input into information that the system can understand and use to process the business event.

   ii) It takes data pertaining to a business event and translates the data for appropriate presentation to the user.

   c. Control classes are those that hold application or business rule logic. Control classes serve as the "traffic cop" containing the application logic or business rules of the event for managing or directing the interaction between the classes.

3. An object-oriented system could be implemented with the above three types of classes. But many methodologists prefer to include two other types of classes.

   a. Persistence classes provide functionality to read and write entity class attributes to a database.
   b. System classes isolate other classes from operating system–specific functionality.

4. In object-oriented design it is necessary to model more advanced relationships in order to accurately specify the software components:

   a. A dependency relationship is used to model the association between two classes in two instances: (1) to indicate that when a change occurs in one class, it may affect the other class, and (2) to indicate the association between a persistent class and a transient class.
   b. By default, associations between classes are bidirectional, meaning that objects of one kind can navigate (send messages) to objects of the other kind. There may be times, though, when you want to limit the message sending to only one direction. You specify navigability by placing an arrowhead on the association in the direction the message will be sent.
   c. How attributes and methods are accessed by other objects is defined by visibility. The UML provides three levels of visibility:

      i) Public—denoted by the symbol "+".
      ii) Protected—denoted by the symbol "#".
      iii) Private—denoted by the symbol "−".

5. Object responsibility is the obligation that an object has to provide a service when requested and thus collaborate with other objects to satisfy the request if required. Object responsibility is closely related to the concept of objects being able to send and/or respond to messages.

6. Object-oriented design includes the following activities:

   a. Refining the use-case model to reflect the implementation environment.
   b. Modeling class interactions, behaviors, and states that support the use-case scenario.
   c. Updating the class diagram to reflect the implementation environment.

7. In OOD, analysis use cases are refined into design use cases to reflect the physical aspects of the implementation environment for the new system.

8. During systems design, use-case descriptions are examined to identify all action-verb phrases.

Action-verb phrases suggest behaviors required to complete a use-case scenario. These behaviors must be associated with a system object.

9. A popular tool for documenting the behaviors and collaborations of an object is the class responsibility collaboration (CRC) card.

10. Another tool for discovering and/or documenting class behaviors and responsibilities is a sequence diagram. A sequence diagram models the logic of a use case by depicting the interaction of messages between classes in time sequence.

11. A state machine diagram models the life cycle of a single object. It depicts the different states a class can have, the events that cause it to change state over time, and the rules that govern its transition between states. In other words, it specifies from which state a class is allowed to transition to another state.

12. A design class diagram represents the software classes in the application. It consists of the following:

   a. Classes.
   b. Associations and gen/spec and aggregation relationships.
   c. Attributes and attribute-type information.
   d. Methods with parameters.
   e. Navigability.
   f. Dependencies.

13. Two overarching goals of object-oriented design are low coupling and high cohesion. Coupling is the degree to which one class is connected to or relies upon other classes. Cohesion is the degree to which the attributes and behaviors of a single class are related to each other. Achieving low coupling and high cohesion promotes object reuse, which lowers the cost of software development.

14. To achieve higher levels of reuse, developers have begun exploiting design patterns. Design patterns are a common solution to a given problem in a given context.

   a. The strategy pattern deals with how to design for varying and changing policy algorithms.
   b. The adapter pattern provides guidance on how to design a stable interface for similar classes with different interfaces.
   c. Many other patterns also exist.

15. Developers also use object frameworks to speed development. A framework is a subsystem of collaborating classes that provide a set of related services.

16. Using components, a developer can package and distribute programming code to others.

17. The UML offers other diagrams for modeling design and implementation aspects of the system:

    a. In design, activity diagrams are very useful for modeling actions that will be performed when an operation is executing as well as the results of those actions—such as modeling the events that cause windows to be displayed or closed.

    b. Communication diagrams model the interaction of objects via messages, focusing on the structural organization of objects in a network format.

    c. Component diagrams are implementation-type diagrams that are used to graphically depict the physical architecture of the software of the system. They can be used to show how programming code is divided into modules (or components) and to depict the dependencies between those components.

    d. Deployment diagrams are implementation-type diagrams that describe the physical architecture of the hardware and software in the system. They depict the software components, processors, and devices that make up the system's architecture.

Review Questions

1. What are the three kinds of objects used in object-oriented design?
2. Why are the three kinds of objects needed in object-oriented design?
3. What is navigability? Please give an example of a navigability relationship.
4. What is visibility in object-oriented design? Explain the different levels of visibility.
5. What is the key reason for object reusability?
6. What are some of the methods developers use to achieve object reusability?
7. What are the main activities of object-oriented design?
8. What is the objective of refining the use-case model in object design? Why is it important?

9. What are some ways that we can use to identify use-cases design objects—namely, interface objects, control objects, and entity objects?
10. What is the goal of constructing object robustness diagrams? What are the components of the diagrams?
11. What should we look for in identifying the object behaviors and responsibilities of the objects?
12. What is the relationship between an object state and state transition event?
13. What are the steps needed to construct the state chart diagram?
14. What are the tools used to document the detailed object interaction for the use cases?
15. What does a design class diagram include?

Problems and Exercises

1. What is the main rationale for using object-oriented methods to develop systems? Why?
2. A project developed in *PL/1* is expected to take 30 months. Assuming the same ratios as those shown in Table 18-1, compare the duration, level of effort, and software size between a project developed in *PL/1* and a project developed in an object-oriented language comparable to *Smalltalk*.

Programming Language	Project Duration (calendar months)	Level of Effort (person-months)	Software size (lines of code)
PL/1	30.0	240.0	41,800
OO Language	5.5	16.4	3,500

3. True or false? Explain your answers as needed.

- A dependency relationship models a two-class association in only two instances.
- To enforce encapsulation, attributes should generally be declared private.
- An object that is supposed to collaborate with other objects when necessary to provide a requested service, but which is unable to do so, is termed an irresponsible object.
- Interface objects are typically persistent.
- During the object-oriented design phase, the object model is updated to reflect the actual implementation environment.

4. What are the interface objects users may find for the following?

    a. Photo printer that doesn't require a computer to print pictures.
    b. Service station gas pump.
    c. Entrance/exit door in retail store.

5. During the design phase in object-oriented design, are any changes made to the use cases created earlier? If so, what are these changes, and what is their overall purpose?

6. Fill in the blanks:

    a. Window _____, e.g., icons, buttons and links, are _____ stated in system design _____.
    b. The term for a set of _____ objects which are _____, have an _____ and, which can act as a single unit is _____.
    c. To be able to _____ objects, they need to _____ correctly by defining them within an appropriate _____ hierarchy so they are _____ enough for easy use in other applications.
    d. During the _____ phase, _____ and _____ are refined to mirror the _____ environment of the solution, rather than an environment based upon a _____ ideal.

7. Match the terms in the first column with the definitions or examples in the second column.

1. Visibility	A. Common reusable solution to given problem in given context
2. Design pattern	B. Execution of software logic in response to message
3. Components	C. Object condition at a specific point during its lifetime
4. State transition event	D. External object's access level to an attribute or method
5. Interface object	E. Obligation to collaborate if needed to provide requested service
6. Control object	F. Collaborating objects subsystem providing set of related services
7. Entity object	G. Model of single object's life cycle states
8. Object state	H. Holds business rule or application logic
9. Object responsibility	I. Acting out use-case scenarios to simulate object behaviors

10. Object framework	J. Representation of business domain's actual data
11. Role playing	K. Change in state caused by occurrence updating attributes' values
12. Method	L. API, screen, window, dialogue box
13. State machine diagram	M. DLL or .exe file

8. Select an application with which you are familiar. Pick one of the processes in the application and create an analysis use case; use the template shown in Figure 18-7. Then, using the guidelines in this chapter, refine the use case and transform it into a design use case. Highlight the areas that you changed or added.

9. After creating the design use case, analyze it in order to identify and classify the use-case design objects; use Figure 18-8 as an example. In general, you will probably have more entity objects than interface objects, and you should have at least one control object. Have a fellow student or co-worker check your work to make sure it is identified and classified correctly.

10. What is the purpose of an object robustness diagram? What are the symbols used in this diagram, and what do they represent? Next, draw an object robustness diagram based upon your use case; use Figure 18-9 as an example.

11. Now go back to the design use case you created. Analyze this use case to identify the required system behaviors; use the matrix shown in Figure 18-10 as an example. After identifying the use case behaviors, determine if each behavior will be automated or manual in the new system. If the behavior is automated, then in the third column assign the object type that will be responsible for executing that behavior.

12. Explain the purpose of the class responsibility collaboration (CRC) card; then create a CRC card for each object type identified in your previous exercises.

13. At this point, take a moment and assess the object-oriented analysis and design techniques you have learned. How do you feel they compare to the other analysis and design approaches taught in this textbook? Do you feel the additional work and complexity will pay off in terms of reduced project development time and other factors? If you were given the choice, which approach would you choose?

Projects and Research

1. As discussed in the textbook, the key element in object-oriented technology is its potential for reusability. Martin Fowler, the chief scientist at ThoughtWorks, is a leader in the use of design patterns and has written numerous articles on the subject. Go to his Web site at *www.martinfowler.com* and read some of the articles and forum entries that are posted there.

   a. Prepare a two- to three-page paper analyzing his early approach and contributions to object reusability.

   b. How do others in the field of design view Fowler's work?

   c. Read some of his most recent papers. What is Fowler currently working on?

   d. Read Fowler's "Agile Manifesto." Do you agree or disagree with it? Explain your answer.

   e. At a theoretical level, what do you feel is the value of Fowler's work in the field of design? What about on a practical level?

2. Fowler has developed a number of approaches to reusability since his initial work on design patterns. Research some of his recent work on the theme of reusability. Select one and write a critical analysis. Compare and contrast his approaches to those of others in the field of design.

3. Now look at Fowler's work on design patterns. Look at the organizational pattern created by Fowler as a way to document the structure of a solution so that its use by others is optimized, and at the adaptation in the textbook. Select an application with which you are familiar. Look for reuse opportunities in the application, then create several design patterns; use the organizational pattern in Figure 18-6. Have someone who is knowledgeable and has experience in object-oriented design review your design patterns for applicability. If this were not a classroom exercise, could your design patterns be used for other actual applications?

4. Take the design use case you created and refined in Problems and Exercises (or another one if you prefer). Create a sequence diagram and class diagram based upon this use case; use Figures 18-16 and 18-17 as examples. Have someone in your organization or school who is knowledgeable in using object-oriented design review your diagrams and modify them as necessary.

5. Envision the implementation for a hypothetical system. Create the component diagrams and deployment diagrams that describe the physical architecture of the system software and hardware; use Figures 18-18 and 18-19 as examples. Have someone in your organization or school who is knowledgeable in using object-oriented implementation techniques review your diagrams and modify them as necessary.

6. Just as object-oriented analysis (OOA) led up to and transitioned into object-oriented design (OOD), so too does OOD lead up to and transition into object-oriented programming. Although this is not a class in programming, understanding the basic concepts and constructs of object oriented programming (OOP) may be beneficial to an overall understanding of the object-oriented approach to systems development. Research object-oriented programming on the Web or in textbooks in order to get an overview. How are the diagrams, use cases, and other artifacts created in the object-oriented analysis and design phases used by object-oriented programming in the construction phase? Does object-oriented programming introduce any new diagrams or other constructs? What are the basic steps and processes used by OOP during the construction phase? What are some of the most popular object-oriented programming languages in use today?

Minicases

1. Get Jim Conallen's book *Building Web Applications with UML* (Boston: Addison-Wesley, 2002):

   a. What are the differences between Web-based UML and traditional UML, shown in Conallen's book?

   b. Wow Munchies, discussed in a previous chapter, has decided to implement an e-commerce site. It would like you to do the UML modeling for the site. What UML modeling techniques will you use? Why?

2. In previous chapters, you interviewed a government department and began designing a new system for it. Return to the notes:

   a. Describe the system that you recommend it have.

b. Using UML modeling, diagram the system you are proposing. (Consider Use Case, Class, Sequence, and State Machine diagrams.)

c. Submit a deliverable to your professor, including a discussion of the previous interviews and parts 2a and 2b of this minicase. You will be graded on correctness, completeness, and professionalism.

3. Using your work from minicase 2 and from previous chapters, create a system prototype for your government client. What language are you using?

Why? Submit your prototype on CD to your professor, as well as a hard copy of your source code, screen shots, a short discussion of any assumptions you made, and a short discussion of the business problem you are solving with the system.

4. Exchange prototypes (from minicase 3) with another group. Evaluate the design and usability of the interface and any output. Document your findings thoroughly and prepare a report. Give one copy of the report to the other group (the one who created the prototype) and one to the professor.

Suggested Readings

Ambler, Scott W. *The Object Primer.* New York: Cambridge University Press, 2001. Very good information about documenting use cases and their use.

Armour, Frank, and Granville Miller. *Advance Use Case Modeling.* Boston: Addison-Wesley, 2001. This book presents excellent coverage of the use-case modeling process.

Booch, G. *Object-Oriented Design with Applications.* Redwood City, CA: Benjamin Cummings, 1994. Many Booch concepts were integrated into the UML.

Coad, P., and E. Yourdon. *Object-Oriented Analysis,* 2nd ed. Englewood Cliffs, NJ: Prentice Hall, 1991. This book provides a very good overview of object-oriented concepts. However, the object model techniques are somewhat limited by comparison to UML and other object-oriented modeling approaches.

Eriksson, Hans-Erik, and Magnus Penker. *UML Toolkit.* New York: John Wiley & Sons, 1998. This book provides detailed coverage of the UML.

Fowler, Martin. *UML Distilled Third Edition, A Brief Guide to the Standard Object Modeling Language.* Reading, MA: Addison-Wesley, 2003. This is a good short guide introducing the concepts and notation of UML 2.0.

Harman, Paul, and Mark Watson. *Understanding UML: The Developer's Guide.* San Francisco: Morgan Kaufmann Publishers, 1997. This is an excellent reference book. The examples were prepared using Popkin's *System Architect.*

Jacobson, Ivar; Magnus Christerson; Patrik Jonsson; and Gunnar Overgaard. *Object-Oriented Software Engineering: A Use Case Driven Approach.* Workingham, England:

Addison-Wesley, 1992. This book presents detailed coverage of how to identify and document use cases.

Larman, Craig. *Applying UML and Patterns: An Introduction to Object-Oriented Analysis and Design.* Englewood Cliffs, NJ: Prentice Hall, 1997. This is an excellent reference book explaining the concepts of OO development utilizing the UML.

Martin, J., and J. Odell. *Object-Oriented Analysis and Design.* Englewood Cliffs, NJ: Prentice Hall, 1992.

Rumbaugh, James; Michael Blaha; William Premerlani; Frederick Eddy; and William Lorensen. *Object-Oriented Modeling and Design.* Englewood Cliffs, NJ: Prentice Hall, 1991. This book presents detailed coverage of the object modeling technique (OMT) and its application throughout the entire systems development life cycle. Many OMT constructs are now in the UML.

Rumbaugh, James; Ivar Jacobson; and Grady Booch. *The Unified Modeling Language Reference Manual.* Reading, MA: Addison-Wesley, 1999. This book presents detailed coverage of the UML by the primary authors who created it.

Rumbaugh, James; Ivar Jacobson; and Grady Booch. *The Unified Modeling Language Users Guide.* Reading, MA: Addison-Wesley, 1999. This book presents detailed coverage of the UML by the primary authors who created it.

Taylor, David A. *Object-Oriented Information Systems: Planning and Implementation.* New York: John Wiley & Sons, 1992. This book is a very good entry-level resource for learning the concepts of object-oriented technology and techniques.

Part Four

Beyond Systems Analysis and Design

Part Four introduces you to the final phases of systems development and the support activities that are ongoing once the system has been placed in operation.

Chapter 19, "Systems Construction and Implementation," presents the process of constructing the system from physical design specifications and the implementation of the constructed system.

Chapter 20, "Systems Operations and Support," discusses four types of systems support for an application. This ongoing maintenance of a system after it has been placed into production consists of correcting errors, recovering the system, assisting users, and adapting the system. Systems support is very important because it is likely that young systems analysts will be responsible for maintaining legacy systems. This chapter concludes our exploration of the systems development life cycle.

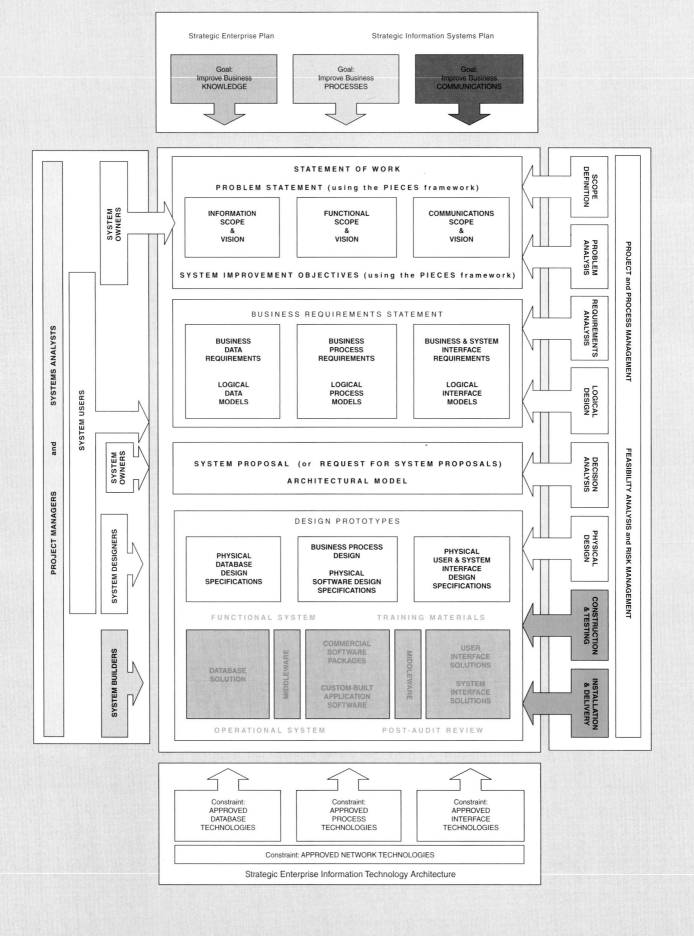

Strategic Enterprise Plan   Strategic Information Systems Plan

Goal:
Improve Business
KNOWLEDGE

Goal:
Improve Business
PROCESSES

Goal:
Improve Business
COMMUNICATIONS

SYSTEM OWNERS

SYSTEMS ANALYSTS

PROJECT MANAGERS and

SYSTEM USERS

SYSTEM OWNERS

SYSTEM DESIGNERS

SYSTEM BUILDERS

STATEMENT OF WORK

PROBLEM STATEMENT (using the PIECES framework)

INFORMATION
SCOPE
&
VISION

FUNCTIONAL
SCOPE
&
VISION

COMMUNICATIONS
SCOPE
&
VISION

SYSTEM IMPROVEMENT OBJECTIVES (using the PIECES framework)

BUSINESS REQUIREMENTS STATEMENT

BUSINESS
DATA
REQUIREMENTS

LOGICAL
DATA
MODELS

BUSINESS
PROCESS
REQUIREMENTS

LOGICAL
PROCESS
MODELS

BUSINESS & SYSTEM
INTERFACE
REQUIREMENTS

LOGICAL
INTERFACE
MODELS

SYSTEM PROPOSAL (or REQUEST FOR SYSTEM PROPOSALS)

ARCHITECTURAL MODEL

DESIGN PROTOTYPES

PHYSICAL
DATABASE
DESIGN
SPECIFICATIONS

BUSINESS PROCESS
DESIGN

PHYSICAL
SOFTWARE DESIGN
SPECIFICATIONS

PHYSICAL
USER & SYSTEM
INTERFACE
DESIGN
SPECIFICATIONS

FUNCTIONAL SYSTEM   TRAINING MATERIALS

DATABASE
SOLUTION

MIDDLEWARE

COMMERCIAL
SOFTWARE
PACKAGES

CUSTOM-BUILT
APPLICATION
SOFTWARE

MIDDLEWARE

USER
INTERFACE
SOLUTIONS

SYSTEM
INTERFACE
SOLUTIONS

OPERATIONAL SYSTEM   POST-AUDIT REVIEW

SCOPE DEFINITION

PROBLEM ANALYSIS

REQUIREMENTS ANALYSIS

LOGICAL DESIGN

DECISION ANALYSIS

PHYSICAL DESIGN

CONSTRUCTION & TESTING

INSTALLATION & DELIVERY

PROJECT and PROCESS MANAGEMENT

FEASIBILITY ANALYSIS and RISK MANAGEMENT

Constraint:
APPROVED
DATABASE
TECHNOLOGIES

Constraint:
APPROVED
PROCESS
TECHNOLOGIES

Constraint:
APPROVED
INTERFACE
TECHNOLOGIES

Constraint: APPROVED NETWORK TECHNOLOGIES

Strategic Enterprise Information Technology Architecture

Systems Construction and Implementation

Chapter Preview and Objectives

In this chapter you will learn more about the construction and implementation phases of systems development. These two phases construct, test, install, and deliver the final system into operation. You will know that you understand the processes of constructing and implementing a system when you can:

▋ Explain the purpose of the construction and implementation phases of the system's life cycle.

▋ Describe the system's construction and implementation phases in terms of your information building blocks.

▋ Describe the system's construction and implementation phases in terms of major tasks, roles, inputs, and outputs.

▋ Explain several application program and system tests.

▋ Identify several system conversion strategies.

▋ Identify the chapters in this textbook that can help you actually perform the tasks of systems construction and implementation.

Although some of the techniques of systems construction and implementation are introduced in this chapter, it is *not* the intent of this chapter to teach the *techniques*. This chapter teaches only the process of construction and implementation.

Introduction

Construction has finally begun on the SoundStage Member Services system. Bob Martinez is an analyst/programmer, which means that he is expected to do some programming as well as systems analysis. Tasked with writing code to implement some of the use cases, Bob is seeing the advantage of all the analysis and design work that has gone on before. From the repository of design documents, Bob can draw essentially everything he needs to know to write his programs. His boss, Sandra, insisted that he write test scripts before he began programming. Again, the use cases told him what alternatives needed to be tested and what the results should be.

Other members of the systems analysis team are working with database programmers, application programmers, Web designers and administrators, software vendors, technical writers, and an outside firm hired to perform systems testing. They are racing to meet the deadline. But it is gratifying to see the system they designed becoming a reality.

What Is Systems Construction and Implementation?

systems construction the development, installation, and testing of system components.

systems implementation the installation and delivery of the entire system into production.

Let's begin with definitions of systems construction and implementation. **Systems construction** is the development, installation, and testing of system components. Unfortunately, *systems development* is a common synonym. (We dislike that synonym since it is more frequently used to describe the *entire* life cycle.) **Systems implementation** is the delivery of that system into production (meaning day-to-day operation).

Relative to the information systems building blocks, systems construction and implementation address IS building blocks primarily from the system builders' perspective (see the chapter home page).

Figure 19-1 illustrates the construction and implementation phases. Notice that the trigger for the systems construction phase is the approval of the physical design specifications resulting from the design phase. Given the design specifications, we can construct and test system components for that design. Eventually we will have built the functional system. The functional system can then be implemented or delivered as an operational system.

This chapter examines each of these phases in detail.

The Construction Phase

The purpose of the construction phase is to develop and test a functional system that fulfills business and design requirements and to implement the interfaces between the new system and existing production systems. Programming is generally recognized as a major aspect of the construction phase. But with the trend toward system solutions that involve acquiring or purchasing software packages, the implementation and integration of software components is becoming an equally, if not more, common and visible aspect of the construction phase.

In this section you will learn about several tasks involved in the construction phase of a typical systems development project. Figure 19-2 depicts the various tasks for the construction phase. Let's examine each construction phase task in greater detail.

> Task 6.1—Build and Test Networks (if Necessary)

Recall that in the requirements analysis phase of systems analysis, we established network requirements. Subsequently, during the design phase we developed distributed

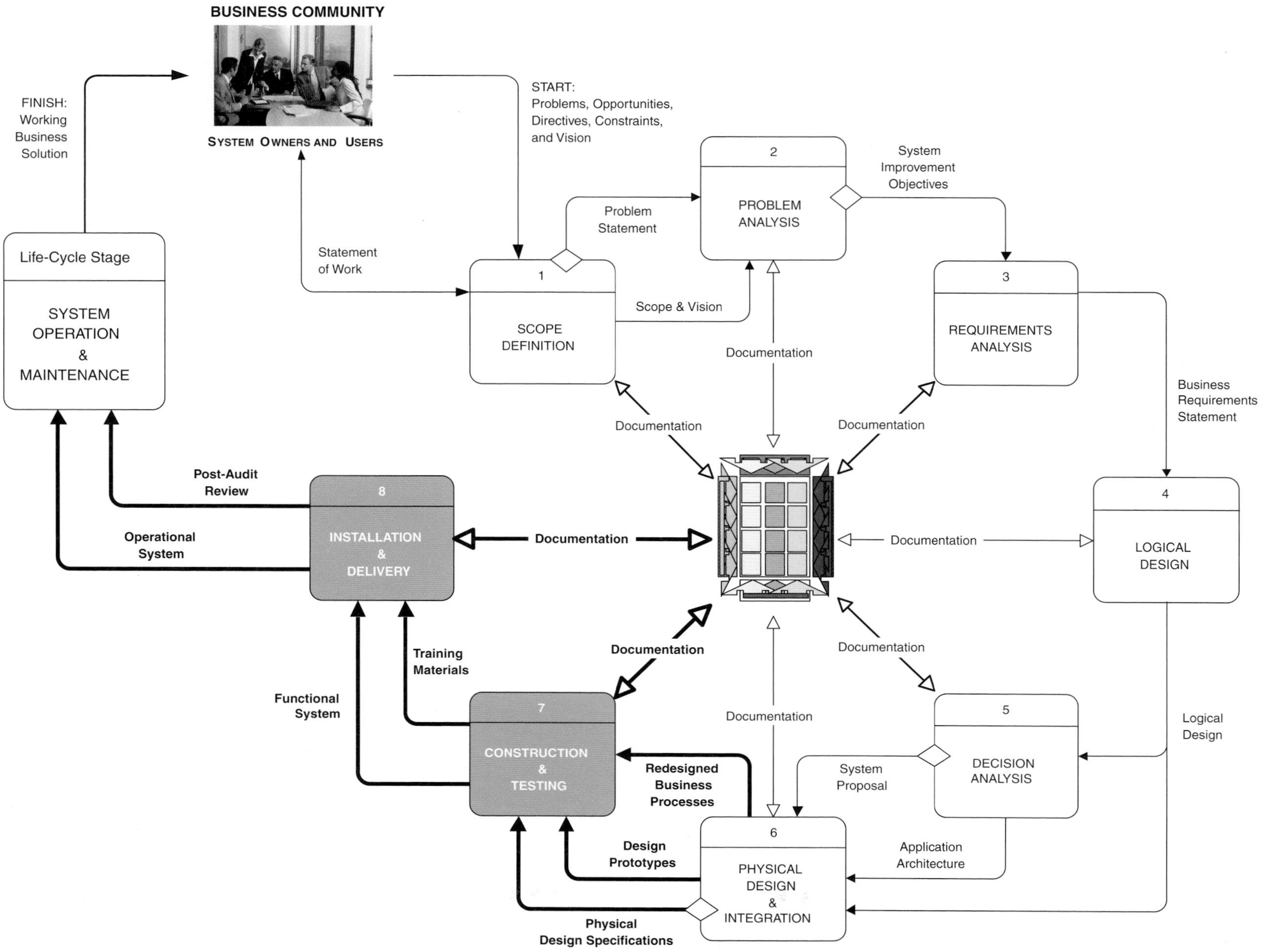

BUSINESS COMMUNITY

SYSTEM OWNERS AND USERS

START: Problems, Opportunities, Directives, Constraints, and Vision

FINISH: Working Business Solution

Life-Cycle Stage

SYSTEM OPERATION & MAINTENANCE

1 SCOPE DEFINITION

2 PROBLEM ANALYSIS

3 REQUIREMENTS ANALYSIS

4 LOGICAL DESIGN

5 DECISION ANALYSIS

6 PHYSICAL DESIGN & INTEGRATION

7 CONSTRUCTION & TESTING

8 INSTALLATION & DELIVERY

Statement of Work

Problem Statement

Scope & Vision

System Improvement Objectives

Business Requirements Statement

Logical Design

Application Architecture

System Proposal

Redesigned Business Processes

Design Prototypes

Physical Design Specifications

Training Materials

Functional System

Operational System

Post-Audit Review

Documentation

FIGURE 19-1 The Context of Systems Construction and Implementation

685

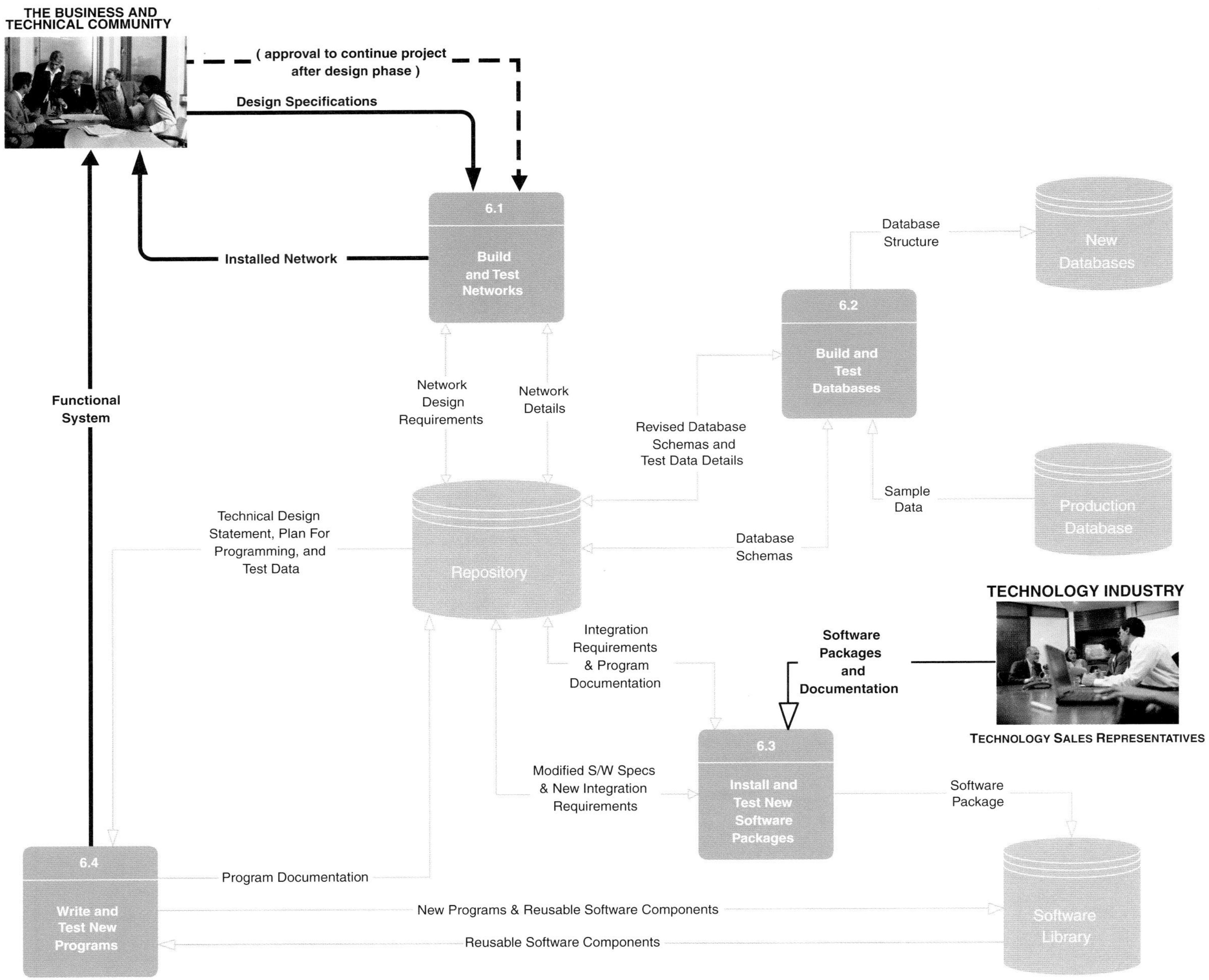

THE BUSINESS AND TECHNICAL COMMUNITY

(approval to continue project after design phase)

Design Specifications

Installed Network

6.1 Build and Test Networks

Network Details

Network Design Requirements

Database Structure

6.2 Build and Test Databases

Sample Data

Revised Database Schemas and Test Data Details

Database Schemas

New Databases

Production Database

Repository

TECHNOLOGY INDUSTRY

Software Packages and Documentation

6.3 Install and Test New Software Packages

Integration Requirements & Program Documentation

Modified S/W Specs & New Integration Requirements

TECHNOLOGY SALES REPRESENTATIVES

Software Package

Software Library

Technical Design Statement, Plan For Programming, and Test Data

Program Documentation

Functional System

6.4 Write and Test New Programs

New Programs & Reusable Software Components

Reusable Software Components

FIGURE 19-2 Systems Construction Tasks

data and process models. Using these technical design specifications to implement the network architecture for an information system is a prerequisite for the remaining construction and implementation activities.

In many cases, new or enhanced applications are built around existing networks. If so, skip this task. However, if the new application calls for new or modified networks, they must normally be implemented before building and testing databases and writing or installing computer programs that will use those networks. Thus, the first task of the construction phase may be to build and test networks.

This phase involves analysts, designers, and builders. A network designer and network administrator assume the primary responsibility for completing this task. The network designer is a specialist in the design of local and wide area networks and their connectivity. The network administrator has the expertise for building and testing network technology for the new system. He or she will also be familiar with network architecture standards that must be adhered to for any possible new networking technology. This person is also responsible for security. (The network designer and network administrator may be the same person.) While the systems analyst may be involved in the completion of this task, the analyst's role is more that of a facilitator and ensures that business requirements are not compromised by the network solution.

> Task 6.2—Build and Test Databases

Building and testing databases are unfamiliar tasks for many students, who are accustomed to having an instructor provide them with the test databases. This task must immediately precede other programming activities because databases are the resources shared by the computer programs to be written. If new or modified databases are required for the new system, we can now build and test those databases.

This task involves systems users, analysts, designers, and builders. The same system specialist that designed the databases will assume the primary responsibility in completing this task. System users may also be involved in this task by providing or approving the test data to be used in the database. When the database to be built is a noncorporate, applications-oriented database, the systems analyst often completes this task. Otherwise, systems analysts mostly ensure business requirements compliance. The database designer will often become the system builder responsible for the completion of this activity. The task may involve database programmers to build and populate the initial database and a database administrator to tune the database performance, add security controls, and provide for backup and recovery.

The primary inputs to this task are the database schema(s) specified during systems design. Sample data from production databases may be loaded into tables for testing the databases. The final product of this task is an unpopulated *database structure* for the new database. The term *unpopulated* means the database structure is implemented but data has not been loaded into the database structure. As you'll soon see, programmers will eventually write programs to populate and maintain those new databases. Revised database schema and test data details are also produced during this task and placed in the project repository for future reference.

> Task 6.3—Install and Test New Software Packages (if Necessary)

Some systems solutions may have required the purchase or lease of software packages. If so, once networks and databases for the new system have been built, we can install and test the new software. This new software will subsequently be placed in the software library.

This activity typically involves systems analysts, designers, builders, and vendors and consultants. This is the first task in the life cycle that is specific to the applications programmer. The systems analyst typically participates in the testing of the software package by clarifying requirements. Likewise, the system designer may be involved in this task to clarify integration requirements and program documentation that is to be used in testing the software. Network administrators may be involved in actually installing the software package on the network server. Finally, this task typically involves participation from the software vendor and consultants who may assist in the installation and testing process.

The main input to this task is the new software packages and documentation that are received from the system vendors. The applications programmer will complete the installation and testing of the package according to integration requirements and program documentation developed during system design. The principal deliverable of this task is the installed and tested software package that is made available in the software library. Any modified software specifications and new integration requirements that were necessary are documented and made available in the project repository to provide a history and serve as future reference.

> Task 6.4—Write and Test New Programs

We are now ready to develop (or complete) any in-house programs for the new system. Recall that prototype programs are frequently constructed in the design phase. These prototypes are included as part of the technical design specifications for completing systems construction and implementation. However, these prototypes are rarely fully functional or complete. Therefore, this activity may involve developing or refining those programs.

This task involves the systems analysts, designers, and builders. The systems analyst typically clarifies business requirements to be implemented by the programs. The designer may have to clarify the program design, integration requirements, and program documentation (developed during systems design) that is used in writing and testing the programs. The system builders will assume the primary responsibility for this activity. The applications programmer (builder) is responsible for writing and testing in-house software. Most large programming projects require a team effort. One popular organization strategy is the use of *chief programmer teams*. The team is managed by the *chief programmer,* a highly proficient and experienced programmer who assumes overall responsibility for the program design strategy, standards, and construction. The chief programmer oversees all coding and testing activities and helps with the most difficult aspects of the programs. Other team members include a *backup chief programmer, program librarian, programmers,* and *specialists.* The applications programmer is often aided by an application or software tester who specializes in building and running *test scripts* that are consistently applied to programs to test all possible events and responses.

The primary inputs to this activity are the technical design statement, plan for programming, and test data developed during systems design. Since any new programs or program components may have already been written and be in use by other existing systems, the experienced applications programmer will know to first check for possible reusable software components available in the software library. The principal deliverables of this activity are the new programs and reusable software components that are placed in the software library. This activity also results in program documentation that may need to be approved by a quality assurance group. Some information systems shops have a quality assurance group staffed by specialists who review the final program documentation for conformity to standards. This group will provide appropriate feedback regarding quality recommendations and requirements. The final program documentation is then placed in the project repository for future reference.

Testing is an important skill that is often overlooked in academic courses on computer programming. Testing should not be deferred until after the entire program has been written! There are three levels of testing to be performed: stub testing, unit or program testing, and systems testing. **Stub testing** is testing performed on individual

stub test a test performed on a subset of a program.

events or modules of a program. In other words, it is the testing of an isolated subset of a program. **Unit** or **program testing** is testing in which all the events and modules that have been coded and stub tested for a program are tested as an integrated unit; it is the testing of an entire program. **Systems testing** ensures that application programs written and tested in isolation work properly when they are integrated into the total system. A system test plan should be developed and followed for testing the system. One or more test scripts are developed for each functional and nonfunctional requirement.

Just because a single program works properly doesn't mean that it works properly with other programs. The integrated set of programs should be run through a systems test to make sure one program properly accepts, as input, the output of other programs. Once the system test is complete and determined to be successful, we can proceed to the implementation of the system.

> **unit** or **program test** a test performed on an entire program.

> **systems test** a test performed on an entire system.

The Implementation Phase

What's left to do? New systems usually represent a departure from the way business is currently done; therefore, the analyst must provide for a smooth transition from the old system to the new system and help users cope with normal start-up problems. Thus, the implementation phase delivers the production system into operation.

The functional system from the construction phase is the key input to the implementation phase (see Figure 19-1). The deliverable of the implementation phase (and the project) is the operational system that will enter the *operation and support* stage of the life cycle.

In your information system framework, the implementation phase considers the same building blocks as does the construction phase (see the chapter home page). In this section you will learn about several tasks involved in the implementation phase for a typical systems development project. Figure 19-3 depicts the various tasks for the implementation phase. Let's examine each implementation phase task in greater detail.

> Task 7.1—Conduct System Test

Now that the software packages and in-house programs have been installed and tested, we need to conduct a final system test. All software packages, custom-built programs, and any existing programs that comprise the new system must be tested to ensure that they all work together.

This task involves analysts, owners, users, and builders. The systems analyst facilitates the completion of this task. The systems analyst typically communicates testing problems and issues with the project team members. The system owners and system users hold the ultimate authority on whether or not a system is operating correctly. System builders, of various specialties, are involved in the systems testing. For example, applications programmers, database programmers, and networking specialists may need to resolve problems revealed during systems testing.

The primary inputs to this task include the software packages, custom-built programs, and any existing programs comprising the new system. The system test is done using the system test data that was developed earlier by the systems analyst. As with previous tests that were performed, the system test may result in required modifications to programs, thus, once again, prompting the return to a construction phase task. This iteration would continue until a successful system test was experienced.

> Task 7.2—Prepare Conversion Plan

Once a successful system test has been completed, we can begin preparations to place the new system into operation. Using the design specifications for the new system, the systems analyst will develop a detailed conversion plan. This plan will

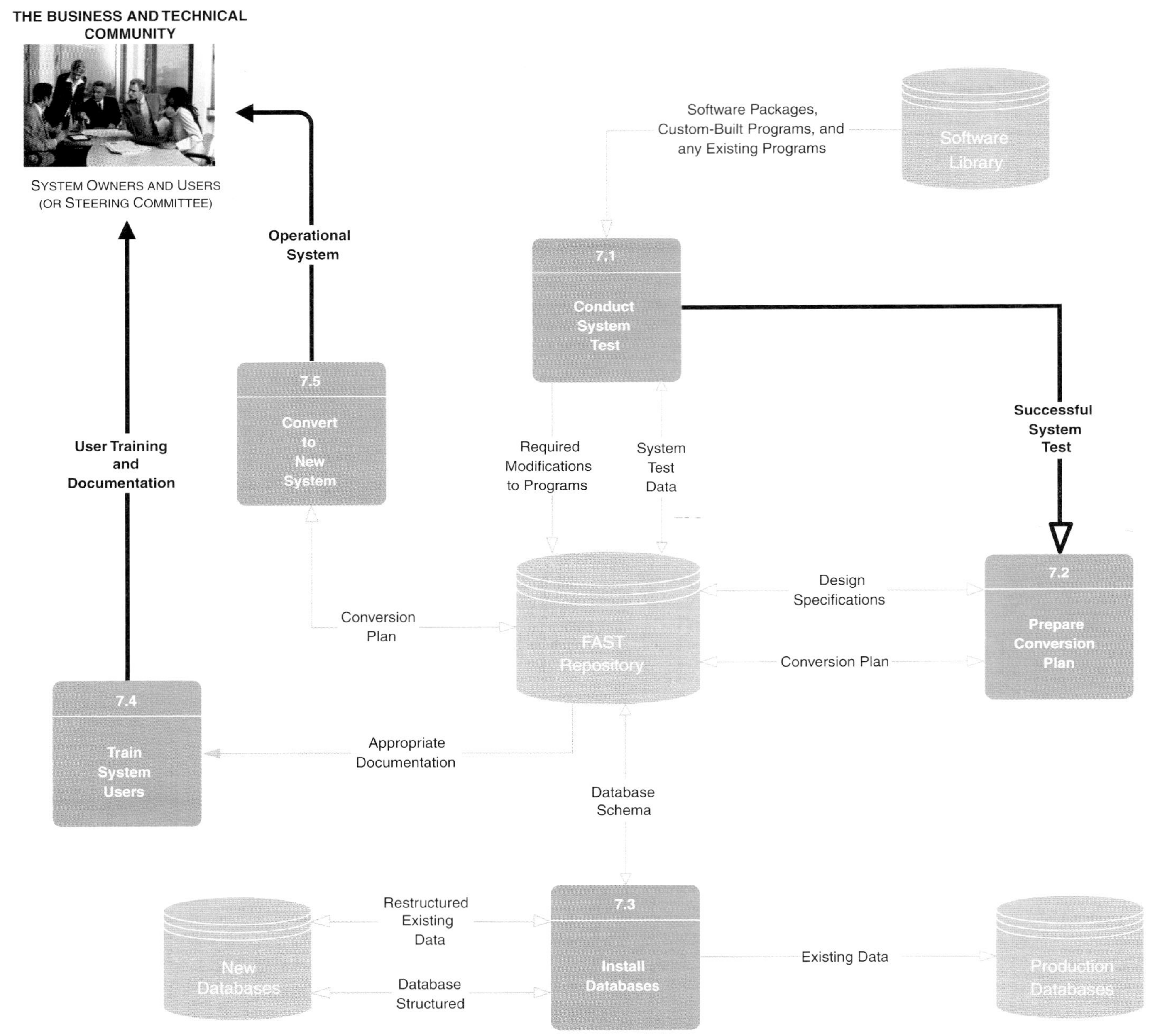

THE BUSINESS AND TECHNICAL COMMUNITY

SYSTEM OWNERS AND USERS
(OR STEERING COMMITTEE)

Software Packages,
Custom-Built Programs, and
any Existing Programs

Software
Library

**Successful
System
Test**

7.1
**Conduct
System
Test**

7.2
**Prepare
Conversion
Plan**

System
Test
Data

Required
Modifications
to Programs

Design
Specifications

Conversion Plan

FAST
Repository

Database
Schema

7.3
**Install
Databases**

Existing Data

Production
Databases

**Operational
System**

7.5
**Convert
to
New
System**

Conversion
Plan

Appropriate
Documentation

Restructured
Existing
Data

Database
Structured

**User Training
and
Documentation**

7.4
**Train
System
Users**

New
Databases

FIGURE 19-3 Systems Implementation Tasks

690

identify databases to be installed, end-user training and documentation that need to be developed, and a strategy for converting from the old system to the new system.

The project manager facilitates the activity. Systems analyst, system designer, and system builder roles are not typically involved unless deemed necessary by the project manager. Finally, many organizations require that all project plans be formally presented to a steering body (sometimes called a *steering committee*) for final approval.

This activity is triggered by the completion of a successful system test. Using the design specifications for the new system, a detailed conversion plan can be assembled. The principal deliverable of this activity is the conversion plan that will identify databases to be installed, end-user training and documentation that need to be developed, and a strategy for converting from the old system to the new system.

The conversion plan may include one of the following commonly used installation strategies:

- *Abrupt cut-over*—On a specific date (usually a date that coincides with an official business period such as month, quarter, or fiscal year), the old system is terminated and the new system is placed into operation. This is a high-risk approach because there may still be major problems that won't be uncovered until the system has been in operation for at least one business period. On the other hand, there are no transition costs. Abrupt cut-over may be necessary if, for instance, a government mandate or business policy becomes effective on a specific date and the system couldn't be implemented before that date.

- *Parallel conversion*—Under this approach, both the old and the new systems are operated for some time period. This ensures that all major problems in the new system have been solved before the old system is discarded. The final cut-over may be either abrupt (usually at the end of one business period) or gradual, as portions of the new system are deemed adequate. This strategy minimizes the risk of major flaws in the new system causing irreparable harm to the business; however, it also means the cost of running two systems over some period must be incurred. Because running two editions of the same system on the computer could place an unreasonable demand on computing resources, this may be possible only if the old system is largely manual.

- *Location conversion*—When the same system will be used in numerous geographical locations, it is usually converted at one location first (using either abrupt or parallel conversion). As soon as that site has approved the system, it can be farmed to the other sites. Other sites can be cut over abruptly because major errors have been fixed. Furthermore, other sites benefit from the learning experiences of the first test site. The first production test site is often called a *beta test site*.

- *Staged conversion*—Like location conversion, staged conversion is a variation on the abrupt and parallel conversions. A staged conversion is based on the version concept introduced earlier. Each successive version of the new system is converted as it is developed. Each version may be converted using the abrupt, parallel, or location strategy.

The conversion plan also typically includes a systems acceptance test plan. The systems acceptance test is the final opportunity for end users, management, and information systems operations management to accept or reject the system. A **systems acceptance test** is a final system test performed by end users using real data over an extended time period. It is an extensive test that addresses three levels of acceptance testing—verification testing, validation testing, and audit testing:

systems acceptance test a test performed on the final system wherein users conduct verification, validation, and audit tests.

- *Verification testing* runs the system in a simulated environment using simulated data. This simulated test is sometimes called *alpha testing*. The simulated test is primarily looking for errors and omissions regarding end-user and design specifications that were specified in the earlier phases but not fulfilled during construction.

- *Validation testing* runs the system in a live environment using real data. This is sometimes called *beta testing*. During this validation, a number of items are tested:

  a. *Systems performance.* Is the throughput and response time for processing adequate to meet a normal processing workload? If not, some programs may have to be rewritten to improve efficiency or processing hardware may have to be replaced or upgraded to handle the additional workload.

  b. *Peak workload processing performance.* Can the system handle the workload during peak processing periods? If not, improved hardware and/or software may be needed to increase efficiency or processing may need to be rescheduled—that is, consider doing some of the less critical processing during nonpeak periods.

  c. *Human engineering test.* Is the system as easy to learn and use as anticipated? If not, is it adequate? Can enhancements to human engineering be deferred until after the system has been placed into operation?

  d. *Methods and procedures test.* During conversion, the methods and procedures for the new system will be put to their first real test. Methods and procedures may have to be modified if they prove to be awkward and inefficient from the end users' standpoint.

  e. *Backup and recovery testing.* All backup and recovery procedures should be tested. This should include simulating a data loss disaster and testing the time required to recover from that disaster. Also, a before-and-after comparison of the data should be performed to ensure that data was properly recovered. It is crucial to test these procedures. Don't wait until the first disaster to find an error in the recovery procedures.

audit test a test performed to ensure a new system is ready to be placed into operation.

- **Audit testing** certifies that the system is free of errors and is ready to be placed into operation. Not all organizations require an audit. But many firms have an independent audit or quality assurance staff that must certify a system's acceptability and documentation before that system is placed into final operation. There are independent companies that perform systems and software certification for end users' organizations.

> Task 7.3—Install Databases

Recall that in a previous phase you built and tested databases. To place the system into operation, you will need fully loaded (or "populated") databases. Therefore, the next task we'll survey is installation of databases. The purpose of this task is to populate the new system's databases with existing data from the old system.

At first, this activity may seem trivial. But consider the implications of loading a typical table, say, MEMBER. Tens or hundreds of thousands of records may have to be loaded. Each must be input, edited, and confirmed before the database table is ready to be placed into operation.

Systems builders play a primary role in this activity. The task will normally be completed by application programmers who will write the special programs to extract data from existing databases and programs to populate the new databases. Systems analysts and designers may play a small role in completing this activity. Their primary involvement will be the calculating of database sizes and estimating of the time required to perform the installation. Finally, data entry personnel or hired help may often be assigned to do data entry.

Special programs will have to be written to populate the new databases. Existing data from the production databases, coupled with the database schema(s) models and database structures for the new databases, will be used to write computer programs to populate the new databases with restructured existing data. The principal deliverable of this task is the restructured existing data that has been populated in the databases for the new system.

> Task 7.4—Train Users

Change may be good, but it's not always easy. Converting to a new system necessitates that system users be trained and provided with documentation (user manuals) that guides them through using the new system.

Training can be performed one on one; however, group training is generally preferred. It is a better use of your time, and it encourages group-learning possibilities. Think about your education for a moment. You really learn more from your fellow students and colleagues than from your instructors. Instructors facilitate learning and instruction, but you master specific skills through practice with large groups where common problems and issues can be addressed more effectively. Take advantage of the ripple effect of education. The first group of trainees can then train several other groups.

The task is completed by the systems analyst and involves system owners and users. Given appropriate documentation for the new system, the systems analysts will provide end-user documentation (typically in the form of manuals) and training for the system users. The system owners must support this activity. They must be willing to approve the release time necessary for people to obtain the training needed to become successful users of the new system. Remember, the system is for the user! User involvement is also important in this activity because the end users will inherit the successes and failures from this effort. Fortunately, users' involvement during this task is rarely overlooked. The most important aspect of their involvement is training and advising the users. They must be trained to use equipment and to follow the procedures required of the new system. But no matter how good the training is, users will become confused at times. Or perhaps they will find mistakes or limitations. Thus, it is the responsibility of the analyst to help the users through the learning period until they become comfortable with the new system.

Given appropriate documentation for the new system, the systems analyst will provide the system users with the documentation and training needed to properly use the new system. The principal deliverable of this task is user training and documentation. Many organizations hire special systems analysts who do nothing but write user documentation and training guides. If you have a skill for writing clearly, the demand for your services is out there! Figure 19-4 is a typical outline for a training manual. The

---

FIGURE 19-4   An Outline for a Training Manual

Training Manual End Users Guide Outline

I. Introduction.

II. Manual.
  A. The manual system (a detailed explanation of people's jobs and standard operating procedures for the new system).
  B. The computer system (how it fits into the overall work flow).
    1. Terminal/keyboard familiarization.
    2. First-time end users.
      a. Getting started.
      b. Lessons.
  C. Reference manual (for nonbeginners).

III. Appendixes.
  A. Error messages.

Golden Rule should apply to user manual writing: "Write unto others as you would have them write unto you." You are not a business expert. Don't expect the reader to be a technical expert. Every possible situation and its proper procedure must be documented.

>Task 7.5—Convert to New System

Conversion to the new system from the old system is a significant milestone. After conversion, the ownership of the system officially transfers from the analysts and programmers to the end users. The analyst completes this task by carefully carrying out the conversion plan. Recall that the conversion plan includes detailed installation strategies to follow for converting from the existing to the new production information system. This task also involves completing a systems audit.

The task involves the systems owners, users, analysts, designers, and builders. The project manager who will oversee the conversion process facilitates it. The system owners provide feedback regarding their experiences with the overall project. They may also provide feedback regarding the new system that has been placed into operation. The system users will provide valuable feedback pertaining to the actual use of the new system. They will be the source of the majority of the feedback used to measure the system's acceptance. The systems analysts, designers, and builders will assess the feedback received from the system owners and users once the system is in operation. In many cases, that feedback may stimulate actions to correct identified shortcomings. Regardless, the feedback will be used to help benchmark new systems projects down the road.

The key input to this activity is the conversion plan that was created in an earlier implementation phase task. The principal deliverable is the operational system that is placed into production in the business.

Learning Roadmap

This chapter provided a detailed overview of the construction and implementation phases of systems development. You are now ready to learn systems operation and support, covered in Chapter 20.

Before proceeding, you may wish to revisit Chapter 3 and its introduction to the systems development process. This review will help you to understand how systems operations support fits into the overall systems development process.

Chapter Review

1. Systems construction is the development, installation, and testing of system components.
2. Systems implementation is the delivery of the system into production (meaning day-to-day operation).
3. The purpose of the construction phase is to develop and test a functional system that fulfills business and design requirements and to implement the interfaces between the new system and existing production systems.
4. The construction phase consists of four tasks: build and test networks, build and test databases, install and test new software packages, and write and test new programs.
5. Three levels of testing are performed on new programs:

    a. Stub testing is testing performed on individual modules, whether they be main program, subroutine, subprogram, block, or paragraph.
    b. Unit or program testing is testing in which all the modules that have been coded and stub tested are tested as an integrated unit.
    c. Systems testing ensures that application programs written in isolation work properly when they are integrated into the total system.

6. The purpose of the implementation phase is to smoothly convert from the old system to the new system.
7. The systems implementation consists of the following activities: conducting a system test, preparing a systems conversion plan, installing databases, training system users, and converting from the old system to the new system.

8. There are several commonly used strategies for converting from an existing to a new production information system, including:

    a. Abrupt cut-over—On a specific date, the old system is terminated and the new system is placed into operation.
    b. Parallel conversion—Both the old and the new systems are operated for some time period to ensure that all major problems in the new system are solved before the old system is discarded.
    c. Location conversion—When the same system will be used in numerous geographical locations, it is usually converted at one location and, following approval, farmed to the other sites.
    d. Staged conversion—Each successive version of the new system is converted as it is developed. Each version may be converted using the abrupt, parallel, or location strategies.

9. The systems acceptance test is the final opportunity for end users, management, and information systems operations management to accept or reject the system. A systems acceptance test is a final system test performed by end users using real data over an extended period. It is an extensive test that addresses three levels of acceptance testing—verification testing, validation testing, and audit testing:

    a. Verification testing runs the system in a simulated environment using simulated data.
    b. Validation testing runs the system in a live environment using real data. This is sometimes called beta testing.
    c. Audit testing certifies that the system is free of errors and is ready to be placed into operation.

Review Questions

1. What is the purpose and the major activity of the construction phase?
2. Who are the network designers and network administrators?
3. What are the tasks needed when building and testing databases?
4. Who are involved in the installation and testing of new software packages? What are their jobs?
5. What are chief programmer teams?
6. What are the three kinds of testing suggested in the textbook?
7. Why is the implementation phase needed?
8. Who are typically involved in conducting system testing in the implementation phase?

9. What are the four common conversion strategies?
10. What are some potential problems of using abrupt cut-over as a conversion strategy?
11. What are some potential problems of using parallel conversion as a conversion strategy?
12. What is the difference between alpha testing and beta testing?
13. Who is the major player in installing databases? What are the responsibilities?
14. What are the responsibilities of the system analysts when training users?
15. Why is feedback essential even though the new system is fully implemented and functional?

Problems and Exercises

1. You are the lead analyst on the system-testing team of a large enterprise system that will touch virtually every business function in the organization. Unfortunately, design and construction ran behind schedule by about two weeks. System testing is scheduled to take four weeks of intensive effort, assuming no major problems are found. Adding resources will not shorten the time required. If you stay on plan, implementation will be delayed by two weeks. The system owner, who is the CEO, finds this unacceptable and tells you: "What do you mean that it is going to take a month to system test? I need this system up in two weeks, not a day later. If you find any problems, they can be fixed later!" What do you do in this situation?

2. Consider a variation of the preceding question. You work as a testing analyst for a software development contractor that has been engaged to develop this enterprise system. If the project is not completed on schedule, your company loses a substantial bonus. Since design and construction ran behind, you will have to cut system testing in half. Your company is putting a great deal of implicit pressure on you to compress testing so the project can finish on schedule and the company will receive its bonus. You have qualms that if testing is compressed, some serious problems may be missed, even with a risk-based testing strategy. What do you do?

3. You are a systems analyst who will be leading a systems-testing team on another project. Your company is adopting a new testing strategy; in the past, the programmers who constructed the system did the system testing themselves. Why was this not a good idea?

4. Who should you select for your systems-testing team? What skills should they have?

5. Are the following statements true or false? Explain your answer as needed.

   a. Building and testing any databases that are needed should occur after programming activities are completed.
   b. Training of users should be done long before actual implementation in order to ensure that everybody receives training without being rushed.
   c. The purpose of parallel conversion is to reduce business risk.
   d. Testing is a highly structured activity that should not be scheduled to commence until the entire application program has been written.

   e. Systems development and systems construction are frequently used as synonyms, but they may not necessarily mean the same thing.

6. As a systems analyst, you have been involved in a project to develop an inventory-tracking system for your business services office. The project is now coming to its final stages and you have been asked to write a training manual. Using the outline shown in Figure 19–4, write a portion of the usual manual (a page or two) describing the manual system or the computer system. Have one of your fellow students or co-workers read and evaluate for clarity the portion you wrote. Did she or he find it understandable and clear? Did it provide the appropriate level of detail that an end user would need?

7. Fill in the blanks:

   a. The final _____ on whether the system is _____ correctly and ready for implementation is the _____.
   b. The key input in to the _____ phase is the _____ system from the _____ phase.
   c. The _____ phase is _____ when the _____ are approved and the design phase is completed.
   d. Once the _____ to the new system is complete, _____ transfers from the project team to the _____.
   e. _____ data from the old database and _____ a new one is a _____ activity that requires careful _____ and execution.

8. During systems construction and implementation, aren't most of the activities technical in nature, so that users don't need to be involved except for system testing?

9. Match the terms in the first column with the definitions or examples in the second column:

1. Beta testing	A. Production database without data loaded in
2. Alpha testing	B. Testing of throughput/ response time under normal load
3. Program testing	C. Migrating completed system into production environment
4. Audit testing	D. Unanticipated sudden system shutdown testing

5. System perfor- E. Application
   mance testing   program–level code
                      testing

6. Unpopulated   F. Independently
   database   performed certification-
                     level testing

7. Backup and   G. Module-level testing of
   recovery testing   code

8. Peak performance H. Extensive verification,
   testing   validation, and audit
                      testing

9. Abrupt cut-off I. Environmental-level
                      testing of application
                      program(s)

10. System   J. Environmental-level
    implementation   testing by users with
                      simulated data

11. Systems   K. Live environmental-
    acceptance testing level testing by users
                        with live data

12. System testing L. Testing of through-
                       put/response time
                       under load spikes

13. Stub testing   M. Installation strategy
                      type

10. "The goal of human interface design is to create a system that is intuitive to use. To require a users manual is an admission of failure." Respond to this statement. Do you agree or disagree with it? Explain why.

11. Many organizations require a postimplementation evaluation report (PIER), usually somewhere between six months and a year after implementation. What purpose(s) does this serve?

12. If a project is poorly designed and constructed, will a well-planned and well-executed implementation effort help the project to succeed? What about the opposite situation? Will a well-designed and well-constructed system overcome a poor implementation effort?

13. One very important final activity should take place after conversion to the new system is successfully completed. What do you think it is?

Projects and Research

1. A number of companies, such as Mercury Interactive of Rational (now owned by IBM), offer automated software-testing packages as stand-alone products or as an integrated part of a larger suite. Research these software packages on the Web and trade journals. Download and try any trial versions you find. In addition, contact the software-testing staff in several local organizations and interview them regarding their software-testing methods.
   a. Describe your research—what products did you find?
   b. Compare and contrast their features and functionality?
   c. Did the software testers you contacted use an automated software testing tool? If so, was it a home-grown tool or a commercial product? Did they indicate any preference as to which one they thought was best?
   d. If you were the testing manager and were given the choice to purchase any automated software-testing tool, which one would you select? Or would you prefer to develop your own home-grown automated testing program? Explain your answer.
   e. What do you see as the primary advantages to using an automated software-testing package?

2. You are a systems analyst working on a major project for an organization that has several hundred employees in its headquarters office, and about a dozen offices located in this country, in Canada, and in Mexico. The objective of the project is to implement an enterprise-wide mission-critical information system. The project is now in the construction and implementation phase, and you have been assigned responsibility for selecting the conversion strategy and for developing the conversion plan. Prepare a summary analysis of the different installation strategies discussed in the book, and recommend the one that you feel would be most appropriate.

3. Assuming that your recommendation is accepted by management, draft a detailed conversion plan that addresses the actual implementation strategy for converting to the new system. After you complete the draft plan, have it reviewed by one or more IT staff members with experience in developing conversion plans. Make any needed changes and repeat the review process until the consensus is that your plan is realistic and doable.

4. The next step is to prepare the systems acceptance test plan. Using the material in the textbook as a general guideline, research on the Web some

of the more detailed components that go into acceptance testing. Select some of the testing templates that should be readily available, and modify them as appropriate. Then draft the test plan and share it with IT testing staff. Make sure that your plan addresses any potential risks. Make any needed changes and repeat the review process as necessary with the testing staff until your plan is ready to put into action.

5. The textbook describes a traditional method for delivering end-user training. Are there other methods, which are Web-based, that may offer more effective and/or efficient methods for delivering end-user training? Research some of the Web-based, training methods that are becoming more widely used. Then use the scenario described in Question 2 to develop a Web-based end-user training plan. After you draft the plan, have some professional trainers review it for completeness and feasibility and make any needed changes. Then have some of your fellow students or co-workers review your training plan from the perspective of an end-user. Were you able to develop a feasible plan?

6. There is an unwritten principle that says that no matter how much you plan for a system implementation, something unanticipated will almost always happen, often at the worst possible moment. Interview several analysts in local organizations who have expertise in implementing systems. Ask them about their experiences, what their worst horror story was, and what they learned from it. Supplement these interviews with research on contingency implementation planning. Then use your anecdotal information and research to put together a set of guidelines on planning for and handling the unexpected during system implementation.

Minicases

1. In minicase 3 of the previous chapter, you created a prototype for a government system. Exchange prototypes with another group (just as you did in minicase 4 of that chapter). This time, *test* each other's prototype. Document your findings thoroughly and prepare a report. Give one copy of the report to the other group (the one who created the prototype) and one to the professor.

2. Wow Munchies has a Web site, *www.wowmunchies. com,* which is currently hosted on server 123coolhost at a Web-hosting company called Cool Hosting. But Wow Munchies has decided to have its Web pages updated and serviced by another hosting company: Reliable Host, using the server 123-reliable. The new hosting company pointed the DNS for *www.wowmunchies.com* to server 123reliable *before* it had the Web pages loaded and tested on its server. It takes 12–72 hours for the DNS change to take place, and Reliable thought it would have the Web pages up in the lag time. It wasn't able to. As a result, the DNS pointed to the new server for several days before the new site was functioning again. Wow Munchies lost an estimated $200,000 in revenues as a result of the sites downtime. Comment on what went wrong, and how it could have been avoided.

3. Use the testing and designing reports given to you by your peer group (from Chapter 18, minicase 4, and Chapter 19, minicase 1) and revise your prototype. Document all of the changes and improvements you have made, and submit a short report to your professor.

4. Prepare a manual for your revised prototype. Remember to orient the manual to the expected user, and not your professor. Bind the manual, and submit it to your professor. You will be graded on clarity, usability, completeness, and professionalism.

Suggested Readings

Bell, P., and C. Evans. *Mastering Documentation.* New York: John Wiley & Sons, 1989.

Brooks, F. P. *The Mythical Man-Month.* Reading, MA: Addison-Wesley, 1995.

Boehm, Barry. "Software Engineering." *IEEE Transactions on Computers, C-25,* December 1976. This classic paper demonstrated the importance of catching errors and omissions before programming begins.

Metzger, Philip W. *Managing a Programming Project,* 2nd ed. Englewood Cliffs, NJ: Prentice Hall, 1981. This is one of the few books to place emphasis solely on systems implementation.

Mosely, D. J. *The Handbook of MIS Application Software Testing.* Englewood Cliffs, NJ: Yourdon Press, 1993.

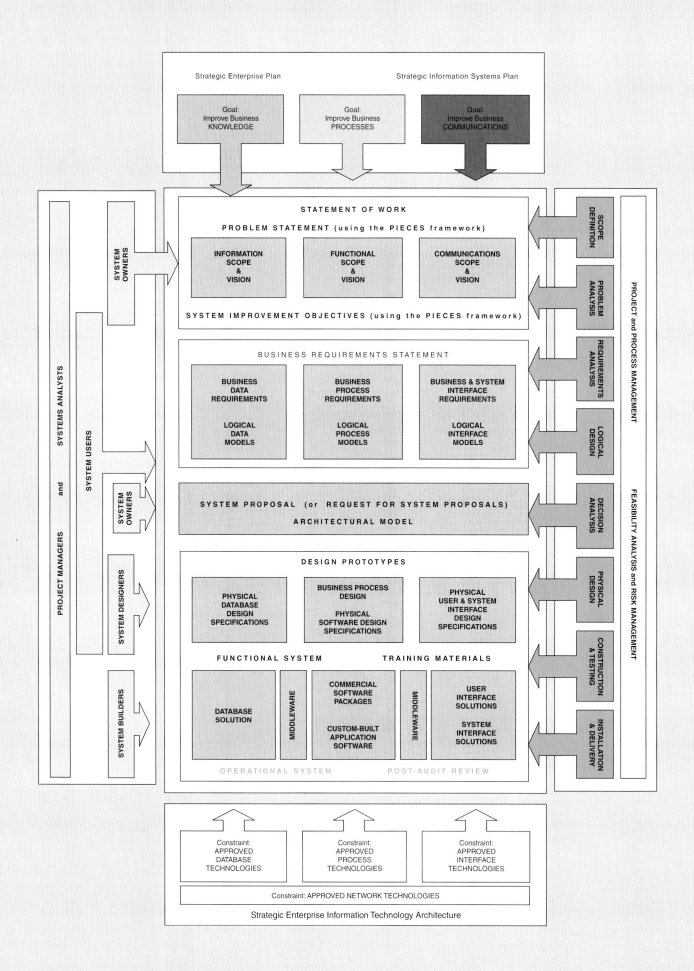

Systems Operations and Support

Chapter Preview and Objectives

Once a system has been implemented, it enters operations and support. Systems operation is the ongoing function in which the system operates until it is replaced. Systems support involves servicing, maintaining, and improving a functional information system through its lifetime. Systems operation and support occur in parallel. In this chapter, you will learn more about systems operation and support. In particular, it is useful to understand the different types of systems support provided for a production system. You will know that you understand the process of systems support when you can:

▊ Define systems operations and support.

▊ Describe the relative roles of a repository, program library, and database in systems operations and support.

▊ Differentiate between maintenance, recovery, technical support, and enhancement as system support activities.

▊ Describe the tasks required to maintain programs in response to bugs.

▊ Describe the role of benchmarking in system maintenance.

▊ Describe the systems analyst's role in system recovery.

▊ Describe forms of technical support provided by a systems analyst for the user community.

▊ Describe the tasks that should be and may be performed in system enhancement and the relationship between the enhancement and the original systems development process.

▊ Describe the role of reengineering in system enhancement. Describe three types of reengineering.

Although some of the techniques of systems support are introduced in this chapter, the chapter does not teach these techniques. This chapter teaches only the *process* of systems support as it relates to the development processes you have been studying throughout this book.

Introduction

It has been one week since SoundStage converted to the new Member Services system. Bob Martinez has been working on a few minor program bugs. As each bug has cropped up, Bob has worked with the users to validate and document the problem. He then passed that information to the programmers, who fixed the problem, tested the revised program, and documented their changes. Finally Bob updated the version control system with information about the fix.

All in all the conversion has gone very well, and Bob was starting to think he was seeing the last of the Member Services system. Then his boss, Sandra, came to him with a list of additional requirements. Most of them were lower-priority use cases that had been intentionally left out of the first iteration. Others were requests that came up during the analysis, design, and implementation phases. There were even requests suggested by users now that they are actually working the system. So Bob is starting back in the requirements analysis phase for iteration two. But now that he is hearing from users how helpful the new system is, Bob is excited to make it even better.

The Context of Systems Operation and Support

Systems support was introduced in Chapter 3. *Systems support* is the ongoing technical support for users, as well as the maintenance required to fix any errors, omissions, or new requirements that may arise. Before an information system can be supported, it must first be in operation. *Systems operation* is the day-to-day, week-to-week, month-to-month, and year-to-year execution of an information system's businesses processes and application programs. An operational system (not to be confused with an operating system) is frequently called a *production system*. Systems operation and support are often ignored in systems analysis and design textbooks. Young analysts are often surprised to learn that half of their duties (or more) are associated with supporting production systems.

In the chapter home page, you can see that systems operation and support use all the information system building blocks since the information system is operational. All the system knowledge and working components of the system are important to its ongoing operation and support. This is reinforced in Figure 20-1, which demonstrates that systems operations and support often require developers to revisit the activities, and hence the building blocks, that were developed during systems analysis, design, construction, and implementation.

Figure 20-2 illustrates systems development, systems operations, and systems support as separate processes. Until this chapter, the entire book has focused on systems development. Systems development responds to problems, opportunities, and directives by developing improved business solutions. To repeat: that process has been the focus of this book. During systems development, we accumulated considerable system knowledge and constructed programs and databases. The figure illustrates these three data stores:

- The *repository* is a data store(s) of accumulated system knowledge—system models, detailed specifications, and any other documentation that has been accumulated during the system's development. This knowledge is reusable and critical to the production system's ongoing support. The repository is implemented with various automated tools, and it is often centralized as an enterprise business and IT resource.
- The *program library* is a data store of all application programs. The source code for these programs must be maintained for the life of the system. Almost always, the software-based librarian will control access (via check-in and

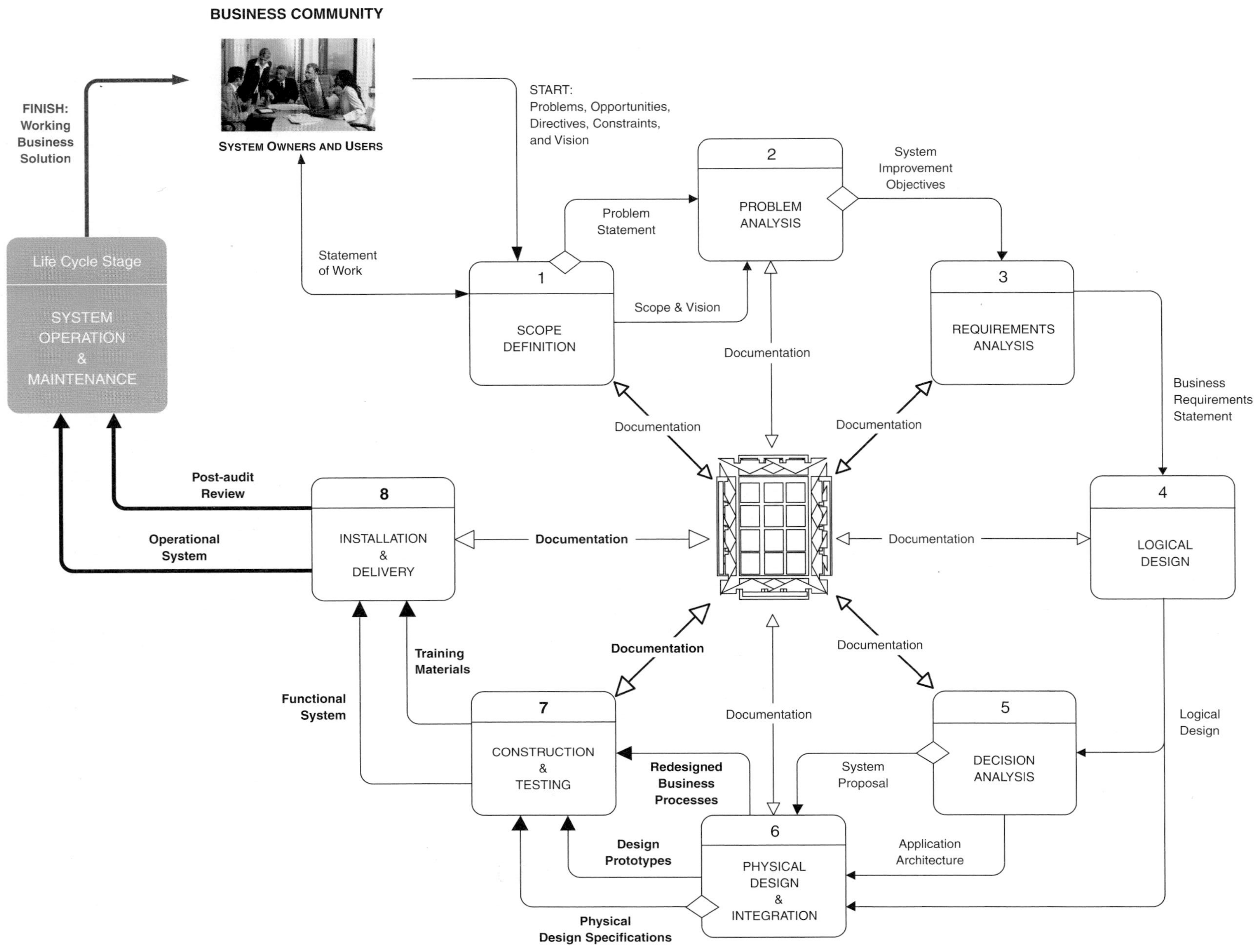

BUSINESS COMMUNITY

SYSTEM OWNERS AND USERS

START:
Problems, Opportunities,
Directives, Constraints,
and Vision

Statement
of Work

FINISH:
Working
Business
Solution

Life Cycle Stage

SYSTEM
OPERATION
&
MAINTENANCE

Operational
System

Post-audit
Review

8

INSTALLATION
&
DELIVERY

Functional
System

Training
Materials

Documentation

7

CONSTRUCTION
&
TESTING

Design
Prototypes

Redesigned
Business
Processes

Physical
Design Specifications

6

PHYSICAL
DESIGN &
INTEGRATION

Documentation

Documentation

Application
Architecture

System
Proposal

5

DECISION
ANALYSIS

Documentation

Logical
Design

4

LOGICAL
DESIGN

Business
Requirements
Statement

Documentation

3

REQUIREMENTS
ANALYSIS

System
Improvement
Objectives

2

PROBLEM
ANALYSIS

Documentation

Problem
Statement

Scope & Vision

1

SCOPE
DEFINITION

Documentation

FIGURE 20-1 The Context of Systems Operation and Support

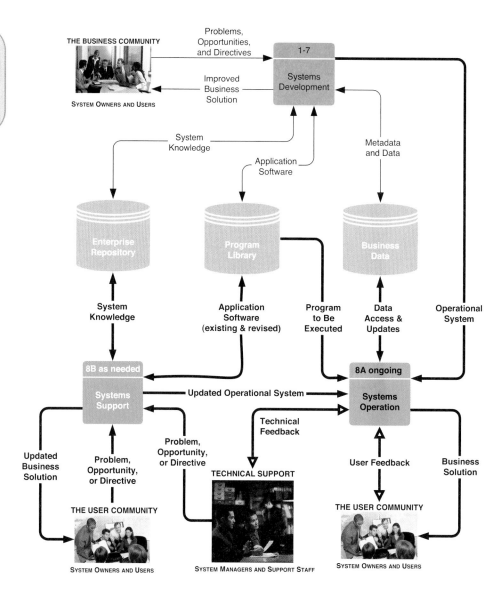

checkout) to the stored programs as well as track changes and maintain several previous versions of the programs in case a problem in a new version forces a temporary use of a prior version. Examples of such *software configuration tools* are Serena's *ChangeMan Professional,* IBM's *CAA Source Code Manager,* and Microsoft's *Visual SourceSafe.*

• The *business data* includes all the actual business data created and maintained by the production application programs. This includes conventional files, relational databases, data warehouses, and any object databases. This data is under the administrative control of the data administrator, who is charged with backup, recovery, security, performance, and the like.

Unlike systems analysis, design, and implementation, systems support cannot sensibly be decomposed into actual phases that a support project must perform. Rather, systems support consists of four ongoing activities. Each activity is a type of support project that is triggered by a particular problem, event, or opportunity encountered with the implemented system.

Figure 20-3 is an activity diagram that illustrates the four types of support activities:

• *Program maintenance*—Unfortunately, most systems suffer from software defects, or bugs—errors that slipped through the testing of software.

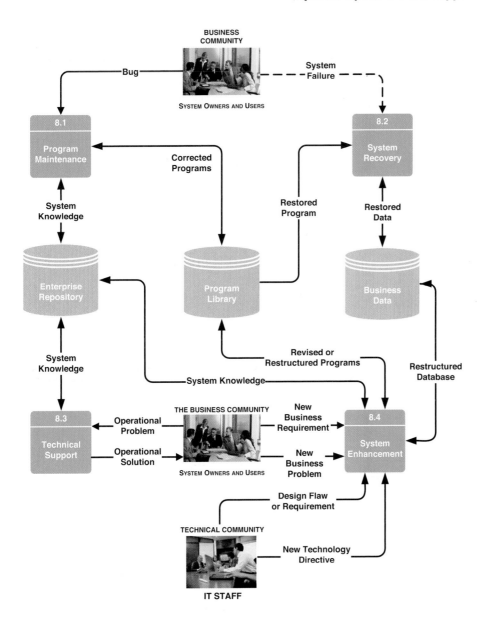

FIGURE 20-3

Systems Support Activities

- *System recovery*—From time to time, a system failure may result in a program "crash" and/or loss of data. Human error or a hardware or software failure may have caused this. The systems analyst or technical support specialists may then be called on to recover the system—that is, to restore a system's files and databases and to restart the system.
- *Technical support*—Regardless of how well the users have been trained and how good the end-user documentation is, users will eventually require additional assistance—unanticipated situations arise, new users are added, and so forth.
- *System enhancement*—New requirements may include new business problems, new business requirements, new technical problems, or new technology requirements.

This chapter examines each of these support activities in various levels of detail, but it abandons the approach used in previous systems analysis, design, and implementation overview chapters. Not all activities will be decomposed into tasks because some of these activities are fairly routine. Also, in some cases, an activity will involve returning to the system development activities that were discussed in earlier chapters.

System Maintenance

Regardless of how well designed, constructed, and tested a system or application may be, errors or bugs will inevitably occur. *Bugs* can be caused by any of the following:

- Poorly validated requirements.
- Poorly communicated requirements.
- Misinterpreted requirements.
- Incorrectly implemented requirements or designs.
- Simple misuse of the programs.

The fundamental objectives of system maintenance are:

- To make predictable changes to existing programs to correct errors that were made during systems design or implementation.
- To preserve those aspects of the programs that were correct and to avoid the possibility that "fixes" to programs cause other aspects of those programs to behave differently.
- To avoid, as much as possible, degradation of system performance. Poor system maintenance can gradually erode system throughput and response time.
- To complete the task as quickly as possible without sacrificing quality and reliability. Few operational information systems can afford to be down for any extended period. Even a few hours can cost millions of dollars.

To achieve these objectives, you need an appropriate understanding of the programs you are fixing and of the applications in which those programs participate. Lack of this understanding is often the downfall of systems maintenance!

How does system maintenance map to your information system building blocks?

- KNOWLEDGE/DATA—System maintenance may improve input data editing or correct a structural problem in the database.
- PROCESSES—Most system maintenance is program maintenance.
- COMMUNICATION—System maintenance may involve correcting problems related to how the application interfaces with the users or another system.

Figure 20-4 illustrates typical system maintenance tasks. Let's examine these tasks.

> Task 8.1.1—Validate the Problem

System maintenance miniprojects are triggered by the identification of the problem, usually called a *bug*. Most such bugs are identified by users when they discover some aspect of the system that does not appear to be working as it should. The first task of the systems analyst or programmer is to *validate the problem.*

Working with the users, the team should attempt to validate the problem by reproducing it. If the problem cannot be reproduced, the project should be suspended until the problem recurs and the user can explain the circumstances under which it occurred. The "as is" program is executed, as closely as possible approximating the circumstances and the data that were present when the problem was first encountered. In most cases, the user who encountered the problem should be the one who re-creates it.

One possible output is an unsubstantiated bug. In this scenario, the user (even with help from the analyst) could not re-create the error. To maintain a productive relationship with the user, the analyst should never make the user feel that it was his or her fault. Instead, the analyst should respect the possibility that the bug is still real and that it will eventually be validated. Should the error recur, the user should be coached to immediately document the exact sequence of events in detail or to call the analyst. Most importantly, the user should be instructed not to perform any subsequent action, if possible, that may prevent the error from being validated.

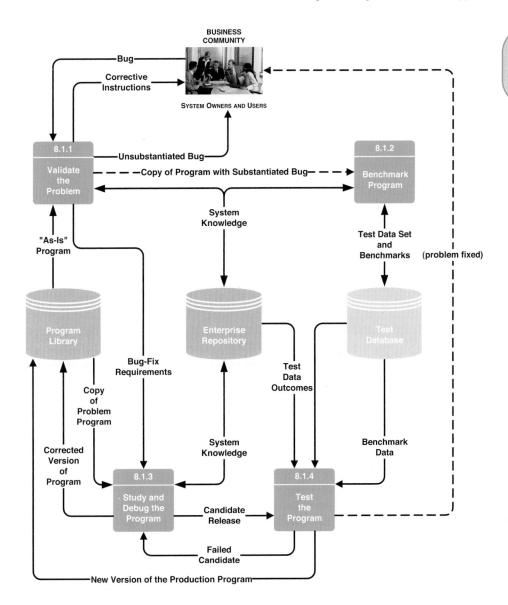

FIGURE 20-4

System Maintenance
Tasks

In some cases the analyst confirms the error but recognizes it as a simple misuse or mistaken use of the program. In such cases, the appropriate output is corrective instructions to the user.

The third possibility is that the bug is real. The analyst should do two things. First, the context of the bug should be examined by studying all relevant documentation (system knowledge) in the repository. In other words, don't try to fix what you don't understand. Second, *all subsequent maintenance should be performed on a copy of the program. The original program remains in the program library and can be used in production systems until it is fixed.*

> Task 8.1.2—Benchmark Program

Given a copy of the program with a substantiated bug, the analyst should *benchmark the program.* System maintenance can result in unpredictable and undesirable side effects that impact the program's or application's overall functionality and performance. In other words, the solution can cause unexpected side effects. For this reason, before any changes are made to programs, the programs should be executed and tested to establish a baseline against which the modified programs and applications can later be measured.

This task can be performed by the systems analyst and/or programmer. Users should also participate to ensure the test is conducted under circumstances that simulate as closely as possible a normal working environment.

Test cases can be defined in either of two ways. First, past test data may exist in the repository as a form of system knowledge. If so, that data should be reexecuted to establish or verify the benchmark. Usually, the test data should also be analyzed for completeness and, if necessary, revised. New test scenarios may have been identified since the system went into operation. Any revised test data should be recorded in the repository for subsequent maintenance projects.

Alternatively, test data can be automatically captured using a software-testing tool. As users enter test data, that data is recorded in a special type of repository as a *test script.* Later the analyst and user can document each test case in the same repository. Ultimately, the test script is executed against the program to test that the program executes properly and also to measure the program's response time and/or throughput. As shown in Figure 20-4, the test data and benchmarks are stored in a test database (not the production database) for the next task.

The analyst or programmer needs to have good testing skills (usually taught in programming courses) and may require training in test tools. Neither is explicitly taught in this textbook.

> Task 8.1.3—Study and Debug the Program

The primary task in system maintenance is to make the required changes to the programs. This task, performed by the application programmer, is not dissimilar from that described in the previous chapter on systems implementation. Essentially, the programmer responds to "bug-fix" requirements that establish the expectation for fixing the problem. The programmer debugs (edits) a copy of the problem program. Changes are not made to the production program. The result is a corrected version of the program. This is a candidate release, meaning a candidate to become the next production version of the program.

Usually, the original programmer is not making these changes. In fact, several programmers may have written parts of any given program that is now being debugged. Those programmers may no longer be available for clarification. Even if they are available, their memory of the application may not be sufficient or accurate. For this reason, the effective maintenance programmer requires system knowledge. Ideally, this knowledge comes from the repository, but that assumes that the knowledge has been properly maintained throughout the system's lifetime. Too often this is not true, especially for older systems. The programmer may need to seek out this knowledge or, in some cases, reconstruct the knowledge through analysis of the program.

Application and program knowledge usually comes from studying the source code. Program understanding can take considerable time. This activity is slowed by some combination of the following limitations:

- Poor program structure—examples include *COBOL* programs written with nonstructured techniques and *Visual Basic* or *C* programs written with nonobject-oriented techniques.
- Unstructured logic (from pre–*structured era* coding styles).
- Prior maintenance (quick fixes and poorly designed extensions).
- Dead code (instructions that cannot be reached or executed—often leftovers from prior testing and debugging).
- Poor or inadequate documentation.

The purpose of application understanding is to see the big picture—that is, how the programs fit into the total application and how they interact with other programs. The purpose of program understanding is to gain insight into how the program works and doesn't work. You need to understand the fields (variables) and where and how they are used, and you need to determine the potential impact of changes throughout the program. Program understanding can also lead to better estimates of the time and resources that will be required to fix the errors.

> Task 8.1.4—Test the Program

There is a big difference between editing a new program and editing an existing program. As the designer and creator of a new program, you are probably intimately familiar with the structure and logic of the program. By contrast, as the editor of the existing program, you are not nearly as familiar (or current) with that program. Changes that you make may have an undesirable ripple effect through other parts of the program or, worse still, other programs in the application and information system.

A candidate release of the program must be tested before it can be placed into operation as the next new version of the production program. The following tests are essential or recommended:

- *Unit testing* (essential) ensures that the stand-alone program fixes the bug <u>without undesirable side effects to the program</u>. The test data and current performance that you recovered, created, edited, or generated when the programs were benchmarked are used here.
- *System testing* (essential) ensures that the entire application, of which the modified program was a part, still works. Again, the test data and current performance are used here.
- *Regression testing* (recommended) extrapolates the impact of the changes on program and application throughput and response time from the before-and-after results using the test data and current performance.

Failed candidates are returned for additional debugging. Successful candidates are released for production. Older versions of the program are retained in the program library for version control. **Version control** is a process whereby a librarian (usually software-based) keeps track of changes made to programs. This allows recovery of prior versions of the programs if new versions cause unexpected problems. In other words, version control allows users to return to a previously accepted version of the system.

version control the tracking of changes made to a program.

The high cost of system maintenance is largely due to failure to update system knowledge (in the repository) and program source code documentation (in the program library). Application documentation is usually the responsibility of the systems analyst who supports the application. Program documentation is usually the responsibility of the programmer who made the program changes.

System Recovery

From time to time a system failure is inevitable. It generally results in an aborted, or "hung," program (also called an *ABEND* or "crash") and may be accompanied by loss of transactions or stored business data. The systems analyst often fixes the system or acts as intermediary between the users and those who can fix the system. This section summarizes the analyst's role in system recovery.

System recovery activities can be summarized as follows:

1. In many cases the analyst can sit at the user's terminal and recover the system. It may be something as simple as pressing a specific key or rebooting the user's personal computer. The systems analyst may need to provide users with corrective instructions to prevent the crash from recurring. In some cases the analyst may arrange to observe the user during the next use of the program or application.
2. In some cases the analyst must contact systems operations personnel to correct the problem. This is commonly required when servers are involved. An appropriate network administrator, database administrator, or Webmaster usually oversees such servers.

3. In some cases the analyst may have to call data administration to recover lost or corrupted data files or databases. In the recovery of business data, it is not only the database that must be restored.

   a. Any transactions that occurred between the last backup and the database's recovery must be reprocessed. This is sometimes called a *roll forward.*

   b. If the crash occurred during a transaction, and that transaction was partially completed, then any transactional updates to the database that occurred before the crash must be undone before reprocessing the complete transaction. This is sometimes called a *roll back.*

   Database management systems and transaction monitors provide facilities for transaction roll forward and roll back. These data backup and recovery techniques are beyond the scope of this book and are deferred to database courses and textbooks.

4. In some cases the analyst may have to call network administration to fix a local, wide, or internetworking problem. Network professionals can usually log out an account and reinitialize programs.

5. In some cases the analyst may have to call technicians or vendor service representatives to fix hardware problems.

6. In some cases the analyst will discover that a possible software bug caused the crash. The analyst attempts to quickly isolate the bug and trap it (automatically or by coaching users to manually avoid it) so that it can't cause another crash. Bugs are then handled as described in the previous section of this chapter.

Technical Support

Another relatively routine ongoing activity of systems support is technical support. No matter how well users have been trained or how well documentation has been written, users will require additional assistance. The systems analyst is generally on call to assist users with the day-to-day use of specific applications. In mission-critical applications, the analyst must be on call day and night.

The most typical tasks include:

- Routinely observing the use of the system.
- Conducting user-satisfaction surveys and meetings.
- Changing business procedures for clarification (written and in the repository).
- Providing additional training as necessary.
- Logging enhancement ideas and requests in the repository.

System Enhancement

Adapting an existing system to new requirements is the norm for all information systems. Business is change! The pace of change in today's economy is accelerated, and rapid response is the expectation. System enhancement requires that the systems analyst evaluate a new requirement to either effect change or direct the change request through an appropriate subset of the original systems development process. In some cases, the analyst may need to recover the system's existing physical structure as a preface to directing the change through systems redevelopment. In this section we will examine two types of adaptive maintenance—system enhancement and systems reengineering.

System enhancement is an adaptive process. Most such enhancement is in response to one of the following events, as shown in Figure 20-5:

- *New business problems*—A new or anticipated business problem will make a portion of the current system unusable or ineffective.
- *New business requirements*—A new business requirement (e.g., new report, transaction, policy, or event) is needed to sustain the value of the current system.

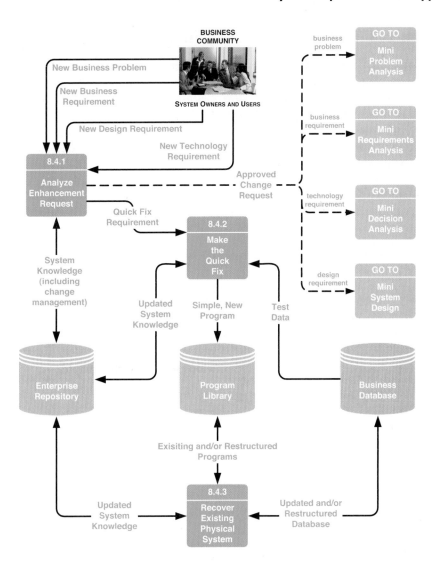

FIGURE 20-5

System Enhancement Tasks

- *New technology requirements*—A decision to consider or use a new technology (e.g., new software or version, or different type of hardware) in an existing system needs to be made.
- *New design requirements*—An element of the existing system needs to be redesigned against the same business requirements (e.g., add new database tables or fields, add or change to a new user interface).

Systems enhancement is reactive in nature—fix it when it breaks or when users or managers request change. System enhancement extends the useful life of an existing system by adapting it to inevitable change. This objective can be linked to your information system building blocks as follows:

- KNOWLEDGE/DATA—Many system enhancements are requests for new information (reports or screens) that can be derived from existing stored data. But some data enhancements call for the restructuring of stored data.
- PROCESSES—Most system enhancements require the modification of existing programs or the creation of new programs to extend the overall application system. But some enhancement requests can be accomplished through careful redesign of existing business processes.
- COMMUNICATION—Many enhancements require modifications to how the users interface with the system and how the system interfaces with other systems.

Figure 20-5 expands on the activities of system enhancement. In this section, we briefly describe each activity, participants and roles, inputs and outputs, and techniques.

> Task 8.4.1—Analyze Enhancement Request

This activity determines the appropriate course of action for achieving a system enhancement requirement. The initial task is to analyze the request against all other outstanding change requests to determine priority. The system knowledge in the repository is invaluable here, especially if it is current. At a minimum, the importance of the requirement must be assessed against the time and cost of a solution.

Requests for change almost always outnumber resources needed to facilitate change. Change management systems formally capture all change requests in the repository so that change can be prioritized. Changes may be batched so that they can be implemented at optimal times.

If immediate change is needed, the approved change request(s) must be directed to solutions according to the type(s) of change required:

- *New business problems* must be directed to a downsized version of the problem analysis phase. From there, the enhancement will be directed through appropriately downsized versions of requirements analysis, decision analysis, design, construction, and implementation.
- *New business requirements* must be directed to downsized requirements analysis, decision analysis, design, construction, and implementation.
- *New technical requirements* must usually be directed to a decision analysis before design, construction, and implementation. The decision analysis determines if the proposed technology will be feasible in the new system. This is exceedingly important because radical technical change may be costly and complex.
- *New design requirements* must obviously be directed to design, construction, and implementation.

To prioritize and plan enhancement projects, the systems analyst should be skilled in project management (Chapter 4) and feasibility techniques (Chapter 11).

> Task 8.4.2—Make the Quick Fix

Some system enhancements can be accomplished quickly by writing new <u>simple</u> programs or making very simple changes to existing programs. Simple programs and changes are those that can be made without restructuring stored data (changing the database structure), without updating stored data, and without inputting new data (for purposes of storing that data). In other words, these programs generate new (or revise existing) reports and outputs. New program requirements represent many of today's enhancement requests.

> NOTE: It is our belief that any new program requirements that exceed our definition of simple should be treated as new business requirements and subjected to systems analysis and design to more fully consider implications within the complete application system's structure.

Most such programs can be easily written with 4GLs or report-writing tools such as *SAS, Brio,* or *Crystal Reports.* With these tools, programs can be completed within hours. Since they generally do not enter or update data stores, testing requirements are not nearly as stringent.

The quick fix might also be as simple as changing business processes so that they can work with existing information system processes. For example, the analyst may suggest ways in which existing reports can be manually adapted to support different needs.

In Figure 20-5, we see that quick fix requirements are identified and studied. The analysts or programmers use existing business data as test data to write simple, new programs that may or may not be added to the program library. Of course, updated system knowledge should be added to the repository.

> Task 8.4.3—Recover Existing Physical System

Sometimes the repository does contain up-to-date or accurate system knowledge. But documentation is frequently out of date. And in some cases, systems were developed without rigorous development processes or enforced documentation standards. In still other cases, systems were purchased; purchased systems are notorious for poor and inadequate documentation. In all of these instances, the analyst may be asked to recover the existing physical structure of a system as a preface to subsequent system enhancement. In some cases, reengineering technology exists to physically restructure and improve system components without altering their functionality. Let's briefly examine some recovery and restructuring possibilities.

Database Recovery and Restructuring   Sometimes systems analysts help in the reengineering of files and databases. Many of today's data stores are still implemented with traditional file structures (such as *VSAM*) or early database structures (such as hierarchical *IMS* structures). Today's database technology of choice is SQL-based relational databases. Tomorrow, object database technology may present yet another paradigm shift. A more common requirement is the changing of versions in an existing database structure (such as *Oracle 9i* to *Oracle 10g*).

A migrating of data structures from one data storage technology or version to another is a major endeavor, filled with opportunities to corrupt essential, existing business data and programs. Thus, reengineering file and database structures has become an important task.

Database reengineering is usually covered more extensively in data and database management courses and textbooks; however, a brief explanation is in order here. The key player in database restructuring is the database administrator. The systems analyst plays a role because of the potential impact on existing applications. Network analysts may also be involved if databases are (to be) distributed across computer networks.

All databases store metadata about their structure. This metadata can be read and transformed into a physical data model. This data model can be stored in the repository (as system knowledge) to assist analysts in the redesign or use of the database. In some cases, an updated and/or restructured database can be generated, but great care must be taken because many programs use the existing database structure. While database technology theoretically separates data structure from program structure, significant restructuring of databases can still cripple programs.

If the database requires such restructuring, it might be better to identify the change in the repository as another new design requirement that should be directed through an appropriate system redevelopment to determine and react to the impact the new structure may have on existing programs.

Program Analysis, Recovery, and Restructuring   Many businesses are questioning the return on investment in corrective and adaptive maintenance of software. If complex and high-cost software can be identified, it might be reengineered to reduce complexity and maintenance costs. One possibility is to analyze program library and maintenance costs. This task almost always requires software capable of performing the analysis.

Software tools such as ASG's *ASG-Recap* measure your software library using a variety of widely accepted software metrics. **Software metrics** are mathematically proven measurements of software quality and productivity. Examples of software metrics applicable to maintenance include:

software metrics mathematically proven measurements of software quality and developer productivity.

- *Control flow knots*—the number of times logic paths cross one another. Ideally, a program should have zero control flow knots. (We have seen knot counts in the thousands on some older, poorly structured programs.)
- *Cycle complexity*—the number of unique paths through a program. Ideally, the fewer, the better.

Software metrics, in combination with cost accounting (on maintenance efforts), can help identify the programs that would benefit from restructuring.

The input to program analysis is existing programs in the program library. The software may generate restructured programs or merely add system knowledge about the programs to be used later for either restructuring or enhancement. Program analysis was a critical first step in solving the year 2000 software problem. Businesses had to analyze programs to determine where dates were used and what impact changes might have as the programs were enhanced to accommodate the millennium rollover.

Program recovery is similar to program analysis. Existing program code is read from the library. It is then transformed into some sort of physical model appropriate to the software. For example, many CASE tools can reverse engineer a *COBOL* program into a structure chart for that program or reverse engineer a *Visual Basic* program into an object model. These models are added to the repository as new system knowledge to assist with enhancement of the system.

Be very careful not to misuse recovery technology. With purchased software applications, the software license usually prohibits reverse engineering. At best, reverse engineering of purchased software may be unethical. In the worst case, reverse engineering of such software may be illegal and a violation of copyright law and trade secrets.

Finally, some reengineering tools actually support the restructuring of programs. There are three distinct types of program restructuring:

- *Code reorganization* restructures the modular organization and/or logic of the program. Logic may be restructured to eliminate control flow knots and reduce cycle complexity.
- *Code conversion* translates the code from one language to another. Typically, this translation is from one language version to another. There is a debate on the usefulness of translators between different languages. If the languages are sufficiently different, the translation may be very difficult. If the translation is easy, the question is, "Why change?"
- *Code slicing* is the most intriguing program-reengineering option. Many programs contain components that could be factored out as subprograms. If factored out, they would be easier to maintain. More importantly, if factored out, they would be reusable. Code slicing cuts out a piece of a program to create a separate program or subprogram. This may sound easy, but it is not! Consider your average *COBOL* program. The code you want to slice out may be located in many paragraphs and have dependent logic in many other paragraphs. Furthermore, you would have to simultaneously slice out a subset of the data division for the new program or subprogram.

The candidate program for restructuring is copied from the program library. It is reengineered using one or more of the preceding methods, it is thoroughly tested (as described earlier in the chapter), and the reengineered program is returned to the program library where it is available for production. Any new data, process, and/or network models are updated in the repository.

System Obsolescence

At some point, it will not be cost-effective to support and maintain an information system. All systems degrade over time. And when support and maintenance become cost-ineffective, a new systems development project must be started to replace the system. At this time we come full circle to Chapters 3–19 of this book.

This chapter provided a detailed overview of the systems support phase of systems development. You learned about the different types of systems support: maintenance, enhancement, reengineering, and design recovery.

If you have been covering the chapters in order, you are now prepared to do systems development. Otherwise, you may wish to return to previous chapters to learn more about the tools and techniques used in systems development. Completion of this book does not guarantee your future success in systems development. Systems development approaches, tools, and techniques continue to evolve. Thus, your learning will be an ongoing process.

 ## Chapter Review

1. Systems support is the ongoing maintenance of a system after it has been placed into operation. This includes program maintenance and system improvements.

2. Systems support involves solving different types of problems. There are several types of systems support: system maintenance, system recovery, end-user assistance, system enhancement, and reengineering.

3. Regardless of how well designed, constructed, and tested a system or application may be, errors or bugs will occur. The corrective action that must be taken is called system maintenance.

4. From time to time a system failure is inevitable. It generally results in an aborted, or "hung," program (also called an ABEND or crash) and possible loss of data. The systems analyst often fixes the system or acts as intermediary between the users and those who can fix the system; this is referred to as system recovery.

5. Another relatively routine ongoing activity of systems support is end-user assistance. No matter how well users have been trained or how well documentation has been written, users will require

additional assistance. The systems analyst is generally on call to assist users with the day-to-day use of specific applications. In mission-critical applications, the analyst must be on call day and night.

6. Most adaptive maintenance is in response to new business problems, new information requirements, or new ideas for enhancement. It is reactive in nature—fix it when it breaks or when users make a request. We call this system enhancement. The objective of system enhancement is to modify or expand the application system in response to constantly changing requirements. Another type of reactive maintenance deals with changing technology. Information system staffs have become increasingly reluctant to wait until systems break. Instead, they choose to analyze their program libraries to determine which applications and programs are costing the most to maintain or which ones are the most difficult to maintain. These systems might be adapted to reduce the costs of maintenance. The preceding examples of adaptive maintenance are classified as reengineering.

 ## Review Questions

1. What is system support?
2. What are the three areas of systems development knowledge suggested in the textbook that are important to system support?
3. What are the four major support activities suggested in the textbook?
4. Why is system maintenance necessary?
5. What are the tasks needed for system maintenance?

6. What does the term *unsubstantiated bug* mean? What should analysts do when an unsubstantiated bug is found?

7. What should analysts do when they have validated that there are bugs in the program?

8. How are test scripts used in the benchmarking of the program?

9. Why is relying on system knowledge important in debugging of the program?

10. Why is the cost of system maintenance high?

11. What are some of the tasks suggested in technical support?

12. Why will system enhancement occur?

13. What are the tasks needed to conduct system enhancement?

14. Why is it necessary to recover the existing physical system in system enhancement?

15. What are the reengineering tools suggested in the textbook?

Problems and Exercises

1. Your company has grown rapidly in the past several years, and its organizational structure has lagged behind. The CIO has asked you to reorganize the systems operation and support section of the IT shop. As you are reorganizing it, what are the four major support activities you need to be aware of, and what is a critical requirement for each of these activities regarding the staff who will be performing them?

2. What are control flow knots, and why should they be avoided?

3. Answer true or false to the following statements:

   a. Microsoft's *SourceSafe* is an example of a software-based librarian.

   b. In a typical IT shop, analysts spend at most about 20 percent of their time on systems operation and support.

   c. A simple program change is one that does not require a change to the database schema or to the data itself.

   d. IT shops no longer need technical IT staff who have expertise in VSAM or IMS because these file and database structures are almost never found in organizations anymore.

   e. The repository is where archival data generated by the application program is stored.

4. All systems eventually grow old and become obsolete. What is the rule of thumb as to when a system should be replaced?

5. Match the terms in the first column with the definitions or examples in the second column:

   1. Version control   A. Code translation to higher version or different language

   2. Program library   B. Software quality and developer productivity measurements

   3. Control flow knots   C. Baseline against which modified program can be measured

   4. Cycle complexity   D. Application source code data store

   5. Roll forward   E. Unique path count in a program

   6. Code slicing   F. Intersecting logic path count

   7. Code reorganization   G. Program change tracking, usually software based

   8. Code conversion   H. Transaction reprocessing as part of recovery process

   9. VIA/Recap   I. Transaction removal as part of recovery process

   10. Benchmark the program   J. Repository for test data and test instructions

   11. Roll back   K. Code removal to create new program or subprogram

   12. Test script   L. Software metric tool

   13. Software metrics   M. Restructuring of program's logic or modular organization

6. One of the common support activities occurs when a software manufacturer, such as Oracle or Microsoft, releases an updated version of one of their software packages. Often the effort to upgrade is significant, while the changes to the software appear minor. This raises the question, why upgrade? Please discuss.

7. Fill in the blanks.

   a. There is usually _____ staff to handle all the _____ requests that are submitted; therefore, they must be _____.

b. If a program contains excessive cycle _____ or an excessive number of _____, a program _____ technique called code _____ can be used.

c. To _____ programs would benefit from _____, reengineering, or _____, software _____ can be used together with _____ techniques.

d. The _____ is responsible for _____ data _____ and maintained by the production application programs.

8. "Quick fixes" can be a very effective way to make simple changes to a system with a minimum amount of time and cost. On the other hand, what is the danger or downside of quick fixes?

9. Explain the concept of adaptive maintenance.

10. Congratulations—you've reached the final chapter in the textbook. Now it is time to reflect on the chapters that you read. Part One covered the context of a systems development project. Reread the introduction at the beginning of Part One regarding the objectives for that part; then for each chapter in Part One, list at least one thing that stuck with you. For example, did Chapter 1 help you to focus on the big picture, and was the concept of information system building blocks, presented in Chapter 2, something that helped you throughout the textbook and that could be a useful tool in your professional career?

11. In Part Two, you covered a wide range of systems analysis methods in seven chapters. Again, reread the introduction at the beginning of Part Two; then list what resonated the most for you in each chapter on systems analysis methods.

12. Please do the same for Part Three as in the preceding two parts.

13. Part Four of the textbook takes you into the final phase of the project, then into ongoing operations and support. Follow the same process as for the preceding parts, but in addition, please take a moment and think about the entire textbook. Where do you feel it is of the greatest value to you in your studies and professional career, and where do you think it can be strengthened? Also, for how long do you think books on systems analysis and systems design principles and methods can be used before they need to be updated or become obsolete? What are the implications of this for your career as a systems analyst or designer?

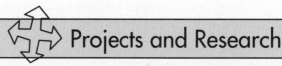

Projects and Research

1. Requests from users for system enhancements or changes are one of the more challenging aspects of systems support, as described in the textbook. Research this issue on the Web, and discuss it with several IT support staff in local organizations. Find out how their organization approaches it. Then:

   a. Write a set of procedures or instructions for your organization to use regarding submitting system enhancement change requests. The procedures should include criteria for determining whether a simple, or "quick fix," program change is appropriate, and the guidelines and processes to be followed when the enhancement doesn't fall under the "quick fix" category. Note: The procedures should be written for the nontechnical business user and should include a sample change request form.

   b. "Pilot" the instructions and form with your fellow students and/or co-workers. Have them evaluate and critique the instructions and form.

   c. Use their feedback to refine your instructions and form. Report the process until there are no further suggestions for improvement.

   d. How many iterations did you go through?

2. At the time that most projects are initiated, an estimate is made of the expected life span of the new system as a part of a cost-benefit evaluation. However, on an anecdotal basis, it appears that most systems remain in production far longer than their original planned life span. Research this topic on the Web and/or in your school library. In addition, conduct an informal survey of system administrators in several local public and private sector agencies.

   a. Describe the articles that you found and their viewpoints on this subject.

   b. Describe the informal survey you conducted, including the nature of organizations that were included in the survey and the types of systems they use.

   c. What was the average difference you found, if any, between expected life span and actual life span of systems?

   d. To what do you attribute the tendency to keep systems in production beyond their expected life span?

   e. What was the oldest system you found still in production? (It is not uncommon to find systems dating back to the 1970s, depending upon the nature of the organization. Finding a system originally implemented in the 1960s would be rare, although some are known to still exist!)

3. The objective of software metrics is to provide a mathematically valid method for measuring software quality and productivity. Research articles on the Web and/or in your school library regarding software metrics, as well as the Web sites of some manufacturers of software metrics tools. If they offer any free trial copies, download them and try them out. In addition, ask the IT staff in several local organizations if they use software metrics tools, and, if so, which one and what their opinion is of it.

a. Describe the research that you conducted.

b. What are some of the different schools of thought and/or controversies regarding the use of software metrics tools?

c. What were the responses from IT staff? Do they use software metric tools? If they do, what are the ones they use and/or prefer? What are the different uses for which the tools are employed? If the IT staff doesn't use software metrics tools, what are the reasons why not?

d. On the basis of your research and perhaps hands-on experience, compare and contrast the different software metrics tools that are available. Which one(s) did you prefer and why?

e. Overall, do you consider the benefits of software metrics tools to be sufficient to justify their cost of purchase and their learning curve?

4. Systems operations and support generally consume an enormous amount of the resources available in the typical IT shop of an organization. Many CIOs are avidly interested in systems support software that may reduce the amount of human effort required. Research the Web and trade journals for information on some of the newly emerging

applications, not discussed in the textbook, that are or soon will be available.

a. Describe the research you conducted.

b. What are some of the new types of systems support applications that you found?

c. Compare and contrast their features.

d. If you were the systems support manager in your organization and you could choose only one of these applications, which one would you choose? Or would you choose any? Explain your answer.

5. Technical support to users is one of the cornerstones of systems support which can consume a tremendous amount of time and staff resources. Research this topic in trade journals and in forums on the Web. Interview support staff in the IT shops of several local organizations regarding the issues they face in providing support. In addition, interview some users in the same organizations regarding *their* perspective on technical support.

a. Describe the research you conducted.

b. What was the range of responses you received regarding technical support?

c. Did the IT staff in the different organizations have the same issues in common?

d. What about the IT staff and the users—what perspectives did they have in common, and where did they differ?

e. Using your research, write a one-page bulleted list of guidelines for providing technical support.

6. Have computers become sophisticated enough to be powerful extensions of how humans think and communicate? Or do computers and humans capture and process data and transform it into information in fundamentally different ways? Explain your answer.

Minicases

1. Create a final exam for this class. Post an electronic copy to the class Web site, as well as a hard copy to your professor. Include solutions to the questions, per your professor's direction.

2. Compile the work you have done throughout this class for the department of the government that you chose. Create two deliverables: one for your contact at the department, and one for your professor. The deliverable to your professor will include ALL of the work you have done: interviews, research, diagrams, testing and design reports and

counter reports, business discussion, manual, and final prototype on CD. The deliverable for your contact in the government will include: diagrams, business discussion, manual, and final prototype on CD. Remember, the manual will be bound separately for both deliverables.

3. Create a portfolio of the work you have done in this class. Include major deliverables that showcase your capabilities as a systems analyst and designer, any letters of team recognition for good teamwork from your group, a description of the

topics you covered in class, and the like. Make sure that this portfolio has a professional appearance and is bound. This is for you to keep as you job-search.

4. Follow up with the contact at the government department. Did they have any questions? Any comments on your work? Does the prototype do everything they wanted and needed? Does it work on their network correctly and without problems? Was the manual clear and easy to follow? Consider this your final system acceptance.

Suggested Readings

Arnold, Robert. *Software Reengineering.* Los Alamitos, CA: IEEE Computer Society Press, 1993. This is a reference book and research compilation on the subject of business process, database, and software reengineering.

Hammer, M., and J. Champy. *Reengineering the Corporation.* New York: Harper Business, 1993. Mike Hammer is widely regarded as the father of business process reengineering.

Martin, E. W.; D. W. DeHayes; J. A. Hoffer; and W. C. Perkins. *Managing Information Technology: What Managers Need to Know.* New York: Macmillan, 1994.

Page 24	Blackberry Wireless Handheld ™ from Research In Motion (RIM)
Page 25	Courtesy of Microsoft
Page 80	Getty/PhotoDisc (DIL)
Page 90	Getty/PhotoDisc (DIL)
Page 95	Getty/PhotoDisc (DIL)
Page 99	Getty/PhotoDisc (DIL)
Page 102	Getty/PhotoDisc (DIL)
Page 102	Getty/PhotoDisc (DIL)
Page 105	Getty/PhotoDisc (DIL)
Page 105	Getty/PhotoDisc (DIL)
Page 105	Getty/PhotoDisc (DIL)
Page 105	Getty/PhotoDisc (DIL)
Page 106	Getty/PhotoDisc (DIL)
Page 129	Getty/PhotoDisc (DIL)
Page 129	Getty/PhotoDisc (DIL)
Page 129	Getty/PhotoDisc (DIL)
Page 161	Getty/PhotoDisc (DIL)
Page 169	Getty/PhotoDisc (DIL)
Page 177	Getty/PhotoDisc (DIL)
Page 187	Getty/PhotoDisc (DIL)
Page 191	Getty/PhotoDisc (DIL)
Page 194	Getty/PhotoDisc (DIL)
Page 194	Getty/EyeWire (DIL)
Page 454	Getty/PhotoDisc (DIL)
Page 455	Getty/PhotoDisc (DIL)
Page 461	Getty/PhotoDisc (DIL)
Page 461	Getty/PhotoDisc (DIL)
Page 461	Getty/PhotoDisc (DIL)
Page 462	Getty/PhotoDisc (DIL)
Page 462	Getty/PhotoDisc (DIL)
Page 585	Blackberry Wireless Handheld ™ from Research In Motion (RIM)
Page 685	Getty/PhotoDisc (DIL)
Page 686	Getty/PhotoDisc (DIL)
Page 686	Getty/PhotoDisc (DIL)
Page 690	Getty/PhotoDisc (DIL)
Page 703	Getty/PhotoDisc (DIL)
Page 704	Getty/PhotoDisc (DIL)
Page 704	Getty/PhotoDisc (DIL)
Page 704	Getty/EyeWire (DIL)
Page 704	Getty/PhotoDisc (DIL)
Page 705	Getty/PhotoDisc (DIL)
Page 705	Getty/PhotoDisc (DIL)
Page 705	Getty/PhotoDisc (DIL)
Page 707	Getty/PhotoDisc (DIL)
Page 711	Getty/PhotoDisc (DIL)

Bold page numbers indicate locations of glossary definitions

override *A technique whereby a subclass (subtype) uses an attribute or behavior of its own instead of an attribute or behavior inherited from the class (supertype).*, **380**

P

paging *Displaying a complete screen of characters at a time*, **619**

parent entity *A data entity that contributes one or more attributes to another entity, called the child. In a one-to-many relationship the parent is the entity on the "one" side.*, **277**

partitioning *The act of determining how to best distribute or duplicate application components across a network.*, **491**, 495

payback analysis *A technique for determining if and when an investment will pay for itself.*, **423**–425

payback period *The period of time that will elapse before accrued benefits overtake accrued costs.*, **423**

persistent class *A class that describes an object that outlives the execution of the program that created it.*, **405**, 649

PERT chart *A graphical network model used to depict the interdependencies between a project's tasks.*, **125**, 127, 147

pessimistic duration (PD) *The estimated maximum amount of time needed to complete a task.*, **133**

physical data flow diagram *A process model used to communicate the technical implementation characteristics of an information system.*, **455**, **477**; *see also* **data flow diagram**

physical design *The translation of business user requirements into a system model that depicts a technical implementation of the users' business requirements. Common synonyms include technical design or, in describing the output, implementation model. The antonym of physical design is logical design (defined earlier in this chapter).*, **86**–87; *see also* **system design**

physical model *A technical pictorial representation that depicts what a system is or does and how the system is implemented. Synonyms are implementation model and technical model.*, **94**, **316**

policy *A set of rules that govern a business process.*, **52**–54

policy *A set of rules that govern how a process is to be completed.*, **357**

polymorphism *Literally meaning "many forms," the concept that different objects can respond to the same message in different ways.*, **380**

present value *The current value of a dollar at any time in the future.*, **424**

presentation *The ongoing activity of communicating findings, recommendations, and documentation for review by interested users and managers. Presentations may be either written or verbal.*, **89**; *see also* **formal presentation**; Reports, written

primary key *A candidate key that will most commonly be used to uniquely identify a single entity instance.*, **274**

primary key *A field or group of fields that uniquely identifies a record.*, **521**, 530, 535

primitive diagram *A data flow diagram that depicts the elementary processes, data stores, and data flows for a single event.*, **335**, 349

U

Unified Modeling Language (UML) *A set of modeling conventions that is used to specify or describe a software system in terms of objects.*, **371**

unit or **program test** *A test performed on an entire program.*, **689**, 709

unstructured interview *An interview that is conducted with only a general goal or subject in mind and with few, if any, specific questions. The interviewer counts on the interviewee to provide a framework and direct the conversation.*, **223**

use case *A business scenario or event for which the system must provide a defined response. Use cases evolved out of object-oriented analysis; however, their use has become common in many other methodologies for systems analysis and design.*, **188**

use case *A behaviorally related sequence of steps (a scenario), both automated and manual, for the purpose of completing a single business task.*, **246**

use case *An analysis tool for finding and identifying business events and responses.*, **342**

use-case dependency diagram *A graphical depiction of the dependencies among use cases.*, **261**–262

use-case diagram *A diagram that depicts the interactions between the system and external systems and users. In other words, it graphically describes who will use the system and in what ways the user expects to interact with the system.*, **246**, 254–256, 384–385

use-case modeling *The process of modeling a system's functions in terms of business events, who initiated the events, and how the system responds to those events.*, **245**

use-case narrative *A textual description of the business event and how the user will interact with the system to accomplish the task.*, **246**

use-case ranking and priority matrix *A tool used to evaluate use cases and determine their priority.*, **260**–261

user dialogue *A specification of how the user moves from window to window or page to page, interacting with the application programs to perform useful work.*, **57**; *see also* Menus